S0-CBI-390

INTERNATIONAL UNION OF PURE AND APPLIED CHEMISTRY

ANALYTICAL CHEMISTRY DIVISION
COMMISSION ON SOLUBILITY DATA

SOLUBILITY DATA SERIES

Volume 13

SCANDIUM, YTTRIUM, LANTHANUM AND LANTHANIDE NITRATES

SOLUBILITY DATA SERIES

Volume 1 H. L. Clever, *Helium and Neon*

Volume 2 H. L. Clever, *Krypton, Xenon and Radon*

Volume 3 M. Salomon, *Silver Azide, Cyanide, Cyanamides, Cyanate, Selenocyanate and Thiocyanate*

Volume 4 H. L. Clever, *Argon*

Volume 5/6 C. L. Young, *Hydrogen and Deuterium*

Volume 7 R. Battino, *Oxygen and Ozone*

Volume 8 C. L. Young, *Oxides of Nitrogen*

Volume 9 W. Hayduk, *Ethane*

Volume 10 R. Battino, *Nitrogen and Air*

Volume 11 B. Scrosati and C. A. Vincent, *Alkali Metal, Alkaline-Earth Metal and Ammonium Halides. Amide Solvents*

Volume 12 C. L. Young, *Sulfur Dioxide, Chlorine, Fluorine and Chlorine Oxides*

Volume 13 S. Siekierski, T. Mioduski and M. Salomon, *Scandium, Yttrium, Lanthanum and Lanthanide Nitrates*

Volume 18 O. Popovych, *Tetraphenylborates*

Volumes in preparation

H. L. Clever and W. Gerrard, *Hydrogen Halides in Non-Aqueous Solvents*

H. Miyamoto and M. Salomon, *Alkaline-Earth Metal Halates*

S. Siekierski and J. K. Navratil, *Actinide Nitrates*

A. L. Horvath and F. W. Getzen, *Halogenated Benzenes, Toluenes, Xylenes and Phenols with Water*

E. M. Woolley, *Aqueous Silver Chloride Systems*

Z. Galus and C. Guminski, *Metals and Intermetallic Compounds in Mercury*

H. Miyamoto and E. M. Woolley, *Silver, Copper and Gold Halates*

J. W. Lorimer, *Beryllium, Strontium, Barium and Radium Sulfates*

H. L. Clever, D. M. Mason and C. L. Young, *Carbon Dioxide*

H. L. Clever and C. L. Young, *Methane*

SOLUBILITY DATA SERIES

Volume 13

SCANDIUM, YTTRIUM, LANTHANUM AND LANTHANIDE NITRATES

Volume Editors

S. SIEKIERSKI
Institute for Nuclear Research
Warsaw, Poland

T. MIODUSKI
Institute for Nuclear Research
Warsaw, Poland

M. SALOMON
US Army, ERADCOM
Fort Monmouth, NJ, USA

Evaluators

S. SIEKIERSKI, T. MIODUSKI and M. SALOMON

Compilers

S. SIEKIERSKI, T. MIODUSKI, M. SALOMON
and
O. POPOVYCH
Brooklyn College of CUNY
Brooklyn, NY, USA

PERGAMON PRESS

OXFORD · NEW YORK · TORONTO · SYDNEY · PARIS · FRANKFURT

U.K.	Pergamon Press Ltd., Headington Hill Hall, Oxford OX3 0BW, England
U.S.A.	Pergamon Press Inc., Maxwell House, Fairview Park, Elmsford, New York 10523, U.S.A.
CANADA	Pergamon Press Canada Ltd., Suite 104, 150 Consumers Road, Willowdale, Ontario M2J 1P9, Canada
AUSTRALIA	Pergamon Press (Aust.) Pty. Ltd., P.O. Box 544, Potts Point, N.S.W. 2011, Australia
FRANCE	Pergamon Press SARL, 24 rue des Ecoles, 75240 Paris, Cedex 05, France
FEDERAL REPUBLIC OF GERMANY	Pergamon Press GmbH, Hammerweg 6, D-6242 Kronberg-Taunus, Federal Republic of Germany

First edition 1983

British Library Cataloguing in Publication Data

Scandium, yttrium, lanthanum and lanthanide nitrates.
—(Solubility data series, ISSN 0191-5622; v.13)
1. Scandium—Solubility—Tables
I. Siekierski, S. II. International Union of
Pure and Applied Chemistry, Commission on
Solubility Data III. Series
546'.401542'0212 QD181.S4

ISBN 0-08-026192-2

In order to make this volume available as economically and as rapidly as possible the authors' typescripts have been reproduced in their original forms. This method unfortunately has its typographical limitations but it is hoped that they in no way distract the reader.

Printed in Great Britain by A. Wheaton & Co. Ltd., Exeter

CONTENTS

SOLUBILITY DATA SERIES

Editor-in-Chief

INTERNATIONAL UNION OF PURE AND APPLIED CHEMISTRY

IUPAC Secretariat: Bank Court Chambers, 2-3 Pound Way,
Cowley Centre, Oxford OX4 3YF, UK

FOREWORD

If the knowledge is undigested or simply wrong, more is not better.

How to communicate and disseminate numerical data effectively in chemical science and technology has been a problem of serious and growing concern to IUPAC, the International Union of Pure and Applied Chemistry, for the last two decades. The steadily expanding volume of numerical information, the formulation of new interdisciplinary areas in which chemistry is a partner, and the links between these and existing traditional subdisciplines in chemistry, along with an increasing number of users, have been considered as urgent aspects of the information problem in general, and of the numerical data problem in particular.

Among the several numerical data projects initiated and operated by various IUPAC commissions, the *Solubility Data Project* is probably one of the most ambitious ones. It is concerned with preparing a comprehensive critical compilation of data on solubilities in all physical systems, of gases, liquids and solids. Both the basic and applied branches of almost all scientific disciplines require a knowledge of solubilities as a function of solvent, temperature and pressure. Solubility data are basic to the fundamental understanding of processes relevant to agronomy, biology, chemistry, geology and oceanography, medicine and pharmacology, and metallurgy and materials science. Knowledge of solubility is very frequently of great importance to such diverse practical applications as drug dosage and drug solubility in biological fluids, anesthesiology, corrosion by dissolution of metals, properties of glasses, ceramics, concretes and coatings, phase relations in the formation of minerals and alloys, the deposits of minerals and radioactive fission products from ocean waters, the composition of ground waters, and the requirements of oxygen and other gases in life support systems.

The widespread relevance of solubility data to many branches and disciplines of science, medicine, technology and engineering, and the difficulty of recovering solubility data from the literature, lead to the proliferation of published data in an ever increasing number of scientific and technical primary sources. The sheer volume of data has overcome the capacity of the classical secondary and tertiary services to respond effectively.

While the proportion of secondary services of the review article type is generally increasing due to the rapid growth of all forms of primary literature, the review articles become more limited in scope, more specialized. The disturbing phenomenon is that in some disciplines, certainly in chemistry, authors are reluctant to treat even those limited-in-scope reviews exhaustively. There is a trend to preselect the literature, sometimes under the pretext of reducing it to manageable size. The crucial problem with such preselection - as far as numerical data are concerned - is that there is no indication as to whether the material was excluded by design or by a less than thorough literature search. We are equally concerned that most current secondary sources, critical in character as they may be, give scant attention to numerical data.

On the other hand, tertiary sources - handbooks, reference books, and other tabulated and graphical compilations - as they exist today, are comprehensive but, as a rule, uncritical. They usually attempt to cover whole disciplines, thus obviously are superficial in treatment. Since they command a wide market, we believe that their service to advancement of science is at least questionable. Additionally, the change which is taking place in the generation of new and diversified numerical data, and the rate at which this is done, is not reflected in an increased third-level service. The emergence of new tertiary literature sources does not parallel the shift that has occurred in the primary literature.

With the status of current secondary and tertiary services being as briefly stated above, the innovative approach of the *Solubility Data Project* is that its compilation and critical evaluation work involve consolidation and reprocessing services when both activities are based on intellectual and scholarly reworking of information from primary sources. It comprises compact compilation, rationalization and simplification, and the fitting of isolated numerical data into a critically evaluated general framework.

The *Solubility Data Project* has developed a mechanism which involves a number of innovations in exploiting the literature fully, and which contains new elements of a more imaginative approach for transfer of reliable information from primary to secondary/tertiary sources. *The fundamental trend of the Solubility Data Project is toward integration of secondary and tertiary services with the objective of producing in-depth critical analysis and evaluation which are characteristic to secondary services, in a scope as broad as conventional tertiary services.*

Fundamental to the philosophy of the project is the recognition that the basic element of strength is the active participation of career scientists in it. Consolidating primary data, producing a truly critically-evaluated set of numerical data, and synthesizing data in a meaningful relationship are demands considered worthy of the efforts of top scientists. Career scientists, who themselves contribute to science by their involvement in active scientific research, are the backbone of the project. The scholarly work is commissioned to recognized authorities, involving a process of careful selection in the best tradition of IUPAC. This selection in turn is the key to the quality of the output. These top experts are expected to view their specific topics dispassionately, paying equal attention to their own contributions and to those of their peers. They digest literature data into a coherent story by weeding out what is wrong from what is believed to be right. To fulfill this task, the evaluator must cover *all* relevant open literature. No reference is excluded by design and every effort is made to detect every bit of relevant primary source. Poor quality or wrong data are mentioned and explicitly disqualified as such. In fact, it is only when the reliable data are presented alongside the unreliable data that proper justice can be done. The user is bound to have incomparably more confidence in a succinct evaluative commentary and a comprehensive review with a complete bibliography to both good and poor data.

It is the standard practice that any given solute-solvent system consists of two essential parts: I. Critical Evaluation and Recommended Values, and II. Compiled Data Sheets.

The Critical Evaluation part gives the following information:
(i) a verbal text of evaluation which discusses the numerical solubility information appearing in the primary sources located in the literature. The evaluation text concerns primarily the quality of data after consideration of the purity of the materials and their characterization, the experimental method employed and the uncertainties in control of physical parameters, the reproducibility of the data, the agreement of the worker's results on accepted test systems with standard values, and finally, the fitting of data, with suitable statistical tests, to mathematical functions;
(ii) a set of recommended numerical data. Whenever possible, the set of recommended data includes weighted average and standard deviations, and a set of smoothing equations derived from the experimental data endorsed by the evaluator;
(iii) a graphical plot of recommended data.

The compilation part consists of data sheets of the best experimental data in the primary literature. Generally speaking, such independent data sheets are given only to the best and endorsed data covering the known range of experimental parameters. Data sheets based on primary sources where the data are of a lower precision are given only when no better data are available. Experimental data with a precision poorer than considered acceptable are reproduced in the form of data sheets when they are the only known data for a particular system. Such data are considered to be still suitable for some applications, and their presence in the compilation should alert researchers to areas that need more work.

The typical data sheet carries the following information:
(i) components - definition of the system - their names, formulas and Chemical Abstracts registry numbers;
(ii) reference to the primary source where the numerical information is reported. In cases when the primary source is a less common periodical or a report document, published though of limited availability, abstract references are also given;
(iii) experimental variables;
(iv) identification of the compiler;
(v) experimental values as they appear in the primary source. Whenever available, the data may be given both in tabular and graphical form. If auxiliary information is available, the experimental data are converted also to SI units by the compiler.

Under the general heading of Auxiliary Information, the essential experimental details are summarized:

(vi) experimental method used for the generation of data;
(vii) type of apparatus and procedure employed;
(viii) source and purity of materials;
(ix) estimated error;
(x) references relevant to the generation of experimental data as cited in the primary source.

This new approach to numerical data presentation, developed during our four years of existence, has been strongly influenced by the diversity of background of those whom we are supposed to serve. We thus deemed it right to preface the evaluation/compilation sheets in each volume with a detailed discussion of the principles of the accurate determination of relevant solubility data and related thermodynamic information.

Finally, the role of education is more than corollary to the efforts we are seeking. The scientific standards advocated here are necessary to strengthen science and technology, and should be regarded as a major effort in the training and formation of the next generation of scientists and engineers. Specifically, we believe that there is going to be an impact of our project on scientific-communication practices. The quality of consolidation adopted by this program offers down-to-earth guidelines, concrete examples which are bound to make primary publication services more responsive than ever before to the needs of users. The self-regulatory message to scientists of 15 years ago to refrain from unnecessary publication has not achieved much. The literature is still, in 1983, cluttered with poor-quality articles. The Weinberg report (in "Reader in Science Information", Eds. J. Sherrod and A. Hodina, Microcard Editions Books, Indian Head, Inc., 1973, p.292) states that "admonition to authors to restrain themselves from premature, unnecessary publication can have little effect unless the climate of the entire technical and scholarly community encourages restraint..." We think that projects of this kind translate the climate into operational terms by exerting pressure on authors to avoid submitting low-grade material. The type of our output, we hope, will encourage attention to quality as authors will increasingly realize that their work will not be suited for permanent retrievability unless it meets the standards adopted in this project. It should help to dispel confusion in the minds of many authors of what represents a permanently useful bit of information of an archival value, and what does not.

If we succeed in that aim, even partially, we have then done our share in protecting the scientific community from unwanted and irrelevant, wrong numerical information.

A. S. Kertes

PREFACE

SCOPE OF THE VOLUME

This volume deals with the solubilities of the nitrates of scandium, yttrium, lanthanum, and the lanthanides, commonly referred to as rare earths. As is the usual practice, lanthanium is included as a member of the lanthanides. All solubility data for binary and multicomponent systems (except for molten systems) are included in this volume. Also included are the solubility data for lanthanide double nitrates. Double salts involving a lanthanide nitrate and a salt of an anion other than nitrate (e.g. a metal halide) are not included in this volume.

A number of papers have been rejected and therefore not compiled, and these have been carefully noted in the critical evaluations. Several papers which have been rejected and not discussed in the critical evaluations are: Holmberg (1) who reported the solubility of the double salt didymium ammonium nitrate in water at 288 K, but didymium is now known to be a mixture of rare earths, mainly neodymium and praseodymium; Rothschild et al. (2) who reported the solubility of a rare earth mixture containing 15 rare earths in 3-methyl-1-butanol (isoamyl alcohol) and 2-octanone (methyl hexyl ketone). The initial composition of the mixture was given, but only a single overall solubility value was reported; Medoks (3) who studied the double salts of La, Ce, Pr, and Nd nitrates with triphenylbenzylarsonium nitrate, but reports the solubilities in a number of organic solvents in terms of "readily soluble, insoluble, etc.," and numerical solubility data are not given.

NATURE OF THE SOLID PHASES

Depending upon temperature and composition of the saturated solutions (e.g. in ternary systems), lanthanide nitrates crystallize with different numbers of waters of hydration ranging from 6 to probably 3 per mole of salt. Lower hydrates predominate at higher temperatures, and when crystallization is carried out in the presence of foreign electrolytes such as HNO_3. A common procedure used to prepare the initial lanthanide nitrate has been to crystallize the salt from aqueous nitric acid followed by desiccation over H_2SO_4 or $CaCl_2$. The nature of the hydrate produced by this method has led to considerable disagreement between various authors concerning both the nature of the equilibrated solid phase and the solubility value. It appears that the stable solid phase for all lanthanides at 298.2 K is the hexahydrate except for Yb and Lu for which the stable solid phase is the pentahydrate. The hexahydrate $\longrightarrow$ pentahydrate transition temperatures are generally higher for the light lanthanides, and for the heavy lanthanides starting at around $Er(NO_3)_3$, this transition temperature is close to 298.2 K.

Metastable equilibria is readily achieved with saturated solutions of the lanthanide nitrates, and this is the main reason for discrepancies between solubilities reported by various authors. The solubility of $Ln(NO_3)_3$ in the various $Ln(NO_3)_3 \cdot nH_2O$ - H_2O systems can be described by a general smoothing equation based upon the treatments in (4,5) and in the INTRODUCTION to this volume:

$$Y = \ln(m/m_o) - nM_2(m - m_o) = a + b/(T/K) + c\,\ln(T/K) \qquad [1]$$

In this smoothing equation, n is the hydration number of the solid phase hydrate, m is the molality of the saturated solution at temperature T, m_o is an arbitrarily selected reference molality (usually the solubility at 298.2 K), M_2 is the molar mass of the solvent, and a, b, c are constants. At the congruent melting point, the solubility of the hydrate in its own waters of hydration is $1000/nM_2 = 55.508/n$ mol kg^{-1}, and eq.[1] thus enables us to calculate the temperature at the congruent melting point. The calculated congruent melting point temperatures when compared to experimental values have enabled the evaluators to confirm the value of n for both stable and metastable systems over a wide range of temperatures. We have not made an exhaustive literature search on congruent melting points, but relied heavily on the citations given by Mellor (6). All references on melting points determined prior to the year 1930 were taken directly from Mellor's treatise.

The readiness to form metastable equilibria results from small differences in lattice structures and, consequently, lattice energies. The light lanthanide nitrate hexahydrates from La to Sm belong to the same $P\bar{1}$ space group. However, two isotypic subgroups can be distinguished, one consisting of La and Ce, and the other of Pr, Nd, and Sm nitrate hexahydrates (7). For $La(NO_3)_3 \cdot 6H_2O$ the coordination polyhedron is formed by three bidentate nitrate groups and five oxygen atoms from the water molecules: thus the structural formula is $[La(NO_3)_3(H_2O)_5] \cdot H_2O$ with coordination number 11 (8). The La-O(nitrate) bonds are somewhat asymmetric, and the uncoordinated water molecule is in close contact with two of the coordinated water molecules. This indicates that the coordination polyhedra are held together by hydrogen bonds. In $Pr(NO_3)_3 \cdot 6H_2O$ the coordination polyhedron is also formed by three bidentate nitrate groups, but only four water oxygen atoms: thus the structural formula is $[Pr(NO_3)_3(H_2O)_4] \cdot 2H_2O$ (9,10). Within each isotypic subgroup unlimited miscibility of the solid nitrates of the various nitrates is observed, whereas with lanthanide nitrates belonging to different subgroups, a miscibility gap occurs such as for La and Sm nitrate hexahydrates (11). There are indications (12) that the light lanthanide nitrate pentahydrates and tetrahydrates also form two isotypic subgroups, one consisting of La and Ce, and the other of Pr, Nd, and Sm. This may explain why the change in solubiity with atomic number within the light lanthanides is almost independent of temperature (Fig. 1). The structural formula of lanthanum nitrate pentahydrate which also belongs to the P1 space group is $[La(NO_3)_3(H_2O)_4]\cdot H_2O$ (13). Thus as in the case with the hexahydrate, the lattice is held together by one water molecule, and the difference concerns only the nearest neighbors of the lanthanum ion. This difference however appears to be small since the lower coordination number of 10 for the pentahydrate as compared to 11 for the hexahydrate is partially compensated by smaller La-O distances, and less asymmetry in the La-O(nitrate) bonds.

SOLUBILITY AS A FUNCTION OF LANTHANIDE ATOMIC NUMBER

The dependence of solubility of lanthanide nitrates on lanthanide atomic number at various temperatures is shown in Fig. 1. The solubility data used to construct this figure were taken from the critical evaluations. From the figure it is seen that the solubility increases between La and Ce, decreases in the interval from Ce to Sm, and then increases monotonically between Gd and Lu with Lu exhibiting an almost singular position. It is noted that changes in the solubility as a function of lanthanide atomic number are almost independent of temperature in the interval 273.2 K to 323.2 K and, presumably, also at higher temperatures.

The solubility of a lanthanide nitrate hydrate in water can be described by the following schematic reaction:

$$Ln(NO_3)_3\cdot nH_2O(s) \longrightarrow Ln(NO_3)_3(aq) + nH_2O(aq) = Ln^{3+}(aq) + 3NO_3^-(aq) + nH_2O(aq) \qquad [2]$$

At equilibrium, the following relation holds:

$$\mu^*_{AB} = \mu^{\ominus}_B + RT\ell n(27m^4\gamma^4_{\pm}) + n\,\mu^*_A + RT\ell n\, a^n_A \qquad [3]$$

where μ^*_{AB}, $\mu^{\ominus}_B$, and μ^*_A are standard chemical potentials of the pure solid hydrate (AB), of the salt (B) at infinite dilution, and of pure solvent (A), respectively. m is the molality of the saturated solution, $\gamma_{\pm}$ the stoichiometric mean molal activity coefficient for the salt, and a_A is the activity of water in the saturated solution. The standard Gibbs energy of solution, ΔG^o_{sln}, for the lanthanide nitrate hydrate follows from eq. [3]:

$$\Delta G^o_{sln} = \mu^{\ominus}_B - \mu^*_{AB} + n\,\mu^*_A \qquad [4]$$

On the other hand, ΔG^o_{sln} for a hydrated salt is also given by

$$\Delta G^o_{sln} = \Delta G^o_h - \Delta G^o_{lat} - n\Delta G^o(H_2O,\ vap) \qquad [5]$$

where ΔG^o_h is the standard Gibbs energy of hydration of $Ln(NO_3)_3$, ΔG^o_{lat} is the standard Gibbs energy of lattice formation of the hydrate, and $\Delta G^o(H_2O,vap)$ is the standard Gibbs energy of vaporization of water. By combining eqs. [3-5], the following exression for the solubility m is obtained:

$$-4RT\,\ell n\, m = \Delta G^o_h - \Delta G^o_{lat} - n\Delta G^o(H_2O,vap) + RT\,\ell n\,\gamma^4_{\pm} + nRT\,\ell n\, a_A + RT\,\ell n\, 27 \qquad [6]$$

Provided n is constant, eq. [6] can be used to explain the observed solubility-atomic number (Z) behavior by variations in the terms ΔG^o_h, ΔG^o_{lat}, and $\gamma_{\pm}$. The activity coefficient term reflects changes in ionic hydration, long range interactions, and complex formation as a function of concentration and lanthanide ion (changes in the activity of water are negligibly small). According to some estimates (14), ΔG^o_h for lanthanide ions decrease from about -3.180 kJ mol^{-1} for La to about -3.598 kJ mol^{-1} for Lu which suggests that both the absolute values and the (relative) changes within the series are substantial. The Gibbs energy of hydration exhibits a well marked double-double effect, i.e. the relative stabilization of f^0, f^7, f^{14}, and f^3, f^4, f^{10}, f^{11} configurations. Apparently nothing is known about the Gibbs energy of lattice formation of lanthanide nitrates, but there is indirect evidence that it must change with atomic number almost the same as ΔG^o_h. It is known (15) that the standard enthalpy of solution of lanthanide nitrates, ΔH^o_{sln} which is equal to $\Delta H^o_h - \Delta H^o_{lat} - n\Delta H^o(H_2O,vap)$, increases from -18.702 kJ mol^{-1} for La to 11.632 kJ mol^{-1} for Lu. Since for certain lanthanide properties such as complexation, the entropy term almost compensates the enthalpy term, it is expected that changes in ΔG^o_{sln} (i.e. changes in the difference $\Delta G^o_h - \Delta G^o_{lat}$) with Z will be much smaller than changes in ΔH^o_{sln}, but that these changes will still be positive. Therefore by considering only the differences in $\Delta G^o_h - \Delta G^o_{lat}$, we would expect a very small decrease in solubility from La to Lu whereas in fact, the solubility increases with Z over the Gd-Lu or Eu-Lu interval. Apparently the contribution from the activity coefficient term is of importance. Since at high lanthanide nitrate concentrations changes in $\gamma_{\pm}$ with Z reflect mainly changes in complex formation between Ln^{3+} and NO_3^- ions (16), the observed solubility behavior as a function of Z may be significantly influenced by differences in complex formation. This conclusion is supported by a partial similarity between the plot of $-4RT \ln m$ against Z (Fig. 2), and changes in the Gibbs energy of complex formation (ΔG_c^o) with Z for many ligands (e.g. for glycolic acid (17)). It is frequently observed that with increasing Z, ΔG_c^o decreases at the beginning of the series, then remains constant or increases slightly, and decreases again in the region of the heavy lanthanides. Thus at least in the Gd-Lu interval, the change in solubility with Z may be due mainly to changes in complex formation. Within the light lanthanides, the appearance of a maximum for the solubility at Ce may be due to the difference in structure between La-Ce and Pr-Sm nitrate hexahydrates as discussed above (11). For La and Ce nitrate hexahydrates, one water molecule is outside the cation coordination sphere, whereas for Pr and probably also the heavier lanthanides, two water molecules hold the lattice together by hydrogen bonds. Thus a discontinuous decrease in lattice enthalpy, and probably also Gibbs energy, should occur between Ce and Pr. It follows from eq. [6] that this change in lattice enthalpies and Gibbs energies will result in a decrease in solubility between Ce and Pr. However since the maximum at Ce also appears at high temperatures, an analogous change in coordination number between Ce and Pr must also occur for lower hydrates, i.e. at least for the pentahydrates which can be seen from the data in the critical evaluations which show a solubility maximum at Ce in the pentahydrate system over the temperature range 298.2 K to 343.2 K. Experimental data on the solubilities in the tetrahydrate system are not sufficiently complete to extend this analysis to the tetrahydrate systems.

It should also be noted that the change in solubility in the La-Nd interval exhibits a conspicuous similarity to the pattern expected for the double-double effect which is greatest for the first segment (*tetrad*). The slightly lower solubility of lutetium nitrate than that expected from the trend in the Gd-Yb interval may also be a manifestation of the effect in the last segment. Since the occurence of the double-double effect in Gibbs energy of hydration of lanthanide ions and in lattice parameters of lanthanide compounds is well established, its influence on solubility behavior is expected, and its direction (convex downwards) is in accordance with the higher coordination number of lanthanide ions in the solid nitrates than in aqueous solution.

POSITION OF YTTRIUM WITHIN THE LANTHANIDES WITH RESPECT TO SOLUBILITY

With respect to lattice parameters containing oxygen atoms as coordinating atoms (18), and with respect to enthalpy and entropy of complex formation, Y lies between Ho and Er. Since for most ligands, yttrium behaves as a slightly heavier quasilanthanide with respect to enthalpy than with respect to entropy of complex formation (19), its position with respect to Gibbs energy of complex formation is that amongst the light lanthanides. However with respect to solubility of the nitrates, yttrium appears almost at the exact place as erbium, i.e. amongst the heavy lanthanides. This means that with respect to enthalpy and entropy of solutions, yttrium behaves as a heavy lanthanide, and that there is little change in the cation's environment in passing from the solid phase to aqueous solution. Indeed lanthanide and yttrium ions in the solid (hexahydrate) phase are surrounded by three NO_3^- ions and 5 or 4 water molecules, whereas in concentrated aqueous solution the environment probably consists of one or two nitrate ions and a number of water molecules completing the coordination number to 9 or 8. Moreover, since donor properties of nitrate ions and water molecules do not differ appreciably, there is little change in the environment upon dissolution which is probably another major reason for the high solubilities of yttrium and lanthanide nitrate hydrates.

GENERAL COMMENTS ON THE PREPARATION OF THIS VOLUME

The literature on the solubilities of Sc, Y and the lanthanide nitrates were covered through 1982. So far as we are aware, the entire literature has been covered in our survey, but some omission may still have occurred. The editors will therefore be grateful to readers who will bring these omissions to their attention so that they may be included in future updates to this volume.

Most of the solubility data reported in the literature are either in mass % units or in units of molality, and the compilers and evaluators have converted all mass % data into mol kg^{-1} units using 1977 IUPAC recommended atomic masses. In several cases such as the ternary systems of high nitric acid content, the conversions to molality led to trivial values: e.g. for nitric acid compositions between 20-99 mass %, the molality of nitric acid is between 10 and 1000 mol kg^{-1}. For cases such as these, conversions to mole fraction units or molality where HNO_3 is taken as part of the solvent would have been preferred. However since this problem was relatively rare, and since molality units were used as the basis for critical evaluation, conversions to mole fraction units were not carried out, and we retained the conversions to mol kg^{-1} even though some of the values were trivial.

Phase diagrams are given for selected multicomponent systems. The basis for selection included clarity of the original diagram, ease of reproduction, and whether in the judgment of the evaluators, they were informative. Thus a number of phase diagrams reported in the original publications are not reproduced in this volume.

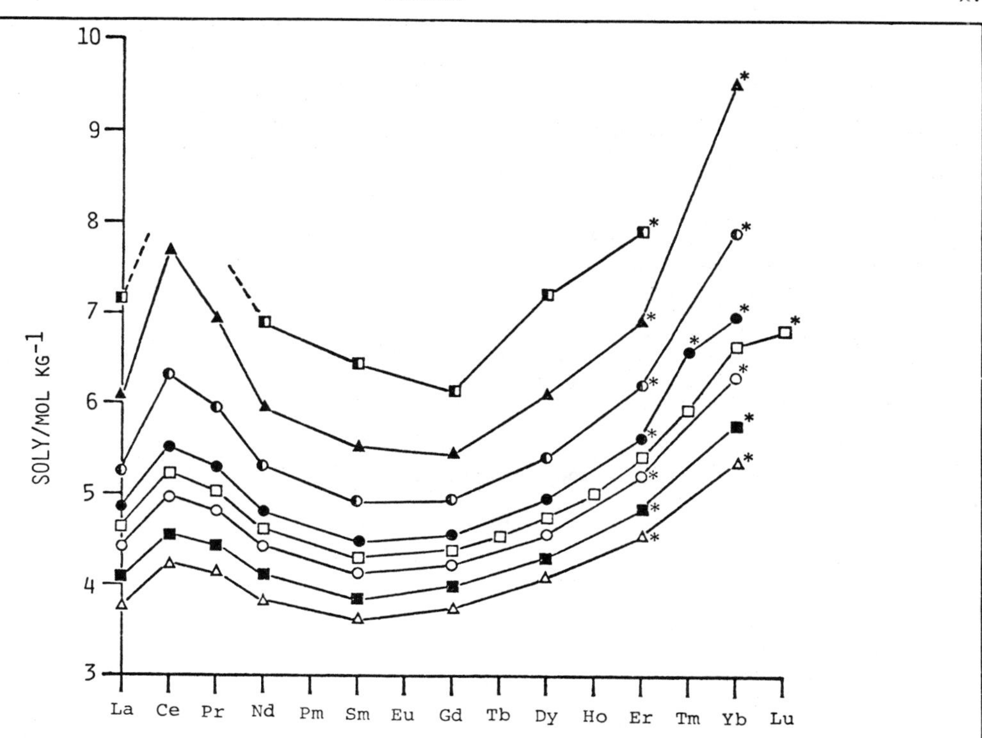

Figure 1. Solubility as a function of atomic number. Solid phase is the hexahydrate except for points with an asterisk (*) which denotes a pentahydrate solid phase. Temperatures are: △ 273.2 K, ■ 283.2 K, ○ 293.2 K, □ 298.2 K, ● 303.2 K, ◐ 313.2 K, ▲ 323.2 K, ◧ 333.2 K.

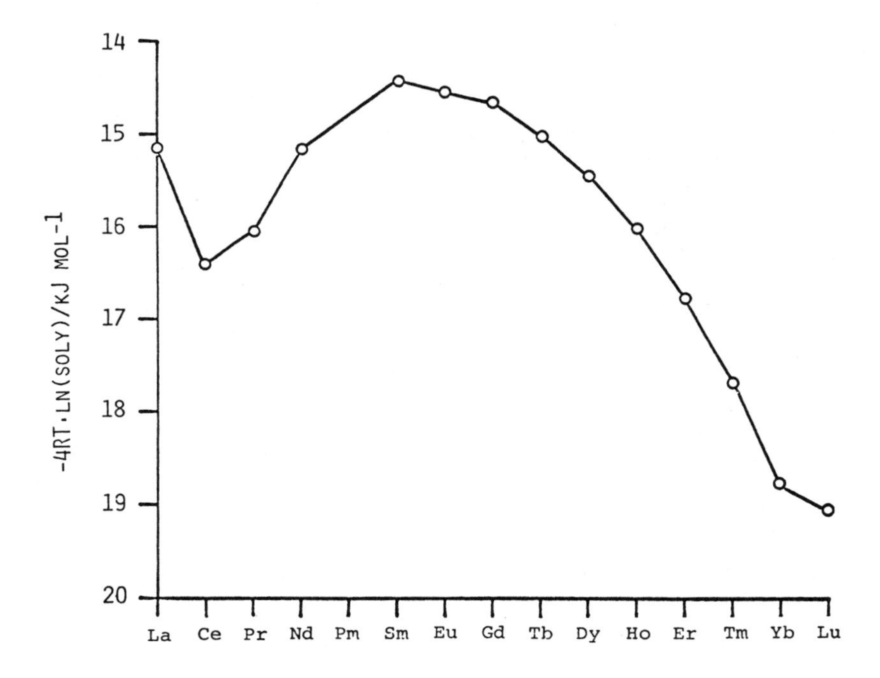

Figure 2. $-4RT\,\ell n\,m_1$ as a function of atomic number (T = 298.2 K).

Chemical Abstract Registry Numbers are given for most of the compounds included in this volume. When possible, compounds which were not assigned Registry Numbers were submitted to Chemical Abstracts Service, and the numbers supplied by Dr. K.L. Loening.

The editors gratefully acknowledge the advice and comments from members of the IUPAC Commission on Solubility Data, and from reviewers who made important suggestions during the preparation of this manuscript. In particular we would like to thank Prof. J. W. Lorimer, Chemistry Department, University of Western Ontario for his many valuable suggestions over the entire course of production of this volume. We also acknowledge the important inputs by Dr. K.L. Loening of Chemical Abstracts Service; Dr. C.F. Coleman, Oak Ridge National Laboratory; Dr. J.A. Rard, Lawrence Livermore Laboratory; Prof. G.H. Brittain, Chemistry Department, Seton Hall University; Prof. F.H. Spedding, Chemistry Department, Iowa State University; and Prof. G. Brunisholz, Chemistry Department, l'Université de Lausanne. Acknowledgements are also due Ms. Brenda Shanholtz, Reference Librarian, ERADCOM Technical Library, and Ms. Susanne Redalje, Assistant Chemistry Librarian, University of Illinois.

The editors would also like to acknowledge the cooperation of VAAP, the Copyright Agency of the USSR, for permission to reproduce phase diagrams directly from their publications.

REFERENCES

1. Holmberg, O. *Z. Anorg. Chem.* 1907, *53*, 83.
2. Rothschild, B.F.; Templeton, C.C.; Hall, N.F. *J. Phys. Colloid Chem.* 1948, *52*, 1006.
3. Medoks, G.V. *Dokl. Akad. Nauk SSSR* 1957, *117*, 421.
4. Williamson, A.T. *Trans. Faraday Soc.* 1944, *40*, 421.
5. Counioux, J.-J.; Tenu, R. *J. Chim. Phys.* 1981, *78*, 816 and 823.
6. Mellor, J.W. *A Comprehensive Treatise on Inorganic and Theoretical Chemistry: Vol. V.* Longmans, Green and Co. London. 1940.
7. Ivernova, V.I.; Tarasova, V.P.; Zolina, Z.K.; Markhasin, G.V.; Sukhodreva, I.M. *Zh. Neorg. Khim.* 1955, *29*, 3145.
8. Eriksson, B.; Larsson, L.O.; Niinisto, L.; Valkonen, J. *Inorg. Chem.* 1980, *19*, 1207.
9. Volodina, G.I.; Rumanova, I.M.; Belov, N.V. *Kristallografiya* 1961, *6*, 919.
10. Fuller, C.C.; Jacobson, R.A. *Cryst. Struct. Commun.* 1975, *5*, 349.
11. Brunisholz, G.; Quinche, J.P.; Kalo, A.M. *Helv. Chim. Acta* 1964, *47*, 14.
12. Mironov, K.E.; Popov, A.P.; Moroz, E.M. *Zh. Neorg. Khim.* 1969, *14*, 320.
13. Eriksson, B.; Larsson, L.D.; Niinisto, L.: Valkonen, J. *Euchem. Conf. Chem. Rare Earths.* Helsinki. 1975.
14. Guillaumont, R.; David, F. *Radiochem. Radioanal. Lett.* 1974, *17*, 25.
15. Spedding, F.G.; Derer, J.L.; Mohs, M.A.; Rard, J.A. *J. Chem. Eng. Data* 1976, *21*, 474.
16. Rard, J.A.; Spedding, F.H. *J. Chem. Eng. Data* 1981, *26*, 391.
17. Grenthe, I. *Acta Chem. Scand.* 1964, *18*, 283.
18. Siekierski, S. *J. Solid State Chem.* 1981, *37*, 279.
19. Siekierski, S. *Radiochem. Radioanal. Lett.* 1981, *48*, 201.

S. Siekierski
Institute for Nuclear Research
Warsaw, Poland

T. Mioduski
Institute for Nuclear Research
Warsaw, Poland

M. Salomon
U.S. Army, ERADCOM
Ft. Monmouth, NJ, USA

February 1983

INTRODUCTION TO THE SOLUBILITY OF SOLIDS IN LIQUIDS

Nature of the Project

The Solubility Data Project (SDP) has as its aim a comprehensive search of the literature for solubilities of gases, liquids, and solids in liquids or solids. Data of suitable precision are compiled on data sheets in a uniform format. The data for each system are evaluated, and where data from different sources agree sufficiently, recommended values are proposed. The evaluation sheets, recommended values, and compiled data sheets are published on consecutive pages.

This series of volumes includes solubilities of solids of all types in liquids of all types.

Definitions

A *mixture* (1,2) describes a gaseous, liquid, or solid phase containing more than one substance, when the substances are all treated in the same way.

A *solution* (1,2) describes a liquid or solid phase containing more than one substance, when for convenience one of the substances, which is called the *solvent* and may itself be a mixture, is treated differently than the other substances, which are called *solutes*. If the sum of the mole fractions of the solutes is small compared to unity, the solution is called a *dilute solution*.

The *solubility* of a substance B is the relative proportion of B (or a substance related chemically to B) in a mixture which is saturated with respect to solid B at a specified temperature and pressure. *Saturated* implies the existence of equilibrium with respect to the processes of dissolution and precipitation; the equilibrium may be stable or metastable. The solubility of a metastable substance is usually greater than that of the corresponding stable substance. (Strictly speaking, it is the activity of the metastable substance that is greater.) Care must be taken to distinguish true metastability from supersaturation, where equilibrium does not exist.

Either point of view, mixture or solution, may be taken in describing solubility. The two points of view find their expression in the quantities used as measures of solubility and in the reference states used for definition of activities and activity coefficients.

The qualifying phrase "substance related chemically to B" requires comment. The composition of the saturated mixture (or solution) can be described in terms of any suitable set of thermodynamic components. Thus, the solubility of a salt hydrate in water is usually given as the relative proportion of anhydrous salt in solution, rather than the relative proportions of hydrated salt and water.

Quantities Used as Measures of Solubility

1. *Mole fraction* of substance B, x_B:

$$x_B = n_B / \sum_{i=1}^{c} n_i \qquad (1)$$

where n_i is the amount of substance of substance i, and c is the number of distinct substances present (often the number of thermodynamic components in the system). *Mole per cent* of B is 100 x_B.

2. *Mass fraction* of substance B, w_B:

$$w_B = m'_B / \sum_{i=1}^{c} m'_i \qquad (2)$$

where m'_i is the mass of substance i. *Mass per cent* of B is 100 w_B. The equivalent terms weight fraction and weight per cent are not used.

3. *Solute mole (mass) fraction* of solute B (3,4):

$$x_{S,B} = n_B / \sum_{i=1}^{c'} n_i = x_B / \sum_{i=1}^{c'} x_i \qquad (3)$$

where the summation is over the solutes only. For the solvent A, $x_{S,A} = x_A$. These quantities are called *Jänecke mole (mass) fractions* in many papers.

4. *Molality* of solute B (1,2) in a solvent A:

$$m_B = n_B/n_A M_A \qquad \text{SI base units: mol kg}^{-1} \qquad (4)$$

where M_A is the molar mass of the solvent.

5. *Concentration* of solute B (1,2) in a solution of volume V:

$$c_B = [B] = n_B/V \qquad \text{SI base units: mol m}^{-3} \qquad (5)$$

The terms molarity and molar are not used.

Mole and mass fractions are appropriate to either the mixture or the solution points of view. The other quantities are appropriate to the solution point of view only. In addition of these quantities, the following are useful in conversions between concentrations and other quantities.

6. *Density*: $\rho = m/V$ SI base units: kg m^{-3} (6)

7. *Relative density*: d; the ratio of the density of a mixture to the density of a reference substance under conditions which must be specified for both (1). The symbol $d_{t'}^{t}$ will be used for the density of a mixture at t°C, 1 atm divided by the density of water at t'°C, 1 atm.

Other quantities will be defined in the prefaces to individual volumes or on specific data sheets.

Thermodynamics of Solubility

The principal aims of the Solubility Data Project are the tabulation and evaluation of: (a) solubilities as defined above; (b) the nature of the saturating solid phase. Thermodynamic analysis of solubility phenomena has two aims: (a) to provide a rational basis for the construction of functions to represent solubility data; (b) to enable thermodynamic quantities to be extracted from solubility data. Both these aims are difficult to achieve in many cases because of a lack of experimental or theoretical information concerning activity coefficients. Where thermodynamic quantities can be found, they are not evaluated critically, since this task would involve critical evaluation of a large body of data that is not directly relevant to solubility. The following discussion is an outline of the principal thermodynamic relations encountered in discussions of solubility. For more extensive discussions and references, see books on thermodynamics, e.g., (5-10).

Activity Coefficients (1)

(a) *Mixtures*. The activity coefficient f_B of a substance B is given by

$$RT\, \ln(f_B x_B) = \mu_B - \mu_B^* \qquad (7)$$

where μ_B is the chemical potential, and μ_B^* is the chemical potential of pure B at the same temperature and pressure. For any substance B in the mixture,

$$\lim_{x_B \to 1} f_B = 1 \qquad (8)$$

(b) *Solutions*.

(i) *Solute substance, B*. The molal activity coefficient γ_B is given by

$$RT\, \ln(\gamma_B m_B) = \mu_B - (\mu_B - RT\, \ln m_B)^{\infty} \qquad (9)$$

where the superscript $^{\infty}$ indicates an infinitely dilute solution. For any solute B,

$$\gamma_B^{\infty} = 1 \qquad (10)$$

Activity coefficients y_B connected with concentration c_B, and $f_{x,B}$ (called the *rational activity coefficient*) connected with mole fraction x_B are defined in analogous ways. The relations among them are (1,9):

$$\gamma_B = x_A f_{x,B} = V_A^*(1 - \sum_s c_s) y_B \qquad (11)$$

or

$$f_{x,B} = (1 + M_A \sum_s m_s)\gamma_B = V_A^* y_B / V_m \qquad (12)$$

or

$$y_B = (V_A + M_A \sum_s m_s V_s)\gamma_B / V_A^* = V_m f_{x,B} / V_A^* \qquad (13)$$

where the summations are over all solutes, V_A^* is the molar volume of the pure solvent, V_i is the partial molar volume of substance i, and V_m is the molar volume of the solution.

For an electrolyte solute $B \equiv C_{\nu+}A_{\nu-}$, the molal activity is replaced by (9)

$$\gamma_B m_B = \gamma_\pm^{\nu} m_B^{\nu} Q^{\nu} \qquad (14)$$

where $\nu = \nu_+ + \nu_-$, $Q = (\nu_+^{\nu+}\nu_-^{\nu-})^{1/\nu}$, and $\gamma_\pm$ is the mean ionic molal activity coefficient. A similar relation holds for the concentration activity $y_B c_B$. For the mol fractional activity,

$$f_{x,B}\, x_B = \nu_+^{\nu_+} \nu_-^{\nu_-} f_\pm^{\nu} x_\pm^{\nu} \qquad (15)$$

The quantities x_+ and x_- are the ionic mole fractions (9), which for a single solute are

$$x_+ = \nu_+ x_B / [1+(\nu-1)x_B]; \qquad x_- = \nu_- x_B / [1+(\nu-1)x_B] \qquad (16)$$

(ii) *Solvent, A:*

The *osmotic coefficient*, ϕ , of a solvent substance A is defined as (1):

$$\phi = (\mu_A^* - \mu_A)/RT\, M_A \sum_s m_s \qquad (17)$$

where μ_A^* is the chemical potential of the pure solvent.

The *rational osmotic coefficient*, ϕ_x, is defined as (1):

$$\phi_x = (\mu_A - \mu_A^*)/RT \ln x_A = \phi M_A \sum_s m_s / \ln(1 + M_A \sum_s m_s) \qquad (18)$$

The activity, a_A, or the activity coefficient f_A is often used for the solvent rather than the osmotic coefficient. The activity coefficient is defined relative to pure A, just as for a mixture.

The Liquid Phase

A general thermodynamic differential equation which gives solubility as a function of temperature, pressure and composition can be derived. The approach is that of Kirkwood and Oppenheim (7). Consider a solid mixture containing c' thermodynamic components i. The Gibbs-Duhem equation for this mixture is:

$$\sum_{i=1}^{c'} x_i'(S_i'dT - V_i'dp + d\mu_i) = 0 \qquad (19)$$

A liquid mixture in equilibrium with this solid phase contains c thermodynamic components i, where, usually, c $\geq$ c'. The Gibbs-Duhem equation for the liquid mixture is:

$$\sum_{i=1}^{c'} x_i(S_i dT - V_i dp + d\mu_i) + \sum_{i=c'+1}^{c} x_i(S_i dT - V_i dp + d\mu_i) = 0 \qquad (20)$$

Eliminate $d\mu_1$ by multiplying (19) by x_1 and (20) x_1'. After some algebra, and use of:

$$d\mu_i = \sum_{j=2}^{c} G_{ij} dx_j - S_i dT + V_i dp \qquad (21)$$

where (7)

$$G_{ij} = (\partial\mu_i/\partial x_j)_{T,P,x_i \neq x_j} \qquad (22)$$

it is found that

$$\sum_{i=2}^{c'} \sum_{j=2}^{c} (x_i' - x_i x_1'/x_1) G_{ij} dx_j - (x_1'/x_1) \sum_{i=c'+1}^{c} \sum_{j=2}^{c} x_i G_{ij} dx_j$$

$$= \sum_{i=1}^{c'} x_i'(H_i - H_i')dT/T - \sum_{i=1}^{c'} x_i'(V_i - V_i')dp \qquad (23)$$

where

$$H_i - H_i' = T(S_i - S_i') \tag{24}$$

is the enthalpy of transfer of component i from the solid to the liquid phase, at a given temperature, pressure and composition, and H_i, S_i, V_i are the partial molar enthalpy, entropy, and volume of component i. Several special cases (all with pressure held constant) will be considered. Other cases will appear in individual evaluations.

(a) Solubility as a function of temperature.
Consider a binary solid compound A_nB in a single solvent A. There is no fundamental thermodynamic distinction between a binary compound of A and B which dissociates completely or partially on melting and a solid mixture of A and B; the binary compound can be regarded as a solid mixture of constant composition. Thus, with c = 2, c' = 1, $x_A' = n/(n+1)$, $x_B' = 1/(n+1)$, eqn (23) becomes

$$(1/x_B - n/x_A)\{1 + (\frac{\partial \ln f_B}{\partial \ln x_B})_{T,P}\}dx_B = (nH_A + H_B - H_{AB}^*)dT/RT^2 \tag{25}$$

where the mole fractional activity coefficient has been introduced. If the mixture is a non-electrolyte, and the activity coefficients are given by the expression for a simple mixture (6):

$$RT \ln f_B = wx_A^2 \tag{26}$$

then it can be shown that, if w is independent of temperature, eqn (25) can be integrated (cf. (5), Chap. XXIII, sect. 5). The enthalpy term becomes

$$nH_A + H_B - H_{AB}^* = \Delta H_{AB} + n(H_A - H_A^*) + (H_B - H_B^*)$$

$$= \Delta H_{AB} + w(nx_B^2 + x_A^2) \tag{27}$$

where ΔH_{AB} is the enthalpy of melting and dissociation of one mole of pure solid A_nB, and H_A^*, H_B^* are the molar enthalpies of pure liquid A and B. The differential equation becomes

$$R\ d\ \ln\{x_B(1-x_B)^n\} = -\Delta H_{AB}\ d(\frac{1}{T}) - w\ d(\frac{x_A^2 + nx_B^2}{T}) \tag{28}$$

Integration from x_B,T to $x_B = 1/(1+n)$, T = T*, the melting point of the pure binary compound, gives:

$$\ln\{x_B(1-x_B)^n\} \simeq \ln\{\frac{n^n}{(1+n)^{n+1}}\} - \{\frac{\Delta H_{AB}^* - T^*\Delta C_p^*}{R}\}(\frac{1}{T} - \frac{1}{T^*}) + \frac{\Delta C_p^*}{R}\ln(\frac{T}{T^*}) - \frac{w}{R}\{\frac{x_A + nx_B}{T} - \frac{n}{(n+1)T^*}\} \tag{29}$$

where ΔC_p^* is the change in molar heat capacity accompanying fusion plus decomposition of the compound at temperature T*, (assumed here to be independent of temperature and composition), and ΔH_{AB}^* is the corresponding change in enthalpy at T = T*. Equation (29) has the general form

$$\ln\{x_B(1-x_B)^n\} = A_1 + A_2/T + A_3\ln T + A_4(x_A^2 + nx_B^2)/T \tag{30}$$

If the solid contains only component B, n = 0 in eqn (29) and (30).
If the infinite dilution standard state is used in eqn (25), eqn (26) becomes

$$RT \ln f_{x,B} = w(x_A^2 - 1) \tag{31}$$

and (27) becomes

$$nH_A + H_B - H_{AB} = (nH_A^* + H_B^\infty - H_{AB}^*) + n(H_A - H_A^*) + (H_B - H_B^\infty) = \Delta H_{AB}^\infty + w(nx_B^2 + x_A^2 - 1) \tag{32}$$

where the first term, ΔH_{AB}^∞, is the enthalpy of melting and dissociation of solid compound A_nB to the infinitely dilute state of solute B in solvent A; H_B^∞ is the partial molar enthalpy of the solute at infinite dilution. Clearly, the integral of eqn (25) will have the same form as eqn (29), with $\Delta H_{AB}^\infty(T^*)$, $\Delta C_p^\infty(T^*)$ replacing ΔH_{AB}^* and ΔC_p^* and $x_A^2 - 1$ replacing x_A^2 in the last term.

If the liquid phase is an aqueous electrolyte solution, and the solid is a salt hydrate, the above treatment needs slight modification. Using rational mean activity coefficients, eqn (25) becomes

$$R\nu(1/x_B-n/x_A)\{1+(\partial\ell nf_\pm/\partial\ell nx_\pm)_{T,P}\}dx_B/\{1+(\nu-1)x_B\}$$

$$= \{\Delta H_{AB}^\infty + n(H_A-H_A^*) + (H_B-H_B^\infty)\}d(1/T) \quad (33)$$

If the terms involving activity coefficients and partial molar enthalpies are negligible, then integration gives (cf. (11)):

$$\ell n\{\frac{x_B^\nu(1-x_B)^n}{1+(\nu-1)x_B{}^{n+\nu}}\} = \ell n\{\frac{n^n}{(n+\nu)^{n+\nu}}\} - \{\frac{\Delta H_{AB}^\infty(T^*)-T^*\Delta C_p^*}{R}\}\left(\frac{1}{T}-\frac{1}{T^*}\right)+\frac{\Delta C_p^*}{R}\ell n(T/T^*) \quad (34)$$

A similar equation (with $\nu=2$ and without the heat capacity terms) has been used to fit solubility data for some $MOH=H_2O$ systems, where M is an alkali metal; the enthalpy values obtained agreed well with known values (11). In many cases, data on activity coefficients (9) and partial molal enthalpies (8,10) in concentrated solution indicate that the terms involving these quantities are not negligible, although they may remain roughly constant along the solubility temperature curve.

The above analysis shows clearly that a rational thermodynamic basis exists for functional representation of solubility-temperature curves in two-component systems, but may be difficult to apply because of lack of experimental or theoretical knowledge of activity coefficients and partial molar enthalpies. Other phenomena which are related ultimately to the stoichiometric activity coefficients and which complicate interpretation include ion pairing, formation of complex ions, and hydrolysis. Similar considerations hold for the variation of solubility with pressure, except that the effects are relatively smaller at the pressures used in many investigations of solubility (5).

(b) *Solubility as a function of composition.*
At constant temperature and pressure, the chemical potential of a saturating solid phase is constant:

$$\mu_{A_nB}^* = \mu_{A_nB}(sln) = n\mu_A + \mu_B \quad (35)$$

$$= (n\mu_A^* + \nu_+\mu_+^\infty+\nu_-\mu_-^\infty) + nRT\ \ell nf_Ax_A$$

$$+ \nu RT\ \ell n\gamma_\pm m_\pm Q_\pm \quad (36)$$

for a salt hydrate A_nB which dissociates to water, (A), and a salt, B, one mole of which ionizes to give ν_+ cations and ν_- anions in a solution in which other substances (ionized or not) may be present. If the saturated solution is sufficiently dilute, $f_A = x_A = 1$, and the quantity K_{s0}^0 in

$$\Delta G^\infty \equiv (\nu_+\mu_+^\infty+\nu_-\mu_-^\infty+n\mu_A^*-\mu_{AB}^*)$$

$$= -RT\ \ell n\ K_{s0}^0$$

$$= -RT\ \ell n\ Q^\nu\gamma_\pm^\nu m_+^{\nu_+}m_-^{\nu_-} \quad (37)$$

is called the *solubility product* of the salt. (It should be noted that it is not customary to extend this definition to hydrated salts, but there is no reason why they should be excluded.) Values of the solubility product are often given on mole fraction or concentration scales. In dilute solutions, the theoretical behaviour of the activity coefficients as a function of ionic strength is often sufficiently well known that reliable extrapolations to infinite dilution can be made, and values of K_{s0}^0 can be determined. In more concentrated solutions, the same problems with activity coefficients that were outlined in the section on variation of solubility with temperature still occur. If these complications do not arise, the solubility of a hydrate salt $C_{\nu_+}A_{\nu_-}\cdot nH_2O$ in the presence of other solutes is given by eqn (36) as

$$\nu\ \ell n\{m_B/m_B(0)\} = -\nu\ell n\{\gamma_\pm/\gamma_\pm(0)\} - n\ \ell n(a_{H_2O}/a_{H_2O}(0)) \quad (38)$$

where a_{H_2O} is the activity of water in the saturated solution, m_B is the molality of the salt in the saturated solution, and (0) indicates absence of other solutes. Similar considerations hold for non-electrolytes.

The Solid Phase

The definition of solubility permits the occurrence of a single solid phase which may be a pure anhydrous compound, a salt hydrate, a non-stoichiometric compound, or a solid mixture (or solid solution, or "mixed crystals"), and may be stable or metastable. As well, any number of solid phases consistent with the requirements of the phase rule may be present. Metastable solid phases are of widespread occurrence, and may appear as polymorphic (or allotropic) forms or crystal solvates whose rate of transition to more stable forms is very slow. Surface heterogeneity may also give rise to metastability, either when one solid precipitates on the surface of another, or if the size of the solid particles is sufficiently small that surface effects become important. In either case, the solid is not in stable equilibrium with the solution. The stability of a solid may also be affected by the atmosphere in which the system is equilibrated.

Many of these phenomena require very careful, and often prolonged, equilibration for their investigation and elimination. A very general analytical method, the "wet residues" method of Schreinemakers (12) (see a text on physical chemistry) is usually used to investigate the composition of solid phases in equilibrium with salt solutions. In principle, the same method can be used with systems of other types. Many other techniques for examination of solids, in particular X-ray, optical, and thermal analysis methods, are used in conjunction with chemical analyses (including the wet residues method).

COMPILATIONS AND EVALUATIONS

The formats for the compilations and critical evaluations have been standardized for all volumes. A brief description of the data sheets has been given in the FOREWORD; additional explanation is given below.

Guide to the Compilations

The format used for the compilations is, for the most part, self-explanatory. The details presented below are those which are not found in the FOREWORD or which are not self-evident.

Components. Each component is listed according to IUPAC name, formula, and Chemical Abstracts (CA) Registry Number. The formula is given either in terms of the IUPAC or Hill (13) system and the choice of formula is governed by what is usual for most current users: i.e. IUPAC for inorganic compounds, and Hill system for organic compounds. Components are ordered according to:

(a) saturating components;
(b) non-saturating components in alphanumerical order;
(c) solvents in alphanumerical order.

The saturating components are arranged in order according to a 18-column, 2-row periodic table:

Columns 1,2: H, groups IA, IIA;
3,12: transition elements (groups IIIB to VIIB, group VIII, groups IB, IIB);
13-18: groups IIIA-VIIA, noble gases.
Row 1: Ce to Lu;
Row 2: Th to the end of the known elements, in order of atomic number.

Salt hydrates are generally not considered to be saturating components since most solubilities are expressed in terms of the anhydrous salt. The existence of hydrates or solvates is carefully noted in the texts, and CA Registry Numbers are given where available, usually in the critical evaluation. Mineralogical names are also quoted, along with their CA Registry Numbers, again usually in the critical evaluation.

Original Measurements. References are abbreviated in the forms given by *Chemical Abstracts Service Source Index (CASSI)*. Names originally in other than Roman alphabets are given as transliterated by *Chemical Abstracts*.

Experimental Values. Data are reported in the units used in the original publication, with the exception that modern *names* for units and quantities are used; e.g., mass per cent for weight per cent; mol dm^{-3} for molar; etc. Both mass and molar values are given. Usually, only one type of value (e.g., mass per cent) is found in the original paper, and the compiler has added the other type of value (e.g., mole per cent) from computer calculations based on 1976 atomic weights (14). Errors in calculations and fitting equations in original papers have been noted and corrected, by computer calculations where necessary.

Method. Source and Purity of Materials. Abbreviations used in *Chemical Abstracts* are often used here to save space.

Estimated Error. If these data were omitted by the original authors, and if relevant information is available, the compilers have attempted to estimate errors from the internal consistency of data and type of apparatus used. Methods used by the compilers for estimating and reporting errors are based on the papers by Ku and Eisenhart (15).

Comments and/or Additional Data. Many compilations include this section which provides short comments relevant to the general nature of the work or additional experimental and thermodynamic data which are judged by the compiler to be of value to the reader.

References. See the above description for Original Measurements.

Guide to the Evaluations

The evaluator's task is to check whether the compiled data are correct, to assess the reliability and quality of the data, to estimate errors where necessary, and to recommend "best" values. The evaluation takes the form of a summary in which all the data supplied by the compiler have been critically reviewed. A brief description of the evaluation sheets is given below.

Components. See the description for the Compilations.

Evaluator. Name and date up to which the literature was checked.

Critical Evaluation

(a) Critical text. The evaluator produces text evaluating *all* the published data for each given system. Thus, in this section the evaluator review the merits or shortcomings of the various data. Only published data are considered; even published data can be considered only if the experimental data permit an assessment of reliability.

(b) Fitting equations. If the use of a smoothing equation is justifiable, the evaluator may provide an equation representing the solubility as a function of the variables reported on all the compilation sheets.

(c) Graphical summary. In addition to (b) above, graphical summaries are often given.

(d) Recommended values. Data are *recommended* if the results of at least two independent groups are available and they are in good agreement, and if the evaluator has no doubt as to the adequacy and reliability of the applied experimental and computational procedures. Data are reported as *tentative* if only one set of measurements is available, or if the evaluator considers some aspect of the computational or experimental method as mildly undesirable but estimates that it should cause only minor errors. Data are considered as *doubtful* if the evaluator considers some aspect of the computational or experimental method as undesirable but still considers the data to have some value in those instances where the order of magnitude of the solubility is needed. Data determined by an inadequate method or under ill-defined conditions are *rejected*. However references to these data are included in the evaluation together with a comment by the evaluator as to the reason for their rejection.

(e) References. All pertinent references are given here. References to those data which, by virtue of their poor precision, have been rejected and not compiled are also listed in this section.

(f) Units. While the original data may be reported in the units used by the investigators, the final recommended values are reported in S.I. units (1,16) when the data can be accurately converted.

References

1. Whiffen, D. H., ed., *Manual of Symbols and Terminology for Physicochemical Quantities and Units*. *Pure Applied Chem.* 1979, *51*, No. 1.
2. McGlashan, M.L. *Physicochemical Quantities and Units*. 2nd ed. Royal Institute of Chemistry. London. 1971.
3. Jänecke, E. *Z. Anorg. Chem.* 1906, *51*, 132.
4. Friedman, H.L. *J. Chem. Phys.* 1960, *32*, 1351.
5. Prigogine, I.; Defay, R. *Chemical Thermodynamics*. D.H. Everett, transl. Longmans, Green. London, New York, Toronto. 1954.
6. Guggenheim, E.A. *Thermodynamics*. North-Holland. Amsterdam. 1959. 4th ed.
7. Kirkwood, J.G.; Oppenheim, I. *Chemical Thermodynamics*. McGraw-Hill, New York, Toronto, London. 1961.
8. Lewis, G.N.; Randall, M. (rev. Pitzer, K.S.; Brewer, L.). *Thermodynamics*. McGraw Hill. New York, Toronto, London. 1961. 2nd ed.
9. Robinson, R.A.; Stokes, R.H. *Electrolyte Solutions*. Butterworths. London. 1959, 2nd ed.
10. Harned, H.S.; Owen, B.B. *The Physical Chemistry of Electrolytic Solutions*. Reinhold. New York. 1958. 3rd ed.
11. Cohen-Adad, R.; Saugier, M.T.; Said, J. *Rev. Chim. Miner.* 1973, *10*, 631.
12. Schreinemakers, F.A.H. *Z. Phys. Chem., stoechiom. Verwandschaftsl.* 1893, *11*, 75.
13. Hill, E.A. *J. Am. Chem. Soc.* 1900, *22*, 478.
14. IUPAC Commission on Atomic Weights. *Pure Appl. Chem.*, 1976, *47*, 75.

15. Ku, H.H., p. 73; Eisenhart, C., p. 69; in Ku, H.H., ed. *Precision Measurement and Calibration*. NBS Special Publication 300. Vol. 1. Washington. 1969.
16. *The International System of Units*. Engl. transl. approved by the BIPM of *Le Système International d'Unités*. H.M.S.O. London. 1970.

R. Cohen-Adad, Villeurbanne, France
J.W. Lorimer, London, Canada
M. Salomon, Fair Haven, New Jersey, U.S.A.

COMPONENTS:	EVALUATOR:
(1) Scandium nitrate; $Sc(NO_3)_3$; [13465-60-6] (2) Water ; H_2O ; [7732-18-5]	Mark Salomon U.S. Army Electronics Technology and Devices Laboratory Fort Monmouth, NJ, USA November , 1982

CRITICAL EVALUATION:

Sc^{3+} is the most extensively hydrolysed ion of the group Y^{3+}, La^{3+}, and the lanthanides. In aqueous solutions the nature of the complex formed by hydrolysis varies with pH, but the major species appear to be (1-5) $ScOH^{2+}$, $Sc_2(OH)_2^{4+}$, $Sc_3(OH)_4^{5+}$, and $Sc_3(OH)_5^{4+}$. Aqueous systems also have a tendency to polymerize (3-5). The degree of hydrolysis of $Sc(NO_3)_3$ is about 6-8 % at ambient temperatures (4), but decreases considerably as the concentration is increased particularly at and above 1 mol dm^{-3} (2). Due to the tendency to hydrolyse, the composition of the solid phases will also depend upon pH as $ScOH(NO_3)_2$ or $Sc(OH)_2NO_3$ tend to precipitate at high pH's (2,4,6). In this critical evaluation only those systems are considered in which saturated $Sc(NO_3)_3$ solutions are in equilibrium with its hydrated solid $Sc(NO_3)_3 \cdot nH_2O$.

This criteria limits the number of publications on the solubility of $Sc(NO_3)_3$ to two (2,7). Only two solid phases have been identified

$Sc(NO_3)_3 \cdot 4H_2O$ [16999-44-3] and $Sc(NO_3)_3 \cdot 2H_2O$ [16999-46-5]

The tetrahydrate can be prepared by crystallization from solutions of high nitric acid concentration (2,8), and the dihydrate can be prepared from the tetrahydrate by heating in air at 323 K or by prolonged dehydration. Both salts are extremely hydroscopic and cannot be completely dehydrated due to the formation of oxide nitrates. Wendlandt (9) states that the hexahydrate was produced by crystallization from concentrated HNO_3, but this result has not been confirmed. Wendlandt also reported thermal studies on $Sc(NO_3)_3 \cdot 6H_2O$, [13759-83-6], and found only decomposition and formation of oxide nitrates. Somewhat similar results were found by Komissarova et al. (8) in their thermal studies of the tetra- and dihydrates. However the latter authors (8) report that the tetrahydrate melts congruently at 323 K, and that the dihydrate melts congruently at 348 K.

An attempt was made to fit the solubility data of Pushkina and Komissarova (2) to the general solubility equation (see INTRODUCTION and reference 10)

$$Y = \ln(m/m_\circ) - nM_2(m - m_\circ) = a + b/(T/K) + c \ln(T/K) \qquad [1]$$

where m is the solubility in mol kg^{-1}, $m_\circ$ an arbitrarily selected reference molality, n is the hydration number, and M_2 is the molar mass of the solvent. Using all six data points from (2) over the temperature range of 273 K to 323 K results in an acceptable fit Y to T/K, but a poor fit with regard to the calculated heat of solution, ΔH_{sln}, and the predicted congruent melting point: i.e. ΔH_{sln} is positive which is opposite to values found for all other saturated lanthanide nitrate solutions, and the congruent melting point could not be calculated due to failure of the calculation to converge. If the solubility at 323 K is neglected, ΔH_{sln} is still positive and an incorrect value of 342 K is obtained for the congruent melting point. A reasonable fit of the data in (2) to eq. [1] is obtained by using only the four data points for the temperatures between 288 K and 313 K. For these four data points it is found that

$$Y = -123.26 + 5178/(T/K) + 18.588 \ln(T/K) \qquad [2]$$

The standard deviation of residuals for the solubility, σ_m, is 0.18 mol kg^{-1}, ΔH_{sln} = -172 kJ mol^{-1}, and the predicted congruent melting point is 325.1 K which is in agreement with the experimental value discussed above. The solubility result of 6.259 mol kg^{-1} for a solution in equilibrium with the dihydrate solid phase (7) seems incorrect because this result suggests that the temperature for the tetrahydrate → dihydrate transition lies well below 303 K which is contrary to expected based on comparison with all other lanthanide nitrates. Thus whereas the solubility data of Pushkina and Komissarova (2) can be designated as *tentative* values, the value for the solubility in the binary system reported by Zholalieva, Sulaimankulov and Ismailov (7) must be designated as *doubtful*.

REFERENCES

1. Baes, C.F.; Mesmer, R.E. *The Hydrolysis of Cations*. Wiley. New York. 1976.
2. Pushkina, G.Ya.; Komissarova, L.N. *Zh. Neorg. Khim.* 1963, *8*, 1498.
3. Komissarova, L.N.; Pushkina, G.Ya.; Ebert, M. *Zh. Neorg. Khim.* 1969, *14*, 2925.
4. Komissarova, L.N. *Zh. Neorg. Khim.* 1980, *25*, 143.
5. Davydov, Yu.P.; Glazacheva, G.I. *Zh. Neorg. Khim.* 1980, *25*, 1462.
6. Mironov, N.N.; Mal'kevich, N.V. *Zh. Neorg. Khim.* 1970, *15*, 599.
7. Zholalieva, Z.; Sulaimankulov, K.; Ismailov, M. *Zh. Neorg. Khim.* 1978, *23*, 860.
8. Komissarova, L.N.; Pushkina, G.Ya.; Spitsyn, V.I. *Zh. Neorg. Khim.* 1963, *8*, 1384.
9. Wendlandt, W.W. *Anal. Chim. Acta* 1956, *15*, 435.
10. Counioux, J.-J.; Tenu, R. *J. Chim. Phys.* 1981, *78*, 816 and 823.

COMPONENTS:

(1) Scandium nitrate; $Sc(NO_3)_3$; [13465-60-6]

(2) Water ; H_2O ; [7732-18-5]

ORIGINAL MEASUREMENTS:

Pushkina, G. Ya.; Komissarova, L.N. *Zh. Neorg. Khim.* 1963, *8*, 1498-504; *Russ. J. Inorg. Chem. Engl. Transl.* 1963, *8*, 777-81

VARIABLES:

Temperature: range $0^{\circ}C$ to $50^{\circ}C$

PREPARED BY:

T. Mioduski, S. Siekierski, M. Salomon

EXPERIMENTAL VALUES:

	Sc_2O_3 [a]	$Sc(NO_3)_3$ [b]		
$t/^{\circ}C$	mass %	mass %	mol kg^{-1}	solid phase
0	16.83	56.37	5.595	$Sc(NO_3)_3 \cdot 4H_2O$
15	18.30	61.30	6.857	"
25	18.62	62.37	7.176	"
30	19.19	64.28	7.791	"
40	20.00	66.99	8.787	"
50	20.19	67.63	9.045	"
60[c]	∞			

a. Original experimental data are mass % Sc_2O_3. The authors also reported calculated values for the solubilities of $Sc(NO_3)_3 \cdot 4H_2O$. These values are not given in this compilation because they are too low by about 0.1% due to the authors' use of old values of atomic weights, and because the solubilities in terms of the *anhydrous* salt are the more relevant quantities.

b. Calculated by the compilers.

c. Authors state that since the melting point of the tetrahydrate is $50^{\circ}C$ (1), the salt is infinitely soluble above this temperature.

AUXILIARY INFORMATION

METHOD/APPARATUS/PROCEDURE:

Isothermal method used. The initial materials were placed in glass vessels and thermostated. For high HNO_3 concentrations, the vessels were sealed with liquid paraffin. Equilibrium was reached within 2-3 days as ascertained by successive analyses. 1-2 g of saturated solution was removed for each analyses with a pipet fitted with a detachable No. 2 or No. 3 Schott filter. 0.5-1 g of solid was also removed for analysis.

Scandium was determined gravimetrically by precipitation as the hydroxide and ignition to the oxide. Nitrogen was determined by Devarda's method, and water was calculated from the loss in weight upon heating to $900^{\circ}C$. The results given in the above table are means of three determinations.

SOURCE AND PURITY OF MATERIALS:

$Sc(NO_3)_3 \cdot 4H_2O$ was crystallized from nitric acid solution containing at least 44.26 mass % N_2O_5. Chemical analysis resulted in the following (mean of 3 detns): Sc 14.84 mass %, NO_3 61.33 mass %. Impurities in the salt stated not to exceed 0.01 %.

A.R. grade conc HNO_3 used (sp. gr. 1.35) or 100 % HNO_3 used obtained by distn from a mixt of nitric and sulfuric acids.

ESTIMATED ERROR:

Soly: reproducibility ± 1% or better (compilers).

Temp: accuracy ± 0.1 K (authors).

REFERENCES:

1. Komissarova, L.N.; Pushkina, G. Ya.; Spitsyn, V.I. *Zh. Neorg. Khim.* 1963, *8*, 1384; *Russ. J. Inorg. Chem. Engl. Transl.* 1963, *8*, 719.

COMPONENTS:

(1) Scandium nitrate; $Sc(NO_3)_3$; [13465-60-6]

(2) Nitric acid; HNO_3; [7697-37-2]

(3) Water; H_2O; [7732-18-5]

ORIGINAL MEASUREMENTS:

Pushkina, G. Ya.; Komissarova, L.N. *Zh. Neorg. Khim.* 1963, *8*, 1498-504; *Russ. J. Inorg. Chem. Engl. Transl.* 1963, *8*, 777-81.

VARIABLES:

Composition at 25°C

PREPARED BY:

Mark Salomon

EXPERIMENTAL VALUES:

Solubility of $Sc(NO_3)_3$ in aqueous HNO_3 solutions at 25°C [a]

Sc_2O_3 mass %	N_2O_5 mass %	H_2O mass %	Sc_2O_3/N_2O_5 mole ratio	$Sc(NO_3)_3$ [b] mass %	$Sc(NO_3)_3$ [b] mol kg^{-1}	HNO_3 [b,c] mass %	HNO_3 [b,c] mol kg^{-1}
18.62	44.20	37.18	1/3.03	62.37 [d]	7.176	0.0 [d]	———
18.51	44.40	37.10	1/3.07	62.00	7.235	0.90	0.385
18.12	45.30	36.58	1/3.20	60.64	7.183	2.73	1.184
17.52	46.30	36.18	1/3.38	58.68	7.022	5.14	2.255
17.00	46.50	36.50	1/3.50	56.94	6.754	6.56	2.852
16.75	46.83	36.42	1/3.57	56.11	6.670	7.47	3.255
16.45	47.34	36.24	1/3.68	55.10	6.583	8.66	3.792
15.67	48.65	35.68	1/3.95	52.49	6.369	11.83	5.262
13.94	50.02	36.04	1/4.59	46.69	5.609	17.27	7.605
12.85	52.48	34.67	1/5.24	43.04	5.375	22.29	10.203
12.40	55.10	32.50	1/5.67	41.53	5.533	25.97	12.681
11.10	56.52	32.38	1/6.50	37.18	4.971	30.44	14.919
10.78	57.70	31.52	1/6.85	36.11	4.960	32.37	16.298
9.36	61.15	29.49	1/8.35	31.35	4.603	39.16	21.074
8.68	65.20	25.12	1/9.57	29.07	5.010	45.81	28.941
7.71	69.10	23.19	1/11.4	25.83	4.822	50.98	34.887
7.04	71.22	22.24	1/12.8	23.58	4.590	54.18	38.661

a. Solid phase is the tetrahydrate, $Sc(NO_3)_3 \cdot 4H_2O$. b. Compiler's calculations.

c. Calculated from mass % HNO_3 = 100 - mass % $Sc(NO_3)_3$ - mass % H_2O.

d. Authors state this data point to be the soly of $Sc(NO_3)_3$ in water: i.e., mass % HNO_3 = 0. However, using the above equation, the compiler calculates mass % HNO_3 = 0.45 for this point: presumably this is due to experimental errors in the detn of Sc_2O_3 and H_2O contents.

AUXILIARY INFORMATION

METHOD/APPARATUS/PROCEDURE:

Isothermal method used. The initial materials were placed in glass vessels and thermostated. For high HNO_3 concentrations, the vessels were sealed with liquid paraffin. Equilibrium was reached within 4-5 days as ascertained by successive analyses. 1-2 g of saturated solution was removed for each analysis with a pipet fitted with a detachable No. 2 or No. 3 Schott filter. 0.5-1 g of solid was also removed for analysis.

Scandium was determined gravimetrically by precipitation as the hydroxide and ignition to oxide. Nitrogen was determined by Devarda's method, and water was calculated from the loss in weight upon heating to 900°C. The results given in the above table are means of two or three determinations.

SOURCE AND PURITY OF MATERIALS:

$Sc(NO_3)_3 \cdot 4H_2O$ was crystallized from nitric acid solution containing at least 44.26 mass % N_2O_5. Chemical analysis resulted in the following (mean of 3 detns): Sc 14.84 mass %, NO_3 61.33 mass %. Impurities in the salt stated not to exceed 0.01 %.

A.R. grade conc HNO_3 used (sp. gr. 1.35) or 100% HNO_3 used obtained by distn from a mixt of nitric and sulfuric acids.

ESTIMATED ERROR:

Soly: reproducibility ± 1 % or better (compiler).

Temp: accuracy ± 0.1 K (authors).

REFERENCES:

COMPONENTS:

(1) Scandium nitrate; $Sc(NO_3)_3$; [13465-60-6]

(2) Urea; CH_4N_2O; [57-13-6]

(3) Water; H_2O; [7732-18-5]

ORIGINAL MEASUREMENTS:

Zholaheva, Z.; Sulaimankulov, K.; Ismailov, M. *Zh. Neorg. Khim.* 1978, *23*, 860-1; *Russ. J. Inorganic Chem. Engl. Transl.* 1978, *23*, 477-8.

VARIABLES:

Composition at 30°C

PREPARED BY:

Mark Salomon

EXPERIMENTAL VALUES:

$Sc(NO_3)_3 - CO(NH_2)_2 - H_2O$ system at 30°C

Compostion of saturated solutions[a]

$Sc(NO_3)_3$ mass %	$Sc(NO_3)_3$ mol kg^{-1}	$CO(NH_2)_2$ mass %	$CO(NH_2)_2$ mol kg^{-1}	nature of the solid phase
---	---	57.5	22.53	$CO(NH_2)_2$
4.36	0.517	59.11	26.944	"
8.54	1.113	58.23	29.178	"
10.82	1.573	59.39	33.196	"
19.09	4.219	61.32	52.121	"
21.99	6.068	62.32	66.138	"
21.60	5.911	62.58	65.868	$CO(NH_2)_2$ + $Sc(NO_3)_3 \cdot 7CO(NH_2)_2$
23.84	7.490	62.38	75.377	$Sc(NO_3)_3 \cdot 7CO(NH_2)_2$
22.65	5.597	59.83	56.863	"
29.95	5.978	48.36	37.125	"
31.71	6.591	47.46	37.938	"
35.98	8.380	45.43	40.692	"
35.49	7.718	44.60	37.300	$Sc(NO_3)_3 \cdot 7CO(NH_2)_2$ + $Sc(NO_3)_3 \cdot 4CO(NH_2)_2$
34.95	7.285	44.28	35.499	$Sc(NO_3)_3 \cdot 4CO(NH_2)_2$
35.01	6.192	40.51	27.555	"
37.31	5.004	30.41	15.687	"
48.12	6.952	21.91	12.173	"

continued.....

AUXILIARY INFORMATION

METHOD/APPARATUS/PROCEDURE:

Isothermal method used. Solutions had high viscosities and required 12-15 hours to reach equilibrium. The liquid phase was separated from the solid residue by filtering with a Schott No. 3 filter. The scandium ion was determined with Trilon B (disodium salt of ethylenediamine tetra-acetic acid). No other information given.

SOURCE AND PURITY OF MATERIALS:

No information given.

ESTIMATED ERROR:

Nothing specified.

REFERENCES:

COMPONENTS:	ORIGINAL MEASUREMENTS:
(1) Scandium nitrate; $Sc(NO_3)_3$; [13465-60-6]	Zholaheva, Z; Sulaimankulov, K; Ismailov, M. *Zh. Neorg. Khim.* 1978, *23*, 860 - 1;
(2) Urea; CH_4N_2O; [57-13-6]	*Russ. J. Inorganic Chem. Engl. Transl.* 1978, *23*, 477 - 8.
(3) Water ; H_2O ; [7732-18-5]	

EXPERIMENTAL VALUES: (continued)

$Sc(NO_3)_3$		$CO(NH_2)_2$		
mass %	mol kg^{-1}	mass %	mol kg^{-1}	nature of the solid phase
47.52	6.726	21.89	11.915	$Sc(NO_3)_3 \cdot 4CO(NH_2)_2 + Sc(NO_3)_3 \cdot CO(NH_2)_2 \cdot 3H_2O$
51.34	7.226	17.90	9.690	$Sc(NO_3)_3 \cdot CO(NH_2)_2 \cdot 3H_2O$
52.07	7.173	16.50	8.741	"
54.04	7.114	13.07	6.617	"
59.43	8.773	11.24	6.381	$Sc(NO_3)_3 \cdot CO(NH_2)_2 \cdot 3H_2O + Sc(NO_3)_3 \cdot 2H_2O$
60.32	9.112	11.02	6.403	$Sc(NO_3)_3 \cdot 2H_2O$
58.44	6.859	4.67	2.108	"
59.33	6.638	1.97	0.848	"
59.11	6.259	---	---	"

a. Molalities calculated by the compiler.

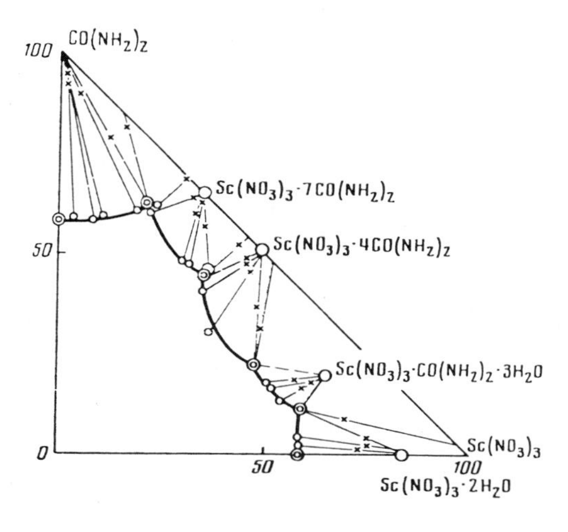

Solubility diagram for the $Sc(NO_3)_3$ - $CO(NH_2)_2$ - H_2O system at 30°C
Concentration units are mass %.

COMPONENTS:

(1) Scandium nitrate; $Sc(NO_3)_3$; [13465-60-6]

(2) 3-Methyl-1-butanol (isoamyl alcohol, isopentyl alcohol); $C_5H_{12}O$; [123-51-3]

(3) Water ; H_2O ; [7732-18-5]

ORIGINAL MEASUREMENTS:

Pushkina, G. Ya.; Komissarova, L.N. *Zh. Neorg. Khim.* 1963, *8*, 1498-504; *Russ. J. Inorg. Chem. Engl. Transl.* 1963, *8*, 777-81

VARIABLES:

One temperature: 25°C

PREPARED BY:

T. Mioduski

EXPERIMENTAL VALUES:

Composition of the saturated solution at 25°C

Sc_2O_3	Sc	NO_3	H_2O [a]	$(CH_3)_2CHCH_2CH_2OH$ [a]	$Sc(NO_3)_3$ [a]	
mass %	mass %	mass %	mass %	mass %	mass %	mol kg^{-1}
14.81	9.67	39.10	15.48	34.91	49.61	4.262

a. Calculated by the compiler based on Sc_2O_3 mass %, and using 1977 IUPAC recommended atomic weights.

AUXILIARY INFORMATION

METHOD/APPARATUS/PROCEDURE:

Isothermal method used. The initial materials were placed in glass vessels and thermostated. Equilibrium was reached within 48 days as ascertained by successive analyses. 0.1 - 0.5 gram samples were removed for analyses using a No. 2 or No. 3 Schott filter fitted to a pipet. 0.5-1 g of solid was also removed for analyses.

Scandium was determined gravimetrically as the oxide. The organic material was first destroyed by heating with HNO_3 to dryness; the residue was then dissolved in HCl, $Sc(OH)_3$ precipitated and ignited to the oxide. Nitrogen was determined by Devarda's method, and water was calculated from the loss in weight upon heating to 900°C. The results given above are the mean of 3-4 determinations. The Sc_2O_3:N_2O_5 molar ratio in the liquid and solid phases is constant at approximately 1:3.

SOURCE AND PURITY OF MATERIALS:

$Sc(NO_3)_3 \cdot 4H_2O$ was crystallized from HNO_3 sln containing at least 44.26 mass % N_2O_5. Chemical analyses resulted in the following (mean of 3 detns): Sc 14.84 mass %, NO_3 61.33 mass %. Impurities in the salt stated not to exceed 0.01%. Prior to use the salt was dried in a desiccator over H_2SO_4 to remove hydroscopic water.

Source of the organic solvent not specified, but authors state that its purity was checked optically.

ESTIMATED ERROR:

Soly: reproducibility around ± 1% (compilers).

Temp: accuracy ± 0.1 K (authors).

REFERENCES:

COMPONENTS:

(1) Scandium nitrate; $Sc(NO_3)_3$; [13465-60-6]

(2) 1-Heptanol; $C_7H_{16}O$; [111-70-6]

(3) Water ; H_2O ; [7732-18-5]

ORIGINAL MEASUREMENTS:

Pushkina, G. Ya.; Komissarova, L.N. *Zh. Neorg. Kim.* 1963, *8*, 1498-504; *Russ. J. Inorg. Chem. Engl. Transl.* 1963, *8*, 777-81

VARIABLES:

One temperature: 25°C

PREPARED BY:

T. Mioduski

EXPERIMENTAL VALUES:

Composition of the saturated solution at 25°C

Sc_2O_3	Sc	NO_3	H_2O [a]	$CH_3(CH_2)_5CH(OH)CH_3$ [a]	$Sc(NO_3)_3$ [a]	
mass %	mass %	mass %	mass %	mass %	mass %	mol kg^{-1}
9.83	6.41	25.94	10.27	56.80	32.93	2.125

a. Calculated by the compiler based on Sc_2O_3 mass %, and using 1977 JUPAC recommended atomic weights.

AUXILIARY INFORMATION

METHOD/APPARATUS/PROCEDURE:

Isothermal method used. The initial materials were placed in glass vessels and thermostated. Equilibrium was reached within 48 days as ascertained by successive analyses. 0.1 - 0.5 gram samples were removed for analyses using a No. 2 or No. 3 Schott filter fitted to a pipet. 0.5-1 g of solid was also removed for analyses.

Scandium was determined gravimetrically as the oxide. The organic material was first destroyed by heating with HNO_3 to dryness; the residue was then dissolved in HCl, $Sc(OH)_3$ precipitated and ignited to the oxide. Nitrogen was determined by Devarda's method, and water was calculated from the loss in weight upon heating to 900°C. The results given above are the mean of 3-4 determinations. The $Sc_2O_3:N_2O_5$ molar ratio in the liquid and solid phases is constant at approximately 1:3.

SOURCE AND PURITY OF MATERIALS:

$Sc(NO_3)_3\cdot4H_2O$ was crystallized from HNO_3 sln containing at least 44.26 mass % N_2O_5. Chemical analyses resulted in the following (mean of 3 detns): Sc 14.84 mass %, NO_3 61.33 mass %. Impurities in the salt stated not to exceed 0.01 %. Prior to use the salt was dried in a desiccator over H_2SO_4 to remove hydroscopic water.

Source of the organic solvent not specified, but authors state that its purity was checked optically.

ESTIMATED ERROR:

Soly: reproducibility around ± 1% (compilers).

Temp: accuracy ± 0.1 K (authors).

REFERENCES:

COMPONENTS:

(1) Scandium nitrate; $Sc(NO_3)_3$; [13465-60-6]

(2) 2-Octanol (sec caprylic alcohol); $C_8H_{18}O$; [123-96-6]

(3) Water ; H_2O ; [7732-18-5]

ORIGINAL MEASUREMENTS:

Pushkina, G. Ya.; Komissarova, L.N. *Zh. Neorg. Khim.* 1963, *8*, 1498-504; *Russ. J. Inorg. Chem. Engl. Transl.* 1963, *8*, 777-81.

VARIABLES:

One temperature: 25°C

PREPARED BY:

T. Mioduski and M. Salomon

EXPERIMENTAL VALUES:

Composition of saturated solution at 25°C

Sc_2O_3	Sc	NO_3	H_2O [a]	$CH_3(CH_2)_5CH(OH)CH_3$ [a]	$Sc(NO_3)_3$ [a]	
mass %	mass %	mass %	mass %	mass %	mass %	mol kg^{-1}
9.06	5.91	24.30	9.47	60.18	30.35	1.886

a. Calculated by the compilers based on Sc_2O_3 mass %, and using 1977 IUPAC recommended atomic weights.

AUXILIARY INFORMATION

METHOD/APPARATUS/PROCEDURE:

Isothermal method used. The initial materials were placed in glass vessels and thermostated. Equilibrium was reached within 48 days as ascertained by successive analyses. 0.1 - 0.5 gram samples were removed for analyses using a No. 2 or No. 3 Schott filter fitted to a pipet. 0.5-1 g of solid was also removed for analyses.

Scandium was determined gravimetrically as the oxide. The organic material was first destroyed by heating HNO_3 to dryness; the residue was then dissolved in HCl, $Sc(OH)_3$ precipitated and ignited to the oxide. Nitrogen was determined by Devarda's method, and water was calculated from the loss in weight upon heating to 900°C. The results given above are the mean of 3-4 determinations. The Sc_2O_3:N_2O_5 molar ratio in the liquid and solid phases is constant at approximately 1:3.

SOURCE AND PURITY OF MATERIALS:

$Sc(NO_3)_3 \cdot 4H_2O$ was crystallized from HNO_3 sln containing at least 44.26 mass % N_2O_5. Chemical analyses resulted in the following (mean of 3 detns): Sc 14.84 mass %, NO_3 61.33 mass %. Impurities in the salt stated not to exceed 0.01 %. Prior to use the salt was dried in a desiccator over H_2SO_4 to remove hydroscopic water.

Source of the organic solvent not specified, but authors state that its purity was checked optically.

ESTIMATED ERROR:

Soly: reproducibility around ± 1% (compilers).

Temp: accuracy ± 0.1 K (authors).

REFERENCES:

COMPONENTS:

(1) Scandium nitrate; $Sc(NO_3)_3$; [13465-60-6]

(2) 2,2'-Oxybis-propane (isopropyl ether); $C_6H_{14}O$; [108-20-3]

(3) Water ; H_2O ; [7732-18-5]

ORIGINAL MEASUREMENTS:

Pushkina, G. Ya.; Komissarova, L.N. *Zh. Neorg. Khim.* 1963, *8*, 1498-504; *Russ. J. Inorg. Chem. Engl. Transl.* 1963, *8*, 777-81.

VARIABLES:

One temperature: 25°C

PREPARED BY:

T. Mioduski

EXPERIMENTAL VALUES:

The solution was very viscous and very slow in settling. Because of this, equilibrium could not be established. Analysis of this liquid phase gave 3.1 mass % Sc_2O_3.

This corresponds to a $Sc(NO_3)_3$ concentration of 10.4 mass % or 0.50 mol kg^{-1} (compiler). The compositions of water and $[(CH_3)_2CH]_2O$ are, respectively, 3.2 mass % and 86.4 mass %.

AUXILIARY INFORMATION

METHOD/APPARATUS/PROCEDURE:

Isothermal method used. The initial materials were placed in glass vessels and thermostated. Equilibrium was reached within 48 days as ascertained by successive analyses. 0.1 - 0.5 gram samples were removed for analyses using a No. 2 or No. 3 Schott filter to a pipet. 0.5-1 g of solid was also removed for analyses.

Scandium was determined gravimetrically as the oxide. The organic material was first destroyed by heating with HNO_3 to dryness; the residue was then dissolved in HCl, $Sc(OH)_3$ precipitated and ignited to the oxide. Nitrogen was determined by Devarda's method, and water was calculated from the loss in weight upon heating to 900°C. The results given above are the mean 3-4 determinations. The $Sc_2O_3:N_2O_5$ molar ratio in the liquid and solid phases is constant at approximately 1:3.

SOURCE AND PURITY OF MATERIALS:

$Sc(NO_3)_3 \cdot 4H_2O$ was crystallized from HNO_3 sln containing at least 44.26 mass % N_2O_5. Chemical analyses resulted in the following (mean of 3 detns): Sc 14.84 mass %, NO_3 61.33 mass %. Impurities in the salt stated not to exceed 0.01 %. Prior to use the salt was dried in a desiccator over H_2SO_4 to remove hydroscopic water.

Source of the organic solvent not specified but authors state that its purity was checked optically.

ESTIMATED ERROR:

Soly: reproducibility around ± 1% (compilers).

Temp: accuracy ± 0.1 K (authors).

REFERENCES:

COMPONENTS:

(1) Scandium nitrate; $Sc(NO_3)_3$; [13465-60-6]

(2) 1,1'-Oxybis-butane (di-n-butyl ether); $C_8H_{18}O$; [142-96-1]

(3) Water ; H_2O ; [7732-18-5]

ORIGINAL MEASUREMENTS:

Pushkina, G. Ya.; Komissarova, L.N. *Zh. Neorg. Khim.* 1963, *8*, 1498-504; *Russ. J. Inorg. Chem. Engl. Transl.* 1963, *8*, 777-81

VARIABLES:

One temperature: 25°C

PREPARED BY:

T. Mioduski

EXPERIMENTAL VALUES:

Composition of the saturated solution at 25°C

Sc_2O_3	Sc	NO_3	H_2O[a]	$[CH_3(CH_2)_3]_2O$[a]	$Sc(NO_3)_3$[a]	
mass %	mass %	mass %	mass %	mass %	mass %	mol kg^{-1}
4.91	3.20	12.89	5.13	78.42	16.45	0.852

a. Calculated by the compiler based on Sc_2O_3 mass %, and using 1977 IUPAC recommended atomic weights.

AUXILIARY INFORMATION

METHOD/APPARATUS/PROCEDURE:

Isothermal method used. The initial materials were placed in glass vessels and thermostated. Equilibrium was reached within 48 days as ascertained by successive analyses. 0.1 - 0.5 gram samples were removed for analyses using a No. 2 or No. 3 Schott filter fitted to a pipet. 0.5-1 g of solid was also removed for analyses.

Scandium was determined gravimetrically as the oxide. The organic material was first destroyed by heating with HNO_3 to dryness; the residue was then dissolved in HCl, $Sc(OH)_3$ precipitated and ignited to the oxide. Nitrogen was determined by Devarda's method, and water was calculated from the loss in weight upon heating to 900°C. The results given above are the mean of 3-4 determinations. The Sc_2O_3:N_2O_5 molar ratio in the liquid and solid phases is constant at approximately 1:3.

SOURCE AND PURITY OF MATERIALS:

$Sc(NO_3)_3 \cdot 4H_2O$ was crystallized from HNO_3 sln containing at least 44.26 mass % N_2O_5. Chemical analyses resulted in the following (mean of 3 detns): Sc 14.84 mass %, NO_3 61.33 mass %. Impurities in the salt stated not to exceed 0.01 %. Prior to use the salt was dried in a desiccator over H_2SO_4 to remove hydroscopic water.

Source of the organic solvent not specified, but authors state that its purity was checked optically.

ESTIMATED ERROR:

Soly: reproducibility around ± 1% (compilers).

Temp: accuracy ± 0.1 K (authors).

REFERENCES:

COMPONENTS:

(1) Scandium nitrate; $Sc(NO_3)_3$; [13465-60-6]

(2) 2-Pentanone (methyl n-propyl ketone); $C_5H_{10}O$; [107-87-9]

(3) Water ; H_2O ; [7732-18-5]

ORIGINAL MEASUREMENTS:

Pushkina, G. Ya.; Komissarova, L.N. *Zh. Neorg. Khim.* 1963, *8*, 1498-504; *Russ. J. Inorg. Chem. Engl. Transl.* 1963, *8*, 777-81.

VARIABLES:

One temperature: 25°C

PREPARED BY:

T. Mioduski

EXPERIMENTAL VALUES:

Composition of the saturated solution at 25°C

Sc_2O_3	Sc	NO_3	H_2O[a]	$CH_3CH_2CH_2COCH_3$[a]	$Sc(NO_3)_3$[a]	
mass %	mass %	mass %	mass %	mass %	mass %	mol kg^{-1}
18.75	12.25	49.40	19.59	17.61	62.80	7.311

a. Calculated by the compiler based on Sc_2O_3 mass %, and using 1977 IUPAC recommended atomic weights.

AUXILIARY INFORMATION

METHOD/APPARATUS/PROCEDURE:

Isothermal method used. The initial materials were placed in glass vessels and thermostated. Equilibrium was reached within 48 days as ascertained by successive analyses. 0.1 - 0.5 gram samples were removed for analyses using a No. 2 or No. 3 Schott filter fitted to a pipet. 0.5-1 g of solid was also removed for analyses.

Scandium was determined gravimetrically as the oxide. The organic material was first destroyed by heating with HNO_3 to dryness; the residue was then dissolved in HCl, $Sc(OH)_3$ precipitated and ignited to the oxide. Nitrogen was determined by Devarda's method, and water was calculated from the loss in weight upon heating 900°C. The results given above are the mean of 3-4 determinations. The Sc_2O_3:N_2O_5 molar ratio in the liquid and solid phases is constant at approximately 1:3.

SOURCE AND PURITY OF MATERIALS:

$Sc(NO_3)_3 \cdot 4H_2O$ was crystallized from HNO_3 sln containing at least 44.26 mass % N_2O_5. Chemical analyses resulted in the following (mean of 3 detn): Sc 14.84 mass %, NO_3 61.33 mass %. Impurities in the salt stated not to exceed 0.01%. Prior to use the salt was dried in a desiccator over H_2SO_4 to remove hydroscopic water.

Source of the organic solvent not specified, but authors state that its purity was checked optically, The ketone was dried prior to use.

ESTIMATED ERROR:

Soly: reproducibility around ± 1% (compilers).

Temp: accuracy ± 0.1 K (authors).

REFERENCES:

COMPONENTS:

(1) Scandium nitrate: $Sc(NO_3)_3$; [13465-60-6]

(2) 2-Hexanone (methyl butyl ketone); $C_6H_{12}O$; [591-78-6]

(3) Water ; H_2O ; [7732-18-5]

ORIGINAL MEASUREMENTS:

Pushkina, G. Ya.; Komissarova, L.N. *Zh. Neorg. Khim.* 1963, *8*, 1498-504; *Russ. J. Inorg. Chem. Engl. Transl.* 1963, *8*, 777-81.

VARIABLES:

One temperature: 25°C

PREPARED BY:

T. Mioduski

EXPERIMENTAL VALUES:

Composition of the saturated solution at 25°C

Sc_2O_3	Sc	NO_3	H_2O[a]	$CH_3(CH_2)_3COCH_3$[a]	$Sc(NO_3)_3$[a]	
mass %	mass %	mass %	mass %	mass %	mass %	mol kg^{-1}
16.89	11.01	47.25	17.65	25.78	56.57	5.639

a. Calculated by the compiler based on Sc_2O_3 mass %, and using 1977 IUPAC recommended atomic weights.

AUXILIARY INFORMATION

METHOD/APPARATUS/PROCEDURE:

Isothermal method used. The initial materials were placed in glass vessels and thermostated. Equilibrium was reached within 48 days as ascertained by successive analyses. 0.1 - 0.5 gram samples were removed for analyses using a No. 2 or No. 3 Schott filter fitted to a pipet. 0.5-1 g of solid was also removed for analyses.

Scandium was determined gravimetrically as the oxide. The organic material was first destroyed by heating with HNO_3 to dryness; the residue was then dissolved in HCl, $Sc(OH)_3$ precipitated and ignited to the oxide. Nitrogen was determined by Devarda's method, and water was calculated from the loss in weight upon heating to 900°C. The results given above are the mean of 3-4 determinations. The Sc_2O_3:N_2O_5 molar ratio in the liquid and solid phases is constant at approximately 1:3.

SOURCE AND PURITY OF MATERIALS:

$Sc(NO_3)_3 \cdot 4H_2O$ was crystallized from HNO_3 sln containing at least 44.26 mass % N_2O_5. Chemical analyses resulted in the following (mean of 3 detns): Sc 14.84 mass %, NO_3 61.33 mass %. Impurities in the salt stated not to exceed 0.01 %. Prior to use the salt was dried in a desiccator over H_2SO_4 to remove hydroscopic water.

Source of the organic solvent not specified, but authors state that its purity was checked optically. The ketone was dried prior to use.

ESTIMATED ERROR:

Soly: reproducibility around ± 1% (compilers).

Temp: accuracy ± 0.1 K (authors).

REFERENCES:

COMPONENTS:

(1) Scandium nitrate; $Sc(NO_3)_3$; [13465-60-6]

(2) Acetophenone (methyl phenyl ketone); C_8H_8O; [98-86-2]

(3) Water ; H_2O ; [7732-18-5]

ORIGINAL MEASUREMENTS:

Pushkina, G. Ya.; Komissarova, L.N. *Zh. Neorg. Khim.* 1963, *8*, 1498-504; *Russ. J. Inorg. Chem. Engl. Transl.* 1963, *8*, 777-81.

VARIABLES:

One temperature: 25°C

PREPARED BY:

T. Mioduski and M. Salomon

EXPERIMENTAL VALUES:

The solution was very viscous and very slow in settling. Because of this, equilibrium could not be established. Analysis of this liquid phase gave 9.8 mass % Sc_2O_3.

This corresponds to a $Sc(NO_3)_3$ concentration of 32.8 mass % or 2.12 mol kg^{-1} (compilers). The compositions of water and $C_6H_5COCH_3$ were calculated by the compilers and are, respectively, 10.2 mass % and 56.9 mass %.

AUXILIARY INFORMATION

METHOD/APPARATUS/PROCEDURE:

Isothermal method used. The initial materials were placed in glass vessels and thermostated. Equilibrium was reached within 48 days as ascertained by successive analyses. 0.1 - 0.5 gram samples were removed for analyses using a No. 2 or No. 3 Schott filter fitted to a pipet. 0.5-1 g of solid was also removed for analyses.

Scandium was determined gravimetrically as the oxide. The organic material was first destroyed by heating with HNO_3 to dryness; the residue was then dissolved in HCl, $Sc(OH)_3$ precipitated and ignited to the oxide. Nitrogen was determined by Devarda's method, and water was calculated from the loss in weight upon heating to 900°C. The results given above are the mean of 3-4 determinations. The Sc_2O_3:N_2O_5 molar ratio in the liquid and solid phase is constant at approximately 1:3.

SOURCE AND PURITY OF MATERIALS:

$Sc(NO_3)_3 \cdot 4H_2O$ was crystallized from HNO_3 sln containing at least 44.26 mass % N_2O_5. Chemical analyses resulted in the following (mean of 3 detns): Sc 14.84 mass %, NO_3 61.33 mass %. Impurities in the salt stated not to exceed 0.01 %. Prior to use the salt was dried in a desiccator over H_2SO_4 to remove hydroscopic water.

Source of the organic solvent not specified, but authors state that its purity was checked optically. The ketone was dried prior to use.

ESTIMATED ERROR:

Soly: reproducibility around ± 1% (compilers).

Temp: accuracy ± 0.1 K (authors).

REFERENCES:

COMPONENTS:

(1) Scandium nitrate; $Sc(NO_3)_3$; [13465-60-6]

(2) n-Butyl propionate ; $C_7H_{14}O_2$; [590-01-2]

(3) Water ; H_2O ; [7732-18-5]

ORIGINAL MEASUREMENTS:

Pushkina, G. Ya.; Komissarova, L.N. *Zh. Neorg. Khim.* 1963, *8*, 1498-504; *Russ. J. Inorg. Chem. Engl. Transl.* 1963, *8*, 777-81.

VARIABLES:

One temperature: 25°C

PREPARED BY:

T. Mioduski

EXPERIMENTAL VALUES:

Composition of the saturated solution at 25°C

Sc_2O_3	Sc	NO_3	H_2O [a]	$CH_3CH_2COO(CH_2)_3CH_3$ [a]	$Sc(NO_3)_3$ [a]	
mass %	mass %	mass %	mass %	mass %	mass %	mol kg^{-1}
8.77	5.72	22.77	9.16	61.46	29.38	1.801

a. Calculated by the compiler based on Sc_2O_3 mass %, and using 1977 IUPAC recommended atomic weights.

AUXILIARY INFORMATION

METHOD/APPARATUS/PROCEDURE:

Isothermal method used. The initial materials were placed in glass vessels and thermostated. Equilibrium was reached within 48 days as ascertained by successive analyses. 0.1 - 0.5 gram samples were removed for analyses using a No. 2 or No. 3 Schott filter fitted to a pipet. 0.5-1 g of solid was also removed for analyses.

Scandium was determined gravimetrically as the oxide. The organic material was first destroyed by heating with HNO_3 to dryness; the residue was then dissolved in HCl, $Sc(OH)_3$ precipitated and ignited to the oxide. Nitrogen was determined by Devarda's method, and water was calculated from the loss in weight upon heating to 900°C. The results given above are the mean of 3-4 determinations. The Sc_2O_3:N_2O_5 molar ratio in the liquid and solid phases is constant at approximately 1:3.

SOURCE AND PURITY OF MATERIALS:

$Sc(NO_3)_3 \cdot 4H_2O$ was crystallized form HNO_3 sln containing at least 44.26 mass % N_2O_5. Chemical analyses resulted in the following (mean of 3 detns): Sc 14.84 mass %, NO_3 61.33 mass %. Impurities in the salt stated not to exceed 0.01 %. Prior to use the salt was dried in a desiccator over H_2SO_4 to remove hydroscopic water.
Source of the organic solvent not specified, but authors state that its purity was checked optically.

ESTIMATED ERROR:

Soly: reproducibility around ± 1% (compilers).

Temp: accuracy ± 0.1 K (authors).

REFERENCES:

COMPONENTS:	EVALUATOR:
(1) Yttrium nitrate; $Y(NO_3)_3$; [10361-93-0] (2) Water ; H_2O ; [7732-18-5]	S. Siekierski, T. Mioduski Institute for Nuclear Research Warsaw, Poland and M. Salomon U.S. Army ET & DL Ft. Monmouth, NJ, USA May 1982

CRITICAL EVALUATION:

INTRODUCTION

The solubility of $Y(NO_3)_3$ has been reported in thirteen publications (1-13), and generally there has been poor agreement for the solubilities in the binary $Y(NO_3)_3$-H_2O system. In addition to the possibility of experimental imprecision (1,3,9,10) and questionable purities of starting materials (1,3,5,6,8,10), much of the disagreement can probably be traced to poor identification of the solid phases. Early studies by Cleve and Hoglund (14) and Demarcay (15) state that the hexahydrate, [13494-98-9], crystallizes upon evaporation of the binary solution, and that the trihydrate, [13470-40-1], is produced by drying over H_2SO_4(14). The trihydrate was also claimed to be produced by crystallization from concentrated HNO_3 (15). Perel'man et al. prepared either the tetrahydrate (7), [13773-69-8], or hexahydrate (9,11,12) by crystallization from HNO_3 solutions, and the trihydrate was obtained by drying the hexahydrate at 373 K or over P_2O_5 (the latter method required a period of six months). Marsh (16) prepared the pentahydrate, [57584-28-8], by evaporating a neutral $Y(NO_3)_3$ solution to a syrup followed by seeding with $Bi(NO_3)_3.5H_2O$. The mother-liquor was seeded with $Dy(NO_3)_3.6H_2O$ to crystallize the hexahydrate $Y(NO_3)_3.6H_2O$. Moret (2) and Kuznetsova et al. (4) prepared the hexahydrate presumably by recrystallization from dilute HNO_3 or from pure water. Crew, Steinert, and Hopkins (1) and James and Pratt (10) recrystallized the salt from HNO_3 solutions but were unable to identify the nature of the hydrate produced. Based on these reports it would appear that the solid phase produced upon crystallization from pure water and dilute HNO_3 solutions is the hexahydrate, and that lower hydrates are easily obtained by desiccation. The stable solid phase at 298.2 K therefore appears to be the hexahydrate.

The three solubility studies which report the stable solid phase to be the hexahydrate are those of Moret (2), Afanas'ev, Azhipa and Sal'nik (3), and Kuznetsova, Yakimova, Yastrebova, and Stepin (4). Based on the chemical similarities between yttrium and erbium (see the INTRODUCTION and critical evaluation for erbium), and on the agreement between the solubility values from (2) and (4), and based upon the above discussions it is concluded by the evaluators that the stable solid phase at 298.2 K is indeed the hexahydrate.

Odent and Duperray (8) state that the stable solid phase at 298.2 K is the pentahydrate whereas Crew et al. (1) and James and Pratt (10) do not specify the nature of the stable hydrate. Perel'man et al. report a solubility value at 298.2 K for the tetrahydrate system which probably represents a metastable system (see below), but Khudaibergenova and Sulaimankulov (13) do state that at 303.2 K the stable solid phase is the tetrahydrate. This latter result of 5.534 mol kg^{-1} for the solubility at 303.2 K is smaller than the solubility value of 5.745 mol kg^{-1} for the hexahydrate at this temperature (2), and it is considerably smaller than the value of 6.217 mol kg^{-1} in the tetrahydrate system at 298.2 K reported by Perel'man et al. (9). In the study by Perel'man et al. (9) the starting material was the hexahydrate, but in the presence of NH_4NO_3 (up to 25 mol kg^{-1}) the solid phase is the tetrahydrate, and it appears that the solubility of $Y(NO_3)_3$ in the absense of NH_4NO_3 was determined indirectly by extrapolation to zero NH_4NO_3 concentration.

TENTATIVE SOLUBILITY VALUES

For saturated solutions in equilibrium with hexahydrate solid phase, *tentative* solubility values are assigned to those data reported by Moret (2) over the temperature range 273-308 K. The data were fitted to the general solubility equation (see INTRODUCTION and references 17, 18)

$$Y = \ln(m/m_o) - nM_2(m - m_o) = a + b/(T/K) + c\ \ln(T/K) \qquad [1]$$

COMPONENTS:	EVALUATOR:
(1) Yttrium nitrate; $Y(NO_3)_3$; [10361-93-0] (2) Water ; H_2O ; [7732-18-5]	S. Siekierski, T. Mioduski Institute for Nuclear Research Warsaw, Poland and M. Salomon U.S. Army ET & DL Ft. Monmouth, NJ USA May 1982

CRITICAL EVALUATION: continued

In this equation m is the solubility in mol kg^{-1}, m_o an arbitrarily selected reference molality, n is the hydrate number, and M_2 is the molar mass of the solvent. Enthalpies and heat capacities of solution, ΔH_{sln} and ΔC_p, can be estimated from the constants a, b, c. The results of fitting Moret's data to eq. [1] are given in Table 1 and in Figure 1. Also shown in Fig. 1 are data points from other investigations (see discussion below). The solubility result of 5.525 mol kg^{-1} by Kuznetsova et al. (4) at 298.2 K was not included in this calculation because inclusion of this data point gives rise to slightly larger standard deviations to the fit of eq. [1].

Moret's results for 313.15 K and 323.15 K which he assigns to the pentahydrate system appear to be either imprecise or that they are precise but are inconsistent with the assignment of equilibrium solid phases. The latter appears to be more probable since these two data points can be fitted to eq. [1] assuming the solid phase is the tetrahydrate. By combining Moret's two data points at 313.35 K and 323.15 K with the value of 6.217 mol kg^{-1} at 298.2 K for the metastable tetrahydrate (9) and the value of 13.88 mol kg^{-1} corresponding to the concentration at the experimental congruent melting point of 349 K (19), an acceptable polytherm results as shown in Figure 1 and in Table 1. Thus there appears to be sufficient justification at this time to designate these solubility values as *tentative*.

The tentative solubility data calculated from the smoothing eq. [1] are given in Table 2. All other solubility data reported in the compilations have been *rejected* for reasons discussed below.

DISCUSSION

The data from references 1, 3, 8, 10 and 13 are rejected because they show large negative deviations from the tentative solubility values for the hexahydrate system. An attempt to fit these data to the smoothing equation [1] yielded inconsistent results for any assumed value of the hydration number n = 4, 5, 6 : i.e. σ values are vary large and the predicted congruent melting points are above 351 K which must be incorrect since the experimental congruent melting point for the tetrahydrate is 349 K (19). It is concluded that the data of Crew et al. (1) contain a large negative systematic error probably due to a combination of experimental error and mixed solid phases of unknown composition.

It is interesting to note that there is a significant agreement for those rejected data points at 298.2 K. Table 3 compares the results for the solubility of $Y(NO_3)_3$ at 298.2 K as reported in (1, 3, 8 and 10). For a Student's t of 3.182 at the 95% level of confidence, the average solubility for the data in this table is 5.10 ± 0.09 mol kg^{-1}. In spite of this surprising agreement, it is concluded that these data are in error due to a common systematic error such as the failure to reach equilibrium. If, as suggested by Odent and Duperray (8), the solid phase in this system is the pentahydrate, then these results are certainly incorrect because at 298.2 K the stable solid phase is the hexahydrate, and the solubility in the metastable pentahydrate system would have to lie between the values of 5.42 mol kg^{-1} for the hexahydrate system and 6.25 mol kg^{-1} for the tetrahydrate system (see Fig 1 and Table 2). Finally as pointed out above, these rejected data points lie on a polytherm which falls to the left of the hexahydrate polytherm (i.e. negative deviations), and rises much too rapidly resulting in a predicted congruent melting point greater than the experimental value of 349 K for the tetrahydrate.

Note added in proof. Rard and Spedding (20) recently reported density, osmotic coefficient, and activity coefficient data for dilute $Y(NO_3)_3$ solutions to supersaturation at 298.15 K. The solubility at 298.15 K was not determined, but it was estimated to be 5.6 ± 0.8 mol kg^{-1}. Based on the smoothing equations given in the source paper (20), the evaluators calculated the following values corresponding to a concentration of 5.424 mol kg^{-1} at 298.2 K (i.e. the *tentative* solubility value at 298.2 K):

apparent molal volume ϕ_V = 0.078706 dm^3 mol^{-1} density = 1.7422 kg m^{-3}

mean molal activity coeff $y_\pm$ = 1.9846 c_1 = 3.793 mol dm^{-3}

osmotic coefficient Φ = 2.1348

activity of water a_2 = 0.4341

COMPONENTS:	EVALUATOR:
(1) Yttrium nitrate; $Y(NO_3)_3$; [10361-93-0] (2) Water ; H_2O ; [7732-18-5]	S. Siekierski, T. Mioduski Institute for Nuclear Research Warsaw, Poland and M. Salomon U.S. Army ET & DL Ft. Monmouth, NJ, USA May 1982

CRITICAL EVALUATION: continued

Table 1. Parameters and Standard Deviations for the Smoothing Equation [1][a]

quantity	hexahydrate	tetrahydrate
a	-48.066	-198.24
b	1835	9000
c	7.3603	29.494
σ_a	0.004	0.01
σ_b	1.2	4.7
σ_c	0.0007	0.003
σ_Y	0.004	0.01
σ_m	0.048	0.75
ΔH_{sln}/kJ mol^{-1}	-60.8	-298
ΔC_p /JK^{-1} mol^{-1}	244.8	981
congruent melting point	323.9 K	349.1 K
concentration at the congruent melting point	9.251 mol kg^{-1}	13.877 mol kg^{-1}

a. Data for the hexahydrate system from (2) over the range 273-308 K. σ_Y and σ_m are the standard deviations of residuals for the quantity Y in eq. [1] and the molality, respectively.

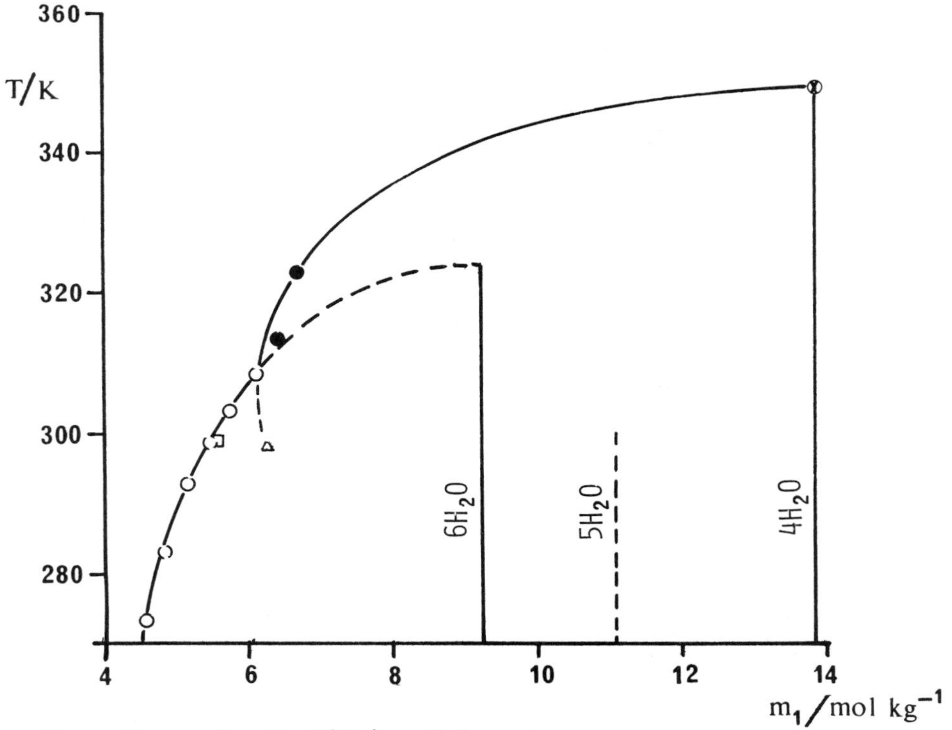

Figure 1. Phase diagram for the $Y(NO_3)_3$ - H_2O system.

○ hexahydrate solid phase from (2).
● pentahydrate solid phase from (2), but probably the tetrahydrate (see text).
□ hexahydrate solid phase from (4).
△ tetrahydrate solid phase from (9).
⊗ congruent melting point of the tetrahydrate from (18).

COMPONENTS:	EVALUATOR:
(1) Yttrium nitrate; $Y(NO_3)_3$; [10361-93-0] (2) Water ; H_2O ; [7732-18-5]	S. Siekierski, T. Mioduski Institute for Nuclear Research Warsaw, Poland and M. Salomon U.S. Army ET & DL Ft. Monmouth, NJ, USA May 1982

CRITICAL EVALUATION: continued

Table 2. Tenative solubilities for $Y(NO_3)_3$ calculated from the smoothing equation [1]. All solubilities are given in units of mol kg^{-1}.

T/K	hexahydrate system[a]	tetrahydrate system[b]
273.2	4.549	
278.2	5.666	
283.2	4.808	
288.2	4.979	
293.2	5.181	
298.2	5.424	6.25[c]
303.2	5.716	6.17[c]
308.2	6.077	6.18[c]
309.5[d]	6.19	6.19
313.2	6.541[c]	6.27
318.2	7.191[c]	6.46
323.2	8.481[c]	6.74
333.2		7.68
343.2		9.61
348.2		12.02

a. At the 95% level of confidence, total uncertainty is ± 0.025 mol kg^{-1} (Student's t = 3.182).

b. At the 95% level of confidence, total uncertainty is ± 2.3 mol kg^{-1} (Student's t = 12.706).

c. Metastable solubilities.

d. Hexahydrate → tetrahydrate transition temperature.

Table 3. Comparison of Rejected Solubility Data for $Y(NO_3)_3$ at 298.2 K.

solubility/mol kg^{-1}	assigned solid phase	reference
5.035[a]	?	1
5.13	hexahydrate	3
5.065	pentahydrate	8
5.151	?	10

a. Calculated by the compilers from a smoothing equation (see compilation).

COMPONENTS:	EVALUATOR:
(1) Yttrium nitrate; $Y(NO_3)_3$; [10361-93-0] (2) Water ; H_2O ; [7732-18-5]	S. Siekierski, T. Mioduski Institute for Nuclear Research Warsaw, Poland and M. Salomon U.S. Army ET & DL Ft. Monmouth, NJ, USA May 1982

CRITICAL EVALUATION:

REFERENCES

1. Crew, M.C.; Steinert, H.E.; Hopkins, B.S. *J. Phys. Chem.* 1925, *29*, 34.
2. Moret, R. *Thèse.* l'Université de Lausanne. 1963.
3. Afanas'ev, Yu.A.; Azhipa, L.T.; Sal'nik, L.V. *Zh. Neorg. Khim.* 1982, *27*, 769.
4. Kuznetsova, G.P.; Yakimova, Z.P.; Yastrebova, L.F.; Stepin, B.D. *Zh. Neorg. Khim.* 1981, *26*, 3161.
5. Perel'man, F.M.; Fedoseeva, E.I. *Zh. Neorg. Khim.* 1963, *8*, 1255.
6. Perel'man, F.M.; Demina, G.A. *Zh. Neorg. Khim.* 1964, *9*, 1772.
7. Perel'man, F.M.; Babievskava, I.Z. *Zh. Neorg. Khim.* 1964, *9*, 986.
8. Odent, G.; Duperray, M.H. *C.R. Hebd. Seances Acad. Sci. Ser. C.* 1974, *279*, 451.
9. Perel'man, F.M.; Zvorykin, A.Ya.; Demina, G.A. *Zh. Neorg. Khim.* 1960, *5*, 960.
10. James, C.; Pratt, L.A. *J. Am. Chem. Soc.* 1910, *32*, 873.
11. Perel'man, F.M.; Babievskaya, I.Z. *Zh. Neorg. Khim.* 1962, *7*, 1479.
12. Perel'man, F.M. *Rev. Chim. Miner.* 1970, *7*, 635.
13. Khudaibergenova, N.; Sulaimankulov, K. *Zh. Neorg. Khim.* 1981, *26*, 1156.
14. Cleve, P.T.; Hoglund, O. *Bihang. Svenska. Akad. Handl.* 1873, 8.
15. Demarcay, E. *Spectres Electriques.* Paris, 1895.
16. Marsh, J.K. *J. Chem. Soc.* 1941, 561.
17. Williamson, A.T. *Trans. Faraday Soc.* 1944, *40*, 421.
18. Counioux, J.-J.; Tenu, R. *J. Chim. Phys.* 1981, *78*, 816 and 823.
19. Babievskaya, I.Z.; Perel'man, F.M. *Zh. Neorg. Khim.* 1966, *11*, 1817.
20. Rard, J.A.; Spedding, F.H. *J. Chem. Eng. Data* 1982, *27*, 454 (this paper was not compiled because the authors did not experimentally determine the solubility of $Y(NO_3)_3$ at 298.15 K).

COMPONENTS:

(1) Yttrium nitrate; $Y(NO_3)_3$; [10361-93-0]

(2) Water; H_2O; [7732-18-5]

ORIGINAL MEASUREMENTS:

Crew, M.C.; Steinert, H.E.; Hopkins, B.S.

J. Phys. Chem. 1925, *29*, 34-8.

VARIABLES:

Temperature

PREPARED BY:

T. Mioduski, S. Siekierski, M. Salomon

EXPERIMENTAL VALUES:

t/°C	mass satd sln/g	mass Y_2O_3/g	solubility of $Y(NO_3)$ g(1)/100 g(2)[a]	g(1)/100 g(2)[b]	mol kg^{-1} [c]
0	1.3078	0.2596	93.1	93.55	3.403
22.5	1.2234	0.2888	136	135.2	4.917
22.5	1.2721	0.2988	133	133.6	4.860
35	0.7403	0.1853	155	156.1	5.677
60.2	0.5738	0.1561	197	196.2	7.138
60.2	0.7974	0.2193	203.1	202.7	7.374
66.5	0.9248	0.2585	211	213.1	7.752

a. Authors' original calculations based on 1925 atomic masses.
b. Compilers' calculations based on atomic masses recommended in the 1977 biennial report of the IUPAC Commission on Atomic Weights.
c. Molalities calculated by the compilers based on 1977 IUPAC recommended atomic masses.

The molalities were fitted to the following smoothing equation:

$$\text{soly/mol kg}^{-1} = -15.126 + 0.070700(T/K) - 1.0360 \times 10^{-5}(T/K)^2$$

The standard deviation for this fit is $\sigma = 0.078$ mol kg^{-1}, and the correlation coefficient is 0.998. Using this smoothing equation, the solubility of $Y(NO_3)_3$ in water at 25°C is calculated to be 5.035 mol kg^{-1}.

AUXILIARY INFORMATION

METHOD/APPARATUS/PROCEDURE:

Isothermal method used. Flasks of the solid and liquid were placed in a thermostat and equilibrated for at least 5 hours with periodic shaking. Equilibrium was approached from above. Samples of saturated solutions were removed from the flasks through a filter, and the funnel used in this procedure was "brought as close as possible" to the temperature of the saturated solutions. The saturated solutions were transfered to a weighing bottle and weighed. The saturated solutions were then evaporated to dryness and ignited to the oxide in a Pt dish.

The nature of the solid phase was not specified.

SOURCE AND PURITY OF MATERIALS:

Yttrium "material slightly short of atomic weight purity" contained traces of Ho and Er. It was twice pptd as the hydroxide and then twice as the oxalate. The oxalate was ignited to Y_2O_3 and dissolved in HNO_3, and the sln evaporated to crystallization. The crystals were washed (presumably with water) and centrifuged to remove excess liquid. Freshly distilled water "protected from the air" was used. HNO_3 was freshly redistilled from a quartz apparatus. Oxalic acid was recrystallized several times.

ESTIMATED ERROR:

Soly: Authors state mean error is ± 1.5%, but in some cases it is ± 3% (compilers).

Temp: precision no better than ± 0.2 K (compilers).

COMPONENTS:

(1) Yttrium nitrate; $Y(NO_3)_3$; [10361-93-0]

(2) Water ; H_2O ; [7732-18-5]

ORIGINAL MEASUREMENTS:

Moret, R. *Thèse*, l'Université de Lausanne. 1963.

VARIABLES:

Temperature: range 0°C to 50°C

PREPARED BY:

T. Mioduski and S. Siekierski

EXPERIMENTAL VALUES:

Solubility[a]

t/°C	mass %	moles of H_2O per 100 moles salt	mol kg^{-1}	solid phase
0	55.51	1223	4.538	$Y(NO_3)_3 \cdot 6H_2O$
10	57.12		4.845	"
20	58.45	1085	5.117	"
25	59.92		5.438	"
30	61.23		5.745	"
35	62.49		6.060	"
40	63.76		6.400	$Y(NO_3)_3 \cdot 5H_2O$
50	64.51		6.612	"

a. Molalities calculated by compilers from mass % values.

AUXILIARY INFORMATION

METHOD/APPARATUS/PROCEDURE:

The isothermal mothod was used as described in (1). Y was determined by complexometric titration using Xylenol Orange indicator in the presence of a small amount of urotropine buffer. Water was determined by difference.

COMMENTS AND/OR ADDITIONAL DATA:

The author states that the temperature for the hexahyrate to pentahydrate transition is 38.5°C. This temperature was determined graphically.

SOURCE AND PURITY OF MATERIALS:

Yttrium nitrate was prepared from Y_2O_3 of purity better than 99.7% (obtained by the ion exchange chromatographic method).

ESTIMATED ERROR:

Soly: precision about ± 0.1% (compilers).

Temp: precision at least ± 0.05 K (compilers).

REFERENCES:

1. Brunisholz, G.; Quinche. J.P.; Kalo, A.M. *Helv. Chim. Acta* 1964. *47*, 14.

COMPONENTS:

(1) Yttrium nitrate; $Y(NO_3)_3$; [10361-93-0]

(2) Nitric acid; HNO_3; [7697-37-2]

(3) Water; H_2O; [7732-18-5]

ORIGINAL MEASUREMENTS:

Kuznetsova, G.P.; Yakimova, Z.P.; Yastrebova, L.F.; Stepin, B.D. *Zh. Neorg. Khim.* 1981, *26*, 3161-4;

Russ. J. Inorg. Chem. Engl. Transl. 1981, *26*, 1692-3.

VARIABLES:

Composition at 25°C

PREPARED BY:

M. Salomon, T. Mioduski, and S. Siekierski

EXPERIMENTAL VALUES:

Composition of saturated solutions [a]

$Y(NO_3)_3$		HNO_3		
mass %	mol kg^{-1}	mass %	mol kg^{-1}	nature of the solid phase
60.30	5.525	——	——	$Y(NO_3)_3 \cdot 6H_2O$
13.98	1.783	57.50	31.995	$Y(NO_3)_3 \cdot 6H_2O + Y(NO_3)_3 \cdot 4H_2O$

a. Molalities calculated by the compilers.

AUXILIARY INFORMATION

METHOD/APPARATUS/PROCEDURE:

The isothermal method was used. Yttrium was determined by complexometric titration. Method of analysis of nitric acid not specified.

COMMENTS AND/OR ADDITIONAL DATA:

The authors report only two numerical values for the soly of $Y(NO_3)_3$ although in the phase diagram, 11 data points were given. Above 56 mass % HNO_3, the stable solid phase is the tetrahydrate.

The caption to the phase diagram in the source paper specifies the temperature as 0°C which is probably a typographical error. All discussion in the text of this paper refers to 25°C.

SOURCE AND PURITY OF MATERIALS:

Yttrium nitrate was prepared by dissolving high purity Y_2O_3 in HNO_3. The nitric acid was c.p. grade which was distilled.

ESTIMATED ERROR:

Soly: precision probably ± 0.1 % (compilers).

Temp: nothing specified.

REFERENCES:

COMPONENTS:

(1) Yttrium nitrate; $Y(NO_3)_3$; [10361-93-0]

(2) Nitric acid; HNO_3; [7697-37-2]

(3) Water ; H_2O ; [7732-18-5]

ORIGINAL MEASUREMENTS:

Afanas'ev, Yu.A.; Azhipa, L.T.; Sal'nik, L.V. *Zh. Neorg. Khim.* 1982, *27*, 769-73; *Russ. J. Inorg. Chem. Engl. Transl.* 1982, *27*, 431-4.

VARIABLES:

HNO_3 concentration at 25°C

PREPARED BY:

T. Mioduski and S. Siekierski

EXPERIMENTAL VALUES:

Solubility of $Y(NO_3)_3$ in HNO_3 solutions at 25°C[a]

$Y(NO_3)_3$ mass %	$Y(NO_3)_3$ mol kg^{-1}	HNO_3 mass %	HNO_3 mol kg^{-1}	solid phase[b]
58.5	5.13	0	———	A
55.1	4.84	3.5	1.34	"
50.9	4.62	9.0	3.56	"
47.7	4.38	12.7	5.09	"
45.3	4.28	16.2	6.68	"
37.5	3.47	23.2	9.37	"
35.2	3.39	27.0	11.34	"
30.2	2.91	32.0	13.43	"
26.4	2.62	36.9	15.96	"
23.6	2.39	40.5	17.90	"
21.5	2.17	42.4	18.64	"
20.5	2.12	44.3	19.97	"
19.8	2.11	46.0	21.35	A+B
13.8	1.38	49.8	21.71	B
8.6	0.96	58.9	28.76	"
9.6	1.20	61.3	33.43	"
9.9	1.32	62.9	36.70	B+C
8.0	1.10	65.6	39.43	C
7.1	1.06	68.6	44.80	"
8.5	1.52	71.2	55.66	"
10.0	2.56	75.8	84.71	C+D
10.4	3.18	77.7	103.6	D
9.3	3.38	80.7	128.1	"
6.6	2.42	83.5	133.9	"
4.5	1.95	87.1	164.6	"
1.8	1.01	91.7	223.9	"

a. Molalities calculated by M. Salomon.

b. Solid phases: A = $Y(NO_3)_3 \cdot 6H_2O$; B = $Y(NO_3)_3 \cdot 5H_2O$

C = $Y(NO_3)_3 \cdot 4H_2O$; D = $Y(NO_3)_3 \cdot nH_2O$

AUXILIARY INFORMATION

METHOD/APPARATUS/PROCEDURE:

The isothermal method was used. The composition of the solutions was changed by addition of 100% HNO_3 to a saturated solution or by addition of the salt to the acid solution. Equilibrium was reached within 3-4 hours. The yttrium content in the saturated solutions and solid phases was determined by complexometric titration using Xylenol Orange indicator. The HNO_3 content was determined by titration with NaOH using methyl red indicator. The compositions of the solid phases were determined by the Schreinemakers' method. The hydrated solid phases were separated and their infrared spectra recorded. Details are given in the source paper.

SOURCE AND PURITY OF MATERIALS:

C.p. grade yttrium nitrate was used.

Nitric acid (source and purity not specified) was concentrated by method recommended in the well-known Brauer's Handbood (the Russian edition was cited by the authors).

ESTIMATED ERROR:

Soly: nothing specified.

Temp: precision within ± 0.1 K.

REFERENCES:

COMPONENTS:

(1) Potassium chromate; K_2CrO_4; [7789-00-6]

(2) Yttrium nitrate; $Y(NO_3)_3$; [10361-93-0]

(3) Water; H_2O; [7732-18-5]

ORIGINAL MEASUREMENTS:

Perel'man, F.M.; Fedoseeva, E.I.

Zh. Neorg. Khim. 1963, *8*, 1255-8; *Russ. J. Inorg. Chem. Engl. Transl.* 1963, *8*, 650-1.

VARIABLES:

Concentration of K_2CrO_4 at 25°C

PREPARED BY:

T. Mioduski and S. Siekierski

EXPERIMENTAL VALUES:

The $Y(NO_3)_3$ - K_2CrO_4 - H_2O system at 25.0°C

Composition of saturated solutions[a]

K_2CrO_4		$Y_2(CrO_4)_3$		KNO_3		$Y(NO_3)_3$		
mass %	mol kg^{-1}	mass %	mol kg^{-1}	mass %	mol kg^{-1}	mass %	mol kg^{-1}	
1.47	0.0816	0.03	0.00062	5.73	0.611	---	---	A
---	---	0.74	0.0154	4.55	0.491	3.09	0.123	A
---	---	0.48	0.0099	6.24	0.670	1.11	0.0438	A
2.62	0.146	0.05	0.00103	4.78	0.511	---	---	A
---	---	0.58	0.0120	5.00	0.537	2.37	0.0937	A
5.68	0.320	---	---	3.00	0.325	---	---	B
8.65	0.496	0.12	0.00254	1.50	0.165	---	---	B
1.00	0.0546	0.07	0.00141	4.60	0.482	---	---	C
7.11	0.406	0.12	0.00253	2.52	0.276	---	---	C

a. Molalities were calculated by the compilers.

Solid Phases:

A = $Y_2(CrO_4)_3 \cdot K_2CrO_4 \cdot 6H_2O$

B = $Y_2(CrO_4)_3 \cdot 4K_2CrO_4 \cdot nH_2O$

C = $Y_2(CrO_4)_3 \cdot 3K_2CrO_4 \cdot nH_2O$

AUXILIARY INFORMATION

METHOD/APPARATUS/PROCEDURE:

The isothermal method was used. Both the liquid and solid phases were analysed after equilibration for 3 days. Potassium was determined by the perchlorate method. Yttrium was determined by the oxalate method. CrO_4^{2-} was determined by reduction with Mohr's salt followed titration of excess Mohr's salt with $KMnO_4$. Water was determined by difference. The composition of the solid residues was determined by Schreinemakers method.

SOURCE AND PURITY OF MATERIALS:

Nothing specified.

ESTIMATED ERROR:

Soly: nothing specified.

Temp: precision is ± 0.1 K.

REFERENCES:

COMPONENTS:

(1) Rubidium nitrate; $RbNO_3$; [13126-12-0]

(2) Yttrium nitrate; $Y(NO_3)_3$; [10361-93-0]

(3) Nitric acid; HNO_3; [7697-37-2]

(4) Water; H_2O; [7732-18-5]

ORIGINAL MEASUREMENTS:

Perel'man, F.M.; Demina, G.A.

Zh. Neorg. Khim. 1964, *9*, 1772-3;

Russ. J. Inorg. Chem. Engl. Transl. 1964, *9*, 960-1.

VARIABLES:

Composition at 25°C

PREPARED BY:

T. Mioduski and S. Siekierski

EXPERIMENTAL VALUES:

The $Y(NO_3)_3$ - $RbNO_3$ - HNO_3 - H_2O system at 25.0°C

Composition of saturated solutions[a]

$Y(NO_3)_3$		$RbNO_3$		HNO_3		
mass %	mol kg^{-1}	mass %	mol kg^{-1}	mass %	mol kg^{-1}	solid phase[b]
28.07	2.818	---	---	35.7	15.64	A
29.62	3.140	3.47	0.685	32.6	15.08	A
29.68	3.919	6.87	1.691	35.9	20.68	A
26.47	3.333	11.14	2.615	33.5	18.40	A
30.71	11.783	26.01	18.605	33.8	56.58	A
30.27	26.029	32.22	51.650	33.28	124.86	A
28.76[c]		36.70		36.2		A + B
27.30	13.811	35.44	33.424	30.07	66.370	B
26.64	14.506	37.15	37.711	29.53	70.155	B
22.52	42.224	43.54	152.19	32.0	261.8	B
20.96	95.300	48.24	408.89	30.0	595.1	B + C
13.97	7.187	49.78	47.745	29.18	65.499	C
13.79	7.850	49.62	52.656	30.2	75.00	C
---	---	51.0	18.40	30.2	25.49	C

a. Molalities calculated by compilers.

b. Solid phases: A = $Y(NO_3)_3 \cdot 4H_2O$ B = $Y(NO_3)_3 \cdot 2RbNO_3$ C = $RbNO_3 \cdot HNO_3$

c. Total mass % of solutes is 101.6 %. It cannot be determined whether this is an experimental error or a typographical error.

AUXILIARY INFORMATION

METHOD/APPARATUS/PROCEDURE:

Nothing specified. Authors state that the method used has been described in previous publications. Compilers assume that ref (1) contains the essential information. This publication on the $Nd(NO_3)_3$ - $RbNO_3$ - HNO_3 - H_2O system has been compiled elsewhere in this volume.

SOURCE AND PURITY OF MATERIALS:

No details given.

ESTIMATED ERROR:

Nothing specified.

REFERENCES:

1. Perel'man, F.M.; Zvorykin, A. Ya.; Demina, G.A. *Z. Neorg. Khim.* 1963, *8*, 1753; *Russ. J. Inorg. Chem. Engl. Tranls.* 1963 *8*, 909.

COMPONENTS:

(1) Yttrium nitrate; $Y(NO_3)_3$; [10361-93-0]

(2) Yttrium hydroxide; $Y(OH)_3$; [16469-22-0]

(3) Water; H_2O; [7732-18-5]

ORIGINAL MEASUREMENTS:

James, C.; Pratt, L.A. *J. Am. Chem. Soc.* 1910, *32*, 873-9.

VARIABLES:

Composition at 25°C

PREPARED BY:

Mark Salomon

EXPERIMENTAL VALUES:

Compostion of saturated solutions[a]

$Y(NO_3)_3$ g/100 g H_2O	$Y(NO_3)_3$ mol kg^{-1}	$Y(OH)_3$ g/100 g H_2O	$Y(OH)_3$ 10^3mol kg^{-1}	density kg m^{-3}	nature of solid[b]
3.13	0.114	0.014	1.00	1.0260	A
8.37	0.304	0.022	1.57	1.0675	A
13.87	0.504	0.034	2.43	1.1106	A
19.05	0.693	0.048	3.43	1.1506	A
24.94	0.907	0.063	4.50	1.1907	A
30.46	1.108	0.091	6.50	1.2350	A
33.02	1.201	0.160	11.43	1.2517	A + B[c]
38.71	1.408	0.122	8.72	1.2897	B
44.35	1.613	0.114	8.15	1.3268	B
51.87	1.887	0.103	7.36	1.3698	B
58.61	2.132	0.095	6.79	1.4104	B
65.89	2.397	0.090	6.43	1.4484	B
73.03	2.656	0.078	5.57	1.4867	B
80.67	2.934	0.072	5.15	1.5231	B
89.06	3.239	0.074	5.29	1.5587	B
95.98	3.491	0.074	5.29	1.5923	B
103.80	3.776	0.075	5.36	1.6259	B
113.40	4.125	0.079	5.65	1.6603	B
122.40	4.452	0.080	5.72	1.6931	B

continued.....

AUXILIARY INFORMATION

METHOD/APPARATUS/PROCEDURE:

Isothermal method. $Y(NO_3)_3$ and excess Y_2O_3 together with water were placed in bottles of 100 cc capacity and rotated in a thermostat at 25°C for 4.5 months. Equilibrium was ascertained by analyses of the solutions beginning at 3 months. Solid phases were permitted to settle and the supernatant was drawn off for analysis. Saturated solutions were analysed by titration with standard 0.1 mol dm^{-3} HNO_3 with methyl orange indicator which yielded the $Y(OH)_3$ concentration. The total Y content was determined gravimetrically by pptn as the oxalate and ignition to the oxide. Solid phases were analysed by pressing samples between filter paper and weighing. Part of each sample was ingited to the oxide and weighed to obtain the total Y, and part was placed in excess HNO_3 and back titrated with $NaCO_3$ solution to obtain the $Y(OH)_3$ content.

SOURCE AND PURITY OF MATERIALS:

"Crude yttria material" was fractionally crystallized by the bromate method (1). The middle fraction was pptd as the hydroxide, washed with boiling water, and converted to the nitrate. The nitrate was dissolved in water, and the oxalate pptd. The oxalate was ignited to the oxide which was then dissolved in excess HNO_3 solution and twice recrystallized from this solvent. Spectroscopic analysis of satd $Y(NO_3)_3$ slns showed very faint absorptions due to Ho and Er. Source and purity of water not stated.

ESTIMATED ERROR:

Nothing specified.

REFERENCES:

(1) *J. Am. Chem. Soc.* 1908, *30*, 182.

COMPONENTS:

(1) Yttrium nitrate; $Y(NO_3)_3$; [10361-93-0]

(2) Yttrium hydroxide; $Y(OH)_3$; [16469-22-0]

(3) Water; H_2O; [7732-18-5]

ORIGINAL MEASUREMENTS:

James, C.; Pratt, L.A. *J. Am. Chem. Soc.* 1910, *32*, 873-9.

EXPERIMENTAL VALUES: continued....

$Y(NO_3)_3$		$Y(OH)_3$		density	
g/100 g H_2O	mol kg^{-1}	g/100 g H_2O	10^3mol kg^{-1}	kg m^{-3}	nature of the solid[b]
132.10	4.805	0.074	5.29	1.7260	B
137.10	4.987	0.083	5.93	1.7440	B + C[c]
141.60	5.151	---	---	1.7446	C

a. Molalities calculated by the compiler.

b. Solid phases are not described in sufficient detail. Authors state that solid A is a solid solution which appears to show a great resemblance to $Y(OH)_3$. If solid A is the hydroxide, which it probably is, then presumably it would be hydrated.

Solid B was stated to be $3Y_2O_3 \cdot 4N_2O_5 \cdot 20H_2O$ as determined by analysis of the wet residues.

Solid C is presumed by the compiler to be $Y(NO_3)_3 \cdot nH_2O$.

c. These compositions were assumed by the compiler.

For the binary $Y(NO_3)_3$ - H_2O system at 25°C, the reported density of the saturated solution permits the calculation of the solubility in volume units:

$$\text{solubility } Y(NO_3)_3 = 3.719 \text{ mol dm}^{-3}$$

COMPONENTS:

(1) Yttrium nitrate; $Y(NO_3)_3$; [10361-93-0]
(2) Lanthanum nitrate; $La(NO_3)_3$; [10099-59-9]
(3) Nitric acid; HNO_3; [7697-37-2]
(4) Water; H_2O; [7732-18-5]

ORIGINAL MEASUREMENTS:

Perel'man, F.M.; Babievskaya, I.Z. *Zh. Neorg. Khim.* 1964, *9*, 986-90; *Russ. J. Inorg. Chem. Engl. Transl.* 1964, *9*, 538-41.

VARIABLES:

Composition at 25°C

PREPARED BY:

T. Mioduski, S. Siekierski and M. Salomon

EXPERIMENTAL VALUES:

The $Y(NO_3)_3$ - $La(NO_3)_3$ - HNO_3 - H_2O system at 25.0°C

Composition of saturated solutions[a]

HNO_3		$Y(NO_3)_3$		$La(NO_3)_3$		
mass %	mol kg^{-1}	mass %	mol kg^{-1}	mass %	mol kg^{-1}	nature of the solid phase[b]
32.8	12.8	26.6	2.38	0	0	$Y(NO_3)_3 \cdot 4H_2O$
31.8	13.9	24.9	2.49	6.88	0.581	"
31.2	14.9	19.8	2.17	15.8	1.46	"
33.4	17.6	15.2	1.84	21.3	2.18	$Y(NO_3)_3 \cdot 4H_2O + La(NO_3)_3 \cdot 6H_2O$
33.0	18.4	15.3	1.96	23.3	2.52	"
30.6	13.8	9.9	1.02	24.2	2.11	$La(NO_3)_3 \cdot 6H_2O$
32.6	14.3	4.5	0.45	26.6	2.26	"
32.9	14.1	0	0	30.0	2.49	"

a. Molalities calculated by the compilers.

b. The eutonic point for HNO_3 = 32.2 mass % corresponds to a saturated solution with $Y(NO_3)_3$ = 15.25 mass % and $La(NO_3)_3$ = 22.3 mass % (these compositions converted to weight units are: HNO_3 = 18.0 mol kg^{-1}, $Y(NO_3)_3$ = 1.90 mol kg^{-1} and $La(NO_3)_3$ = 2.35 mol kg^{-1} (compilers)), and the solid phases are $Y(NO_3)_3 \cdot 4H_2O$ and $La(NO_3)_3 \cdot 6H_2O$.

AUXILIARY INFORMATION

METHOD/APPARATUS/PROCEDURE:

Isothermal method. Slns equilibrated 2-3 d. Both satd slns and residues analysed. HNO_3 in sln detd acidimetrically. Total Y + La nitrates detd by two methods: first by pptn of the oxalate which was ignited to the oxide, and second by titrn of the oxalate with stnd $KMnO_4$. The mass % of each nitrate was calcd from

$$\% \; Y(NO_3)_3 = 4.166(162.9a_2 - 132a_1)/b$$

$$\% \; La(NO_3)_3 = 4.923(132a_1 - 112.9a_2)/b$$

where b = total weight of sample consisting of x % $Y(NO_3)_3$ and y % $La(NO_3)_3$, a_1 = total weight of oxide and a_2 = total oxalate (ion) weight. From mass balance considerations, the authors solved for x and y by method of simultaneous equations. The compilers consider this method of determining the individual nitrate contents to be approximate. The following molecular weights were used by the authors: $Y(NO_3)_3$ = 274.9, $La(NO_3)_3$ = 324.9, $Y_2O_3/2$ = 112.9, $La_2O_3/2$ = 162.9, and $3C_2O_4/2$ = 132.

SOURCE AND PURITY OF MATERIALS:

The nitrate salts were prepared by dissolving "chemically pure" oxides (99.9%) in nitric acid and evaporating until crystallization started. The refractive indices of the crystals were determined by the immersion method and are:

for $Y(NO_3)_3 \cdot 4H_2O$, n_p = 1.420, n_m = 1.528, n_g = 1.570.

for $La(NO_3)_3 \cdot 6H_2O$, n_p = 1.464, n_m = 1.579, n_g = 1.591.

ESTIMATED ERROR:

Soly: analysis checked with standard mixtures showed accuracy is from 2 % to 4 % providing both components do not fall below 10-15 % of the total concn. At ratios of the two nitrates of 90:10 the accuracy reaches 12 % for the nitrate present in the least amount, and 1.0 % for the preponderant nitrate.

Temp: precision ± 0.1 K.

COMPONENTS:

(1) Yttrium nitrate; $Y(NO_3)_3$; [10361-93-0]

(2) Gadolinium nitrate; $Gd(NO_3)_3$; [10168-81-7]

(3) Nitric acid; HNO_3; [7697-37-2]

(4) Water; H_2O; [7732-18-5]

ORIGINAL MEASUREMENTS:

Perel'man, F.M.; Babievskaya, I.Z. *Zh. Neorg. Khim.* 1964, *9*, 986-90: *Russ. J. Inorg. Chem.* 1964, *9*, 538-41.

Perel'man, F.M. *Rev. Chim. Miner.* 1970, *7*, 635-45.

VARIABLES:

Composition at 25°C

PREPARED BY:

T. Mioduski, S. Siekierski and M. Salomon

EXPERIMENTAL VALUES:

The $Y(NO_3)_3$ - $Gd(NO_3)_3$ - HNO_3 - H_2O system at 25.0°C

Composition of saturated solutions[a]

HNO_3		$Y(NO_3)_3$		$Gd(NO_3)_3$		
mass %	mol kg^{-1}	mass %	mol kg^{-1}	mass %	mol kg^{-1}	nature of the solid phase
31.8	12.6	28.3	2.58	0	0	$Y(NO_3)_3 \cdot nH_2O$
31.4	13.1	28.6	2.73	1.95	0.149	solid solutions
34.3	14.1	25.4	2.40	1.8	0.14	"
31.9	13.2	24.4	2.31	5.28	0.340	"
31.7	12.6	22.7	2.06	5.6	0.41	"
35.3	16.0	24.5	2.55	5.22	0.435	"
33.6	14.8	23.9	2.41	6.5	0.53	"
34.0	13.7	17.8	1.64	8.8	0.65	"
34.7	14.3	15.3	1.44	11.4	0.860	"
28.5	11.0	13.0	1.15	17.4	1.23	"
32.9	13.0	5.1	0.46	21.8	1.58	"
32.0	12.1	0	0	26.0	1.80	$Gd(NO_3)_3 \cdot nH_2O$

a. Molalities calculated by the compilers.

AUXILIARY INFORMATION

METHOD/APPARATUS/PROCEDURE:

Isothermal method. Slns equilibrated 2-3 d. Both satd slns and residues analysed. HNO_3 in sln detd acidimetrically. Total Y + Gd nitrates detd by two methods: first by pptn of the oxalate which was ignited to the oxide, and second by titrn of the oxalate with stnd $KMnO_4$. The mass % of each nitrate was calcd from

$$\% \; Y(NO_3)_3 = 3.062(180.9a_2 - 132a_1)/b$$

$$\% \; Gd(NO_3)_3 = 3.82(132a_1 - 112.9a_2)/b$$

where b = total weight of sample consisting of x % $Y(NO_3)_3$ and y % $Gd(NO_3)_3$, a_1 = total weight of oxide, and a_2 = total oxalate (ion) weight. From mass balance considerations, the authors solved for x and y by method of simultaneous equations. The compilers consider this method of determining the individual metal nitrate contents to be approximate. The following molecular weights were used by the authors: $Y(NO_3)_3$ = 274.9, $Gd(NO_3)$ = 342.9, $Y_2O_3/2$ = 112.9, $Gd_2O_3/2$ = 180.9, and $3C_2O_4/2$ = 132.

SOURCE AND PURITY OF MATERIALS:

$Y(NO_3)_3 \cdot nH_2O$ and $Gd(NO_3)_3 \cdot nH_2O$ were prepared by dissolving chemically pure oxides (99.9%) in nitric acid and evaporating until crystallization started. The refractive indices for $Y(NO_3)_3 \cdot 4H_2O$ were determined by the immersion method and are: n_p = 1.420, n_m = 1.528, and n_g = 1.570.

ESTIMATED ERROR:

Soly: analysis checked with standard mixtures showed accuracy is from 2 % to 4 % providing both components do not fall below 10-15 % of the total concn. At ratios of the two nitrates of 90:10 the accuracy reaches 12 % for the nitrate present in the least amount, and 1.0% for the preponderant nitrate.

Temp: precision ± 0.1 K.

COMPONENTS:

(1) Yttrium nitrate; $Y(NO_3)_3$; [10361-93-0]

(2) Cobalt nitrate; $Co(NO_3)_2$; [10141-05-6]

(3) Water; H_2O; [7732-18-5]

ORIGINAL MEASUREMENTS:

Odent, G.; Duperray, M.H. *C.R. Hebd. Seances Acad. Sci., Ser. C* 1974, *279*, 451-3.

VARIABLES:

Composition at 25°C

PREPARED BY:

T. Mioduski and S. Siekierski

EXPERIMENTAL VALUES:

The $Y(NO_3)_3$ - $Co(NO_3)_2$ - H_2O system at 25.0°C

Composition of saturated solutions[a]

$Y(NO_3)_3$		$Co(NO_3)_2$		
mass %	mol kg^{-1}	mass %	mol kg^{-1}	nature of the solid phase
58.20 [b]	5.065	---	---	$Y(NO_3)_3 \cdot 5H_2O$
54.60	4.799	4.02	0.531	"
50.10	4.295	7.47	0.962	"
48.04	4.274	11.08	1.482	"
42.07	3.607	15.51	1.999	"
38.82	3.431	20.03	2.661	"
36.02	3.083	21.48	2.763	$Y(NO_3)_3 \cdot 5H_2O$ + $Co(NO_3)_2 \cdot 6H_2O$
37.36	3.244	20.75	2.708	$Co(NO_3)_2 \cdot 6H_2O$
37.46	3.424	22.75	3.125	"
36.25	3.197	22.50	2.982	"
34.48	2.895	22.20	2.801	"
31.96	2.684	24.72	3.119	"
24.62	1.990	30.38	3.690	"
15.18	1.186	38.27	4.494	"
9.32	0.707	42.72	4.869	"
2.39	0.179	48.98	5.506	"
---	---	50.62	5.603	"

a. Molalities calculated by the compilers.
b. Authors report the soly of $Y(NO_3)_3$ in water as 139.23 g per 100 g water. The corresponding molality calculated by the compilers is 5.0645 mol kg^{-1}.

AUXILIARY INFORMATION

METHOD/APPARATUS/PROCEDURE:

The isothermal method was used. The solutions were equilibrated until their densities, measured with a Cornec-Cottet pipet, remained constant for three successive measurements performed at 24 hour intervals. Both the compositions of the saturated solutions and the solid residues were analysed. The total Y + Co was determined by titration with EDTA and back titrating with standard Zn^{2+} solutions using Eriochrom Black indicator at pH 9. Co was determined separately by electrolysis, and Y was then obtained by difference. The composition of the solid phases was determined by Schreinemakers' method, and by X-ray diffraction of the dried solid phases using the K_α ray of Cu.

SOURCE AND PURITY OF MATERIALS:

No information given.

ESTIMATED ERROR:

Soly: nothing specified.

Temp: precision of ± 0.1 K (compilers).

REFERENCES:

COMPONENTS:

(1) Yttrium nitrate; $Y(NO_3)_3$; [10361-93-0]

(2) Aluminum nitrate; $Al(NO_3)_3$; [13473-90-0]

(3) Water; H_2O; [7732-18-5]

ORIGINAL MEASUREMENTS:

Kuznetsova, G.P.; Yakimova, Z.P.; Yastrebova, L.F.; Stepin, B.D. *Zh. Neorg. Khim.* 1981, *26*, 3161-4;

Russ. J. Inorg. Chem. Engl. Transl. 1981, *26*, 1692-3.

VARIABLES:

Composition at 25°C

PREPARED BY:

M. Salomon, T. Mioduski, and S. Siekierski

EXPERIMENTAL VALUES:

Composition of saturated solutions [a]

$Y(NO_3)_3$		$Al(NO_3)_3$		
mass %	mol kg^{-1}	mass %	mol kg^{-1}	nature of the solid phase
60.30	5.525	———	———	$Y(NO_3)_3\cdot 6H_2O$
50.87	4.454	7.59	0.858	$Y(NO_3)_3\cdot 6H_2O + Al(NO_3)_3\cdot 9H_2O$
———	———	40.73	3.226	$Al(NO_3)_3\cdot 9H_2O$

a. Molalities calculated by the compilers.

AUXILIARY INFORMATION

METHOD/APPARATUS/PROCEDURE:

The isothermal method was used. Yttrium was determined complexometrically using sulfosalicylate as a masking agent for aluminum. Aluminum was determined by back-titration of excess EDTA with standard $ZnSO_4$ solution using dithizone indicator (1).

COMMENTS AND/OR ADDITIONAL DATA:

The authors report a phase diagram in the source paper indicating 9 data points. However, only the three data points given above were listed in the paper. The caption to the phase diagram in the source paper specifies the temperature to be 0°C which is probably a typographical error. Discussion in the text refers to 25°C.

SOURCE AND PURITY OF MATERIALS:

Yttrium nitrate was prepared by dissolving high purity Y_2O_3 in HNO_3. The nitric acid was c.p. grade which was distilled.

A.R. grade $Al(NO_3)_3\cdot 9H_2O$ was recrystallized before use.

ESTIMATED ERROR:

Soly: precision probably ± 0.1 % (compilers).

Temp: nothing specified.

REFERENCES:

1. Grosskreutz, W.; Schultze, D.; Wilke, K.T. *Z. Anal. Chem.* 1967, *232*, 278.

COMPONENTS:

(1) Yttrium nitrate; $Y(NO_3)_3$; [10361-93-0]

(2) Ammonium nitrate; NH_4NO_3; [6484-52-2]

(3) Water; H_2O; [7732-18-5]

ORIGINAL MEASUREMENTS:

Perel'man, F.M.; Zvorykin, A. Ya.; Demina, G.A. *Zh. Neorg. Khim.* 1960, *5*, 960-3; *Russ. J. Inorg. Chem. Engl. Transl.* 1960, *5*, 460-2.

VARIABLES:

Composition at 25°C and 50°C

PREPARED BY:

T. Mioduski and S. Siekierski

EXPERIMENTAL VALUES: The system $Y(NO_3)_3 - NH_4NO_3 - H_2O$ at 25°C

NH_4NO_3 in satd sln[a]		$Y(NO_3)_3$ in satd sln[a]		
mass %	mol kg^{-1}	mass %	mol kg^{-1}	nature of the solid phase
---	---	63.09	6.217	$Y(NO_3)_3 \cdot 4H_2O$
11.43	4.774	58.66	7.134	
18.74	9.217	55.86	7.999	
31.31[b]	24.266	52.57	11.862	$Y(NO_3)_3 \cdot 4H_2O + Y(NO_3)_3 \cdot 2NH_4NO_3 \cdot nH_2O$
32.23	15.809	42.30	6.041	$Y(NO_3)_3 \cdot 2NH_4NO_3 \cdot nH_2O$[c]
36.40	14.323	31.85	3.649	
43.77	17.847	25.59	3.038	
44.86[b]	18.813	25.35	3.095	$Y(NO_3)_3 \cdot 2NH_4NO_3 \cdot nH_2O + NH_4NO_3$
44.86	18.813	25.35	3.095	NH_4NO_3
47.66	19.158	21.26	2.488	
52.46	20.571	15.68	1.790	
56.04	20.340	9.54	1.008	
66.75	25.080	---	---	

a. Molalities calculated by the compilers.
b. Eutonic points were determined graphically.
c. Authors state this solid phase to be contaminated with NH_4NO_3.

continued............

AUXILIARY INFORMATION

METHOD/APPARATUS/PROCEDURE:

Isothermal method. Samples of saturated solutions with an excess of solids were isothermally equilibrated with constant agitation for 2-3 days at 50°C, 5 days at 25°C and 2-3 weeks when the solutions had high viscosities. Both the saturated solutions and wet residues were analysed. Ammonia was determined by the Kjeldahl method. Y was determined by precipitation with ammonia and the precipate washed and filtered. The precipitate was ignited to Y_2O_3 which was then weighed. The composition of the dry solid residues was determined graphically by Schreinemakers' method.

SOURCE AND PURITY OF MATERIALS:

$Y(NO_3)_3 \cdot 6H_2O$ was prepared by dissolving 99.28 % Y_2O_3 in aqueous HNO_3 (1:1) followed by crystallization. The hexahydrate contained 30.6 mass % Y_2O_3, and the pH of its aqueous solution was 5.5.

CP grade NH_4NO_3 was recrystallized twice before use.

ESTIMATED ERROR:

Soly: nothing specified.

Temp: precision ± 0.1 K.

REFERENCES:

COMPONENTS:

(1) Yttrium nitrate; $Y(NO_3)_3$; [10361-93-0]

(2) Ammonium nitrate; NH_4NO_3; [6484-52-2]

(3) Water; H_2O; [7732-18-5]

ORIGINAL MEASUREMENTS

Perel'man, F.M.; Zvorykin, A. Ya. ; Demina, G.A. *Zh. Neorg. Khim.* 1960, *5*, 960-3; *Russ. J. Inorg. Chem. Engl. Transl.* 1960, *5*, 460-2.

EXPERIMETNAL VALUES: (continued) The system $Y(NO_3)_3$ - NH_4NO_3 - H_2O at 50°C

NH_4NO_3 in satd sln[a]		$Y(NO_3)_3$ in satd sln[a]		
mass %	mol kg^{-1}	mass %	mol kg^{-1}	nature of the solid phase
14.00	8.165	64.58	10.967	$Y(NO_3)_3 \cdot 4H_2O$
14.00	7.446	62.51	9.680	
17.37	13.216	66.21	14.667	
17.4[b]	13.26	66.2	14.68	$Y(NO_3)_3 \cdot 4H_2O + Y(NO_3)_3 \cdot 2NH_4NO_3$
27.25	27.213	60.24	17.515	$Y(NO_3)_3 \cdot 2NH_4NO_3$
30.20	32.056	58.03	17.934	
33.88	45.464	56.81	22.196	
40.6	61.04	51.09	22.363	
44.5[b]	74.13	48.00	23.279	$Y(NO_3)_3 \cdot 2NH_4NO_3 + NH_4NO_3$
44.30	71.598	47.97	22.573	NH_4NO_3
50.70	31.813	29.39	5.369	
54.85	32.756	24.23	4.213	
62.40	34.756	15.17	2.460	
75.83	39.196	---	---	

a and b: see previous page.

COMMENTS AND/OR ADDITIONAL DATA:

Analyses of the solubility data in the tables and the phase diagram show that at 50°C the solubility curve has three branches corresponding to crystallization of $Y(NO_3)_3 \cdot 4H_2O$, $Y(NO_3)_3 \cdot 2NH_4NO_3$, and NH_4NO_3. Anhydrous $Y(NO_3)_3 \cdot 2NH_4NO_3$ is congruently soluble and crystallizes in the NH_4NO_3 concentration range of 18 to 44 mass %, and in the $Y(NO_3)_3$ concentration range of 66 to 48 mass %: its solubility in water is 88 mass % at 50°C. At 25°C there is insufficient proof that the double salt is actually formed, and it was impossible to separate the salt. Based on the middle branch of the 25°C isotherm the authors believe that the double salt is $Y(NO_3)_3 \cdot 2NH_4NO_3 \cdot nH_2O$ and that its solubility in water reaches 65 mass % (based on the anhydrous salt). The authors state that the tetrahydrate $Y(NO_3)_3 \cdot 4H_2O$ forms at both 25°C and 50°C.

COMPONENTS:

(1) Yttrium nitrate; $Y(NO_3)_3$; [10361-93-0]

(2) Ammonium nitrate; NH_4NO_3; [6484-52-2]

(3) Nitric acid; HNO_3; [7697-37-2]

(4) Water; H_2O; [7732-18-5]

ORIGINAL MEASUREMENTS:

Perel'man, F.M.; Babievskaya, I.Z. *Zh. Neorg. Khim.* 1962, 7, 1479-81; *Russ. J. Inorg. Chem. Engl. Transl.* 1962, 7, 762-3:
Perel'man, F.M. *Rev. Chim. Miner.* 1970, 7, 635-45.

VARIABLES:

Composition at 25°C

PREPARED BY:

T. Mioduski and S. Siekierski

EXPERIMENTAL VALUES:

The $Y(NO_3)_3 - NH_4NO_3 - HNO_3 - H_2O$ system at 25°C

Composition of saturated solutions[a]

HNO_3		$Y(NO_3)_3$		NH_4NO_3		
mass %	mol kg^{-1}	mass %	mol kg^{-1}	mass %	mol kg^{-1}	solid phase[b]
32.67	13.633	29.3	2.80	---	---	A
34.4	16.15	26.6	2.86	5.20	1.92	B
34.05	19.903	27.4	3.67	11.4	5.25	B
34.95	37.603	32.7	8.06	17.6	14.91	B
31.0	60.36	41.8	18.66	19.05	29.202	C
32.8	59.83	34.2	14.30	24.3	34.89	C
33.96	118.97	31.07	24.948	30.44	83.950	C
34.02	90.283	19.2	11.68	40.8	85.24	C
33.02	159.76	13.8	15.30	49.9	190.1	C + D
34.04	126.81	12.2	10.42	49.5	145.2	D
32.1	30.86	5.69	1.254	45.7	34.58	D
32.4	26.11	3.71	0.685	44.2	28.04	D
32.4	22.89	1.24	0.201	43.9	24.42	D
33.23	26.621	---	---	46.96	29.615	D

a. Molalities calculated by the compilers.

b. A = $Y(NO_3)_3 \cdot 6H_2O$ B = $Y(NO_3)_3 \cdot 4H_2O$ C = $Y(NO_3)_3 \cdot 2NH_4NO_3$ D = NH_4NO_3

The solubility isotherm consists of three branches corresponding to crystallization of $Y(NO_3)_3 \cdot 4H_2O$, $Y(NO_3)_3 \cdot 2NH_4NO_3$ (congruently soluble), and NH_4OH. The eutonic points between $Y(NO_3) \cdot 2NH_4NO_3$ and $Y(NO_3)_3 \cdot 4H_2O$ are given below.

AUXILIARY INFORMATION

METHOD/APPARATUS/PROCEDURE:

Isothermal method. Equilibrium stated to be reached in 2 days. Nitrogen was determined by the Kjeldahl method. Yttrium was determined by precipitation of $Y(OH)_3$ with ammonia followed by ignition at 800-900°C to form the oxide Y_2O_3. Composition of the solid phases were determined by Schreinemakers method. The solid phases were subjected to thermographical and crystal-optical analysis.

COMMENTS AND/OR ADDITIONAL DATA:

The eutonic points between $Y(NO_3)_3 \cdot 2NH_4NO_3$ and $Y(NO_3)_3 \cdot 4H_2O$ and vice versa are, respectively:

1. $Y(NO_3)_3$ = 42% and NH_4NO_3 = 18.8 mass %
2. $Y(NO_3)_3$ = 13.8 % and NH_4NO_3 = 49.9 mass %.

SOURCE AND PURITY OF MATERIALS:

The hexahydrate $Y(NO_3)_3 \cdot 6H_2O$ was prepared by dissolving 99.23 % Y_2O_3 in an excess of 58.5 % nitric acid followed by crystallization. C.p. grade NH_4NO_3 was recrystallized before use.

ESTIMATED ERROR:

Soly: Nothing specified.

Temp: precision ± 0.1 K.

REFERENCES:

COMPONENTS:

(1) Yttrium nitrate; $Y(NO_3)_3$; [10361-93-0]

(2) Urea; CH_4N_2O; [57-13-6]

(3) Water ; H_2O ; [7732-18-5]

ORIGINAL MEASUREMENTS:

Khudaibergenova, N.; Sulaimankulov, K.

Zh. Neorg. Khim. 1981, *26*, 1156-9;

Russ. J. Inorg. Chem. Engl. Transl. 1981, *26*, 627-8.

VARIABLES:

Composition at 30°C

PREPARED BY:

M. Salomon

EXPERIMENTAL VALUES:

$Y(NO_3)_3$ - $CO(NH_2)_2$ - H_2O system at 30°C

Composition of saturated solutions[a]

$Y(NO_3)_3$		$CO(NH_2)_2$		
mass %	mol kg^{-1}	mass %	mol kg^{-1}	nature of the solid phase
——	——	57.56	22.583	$CO(NH_2)_2$
13.47	1.617	56.23	30.901	"
27.08	6.088	56.74	58.392	"
32.92	10.278	55.43	79.225	"
34.58	12.224	55.13	89.211	$Y(NO_3)_3 \cdot 4CO(NH_2)_2$
35.55	10.150	51.71	67.585	"
36.64	7.454	45.48	42.354	"
39.58	6.482	38.21	28.647	"
42.61	6.346	34.54	24.115	"
45.29	7.978	34.06	27.464	"
47.64	7.634	29.66	21.757	"
47.40	7.349	29.14	20.683	$Y(NO_3)_3 \cdot 3CO(NH_2)_2$
48.79	7.488	27.51	19.328	"
49.26	6.090	21.32	12.067	"
50.60	——	49.61[b]	——	"
56.10	7.195	15.54	9.124	"
58.09	7.592	14.08	8.424	"
59.46	8.052	13.68	8.481	"

continued ....

AUXILIARY INFORMATION

METHOD/APPARATUS/PROCEDURE:

Isothermal method used. Equilibrium was reached after 7-9 h. After the liquid and solid phases had been separated, their nitrogen content was determined by the Kjeldahl method, and yttrium was determined as described previously (1).

COMMENTS AND/OR ADDITIONAL DATA:

The complex $Y(NO_3)_3 \cdot 3CO(NH_2)_2$ is incongruently soluble, and $Y(NO_3)_3 \cdot 4CO(NH_2)_2$ is congruently soluble. The compositions of these salts were confirmed by chemical analyses.

SOURCE AND PURITY OF MATERIALS:

"Chemically pure" grade urea and crystalline yttrium nitrate hydrate were used. No other information given.

ESTIMATED ERROR:

Soly: accuracy for the $Y(NO_3)_3$ - H_2O binary system very poor (see critical evaluation).

Temp: nothing specified.

REFERENCES:

1. Khudaibergenova, N.; Sulaimankulov, K. *Zh. Neorg. Khim.* 1979, *24*, 2005.

COMPONENTS:	ORIGINAL MEASUREMENT:
(1) Yttrium nitrate; $Y(NO_3)_3$; [10361-93-0]	Khudaibergenova, N.; Sulaimankulov, K.
(2) Urea; CH_4N_2O; [57-13-6]	*Zh. Neorg. Khim.* 1981, *26*, 1156-9; *Russ. J. Inorg. Chem. Engl. Transl.* 1981, *26*, 627-8.
(3) Water ; H_2O ; [7732-18-5]	

EXPERIMENTAL VALUES: continued ....

$Y(NO_3)_3$		$CO(NH_2)_2$		
mass %	mol kg^{-1}	mass %	mol kg^{-1}	nature of the solid phase
60.08	7.872	12.16	7.294	$Y(NO_3)_3 \cdot 3CO(NH_2)_2 + Y(NO_3)_3 \cdot 4H_2O$
60.98	7.635	9.97	5.714	$Y(NO_3)_3 \cdot 4H_2O$
60.38	7.080	8.60	4.616	"
60.08	5.772	2.06	0.906	"
60.34	5.534	———	———	"

a. Molalities calculated by the compiler.

b. This appears to be a typographical error. The correct value may be 19.61 mass % $CO(NH_2)_2$.

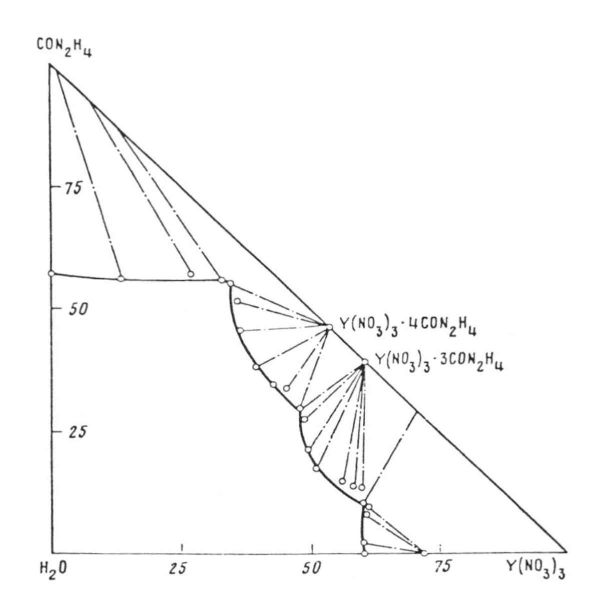

30°C isotherm for the $Y(NO_3)_3 - CO(NH_2)_2 - H_2O$ system. Concentrations in units of mass %.

COMPONENTS:

(1) Yttrium nitrate; $Y(NO_3)_3$; [10361-93-0]

(2) Diethyl ether; $C_4H_{10}O$; [60-29-7]

ORIGINAL MEASUREMENTS:

Wells, R.C. *J. Wash. Acad. Sci.* 1930, *20*, 146-8.

VARIABLES:

Room temperature (about 20°C)

PREPARED BY:

T. Mioduski, S. Siekierski, M. Salomon

EXPERIMENTAL VALUES:

Experiment 1. This experiment involves the hydrated yttrium nitrate as the initial solid, and which the compilers assume to be the hexahydrate.

Authors report the solubility as 0.0792 g Y_2O_3 in 10 ml ether.

This is equivalent to a $Y(NO_3)_3$ soly of 0.0701 mol dm^{-3} (compilers).

Experiment 2. This experiment involved yttrium nitrate dehydrated as described in the METHOD/APPARATUS/PROCEDURE box below.

Authors report the solubility as 0.0803 Y_2O_3 in 10 ml ether.

This is equivalent to a $Y(NO_3)_3$ soly of 0.0711 mol dm^{-3} (compilers).

AUXILIARY INFORMATION

METHOD/APPARATUS/PROCEDURE:

The isothermal method was used. The soly of yttrium nitrate was determined in two experiments in which the nature of the initial solid phase differs.

Experiment 1. A few grams of yttrium nitrate (presumably the hexahydrate, compilers) were added to about 20 ml of ether in small stoppered flasks. The flasks were periodically agitated and permitted to stand at about 20°C overnight. A 10 ml sample was removed, filtered, the solvent evaporated and the salt ignited to the oxide and weighed.

Experiment 2. The remaining salt in the flask was freed from ether, dissolved in water and a few drops of HNO_3 added. The solution was evaporated to dryness and heated to 150°C. The solubility in ether was then determined again with this "dehydrated" salt.

SOURCE AND PURITY OF MATERIALS:

Nothing specified.

ESTIMATED ERROR:

Soly: precision probably around ± 10 % (compilers).

Temp: precision probably ± 4 K (compilers).

REFERENCES:

COMPONENTS:

(1) Lanthanum nitrate; $La(NO_3)_3$; [10099-59-9]

(2) Water ; H_2O ; [7732-18-5]

EVALUATOR: S. Siekierski, T. Mioduski
Institute for Nuclear Research
Warsaw, Poland
and
M. Salomon
U.S. Army ET & DL
Ft. Monmouth, NJ, USA
May 1982

CRITICAL EVALUATION:

THE BINARY SYSTEM

INTRODUCTION

Data for the solubility of $La(NO_3)_3$ in water has been reported in 35 publications (1-23, 25-32, 34-37). Many of these studies deal with ternary systems, and the solubility in the binary system is given as one point on a phase diagram. It appears that for many of these studies involving ternary systems, the solubility in the binary system was measured only once by a given research group, and that this value was reported in subsequent publications. It thus appears that of these 35 publications, only 25 report independently determined solubilities in water.

Only one study (3) reports the solubility determined by the synthetic method, and all other studies employ the isothermal method. With very few exceptions, most papers do not report experimental errors, and the compilers and evaluators have estimated either the precision or accuracy when possible.

Depending upon temperature and composition (e.g. in ternary systems which are discussed below), equilibrated solid phases of varying degrees of hydration have been reported. The following solid phases have been identified:

$La(NO_3)_3.6H_2O$	[10277-43-7]	$La(NO_3)_3.3H_2O$	[80573-05-3]
$La(NO_3)_3.5H_2O$	[15878-72-5]	$La(NO_3)_3.2H_2O$	[80573-06-4]
$La(NO_3)_3.4H_2O$	[15878-75-8]	$La(NO_3)_3$	

It is interesting to note that no evidence has been reported indicating the existence of a stable or metastable dihydrate.

The temperature dependence of the solubility of $La(NO_3)_3$ in pure water has been studied by Friend (1) from 273-239 K, by Brunisholz et al. (2) over the range of 273-323 K, and by Mironov and Popov (3) over the range of 247-395 K. The results of Friend are inconsistent with most other data, and they are rejected. This author claims the existence of α and β forms of the hexahydrate, $La(NO_3)_3.6H_2O$, with the α form being more stable at lower temperatures. The existence of α and β hexahydrate phases has not been confirmed by other investigators. Although Friend's data are in error by what appears to be a large systematic error, they are still of interest for several reasons. Friend reports the $\alpha \rightarrow \beta$ transition temperature as approximately 316 K which is very close to the value of 316.4 K for the hexahydrate to pentahydrate transition determined graphically by the evaluators below. It thus appears that Friend's α - phase corresponds to the hexahydrate and his β - phase corresponds to the pentahydrate as originally suggested by Mironov and Popov (3). Friend also reported the melting point of the hexahydrate as approximately 338.6 K which is very close to the value of 339.7 K for the congruently melting hydrate determined theoretically by the evaluators below. Based on these results and the quantitative results for the solubilities mainly from references (2-8, 27) we conclude that the stable solid phase below 316.4 K is the hexahydrate, and above 316.4 K the stable solid phase is the pentahydrate. Below we present evidence which suggests that above 343 K, the stable solid phase is a lower hydrate or a mixture of lower hydrates.

EVALUATION PROCEDURE

The data reported in the compilations were examined and either rejected immediately because of large obvious errors, or were analysed by a weighted least squares fit to a smoothing equation. It should be noted that only experimental solubility values were used in the least squares analyses: smoothed or extrapolated data were not used. The data were fitted to the general solubility equation (see INTRODUCTION and references 42, 43)

$$Y = \ln(m/m_o) - nM_2(m - m_o) = a + b/(T/K) + c\ln(T/K)$$

In this equation m is the solubility in mol kg^{-1} at temperature T/K, m_o is an arbitrarily selected reference molality (usually for 298.2 K), M_2 the molar mass of the solvent (18.0153 g mol^{-1}), and a, b and c are constants from which the enthalpy of solution, $\Delta Hsln$, and heat capacity of solutions, ΔCp, can be estimated (42, 42). In fitting the solubility data to eq. [1], weight factors of 0, 1, 2, 3 were assigned to each

COMPONENTS:	EVALUATOR:
(1) Lanthanum nitrate; $La(NO_3)_3$; [10099-59-9] (2) Water ; H_2O ; [7732-18-5]	S. Siekierski, T. Mioduski Institute of Nuclear Research Warsaw, Poland and M. Salomon U.S. Army ET & DL Ft. Monmouth, NJ, USA May 1982

CRITICAL EVALUATION: (continued)

solubility value depending upon the precision of the solubility determination and the temperature.

In the fitting of the data to eq. [1], if the calculated error between the observed and calculated molalities, Δm, was larger than twice the standard error of estimate, σ_m, the data point was either rejected or its weight factor decreased. The fitting of the data was repeated until all Δm values were equal to or less than $\pm 2\sigma_m$.

SOLUBILITIES IN THE $La(NO_3)_3 \cdot 6H_2O - H_2O$ SYSTEM

A number of papers have been rejected. The papers by Kolesnikov et al. (24, 33) report the solubility of lanthanum nitrate at 293.2 K and 323.2 K with trace amounts of $Nd(NO_3)_3$ (24) or $Sm(NO_3)_3$ (33). It is not possible to extrapolate the $La(NO_3)_3$ solubility to zero $Nd(NO_3)_3$ or $Sm(NO_3)_3$ concentration because of too much scatter in the data, and inconsistencies in some results (e.g. the solubility at 293 K is greater than that at 298 K). Data from the ternary study of the $La(NO_3)_3$-$Sm(NO_3)_3$-H_2O system by Petelina et al. (38) were not compiled because of incompleteness: these investigators report mass % values for $La(NO_3)_3$ along 5 sections of the phase diagram. Most of the data are presented graphically and some numerical data are given in the text of the original publication. The data of Templeton and Daly (39) were also rejected because the aqueous phase must contain an unknown but significant amount of 1-hexanol (as discussed in the compliation). The data of Friend (1) were rejected for the reasons discussed above. One paper (7) was not compiled because the experimental details are the same as in (4-6) which have been compiled.

The data considered are given in Table 1 along with the initial and final weight factors used in the least squares fitting to the smoothing equation [1]. In assigning the initial weight factors, we first consider the solubility data at 298.2 K.

Table 1. Solubility of $La(NO_3)_3$ in the $La(NO_3)_3 \cdot 6H_2O - H_2O$ system.

T/K	solubility mol kg^{-1}	ref	weight initial/final	T/K	solubility mol kg^{-1}	ref	weight initial/final
258.2	3.39	3	2/2	298.2	4.671	9,25	1/0
273.15	3.760	2	3/3	298.2	4.688	11,32	1/0
273.2	3.14	15	0/0	298.2	3.985	12	0/0
275.9	3.67	3	2/0	298.2	4.704	13	0/0
278.15	3.898	2	3/3	298.2	4.403	14	0/0
278.5	3.81	3	2/0	298.2	4.58	19,20	1/1
283.15	4.095	2	3/3	298.2	4.649	21	1/1
289.0	4.23	3	2/2	298.2	4.411	23	1/0
293.15	4.423	2	3/3	298.2	4.544	37	1/0
293.15	4.405	36	3/3	300.9	4.81	3	2/0
293.2	4.642	22	0/0	308.15	5.095	2	3/3
293.2	4.25	28,29,34	0/0	309.5	5.15	3	2/2
293.2	4.52	35	0/0	313.2	4.864	14	0/0
294.6	4.47	3	2/2	323.15	6.052	2	3/3
298.15	4.613	2	3/3	323.2	6.297	9.25	0/0
298.15	4.610	4	3/3	323.2	6.11	16,18,28	1/1
298.15	4.610	5	3/3	323.2	5.84	10,15,17 18,30,31	0/0
298.15	4.608	6	3/3	323.2	6.151	26	2/0
298.15	4.610	7	3/3	323.2	5.59	29,34	0/0
298.2	4.669	27	1/0	333.2	7.013	12	1/0
298.2	4.34	10,15,18 28,30,31	0/0				

COMPONENTS:	EVALUATOR:
(1) Lanthanum nitrate; $La(NO_3)_3$; [10099-59-9] (2) Water ; H_2O ; [7732-18-5]	S. Siekierski, T. Mioduski Institute for Nuclear Research Warsaw, Poland and M. Salomon U.S. Army ET & DL Ft. Monmouth, NJ, USA May 1982

CRITICAL EVALUATION:

The solubility of $La(NO_3)_3$ at 298.2 K has been reported in 27 publications as indicated in Table 1 (excluding those rejected papers). A number of publications are considered by the evaluators not to be reports of independently determined solubilities: i.e. it appears that the solubility was measured once by a given research group and the same value reported in two or more publications. The data in (9,25) appear to represent a single measurement and we list the given value once in Table 1. Similarly the results from (11,31) are identical and are treated as a single determination in the table. The solubility of 4.34 mol kg^{-1} has been reported in 8 separate publications by Gorshunova and Zhuravlev (10,15-18,28,30,31) and again this value is treated as a single determination in Table 1. Even if we were to assume that each of these publications report an independently measured solubility, the final results for the critically evaluated data would remain unchanged because in the final analyses these data are assigned zero weights. In the initial assignments of weights all solubilities at 298.2 K falling below 4.41 mol kg^{-1} were assigned zero weight. In addition the highest value of 4.704 mol kg^{-1} was assigned an initial weight of zero. The mean of the remaining 12 values is 4.61 mol kg^{-1} with a standard deviation of σ = 0.07 mol kg^{-1}. The mean of the most precise results (2,4-7) is 4.610 mol kg^{-1} with σ = 0.002 mol kg^{-1}. At the 95% confidence level and a Student's t = 2.751, the uncertainly in this average value is ± 0.002 mol kg^{-1}. Combining this uncertainly with the experimental precision of ± 0.1%, the *recommended* solubility of $La(NO_3)_3$ at 298.15 K is 4.610 mol kg^{-1} with an estimated overall uncertainty of ± 0.005 mol kg^{-1}.

In the initial weight assignments, all values reported by Brunisholz et al. (2,36) and by Spedding and co-workers (4-7) are assigned weights of 3. The data of Mironov and Popov (3) must be considered to be of high accuracy since the solubilities were determined by the synthetic method, and the accuracy is determined by the experimental accuracy in weighing the components which is probably better than ± 0.1%. The major source of error in Mironov and Popov's results is due to the visual observation of the temperatures of crystallization, and there are a number of cases in which the solubilities reported by Mironov and Popov deviate significantly from the smoothed solubility curve (see Table 1 and Figure 1). The precision in the visually determined temperatures of crystallization was given as ± 0.2 K, and the evaluators estimate an overall uncertainly of around ± 0.02 mol kg^{-1} due mainly to this imprecision in temperature: all of Mironov and Popov's data were assigned initial weights of 2. The remaining data were assigned weights of 0 or 1 depending upon the estimated precision.

Initial weight factors for solubilities at temperatures other than 298.2 K were assigned in a similar manner. The data reported by Brunisholz and co-workers (2,36) were assigned weights of 3, those of Mironov and Popov (3) were assigned weights of 2. The remaining data were assigned weights from 0-2 depending upon the reported or compilers' estimates of experimental precision. It is noted that in one paper by Gorshunova and Zhuravlev (18), the solubility of $La(NO_3)_3$ at 323.2 K is reported as 5.84 mol kg^{-1} and 6.11 mol kg^{-1}. While the lower solubility value must be rejected, it is interesting to note that this value lies very close to the curve for the stable pentahydrate system (see Figure 1 and Table 4).

The results of the analyses for the solubility of $La(NO_3)_3$ in the $La(NO_3)_3.6H_2O$ - H_2O system are given in Tables 3 and 4, and in Figure 1. Table 3 lists the parameters of the smoothing equation, and Table 4 gives recommended solubilities obtained from the smoothing equation.

The uncertainly in the calculated *recommended* solubilities is ± 0.005 mol kg^{-1} at a 95% level of confidence (Student's t = 2.00), and combining this with an average experimental precision of about 0.2%, the overall uncertainly in the *recommended* values in Table 4 is ± 0.011 mol kg^{-1}. Figure 1 is the phase diagram showing the domains of the stable and metastable hydrates.

SOLUBILITIES IN THE $La(NO_3)_3.nH_2O$ - H_2O SYSTEM: $n \leq 5$

The solubility of the pentahydrate at 298.2 K has been reported in two papers by the same research group (8,27). As pointed out in the compilation of (8), it is not clear whether the value of 5.13 mol kg^{-1} represents excellent agreement of two independent measurements or if the more recent paper (8) is simply reporting the results of the earlier study (27). In (8) the data from the ternary $La(NO_3)_3$-HNO_3-H_2O system reported in (27) were used in Kirgintsev's equation (41) to reproduce the solubility branch for the pentahydrate:

$$\log(\text{soly/mol kg}^{-1}) = \log 5.13 - 0.637 \log y_1$$

COMPONENTS:	EVALUATOR:
(1) Lanthanum nitrate; $La(NO_3)_3$; [10099-59-9] (2) Water ; H_2O ; [7732-18-5]	S. Siekierski, T. Mioduski Institute for Nuclear Research Warsaw, Poland and M. Salomon U.S. Army ET & DL Ft. Monmouth, NJ, USA May 1982

CRITICAL EVALUATION: (continued)

where y_1 is the mole fraction of $La(NO_3)_3$ in the saturated solution (the solute mole fraction is defined by mole fraction $La(NO_3)_3$ + mole fraction HNO_3 = 1). The solubility of the metastable tetrahydrate could not be measured at 298.2 K, but a value could be obtained (8) by extrapolation using Kirgintsev's equation. The relation obtained in (8) for the solubility branch of the tetrahydrate in the $La(NO_3)_3$-HNO_3-H_2O system is

$$\log (\text{soly/mol kg}^{-1}) = \log 7.35 - 0.609 \log y_1$$

Thus the solubility of the metastable tetrahydrate in pure water at 298.2 K is calculated to be 7.35 mol kg^{-1}.

All other data for the solubilities of the lower hydrates as a function of temperature are reported in one publication by Mironov and Popov (3). The data are given in Table 2 from which it is seen that there is no definitive assignment of the nature of the solid phases above 314.5 K. The evaluators were however able to assign reasonable solid phase compositions for several of these points which permitted the analyses of the pentahydrate solubility branch by the smoothing equation. The data fitted to eq. [1] were all assigned equal weights of unity because they represent the results of one research group, and there are no other publications available with which a comparative study can be made. The results of the analyses are given in Tables 3 and 4, and in Figure 1.

The uncertainty in the calculated *tentative* solubilities in Table 4 for the pentahydrate system is ± 0.20 mol kg^{-1} at the 95% level of confidence (Student's t = 2.78). Combining this uncertainty with the estimated experimental precision of 0.5%, the overall uncertainty in the *tentative* values in Table 4 is ± 0.21 mol kg^{-1}.

Table 2. Solubility in the $La(NO_3)_3 \cdot nH_2O$ system: $n \leqslant 5$

T/K	solubility mol kg^{-1}	ref	value of n in $La(NO_3)_3 \cdot nH_2O$ author's assignment	evaluator's assignment
298.2	5.123	8,27	5	
298.2	7.35[a]	8	4	
314.5	5.47	3	5 + 4	
321.8	5.79	3	5 + 4 + 3[b]	5
325.9	6.11	3	5 + 4 + 3	5
328.6	6.42	3	5 + 4 + 3	5
334.1	7.25	3	5 + 4 + 3	5
343.1	8.49	3	5 + 4 + 3	5
353.1	10.07	3	5 + 4 + 3	
371.6	11.44	3	5 + 4 + 3	
395.2	12.23	3	5 + 4 + 3	

a. Extrapolated value: see text for discussion.

b. Note that for a 2-component system at constant pressure, the phase rule specifies that only three phases can coexist in equilibrium. Thus the assignment of three solid phases in equilibrium with the saturated solutions cannot be correct.

RECOMMENDED AND TENTATIVE SOLUBILITIES

At 298.15 K the *recommended* solubility of $La(NO_3)_3$ in the hexahydrate system is 4.610 mol kg^{-1} with an overall uncertainty of ± 0.005 mol kg^{-1} at the 95% confidence level. The recommended and tentive solubilities at other temperatures were obtained from the smoothing equation, and the results are given in Tables 3 and 4, and in the phase diagram of Figure 1. The solubilities in the hexahydrate system are designated as *recommended* values, and at the 95% confidence level, the total uncertainty in these solubilities is ± 0.011 mol kg^{-1}. The pentahydrate data are designated as *tentative*, and at the 95% confidence level, the overall uncertainty in these values is ± 0.21 mol kg^{-1}. These recommended and tentative solubilities were used to construct the phase diagram of Figure 1.

COMPONENTS:	EVALUATOR:
(1) Lanthanum nitrate; $La(NO_3)_3$; [10099-59-9] (2) Water ; H_2O ; [7732-18-5]	S. Siekierski, T. Mioduski Institute for Nuclear Research Warsaw, Poland and M. Salomon U.S. Army ET & DL Ft. Monmouth, NJ, USA May 1982

CRITICAL EVALUATION: (continued)

In Figure 1 the lines were calculated from eq. [1], and the experimental data were selected from those publications reporting the most precise solubilities as discussed above. Support of the assignment of domains of existence of the hexahydrate and pentahydrate comes from direct chemical analyses and from the temperature dependence of the solubilities. The temperature of the transition of the stable hexahydrate to the stable pentahydrate is found graphically to be 316.5 K. This is in excellent agreement with Mironov and Popov's value (3) of 314.5 K, and Friend's value (1) of 316 K. The predicted values of the congruent melting points of the hexahydrate and penthydrate are 339.7 K and 346.3 K, respectively. For the hexahydrate, the experimental values for the congruent melting of the solid are 338.6 K (1), 338 K (28), 339.7 ± 0.5 K (44, 45) , 343 ± 3 K (27), and 343 ± 1 K (46). The only reported value for the congruent melting point of the pentahydrate is that of Mironov, Popov, and Khirpin (27) who reported a value of 343 ± 3 K (i.e. almost identical to the melting point of the hexahydrate).

We were not able to fit the solubility data above 343.1 K to the smoothing equation because of the lack of sufficient knowledge of the solid phase compositions, and the lack of sufficient data points to permit a statistical analysis. Noting that the congruent melting point of the tetrahydrate is 368 ± 3 K (27), it is possible that the two solubility values at 343.1 K and 353.1 K fall on a tetrahydrate solubility isotherm.

To summarize the analyses, the hexahydrate is the stable solid phase up to 316.4 K, and the pentahydrate is the stable solid phase up to about 343.1 K. The metastable pentahydrate is however very stable below 316.4 K as indicated by the failure to convert the metastable pentahydrate to the stable hexahydrate by seeding (27). According to Mironov, Popov, and Khripin, different hydrated forms of $La(NO_3)_3$ can be obtained from identical aqueous solutions, and that discrepancies in the solubilities reported in the literature can often be attributed to the uncertainty in the nature of the hydrate involved. As an example the data of Gorshunova and Shuravlev are cited (see ref 18 in particular and the discussion above): the solubility of 5.84 mol kg^{-1} in the hexahydrate system at 323.2 K is probably in error due to the probability that the solutions were in equilibrium with the pentahydrate solid phase. The solid phase (or phases) which exists above 343.1 K is probably a lower hydrate such as the tetra or trihydrates, or a mixture of lower hydrates.

Table 3. Parameters and standard deviations for the smoothing equation [1]

	hexahydrate	pentahydrate
a	-20.980	-131.817
b	586.81	5815.5
c	3.3369	19.967
σ_a	0.002	0.009
σ_b	0.53	2.9
σ_c	0.0003	0.001
σ_Y	0.002	0.009
σ_{soly}	0.02	0.2
ΔH_{sln}/kJ mol^{-1}	-19.4	-192.8
ΔC_p/J K^{-1} mol^{-1}	111.0	655.1
congruent melting point	339.7 K	346.3 K
concn at the melting point/mol kg^{-1}	9.251	11.102

COMPONENTS:	EVALUATOR:
(1) Lanthanum nitrate; $La(NO_3)_3$; [10099-59-9] (2) Water ; H_2O ; [7732-18-5]	S. Siekierski, T. Mioduski Institute for Nuclear Research Warsaw, Poland and M. Salomon U.S. Army ET & DL Ft. Monmouth, NJ, USA May 1982

CRITICAL EVALUATION: (continued)

Table 4. Recommended and Tentative Solubilities obtained from the smoothing equation All solubilities given in units of mol kg^{-1}.

T/K	hexahydrate[a]	pentahydrate[b,c]
273.2	3.76	
278.2	3.91	
283.2	4.06	
288.2	4.23	
293.2	4.41	5.10
298.2	4.61	5.11
303.2	4.84	5.16
308.2	5.08	5.27
313.2	5.37	5.44
318.2	5.69	5.66
323.2	6.06	5.96
333.2	7.15	6.86
343.2		8.79

a. Recommended values.
b. Tentative values.
c. For solubilities of the lower hydrates, see Table 2.

COMPONENTS:	EVALUATOR:
(1) Lanthanum nitrate; $La(NO_3)_3$; [10099-59-9] (2) Water ; H_2O ; [7732-18-5]	S. Siekierski, T. Mioduski Institute for Nuclear Research Warsaw, Poland and M. Salomon U.S. Army ET & DL Ft. Monmouth, NJ, USA May 1982

CRITICAL EVALUATION:

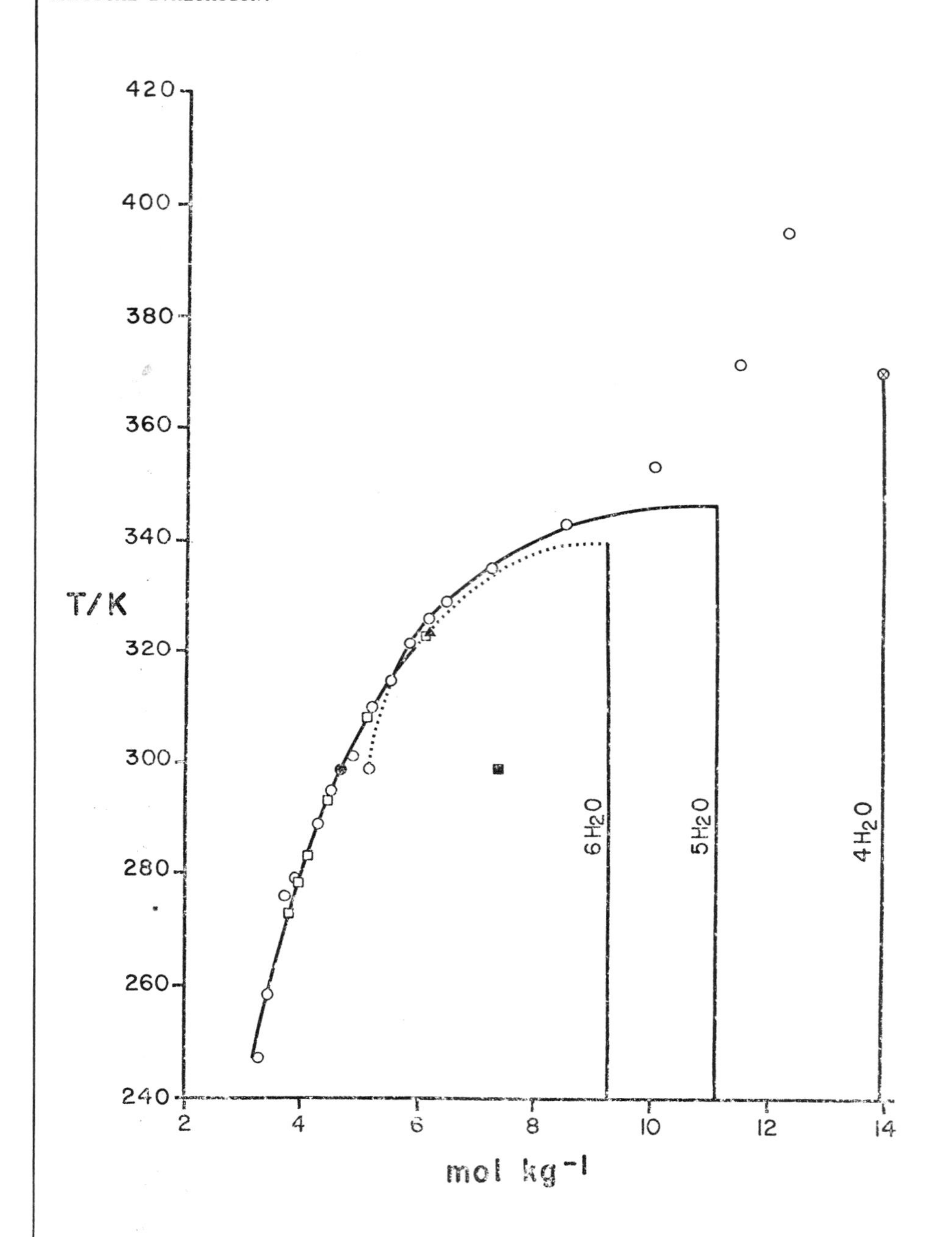

Figure 1. Phase diagram for the $La(NO_3)_3$-H_2O system.
Solid lines for stable phases and dashed lines for metastable phases were calculated from the smoothing equation. Experimental data: ○ (3); □ (2); ● (2,4-7); ▲ (16.18,28); ■ (8,27); ⊗ melting point of the tetrahydrate (27).

COMPONENTS:

(1) Lanthanum nitrate; $La(NO_3)_3$; [10099-59-9]

(2) Water ; H_2O ; [7732-18-5]

EVALUATOR: S. Siekierski, T. Mioduski
Institute for Nuclear Research
Warsaw, Poland
and
M. Salomon
U.S. Army ET & DL
Ft. Monmouth, NJ, USA
May 1982

CRITICAL EVALUATION:

TERNARY SYSTEMS

Systems with Nitric Acid. The two comprehensive studies available (26, 27) are in basic agreement that all hydrates (except the dihydrate) and the anhydrous salt are all stable at 298.2 K depending upon composition. At high nitric acid concentrations (over 80 mass %), the anhydrous salt in stable at 298.2 and 323.2 K. One study (27) reports the existence of the monohydrate at nitric acid concentrations between 65 and 88 mass %. The existence of a double salt with HNO_3 was not reported. A third study (40) reports solubilities in the four component systen $Y(NO_3)_3$-$La(NO_3)_3$-HNO_3-H_2O.

Systems with Alkali Metal and Ammonium Nitrates. The dominant feature in these studies is the existence of double salts of the type $M_2La(NO_3)_5.nH_2O$ where M is the alkali metal. Those double salts which were experimentally verified are:

$2KNO_3.La(NO_3)_3.2H_2O$	[77076-82-5]	(29)
$2RbNO_3.La(NO_3)_3.3H_2O$	[61391-42-2]	(32)
$2CsNO_3.La(NO_3)_3.3H_2O$	[61192-73-2]	(13)
$La(NO_3)_3.2NH_4NO_3.4H_2O$	[80573-07-5]	(9)

A study of the $NaNO_3$-$La(NO_3)_3$-H_2O system (34) did not report the existence of a double salt.

Systems with Alkaline Earth and Transition Metal Nitrates. The dominant feature in all these studies is the existence of a tetracosahydrate double salt with general formula $La_2M_3(NO_3)_{12}.24H_2O$. The following tetracosahydrates have been reported:

$3Mg(NO_3)_2.2La(NO_3)_3.24H_2O$	[13984-18-4]	(9, 23)
$2La(NO_3)_3.3Co(NO_3)_2.24H_2O$	[18851-80-4]	(14)
$2La(NO_3)_3.3Ni(NO_3)_2.24H_2O$	[18851-79-1]	(25)
$2La(NO_3)_3.3Zn(NO_3)_2.24H_2O$	[14520-95-7]	(23, 25, 36)

Although the tetracosahydrate $2La(NO_3)_3.3Mn(NO_3)_2.24H_2O$, [18851-78-0], is known to exist, it was not reported by di Capua (di Capua (22) did not analyse the solid phase in any of his work). No double salt was reported in the study of the $La(NO_3)_3$-$Ba(NO_3)_2$-H_2O system (11). In addition, the existence of solid solutions have been reported for systems containing $Mg(NO_3)_2$ (9, 23), $Ni(NO_3)_2$ (25), and $Zn(NO_3)_2$ (25).

Systems with Other Lanthanide Salts. The dominant feature in those studies concerning other lanthanide nitrates is the existence of solid solutions. The work of Brunisholz et al. (2) suggest that these systems contain miscibility gaps. In the rejected paper (38), the existence of solid solutions in the $La(NO_3)_3$-$Sm(NO_3)_3$-H_2O was not reported whereas Brunisholz's study on this system does reveal the presence of solid solutions.

For systems with $LaCl_3$, no double salts were reported and the authors report a phase diagram of the simple eutonic type (20) (note that the data for the ternary system in this study were rejected for reasons discussed in the compilation). For the mixed lanthanum nitrate/acetate system (19) a double salt was found with the composition:

$$2La(NO_3)_3.La(CH_3COO)_3.13H_2O$$

COMPONENTS:	EVALUATOR:
(1) Lanthanum nitrate; $La(NO_3)_3$; [10099-59-9] (2) Water ; H_2O ; [7732-18-5]	S. Siekierski, T. Mioduski Institute for Nuclear Research Warsaw, Poland and M. Salomon U.S. Army ET & DL Ft. Monmouth, NJ USA May 1982

CRITICAL EVALUATION:

Other Systems. Most other studies on ternary systems involve hydrazine nitrates and organic nitrogen compounds. For those ternary systems in which double salts are not formed, the phase diagrams are generally of the simple eutonic type: these are the studies involving guanidine monontrite (17), diethylamine nitrate (15), dimethylamine nitrate (15), and triethylamine (35). Double salts were found in the remaining systems:

$La(NO_3)_3.2[N_2H_4.HNO_3]$	[33412-01-0]	(18)
$La(NO_3)_3.3[N_2H_4.2HNO_3]$	[33412-02-1]	(18)
$La(NO_3)_3.H_2NCH_2CH_2NH_2.2HNO_3.6H_2O$	[27099-40-7]	(16)
$La(NO_3)_3.2[C_5H_5N.HNO_3].4H_2O$	[51537-72-5]	(30)
$La(NO_3)_3.4[C_5H_{10}NH.HNO_3]$	[36354-70-8]	(28)
$La(NO_3)_3.3C_6H_5CONH_2.3H_2O$	[80573-10-0]	(12)
$La(NO_3)_3.C_9H_7.HNO_3.4H_2O$	[80573-09-7]	(31)

COMPONENTS:	EVALUATOR:
(1) Lanthanum nitrate; $La(NO_3)_3$; [10099-59-9] (2) Water ; H_2O ; [7732-18-5]	S. Siekierski, T. Mioduski Institute for Nuclear Research Warsaw, Poland and M. Salomon U.S. Army ET & DL Ft. Monmouth, NJ, USA May 1982

CRITICAL EVALUATION: (continued)

REFERENCES

1. Friend, J.N. *J. Chem. Soc.* 1935, 824.
2. Brunisholz, G.; Quinche, J.P.; Kalo, A.M. *Helv. Chim. Acta* 1964, *47*; 14.
3. Mironov, K.E.; Popov, A.P. *Rev. Roum. Chim.* 1966, *11*, 1373.
4. Spedding. F.H.; Shiers, L.E.; Rard, J.A. *J. Chem. Eng. Data* 1975, *20*, 88.
5. Rard, J.A.; Spedding, F.H. *J. Phys. Chem.* 1975, *79*, 257.
6. Spedding, F.H.; Derer, J.L.; Mohs, M.A.; Rard, J.A. *J. Chem. Eng. Data* 1976, *21*, 474.
7. Spedding, F.H.; Shiers, L.E.; Brown, M.A.; Baker, J.L.; Guitierrez, L; McDowell, L.S.; Habenschuss, A. *J. Phys. Chem.* 1975, *79*, 1087.
8. Popov, A.P.; Mironov, K.E. *Zh. Neorg. Khim.* 1971, *16*, 464.
9. Urazov, G.G.; Shevtsova, Z.N. *Zh. Neorg. Khim.* 1957, *2*, 655.
10. Gorshunova, V.P.; Zhuravlev, E.F. *Zh. Neorg. Khim.* 1970, *15*, 1422.
11. Molodkin, A.K.; Odinets, Z.K.; Vargas Ponse, O. *Zh. Neorg. Khim.* 1977, *22*, 3388.
12. Runov, N.N.; Shchevev, A.V. *Zh. Neorg. Khim.* 1980, *25*, 721.
13. Molodkin, A.K.; Odinets, Z.K.; Pereira Paveze, V. *Zh. Neorg. Khim.* 1976, *21*, 2792.
14. Odent, G.; Venot, A. *C.R. Hebd. Seances Acad. Sci., Ser. C.* 1975, *280*, 377.
15. Gorshunova, V.P.; Zhuravlev, E.F. *Zh. Neorg. Khim.* 1970, *15*, 3355.
16. Zhuravlev, E.F.; Gorshunova, V.P. *Zh. Neorg. Khim.* 1970, *15*, 195.
17. Gorshunova, V.P.; Zhuravlev, E.F. *Zh. Neorg. Khim.* 1971, *16*, 1739.
18. Gorshunova, V.P.; Zhuravlev, E.F. *Zh. Neorg. Khim.* 1971, *16*, 1700.
19. Petelina, V.S.; Nikurashina, N.I.; Bakhtiarova, G.A. *Izv. Vysch. Ucheb. Zaved. Khim. Khim. Tekhnol.* 1971, *14*, 1611.
20. Petelina, V.S.; Mertslin, R.V.; Nikurashina, N.I.; Sedova, L.K. *Issled. v. Obl. Khim. Redkozem. Elementov* 1969, 85.
21. James, C.; Whittemore, C.F. *J. Am. Chem. Soc.* 1912, *34*, 1168.
22. di Capua, C. *Gazz. Chim. Ital.* 1929, *59*, 164.
23. Yakimov, M.A.; Gizhavina, E.I. *Zh. Neorg. Khim.* 1971, *16*, 507.
24. Kolosnikov, A.A.; Korotkevich, I.B.; Bui Van Tuan; Stepin, B.D. *Zh. Neorg. Khim.* 1978, *23*, 2833.
25. Urazov, G.G.; Shevtsova, Z.N. *Zh. Neorg. Khim.* 1957, *2*, 659.
26. Quill, L.L.; Robey, R.F. *J. Am. Chem. Soc.* 1937, *59*, 2591.
27. Mironov, K.E.; Popov, A.P.; Khripin, L.A. *Zh. Neorg. Khim.* 1966, *11*, 2789.
28. Gorshunova, V.P.; Zhuravlev, E.F. *Zh. Neorg. Khim.* 1972, *17*, 231.
29. Bogdanovskaya, R.L.; Armisheva, L.N. *Uch. Zap. Perm. Univ.* 1973, *289*, 32.
30. Gorshunova, V.P.; Zhuravlev, E.F. *Zh. Neorg. Khim.* 1974, *19*, 249.
31. Gorshunova, V.P.; Zhuravlev, E.F. *Zh. Neorg. Khim.* 1973, *18*, 1688.
32. Molodkin, A.K.; Odinets, Z.K.; Vargas Ponse, O. *Zh. Neorg. Khim.* 1976, *21*, 2590.
33. Kolesnikov, A.A.; Korotkench, I.B.; Shakhaleeva, N.N.; Stepin, B.D. *Zh. Neorg. Khim.* 1978, *23*, 2524.
34. Bogdanovskaya, R.L. *Uch. Zap. Perm. Univ.* 1970, *229*, 32.
35. Shabikova, G. Kh.; Sergeeva, V.F.; Izmeleuova, M.B. *Zh. Obshch. Khim.* 1975, *45*. 990.
36. Brunisholz, G.; Klipfel, K. *Rec. Chim. Miner.* 1970, *7*, 349.
37. Zwietasch, K.-J.; Kirmse, E.M. *Wiss. Hefte d. Päd. Hochschule "W. Ratke" Köthen.* 1977, *4*, 133.
38. Petelina, V.S.; Guzhvina, O.V. *Issled. Mnogokomponent. Sistem s Razl. Vzaimodeistviem Komponentov (Saratov)* 1977, 86.
39. Templeton, C.C.; Daly, L.K. *J. Am. Chem. Soc.* 1951, *73*, 3989.
40. Perel'man, F.M.; Babievskaya, I.Z. *Zh. Neorg. Khim.* 1964, *9*, 986 (see the section on yttrium nitrates for this compilation).
41. Kirgintsev, A.N. *Izv. Akad. Nauk. SSSR, Ser. Khim. Nauk.* 1965, *No. 8*, 1591.
42. Williamson, A.T. *Trans. Faraday Soc.* 1944, *40*, 421.
43. Counioux, J.-J.; Tenu, R. *J. Chim. Phys.* 1981, *78*, 816 and 823.
44. Quill, L.L.; Robey, R. *J. Am. Chem. Soc.* 1937, *59*, 1071.
45. Quill, L.L.; Robey, R.; Seifter, S. *Ind. Eng. Chem. Anal. Ed.* 1937, *9*, 389.
46. Wendlandt, W.W.; Sewell, R.G. *Texas J. Sci.* 1961, *13*, 231.

COMPONENTS:

(1) Lanthanum nitrate; $La(NO_3)_3$; [10099-59-9]

(2) Water; H_2O; [7732-18-5]

ORIGINAL MEASUREMENTS:

Friend, J.N. *J. Chem. Soc.* 1935, 824-6.

VARIABLES:

Temperature

PREPARED BY:

T. Mioduski, S. Siekierski, M. Salomon

EXPERIMENTAL VALUES:

Solubility of $La(NO_3)_3$ in water as a function of temperature:[a]

t/°C	mass %	mol kg^{-1}	solid phase
0	50.03	3.081	α-$La(NO_3)_3\cdot 6H_2O$
18.4	54.16	3.636	"
21.2	55.03	3.766	"
25.4	55.80	3.885	"
35.4	59.12	4.451	"
42.4	63.84	5.434	"
44.2	65.13	5.748	"
14.4	56.27	3.960	β-$La(NO_3)_3\cdot 6H_2O$
15.2	56.94	4.070	"
15.8	56.85	4.055	"
16.0	56.74	4.037	"
23.2	58.7	4.374	"
29.6	60.08	4.632	"
32.2	61.34	4.883	"
40.0	62.71	5.176	"
46.4	64.55	5.604	"
49.4	65.17	5.759	"
56.0	68.30	6.631	"
64.5[b]	75.04	9.253	"

[a] The transition temperature for the $\alpha \rightarrow \beta$ was reported as approximately 43°C as read from the intersection of the plots of solubility vs t/°C for the α and β hexahydrates.

[b] Melting point of the hexahydrate (theoretical concentration at the congruent melting point is 9.251 mol kg^{-1}, compilers).

AUXILIARY INFORMATION

METHOD/APPARATUS/PROCEDURE:

The isothermal method was used as described earlier (1). Aliquots of around 50 cc were taken from 250 cc "saturation bottles" and La precipitated as the oxalate. La was determined gravimetrically by ignition of the oxalate to the oxide. As a check of the method, 5 samples were evaporated to dryness and directly ignited to the oxide, and the results were always higher by 0.1 to 0.4 %.

COMMENTS AND/OR ADDITIONAL DATA:

This paper is the only study that reports α and β phases for the hexahydrate. This is discussed further in the critical evaluation.

The molalities in the table were calculated by the compilers.

SOURCE AND PURITY OF MATERIALS:

Lanthanum nitrate was prepared by dissolving the oxide in dilute nitric acid and inducing crystallization by seeding with $Bi(NO_3)_3\cdot 5H_2O$. The author claims that the initial precipitate is β-$La(NO_3)_3\cdot 6H_2O$ which is transformed into the more stable α form upon prolonged standing at room temperature. The α-salt is also said to form from the supersaturated solution "upon appropriate seeding." No details were given.

ESTIMATED ERROR:

Soly: Accuracy around ± 30 % (compilers).

Temp: Probably around ± 0.2 K as in (1).

REFERENCES:

1. Friend, J.N. *J. Chem. Soc.* 1930, 1633.

COMPONENTS:

(1) Lanthanum nitrate; $La(NO_3)_3$; [10099-59-9]

(2) Water ; H_2O ; [7732-18-5]

ORIGINAL MEASUREMENTS:

Brunisholz, G.; Quinche, J.P.; Kalo, A.M. *Helv. Chim. Acta* 1964, *47*, 14-27.

VARIABLES:

Temperature

PREPARED BY:

T. Mioduski and S. Siekierski

EXPERIMENTAL VALUES:

Solubility of $La(NO_3)_3$ in water[a]

t/°C	mass %	mol kg^{-1}
0	54.99	3.760
5	55.88	3.898
10	57.09	4.095
20	58.97	4.423
25	59.98	4.613
35	62.34	5.095
50	66.29	6.052

a. Molalities calculated by M. Salomon.

Authors report that the solid phase is $La(NO_3)_3 \cdot 6H_2O$

AUXILIARY INFORMATION

METHOD/APPARATUS/PROCEDURE:

The isothermal method was used. Lanthanum determined by complexometric titration using xylenol orange indicator in the presence of a small quantity of urotropine buffer. Water was determined by difference. No additional information given.

SOURCE AND PURITY OF MATERIALS:

La_2O_3 purified by the ion exchange method had a purity better than 99.7 %. The nitrate salt was presumably prepared from the oxide by addition of nitric acid followed by crystallization. No additional details were given.

ESTIMATED ERROR:

Soly: precision ± 0.2 % (compilers).

Temp: precision about ± 0.05 K (compilers).

REFERENCES:

COMPONENTS:

(1) Lanthanum nitrate; $La(NO_3)_3$; [10099-59-9]

(2) Water ; H_2O ; [7732-18-5]

ORIGINAL MEASUREMENTS:

Mironov, K.E.; Popov, A.P. *Rev. Roum. Chim.* 1966, *11*, 1373 - 81.

VARIABLES:

Temperature

PREPARED BY:

T. Mioduski and S. Siekierski

EXPERIMENTAL VALUES:

Solubility of $La(NO_3)_3$ in water[a]

t/°C	mass %	mol kg^{-1}	solid phase[b]	t/°C	mass %	mol kg^{-1}	solid phase[b]
-0.3	8.6	0.29	A	21.4	59.2	4.47	B
-2.6	18.5	0.699	A	27.7	61.0	4.81	B
-4.9	23.3	0.935	A	36.3	62.6	5.15	B
-9.5	35.6	1.70	A	41.3	64.0	5.47	B + C
-12.8	40.2	2.07	A				
-18.1	45.9	2.61	A	48.6	65.3	5.79	C + D + E
-22.9	48.9	2.95	A	52.7	66.5	6.11	C + D + E
-25.9	51.3	3.24	A + B	55.4	67.6	6.42	C + D + E
				60.9	70.2	7.25	C + D + E
-15.0	52.4	3.39	B	69.9	73.4	8.49	C + D + E
2.7	54.4	3.67	B	79.9	76.6	10.07	C + D + E
5.3	55.3	3.81	B	98.4	78.8	11.44	C + D + E
15.8	57.9	4.23	B	122.0	79.9	12.23	C + D + E

a. Molalities calculated by M. Salomon.

b. Solid phases: A = ice

B = $La(NO_3)_3 \cdot 6H_2O$

C = $La(NO_3)_3 \cdot 5H_2O$

D = $La(NO_3)_3 \cdot 4H_2O$

E = $La(NO_3)_3 \cdot 3H_2O$

AUXILIARY INFORMATION

METHOD/APPARATUS/PROCEDURE:

The synthetic method was used. The temperature of crystallization was determined visually. Authors state that only stable equilibrium data are given. No additional information given.

SOURCE AND PURITY OF MATERIALS:

$La(NO_3)_3 \cdot 6H_2O$ prepared by dissolving 99.9 % pure La_2O_3 in nitric acid followed by crystallization. Double distilled water was used.

ESTIMATED ERROR:

Soly: precision probably ± 0.2 to ± 0.5% (compilers).

Temp: precision ± 0.2 K (authors).

REFERENCES:

COMPONENTS:

(1) Lanthanum nitrate; $La(NO_3)_3$; [10099-59-9]

(2) Water; H_2O; [7732-18-5]

ORIGINAL MEASUREMENTS:

Popov, A.P.; Mironov, K.E. *Zh. Neorg. Khim.* 1971, *16*, 464-6; *Russ. J. Inorg. Chem. Engl. Transl.* 1971, *16*, 464-6.

VARIABLES:

One temperature: 25°C

PREPARED BY:

T. Mioduski, S. Siekierski, and M. Salomon

EXPERIMENTAL VALUES:

The solubility of metastable $La(NO_3)_3 \cdot 5H_2O$ in water at 25°C was reported to be

62.5 mass %

The corresponding value in molality was calculated by the compilers:

5.13 mol kg^{-1}

COMMENTS AND/OR ADDITIONAL DATA:

It is not clearly stated whether the solubility of the pentahydrate was actually measured in the present work or whether the authors are reporting the value obtained from earlier work (1). In reference (1) the solubility of the pentahydrate at 25°C was given as 62.47 mass % (see the compilation of the data from this paper).

By extrapolation of the solubility branch for $La(NO_3)_3 \cdot 4H_2O$ from data in the ternary system $La(NO_3)_3$-HNO_3-H_2O, the authors calculated the solubility of the metastable tetrahydrate in pure water as 70.5 mass % (7.35 mol kg^{-1}). See the critical evaluation for additional details.

AUXILIARY INFORMATION

METHOD/APPARATUS/PROCEDURE:

Experimental method not specified. Metastable (at 25°C) $La(NO_3)_3 \cdot 5H_2O$ was probably obtained by slow cooling of a saturated $La(NO_3)_3$ solution from 40-50°C to 25°C (1).

SOURCE AND PURITY OF MATERIALS:

No details given.

ESTIMATED ERROR:

Soly: Not specified, but precision probably ± 0.5 mass % units (compilers).

Temp: precision around ± 0.2 K (compilers).

REFERENCES:

1. Mironov, K.E.: Popov, A.P.; Khripin, L.A. *Zh. Neorg. Khim.* 1966, *11*, 2789: *Russ. J. Inorg. Chem. Engl. Transl.* 1966, *11*, 1499.

COMPONENTS:

(1) Lanthanum nitrate; $La(NO_3)_3$; [10099-59-9]

(2) Water; H_2O; [7732-18-5]

ORIGINAL MEASUREMENTS:

1. Spedding, F.H.; Shiers, L.E.; Rard, J.A. *J. Chem. Eng. Data* 1975, *20*, 88-93.
2. Rard, J.A.; Spedding, F.H. *J. Phys. Chem.* 1975, *79*, 257-62.
3. Spedding, F.H.: Derer, J.L.; Mohs, M.A.; Rard, J.A. *J. Chem. Eng. Data* 1976, *21*, 474-88.

VARIABLES:

One temperature: 25°C

PREPARED BY:

T. Mioduski, S. Siekierski, and M. Salomon

EXPERIMENTAL VALUES:

The solubility of $La(NO_3)$ in water at 25.00°C has been reported by Spedding and co-workers in three publications. Source paper [3] reports the solubility to be 4.608 mol kg^{-1}, but the preferred value is given in source papers [1] and [2] as 4.610 mol kg^{-1}.

COMMENTS AND/OR ADDITIONAL DATA:

Source paper [1] reports the relative viscosity, η_R, of a saturated solution to be 20.078. Taking the viscosity of water to be 0.008903 P at 25°C, the viscosity of a saturated $La(NO_3)_3$ solution at 25°C is 0.17875 P.

Supplementary data available in the microfilm edition to *J. Phys. Chem.* 1975, *79* have enabled the compilers to provide the following additional data on the $La(NO_3)_3$ - H_2O system.

The density of the saturated solution was calculated by the compilers from the smoothing equation in source paper [2], and the value at 25°C is 1.8097 kg m^{-3}. Using this density, the solubility of $La(NO_3)_3$ in volume units (based on m_{satd} = 4.6100 mol kg^{-1}) is

$$c_{satd} = 3.340 \text{ mol dm}^{-3}$$

Source paper [2] reports the electrolytic conductivity of the saturated solution to be (corrected for the electrolytic conductivity of the solvent) κ = 0.020369 S cm^{-1}.

The molar conductivity of the saturated solution as calculated from $1000\kappa/3c_{satd}$ is

$$\Lambda(\tfrac{1}{3}La(NO_3)_3) = 2.033 \text{ S cm}^2 \text{ mol}^{-1}$$

AUXILIARY INFORMATION

METHOD/APPARATUS/PROCEDURE:

Isothermal method used. Solutions were prepared as described in (1) and (2). The concentration of the saturated solution was determined by both EDTA (1) and sulfate (2) methods which is said to be reliable to ± 0.1 % or better. In the sulfate analysis, the salt was first decomposed with HCl followed by evaporation to dryness before H_2SO_4 additions were made. This eliminated the possibility of nitrate ion coprecipitation.

SOURCE AND PURITY OF MATERIALS:

$La(NO_3)_3 \cdot 6H_2O$ was prepared by addition of HNO_3 to the oxide. The oxide was purified by an ion exchange method and rare earth impurities were less than 0.1%. Ca and Fe impurities were also less than 0.1 %.

In source paper [3] the salt was analysed for water of hydration and found to be within ± 0.016 water molecules of the hexahydrate.

Water was distilled from alkaline permanganate.

ESTIMATED ERROR:

Soly: Duplicate analyses agreed to at least ± 0.1 %.

Temp: Not specified, but probably accurate to at least ± 0.01 K as in (3).

REFERENCES:

1. Spedding, F.H.; Cullen, P.F.; Habenschuss, A. *J. Phys. Chem.* 1974, *78*, 1106.
2. Spedding, F.H.; Pikal, M.J. Ayers, B.O. *J. Phys. Chem.* 1966, *70*, 2440.
3. Spedding, F.H.; et. al. *J. Chem. Eng. Data* 1975, *20*, 72.

COMPONENTS:

(1) Lanthanum nitrate; $La(NO_3)_3$; [10099-59-9]

(2) Nitric acid; HNO_3; [7697-37-2]

(3) Water ; H_2O ; [7732-18-5]

ORIGINAL MEASUREMENTS:

Quill, L.L.; Robey, R.F. *J. Am. Chem. Soc.* 1937, *59*, 2591-5.

VARIABLES:

Composition at 25°C and 50°C

PREPARED BY:

T. Mioduski, S. Siekierski, M. Salomon

EXPERIMENTAL VALUES:

Composition of saturated solutions[a]

	$La(NO_3)_3$		HNO_3		density	
t/°C	mass %	mol kg^{-1}	mass %	mol kg^{-1}	kg m^{-3}	nature of the solid phase
25	56.42	4.285	3.06	1.198	1.771	$La(NO_3)_3 \cdot 6H_2O$
	46.42	3.432	11.95	4.555	---	"
	29.10	2.473	34.69	15.204	---	"
	29.62	3.018	40.17	21.102	---	$La(NO_3)_3 \cdot 4H_2O$
	28.73	2.935	41.14	21.669	---	"
	25.18	3.797	54.41	42.306	---	"
	21.61	3.320	58.36	46.239	---	"
	2.79	0.324	70.70	42.323	1.489	$La(NO_3)_3 \cdot 3H_2O$
	0.56	0.15	87.85	120.29	---	$La(NO_3)_3$
50	66.65	6.151	---	---	1.929	$La(NO_3)_3 \cdot 6H_2O$
	64.21	5.952	2.59	1.238	1.912	"
	61.85	5.793	5.29	2.555	1.892	"
	56.70	5.584	12.05	6.119	1.852	"
	56.34	6.035	14.93	8.247	1.880	$La(NO_3)_3 \cdot 6H_2O$ + $La(NO_3)_3 \cdot 4H_2O$
	44.23	4.855	27.73	15.694	1.755	$La(NO_3)_3 \cdot 4H_2O$
	30.31	3.700	44.48	28.000	1.645	"
	29.58	4.577	50.53	40.317	---	"
	5.48	1.040	78.31	76.666	---	$La(NO_3)_3$?
	1.38	0.324	85.51	103.51	---	$La(NO_3)_3$?
	0.41	0.15	91.15	171.39	1.419	$La(NO_3)_3$?

a. Molalities calculated by the compilers.

AUXILIARY INFORMATION

METHOD/APPARATUS/PROCEDURE:

Isothermal method. Appropriate quantities of $La(NO_3)_3$ and HNO_3 were placed in Pyrex tubes, heated to induce supersaturation, and thermostated. The Pyrex tubes were sealed after a small crystal of $Bi(NO_3)_3 \cdot 5H_2O$ was added to "seed" crystallization. The sealed tubes were shaken in the thermostat for at least 8 hours (equilibrium was reached in 4 hours). Authors state that approach to equilibrium from undersaturation gave identical results within experimental error. All data reported in the above table are the results obtained by approach from supersaturation.

A "filtering pipet" maintained at a temperature slightly higher than the thermostat temperature was used to withdraw samples for analyses. Weighed samples of liquid and solid phases were analysed for HNO_3 by titrn with 0.1 mol dm^{-3} NaOH with methyl red indicator. La was pptd as the oxalate, filtered, washed with hot dilute oxalic acid, and ignited to the oxide.

SOURCE AND PURITY OF MATERIALS:

HNO_3 prepd from c.p. grade by distillation in an all Pyrex still and retaining the middle fraction. For very high HNO_3 concs, reagent grade fuming HNO_3 used as received. $La(NO_3)_3 \cdot 6H_2O$ prepared by dissolving the oxide in pure nitric acid, recrystalling twice, and dried over 55% sulfuric acid in a desiccator. No trace of any other rare earth was found in the oxide by arc emission spectroscopy.
Distilled water was used which had a conductivity of 2 x 10^{-6} S cm^{-1}.

ESTIMATED ERROR:

Soly: precision about ± 1 % (compilers).
Temp: at 25°C, accuracy was ± 0.03 K.
at 50°C, accuracy was ± 0.1 K.

REFERENCES:

COMPONENTS:

(1) Lanthanum nitrate; $La(NO_3)_3$; [10099-59-9]

(2) Nitric acid; HNO_3; [7697-37-2]

(3) Water ; H_2O ; [7732-18-5]

ORIGINAL MEASUREMENTS:

Mironov, K.E.; Popov, A.P.; Khripin, L.A. *Zh. Neorg. Khim.* 1966, *11*, 2789-96; *Russ. J. Inorg. Chem. Engl. Transl.* 1966, *11*, 1499-1503.

VARIABLES:

Composition

PREPARED BY:

T. Mioduski and S. Siekierski

EXPERIMENTAL VALUES:

The $La(NO_3)_3$ - HNO_3 - H_2O system at 25.0°C

Composition of saturated solutions for conditions of stable equilibrium[a]

$La(NO_3)_3$ mass %	$La(NO_3)_3$ mol kg^{-1}	HNO_3 mass %	HNO_3 mol kg^{-1}	nature of the solid phase
60.2_7	4.669	---	---	$La(NO_3)_3 \cdot 6H_2O$
59.3	4.57	0.8	0.32	"
43.3	3.23	15.5	5.97	"
36.4	2.72	22.4	8.63	"
29.1	2.50	35.1	15.56	"
28.2	2.48	36.8	16.69	"
27.7	2.52	38.5	18.08	"
28.8 [b]	2.74	38.8	19.00	$La(NO_3)_3 \cdot 6H_2O + La(NO_3)_3 \cdot 5H_2O$
28.8	2.80	39.6	19.89	$La(NO_3)_3 \cdot 5H_2O$
27.8	2.70	40.5	20.28	"
28.1	2.81	41.1	21.18	"
20.3	2.30	52.5	30.63	$La(NO_3)_3 \cdot 4H_2O$
18.6	2.35	57.0	37.07	"
18.9 [b]	2.56	58.4	40.83	$La(NO_3)_3 \cdot 4H_2O + La(NO_3)_3 \cdot 3H_2O$
17.1	2.32	60.2	42.09	$La(NO_3)_3 \cdot 3H_2O$

continued.....

AUXILIARY INFORMATION

METHOD/APPARATUS/PROCEDURE:

The isothermal method was used. Equilibrium at 25°C was approached from both above and below with 4-6 hours being required for equilibrium in both approaches. Both the saturated solutions and solid phases were analysed. Lanthanum was determined complexometrically, nitric acid was determined by potentiometric titration, and for several crystal hydrates water was determined by the Karl Fischer method. Solid phases were identified by Schreinemakers' method of residues and by chemical analyses of the precipitates washed free of mother-liquor with anhydrous dichloroethane. Only the monohydrate and anhydrous salts could not be completely freed of HNO_3 in this way (water of hydration is not effected by this treatment).

SOURCE AND PURITY OF MATERIALS:

Lanthanum nitrate was made from spec pure lanthanum oxide. Analysis of the salt showed it to be the hexahydrate. Completely colorless 100% nitric acid was made by the Brauer method (1). Double distilled water was used.

ESTIMATED ERROR:

Soly: precision about ± 0.5 % (compilers).

Temp: precision ± 0.1 K.

REFERENCES:

1. Brauer, G. (ed.) *Handbuch der praparativen anorganischen Chemie. Russ. Transl.* Inostr. Lit. Moscow. 1956. p. 243.

COMPONENTS:

(1) Lanthanum nitrate; $La(NO_3)_3$; [10099-59-9]

(2) Nitric acid; HNO_3; [7697-37-2]

(3) Water ; H_2O ; [7732-18-5]

ORIGINAL MEASUREMENTS:

Mironov, K.E.; Popov, A.P.; Khripin, L.A.

Zh. Neorg. Khim. 1966, *11*, 2789-96; *Russ. J. Inorg. Chem. Engl. Transl.* 1966, *11*, 1499-1503.

VARIABLES: continued.....

Composition of saturated solutions for conditions of stable equilibrium (cont.)

$La(NO_3)_3$ mass %	$La(NO_3)_3$ mol kg^{-1}	HNO_3 mass %	HNO_3 mol kg^{-1}	nature of the solid phase
17.0	3.39	62.1	44.28	$La(NO_3)_3 \cdot 3H_2O$
17.3	2.69	62.9	50.41	"
19.5	3.33	62.5	55.10	"
20.4[b]	3.63	62.3	57.15	$La(NO_3)_3 \cdot 3H_2O + La(NO_3)_3 \cdot H_2O$
18.2	3.29	64.8	60.49	$La(NO_3)_3 \cdot H_2O$
13.0	2.56	71.4	72.63	"
7.6	1.62	78.0	85.96	"
4.9	1.08	81.1	91.93	"
2.0	0.48	85.2	105.6	"
1.1[b]	0.30	87.8	125.5	$La(NO_3)_3 \cdot H_2O + La(NO_3)_3$
0.4	0.14	90.5	157.8	$La(NO_3)_3$
0.3	0.12	92.0	189.6	"
0.3	0.16	93.8	252.3	"
0.2	0.13	95.1	321.1	"
0.2	0.18	96.3	436.6	
solubilities for metastable conditions				
62.4_7	5.123	---	---	$La(NO_3)_3 \cdot 5H_2O$
61.1	5.00	1.3	0.55	"
59.5	4.82	2.5	1.04	"
43.6	3.47	17.7	7.26	"
40.5	3.21	20.7	8.47	"
37.6	3.00	23.8	9.78	"
36.0	2.89	25.7	10.65	"
28.9	3.07	42.1	23.04	"
29.2	3.10	41.8	22.87	$La(NO_3)_3 \cdot 4H_2O$
29.0	3.16	42.8	24.09	$La(NO_3)_3 \cdot 3H_2O$
26.4	2.90	45.6	25.85	"
25.0	2.88	48.3	28.71	"
23.6	2.70	49.5	29.20	"
22.8	2.77	51.9	32.55	"
22.5	2.72	52.0	32.36	"
22.6	2.89	53.3	35.10	"
19.6	2.59	57.1	38.89	"
21.2	3.71	61.2	55.18	"
Additional eutonic points not listed above (stable slns)				
28.5	2.93	41.6	22.08	$La(NO_3)_3 \cdot 5H_2O + La(NO_3)_3 \cdot 4H_2O$
29.4	3.23	42.6	24.14	$La(NO_3)_3 \cdot 5H_2O + La(NO_3)_3 \cdot 3H_2O$
30.4	3.11	39.5	20.83	$La(NO_3)_3 \cdot 6H_2O + La(NO_3)_3 \cdot 4H_2O$

a. Molalities calculated by M. Salomon.
b. Eutonic points.

COMMENTS AND/OR ADDITIONAL DATA:

The authors state that metastable hydrates do not change into the stable forms when introducing seeds of the stable phases. Relative densities (d_4^{25}) and refractive indices are given for each of the hydrates and for the anhydrous salt in the source paper.

COMPONENTS:

(1) Lanthanum nitrate; $La(NO_3)_3$; [10099-59-9]

(2) Acetone; C_3H_6O; [67-64-1]

(3) Water ; H_2O ; [7732-18-5]

ORIGINAL MEASUREMENTS:

Zwietasch, K.-J.; Kirmse, E.M. *Wiss. Hefte d. Päd. Hochschule "W. Ratke" Köthen* 1977, *4*, 133-5

VARIABLES:

Acetone concentration

PREPARED BY:

T. Mioduski, S. Siekierski and M. Salomon

EXPERIMENTAL VALUES:

Composition of saturated solutions at 25°C

acetone	$La(NO_3)_3 \cdot 6H_2O$	anhydrous $La(NO_3)_3$ [a]	
mass %	mass %	mass %	mol kg^{-1}
0	79.45	59.62	4.544
1.63	79.04	59.31	4.486
2.93	79.64	59.76	4.571
3.70	79.12	59.37	4.496
5.40	78.94	59.23	4.472
7.72	79.04	59.31	4.486
11.75	78.50	58.90	4.411
14.54	77.52	58.17	4.280
20.66	76.19	57.17	4.108
21.82	75.85	56.92	4.066
24.61	75.39	56.57	4.009

a. Mass % of anyhr salt and molalities calculated by the compilers.

In all cases the authors state that the solid phase is $La(NO_3)_3 \cdot 6H_2O$

AUXILIARY INFORMATION

METHOD/APPARATUS/PROCEDURE:

The isothermal method was used. The reaction mixtures were agitated in a thermostat at 25°C for 24 hours. Lanthanum was determined by complexometric titration with EDTA and xylenol orange indicator. Acetone was determined by iodometric titration, and water content was determined by difference. The composition of the solid phases were determined by Schreinemakers' method of wet residues.

SOURCE AND PURITY OF MATERIALS:

$La(NO_3)_3 \cdot 6H_2O$ was used as the starting solid. No other information given.

ESTIMATED ERROR:

Soly: accuracy around 1-2 % (compilers).

Temp: nothing specified.

REFERENCES:

COMPONENTS:

(1) Sodium nitrate; $NaNO_3$; [7631-99-4]

(2) Lanthanum nitrate; $La(NO_3)_3$; [10099-59-9]

(3) Water ; H_2O ; [7732-18-5]

ORIGINAL MEASUREMENTS:

Bogdanovskaya, R.L. *Uch. Zap. Perm. Univ.* 1970, *229*, 32-4.

VARIABLES:

Composition at 20°C and 50°C

PREPARED BY:

T. Mioduski and S. Siekierski

EXPERIMENTAL VALUES:

Composition of saturated solutions[a]

	$La(NO_3)_3$		$NaNO_3$		
t/°C	mass %	mol kg^{-1}	mass %	mol kg^{-1}	nature of the solid phase
20	---	----	46.8	10.35	$NaNO_3$
	12.4	0.77	38.0	9.01	"
	30.8	2.05	23.0	5.86	"
	36.9	2.52	18.0	4.70	"
	50.1	3.89	10.3	3.06	"
	55.8	4.62	7.0	2.21	"
	57.0	4.68	5.5	1.73	$NaNO_3 + La(NO_3)_3 \cdot 6H_2O$
	57.5	4.68	4.7	1.46	$La(NO_3)_3 \cdot 6H_2O$
	57.8	4.46	2.3	0.68	"
	58.0	4.25	---	----	"
50	---	----	53.0	13.27	$NaNO_3$
	11.4	0.77	43.0	11.09	"
	28.4	2.05	29.0	8.01	"
	39.5	3.07	20.9	6.21	"
	51.6	4.37	12.1	3.92	"
	58.0	5.25	8.0	2.77	"
	63.6	6.14	4.5	1.66	$NaNO_3 + La(NO_3)_3 \cdot 6H_2O$
	63.8	5.91	3.0	1.06	$La(NO_3)_3 \cdot 6H_2O$
	64.0	5.71	1.5	0.51	"
	64.5	5.59	---	----	"

a. Molalities calculated by M. Salomon.

AUXILIARY INFORMATION

METHOD/APPARATUS/PROCEDURE:

The method of isothermal sections was used with refractometric analysis (1). Solutions were prepared by mixing known amounts of solids + liquid and equilibrated until their refractive indices were constant. Equilibrium was reached in 5 hours at 50°C and in 8 hours at 20°C. The composition of the satd slns and solid phases were found as break points on the composition-refractive index diagram (see the compilation for reference 2). The refractive indices were determined at 50°C using an IRF-22 refractometer.

COMMENTS AND/OR ADDITIONAL DATA:

The isotherms at 20°C and 50°C are identical and consist of two branches corresponding to the crystallization of $La(NO_3)_3 \cdot 6H_2O$ and $NaNO_3$. The eutonic sln is strongly enriched in $La(NO_3)_3$: i.e. this component shows a strong salting out effect with respect to $NaNO_3$. At higher temperatures the salting out effect of $La(NO_3)_3$ increases. The influence of $NaNO_3$ on the soly of $La(NO_3)_3$ is much weaker.

SOURCE AND PURITY OF MATERIALS:

Pure grade $La(NO_3)_3 \cdot 6H_2O$ was used.

Analar grade $NaNO_3$ was used.

No other information given.

ESTIMATED ERROR:

Soly: nothing specified.

Temp: precision reported as ± 0.1 to 0.2 K.

REFERENCES:

1. Zhuravlev, E.F.; Sheveleva, A.D. *Zh. Neorg. Khim.* 1960, *5*, 2630.

2. Bogdanovskaya, R.L.: Armisheva, L.N. *Uch. Zap. Perm. Univ.* 1973, *289*, 32.

COMPONENTS:

(1) Potassium nitrate; KNO_3; [7757-79-1]

(2) Lanthanum nitrate; $La(NO_3)_3$; [10099-59-9]

(3) Water ; H_2O ; [7732-18-5]

ORIGINAL MEASUREMENTS:

Bogdanovskaya, R.L.; Armisheva, L.N.

Uch. Zap. Perm. Univ. 1973, *289*, 32-5.

VARIABLES:

Composition at 20°C and 50°C

PREPARED BY:

T. Mioduski and S. Siekierski

EXPERIMENTAL VALUES:

The $La(NO_3)_3$ - KNO_3 - H_2O system at 20.0°C

Composition of saturated solutions[a]

$La(NO_3)_3$ mass %	$La(NO_3)_3$ mol kg^{-1}	KNO_3 mass %	KNO_3 mol kg^{-1}	nature of the solid phase[b]
---	---	24.0	3.12	KNO_3
8.0	0.34	20.0	2.75	"
16.4	0.77	18.0	2.71	"
18.0	0.87	18.0	2.78	"
23.0	1.19	17.5	2.91	"
24.8	1.32	17.5	3.00	"
29.0	1.67	17.7	3.28	"
32.6	2.05	18.5	3.74	"
39.6	3.13	21.5	5.47	"
41.0	3.46	22.5	6.10	KNO_3 + $2KNO_3 \cdot La(NO_3)_3 \cdot 2H_2O$
42.0	3.49	21.0	5.61	$2KNO_3 \cdot La(NO_3)_3 \cdot 2H_2O$
45.0	3.64	17.0	4.42	"
47.5	3.95	15.5	4.14	"
56.5	5.10	9.4	2.73	"
60.0	5.43	6.0	1.75	$2KNO_3 \cdot La(NO_3)_3 \cdot 2H_2O$ + $La(NO_3)_3 \cdot 6H_2O$
58.5	4.91	4.8	1.29	$La(NO_3)_3 \cdot 6H_2O$
58.3	4.55	2.3	0.58	"
58.0	4.25	---	---	"

continued.....

AUXILIARY INFORMATION

METHOD/APPARATUS/PROCEDURE:

The solubility was studied by the method of isothermal sections (1) by measuring the refractive indices of saturated solutions along directed sections of the phase diagram. Equilibrium was checked by repeated measurements of the refractive indices as a function of time. At 50°C equilibrium was reached in 6 hours, and at 20°C equilibrium was reached in 10 hours. The results were used to graph the relation between the refractive indices and the compositions of the components for each of the sections studied. The graphs were used to find the break points corresponding to the composition of the saturated solutions.

The composition of $2KNO_3.La(NO_3)_3 \cdot 2H_2O$ was confirmed by chemical analysis, crystal-optical and microscopic analysis. La was determined by titration with Trilon, K was determined as the tetraphenylborate, and water by the Karl Fischer method. Nitrate was determined by difference.

SOURCE AND PURITY OF MATERIALS:

$La(NO_3)_3 \cdot 6H_2O$ was recrystallized.

Anhydrous c.p. grade KNO_3 was presumably used as received.

No other information given.

ESTIMATED ERROR:

Soly: nothing specified.

Temp: precision ± 0.2 K.

REFERENCES:

1. Zhuravlev, E.F.; Sheveleva, A.D. *Zh. Neorg. Khim.* 1960, *5*, 2630.

COMPONENTS:

(1) Potassium nitrate; KNO_3; [7757-79-1]

(2) Lanthanum nitrate; $La(NO_3)_3$; [10099-59-9]

(3) Water ; H_2O ; [7732-18-5]

ORIGINAL MEASUREMENTS:

Bogdanovskaya, R.L.; Armisheva, L.N.

Uch. Zap. Perm. Univ. 1973, *289*, 32-5.

EXPERIMENTAL VALUES: continued.....

The $La(NO_3)_3$ - KNO_3 - H_2O system at 50.0°C

Composition of saturated solutions[a]

$La(NO_3)_3$		KNO_3		
mass %	mol kg^{-1}	mass %	mol kg^{-1}	nature of the solid phase[b]
---	---	44.0	7.77	KNO_3
6.0	0.35	40.5	7.49	"
12.6	0.77	37.0	7.26	"
20.1	1.32	33.0	6.96	"
23.5	1.63	32.0	7.11	"
24.6	1.71	31.0	6.91	"
30.5	2.41	30.5	7.74	"
35.5	3.08	29.0	8.08	"
38.0	3.49	28.5	8.41	"
42.5	4.51	28.5	9.72	KNO_3 + $2KNO_3 \cdot La(NO_3)_3 \cdot 2H_2O$
43.4	4.44	26.5	8.71	$2KNO_3 \cdot La(NO_3)_3 \cdot 2H_2O$
45.0	4.62	25.0	8.24	"
46.0	4.57	23.0	7.34	"
51.3	4.98	17.0	5.30	"
59.0	5.82	9.8	3.11	"
67.0	7.24	4.5	1.56	$2KNO_3 \cdot La(NO_3)_3 \cdot 2H_2O$ + $La(NO_3)_3 \cdot 6H_2O$
66.0	6.47	2.6	0.82	$La(NO_3)_3 \cdot 6H_2O$
66.5	6.36	1.3	0.40	"
64.5	5.59	---	---	"

a. Molalities calculated by M. Salomon.

b. The solubility isotherms at 20° and 50°C are identical and consist of three branches corresponding to the crystallization of KNO_3, the double nitrate, and $La(NO_3)_3 \cdot 6H_2O$. With increasing temperature, the crystallization field of the double nitrate increases, and at 50°C it is congruently soluble in water.

COMMENTS AND/OR ADDITIONAL DATA:

The crystals of the double nitrate $2KNO_3 \cdot La(NO_3)_3 \cdot 2H_2O$ are white and are stable in air. Refractive indices were measured by the immersion method, and the density was measured pycnometrically with CCl_4. The results of these measurements follow:

$$d^{20} = 2.47 \pm 0.01 \text{ kg m}^{-3}$$

$$n_q = 1.560 \pm 0.001$$

$$n_p = 1.505 \pm 0.001$$

COMPONENTS:

(1) Rubidium nitrate; $RbNO_3$; [13126-12-0]

(2) Lanthanum nitrate; $La(NO_3)_3$; [10099-59-9]

(3) Water ; H_2O ; [7732-18-5]

ORIGINAL MEASUREMENTS:

Molodkin, A.K.; Odinets, Z.K.; Vargas Ponse, O. *Zh. Neorg. Khim.* 1976, *21*, 2590-3; *Russ. J. Inorg. Chem. Engl. Transl.* 1976, *21*, 1425-7.

VARIABLES:

Composition at 25°C

PREPARED BY:

T. Mioduski and S. Siekierski

EXPERIMENTAL VALUES:

Composition of saturated solutions at 25°C[a]

$La(NO_3)_3$		$RbNO_3$		
mass %	mol kg^{-1}	mass %	mol kg^{-1}	nature of the solid phase[b]
---	---	40.65	4.644	$RbNO_3$
7.4	0.42	37.97	4.713	"
11.7	0.668	33.99	4.249	"
19.66	1.221	30.77	4.209	"
30.13	2.396	31.16	5.458	"
31.99	2.829	33.21	6.471	"
33.09	3.029	33.29	6.714	"
35.48	3.396	32.37	6.827	$RbNO_3$ + $2RbNO_3 \cdot La(NO_3)_3 \cdot 3H_2O$
36.10	3.675	33.67	7.553	$2RbNO_3 \cdot La(NO_3)_3 \cdot 3H_2O$
38.01	3.481	28.38	5.726	"
46.72	3.946	16.84	3.134	"
49.89	4.215	13.68	2.546	"
51.90	4.452	12.22	2.309	"
54.60	4.767	10.15	1.953	$2RbNO_3 \cdot La(NO_3)_3 \cdot 3H_2O$ + $La(NO_3)_3 \cdot 6H_2O$
58.44	5.387	8.17	1.659	$La(NO_3)_3 \cdot 6H_2O$
57.72	5.018	6.88	1.318	"
58.02	4.754	4.42	0.798	"
60.37	4.688	---	----	"

a. Molalities calculated by M. Salomon.

b. In the phase diagram, A is the soly of $RbNO_3$ and B is the soly of $La(NO_3)_3$. AP is the soly curve of $RbNO_3$, PE that of the double salt, and EB that of $La(NO_3)_3 \cdot 6H_2O$. H_2O-APEB is the region of unsatd slns, below the line APEB is the region of satd slns. Point C was found using Schreinemakers' method.

AUXILIARY INFORMATION

METHOD/APPARATUS/PROCEDURE:

Authors state method of additions used. This is probably the isothermal method. Equilibrium reached after 6-7 h as detd by constancy in refractive indices. Both solid and liquid phases analysed. La detmd by pptn with NH_4OH followed by ignition to the oxide. After separation and filtration of $La(OH)_3$, the filtrate and wash waters were evaporated to dryness and the residue calcined at 350°C to constant weight. DTA studies showed $RbNO_3$ to be stable up to 536°C. The method of analysis was checked by analysis of standard mixtures of the nitrates. Total nitrate was detd gravimetrically with nitron. Composition of the residues detd by Schreinmakers' method and checked by crystal-optical analysis, IR spectroscopy, and differential thermal analysis (DTA). The crystals of the double salt, $2RbNO_3 \cdot La(NO_3)_3 \cdot 3H_2O$ were white, anisotropic with n = 1.625, and readily soluble in water. The refractive index was detd by the immersion method.

SOURCE AND PURITY OF MATERIALS:

Nothing specified.

ESTIMATED ERROR:

Soly: nothing specified.

Temp: precision about ± 0.2 K (compilers).

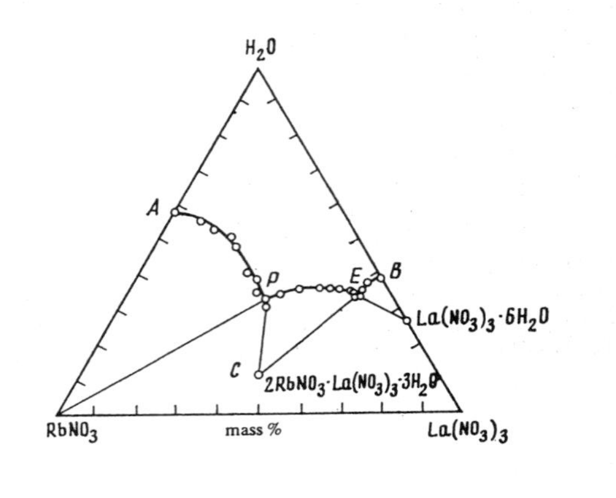

COMPONENTS:

(1) Cesium nitrate; $CsNO_3$; [7789-18-6]

(2) Lanthanum nitrate; $La(NO_3)_3$; [10099-59-9]

(3) Water; H_2O; [7732-18-5]

ORIGINAL MEASUREMENTS:

Molodkin, A.K.; Odinets, Z.K.; Pereira Paveze, V. *Zh. Neorg. Khim.* 1976, *21*, 2792-4: *Russ. J. Inorg. Chem. Engl. Transl.* 1976, *21*, 1540-1.

VARIABLES:

Composition at 25°C

PREPARED BY:

T. Mioduski and S. Siekierski

EXPERIMENTAL VALUES:

Composition of saturated solutions at 25°C[a]

$La(NO_3)_3$		$CsNO_3$		
mass %	mol kg^{-1}	mass %	mol kg^{-1}	nature of the solid phase[b]
---	---	22.03	1.450	$CsNO_3$
5.12	0.210	19.77	1.350	"
11.58	0.493	16.14	1.146	"
18.61	0.874	15.82	1.238	"
24.87	1.321	17.17	1.520	"
36.14	2.605	21.16	2.542	"
41.38	3.937	26.27	4.166	"
46.73	5.370	26.49	5.075	$CsNO_3$ + $2CsNO_3\cdot La(NO_3)_3\cdot 3H_2O$
48.11	5.500	24.97	4.759	$2CsNO_3\cdot La(NO_3)_3\cdot 3H_2O$
50.23	5.551	21.92	4.038	"
52.40	6.033	20.87	4.006	$2CsNO_3\cdot La(NO_3)_3\cdot 3H_2O$ + $La(NO_3)_3\cdot 6H_2O$
54.31	4.818	11.00	1.627	$La(NO_3)_3\cdot 6H_2O$
57.71	4.806	5.33	0.740	"
60.45	4.704	---	---	"

a. Molalities calculated by M. Salomon.

b. In the phase diagram below, point A is the soly of $CsNO_3$ in water and B is the soly of $La(NO_3)_3$ in water. P is a transformation point, E a eutonic, AP the crystallization branch for $CsNO_3$, PE that for the incongruently soluble compound, and EB that of $La(NO_3)_3\cdot 6H_2O$. Unsaturated solutions correspond to the region H_2O-APEB, and below the segmented line APEB there is a region of saturated solutions and solid phases. Point C corresponds to the composition of the incongruently soluble salt which was determined by Schreinemakers' method.

AUXILIARY INFORMATION

METHOD/APPARATUS/PROCEDURE:

Isothermal method used, and equilibrium was reached after 6-7 h as determined by constancy in the refractive index. Both solid and liquid phases analysed. La detd by pptn from sln with NH_4OH followed by ingition to the oxide. After separation and filtration, the filtrate and wash waters were evaporated to dryness and the residue ignited to constant weight at 350°C. DTA studies showed $CsNO_3$ to be stable up to 580°C. Total NO_3 detd gravimetrically with nitron. Composition of the solid phases found by Schreinemakers method and checked by crystal-optical analysis, IR spectroscopy, and differential thermal analysis (DTA).

For $2CsNO_3\cdot La(NO_3)_3\cdot 3H_2O$, the refractive index was given as 1.568 and chemical analysis gave the following: found, % Cs 34.36, La 18.74, NO_3 40.27; Calcd, Cs 34.59, La 18.08, NO_3 40.31.

SOURCE AND PURITY OF MATERIALS:

No details given.

ESTIMATED ERROR:

Soly: nothing specified.

Temp: precision ± 0.2 K (compilers).

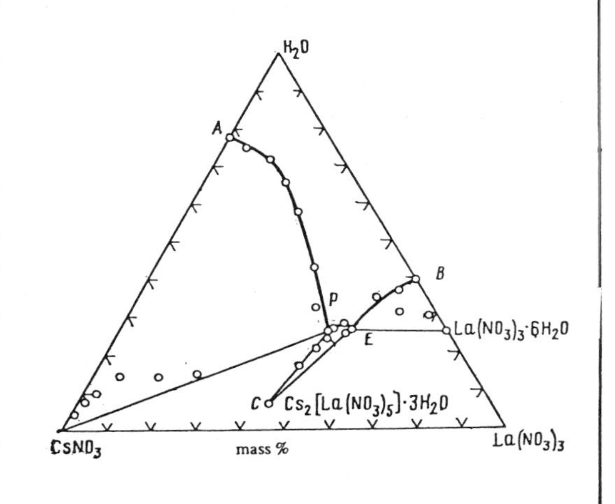

COMPONENTS:

(1) Magnesium nitrate; $Mg(NO_3)_2$; [10377-60-3]

(2) Lanthanum nitrate; $La(NO_3)_3$; [10099-59-9]

(3) Water ; H_2O ; [7732-18-5]

ORIGINAL MEASUREMENTS:

di Capua, C. *Gazz. Chim. Ital.* 1929, *59*, 164 - 9.

VARIABLES:

Composition at 20°C

PREPARED BY:

T. Mioduski and S. Siekierski

EXPERIMENTAL VALUES:

The $La(NO_3)_3$ - $Mg(NO_3)_2$ - H_2O system at 20°C

Composition of saturated solutions[a]

$La(NO_3)_3$ mass %	$La(NO_3)_3$ mol kg^{-1}	$Mg(NO_3)_2$ mass %	$Mg(NO_3)_2$ mol kg^{-1}
60.13	4.642	---	---
59.65	4.687	1.18	0.203
58.91	4.662	2.20	0.381
57.46	4.578	3.91	0.682
55.06	4.375	6.21	1.081
50.11	3.850	9.83	1.654
48.72	3.873	12.56	2.187
42.98	3.339	17.40	2.961
37.02	2.605	19.25	2.968
27.96	1.767	23.34	3.231
26.00	1.600	24.00	3.236
22.90	1.365	25.48	3.328
16.78	0.955	29.15	3.635
15.35	0.879	30.93	3.882
6.89	0.387	38.30	4.711
3.20	0.171	39.25	4.598
---	---	43.68	5.229

a. Molalities calculated by M. Salomon.

Nature of the solid phases not specified.

AUXILIARY INFORMATION

METHOD/APPARATUS/PROCEDURE:

The isothermal method was used. The total $La(NO_3)_3$ + $Mg(NO_3)_2$ content was determined by evaporation of the saturated solutions. La was determined by the oxalate method, and Mg and water were determined by difference. No other details given.

SOURCE AND PURITY OF MATERIALS:

Nothing specified.

ESTIMATED ERROR:

Soly: precision about ± 0.5 % (compilers).

Temp: precision probably ± 0.5 K (compilers).

REFERENCES:

COMPONENTS:

(1) Magnesium nitrate; $Mg(NO_3)_2$; [10377-60-3]

(2) Lanthanum nitrate; $La(NO_3)_3$; [10099-59-9]

(3) Water ; H_2O ; [7732-18-5]

ORIGINAL MEASUREMENTS:

Urazov, G.G.; Shevtsova, Z.N. *Zh. Neorg. Khim.* 1957, *2*, 655-8; *J. Inorg. Chem. (USSR)* 1957, *2*, 288-94.

VARIABLES:

Composition at 25°C and 50°C

PREPARED BY:

T. Mioduski and S. Siekierski

EXPERIMENTAL VALUES:

Composition of saturated slns at 25°C[a]

$La(NO_3)_3$ mass %	$La(NO_3)_3$ mol kg^{-1}	$Mg(NO_3)_2$ mass %	$Mg(NO_3)_2$ mol kg^{-1}	solid phase[b]
60.28	4.671	---	----	A
59.22	4.644	1.53	0.263	B
58.92	4.697	2.47	0.431	B
57.94	4.620	3.46	0.604	B
53.99	4.318	7.53	1.319	C
44.67	3.068	10.52	1.583	C
37.46	2.362	13.73	1.897	C
29.20	1.739	19.12	2.494	C
23.66	1.377	23.47	2.993	C
15.64	0.863	28.58	3.455	C
10.35	0.553	32.00	3.743	C
8.35	0.454	35.00	4.166	C
4.49	0.243	38.69	4.591	C
2.52	0.134	39.54	4.601	C
1.65	0.087	40.00	4.622	D
---	----	43.06	5.099	D
---	----	43.55	5.202	E

Composition of saturated slns at 50°C

$La(NO_3)_3$ mass %	$La(NO_3)_3$ mol kg^{-1}	$Mg(NO_3)_2$ mass %	$Mg(NO_3)_2$ mol kg^{-1}	solid phase
67.17	6.297	---	----	A
62.00	5.659	4.28	0.856	B
59.83	5.266	5.20	1.003	C
52.42	4.298	10.04	1.803	C
45.00	3.397	14.23	2.353	C
41.05	2.999	16.82	2.692	C
35.86	2.440	18.91	2.819	C
33.50	2.251	20.70	3.047	C
31.00	2.054	22.54	3.271	C
26.68	1.699	25.00	3.488	C
23.15	1.443	27.47	3.751	C
12.61	0.727	34.00	4.294	C
10.16	0.585	36.37	4.586	C
5.10	0.286	40.00	4.913	C
3.03	0.171	42.40	5.239	C
3.00	0.169	42.52	5.262	C + D
---	----	44.73	5.457	D
---	----	45.74	5.684	E

a. Molalities calculated by M. Salomon.

b. A = $La(NO_3)_3 \cdot 6H_2O$ B,D = solid solutions C = $3Mg(NO_3)_2 \cdot 2La(NO_3)_3 \cdot 24H_2O$

E = $Mg(NO_3)_2 \cdot 6H_2O$

AUXILIARY INFORMATION

METHOD/APPARATUS/PROCEDURE:

The isothermal method was used. Equilibrium was verified and required 20 days at 25°C and 5 days at 50°C. The compositions of the solid phases were determined by chemical analyses, thermographical studies, crystal-optical and graphical methods. Schreinmakers method of residues was used to determine the composition of the solid phases.

No other information given.

SOURCE AND PURITY OF MATERIALS:

Sources not specified, but the salts were analysed. The compilers assume the analyses showed the salts to be of acceptable purity.

ESTIMATED ERROR:

Soly: Analysis agreed to 0.1-0.2% for each point on the soly curve (authors). The compilers assume this is mass %.

Temp: precision ± 0.1 K.

REFERENCES:

COMPONENTS:

(1) Magnesium nitrate; $Mg(NO_3)_2$; [10377-60-3]

(2) Lanthanum nitrate; $La(NO_3)_3$; [10099-59-9]

(3) Water; H_2O; [7732-18-5]

ORIGINAL MEASUREMENTS:

Yakimov, M.A.; Gizhavina, E.I. *Zh. Neorg. Khim.* 1971, *16*, 507-9; *Russ. J. Inorg. Chem. Engl. Transl.* 1971, *16*, 268-9.

VARIABLES:

Composition at 25°C

PREPARED BY:

T. Mioduski and S. Siekierski

EXPERIMENTAL VALUES:

Composition of saturated solutions[a]

$La(NO_3)_3$ mass %	$La(NO_3)_3$ mol kg^{-1}	$Mg(NO_3)_2$ mass %	$Mg(NO_3)_2$ mol kg^{-1}	solid phase[b]
58.90	4.411	---	---	A
57.20	4.374	2.55	0.427	A
57.20	4.396	2.75	0.463	A + B
55.84	4.171	2.96	0.484	B
54.12	3.968	3.90	0.626	B
50.58	3.514	5.12	0.779	B
49.00	3.334	5.77	0.860	B
46.80	3.132	7.21	1.057	B
42.36	2.731	9.90	1.398	B
37.93	2.336	12.10	1.633	B
32.43	1.908	15.26	1.967	B
29.30	1.695	17.51	2.220	B
27.28	1.560	18.91	2.369	B

Composition of saturated solutions[a]

$La(NO_3)_3$ mass %	$La(NO_3)_3$ mol kg^{-1}	$Mg(NO_3)_2$ mass %	$Mg(NO_3)_2$ mol kg^{-1}	solid phase
24.58	1.386	20.85	2.576	B
23.08	1.284	21.61	2.634	B
22.35	1.246	22.46	2.744	B
17.66	0.948	24.98	2.936	B
17.50	0.939	25.17	2.960	B
13.26	0.697	28.16	3.241	B
8.53	0.439	31.70	3.576	B
4.55	0.232	35.13	3.927	B
2.60	0.131	36.20	3.988	B
1.05	0.054	39.10	4.405	B
0.42	0.022	41.44	4.806	B
0.50	0.027	41.50	4.824	B + C
---	---	41.80	4.842	C

a. Molalities calculated by M. Salomon.

b. Solid phases: A = $La(NO_3)_3 \cdot 6H_2O$

B = $2La(NO_3)_3 \cdot 3Mg(NO_3)_2 \cdot 24H_2O$

C = $Mg(NO_3)_2 \cdot 6H_2O$

AUXILIARY INFORMATION

METHOD/APPARATUS/PROCEDURE:

The isothermal method was used. The compositions of the saturated solutions were determined by chemical analysis. Lanthanum was precipitated as the oxalate followed by ignition to La_2O_3. The total lanthanum and magnesium content was determined by back titration with Trilon B, and the magnesium was then found by difference.

Composition of the solid phases were determined by Schreinemakers' method.

No other information given.

SOURCE AND PURITY OF MATERIALS:

No information given.

ESTIMATED ERROR:

Soly: precision about ± 0.5 % (compilers).

Temp: precision probably ± 0.2 K (compilers).

REFERENCES:

COMPONENTS:

(1) Barium nitrate; $Ba(NO_3)_2$; [10022-31-8]

(2) Lanthanum nitrate; $La(NO_3)_3$; [10099-59-9]

(3) Water; H_2O; [7732-18-5]

ORIGINAL MEASUREMENTS:

Molodkin, A.K.; Odinets, Z.K.; Vargas Ponse, O. *Zh. Neorg. Khim.* 1977, *22*, 3388-9; *Russ. J. Inorg. Chem. Engl. Transl.* 1977, *22*, 1852-3.

VARIABLES:

Composition at 25°C

PREPARED BY:

T. Mioduski and S. Siekierski

EXPERIMENTAL VALUES:

The $La(NO_3)_3 - Ba(NO_3)_2 - H_2O$ system at 25.0°C

Composition of saturated solutions.[a]

$La(NO_3)_3$ mass %	$La(NO_3)_3$ mol kg^{-1}	$Ba(NO_3)_2$ mass %	$Ba(NO_3)_2$ mol kg^{-1}	nature of the solid phase
---	---	9.53	0.403	$Ba(NO_3)_2$
1.70	0.058	7.59	0.320	"
9.24	0.329	4.33	0.192	"
20.78	0.837	2.80	0.140	"
31.62	1.452	1.38	0.079	"
43.24	2.385	0.97	0.067	"
51.30	3.301	0.87	0.070	"
58.83	4.467	0.64	0.060	"
59.20	4.505	0.36	0.034	$Ba(NO_3)_2 + La(NO_3)_3 \cdot 6H_2O$
60.37	4.688	---	----	$La(NO_3)_3 \cdot 6H_2O$

a. Molalities calculated by M. Salomon.

b. The system is of the simple eutonic type with a well developed crystallization field of $Ba(NO_3)_2$. The composition of the eutonic solution is practically the same as that of the solution saturated with $La(NO_3)_3$.

AUXILIARY INFORMATION

METHOD/APPARATUS/PROCEDURE:

The method of isothermal sections with refractometric analysis was used. Equilibrium was recorded after 48 hours and at which time the refractive index of the saturated solution was constant. The composition of the solid phases was determined graphically by Schreinemakers' method and confirmed by crystal-optical, IR, and thermogravimetrical studies. An IRF-22 refractometer and a UR-20 spectrophotometer were used.

SOURCE AND PURITY OF MATERIALS:

$La(NO_3)_3 \cdot 6H_2O$ prepd by dissolving La_2O_3 in HNO_3. Gravimetric analysis yielded the following (NO_3^- detd as nitron nitrate): 32.17% La, 42.12% NO_3, and 24.71% H_2O.

C.p. grade $Ba(NO_3)_2$ was recrystallized twice from bidistilled water prior to use.

Both salts were analysed by IR spectroscopy and derivatography and presumed to be of acceptable purity.

ESTIMATED ERROR:

Soly: precision probably around 0.5 mass % (compilers).

Temp: precision of ± 0.1 to 0.2 K.

REFERENCES:

COMPONENTS:

(1) Lanthanum acetate; $La(CH_3COO)_3$; [917-70-4]

(2) Lanthanum nitrate; $La(NO_3)_3$; [10099-59-9]

(3) Water; H_2O; [7732-18-5]

ORIGINAL MEASUREMENTS:

Petelina, V.S.; Nikurashina, N.I.; Bakhtiarova, G.A. *Izv. Vyssh. Ucheb. Zaved. Khim. Khim. Tekhnol.* 1971, *14*, 1611-4.

VARIABLES:

Composition

PREPARED BY:

T. Mioduski and S. Siekierski

EXPERIMENTAL VALUES:

The $La(NO_3)_3$ - $La(CH_3COO)_3$ - H_2O system at 25°C

Composition of saturated solutions[a]

$La(NO_3)_3$		$La(CH_3COO)_3$		
mass %	mol kg^{-1}	mass %	mol kg^{-1}	nature of the solid phase
59.8	4.58	---	---	$La(NO_3)_3 \cdot 6H_2O$
56.6[b]	4.51	4.8	0.39	$2La(NO_3)_3 \cdot La(CH_3COO)_3 \cdot 13H_2O$
32.6[b]	2.33	24.4	1.80	"
---	---	16.6	0.630	$La(CH_3COO)_3 \cdot H_2O$

a. Molalities calculated by M. Salomon.

b. Eutonic points.

AUXILIARY INFORMATION

METHOD/APPARATUS/PROCEDURE:

The solubilities were studied by the method of isothermal sections as described by Mochalov (1). Analyses were made by refractive index measurements along directed sections of the phase diagram.

The reaction mixtures were periodically agitated at room temperature for a period of 20 days. Before the refractive indices were measured, the saturated solutions were centrifuged to separate the precipitate and then thermostated at 25°C for 3 hours.

SOURCE AND PURITY OF MATERIALS:

$La(NO_3)_3 \cdot 6H_2O$ was prepared by dissolving c.p. grade La_2O_3 in dilute HNO_3. The salt was dried and analysed for La by the oxalate method.
$La(CH_3COO)_3 \cdot H_2O$ was prepared by dissolving freshly precipitated $La_2(CO_3)_3$ in dilute CH_3COOH. The salt was dried and analysed for La by the oxalate method.

The compilers assume that the results of the analyses showed the salts to be of acceptable purity.

ESTIMATED ERROR:

Soly: nothing specified.

Temp: precision probably ± 0.2 K (compilers).

REFERENCES:

1. Mochalov, K.I. *Zh. Obshch. Khim.* 1939, *9*, 1701.

COMPONENTS:

(1) Lanthanum oxalate; $La(C_2O_4)_3$; [537-03-1]

(2) Lanthanum nitrate; $La(NO_3)_3$; [10099-59-9]

(3) Water ; H_2O ; [7732-18-5]

ORIGINAL MEASUREMENTS:

James, C.; Whittemore, C.F. *J. Am. Chem. Soc.* 1912, *34*, 1168-71.

VARIABLES:

Composition at 25°C

PREPARED BY:

T. Mioduski, S. Siekierski and M. Salomon

EXPERIMENTAL VALUES:

Composition of saturated solutions[a]

$La(NO_3)_3$		$La_2(C_2O_4)_3$		
mass %	mol kg^{-1}	mass %	mol kg^{-1}	nature of the solid phase[b]
5.06	0.165	0.28	0.0055	$La_2(C_2O_4)_3 \cdot 8H_2O$
9.89	0.340	0.50	0.010	"
14.04	0.507	0.88	0.019	"
17.99	0.685	1.18	0.0269	"
22.15	0.892	1.46	0.0353	"
25.17	1.060	1.73	0.0437	"
28.63	1.270	2.01	0.0535	"
31.53	1.465	2.21	0.0616	"
34.61	1.691	2.41	0.0706	"
35.37	1.768	2.51	0.0748	"
36.24	1.823	2.59	0.0781	"
37.18	1.901	2.63	0.0806	"
37.42	1.922	2.67	0.0822	"
38.50	2.019	2.80	0.088	"
39.89	2.150	3.00	0.0969	"
40.83	2.241	3.09	0.102	"
42.27	2.391	3.32	0.113	"
45.26	2.676	2.68	0.0950	$La_2(C_2O_4)_3 \cdot 3H_2O$
45.06	2.654	2.68	0.0946	"
46.39	2.801	2.46	0.0956	"
49.84	3.225	2.59	0.100	"
51.30	3.419	2.52	0.101	"

continued........

AUXILIARY INFORMATION

METHOD/APPARATUS/PROCEDURE:

The isothermal method was used. Varying amounts of lanthanum nitrate, lanthanum oxalate and sufficient water to make 50 cc were placed in bottles which were rotated in a thermostat at 25°C until equilibrium was reached (time required to reach equilibrium not specified). Solutions were then permitted to settle and aliquots removed for analyses. Total lanthanum was determined gravimetrically by precipitation with excess oxalic acid followed by ignition to the oxide. Oxalate in the saturated solutions was determined by titration with standard $KMnO_4$ solution.

Samples of the solid phases were pressed between filter papers in a screw press and analysed for total lanthanum by simple ignition to La_2O_3. In order to estimate the oxalate by titration with $KMnO_4$, it was found necessary to add dilute sulfuric acid to the solid since under these conditions the solid dissolved rapidly upon heating.

SOURCE AND PURITY OF MATERIALS:

Lanthanum oxalate was prepared "at a fairly low temperature since that obtained at 100°C 100°C " was composed of a lower hydrate which required long equilibrium times.

No other information given.

ESTIMATED ERROR:

Soly: precision probably ± 0.5 % (compilers).

Temp: nothing specified.

REFERENCES:

COMPONENTS:

(1) Lanthanum oxalate; $La(C_2O_4)_3$; [537-03-1]

(2) Lanthanum nitrate; $La(NO_3)_3$; [10099-59-9]

(3) Water ; H_2O ; [7732-18-5]

ORIGINAL MEASUREMENTS:

James, C.; Whittemore, C.F. *J. Am. Chem. Soc.* 1912, *34*, 1168-71.

EXPERIMETNAL VALUES: continued......

Composition of saturated solutions at 25°C

mass %	mol kg^{-1}	mass %	mol kg^{-1}	nature of the solid phase
52.74	3.624	2.47	0.102	$La_2(C_2O_4)_3 \cdot 3H_2O$
54.11	3.830	2.41	0.102	"
55.20	4.001	2.34	0.102	"
56.54	4.230	2.32	0.104	"
58.22	4.534	2.26	0.106	"
59.03	4.690	2.23	0.106	$La_2(C_2O_4)_3 \cdot 3H_2O + La(NO_3)_3 \cdot 6H_2O^c$
59.03	4.674	2.10	0.0997	$La(NO_3)_3 \cdot 6H_2O^c$
59.91	4.677	0.67	0.031	"
60.17	4.649	----	----	"

a. Molalities calculated by the compilers.

b. From the phase diagram, it was predicted that the solvate $La_2(C_2O_4)_3 \cdot 5H_2O$ exists, but no solubility data were obtained for solutions in which this solvate is part of the solid phase.

c. The authors did not specify the nature of the hydration of the lanthanum nitrate. The compilers assume that references to lanthanum nitrate as the solid phase involve the hexahydrate.

COMPONENTS:

(1) Lanthanum nitrate; $La(NO_3)_3$; [10099-59-9]

(2) Lanthanum chloride; $LaCl_3$; [10099-58-8]

(3) Water ; H_2O ; [7732-18-5]

ORIGINAL MEASUREMENTS:

Petelina, N.S.; Mertslin, R.V.; Nikurashina, N.I.; Sedova, L.K. *Issled. v. Obl. Khim. Redkozem. Elementov* 1969, 85-9.

VARIABLES:

Composition at 25°C

PREPARED BY:

T. Mioduski and S. Siekierski

EXPERIMENTAL VALUES:

The solubilities for the ternary system at 25°C were given numerically only for $LaCl_3$ along four sections. Neither the $La(NO_3)_3$ or the H_2O composition along these sections are given, the quantitative significance of the numerical data for $LaCl_3$ is questionable. The authors do present a phase diagram, but the % composition for each axis is not indicated, and it is not possible to interpolate these values with sufficient precision. For these reasons the compilers have rejected the data for the ternary system.

The authors do report the solubility of $La(NO_3)_3$ in pure water at 25°C as 59.8 mass %. Converting this value to molality, the compilers obtain a solubity of 4.58 mol kg^{-1}.

AUXILIARY INFORMATION

METHOD/APPARATUS/PROCEDURE:

The isothermal method of sections was used as described in (1). The refractive indices of saturated solutions are measured along directed sections of the phase diagram. The results are used to graph the relation between the refractive index and the composition of the components for of the sections studied. Solutions were equilibrated for 20 days (presumably at room temperature) with periodic agitation. Equilibrium was confirmed by constancy in the refractive indices. Prior to measuring the refractive indices, the solutions were thermostated for 3 hours at 25°C.

SOURCE AND PURITY OF MATERIALS:

Each salt was prepared by dissolving La_2O_3 in the appropriate acid followed by crystallization. The salts were analysed for La content by the oxalate method.

No other information given.

ESTIMATED ERROR:

Soly: nothing specified.

Temp: precision ± 0.1 to 0.2 K.

REFERENCES:

1. Mertslin, R.V. *Zh. Obshch. Khim.* 1936, 7, 1828.

COMPONENTS:

(1) Lanthanum nitrate; $La(NO_3)_3$; [10099-59-9]

(2) Praseodymium nitrate; $Pr(NO_3)_3$; [10361-80-5]

(3) Water; H_2O; [7732-18-5]

ORIGINAL MEASUREMENTS:

Brunisholz, G.; Quinche, J.P.; Kalo, A.M. *Helv. Chim. Acta* 1964, *47*, 14-27.

VARIABLES:

Composition at 20°C

PREPARED BY:

T. Mioduski, S. Siekierski, and M. Salomon

EXPERIMENTAL VALUES:

The $La(NO_3)_3$ - $Pr(NO_3)_3$ - H_2O system at 20°C

Composition of the saturated solutions[a]

mol % Pr of total Pr + La	moles H_2O per 100 mol Pr + La	$La(NO_3)_3$ mol kg^{-1}	$Pr(NO_3)_3$ mol kg^{-1}	nature of the solid phase
0.0	1255	4.42	---	$La(NO_3)_3 \cdot 6H_2O$[b]
8.9	1221	4.14	0.40	solid solution I[c]
24.0	1200	3.52	1.11	"
38.7	1159	2.94	1.85	"
61.3	1105	1.94	3.08	"
70.9	1073	1.51	3.67	"
73.1	1064	1.40	3.81	I + II
74.2	1050	1.36	3.92	solid solution II[d]
80.3	1087	1.01	4.10	"
90.7	1116	0.46	4.51	"
100	1149	---	4.83	$Pr(NO_3)_3 \cdot 6H_2O$[b]

a. Molalities calculated by M. Salomon.

b. Composition not specified, but assumed by compilers.

c. (<u>La</u>,Pr)$(NO_3)_3 \cdot 6H_2O$ solid solution: miscibility limit of Pr = 45 mol %.

d. (La,<u>Pr</u>)$(NO_3)_3 \cdot 6H_2O$ solid solution: miscibility limit of La = 9.0 mol %.

AUXILIARY INFORMATION

METHOD/APPARATUS/PROCEDURE:

The isothermal method was used. The mixed homogeneous crystals were equilibrated for 3-5 weeks using continuous pulverization of the solid (500000 strokes per week). This technique of pulverization was accomplished by placing the solids and liquid into glass tubes into which a small "dumb-bell" shaped pestle was placed. The sealed tubes were placed in larger tubes and rotated in a thermostat at 20°C. La and Pr in the satd solutions were determined by complexometric titration using Xylenol orange indicator and urotropine buffer. Water was determined by difference. The solid phases were identified by X-ray analysis and by Schreinemakers' method of residues.

SOURCE AND PURITY OF MATERIALS:

$La(NO_3)_3 \cdot 6H_2O$ and $Pr(NO_3)_3 \cdot 6H_2O$ were prepared from the oxides. The oxides were purified by ion exchange chromatography, and were at least 99.7 % pure.

ESTIMATED ERROR:

Soly: precision ± 0.2 % (compilers).

Temp: precision at least ± 0.05 K (compilers).

REFERENCES:

COMPONENTS:

(1) Lanthanum nitrate; $La(NO_3)_3$; [10099-59-9]

(2) Neodymium nitrate; $Nd(NO_3)_3$; [10045-95-1]

(3) Water ; H_2O ; [7732-18-5]

ORIGINAL MEASUREMENTS:

Brunisholz, G.; Quinche, J.P.; Kalo, A.M.
Helv. Chim. Acta 1964, *47*, 14-27.

VARIABLES:

Composition at 0°C and 20°C

PREPARED BY:

T. Mioduski, S. Siekierski, and M. Salomon

EXPERIMENTAL VALUES:

The $La(NO_3)_3$ - $Nd(NO_3)_3$ - H_2O system

Composition of saturated solutions at 0°C

mol % Nd of total Nd + La	moles H_2O per 100 mol Nd + La	$La(NO_3)_3$ mol kg^{-1}	$Nd(NO_3)_3$ mol kg^{-1}	nature of the solid phase
0.0	1476	3.76	---	$La(NO_3)_3 \cdot 6H_2O$ [b]
15.5	1434	3.27	0.60	solid solution I [c]
35.0	1380	2.61	1.41	"
51.0	1321	2.06	2.14	"
57.2	1288	1.84	2.47	I + II
57.3	1301	1.82	2.44	"
62.8	1310	1.58	2.66	solid solution II [d]
73.0	1343	1.12	3.02	"
83.7	1387	0.65	3.35	"
100	1446	---	3.84	$Nd(NO_3)_3 \cdot 6H_2O$ [b]

a. Molalities calculated by the compilers.

b. Composition not specified, but assumed by M. Salomon.

c. (<u>La</u>,Nd)$(NO_3)_3 \cdot 6H_2O$ solid solution: miscibility limit of Nd = 12.2 mol %.

d. (La,<u>Nd</u>)$(NO_3)_3 \cdot 6H_2O$ solid solution: miscibility limit of La = 14.7 mol %.

continued......

AUXILIARY INFORMATION

METHOD/APPARATUS/PROCEDURE:

The isothermal method was used. The mixed homogeneous crystals were equilibrated for 3-5 weeks using continuous pulverization of the solid (500000 strokes per week). This technique of pulverization was accomplished by placing the solids and liquid into glass tubes into which a small "dumb-bell" shaped pestle was placed. The sealed tubes were placed in larger tubes and rotated in a thermostat at the required temperature. La and Nd in the saturated solutions were determined by complexometric titration using Xylenol orange indicator and urotropine buffer, and by chromatographic analysis. Water was determined by difference. The solid phases were identified by X-ray diffraction and by Schreinemakers's method of residues.

SOURCE AND PURITY OF MATERIALS:

$La(NO_3)_3 \cdot 6H_2O$ and $Nd(NO_3)_3 \cdot 6H_2O$ were prepared from the oxides. The oxides were purified by ion exchange chromatography, and were at least 99.7 % pure.

ESTIMATED ERROR:

Soly: precision ± 0.2 % (compilers).

Temp: precision at least ± 0.05 K (compilers).

REFERENCES:

COMPONENTS:

(1) Lanthanum nitrate; $La(NO_3)_3$; [10099-59-9]

(2) Neodymium nitrate; $Nd(NO_3)_3$; [10045-95-1]

(3) Water ; H_2O ; [7732-18-5]

ORIGINAL MEASUREMENTS:

Brunisholz, G.; Quinche, J.P.; Kalo, A.M.

Helv. Chim. Acta 1964, *47*, 14-27.

EXPERIMENTAL VALUES: continued......

Composition of saturated solutions at 20°C[a]

mol % Nd of total Nd + La	moles H_2O per 100 mol Nd + La	$La(NO_3)_3$ mol kg^{-1}	$Nd(NO_3)_3$ mol kg^{-1}	nature of the solid phase
0	1255	4.42	---	$La(NO_3)_3 \cdot 6H_2O$[b]
10.1	1227	4.07	0.46	solid solution I[c]
12.2	1214	4.01	0.56	"
27.8	1203	3.33	1.28	"
44.0	1158	2.68	2.11	"
53.4	1119	2.31	2.65	"
55.4	1110	2.23	2.77	"
57.5	1094	2.16	2.92	I + II
57.9	1099	2.13	2.92	solid solution II[d]
61.0	1112	1.95	3.04	"
70.30	1162	1.42	3.36	"
89.2	1221	0.49	4.06	"
100	1259	---	4.41	$Nd(NO_3)_3 \cdot 6H_2O$[b]

a. Molalities calculated by compilers.

b. Composition not specified, but assumed by M. Salomon.

c. (La,Nd)$(NO_3)_3 \cdot 6H_2O$ solid solutions: miscibility limit of Nd = 12.7 mol %.

d. (La,Nd)$(NO_3)_3 \cdot 6H_2O$ solid solutions: miscibility limit of La = 20.0 mol %.

COMPONENTS:

(1) Lanthanum nitrate; $La(NO_3)_3$; [10099-59-9]

(2) Neodymium nitrate; $Nd(NO_3)_3$; [10045-95-1]

(3) Tri-n-butyl phosphate; $C_{12}H_{27}O_4P$; [126-73-8]

(4) Water ; H_2O ; [7732-18-5]

ORIGINAL MEASUREMENTS:

Kolesnikov, A.A.; Korotkevich, I.B.; Bui Van Tuan; Stepin, B.D. *Zh. Neorg. Khim.* 1978, *23*, 2833-8; *Russ. J. Inorg. Chem. Engl. Transl.* 1978, *23*, 1570-3.

VARIABLES:

Composition, temperature

PREPARED BY:

T. Mioduski and S. Siekierski

EXPERIMENTAL VALUES:

	Composition of saturated aqueous phase[a]				Composition of saturated organic phase[a]			
	$La(NO_3)_3$		$Nd(NO_3)_3$		$La(NO_3)_3$		$Nd(NO_3)_3$	
t/°C	mass %	mol kg^{-1}	mass %	mol kg^{-1}	mass %	mol kg^{-1}	mass %	mol kg^{-1}
50	4.90	0.480	63.65	6.128				
20	10.05	0.840	51.02	4.015	1.45	0.063	28.01	1.202
20	8.31	0.656	52.68	4.089				
50	14.08	1.377	54.45	5.239				
20	20.90	1.677	40.75	3.217	2.15	0.094	27.74	1.198
20	15.70	1.280	46.56	3.736				
50	21.15	2.162	48.74	4.901				
20	25.82	2.138	37.02	3.017	2.04	0.089	27.50	1.182
20	23.16	1.933	39.97	3.283				
50	31.74	3.280	38.48	3.913	2.36	0.105	28.44	1.244
40	32.98	3.245	35.74	3.460	2.59	0.115	27.88	1.214
30	32.10	2.840	33.11	2.882	3.04	0.134	27.31	1.187
20	32.86	2.722	29.98	2.443	4.08	0.179	25.84	1.116
50	39.17	4.195	32.09	3.381	3.92	0.173	26.17	1.133
40	40.03	3.935	28.66	2.772	4.12	0.181	25.95	1.124
30	39.67	3.406	24.48	2.068	4.21	0.186	26.04	1.130
20	40.54	3.163	20.01	1.536	4.37	0.225	25.81	1.813

continued.......

AUXILIARY INFORMATION

METHOD/APPARATUS/PROCEDURE:

The system was studied by the extractive crystallization ray (ECR) method similar to that described by Nikolaev (1). An ECR is a line of a phase diagram which reflects the change in compn of the aqueous phase during extractive crystn. Satd slns of both salts in water is prepd at 50°C with varying amts of solid $La(NO_3)_3$ and $Nd(NO_3)_3$. The extractant is added to these slns and the temp reduced. The phases were analysed after equil was reached (2 days for aqueous systems, and 1/4 to 1/5 less when tributyl phosphate was present). Total La + Nd was detd by complexometric titrn. Nd detd spectrophotometrically at a wavelength of 569 nm, but at concns of 1 x 10^{-4} mass % and less, Nd was detd by radioassay using ^{147}Nd. Authors state that the compositions of the solid phases were detd by Schreinemakers' method, but they do not report the results of these analyses. They do report the mass % of anhydrous $La(NO_3)_3$ and $Nd(NO_3)_3$ in the solid phases. both salts are present in all the solid phases for the equilibrated satd slns given in the soly tables. The source paper also lists distribution coefficients for each sln.

SOURCE AND PURITY OF MATERIALS:

C.p. grade lanthanum nitrate was used. When ^{147}Nd was used, spec grade lanthanum nitrate was used.

C.p. grade neodymium nitrate was used.

The extractant, $[CH_3(CH_2)_3O]PO$, was purified as in (2).

Source and purity of water not specified.

ESTIMATED ERROR:

Soly: based on the method, precision is about 1-4 % of the indicated solubilities (compilers).

Temp: precision ± 0.2 K (authors).

REFERENCES:

1. Nikolaev, A.V. *Ekstraktsiya Neorgicheskikh Veshchestv*. Izd. Nauka. Novosibirsk. 1970. p 81.
2. Alcock, K.; Grimley, S.S.; Healy, T.V.; McKay, H.A.C. *Trans. Faraday Soc.* 1956, *52*, 39.

COMPONENTS:

(1) Lanthanum nitrate; $La(NO_3)_3$; [10099-59-9]

(2) Neodymium nitrate; $Nd(NO_3)_3$; [10045-95-1]

(3) Tri-n-butyl phosphate; $C_{12}H_{27}O_4P$; [126-73-8]

(4) Water ; H_2O ; [7732-18-5]

ORIGINAL MEARUREMENTS:

Kolesnikov, A.A.; Korotkevich, I.B.; Bui Van Tuan; Stepin, B.D. *Zh. Neorg. Khim.* 1978, *23*, 2833-8; *Russ. J. Inorg. Chem. Engl. Transl.* 1978, *23*, 1570-3.

EXPERIMENTAL VALUES: continued......

	Composition of saturated aqueous phase[a]				Composition of saturated organic phase[a]			
	$La(NO_3)_3$		$Nd(NO_3)_3$		$La(NO_3)_3$		$Nd(NO_3)_3$	
t/°C	mass %	mol kg^{-1}	mass %	mol kg^{-1}	mass %	mol kg^{-1}	mass %	mol kg^{-1}
50	44.85	4.467	24.25	2.376	4.23	0.185	25.46	1.096
40	45.12	4.192	21.75	1.988	4.41	0.193	25.34	1.092
30	45.00	3.849	19.02	1.601	4.73	0.208	25.42	1.102
20	44.62	3.436	15.41	1.167	5.12	0.225	24.98	1.082
50	54.07	5.320	14.65	1.418				
40	53.01	4.715	12.39	1.084	11.23	0.488	17.95	0.767
30	52.49	4.328	10.18	0.826	12.37	0.533	16.21	0.687
20	51.41	3.910	8.12	0.608	13.82	0.595	14.70	0.623
50	63.11	6.057	4.82	0.455				
20	57.13	4.303	2.01	0.149	14.43	0.622	14.21	0.603
50	66.5	6.14	0.14	0.013				
20	59.5	4.53	0.062	0.0046				
20	59.7	4.56	0.037	0.0028	28.3	1.22	0.1	0.004
50	66.0	5.98	0.0054	0.00048				
20	58.5	4.34	0.0021	0.00015				
20	59.0	4.43	0.0014	0.00010	29.0	1.26	0.004	0.0002
50	66.3	6.06	0.0006	0.00005				
20	59.8	4.58	0.0002	0.00002				
20	59.00	4.43	0.00018	0.00001	29.3	1.28	0.00042	0.000018
50	66.3	6.05	0.00023	0.000021				
20	58.3	4.30	0.00009	0.000007				
20	58.8	4.39	0.00005	0.000004	28.8	1.24	0.000009	0.0000004

a. Molalities calculated by M. Salomon.

COMPONENTS:

(1) Lanthanum nitrate; $La(NO_3)_3$; [10099-59-9]

(2) Samarium nitrate; $Sm(NO_3)_3$; [10361-83-8]

(3) Water ; H_2O ; [7732-18-5]

ORIGINAL MEASUREMENTS:

Brunisholz, G.; Quinche, J.P.; Kalo, A.M.

Helv. Chim. Acta 1964, *47*, 14-27.

VARIABLES:

Composition and temperature

PREPARED BY:

T. Mioduski, S. Siekierski, and M. Salomon

EXPERIMENTAL VALUES:

The $La(NO_3)_3$ - $Sm(NO_3)_3$ - H_2O system

Composition of the saturated solutions[a]

t/°C	mol % Sm of total Sm + La	moles H_2O per 100 mol La + Sm	$La(NO_3)_3$ mol kg^{-1}	$Sm(NO_3)_3$ mol kg^{-1}	nature of the solid phase
0	51.99	1327	2.008	----	I + II[b]
20	0	1255	4.42	----	$La(NO_3)_3 \cdot 6H_2O$[c]
	11.7	1243	3.94	0.52	I
	22.5	1217	3.53	1.03	I
	33.3	1189	3.11	1.55	I
	43.8	1157	2.70	2.10	I
	48.22	1141	2.519	2.346	I + II
	51.8	1153	2.32	2.49	II
	61.6	1197	1.78	2.86	II
	77.8	1268	0.97	3.41	II
	100	1350	---	4.11	$Sm(NO_3)_3 \cdot 6H_2O$[c]
35	44.31	956.5	3.232	2.571	I + II

continued......

AUXILIARY INFORMATION

METHOD/APPARATUS/PROCEDURE:

The isothermal method was used. The mixed homogeneous crystals were equilibrated for 3-5 weeks using continuous pulverization of the solid (500000 strokes per week). This technique of pulverization was accomplished by placing the solids and liquid into glass tubes into which a small "dumb-bell" shaped pestle was placed. The sealed tubes were placed in larger tubes and rotated in a thermostat at the required temperature. La and Pr in the saturated solutions were determined by complexometric titration using Xylenol orange indicator and urotropine buffer, and by chromatographic analysis. Water was determined by difference. The solid phases were identified by X-ray diffraction and by Schreinemakers' method of residues.

SOURCE AND PURITY OF MATERIALS:

$La(NO_3)_3 \cdot 6H_2O$ and $Sm(NO_3)_3 \cdot 6H_2O$ were prepared from the oxides. The oxides were purified by ion exchange chromatography, and were at least 99.7% pure.

ESTIMATED ERROR:

Soly: precision ± 0.2 % (compilers).

Temp: precision at least ± 0.05 K (compilers).

REFERENCES:

COMPONENTS:

(1) Lanthanum nitrate; $La(NO_3)_3$; [10099-59-9]

(2) Samarium nitrate; $Sm(NO_3)_3$; [10361-83-8]

(3) Water ; H_2O ; [7732-18-5]

ORIGINAL MEASUREMENTS:

Brunisholz, G.; Quinche, J.P.; Kalo, A.M.

Helv. Chim. Acta 1964, *47*, 14-27.

EXPERIMENTAL VALUES: continued......

t/°C	mol % Sm of total La + Sm	moles H_2O per 100 mol La + Sm	$La(NO_3)_3$ mol kg^{-1}	$Sm(NO_3)_3$ mol kg^{-1}	nature of the solid phase
50	0	917.2	6.05	---	$La(NO_3)_3 \cdot 6H_2O$[c]
	16.5	865.3	5.36	1.06	I[b]
	32.4	810.8	4.63	2.22	I
	40.72	781.9	4.208	2.891	I + II
	45.3	800.6	3.79	3.14	II
	45.6	786.6	3.84	3.22	II
	52.1	819.9	3.24	3.53	II
	63.9	890.2	2.25	3.98	II
	74.4	921.0	1.54	4.48	II
	100	1020	---	5.44	$Sm(NO_3)_3 \cdot 6H_2O$[c]

a. Molalities were calculated by M. Salomon.

b. Solid phases are solid solutions: see COMMENTS below.

C. Composition not specified, but assumed by compilers.

COMMENTS AND/OR ADDITIONAL DATA:

The solid phases were identified by the authors as follows:

For 0°C

Solid solution I = (<u>La</u>,Sm)$(NO_3)_3 \cdot 6H_2O$, miscibility limit of Sm = 1.7 mol %.

Solid solution II = (La,<u>Sm</u>)$(NO_3)_3 \cdot 6H_2)$, miscibility limit of La = 10.9 mol %.

For 20°C

Solid solution I = (<u>La</u>,Sm)$(NO_3)_3 \cdot 6H_2O$, miscibility limit of Sm = 3.0 mol %.

Solid solution II = (La,<u>Sm</u>)$(NO_3)_3 \cdot 6H_2O$, miscibility limit of La = 14.1 mol %.

For 35°C

Solid solution I = (<u>La</u>,Sm)$(NO_3)_3 \cdot 6H_2O$, miscibility limit of Sm = 7.8 mol %.

Solid solution II = (La,<u>Sm</u>)$(NO_3)_3 \cdot 6H_2O$, miscibility limit of La = 19.2 mol %.

For 50°C

Solid solution I = (<u>La</u>,Sm)$(NO_3)_3 \cdot 6H_2O$, miscibility limit of Sm = 14.5 mol %.

Solid solution II = (La,<u>Sm</u>)$(NO_3)_3 \cdot 6H_2O$, miscibility limit of La = 20.5 mol %.

COMPONENTS:	ORIGINAL MEASUREMENTS:
(1) Lanthanum nitrate; $La(NO_3)_3$; [10099-59-9] (2) Samarium nitrate; $Sm(NO_3)_3$; [10361-83-8] (3) Tri-n-butyl phosphate; $C_{12}H_{27}O_4P$; [126-73-8] (4) Water ; H_2O ; [7732-18-5]	Kolesnikov, A.A.; Korotkevich, I.B.; Shakhaleeva, N.N.; Stepin, B.D. *Zh. Neorg. Khim.* 1978, *23*, 2524-8; *Russ. J. Inorg. Chem. Engl. Transl.* 1978, *23*, 1395-8.
VARIABLES:	PREPARED BY:
Composition and temperature	T. Mioduski and S. Siekierski

EXPERIMENTAL VALUES:

		Composition of saturated aqueous phase[a]				Composition of saturated organic phase[a]			
		$La(NO_3)_3$		$Sm(NO_3)_3$		$La(NO_3)_3$		$Sm(NO_3)_3$	
R[b]	t/°C	mass %	mol kg^{-1}	mass %	mol kg^{-1}	mass %	mol kg^{-1}	mass %	mol kg^{-1}
	50	5.7	0.56	63.1	6.06				
	25	12.0	0.90	47.0	3.41				
0.2	25	7.0	0.57	55.2	4.34	4.0	0.17	24.8	1.04
0.3	25	8.2	0.66	53.7	4.19	3.2	0.14	26.0	1.09
0.5	25	9.5	0.74	51.2	3.87	2.0	0.087	27.5	1.16
	50	20.4	2.73	56.6	7.32				
	25	16.2	1.17	41.3	2.89				
0.2	25	15.6	1.49	52.1	4.80	5.0	0.22	24.0	1.00
0.3	25	18.1	1.52	45.2	3.66	4.2	0.18	24.3	1.01
0.5	25	22.3	1.98	43.1	3.70	3.4	0.14	22.3	0.89
	50	20.1	2.10	50.4	5.08				
	25	19.2	1.67	45.5	3.83				
0.2	25	45.0	4.05	20.8	1.81	8.1	0.36	21.8	0.92
0.3	25	21.9	1.93	43.1	3.66	7.3	0.32	23.0	0.98
0.5	25	21.0	1.65	39.8	3.02	6.9	0.29	21.0	0.87
	50	33.0	3.53	38.2	3.94				
	25	28.1	2.41	36.0	2.98				
0.2	25	31.9	2.82	33.3	2.84	12.5	0.55	17.0	0.72
0.3	25	36.0	2.92	26.1	2.04	11.7	0.52	18.5	0.79
0.5	25	41.0	2.85	14.8	1.00	11.0	0.48	19.0	0.81

continued......

AUXILIARY INFORMATION

METHOD/APPARATUS/PROCEDURE:

The system was studied by the extractive crystallization ray (ECR) method similar to that described by Nikolaev (1). For additional details, see the compilation of ref (2). The extraction rays were based on a series of extractions from a given saturated solution, but at different ratios, R, of the volumes of organic extractant to aqueous solution (V_{org}/V_{aq} = R in the above table).

The extraction crystallization was carried out in Fluoroplast-4 test tubes with screw stoppers (Fluoroplast-4 is a Teflon type of plastic). Lanthanum and the total La + Sm was determined by complexometric titration. Samarium was determined spectophotometrically at a wavelength of 401.3 nm, but at concentrations of 1 x 10^{-4} mass % and less, Sm was determined by radioassay using ^{153}Sm. Composition of the solid phases not specified, but analysis of anhydrous solid phases for $La(NO_3)_3$ and $Sm(NO_3)_3$ are given in the source paper. The source paper also lists distribution coefficients for $La(NO_3)_3$ and $Sm(NO_3)_3$ between the aqueous and organic phases.

SOURCE AND PURITY OF MATERIALS:

C.p. and spec. pure lanthanum nitrate, and c.p. grade samarium nitrate were used. The tri-n-butyl phosphate extractant was purified as in (3). Source and purity of water was not specified.

ESTIMATED ERROR:

Soly: based on the method, precision is about 1-4 % of the indicated solubilities (compilers).

Temp: precision ± 0.2 K (authors).

REFERENCES:

1. Nikolaev, A.V. *Ekstraktsiya Neorgicheskikh Veshchestv.* Izd. Nauka. Novosibirsk. 1970. p 81.
2. Kolesnikov, A.A.; Kortkevich, I.B.; Bui Van Tuan; Stepin, B.D. *Zh. Neorg. Khim.* 1978, *23*, 2833.
3. Alcock, K.; Grimley, S.S.; Healy, T.V.; McKay, H.A.C. *Trans. Faraday Soc.* 1956, *52*, 39.

COMPONENTS:

(1) Lanthanum nitrate; $La(NO_3)_3$; [10099-59-9]

(2) Samarium nitrate; $Sm(NO_3)_3$; [10361-83-8]

(3) Tri-n-butyl phosphate; $C_{12}H_{27}O_4P$; [126-73-8]

(4) Water ; H_2O ; [7732-18-5]

ORIGINAL MEASUREMENTS:

Kolesnikov, A.A.; Korotkevich, I.B.; Shakhaleeva, N.N. Stepin, B.D. *Zh. Neorg. Khim.* 1978, *23*, 2524-8; *Russ. J. Inorg. Chem. Engl. Transl.* 1978, *23*, 1395-8.

EXPERIMENTAL VALUES: continued......

		Composition of saturated aqueous phase[a]				Composition of saturated organic phase[a]			
		$La(NO_3)_3$		$Sm(NO_3)_3$		$La(NO_3)_3$		$Sm(NO_3)_3$	
R[b]	t/°C	mass %	mol kg^{-1}	mass %	mol kg^{-1}	mass %	mol kg^{-1}	mass %	mol kg^{-1}
	50	48.0	4.96	22.2	2.21				
	25	48.2	4.39	18.0	1.58				
0.2	25	48.5	4.38	17.4	1.52	10.4	0.45	18.3	0.76
0.3	25	49.5	4.06	13.0	1.03	6.0	0.26	22.5	0.94
0.5	25	50.6	3.90	9.5	0.71	2.8	0.12	24.7	1.01
	50	53.0	4.80	13.0	1.14				
	25	29.1	2.23	30.8	2.28				
0.2	25	54.0	4.62	10.0	0.83	5.0	0.23	26.7	1.16
0.3	25	52.0	3.90	7.0	0.51	2.0	0.089	29.0	1.25
0.5	25	57.5	4.60	4.0	0.31	1.5	0.066	29.0	1.24
	50	59.7	6.37	11.45	1.18				
	25	37.2	3.09	25.8	2.07				
0.2	25	53.8	4.16	6.4	0.48	9.6	0.43	21.0	0.90
0.5	25	54.2	3.94	3.5	0.25	9.0	0.40	21.6	0.93
	50	60.0	5.77	8.0	0.74				
	25	49.3	3.82	11.0	0.82				
0.2	25	55.0	4.23	5.0	0.37	17.0	0.70	8.5	0.34
0.5	25	56.0	4.10	2.0	0.14	17.5	0.73	9.0	0.36
	50	65.5	6.06	1.25	0.11				
	25	56.0	3.99	0.77	0.053				
0.5	25	56.0	3.95	0.41	0.028	29.0	1.28	1.22	0.052
	50	66.0	6.01	0.21	0.018				
	25	54.0	3.63	0.18	0.012				
0.5	25	54.0	3.62	0.029	0.0019	27.0	1.14	0.18	0.0073
	50	66.0	5.98	0.019	0.0017				
	25	55.0	4.25	0.017	0.0012				
0.5	25	56.0	3.92	0.0043	0.00029	27.0	1.14	0.2	0.008
	50	66.5	6.11	0.0022	0.00020				
	25	57.0	4.08	0.0013	0.000090				
0.5	25	56.0	3.92	0.00059	0.000040	28.0	1.20	0.004	0.0002

a. Molalities calculated by M. Salomon.

b. $R = V_{org}/V_{aq}$, the ratio of volumes of the organic to aqueous phases.

COMPONENTS:

(1) Lanthanum nitrate; $La(NO_3)_3$; [10099-59-9]

(2) Manganese nitrate; $Mn(NO_3)_2$; [10377-66-9]

(3) Water ; H_2O ; [7732-18-5]

ORIGINAL MEASUREMENTS:

di Capua, C. *Gazz. Chim. Ital.* 1929, *59*, 164-9.

VARIABLES:

Composition at 20°C

PREPARED BY:

T. Mioduski and S. Siekierski

EXPERIMENTAL VALUES:

The $La(NO_3)_3$ - $Mn(NO_3)_2$ - H_2O system at 20°C[a]

$La(NO_3)_3$		$Mn(NO_3)_2$		$La(NO_3)_3$		$Mn(NO_3)_2$	
mass %	mol kg^{-1}	mass %	mol kg^{-1}	mass %	mol kg^{-1}	mass %	mol kg^{-1}
60.13	4.642	---	---	24.00	1.495	26.60	3.009
53.90	4.044	5.08	0.692	23.80	1.533	28.41	3.322
51.75	3.826	6.62	0.889	22.95	1.425	27.50	3.101
49.20	3.615	8.91	1.189	22.30	1.392	28.40	3.219
41.45	2.814	13.22	1.630	21.25	1.319	29.18	3.290
38.00	2.536	15.88	1.924	14.15	0.973	41.10	5.132
36.82	2.412	16.19	1.925	12.03	0.833	43.52	5.471
35.60	2.339	17.55	2.093	10.00	0.684	45.00	5.588
29.81	1.865	21.00	2.386	6.22	0.429	49.20	6.167
27.45	1.670	21.95	2.424	---	---	56.81	7.350
27.40	1.567	26.80	3.088				

a. Nature of the solid phases not specified.

Molalities calculated by M. Salomon.

AUXILIARY INFORMATION

METHOD/APPARATUS/PROCEDURE:

The isothermal method was used. Lanthanum was determined gravimetrically by the oxalate method (i.e. ignition of the oxalate to the oxide), and Mn was determined by the Knorre method. Water was determined by difference.

SOURCE AND PURITY OF MATERIALS:

Nothing specified.

ESTIMATED ERROR:

Soly: precision probably ± 0.5 % (compilers).

Temp: precision probably ± 0.5 K (compilers).

REFERENCES:

COMPONENTS:

(1) Lanthanum nitrate; $La(NO_3)_3$; [10099-59-9]

(2) Cobalt nitrate; $Co(NO_3)_2$; [10141-05-6]

(3) Water; H_2O; [7732-18-5]

ORIGINAL MEASUREMENTS:

Odent, G.; Venot, A. *C.R. Hebd. Seances Acad. Sci., Ser. C.* 1975, *280*, 377-80.

VARIABLES:

Composition at 25 and 40°C

PREPARED BY:

T. Mioduski and S. Siekierski

EXPERIMENTAL VALUES:

Composition of saturated solutions.[a]

	$La(NO_3)_3$		$Co(NO_3)_2$		
t/°C	mass %	mol kg^{-1}	mass %	mol kg^{-1}	nature of the solid phase
25.0	58.858	4.403	---	---	$La(NO_3)_3 \cdot 6H_2O$
	51.85[b]	3.740	5.48	0.702	$La_2Co_3(NO_3)_{12} \cdot 24H_2O$
	5.24[b]	0.341	47.52	5.499	"
	---	---	50.29	5.532	$Co(NO_3)_2 \cdot 6H_2O$
40.0	61.160	4.846	---	---	$La(NO_3)_3 \cdot 6H_2O$
	52.16[b]	4.104	8.72	1.218	$La_2Co_3(NO_3)_{12} \cdot 24H_2O$
	0.91[b]	0.057	50.11	5.592	"
	---	---	55.349	6.776	$Co(NO_3)_2 \cdot 6H_2O$

a. Molalities calculated by M. Salomon.
b. Transition points.

The authors also reported the solubility of $La(NO_3)_3$ in units of g(1)/100 g H_2O. At 25.0°C they report a solubility of 143.06 g(1)/100 g H_2O, and at 40.0°C they report the solubility as 160.40 g(1)/100 g H_2O. Similarly for $Co(NO_3)_2$ the solubilities at 25.0° and 40.0°C are, respectively, 101.20 g(2)/100 g H_2O and 123.96 g(2)/100 g H_2O.

AUXILIARY INFORMATION

METHOD/APPARATUS/PROCEDURE:

The isothermal method was used. The solutions were equilibrated until their densities, measured with a Cornec-Cottet pipet, remained constant for three successive measurements performed at 24 hour intervals. Both the compositions of the saturated solutions and the solid phases were determined. The total La + Co was determined by titration with EDTA and back titrating with standard Zn^{2+} solutions using Eriochrome Black indicator at pH 9. Co was determined separately by electrolysis, and La was obtained by difference. The composition of the solid phases were determined by Schreinemakers' method. The densities of the solid phases were determined pycnometrically using xylene as a reference liquid.

SOURCE AND PURITY OF MATERIALS:

Nothing specified.

ESTIMATED ERROR:

Soly: based on the methods used, the compilers estimate a precision of ±0.3 mass%.

Temp: precision probably ± 0.1 K (compilers).

REFERENCES:

COMPONENTS:

(1) Lanthanum nitrate; $La(NO_3)_3$; [10099-59-9]

(2) Nickel nitrate; $Ni(NO_3)_2$; [13138-45-9]

(3) Water ; H_2O ; [7732-18-5]

ORIGINAL MEASUREMENTS:

Urazov, G.G.; Shevtsova, Z.N. *Zh. Neorg. Khim.* 1957, *2*, 659-61 ; *J. Inorg. Chem. (USSR)* 1957, *2*, 295-9.

VARIABLES:

Composition at 25°C

PREPARED BY:

T. Mioduski, S. Siekierski, and M. Salomon

EXPERIMENTAL VALUES:

The $La(NO_3)_3$ - $Ni(NO_3)_2$ - H_2O system at 25.0°C

Composition of saturated solutions[a]

$La(NO_3)_3$		$Ni(NO_3)_2$		
mass %	mol kg^{-1}	mass %	mol kg^{-1}	nature of the solid phase
60.28	4.671	---	---	$La(NO_3)_3 \cdot 6H_2O$
55.26	4.160	3.86	0.517	$2La(NO_3)_3 \cdot 3Ni(NO_3)_2 \cdot 24H_2O$
52.34	3.857	5.90	0.773	"
42.53	2.886	12.12	1.463	"
35.60	2.287	16.50	1.885	"
29.20	1.790	20.60	2.246	"
24.23	1.425	23.45	2.453	"
21.00	1.252	27.38	2.903	"
6.13	0.305	32.00	2.831	"
10.37	0.606	36.98	3.844	"
6.12	0.360	41.50	4.336	"
2.35	0.134	43.54	4.404	"
---	---	48.08	5.068	solid solutions
---	---	50.28	5.535	"
---	---	51.70	5.858	$Ni(NO_3)_2 \cdot 6H_2O$

a. Molalities calculated by M. Salomon.

AUXILIARY INFORMATION

METHOD/APPARATUS/PROCEDURE:

The isothermal method was used. Equilibrium was established in 20 days. The compositions of the solid phases were determined by chemical analyses, and by thermographical, crystal-optical, and graphical methods.

Lanthanum was precipitated first with NH_4OH, and then as the oxalate, The oxalate was ignited to the oxide and lanthanum determined gravimetrically. Nickel was determined by precipitation with dimethylglyoxime, and water was determined by difference. The composition of the solid phases was determined by Schreinemakers' method.

SOURCE AND PURITY OF MATERIALS:

"Pure grade" starting materials were used. No other information given.

ESTIMATED ERROR:

Soly: analyses agreed to 0.1-0.2 % for each point on the soly curve (authors). The compilers assume this error refers to mass %.

Temp: precision ± 0.1 K.

COMPONENTS:

(1) Lanthanum nitrate; $La(NO_3)_3$; [10099-59-9]

(2) Zinc nitrate; $Zn(NO_3)_2$; [7779-88-6]

(3) Water ; H_2O ; [7732-18-5]

ORIGINAL MEASUREMENTS:

Urazov, G.G.; Shevtsova, Z.N. *Zh. Neorg. Khim.* 1957, *2*, 659-61; *J. Inorg. Chem. (USSR)* 1957, *2*, 295-9.

VARIABLES:

Composition at 25°C

PREPARED BY:

T. Mioduski, S. Siekierski, and M. Salomon.

EXPERIMENTAL VALUES:

Composition of saturated solutions at 25.0°C[a]

$La(NO_3)_3$ mass %	$La(NO_3)_3$ mol kg^{-1}	$Zn(NO_3)_2$ mass %	$Zn(NO_3)_2$ mol kg^{-1}	nature of the solid phase
60.28[b]	4.671	---	----	$La(NO_3)_3 \cdot 6H_2O$
56.40	4.288	3.12	0.407	solid solutions
54.13	4.165	5.87	0.775	"
53.12	4.034	6.35	0.827	$2La(NO_3)_3 \cdot 3Zn(NO_3)_2 \cdot 24H_2O$
44.30	3.045	10.92	1.288	"
41.53	2.868	13.90	1.647	"
36.42	2.387	16.62	1.869	"
32.97	2.175	20.38	2.307	"
27.50	1.747	24.05	2.621	"
20.51	1.281	30.22	3.239	"
11.42	0.679	36.81	3.754	"
6.01	0.368	43.72	4.592	"
3.25	0.199	46.48	4.882	"
1.64	0.101	48.18	5.070	"
1.05	0.068	51.13	5.646	solid solutions
0.5	0.033	53.10	6.043	"
---	----	55.98	6.715	$Zn(NO_3)_2 \cdot 6H_2O$

a. Molalities calculated by M. Salomon.

b. The English translation gives 60.27 mass % for this point.

AUXILIARY INFORMATION

METHOD/APPARATUS/PROCEDURE:

The isothermal method was used. Solutions were equilibrated for 20 days. The compositions of the solid phases were determined by chemical analyses, thermographical method, and by crystal-optical and graphical methods. Lanthanum was precipitated with NH_4OH and then as the oxalate: the oxalate was ignited to the oxide and weighed. From the filtrate zinc was determined as the phosphate, and water was determined by difference. The composition of the solid phases were determined by Schreinemakers' method.

SOURCE AND PURITY OF MATERIALS:

"Pure grade" starting materials were used. No other information given.

ESTIMATED ERROR:

Soly: Analyses agreed to 0.1-0.2 % for each point of the soly surve. Presumably the authors mean mass %.

Temp: precision ± 0.1 K.

REFERENCES:

COMPONENTS:

(1) Lanthanum nitrate; $La(NO_3)_3$; [10099-59-9]

(2) Zinc nitrate; $Zn(NO_3)_2$; [7779-88-6]

(3) Water ; H_2O ; [7732-18-5]

ORIGINAL MEASUREMENTS:

Brunisholz, G.; Klipfel, K. *Rec. Chim. Miner.* 1970, *7*, 349-58.

VARIABLES:

Composition at 20°C

PREPARED BY:

T. Mioduski and S. Siekierski

EXPERIMENTAL VALUES:

The $La(NO_3)_3 - Zn(NO_3)_2 - H_2O$ system at 20.00°C

Composition of the saturated solutions[a]

mol % La of total La + Zn	moles H_2O per 100 mol La + Zn	$La(NO_3)_3$ mol kg^{-1}	$Zn(NO_3)_2$ mol kg^{-1}	solid phase[b]
0	892	---	6.22	I
0.26	871	0.017	6.36	I + II
0.66	1052	0.035	5.24	II
6.16	1245	0.275	4.184	II
16.13	1309	0.684	3.557	II
30.00	1317	1.264	2.950	II
48.44	1309	2.054	2.186	II
65.20	1260	2.872	1.533	II
85.00	1180	3.998	0.706	II + III
85.02	1181	3.996	0.704	II + III
87.13	1208	4.004	0.591	III
93.07	1275	4.052	0.302	III
100	1260	4.41	---	III

a. Molalities calculated by M. Salomon.

b. Solid phases: I = $Zn(NO_3)_3 \cdot 6H_2O$

II = $2La(NO_3)_3 \cdot 3Zn(NO_3)_2 \cdot 24H_2O$

III = $La(NO_3)_3 \cdot 6H_2O$

AUXILIARY INFORMATION

METHOD/APPARATUS/PROCEDURE:

The isothermal method was used. The saturated solutions were equilibrated over a period of 2 to 4 weeks with continuous pulverization of the solid. Lanthanum was determined by titration with 0.01 mol dm^{-3} solution of the disodium salt of ethylenediaminetetraacetic acid. Xylenol orange indicator was used with urotropine buffer. Zn was precipitated as ZnS with thioacetamide or was separated by ion exchange chromatography.

The solid phases were identified by X-ray diffraction, and by Schreinemakers' method of residues: the latter method is preferred for identification of the solid phases.

SOURCE AND PURITY OF MATERIALS:

Source and purity of initial materials not specified. Both nitrates were prepared from their oxides by dissolution in nitric acid followed by crystallization. The salts were dried in a desiccator in vacuum over KOH.

ESTIMATED ERROR:

Soly: precision about ± 0.2 % (compilers).

Temp: precision ± 0.005 K (authors).

REFERENCES:

COMPONENTS:

(1) Lanthanum nitrate; $La(NO_3)_3$; [10099-59-9]

(2) Zinc nitrate; $Zn(NO_3)_2$; [7779-88-6]

(3) Water ; H_2O ; [7732-18-5]

ORIGINAL MEASUREMENTS:

Yakimov, M.A.; Gizhavina, E.I. *Zh. Neorg. Khim.* 1971, *16*, 507-9; *Russ. J. Inorg. Chem. Engl. Transl.* 1971, *16*, 268-9.

VARIABLES:

Composition at 25°C

PREPARED BY:

T. Mioduski and S. Siekierski

EXPERIMENTAL VALUES:

The $La(NO_3)_3 - Zn(NO_3)_2 - H_2O$ system at 25°C

Composition of saturated solutions[a]

$La(NO_3)_3$ mass %	$La(NO_3)_3$ mol kg^{-1}	$Zn(NO_3)_2$ mass %	$Zn(NO_3)_2$ mol kg^{-1}	nature of the solid phase
58.90	4.411	----	----	$La(NO_3)_3\cdot 6H_2O$
56.01	4.180	2.75	0.352	"
53.46	3.955	4.94	0.627	"
52.48	3.939	6.52	0.840	$La(NO_3)_3\cdot 6H_2O + 2La(NO_3)_3\cdot 3Zn(NO_3)_2\cdot 24H_2O$
48.63	3.479	8.35	1.025	$2La(NO_3)_3\cdot 3Zn(NO_3)_2\cdot 24H_2O$
44.13	3.041	11.21	1.325	"
39.81	2.655	14.05	1.608	"
37.38	2.465	15.94	1.803	"
36.51	2.437	17.39	1.992	"
29.67	1.875	21.63	2.345	"
25.03	1.540	24.96	2.635	"
20.04	1.197	28.44	2.915	"
15.19	0.896	32.64	3.303	"
12.49	0.731	34.91	3.504	"
8.87	0.515	38.12	3.797	"
5.66	0.330	41.59	4.163	"
1.14	0.068	47.10	4.805	"
<0.2	0.012	50.31	5.368	"
<0.2	0.014	55.33	6.570	$2La(NO_3)_3\cdot 3Zn(NO_3)_2\cdot 24H_2O + Zn(NO_3)_2\cdot 6H_2O$
---	----	56.00	6.720	$Zn(NO_3)_2\cdot 6H_2O$

a. Molalities calculated by M. Salomon.

AUXILIARY INFORMATION

METHOD/APPARATUS/PROCEDURE:

The isothermal method was used. The composition of saturated solutions was determined by chemical analysis. The total La + Zn content was determined by back titration with Trilon B, and then the Zn concentration found using unithiol as a complexing agent (1).

SOURCE AND PURITY OF MATERIALS:

No information given.

ESTIMATED ERROR:

Soly: precision about ± 0.5 % (compilers).

Temp: precision about ± 0.2 K (compilers).

REFERENCES:

1. Morachevskii, Yu. V.; Vol'f, L.A. *Zh. Anal. Khim.* 1960, *15*, 656.

COMPONENTS:

(1) Lanthanum nitrate; $La(NO_3)_3$; [10099-59-9]

(2) Ammonium nitrate; NH_4NO_3; [6484-52-2]

(3) Water; H_2O; [7732-18-5]

ORIGINAL MEASUREMENTS:

Urazov, G.G.; Shevtsova, Z.N. *Zh. Neorg. Khim.* 1957, *2*, 655-8; *J. Inorg. Chem. (USSR)* 1957, *2*, 288-94.

VARIABLES:

Composition

PREPARED BY:

T. Mioduski and S. Siekierski

EXPERIMENTAL VALUES:

The $La(NO_3)_3$ - NH_4NO_3 - H_2O system at 25.0°C

Composition of saturated solutions.[a]

$La(NO_3)_3$ mass %	$La(NO_3)_3$ mol kg^{-1}	NH_4NO_3 mass %	NH_4NO_3 mol kg^{-1}	nature of the solid phase
60.28	4.671	---	---	$La(NO_3)_3 \cdot 6H_2O$
59.75	4.741	1.46	0.470	solid solutions
59.00	4.772	2.95	0.969	"
58.08	4.916	5.56	1.910	"
57.55	5.141	8.00	2.901	"
54.20	5.279	14.20	5.614	$La(NO_3)_3 \cdot 2NH_4NO_3 \cdot 4H_2O$
50.00	4.428	15.25	5.483	"
47.12	4.042	17.00	5.919	"
46.00	3.935	18.02	6.257	"
42.71	3.466	19.37	6.382	"
40.02	3.261	22.21	7.346	$La(NO_3)_3 \cdot 2NH_4NO_3 \cdot 4H_2O + NH_4NO_3$
33.63	2.461	24.31	7.221	NH_4NO_3
31.70	2.293	25.75	7.561	"
22.36	1.470	30.83	8.228	"
14.00	0.907	38.47	10.112	"
6.35	0.437	48.90	13.652	"
4.72	0.352	54.05	16.378	"
4.0	0.30	55.00	16.76	"
2.56	0.215	60.85	20.776	"
1.35	0.124	65.05	24.187	"
0.50	0.048	67.14	25.921	"
---	----	68.00	26.548	"

a. Molalities calculated by M. Salomon.

AUXILIARY INFORMATION

METHOD/APPARATUS/PROCEDURE:

The isothermal method was used. Equilibrium was verified and required 3 days to be reached. The composition of the solid phases was determined by chemical analysis using Schreinemakers' method, by crystal-optical analysis, and by thermographical analysis.

SOURCE AND PURITY OF MATERIALS:

Sources not specified but the salts were analysed (results not given). Since the salts were analysed the compilers assume they were of acceptable purity.

ESTIMATED ERROR:

Soly: Analyses agreed to 0.1 to 0.2 % for each point on the soly curve (authors). The compilers assume that this error refers to mass % units.

Temp: precision ± 0.1 K.

COMPONENTS:

(1) Lanthanum nitrate; $La(NO_3)_3$; [10099-59-9]

(2) Hydrazine mononitrate; $N_2H_4 \cdot HNO_3$; [13464-97-6]

(3) Water; H_2O; [7732-18-5]

ORIGINAL MEASUREMENTS:

Gorshunova, V.P.; Zhuravlev, E.F.
Zh. Neorg. Khim. 1971, *16*, 1700-3: *Russ. J. Inorg. Chem. Engl. Transl.* 1971, *16*, 898-900.

VARIABLES:

Composition at 25°C and 50°C

PREPARED BY:

T. Mioduski and S. Siekierski

EXPERIMENTAL VALUES: The $La(NO_3)_3 - N_2H_4 \cdot HNO_3 - H_2O$ system

Composition of saturated solutions at 25°C [a]

$La(NO_3)_3$ mass %	$La(NO_3)_3$ mol kg^{-1}	$N_2H_4 \cdot HNO_3$ mass %	$N_2H_4 \cdot HNO_3$ mol kg^{-1}	n_D^{50}	nature of the solid phase
——	——	76.6	35.18	1.4680	$N_2H_4 \cdot HNO_3$
4.5	0.58	71.5	32.02	1.4700	"
10.5	1.38	66.0	30.19	1.4742	"
19.0	2.85	60.5	31.72	1.4825	"
29.0	6.87	58.0	47.95	1.5025	$N_2H_4 \cdot HNO_3 + La(NO_3)_3 \cdot 2N_2H_4 \cdot HNO_3$
32.0	6.35	52.5	36.40	1.4975	$La(NO_3)_3 \cdot 2H_2N_4 \cdot HNO_3$
41.5	5.81	36.5	17.83	1.4840	"
51.0	6.04	23.0	9.51	1.4785	"
59.0	6.73	14.0	5.57	1.4825	$La(NO_3)_3 \cdot 2N_2H_4 \cdot HNO_3 + La(NO_3)_3 \cdot 6H_2O$ [b]
58.5	6.21	12.5	4.36	1.4795	$La(NO_3)_3 \cdot 6H_2O$
58.0	5.41	9.0	2.93	1.4655	"
58.0	4.82	5.0	1.45	1.4570	"
58.5	4.34	——	——	1.4520	"

a. Molalities calculated by M. Salomon.

b. Found by extrapolation.

continued........

AUXILIARY INFORMATION

METHOD/APPARATUS/PROCEDURE:

The solubility was studied by the method of isothermal sections (1) by measuring the refractive indices of saturated solutions along directed sections of the phase diagram. Equilibrium was checked by repeated measurements of the refractive index as a function of time. The results were used to graph the relation between the refractive indices and the composition of the components for each of the sections studied. The graphs were used to find the inflection or break points corresponding to the composition of the saturated solutions. At 50°C, equilibrium was reached in 10-12 hours, and at 25°C, equilibrium was reached in two days.

SOURCE AND PURITY OF MATERIALS:

C.p. grade $La(NO_3)_3 \cdot 6H_2O$ was recrystallized. $N_2H_4 \cdot HNO_3$ prepd from "pure" grade aq N_2H_4 and c.p. grade concentrated HNO_3. After evaporation of the solvent, the salt was recrystallized and analysed for NO_3 with nitron. The analysis corresponded to the composition $N_2H_4 \cdot HNO_3$, and its melting point was 73°C (lit. 73°C (2)).

Double distilled water was used.

ESTIMATED ERROR:

Soly: precision ± 1% at best (compilers).

Temp: precision probably ± 0.2 K (compilers).

REFERENCES:

1. Zhuravlev, E.F.; Sheveleva, A.D. *Zh. Neorg. Khim.* 1960, *5*, 2630.

2. Grekov, A.P. *Organicheskaya Khimiya Gidrazina. Kiev.* 1966, p 11.

COMPONENTS:	ORIGINAL MEASUREMENTS:
(1) Lanthanum nitrate; $La(NO_3)_3$; [10099-59-9] (2) Hydrazine mononitrate; $N_2H_4 \cdot HNO_3$; [13464-97-6] (3) Water; H_2O; [7732-18-5]	Gorshunova, V.P. Zhuravlev, E.F. *Zh. Neorg. Khim.* 1971, *16*, 1700-3: *Russ. J. Inorg. Chem. Engl. Transl.* 1971, *16*, 898-900.

EXPERIMENTAL VALUES: continued.......

The $La(NO_3)_3 - N_2H_4 \cdot HNO_3 - H_2O$ system

Composition of saturated solutions at 50°C [a]

$La(NO_3)_3$		$N_2H_4 \cdot HNO_3$			
mass %	mol kg^{-1}	mass %	mol kg^{-1}	n_D^{50}	nature of the solid phase
——	——	91.0	106.4	1.4935	$N_2H_4 \cdot HNO_3$
1.5	0.54	90.0	111.4	1.5000	"
4.0	1.76	89.0	133.8	1.5025	"
6.0	3.08	88.0	154.3	1.5050	"
9.5	5.32	85.0	162.6	1.5125	"
35.0	8.98	53.0	46.46	1.5110	$La(NO_3)_3 \cdot 2N_2H_4 \cdot HNO_3$
36.5	8.32	50.0	38.96	1.5070	"
39.5	7.84	45.0	30.54	1.5030	"
44.0	7.52	38.0	22.21	1.5000	"
53.5	7.32	24.0	11.22	1.4960	"
64.5	7.78	10.0	4.13	1.5000	$La(NO_3)_3 \cdot 3N_2H_4 \cdot 2HNO_3 + La(NO_3)_3 \cdot 6H_2O$ [b]
65.0	7.41	8.0	3.12	1.4850	$La(NO_3)_3 \cdot 6H_2O$
65.0	6.78	5.5	1.96	1.4800	"
65.5	6.40	3.0	1.00	1.4755	"
65.5	6.11	——	——	1.4720	"

a. Molalities calculated by M. Salomon.

b. Found by extrapolation.

COMMENTS AND/OR ADDITIONAL DATA:

The double salt was isolated and analysed for NO_3^- by precipitation with nitron, and for La by the oxalate method. The hydrazinium ion was found by difference. The density of the double salt was measured pycnometrically using benzene, and the refractive indices were measured by the immersion method. For $La(NO_3)_3.2N_2H_4.HNO_3$, the authors report the following additional data:

$$d^{20} = 1.3812 \text{ kg m}^{-3}$$

$$n_g = 1.559 \pm 0.001$$

$$n_p = 1.553 \pm 0.001$$

COMPONENTS:

(1) Lanthanum nitrate; $La(NO_3)_3$; [10099-59-9]

(2) Hydrazine dinitrate; $N_2H_4 \cdot 2HNO_3$; [13464-98-7]

(3) Water ; H_2O ; [7732-18-5]

ORIGINAL MEASUREMENTS:

Gorshunova, V.P.; Zhuravlev, E.F. *Zh. Neorg. Khim.* 1971, *16*, 1700-3; *Russ. J. Inorg. Chem. Engl. Transl.* 1971, *16*, 898-900.

VARIABLES:

Composition at 25°C and 50°C

PREPARED BY:

T. Mioduski and S. Siekierski

EXPERIMENTAL VALUES:

The $La(NO_3)_3$ - $N_2H_4 \cdot 2HNO_3$ - H_2O system.

Composition of saturated solutions at 25°C[a]

$La(NO_3)_3$ mass %	$La(NO_3)_3$ mol kg^{-1}	$N_2H_4 \cdot 2HNO_3$ mass %	$N_2H_4 \cdot 2HNO_3$ mol kg^{-1}	n_D^{50}	nature of the solid phase[b]
---	---	70.0	14.76	1.4665	A
1.0	0.10	69.5	14.90	1.4670	A + B
2.0	0.19	66.0	13.05	1.4630	B
6.5	0.59	59.5	11.07	1.4570	B
17.5	1.44	45.0	7.59	1.4485	B
27.5	2.12	32.5	5.14	1.4450	B
30.0	2.31	30.0	4.74	1.4445	B
37.5	2.81	21.5	3.32	1.4450	B
43.0	3.31	17.0	2.69	1.4465	B
50.5	3.93	10.0	1.60	1.4510	B
57.5	4.78	5.5	0.94	1.4560	B + C[c]
57.0	4.62	5.0	0.83	1.4550	C
58.0	4.52	2.5	0.40	1.4530	C
58.5	4.34	---	---	1.4510	C

a. Molalities calculated by M. Salomon.

b. Solid phases: A = $N_2H_4 \cdot 2HNO_3$ B = $La(NO_3)_3 \cdot 3N_2H_4 \cdot 2HNO_3$ C = $La(NO_3)_3 \cdot 6H_2O$

c. Found by extrapolation.

continued......

AUXILIARY INFORMATION

METHOD/APPARATUS/PROCEDURE:

The solubility was studied by the method of isothermal sections (1) by measuring the refractive indices of saturated solutions along directed sections of the phase diagram. Equilibrium was checked by repeated measurements of the refractive index as a function of time. At 50°C equilibrium was reached in 10-12 hours, and at 25°C equilibrium was reached in two days. The results were used to graph the relation between the refractive indices and the composition of the components for each of the sections studied. The graphs were used to find the break points corresponding to the composition of the saturated solutions. The refractive indices of the saturated solutions, n_D^{50}, are included in the data tables.

SOURCE AND PURITY OF MATERIALS:

C.p. grade $La(NO_3)_3 \cdot 6H_2O$ was recrystallized. $N_2H_4 \cdot 2HNO_3$ prepd by stoichiometric mixing of HNO_3 and N_2H_4, or by mixing of HNO_3 with $N_2H_4 \cdot HNO_3$ (see previous compilation for prep of hydrazine mononitrate). Analysis for NO_3^- by pptn with nitron confirmed the composition $N_2H_4 \cdot 2HNO_3$, and its melting point was 102°C (lit 103°C (2)).

Double distilled water was used.

ESTIMATED ERROR:

Soly: precision ± 1% at best (compilers).

Temp: precision probably ± 0.2 K (compilers).

REFERENCES:

1. Zhuravlev, E.F.; Sheveleva, A.D. *Zh. Neorg. Khim.* 1960, *5*, 2630.

2. Grekov, A.P. *Organicheskaya Khimiya Gidrazina.* Kiev. 1966. p 11.

COMPONENTS:	ORIGINAL MEASUREMENTS:
(1) Lanthanum nitrate; $La(NO_3)_3$; [10099-59-9]	Gorshunova, V.P.; Zhuravlev, E.F. *Zh. Neorg. Khim.* 1971, *16*, 1700-3; *Russ. J. Inorg. Chem. Engl. Transl.* 1971, *16*, 898-900.
(2) Hydrazine dinitrate; $N_2H_4 \cdot 2HNO_3$; [13464-98-7]	
(3) Water; H_2O; [7732-18-5]	

EXPERIMENTAL VALUES: continued.....

Composition of saturated solutions at 50°C[a]

$La(NO_3)_3$		$N_2H_4 \cdot 2HNO_3$			
mass %	mol kg^{-1}	mass %	mol kg^{-1}	n_D^{50}	nature of the solid phase[b]
---	---	75.0	18.98	1.4780	A
0.5	0.06	74.5	18.85	1.4790	A + B
1.5	0.17	71.5	16.75	1.4750	B
6.0	0.58	62.0	12.26	1.4620	B
16.0	1.41	49.0	8.86	1.4555	B
24.5	2.04	38.5	6.58	1.4530	B
35.0	2.91	28.0	4.79	1.4520	B
36.0	2.92	26.0	4.33	1.4520	B
51.0	4.36	13.0	2.28	1.4580	B
59.5	5.31	6.0	1.10	1.4690	B
66.0	6.66	3.5	0.73	1.4760	B + C[c]
66.0	6.45	2.5	0.50	1.4730	C
65.5	6.11	1.5	0.29	1.4715	C
65.5	5.84	---	---	1.4710	C

a. Molalities calculated by M. Salomon.

b. Solid phases: A = $N_2H_4 \cdot 2HNO_3$ B = $La(NO_3)_3 \cdot 3N_2H_4 \cdot 2HNO_3$ C = $La(NO_3)_3 \cdot 6H_2O$

c. Found by extrapolation.

COMMENTS AND/OR ADDITIONAL DATA:

The double salt was isolated and analysed for NO_3^- by precipitation with nitron, and for La by the oxalate method. The hydrazinium ion was found by difference. The density of the double salt was measured pycnometrically using benzene, and the refractive indices were measured by the immersion method. For $La(NO_3)_3 \cdot 3N_2H_4 \cdot 2HNO_3$, the authors report the following additional data:

$$d^{20} = 1.9941 \text{ kg m}^{-3}$$

$$n_g = 1.565 \pm 0.001$$

$$n_p = 1.557 \pm 0.001$$

COMPONENTS:

(1) Lanthanum nitrate; $La(NO_3)_3$; [10099-59-9]

(2) Dimethylamine nitrate; $C_2H_8N_2O_3$; [30781-73-8]

(3) Water ; H_2O ; [7732-18-5]

ORIGINAL MEASUREMENTS:

Gorshunova, V.P. Zhuravlev, E.F. *Zh. Neorg. Khim.* 1970, *15*, 3355-8; *Russ. J. Inorg. Chem. Engl. Transl.* 1970, *15*, 1750-1.

VARIABLES:

Composition at 25°C and 50°C

PREPARED BY:

T. Mioduski and S. Siekierski

EXPERIMENTAL VALUES:

The $La(NO_3)_3$ - $(CH_3)_2NH{\cdot}HNO_3$ - H_2O system at 25°C

Composition of saturated solutions[a]

$La(NO_3)_3$ mass %	$La(NO_3)_3$ mol kg^{-1}	$(CH_3)_2NH{\cdot}HNO_3$ mass %	$(CH_3)_2NH{\cdot}HNO_3$ mol kg^{-1}	n_D^{50}	nature of the solid phase
---	----	91.5	115.7	1.4312	$(CH_3)_2NH{\cdot}HNO_3$
1.5	0.49	89.0	100.7	1.4325	"
5.5	2.42	87.5	134.3	1.4350	"
7.5	3.30	85.5	131.2	1.4360	"
15.5	10.6	80.0	191.0	1.4450	"
58.5	9.48	22.5	12.72	1.4730	$La(NO_3)_3{\cdot}6H_2O$
57.0	7.16	18.5	8.11	1.4700	"
57.5	6.21	14.0	5.28	1.4640	"
57.5	7.53	19.0	8.69	1.4585	"
57.5	4.66	4.5	1.27	1.4549	"
58.5	4.34	----	---	1.4520	"

a. Molalities calculated by M. Salomon.

continued......

AUXILIARY INFORMATION

METHOD/APPARATUS/PROCEDURE:

The solubilities were studied by the method of isothermal sections (1,2) which involves the measurement of refractive indices of saturated solutions along directed sections of the phase diagram. Mixtures of known composition were equilibrated until their refractive indices remained constant after repeated measurements. All refractive indices were measured at 50°C.

The results were used to graph the relation between the refractive indices and the composition of the components for each section studied. The graphs were used to find the break points corresponding to the composition of the saturated solutions.

The phase diagram is of the eutonic type.

SOURCE AND PURITY OF MATERIALS:

C.p grade $La(NO_3)_3{\cdot}6H_2O$; "analytical reagent" dimethylamine, and c.p. grade HNO_3 were used. Dimethylamine nitrate was prepared by neutralizing dimethylamine with HNO_3. The resulting solution was evaporated, the solid product recrystallized twice and dried in a vacuum desicaator over $CaCl_2$ to constant mass.

ESTIMATED ERROR:

Soly: precision probably 1% (compilers).

Temp: precision probably ± 0.2 K (compilers).

REFERENCES:

1. Zhuravlev, E.F.; Gorshunova, V.P. *Zh. Neorg. Khim.* 1970, *15*, 195.
2. Mertslin, R.V. *Izv. Biol. Nauch.-Issled. Inst. pri Perm. Gos. Univ.* 1937, *11*, 1.

COMPONENTS:	ORIGINAL MEASUREMENTS:
(1) Lanthanum nitrate; $La(NO_3)_3$; [10099-59-9] (2) Dimethylamine nitrate; $C_2H_8N_2O_3$; [25238-43-1] (3) Water ; H_2O ; [7732-18-5]	Gorshunova, V.P.; Zhuravlev, E.F. *Zh. Neorg. Khim.* 1970, *15*, 3355-8; *Russ. J. Inorg. Chem. Engl. Transl.* 1970, *15*, 1750-1.

EXPERIMENTAL VALUES: continued.....

The $La(NO_3)_3$ - $(CH_3)_2NH{\cdot}HNO_3$ - H_2O system at 50°C

Composition of saturated solutions[a]

$La(NO_3)_3$		$(CH_3)_2NH{\cdot}HNO_3$			
mass %	mol kg^{-1}	mass %	mol kg^{-1}	n_D^{50}	nature of the solid phase
---	----	95.5	228.0	1.4360	$(CH_3)_2NH{\cdot}HNO_3$
1.0	0.77	95.0	255.2	1.4360	"
4.0	3.52	92.5	284.0	1.4375	"
4.5	3.96	92.0	282.5	1.4392	"
10.0	15.4	88.0	472.8	1.4430	"
66.0	9.03	11.5	5.49	1.4840	$La(NO_3)_3{\cdot}6H_2O$
66.0	8.29	9.5	4.17	1.4825	"
66.5	7.72	7.0	2.84	1.4780	"
66.0	6.89	4.5	1.64	1.4760	"
65.5	6.20	2.0	0.66	1.4730	"
65.5	5.84	---	----	1.4720	"

a. Molalities calculated by M. Salomon.

COMPONENTS:

(1) Lanthanum nitrate; $La(NO_3)_3$; [10099-59-9]

(2) Guanidine mononitrate; $CH_6N_4O_3$; [506-93-4]

(3) Water ; H_2O ; [7732-18-5]

ORIGINAL MEASUREMENTS:

Gorshunova, V.P.; Zhuravlev, E.F. *Zh. Neorg. Khim.* 1971, *16*, 1739-40; *Russ. J. Inorg. Chem. Engl. Transl.* 1971, *16*, 922.

VARIABLES:

Composition at 25°C and 50°C

PREPARED BY:

T. Mioduski and S. Siekierski

EXPERIMENTAL VALUES:

Composition of saturated solutions at 25°C[a]

$La(NO_3)_3$ mass %	$La(NO_3)_3$ mol kg^{-1}	$CH_5N_3\cdot HNO_3$ mass %	$CH_5N_3\cdot HNO_3$ mol kg^{-1}	n_D^{50}	nature of the solid phase
---	----	14.5	1.39	1.3510	$CH_5N_3\cdot HNO_3$
9.0	0.34	9.5	0.95	1.3580	"
28.5	1.32	5.0	0.62	1.3870	"
49.0	3.02	1.0	0.16	1.4270	"
58.0	4.35	1.0	0.20	1.4510	$CH_5N_3\cdot HNO_3 + La(NO_3)_3\cdot 6H_2O$
58.5	4.34	---	----	1.4510	$La(NO_3)_3\cdot 6H_2O$

Composition of saturated solutions at 50°C

$La(NO_3)_3$ mass %	$La(NO_3)_3$ mol kg^{-1}	$CH_5N_3\cdot HNO_3$ mass %	$CH_5N_3\cdot HNO_3$ mol kg^{-1}	n_D^{50}	nature of the solid phase
---	----	28.5	3.26	1.3720	$CH_5N_3\cdot HNO_3$
8.0	0.36	24.0	2.89	1.3780	"
25.0	1.33	17.0	2.40	1.3980	"
45.5	3.08	9.0	1.62	1.4320	"
66.5	6.94	4.0	1.11	1.4725	$CH_5N_3\cdot HNO_3$[b] + $La(NO_3)_3\cdot 6H_2O$
66.0	6.45	2.5	0.65	1.4720	$La(NO_3)_3\cdot 6H_2O$
66.0	6.16	1.0	0.25	1.4710	"
65.5	5.93	0.5	0.12	1.4710	"
65.5	5.84	---	----	1.4705	"

a. Molalities calculated by M. Salomon.

b. Found by extrapolation.

AUXILIARY INFORMATION

METHOD/APPARATUS/PROCEDURE:

The solubilities were studied by the method of isothermal sections (1,2). The refractive indices of the liquid phases were measured at 50°C. No other information given. For details on the isothermal sections method, see the compilations of the data in references 1 and 2.

COMMENTS AND/OR ADDITIONAL DATA:

The phase diagram is characterized by an extensively developed crystallization field for guanidine mononitrate, and a very small crystallization for lanthanum nitrate hexahydrate. This indicates that lanthanum nitrate hexahydrate has a strong salting-out action on guanidine mononitrate.

SOURCE AND PURITY OF MATERIALS:

$La(NO_3)_3\cdot 6H_2O$ was c.p. grade and presumably used as received.

"Pure" grade guanidine mononitrate, $H_2NC(=NH)NH_2\cdot HNO_3$, was twice recrystallized.

Double distilled water was used.

ESTIMATED ERROR:

Soly: precision around 1 % (compilers).

Temp: precision probably ± 0.2 K (compilers).

REFERENCES:

1. Zhuravlev, E.F.; Gorshunova, V.P. *Zh. Neorg. Khim.* 1970, *15*, 195.
2. Gorshunova, V.P.; Zhuravlev, E.F. *Zh. Neorg. Khim.* 1970, *15*, 1422.

COMPONENTS:

(1) Lanthanum nitrate; $La(NO_3)_3$; [10099-59-9]

(2) Diethylamine nitrate; $C_4H_{12}N_2O_3$; [27096-30-6]

(3) Water ; H_2O ; [7732-18-5]

ORIGINAL MEASUREMENTS:

Gorshunova, V.P.; Zhuravlev, E.F. *Zh. Neorg. Khim.* 1970, *15*, 3355-8; *Russ. J. Inorg. Chem. Engl. Transl.* 1970, *15*, 1750-1.

VARIABLES:

Composition at 0°C, 25°C and 50°C

PREPARED BY:

T. Mioduski and S. Siekierski

EXPERIMENTAL VALUES:

The $La(NO_3)_3$ - $(C_2H_5)_2NH{\cdot}HNO_3$ - H_2O system at 0°C

Composition of saturated solutions[a]

$La(NO_3)_3$		$(C_2H_5)_2NH{\cdot}HNO_3$			
mass %	mol kg^{-1}	mass %	mol kg^{-1}	n_D^{50}	nature of the solid phase
---	----	67.5	15.3	1.4100	$(C_2H_5)_2NH{\cdot}HNO_3$
17.0	2.91	65.0	26.5	1.4181	"
16.0	2.10	60.5	18.9	1.4278	"
21.5	3.08	57.0	19.5	1.4340	"
27.0	4.75	55.5	23.3	1.4400	"
33.0	7.25	53.0	27.8	1.4472	"
43.5	5.82	33.5	10.7	1.4530	$La(NO_3)_3{\cdot}6H_2O$
46.0	5.34	27.5	7.62	1.4485	"
49.0	4.57	18.0	4.01	1.4425	"
51.5	3.73	6.0	1.04	1.4380	"
50.5	3.14	---	----	1.4341	"

a. Molalities calculated by M. Salomon.

continued.......

AUXILIARY INFORMATION

METHOD/APPARATUS/PROCEDURE:

The solubilities were studied by the method of isothermal sections (1,2) which involves the measurement of refractive indices of saturated solutions along directed sections of the phase diagram. Mixtures of known composition were equilibrated until their refractive indices remained constant after repeated measurements. All refractive indices were measured at 50°C.

The results were used to graph the relation between the refractive indices and the composition of the components for each section studied. The graphs were used to find the break points corresponding to the composition of the saturated solutions.

The phase diagram is of the eutonic type.

SOURCE AND PURITY OF MATERIALS:

C.p grade $La(NO_3)_3{\cdot}6H_2O$, c.p. grade nitric acid, and "analytical reagent" diethylamine were used. Diethylamine nitrate was prepared by neutralizing diethylamine with nitric acid. The solution was evaporated on a water bath until crystallization: the salt was twice recrystallized and dried in a vacuum desiccator over calcium chloride to constant mass.

ESTIMATED ERROR:

Soly: precision probably 1 % (compilers).

Temp: precision probably ± 0.2 K (compilers).

REFERENCES:

1. Zhuravlev, E.F.; Gorshunova, V.P. *Zh. Neorg. Khim.* 1970, *15*, 195.

2. Mertslin, R.V. *Izv. Biol. Nauch.-Issled. Inst. pri Perm. Gos. Univ.* 1937, *11*, 1.

COMPONENTS:	ORIGINAL MEASUREMENTS:
(1) Lanthanum nitrate; $La(NO_3)_3$; [10099-59-9] (2) Diethylamine nitrate; $C_4H_{12}N_2O_3$; [27096-30-6] (3) Water ; H_2O ; [7732-18-5]	Gorshunova, V.P.; Zhuravlev, E.F. *Zh. Neorg. Khim.* 1970, *15*, 3355-8; *Russ. J. Inorg. Chem. Engl. Transl.* 1970, *15*, 1750-1.

EXPERIMENTAL VALUES: continued.......

The $La(NO_3)_3$ - $(C_2H_5)_2NH{\cdot}HNO_3$ - H_2O system

Composition of saturated solutions at 25°C[a]

$La(NO_3)_3$		$(C_2H_5)_2NH{\cdot}HNO_3$			
mass %	mol kg^{-1}	mass %	mol kg^{-1}	n_D^{50}	nature of the solid phase
---	----	82.0	33.5	1.4245	$(C_2H_5)_2NH{\cdot}HNO_3$
4.0	0.75	79.5	35.4	1.4285	"
10.0	2.37	77.0	43.5	1.4350	"
12.0	3.08	76.0	46.5	1.4380	"
20.5	9.01	72.5	76.1	1.4500	"
51.0	9.51	32.5	14.5	1.4700	$La(NO_3)_3{\cdot}6H_2O$
53.5	7.01	23.0	7.19	1.4650	"
54.0	5.73	17.0	4.31	1.4600	"
57.0	5.32	10.0	2.23	1.4570	"
58.0	4.76	4.5	0.88	1.4540	"
58.5	4.34	---	----	1.4520	"

Compositions of saturated solutions at 50°C

$La(NO_3)_3$		$(C_2H_5)_2NH{\cdot}HNO_3$			
mass %	mol kg^{-1}	mass %	mol kg^{-1}	n_D^{50}	nature of the solid phase
---	----	10.0[b]	?	1.4325	$(C_2H_5)_2NH{\cdot}HNO_3$
3.0	1.09	88.5	76.5	1.4350	"
5.5	1.99	86.0	74.3	1.4385	"
7.5	2.89	84.5	77.6	1.4400	"
15.0	11.51	81.0	148.7	1.4470	"
64.0	9.38	15.0	5.25	1.4810	$La(NO_3)_3{\cdot}6H_2O$
63.5	7.82	11.5	3.38	1.4775	"
64.5	7.22	8.0	2.14	1.4760	"
65.0	6.78	5.5	1.37	1.4745	"
65.0	6.45	4.0	0.95	1.4730	"
65.5	5.84	---	----	1.4720	"

a. Molalities calculated by M. Salomon.

b. This appears to be a typographical error. The value of 10.0 mass % appears in both the original Russian publication and in the English translation of the source paper.

COMPONENTS:

(1) Lanthanum nitrate; $La(NO_3)_3$; [10099-59-9]

(2) Triethylamine; $C_6H_{15}N$; [121-44-8]

(3) Water ; H_2O ; [7732-18-5]

ORIGINAL MEASUREMENTS:

Shabikova, G. Kh.; Sergeeva, V.F.; Izmeleuova, M.B. *Zh. Obshch. Khim.* 1975, *45*, 990-5.

VARIABLES:

Solvent composition at 20°C

PREPARED BY:

T. Mioduski, S. Siekierski and M. Salomon

EXPERIMENTAL VALUES:

The $La(NO_3)_3$ - $(C_2H_5)_3N$ - H_2O system at 20°C

mass % $(C_2H_5)_3$ in original solvent		Composition of equilibrated saturated solutions[a,b]			
		$La(NO_3)_3$		$(C_2H_5)_3N$	
exptl	calcd[c]	mass %	mol kg^{-1}	mass %	mol kg^{-1}
0	---	59.5	4.52	---	---
10	9.0	54.4	4.03	4.1	0.98
20	19.6	48.0	3.53	10.2	2.41
30	30.1	38.2	2.72	18.6	4.25
40	40.1	24.9	1.70	30.1	6.61
50	51.2	3.9	0.26	49.2	10.37

a. Molalities calculated by the compilers.

b. Nature of the solid phase not specified: presumably $La(NO_3)_3 \cdot 6H_2O$ is the major component, particularly at low triethylamine concentrations (compilers).

c. To check on the internal consistency of the analyses, the compilers calculated the mass % $(C_2H_5)_3N$ to total $(C_2H_5)_3N + H_2O$ in the equilibrated saturated solutions. This value should equal the mass % of triethylamine in the original solvent mixture. The mass % triethylamine in the saturated solutions (i.e. neglecting $La(NO_3)_3$) was calculated from

$$\text{mass \% } (C_2H_5)_3N \text{ (in satd sln) x } 100/(\text{mass \% } H_2O + \text{mass \% } (C_2H_5)_3N)$$

where mass % H_2O in the satd sln is obtained from 100-mass % $La(NO_3)_3$ - mass % $(C_2H_5)_3N$. The difference between this calculated mass % for triethylamine and the value reported by the authors for the orignal solvent composition is attributed by the compilers to an error in accuracy in the chemical analyses.

AUXILIARY INFORMATION

METHOD/APPARATUS/PROCEDURE:

The method of isothermal sections was used according to Mertslin and described by Mochalov (1). Analyses and establishment of equilibrium accomplished by measurement of refractive indices. Reaction mixtures were equilibrated for 7-8 hours. No other information given.

SOURCE AND PURITY OF MATERIALS:

C.p grade lanthanum nitrate presumably used as received.

Triethylamine was dried over ignited Al_2O_3 and distilled at reduced pressure. The resulting product had the following properties: d_4^{25}=0.7222, n_D^{20} = 1.4010, b.p. = 86.9°C.

Water was twice distilled from $KMnO_4$ solution.

ESTIMATED ERROR:

Soly: based on internal consistency of data, compilers estimate an accuracy of ± 2-3 %.

Temp: precision probably ± 0.2 K (compilers).

REFERENCES:

1. Mochalov, K.I. *Zh. Obshch. Khim.* 1939, *9*, 1701.

COMPONENTS:

(1) Lanthanum nitrate; $La(NO_3)_3$; [10099-59-9]

(2) Triethylamine nitrate; $C_6H_{16}N_2O_3$; [27096-31-7]

(3) Water ; H_2O ; [7732-18-5]

ORIGINAL MEASUREMENTS:

Gorshunova, V.P.; Zhuravlev, E.F. *Zh. Neorg. Khim.* 1970, *15*, 3355-8; *Russ. J. Inorg. Chem. Engl. Transl.* 1970, *15*, 1750-1.

VARIABLES:

Composition at 25°C and 50°C

PREPARED BY:

T. Mioduski and S. Siekierski

EXPERIMENTAL VALUES:

The $La(NO_3)_3 - (C_2H_5)_3N{\cdot}HNO_3 - H_2O$ system at 25°C

Composition of saturated solutions[a]

$La(NO_3)_3$ mass %	$La(NO_3)_3$ mol kg^{-1}	$(C_2H_5)_3N{\cdot}HNO_3$ mass %	$(C_2H_5)_3N{\cdot}HNO_3$ mol kg^{-1}	n_D^{50}	nature of the solid phase
---	----	91.5	65.6	1.4460	$(C_2H_5)_3N{\cdot}HNO_3$
2.5	0.81	88.0	56.4	1.4470	"
6.0	2.46	86.5	70.2	1.4500	"
8.0	3.28	84.5	68.6	1.4510	"
13.5	10.39	82.5	125.6	1.4552	"
54.0	8.98	27.5	9.05	1.4670	$La(NO_3)_3{\cdot}6H_2O$
55.0	7.20	21.5	5.57	1.4635	"
55.0	5.74	15.5	3.20	1.4610	"
56.5	5.11	9.5	1.70	1.4560	"
57.5	4.66	4.5	0.72	1.4530	"
58.5	4.34	---	----	1.4520	"

a. Molalities calculated by M. Salomon.

continued.........

AUXILIARY INFORMATION

METHOD/APPARATUS/PROCEDURE:

The solubilities were studied by the method of isothermal sections (1,2) which involves the measurement of refractive indices of saturated solutions along directed sections of the phase diagram. Mixtures of known composition were equilibrated until their refractive indices remained constant after repeated measurements. All refractive indices were measured at 50°C.

The results were used to graph the relation between the refractive indices and the composition of the components for each section studied. The graphs were used to find the break points corresponding to the composition of the saturated solutions.

The phase diagram is of the eutonic type.

SOURCE AND PURITY OF MATERIALS:

C.p. grade $La(NO_3)_3{\cdot}6H_2O$, 'analytical reagent' triethylamine, and c.p. grade nitric acid were used. Triethylamine nitrate was prepared by neutralizing triethylamine with nitric acid. The solution was evaporated on a water bath until crystallization: the salt was recrystallized twice and dried in vacuum over calcium chloride to constant mass.

ESTIMATED ERROR:

Soly: precision probably 1 % (compilers).

Temp: precision probably ± 0.2 K (compilers).

REFERENCES:

1. Zhuravlev, E.F.; Gorshunova, V.P. *Zh. Neorg. Khim.* 1970, *15*, 195.

2. Mertslin, R.V. *Izv. Biol. Nauch.-Issled. Inst. pri Perm. Gos. Univ.* 1937, *11*, 1.

COMPONENTS:	ORIGINAL MEASUREMENTS:
(1) Lanthanum nitrate; $La(NO_3)_3$; [10099-59-9] (2) Triethylamine nitrate; $C_6H_{16}N_2O_3$; [27096-31-7] (3) Water ; H_2O ; [7732-18-5]	Gorshunova, V.P.; Zhuravlev, E.F. *Zh. Neorg. Khim.* 1970, *15*, 3355-8; *Russ. J. Inorg. Chem. Engl. Transl.* 1970, *15*, 1750-1.

EXPERIMENTAL VALUES: continued.......

The $La(NO_3)_3 - (C_2H_5)_3N{\cdot}HNO_3 - H_2O$ system at 50°C

Composition of saturated solutions[a]

$La(NO_3)_3$		$(C_2H_5)_3N{\cdot}HNO_3$			
mass %	mol kg^{-1}	mass %	mol kg^{-1}	n_D^{50}	nature of the solid phase
---	----	94.0	95.4	1.4490	$(C_2H_5)_3N{\cdot}HNO_3$
2.5	1.18	91.0	85.3	1.4500	"
4.5	2.13	89.0	83.4	1.4515	"
7.5	4.20	87.0	96.3	1.4530	"
11.0	11.28	86.0	174.6	1.4560	"
65.0	9.30	13.5	3.82	1.4780	$La(NO_3)_3{\cdot}6H_2O$
65.0	8.00	10.0	2.44	1.4760	"
64.5	7.22	8.0	1.77	1.4755	"
65.5	6.72	4.5	0.91	1.4740	"
66.0	6.45	2.5	0.48	1.4730	"
65.5	5.84	---	----	1.4720	"

a. Molalities calculated by M. Salomon.

COMPONENTS:

(1) Lanthanum nitrate; $La(NO_3)_3$; [10099-59-9]

(2) Ethylenediamine dinitrate; $C_2H_{10}N_4O_6$; [20829-66-7]

(3) Water ; H_2O ; [7732-18-5]

ORIGINAL MEASUREMENTS:

Zhuravlev, E.F.; Gorshunova, V.P. *Zh. Neorg. Khim.* 1970, *15*, 195-200; *Russ. J. Inorg. Chem. Engl. Transl.* 1970, *15*, 100-3.

VARIABLES:

Composition at 25°C and 50°C

PREPARED BY:

T. Mioduski and S. Siekierski

EXPERIMENTAL VALUES:

The $La(NO_3)_3 - H_2NCH_2CH_2NH_2.2HNO_3 - H_2O$ system at 25°C

Composition of saturated solutions[a,b]

$La(NO_3)_3$		$En.2HNO_3$		
mass %	mol kg^{-1}	mass %	mol kg^{-1}	nature of the solid phase
---	---	50.5	17.0	$En.2HNO_3$
9.0	0.59	44.0	15.6	"
20.0	1.42	36.5	14.0	"
25.0	1.85	33.5	13.4	"
32.0	2.53	29.0	12.4	$En.2HNO_3 + La(NO_3)_3.En.2HNO_3.6H_2O$
37.0	2.71	21.0	8.32	$La(NO_3)_3.En.2HNO_3.6H_2O$
42.0	2.94	14.0	5.29	"
48.5	3.35	7.0	2.62	"
53.0	3.79	4.0	1.55	"
59.5	4.64	1.0	0.42	$La(NO_3)_3.2En.2HNO_3.6H_2O + La(NO_3)_3.6H_2O$
58.5	4.34	---	---	$La(NO_3)_3.6H_2O$

a. Molalities calculated by M. Salomon.

b. En = ethylenediamine, $H_2NCH_2CH_2NH_2$.

continued......

AUXILIARY INFORMATION

METHOD/APPARATUS/PROCEDURE:

The solubilities were determined by the method of isothermal sections (1) which involves the measurement of refractive indices of saturated solutions along directed sections of the phase diagram. Mixtures of known composition were equilibrated until their refractive indices were constant after repeated measurements. The results were used to graph the relation between the refractive indices and the composition of the components for each of the sections studied. The graphs were used to find the break points corresponding to the composition of the saturated solutions. The refractive indices were not reported in the source paper.

Mixtures were placed in vessels which were closed and kept in a thermostat with periodic shaking for 2-5 days. Part of the liquid phases were removed periodically and the refractive indices measured at 50°C. Apparently the mass % H_2O was obtained by difference.

SOURCE AND PURITY OF MATERIALS:

C.p. grade $La(NO_3)_3.6H_2O$ analysed by the oxalate method and contained 75 mass % $La(NO_3)_3$.

Ethylenediamine dinitrate prepd by mixing solutions of c.p. grade nitric acid and "pure grade" ethylenediamine. In slight excess of nitric acid the dinitrate salt crystallized. The crystals were filtered and dried to constant mass at 50-70°C.

Double distilled water was used.

ESTIMATED ERROR:

Soly: precision of 1 % (compilers).

Temp: precision probably ± 0.2 K (compilers).

REFERENCES:

1. Zhuravlev, E.F.; Sheveleva, A.D. *Zh. Neorg. Khim.* 1960, *5*, 2630.

COMPONENTS:	ORIGINAL MEASUREMENTS:
(1) Lanthanum nitrate; $La(NO_3)_3$; [10099-59-9]	Zhuravlev, E.F.; Gorshunova, V.P. *Zh. Neorg. Khim.* 1970, *15*, 195-200; *Russ. J. Inorg. Chem. Engl. Transl.* 1970, *15*, 100-3.
(2) Ethylenediamine dinitrate; $C_2H_{10}N_4O_6$; [20829-66-7]	
(3) Water ; H_2O ; [7732-18-5]	

EXPERIMENTAL VALUES: continued.......

The $La(NO_3)_3$ - $H_2NCH_2CH_2NH_2$ - H_2O system at 50°C

Composition of saturated solutions[a,b]

$La(NO_3)_3$ mass %	$La(NO_3)_3$ mol kg^{-1}	$En\cdot 2HNO_3$ mass %	$En\cdot 2HNO_3$ mol kg^{-1}	nature of the solid phase
---	---	67.0	33.8	$En\cdot 2HNO_3$
7.0	0.62	58.5	28.2	"
15.0	1.40	52.0	26.2	"
26.5	2.91	45.5	27.0	"
31.0	3.82	44.0	29.3	"
33.0	4.23	43.0	29.8	$En\cdot 2HNO_3$ + $La(NO_3)_3\cdot En\cdot 2HNO_3\cdot 6H_2O$
38.0	4.10	33.5	19.6	$La(NO_3)_3\cdot En\cdot 2HNO_3\cdot 6H_2O$
43.5	4.18	24.5	12.7	"
50.0	4.53	16.0	7.83	"
57.5	5.13	8.0	3.86	"
66.0	6.45	2.5	1.32	$La(NO_3)_3\cdot En\cdot 2HNO_3\cdot 6H_2O$ + $La(NO_3)_3\cdot 6H_2O$
66.0	6.16	1.0	0.50	$La(NO_3)_3\cdot 6H_2O$
66.5	6.11	---	---	"

a. Molalities calculated by M. Salomon.

b. En = ethylenediamine, $H_2NCH_2CH_2NH_2$.

COMPONENTS:

(1) Lanthanum nitrate; $La(NO_3)_3$; [10099-59-9]

(2) Piperidine nitrate; $C_5H_{12}N_2O_3$; [6091-45-8]

(3) Water ; H_2O ; [7732-18-5]

ORIGINAL MEASUREMENTS:

Gorshunova, V.P.; Zhuravlev, E.F. *Zh. Neorg. Khim.* 1972, *17*, 231-3; *Russ. J. Inorg. Chem. Engl. Transl.* 1972, *17*, 121-2.

VARIABLES:

Composition at 25°C and 50°C

PREPARED BY:

T. Mioduski, S. Siekierski and M. Salomon

EXPERIMENTAL VALUES:

The $La(NO_3)_3 - C_5H_{10}NH{\cdot}HNO_3 - H_2O$ system at 25°C

Composition of saturated solutions[a]

mass %	mol kg^{-1}	mass %	mol kg^{-1}	n_D^{50}	nature of the solid phase
---	----	87.0	45.16	1.4600	$C_5H_{10}NH{\cdot}HNO_3$
2.6	0.67	85.5	48.49	1.4635	"
6.0	1.03	76.0	28.50	1.4542	$La(NO_3)_3{\cdot}4C_5H_{10}NH{\cdot}HNO_3$
11.3	1.33	62.5	16.10	1.4426	"
21.5	2.07	46.5	9.81	1.4388	"
25.5	2.73	45.8	10.77	1.4420	"
33.5	3.08	33.0	6.65	1.4441	"
33.6	3.10	33.0	6.67	1.4440	"
40.8	3.78	26.0	5.29	1.4490	"
50.5	4.78	17.0	3.53	1.4594	"
54.5	5.24	13.5	2.85	1.4654	$La(NO_3)_3{\cdot}6H_2O$
56.5	5.16	9.8	1.96	1.4620	"
58.0	4.82	5.0	0.91	1.4565	"
58.5	4.34	---	----	1.4520	"

a. Molalities calculated by the compilers.

continued........

AUXILIARY INFORMATION

METHOD/APPARATUS/PROCEDURE:

The method of isothermal sections was used (1,2). Equilibrium was reached after 2 days at 25°C, and after 10-12 hours at 50°C.

No other information given. For details on the isothermal sections method, see the compilations for references 1 and 2.

SOURCE AND PURITY OF MATERIALS:

C.p. grade $La(NO_3)_3{\cdot}6H_2O$ analysed and found to contain 75% $La(NO_3)_3$: the product was recrystallized and had a m.p. of 65°C, d^{20} = 2.354 g cm^{-3}. "Pure" grade Piperidine and c.p. grade nitric acid sln were used to prepare piperidine nitrate. The resulting sln was evaporated at 70-80°C, the solid separated, recystallized and air dried at 40-50°C to const mass: composition confirmed by analysis for NO_3^- with nitron. The m.p. of this salt was 141°C, d^{20} = 1.1538 g cm^{-3}, and its refractive index was n = 1.485.

ESTIMATED ERROR:

Soly: precision about 1 % (compilers).

Temp: precision probably ± 0.2 K (compilers).

REFERENCES:

1. Zhuravlev, E.F.; Gorshunova, V.P. *Zh. Neorg. Khim.* 1970, *15*, 195.

2. Gorshunova, V.P.; Zhuravlev, E.F. *Zh. Neorg. Khim.* 1970, *15*, 1422.

COMPONENTS:	ORIGINAL MEASUREMENTS:
(1) Lanthanum nitrate; $La(NO_3)_3$; [10099-59-9] (2) Piperidine nitrate; $C_5H_{12}N_2O_3$; [6091-45-8] (3) Water ; H_2O ; [7732-18-5]	Gorshunova, V.P.; Zhuravlev, E.F. *Zh. Neorg. Khim.* 1972, *17*, 231-3; *Russ. J. Inorg. Chem. Engl. Transl.* 1972, *17*, 121-2.

EXPERIMENTAL VALUES: continued.........

The $La(NO_3)_3$ - $C_5H_{10}NH{\cdot}HNO_3$ - H_2O system at 50°C

composition of saturated solutions[a]

$La(NO_3)_3$		$C_5H_{10}NH{\cdot}HNO_3$			
mass %	mol kg^{-1}	mass %	mol kg^{-1}	n_D^{50}	nature of the solid phase
---	----	89.0	54.61	1.4630	$C_5H_{10}NH{\cdot}HNO_3$
2.3	0.69	87.5	57.90	1.4655	"
3.0	0.84	86.0	52.77	1.4665	"
4.5	1.54	86.5	64.87	1.4670	"
6.0	2.05	85.0	64.74	1.4690	"
15.5	3.18	69.5	31.27	1.4637	$La(NO_3)_3{\cdot}4C_5H_{10}NH{\cdot}HNO_3$
23.2	3.61	57.0	19.43	1.4611	"
25.5	3.65	53.0	16.64	1.4600	"
40.8	5.66	37.0	11.25	1.4672	"
57.5	7.66	19.4	5.67	1.4840	"
65.5	7.44	7.4	1.84	1.4830	$La(NO_3)_3{\cdot}6H_2O$
65.5	7.17	6.4	1.54	1.4820	"
65.5	6.83	5.0	1.14	1.4790	"
66.0	6.35	2.0	0.42	1.4740	"
66.5	6.11	---	----	1.4720	"

a. Molalities calculated by the compilers.

COMMENTS AND/OR ADDITIONAL DATA:

At 50°C a considerable part of the phase diagram is occupied by the crystallization field of the congruently soluble double salt. This field shrinks as the temperature is decreased. The crystallization field of $C_5H_{10}NH{\cdot}HNO_3$ is extremely small at both temperatures.

The double salt was isolated and analysed to check its composition. Lanthanum was determined by the oxalate method, nitrate was determined by precipation with nitron, and the piperidinium cation was determiend by difference. The double salt melted and decomposed at 148°C. Its density d^{20} = 1.557 g cm^{-3}, and the refractive indices were found to equal: n_g = 1.524 and n_p = 1.518. The solubility of this double salt in pure water was found to equal:

66.8 mass % at 25°C (4.25 mol kg^{-1}, compilers)

79.4 mass % at 50°C (8.15 mol kg^{-1}, compilers).

COMPONENTS:

(1) Lanthanum nitrate; $La(NO_3)_3$; [10099-59-9]

(2) Pyridine nitrate; $C_5H_6N_2O_3$; [543-53-3]

(3) Water ; H_2O ; [7732-18-5]

ORIGINAL MEASUREMENTS:

Gorshunova, V.P.; Zhuravlev, E.F. *Zh. Neorg. Khim.* 1974, *19*, 249-52; *Russ. J. Inorg. Chem. Engl. Transl.* 1974, *19*, 137-9.

VARIABLES:

Composition at 25°C and 50°C

PREPARED BY:

T. Mioduski, S. Siekierski, and M. Salomon

EXPERIMENTAL VALUES:

The $La(NO_3)_3$ - $C_5H_5N{\cdot}HNO_3$ - H_2O system at 25°C

Composition of saturated solutions[a]

$La(NO_3)_3$ mass %	$La(NO_3)_3$ mol kg^{-1}	$C_5H_5N{\cdot}HNO_3$ mass %	$C_5H_5N{\cdot}HNO_3$ mol kg^{-1}	n_D^{50}	nature of the solid phase
---	----	74.0	20.03	1.4682	$C_5H_5N{\cdot}HNO_3$
6.7	0.97	72.0	23.79	1.4837	"
15.8[b]	?	87.5[b]	?	1.4941	"
26.0	3.08	48.0	12.99	1.4700	$La(NO_3)_3{\cdot}2C_5H_5N{\cdot}HNO_3{\cdot}4H_2O$
29.5	2.84	38.5	8.47	1.4615	"
29.5	2.67	36.5	7.55	1.4595	"
33.5	2.82	30.0	5.78	1.4563	"
41.0	3.08	18.0	3.09	1.4460	"
44.7	3.41	15.0	2.62	1.4480	"
49.0	3.82	11.5	2.05	1.4535	"
57.0	4.87	7.0	1.37	1.4625	$La(NO_3)_3{\cdot}6H_2O$
58.0	4.89	5.5	1.06	1.4595	"
58.5	4.62	2.5	0.45	1.4545	"
58.5	4.34	---	----	1.4520	"

a. Molalities calculated by the compilers.

b. The compilers assume that a typographical error exists because the sum of these mass % values equals 104.2. Both the original Russian publication and the English translation list the same mass % values for $La(NO_3)_3$ and $C_5H_5N{\cdot}HNO_3$.

continued.........

AUXILIARY INFORMATION

METHOD/APPARATUS/PROCEDURE:

The solubilities were studied by the method of isothermal sections (1,2). The refractive indices of the liquid phases were measured at 50°C. Equilibrium was ascertained by constancy in the refractive indices after repeated measurements. No other information was given.

For details on the method of isothermal sections, see the compilations for references 1 and 2.

SOURCE AND PURITY OF MATERIALS:

C.p. grade $La(NO_3)_3$ was recryst: m.p. = 65°C, d^{20} = 2.354 g cm^{-3}, and refractive indices n_g = 1.591, n_p = 1.581.

C.p. grade aq HNO_3 and "pure" grade pyridine used to prepare pyridine nitrate. The resulting sln was evap to crystn, the solid recryst and dried at 40-50°C. Compn confirmed by NO_3^- analysis with nitron: m.p. = 118°C, d^{20} = 1.30 g cm^{-3}, and refractive indices n_g = 1.660, n_p = 1.515.

Source and purity of water not specified.

ESTIMATED ERROR:

Soly: precision around 1 % (compilers).

Temp: precision probably ± 0.2 K (compilers).

REFERENCES:

1. Zhuravlev, E.F.; Gorshunova, V.P. *Zh. Neorg. Khim.* 1970, *15*, 195.

2. Gorshunova, V.P.; Zhuravlev, E.F. *Zh. Neorg. Khim.* 1970, *15*, 1422.

COMPONENTS:

(1) Lanthanum nitrate; $La(NO_3)_3$; [10099-59-9]

(2) Pyridine nitrate; $C_5H_6N_2O_3$; [543-53-3]

(3) Water ; H_2O ; [7732-18-5]

ORIGINAL MEASUREMENTS:

Gorshunova, V.P.; Zhuravlev, E.F. *Zh. Neorg. Khim.* 1974, *19*, 249-52; *Russ. J. Inorg. Chem. Engl. Transl.* 1974, *19*, 137-9.

EXPERIMENTAL VALUES: continued......

The $La(NO_3)_3 - C_5H_5N\cdot HNO_3 - H_2O$ system at 50°C

Composition of saturated solutions[a]

$La(NO_3)_3$		$C_5H_5N\cdot HNO_3$			
mass %	mol kg^{-1}	mass %	mol kg^{-1}	n_D^{50}	nature of the solid phase
---	----	80.0	28.15	1.4850	$C_5H_5N\cdot HNO_3$
5.0	0.93	78.5	33.48	1.4960	"
9.5	2.09	76.5	38.45	1.5050	"
12.3	3.10	75.5	43.55	1.5091	"
16.5	5.35	74.0	54.81	1.5170	"
26.5	4.74	56.3	23.03	1.5150	$La(NO_3)_3\cdot 2C_5H_5N\cdot HNO_3\cdot 4H_2O$
35.5	6.07	46.5	18.18	1.5020	"
37.0	5.42	42.0	14.07	1.4912	"
52.0	4.57	13.0	2.61	1.4470	"
65.5	7.47	7.5	1.95	1.4885	$La(NO_3)_3\cdot 6H_2O$
65.5	6.95	5.5	1.33	1.4830	"
65.5	6.72	4.5	1.06	1.4800	"
66.5	6.71	3.0	0.69	1.4770	"
65.5	6.11	1.5	0.32	1.4730	"
65.5	5.84	---	----	1.4720	"

a. Molalities calculated by the compilers.

COMMENTS AND/OR ADDITIONAL DATA:

The double nitrate $2C_5H_5N\cdot HNO_3\cdot La(NO_3)_3\cdot 4H_2O$ was isolated and its composition confirmed by chemical analysis. Lanthanum was determined by the oxalate method, nitrate was precipitated with nitron, water was determined by the Karl Fischer method, and the pyridinium cation determined by difference. The density of this double nitrate was d^{20} = 1.89 g cm^{-3}, and the refractive indices were n_g = 1.614 and n_p = 1.556. The source paper also discusses the results of differential thermal analysis on this double nitrate. The double salt melts congruently at 75°C.

COMPONENTS:

(1) Lanthanum nitrate; $La(NO_3)_3$; [10099-59-9]

(2) Aniline nitrate; $C_6H_8N_2O_3$; [542-15-4]

(3) Water ; H_2O ; [7732-18-5]

ORIGINAL MEASUREMENTS:

Gorshunova, V.P.; Zhuravlev, E.F. *Zh. Neorg. Khim.* 1970, *15*, 1422-4; *Russ. J. Inorg. Chem. Engl. Transl.* 1970, *15*, 729-30.

VARIABLES:

Composition at 25°C and 50°C

PREPARED BY:

T. Mioduski and S. Siekierski

EXPERIMENTAL VALUES:

The $La(NO_3)_3$ - $C_6H_5NH_2{\cdot}HNO_3$ - H_2O system at 25°C

Composition of saturated solutions[a]

$La(NO_3)_3$ mass %	$La(NO_3)_3$ mol kg^{-1}	$C_6H_5NH_2{\cdot}HNO_3$ mass %	$C_6H_5NH_2{\cdot}HNO_3$ mol kg^{-1}	total moles solutes in 1000 mol H_2O	n_D^{50}	solid phase[b]
---	---	22.5	1.86	33.4	1.3725	A
6.0	0.24	17.0	1.41	29.6	1.3730	A
15.0	0.63	11.5	1.00	29.3	1.3771	A
22.5	1.00	8.0	0.74	31.1	1.3833	A
29.5	1.42	6.5	0.65	37.4	1.3931	A
37.5	1.97	4.0	0.44	43.3	1.4060	A
47.0	2.92	3.5	0.45	60.7	1.4250	A
58.5	4.45	1.0	0.16	82.6	1.4520	A + B
58.5	4.34	---	---	77.9[c]	1.4510	B

a. Molalities calculated by M. Salomon.

b. Solid phases: A = $C_6H_5NH_2{\cdot}HNO_3$ B = $La(NO_3)_3{\cdot}6H_2O$

c. Based on this value, the compilers compute soly of $La(NO_3)_3$ = 4.32 mol kg^{-1}.

continued......

AUXILIARY INFORMATION

METHOD/APPARATUS/PROCEDURE:

The solubilities were determined by the method of isothermal sections (1) which involves the measurement of refractive indices of saturated solutions along directed sections of the phase diagram. Mixtures of known composition were equilibrated until their refractive indices were constant after repeated measurements. Time required to reach equilibrium was 8-12 hours. The results were used to graph the relation between the refractive indices and the composition of the components for each of the sections studied. The graphs were used to find the break points corresponding to the composition of the saturated solutions.

SOURCE AND PURITY OF MATERIALS:

C.p. grade $La(NO_3)_3{\cdot}6H_2O$ was used.

Aniline nitrate, $C_6H_5NH_2{\cdot}HNO_3$, was a "pure" grade and recrystallized twice from aqueous solution.

Water was doubly distilled.

ESTIMATED ERROR:

Soly: precision of 1 % (compilers).

Temp: precision probably ± 0.2 K (compilers).

REFERENCES:

1. Zhuravlev, E.F.; Sheveleva, A.D. *Zh. Neorg. Khim.* 1960, *5*, 2630.

COMPONENTS:

(1) Lanthanum nitrate; $La(NO_3)_3$; [10099-59-9]

(2) Aniline nitrate; $C_6H_8N_2O_3$; [542-15-4]

(3) Water ; H_2O ; [7732-18-5]

ORIGINAL MEASUREMENTS:

Gorshunova, V.P.; Zhuravlev, E.F. *Zh. Neorg. Khim.* 1970, *15*, 1422-4; *Russ. J. Inorg. Chem. Engl. Transl.* 1970, *15*, 729-30.

EXPERIMENTAL VALUES: continued.....

The $La(NO_3)_3$ - $C_6H_5NH_2 \cdot HNO_3$ - H_2O system at 50°C

Composition of saturated solutions[a]

$La(NO_3)_3$		$C_6H_5NH_2 \cdot HNO_3$		total moles solutes in		
mass %	mol kg^{-1}	mass %	mol kg^{-1}	1000 mol H_2O	n_D^{50}	solid phase[b]
---	---	49.5	6.28	113.2	1.4265	A
4.0	0.23	43.0	5.20	97.6	1.4232	A
11.0	0.63	35.0	4.15	86.0	1.4203	A
18.0	1.05	29.0	3.50	81.6	1.4200	A
24.5	].44	23.0	2.81	76.0	1.4207	A
32.5	1.98	17.0	2.16	74.6	1.4260	A
43.0	2.88	11.0	1.53	79.2	1.4358	A
54.0	4.15	6.0	0.96	91.8	1.4500	A
66.0	6.16	1.0	0.19	111.1	1.4711	A + B
66.0	6.06	0.5	0.096	110.6	1.4710	B
65.5	5.84	---	---	105.2	1.4708	B

a. Molalities calculated by M. Salomon.

b. Solid phases: A = $C_6H_5NH_2 \cdot HNO_3$ B = $La(NO_3)_3 \cdot 6H_2O$

COMMENTS AND/OR ADDITIONAL DATA:

The phase diagrams for the 25°C and 50°C isotherms are of the eutonic type. The crystallization field of $C_6H_5NH_2 \cdot HNO_3$ occupies a large concentration range whereas the field of $La(NO_3)_3 \cdot 6H_2O$ is not very extensive. This is shown by the significant enrichment of the eutonic solutions by lanthanum nitrate. The ratio of the salt components at the invariant points are found more precisely by the graphical method: i.e. by plotting the salt composition of the saturated solutions against either the refractive index or the sum of moles of total solutes per 1000 moles of water. The results of these more precise determinations of the composition at the invariant points are:

25°C mol % $La(NO_3)_3$ = 98.5 (63.7 mass % or 5.47 mol kg^{-1}, compilers).

mol % $C_6H_5NH_2 \cdot HNO_3$ = 1.5 (0.47 mass % or 0.083 mol kg^{-1}, compilers).

50°C mol % $La(NO_3)_3$ = 98.3 (63.6 mass % or 5.46 mol kg^{-1}, compilers).

mol % $C_6H_5NH_2 \cdot HNO_3$ = 1.7 (0.53 mass % or 0.094 mol kg^{-1}, compilers).

COMPONENTS:

(1) Lanthanum nitrate; $La(NO_3)_3$; [10099-59-9]

(2) Quinoline nitrate; $C_9H_8N_2O_3$; [21640-15-3]

(3) Water ; H_2O ; [7732-18-5]

ORIGINAL MEASUREMENTS:

Gorshunova, V.P.; Zhuravlev, E.F. *Zh. Neorg. Khim.* 1973, *18*, 1688-90; *Russ. J. Inorg. Chem. Engl. Transl.* 1973, *18*, 890-1.

VARIABLES:

Composition at 25°C and 50°C

PREPARED BY:

T. Mioduski, S. Siekierski, and M. Salomon

EXPERIMENTAL VALUES:

The $La(NO_3)_3 - C_9H_7N{\cdot}HNO_3 - H_2O$ system at 25°C

Composition of saturated solutions[a]

$La(NO_3)_3$ mass %	$La(NO_3)_3$ mol kg^{-1}	$C_9H_7N{\cdot}HNO_3$ mass %	$C_9H_7N{\cdot}HNO_3$ mol kg^{-1}	n_D^{50}	nature of the solid phase
---	----	76.0	16.48	1.5340	$C_9H_7N{\cdot}HNO_3$
2.3	0.31	75.0	17.19	1.5390	"
8.7	0.86	60.0	9.97	1.5072	$La(NO_3)_3{\cdot}C_9H_7N{\cdot}HNO_3{\cdot}4H_2O$
11.5	0.86	47.2	5.95	1.4753	"
13.5	0.92	41.2	4.72	1.4628	"
17.0	1.01	31.0	3.10	1.4391	"
20.0	1.06	22.0	1.97	1.4220	"
26.0	1.32	13.3	1.14	1.4086	"
38.51	2.08	4.5	0.41	1.4100	"
40.0	2.21	4.3	0.40	1.4160	"
58.0	4.35	1.0	0.13	1.4531	$La(NO_3)_3{\cdot}C_9H_7N{\cdot}HNO_3{\cdot}4H_2O$ + $La(NO_3)_3{\cdot}6H_2O$
58.5	4.34	---	----	1.4520	$La(NO_3)_3{\cdot}6H_2O$

a. Molalities calculated by the compilers.

continued........

AUXILIARY INFORMATION

METHOD/APPARATUS/PROCEDURE:

The solubilities were studied by the method of isothermal sections (1,2). The refractive indices of the liquid phases were measured at 50°C. No other information given.

For details on the method of isothermal sections, see the compilations for references 1 and 2.

SOURCE AND PURITY OF MATERIALS:

C.p. grade $La(NO_3)_3{\cdot}6H_2O$ was recrystallized.

Commercial "pure" grade quinoline nitrate was recrystallized.

Double distilled water was used.

ESTIMATED ERROR:

Soly: precision about 1 % (compilers).

Temp: precision probably ± 0.2 K (compilers).

REFERENCES:

1. Zhuravlev, E.F.; Gorshunova, V.P. *Zh. Neorg. Khim.* 1970, *15*, 195.
2. Gorshunova, V.P.; Zhuravlev, E.F. *Zh. Neorg. Khim.* 1970, *15*, 1422.

COMPONENTS:

(1) Lanthanum nitrate; $La(NO_3)_3$; [10099-59-9]

(2) Quinoline nitrate; $C_9H_8N_2O_3$; [21640-15-3]

(3) Water ; H_2O ; [7732-18-5]

ORIGINAL MEASUREMENTS:

Gorshunova, V.P.; Zhuravlev, E.F. *Zh. Neorg. Khim.* 1973, *18*. 1688-90; *Russ. J. Inorg. Chem. Engl. Transl.* 1973, *18*, 890-1.

EXPERIMENTAL VALUES: continued........

The $La(NO_3)_3$ - $C_9H_7N{\cdot}HNO_3$ - H_2O system at 50°C

Composition of saturated solutions[a]

$La(NO_3)_3$		$C_9H_7N{\cdot}HNO_3$			
mass %	mol kg^{-1}	mass %	mol kg^{-1}	n_D^{50}	nature of the solid phase
---	----	83.5	26.33	1.5600	$C_9H_7N{\cdot}HNO_3$
1.6	0.31	82.5	27.00	1.5620	"
4.0	0.82	81.0	28.10	1.5640	"
6.5	1.48	80.0	30.84	1.5681	"
15.0	1.88	60.5	12.85	1.5290	$La(NO_3)_3{\cdot}C_9H_7N{\cdot}HNO_3{\cdot}4H_2O$
25.6	2.11	37.0	5.15	1.4792	"
27.5	2.09	32.0	4.11	1.4721	"
32.0	2.19	23.0	2.66	1.4538	"
37.0	2.39	15.4	1.68	1.4442	"
48.0	3.28	7.0	0.81	1.4449	"
65.5	6.07	1.3	0.20	1.4720	$La(NO_3)_3{\cdot}C_9H_7N{\cdot}HNO_3{\cdot}4H_2O$ + $La(NO_3)_3{\cdot}6H_2O$
65.5	5.93	0.5	0.08	1.4720	
65.5	5.84	---	----	1.4720	$La(NO_3)_3{\cdot}6H_2O$

a. Molalities calculated by the compilers.

COMMENTS AND/OR ADDITIONAL DATA:

The crystallization field for the double salt $La(NO_3)_3{\cdot}C_9H_7N{\cdot}HNO_3{\cdot}4H_2O$ dominates the phase diagram. This double salt was isolated and its composition checked by chemical analyses: La was determined by the oxalate method, nitrate was determined by precipitation with nitron, was was determined by the Karl Fischer method, and the quinolinium ion was determined by difference. The density was measured pycnometrically in absolute benzene, and the refractive indices were determined by the immersion method. The results are:

$$d^{20} = 1.96 \text{ g cm}^{-3}$$

$$n_g = 1.575$$

$$n_p = 1.536$$

The solubility of this double salt in pure water was measured with the following results:

At 25°C, soly = 48.9 mass % (1.62 mol kg^{-1}, compilers).

At 50°C, soly = 64.8 mass % (3.12 mol kg^{-1}, compilers).

COMPONENTS:

(1) Lanthanum nitrate; $La(NO_3)_3$; [10099-59-9]

(2) Benzamide; C_7H_7NO; [55-21-0]

(3) Water ; H_2O ; [7732-18-5]

ORIGINAL MEASUREMENTS:

Runov, N.N.; Shchenev, A.V. *Zh. Neorg. Khim.* 1980, *25*, 721-4; *Russ. J. Inorg. Chem. Engl. Transl.* 1980, *25*, 394-6.

VARIABLES:

Composition at 25°C and 60°C

PREPARED BY:

T. Mioduski and S. Siekierski

EXPERIMENTAL VALUES:

25°C Isotherm[a]

$La(NO_3)_3$ soly		$C_6H_5CONH_2$ soly		nature of the solid phase
mass %	mol kg^{-1}	mass %	mol kg^{-1}	
---	---	1.36	0.114	$C_6H_5CONH_2$
13.37	0.483	1.44	0.140	"
21.43	0.857	1.62	0.174	"
29.37	1.310	1.65	0.197	"
41.49	2.247	1.67	0.243	"
47.48	2.881	1.80	0.293	"
60.35	4.861	1.44	0.311	$C_6H_5CONH_3 + La(NO_3)_3 \cdot 6H_2O$
58.76	4.385	---	---	$La(NO_3)_3 \cdot 6H_2O$

60°C Isotherm

$La(NO_3)_3$ soly		$C_6H_5CONH_2$ soly		nature of the solid phase
mass %	mol kg^{-1}	mass %	mol kg^{-1}	
		5.26	0.458	$C_6H_5CONH_2$
25.54	1.164	6.92	0.846	"
33.14	1.795	10.04	1.459	"
34.39	2.431	22.07	4.184	"
29.48	3.801	46.65	16.133	"
31.68	6.011	52.10	26.515	"
40.12	6.083	39.58	16.095	$La(NO_3)_3 \cdot 3C_6H_5CONH_2 \cdot 3H_2O$
46.32	6.462	31.62	11.832	"
48.17	6.271	28.19	9.844	"
56.43	6.947	18.57	6.132	"
69.45	7.495	2.03	0.588	$La(NO_3)_3 \cdot 3C_6H_5CONH_2 \cdot 3H_2O + La(NO_3)_3 \cdot 6H_2O$
69.50	7.013	---	---	$La(NO_3)_3 \cdot 6H_2O$

a. Molalities calculated by M. Salomon.

AUXILIARY INFORMATION

METHOD/APPARATUS/PROCEDURE:

Isothermal method used (1). Benzamide was determined by the Kjeldahl method, and La by complexometric titration using xylenol orange indicator. The equilibrated solid phases were presumably analysed, but no details given. These solids were studied by IR-spectroscopy (solid samples were powdered in liquid paraffin or fluorinated oils).

SOURCE AND PURITY OF MATERIALS:

Nothing specified.

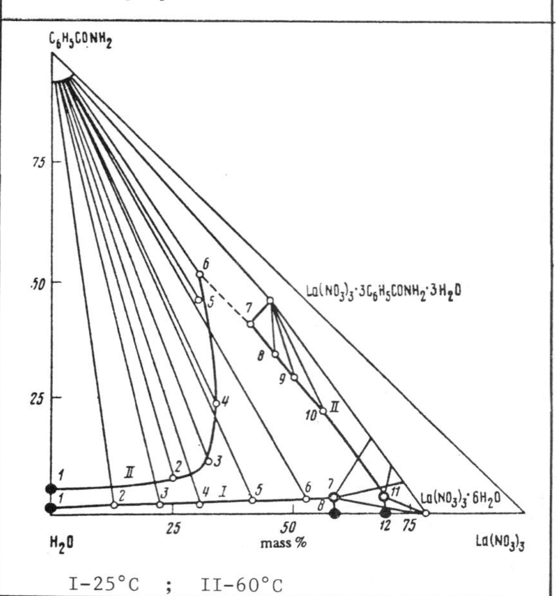

I-25°C ; II-60°C

ESTIMATED ERROR:

Soly: precision probably about ± 0.3 mass % units, but accuracy appears to be poor. For the binary $La(NO_3)_3$-H_2O system at 25°C and 60°C, the soly of the nitrate appears to be too low by about 5%.

Temp: precision probably around 0.1 to 0.2 K (compilers).

REFERENCES:

1. Tarakanov, V.F.; Shchenev, A.V. *V. sb. Fiz. Khim. Issled. Ravnovesii v Rastvorach* 1978, *169*, 52 (*Republican Conference: Physicochemical Investigations of Equilibria in Solutions*).

COMPONENTS:	EVALUATOR:
(1) Lanthanum nitrate; $La(NO_3)_3$; [1099-59-9] (2) Organic solvents	Mark Salomon U.S. Army Electronics Technology and Devices Laboratory Fort Monmouth, NJ, USA November 1982

CRITICAL EVALUATION:

The solubility of $La(NO_3)_3$ or $La(NO_3)_3 \cdot 6H_2O$ in thirty-six organic solvents has been reported in seven publications (1-7). With the exception of the data reported by Moeller et al. (2,3), all data from (1,4-7) are of poor accuracy due to a combination of poor experimental technique, impure materials, and ill defined conditions. Even when comparisons can be made, the agreement is so poor that none of the data can be critically evaluated. For example in the system of 1-hexanol equilibrated with $La(NO_3)_3 \cdot 6H_2O$, the solubility of $La(NO_3)_3$ reported in (1) is 0.418 mol kg^{-1} (compiler's calculation) which differs significantly from the value of 0.264 mol kg^{-1} reported in (4). The results for diethyl ether solutions reported in (6) and (7) differ by 1 to 2 orders of magnitude and is probably due to a combination of poor experimental techniques and water impurity.

A number of the systems studied are ill defined since it appears to the evaluator that phase separation could have occurred, but was not reported. For example in the system of 1-hexanol equilibrated with $La(NO_3)_3 \cdot 6H_2O$ reported by Stewart and Wendlandt (1), the compiler calculated the water content of the solutions to be 4.0 mass % H_2O based on the solubility of 1.468 mol kg^{-1} $La(NO_3)_3$ at 298 K. At 293 K the solubility of water in 1-hexanol is 0.71 mass % (8) which thus raises the question of possible phase separation. Templeton and Daly (4) report the water content of this identical system to be 4.8 mass %. Similar problems exist with an ill defined water content of $La(NO_3)_3$ saturated solutions in o-toluidine (1), cyclohexanone (1), and methylacetate (1) (see the compilations for details).

One publication (9) was completely rejected (i.e. not compiled) because of its questionable significance. This paper reports the mutual solubilities of $Th(NO_3)_4$, $ZrO(NO_3)_2$, $Ca(NO_3)_2$, and a mixture of Y, La and the thirteen stable lanthanide nitrates in 2-octanone (methyl-n-hexyl ketone) and in 3-methyl-1-butanol (isoamyl alchol). The authors only report the total rare earth nitrate solubility.

At this time the only data which appear to be sufficiently reliable to be classified as *tentative* solubility data are those reported by Moeller et al. (2,3) for the solubility of $La(NO_3)_3$ in ethanolamine, ethylenediamine, and morpholine at 303.2 K (see the compilations for details).

REFERENCES

1. Stewart, D.F.; Wendlandt, W.W. *J. Phys. Chem.* 1959, *63*, 1330.
2. Moeller, T,; Zimmerman, P.A. *J. Am. Chem. Soc.* 1953, *75*, 3950.
3. Moeller, T.; Cullen, G.W. *J. Inorg. Nucl. Chem.* 1959, *10*, 148.
4. Templeton, C.C.; Daly, L.K. *J. Am. Chem. Soc.* 1951, *73*, 3989.
5. Wendlandt, W.W.; Bryant, J.M. *J. Phys. Chem.* 1956, *60*, 1145.
6. Wells, R.C. *J. Wash. Acad. Sci.* 1930, *20*, 146.
7. Hopkins, B.S.; Quill, L.L. *Proc. Natl. Acad. Sci. U.S.A.* 1933, *19*, 64.
8. Addison, C.C. *J. Chem. Soc.* 1945, 98.
9. Rothschild, B.F.; Templeton, C.C.; Hall, N.F. *J. Phys. Colloid.Chem.* 1948, *52*, 1006.

COMPONENTS:

(1) Lanthanum nitrate; $La(NO_3)_3$; [10099-59-9]

(2) Methanol; CH_4O ; [67-56-1]

(3) Water ; H_2O ; [7732-18-5]

ORIGINAL MEASUREMENTS:

Stewart, D.F.; Wendlandt, W.W.

J. Phys. Chem. 1959, *63*, 1330-1.

VARIABLES:

One temperature: 25.0°C

PREPARED BY:

Mark Salomon

EXPERIMENTAL VALUES:

Original data:

$La(NO_3)_3 \cdot 6H_2O$/mass %	ΔH_{sln}/kcal mol^{-1}
87.45	-3.75
87.47	-3.58

Compiler's conversions for anhydrous salt:

average solubility of $La(NO_3)_3$ = 65.63 mass % (5.876 mol kg^{-1})

COMMENTS AND/OR ADDITIONAL DATA:

The solid phase was not analysed and the nature of the solvate formed in the equilibrated solution is unknown. The water content of the equilibrated solution saturated with $La(NO_3)_3$ was calculated by the compiler as 21.83 mass % H_2O.

Based on the duplicate analysis, an average precision of ± 0.02 mass % is indicated. However due to the unknown purity of the solvent and uncertainties associated with the experimental technique (e.g. temperature at which centrifuging was carried out), the compiler estimates an accuracy of no better than ± 1 % for the solubility.

AUXILIARY INFORMATION

METHOD/APPARATUS/PROCEDURE:

Isothermal method used. Excess lanthanum nitrate hexahydrate and 10 ml of CH_3OH were placed in 25 x 160 mm screw-cap vial, sealed with paraffin, and allowed to stand at room temperature (25-35°C) for several hours with periodic shaking. The vial was then placed in a water bath at 25 ± 0.2°C and mechanically rotated end-over-end at 30 r.p.m. Preliminary experiments indicated that equilibrium was established in 4 days, but the solutions was equilibrated for at least 7 days.

The saturated solution was centrifuged and analysed for lanthanum gravimetrically by precipitating as the oxalate and ignition to the oxide.

Heats of solution were measured calorimetrically as described in (1). The solvent used for this determination was dried and fractionated.

SOURCE AND PURITY OF MATERIALS:

$La(NO_3)_3 \cdot 6H_2O$ was prepared from 99.99 % La_2O_3 and aqueous nitric acid. The solution was evaporated and the resulting solid redissolved in water and evaporated to crystallization. The crystals were filtered and excess water removed by drying over conc sulfuric acid followed by drying over partially dehydrated $La(NO_3)_3 \cdot 6H_2O$ in a desiccator.

Methanol was stock material obtained from a commercial supplier.

ESTIMATED ERROR:

See COMMENTS above.

REFERENCES:

1. Van Tassel, J.H.; Wendlandt, W.W. *J. Am. Chem. Soc.* 1959, *81*, 813.

COMPONENTS:

(1) Lanthanum nitrate; $La(NO_3)_3$; [10099-59-9]

(2) Ethanol (ethyl alcohol); C_2H_6O; [64-17-5]

(3) Water; H_2O; [7732-18-5]

ORIGINAL MEASUREMENTS:

Stewart, D.F.; Wendlandt, W.W.
J. Phys. Chem. 1959, *63*, 1330-1.

VARIABLES:

Water content at 25°C

PREPARED BY:

Mark Salomon

EXPERIMENTAL VALUES:

Solubility at 25°C

water content/mass %		$La(NO_3)_3 \cdot 6H_2O$	$La(NO_3)_3$ [a]		ΔH_{sln}
initial	final [a]	mass %	mass %	mol kg^{-1}	kol mol^{-1}
7.59 [b]	25.78	72.86	54.67	3.712	2.58
7.59 [b]	25.88	73.28	54.99	3.760	2.58
0	17.95	71.89	53.94	3.605	1.57
0	17.90	71.69	53.79	3.583	1.64

a. Calculated by the compiler.

b. Authors state the initial solvent was 95% alcohol. The compiler *assumes* this to be 95% by volume which contains 7.59 mass % water (1). The total water content is thus the initial water plus the water of hydration.

COMMENTS AND/OR ADDITIONAL DATA:

The solid phase was not analysed, and the type of solvate formed in the equilibrated solution is unknown.

Based on the duplicate analysis, an average precision of ± 0.2 mass % is indicated. However, due to the unknown purity of the solvent and uncertainties associated with the experimental technique (e.g., temperature at which centrifuging was carried out), the compiler estimates an accuracy of no better than ± 1 % for the solubility.

AUXILIARY INFORMATION

METHOD/APPARATUS/PROCEDURE:

Isothermal method used. Excess lanthanum nitrate hexahydrate and 10 ml of solvent were placed in 25 x 160 mm screw-cap vials, sealed with paraffin, and allowed to stand at room temperature (25-35°C) for several hours with periodic shaking. The vial was then placed in a water bath at 25 ± 0.2°C and mechanically rotated end-over-end at 30 r.p.m. Preliminary experiments indicated that equilibrium was established in 4 days, but the solutions were equilibrated for at least 7 days.

The saturated solution was centrifuged and analysed for lanthanum gravimetrically by precipitating as the oxalate and ignition to the oxide.

Heats of solution were measured calorimetrically as described in (2). The solvent used for this determination was dried and fractionated.

SOURCE AND PURITY OF MATERIALS:

$La(NO_3)_3 \cdot 6H_2O$ was prepared from 99.99 % La_2O_3 and aqueous nitric acid. The solution was evaporated and the remaining solid redissolved in water and evaporated to crystallization. The crystals were filtered and excess water removed by drying over conc sulfuric acid followed by drying over partially dehydrated $La(NO_3)_3 \cdot 6H_2O$ in a desiccator.

The alcohol was stock material obtained from a commercial supplier.

ESTIMATED ERROR:

See COMMENTS above.

REFERENCES:

1. Lange, N.A.; Forker, G.M. *Handbook of Chemistry*. Mc-Graw Hill. N.Y. 1961. Tenth Edition.

2. Van Tassel, J.H.; Wendlandt, W.W. *J. Am. Chem. Soc.* 1959, *81*, 813.

COMPONENTS:

(1) Lanthanum nitrate; $La(NO_3)_3$; [10099-59-9]

(2) 1,2-Ethanediol (ethylene glycol); $C_2H_6O_2$; [107-21-1]

(3) Water; H_2O; [7732-18-5]

ORIGINAL MEASUREMENTS:

Stewart, D.F.; Wendlandt, W.W.
J. Phys. Chem. 1959, *63*, 1330-1.

VARIABLES:

One temperature: 25.0°C

PREPARED BY:

Mark Salomon

EXPERIMENTAL VALUES:

Original data:

$La(NO_3)_3 \cdot 6H_2O$/mass %
84.06
84.07

Compiler's conversions for anhydrous salt:

average solubility of $La(NO_3)_3$ = 63.08 mass % (5.258 mol kg^{-1})

COMMENTS AND/OR ADDITIONAL DATA:

The solid phase was not analysed and the nature of the solvate formed in the equilibrated solution is unknown. The water content of the equilibrated solution saturated with $La(NO_3)_3$ was calculated by the compiler as 20.98 mass % H_2O.

Based on the duplicate analysis, an average precision of ± 0.01 mass % is indicated. However, due to the unknown purity of the solvent and uncertainties associated with the experimental technique (e.g., temperature at which centrifuging was carried out), the compiler estimates an accuracy of no better than ± 1 % for the solubility.

AUXILIARY INFORMATION

METHOD/APPARATUS/PROCEDURE:

Isothermal method used. Excess lanthanum nitrate hexahydrate and 10 ml of solvent were placed in 25 x 160 mm screw cap vials, sealed with paraffin, and allowed to stand at room temperature (25-35°C) for several hours with periodic shaking. The vial was then placed in a water bath at 25 ± 0.2°C and mechanically rotated end-over-end at 30 r.p.m. Preliminary experiments indicated that equilibrium was established in 4 days, but the solutions were equilibrated for at least 7 days.

The saturated solution was centrifuged and analysed for lanthanum gravimetrically by precipitating as the oxalate and ignition to the oxide.

SOURCE AND PURITY OF MATERIALS:

$La(NO_3)_3 \cdot 6H_2O$ was prepared from 99.99 % La_2O_3 and aqueous nitric acid. The solution was evaporated and the resulting solid redissolved in water and evaporated to crystallization. The crystals were filtered and excess water removed by drying over conc sulfuric acid followed by drying over partially dehydrated $La(NO_3)_3 \cdot 6H_2O$ in a desiccator.

1,2-ethanediol was stock material obtained form a commercial supplier.

ESTIMATED ERROR:

See COMMENTS above.

REFERENCES:

COMPONENTS:

(1) Lanthanum nitrate; $La(NO_3)_3$; [10099-59-9]

(2) 2-Methoxyethanol; $C_3H_8O_2$; [109-86-4]

(3) Water; H_2O; [7732-18-5]

ORIGINAL MEASUREMENTS:

Stewart, D.F.; Wendlandt, W.W.
J. Phys. Chem. 1959, *63*, 1330-1.

VARIABLES:

One temperature: 25.0°C

PREPARED BY:

Mark Salomon

EXPERIMENTAL VALUES:

Original data:

$La(NO_3)_3 \cdot 6H_2O$/mass %	ΔH_{sln}/kcal mol^{-1}
77.93	-3.74
78.06	-3.58

Compiler's conversions for anhydrous salt:

average solubility of $La(NO_3)_3$ = 58.53 mass % (4.343 mol kg^{-1})

COMMENTS AND/OR ADDITIONAL DATA:

The solid phase was not analysed and the nature of the solvate formed in the equilibrated solution is unknown. The water content of the equilibrated solution saturated with $La(NO_3)_3$ was calculated by the compiler as 19.47 mass % H_2O.

Based on the duplicate analysis, an average precision of ± 0.07 mass % is indicated. However, due to the unknown purity of the solvent and uncertainties associated with the experimental technique (e.g., temperature at which centrifuging was carried out), the compiler estimates an accuracy of no better than ± 1 % for the solubility.

AUXILIARY INFORMATION

METHOD/APPARATUS/PROCEDURE:

Isothermal method used. Excess lanthanum nitrate hexahydrate and 10 ml of $CH_3OCH_2CH_2OH$ were placed in 25 x 160 mm screw-cap vials, sealed with paraffin, and allowed to stand at room temperature (25-35°C) for several hours with periodic shaking. The vial was then placed in a water bath at 25 ± 0.2°C and mechanically rotated end-over-end at 30 r.p.m. Preliminary experiments indicated that equilibrium was established in 4 days, but the solutions were equilibrated for at least 7 days.

The saturated solution was centrifuged and analysed for lanthanum gravimetrically by precipitating as the oxalate and ignition to the oxide.

Heats of solution were measured calorimetrically as described in (1). The solvent used for this determination was dried and fractionated.

SOURCE AND PURITY OF MATERIALS:

$La(NO_3)_3 \cdot 6H_2O$ was prepared from 99.99 % La_2O_3 and aqueous nitric acid. The solution was evaporated and the resulting solid redissolved in water and evaporated to crystallization. The crystals were filtered and excess water removed by drying over conc sulfuric acid followed by drying over partially dehydrated $La(NO_3)_3 \cdot 6H_2O$ in a desiccator.

The alcohol was stock material obtained from a commercial supplier.

ESTIMATED ERROR:

See COMMENTS above.

REFERENCES:

1. Van Tassel, J.H.; Wendlandt, W.W. *J. Am. Chem. Soc.* 1959, *81*, 813.

COMPONENTS:

(1) Lanthanum nitrate; $La(NO_3)_3$; [10099-59-9]
(2) 2-Ethoxyethanol (cellosolve, ethylene glycol ethyl ether); $C_4H_{10}O_2$; [110-80-5]
(3) Water; H_2O; [7732-18-5]

ORIGINAL MEASUREMENTS:

Stewart, D.F.; Wendlandt, W.W. *J. Phys. Chem.* 1959, *63*, 1330-1.

VARIABLES:

One temperature: 25.0°C

PREPARED BY:

Mark Salomon

EXPERIMENTAL VALUES:

Original data:

$La(NO_3)_3 \cdot 6H_2O$/mass %	ΔH_{sln}/kcal mol^{-1}
70.00	-1.78
70.07	-1.76

Compiler's conversions for anhydrous salt:

average solubility of $La(NO_3)_3$ = 52.55 mass % (3.409 mol kg^{-1})

COMMENTS AND/OR ADDITIONAL DATA:

The solid phase was not analysed and the nature of the solvate formed in the equilibrated solution is unknown. The water content of the equilibrated solution saturated with $La(NO_3)_3$ was calculated by the compiler as 17.48 mass % H_2O.

Based on the duplicate analysis, an average precision of ± 0.04 mass % is indicated. However, due to the unknown purity of the solvent and uncertainties associated with the experimental technique (e.g., temperature at which centrifuging was carried out), the compiler estimates an accuracy of no better than ± 1 % for the solubility.

AUXILIARY INFORMATION

METHOD/APPARATUS/PROCEDURE:

Isothermal method used. Excess lanthanum nitrate hexahydrate and 10 ml of $CH_3CH_2OCH_2CH_2OH$ were placed in 25 x 160 mm screw-cap vials, sealed with paraffin, and allowed to stand at room temperature (25-35°C) for several hours with periodic shaking. The vial was then placed in a water bath at 25 ± 0.2°C and mechanically rotated end-over-end at 30 r.p.m. Preliminary experiments indicated that equilibrium was established in 4 days, but the solutions were equilibrated for at least 7 days.

The saturated solution was centrifuged and analysed for lanthanum gravimetrically by precipitating as the oxalate and ignition to the oxide.

Heats of solution were measured calorimetrically as described in (1). The solvent used for this determination was dried and fractionated.

SOURCE AND PURITY OF MATERIALS:

$La(NO_3)_3 \cdot 6H_2O$ was prepared from 99.99 % La_2O_3 and aqueous nitric acid. The solution was evaporated and the resulting solid redissolved in water and evaporated to crystallization. The crystals were filtered and excess water removed by drying over conc sulfuric acid followed by drying over partially dehydrated $La(NO_3)_3 \cdot 6H_2O$ in a desiccator.
2-Ethoxyethanol was stock material obtained from a commercial supplier.

ESTIMATED ERROR:

See COMMENTS above.

REFERENCES:

1. Van Tassel, J.H.; Wendlandt, W.W. *J. Am. Chem. Soc.* 1959, *81*, 813.

COMPONENTS:

(1) Lanthanum nitrate; $La(NO_3)_3$; [10099-59-9]

(2) 2-Aminoethanol (ethanolamine); C_2H_7NO; [141-43-5]

ORIGINAL MEASUREMENTS:

Moeller, T.; Zimmerman, P.A. *J. Am. Chem. Soc.* 1953, *75*, 3950-3.

VARIABLES:

One temperature: 30°C

PREPARED BY:

T. Mioduski and S. Siekierski

EXPERIMENTAL VALUES:

The solubility of anhydrous $La(NO_3)_3$ in $H_2N{\cdot}CH_2CH_2OH$ at 30°C is

7.75 g/100 g solvent

The compilers have converted this value to molality:

0.239 mol kg^{-1}

AUXILIARY INFORMATION

METHOD/APPARATUS/PROCEDURE:

The isothermal method was used. Reaction mixtures were sealed in 25 x 200 mm test tubes and thermostated for one week at 30 ± 0.05°C, and frequently agitated.
The density of the supernatant liquid was determined pycnometrically, and the lanthanum content determined by precipitating the hydrous hydroxide from measured volumes by adding excess and weighing the ignited oxide.

All anhydrous substances were handled in a dry box through which a current of nitrogen was passed. The nitrogen was freed of CO_2 and moisture by passage through concentrated sulfuric acid, soda lime, and Drierite. All solutions were prepared in the dry box, and all apparatus containing these solutions were sealed before being removed.

SOURCE AND PURITY OF MATERIALS:

Lanthanum oxide was converted to the nitrate by high temperature reaction with NH_4NO_3. Unreacted NH_4NO_3 was removed by heating in N_2 and then in vacuo. The oxide, obtained from University stocks, contained traces of other rare earth metals.
Ethanolamine was purified as in (1). Its boiling point was 168°C (uncor), density of 1.0108 g/ml at 26.5°C (lit. 1.0106 g/ml), and electrolytic conductivity at 20°C was 1.93×10^{-5} S cm^{-1}. The solvent was stored in wax sealed glass stoppered flasks.

ESTIMATED ERROR:

Soly: precision probably around ± 1 % (compilers).
Temp: thermostat control was ± 0.05 K, but uncertainty due to handling during analysis is closer to ± 0.5 K (compilers).

REFERENCES:

1. Dirkse, T.P.; Briscoe, H.T. *Metal Ind.* 1938, *36*, 284.

COMPONENTS:

(1) Lanthanum nitrate; $La(NO_3)_3$; [10099-59-9]

(2) 1-Propanol; C_3H_8O; [71-23-8]

(3) Water; H_2O; [7732-18-5]

ORIGINAL MEASUREMENTS:

Stewart, D.F.; Wendlandt, W.W.
J. Phys. Chem. 1959, *63*, 1330-1.

VARIABLES:

One temperature: 25.0°C

PREPARED BY:

Mark Salomon

EXPERIMENTAL VALUES:

Original data:

$La(NO_3)_3 \cdot 6H_2O$/mass %	ΔH_{sln}/kcal mol^{-1}
45.15	6.30
45.23	6.47

Compiler's conversions for anhydrous salt:

average solubility of $La(NO_3)_3$ = 33.91 mass % (1.579 mol kg^{-1})

COMMENTS AND/OR ADDITIONAL DATA:

The solid phase was not analysed and the nature of the solvate formed in the equilibrated solution is unknown. The water content of the equilibrated solution saturated with $La(NO_3)_3$ was calculated by the compiler as 11.28 mass % H_2O.

Based on the duplicate analysis, an average precision of ± 0.04 mass % is indicated. However, due to the unknown purity of the solvent and uncertainties associated with the experimental technique (e.g., temperature at which centrifuging was carried out), the compiler estimates an accuracy of no better than ± 1 % for the solubility.

AUXILIARY INFORMATION

METHOD/APPARATUS/PROCEDURE:

Isothermal method used. Excess lanthanum nitrate hexahydrate and 10 ml of $CH_3CH_2CH_2OH$ were placed in 25 x 160 mm screw-cap vials, sealed with paraffin, and allowed to stand at room temperature (25-35°C) for several hours with periodic shaking. The vial was then placed in a water bath at 25 ± 0.2°C and mechanically rotated end-over-end at 30 r.p.m. Preliminary experiments indicated that equilibrium was established in 4 days, but the solutions were equilibrated for at least 7 days.

The saturated solution was centrifuged and analysed for lanthanum gravimetrically by precipitating as the oxalate and ignition to the oxide.

Heats of solution were measured calorimetrically as described in (1). The solvent used for this determination was dried and fractionated.

SOURCE AND PURITY OF MATERIALS:

$La(NO_3)_3 \cdot 6H_2O$ was prepared from 99.99 % La_2O_3 and aqueous nitric acid. The solution was evaporated and the resulting solid redissolved in water and evaporated to crystallization. The crystals were filtered and excess water removed by drying over conc sulfuric acid followed by drying over partially dehydrated $La(NO_3)_3 \cdot 6H_2O$ in a desiccator.
The alcohol was stock material obtained from a commercial supplier.

ESTIMATED ERROR:

See COMMENTS above.

REFERENCES:

1. Van Tassel, J.H.; Wendlandt, W.W. *J. Am. Chem. Soc.* 1959, *81*, 813.

COMPONENTS:

(1) Lanthanum nitrate; $La(NO_3)_3$; [10099-59-9]

(2) 2-Propanol (isopropanol); C_3H_8O; [67-63-0]

(3) Water; H_2O; [7732-18-5]

ORIGINAL MEASUREMENTS:

Stewart, D.F.; Wendlandt, W.W.
J. Phys. Chem. 1959, *63*, 1330-1.

VARIABLES:

One temperature: 25.0°C

PREPARED BY:

Mark Salomon

EXPERIMENTAL VALUES:

Original data:

$La(NO_3)_3 \cdot 6H_2O$/mass %	ΔH_{sln}/kcal mol^{-1}
42.06	12.0
42.53	12.5

Compiler's conversions for anhydrous salt:

average solubility of $La(NO_3)_3$ = 31.74 mass % (1.431 mol kg^{-1})

COMMENTS AND/OR ADDITIONAL DATA:

The solid phase was not analysed and the nature of the solvate formed in the equilibrated solution is unknown. The water content of the equilibrated solution saturated with $La(NO_3)_3$ was calculated by the compiler as 10.56 mass % H_2O.

Based on the duplicate analysis, an average precision of ± 0.2 mass % is indicated. However, due to the unknown purity of the solvent and the uncertainties associated with the experimental technique (e.g., temperature at which centrifuging was carried out), the compiler estimates an accuracy of no better than ± 1 % for the solubility.

AUXILIARY INFORMATION

METHOD/APPARATUS/PROCEDURE:

Isothermal method used. Excess lanthanum nitrate hexahydrate and 10 ml of $CH_3CHOHCH_3$ were placed in 25 x 160 mm screw-cap vials, sealed with paraffin, and allowed to stand at room temperature (25-35°C) for several hours with periodic shaking. The vial was then placed in a water bath at 25 ± 0.2°C and mechanically rotated end-over-end at 30 r.p.m. Preliminary experiments indicated that equilibrium was established in 4 days, but the solutions were equilibrated for at least 7 days.

The saturated solution was centrifuged and analysed for lanthanum gravimetrically by precipitating as the oxalate and ignition to the oxide.

Heats of solution were measured calorimetrically as described in (1). The solvent used for this determination was dried and fractionated.

SOURCE AND PURITY OF MATERIALS:

$La(NO_3)_3 \cdot 6H_2O$ was prepared from 99.99 % La_2O_3 and aqueous nitric acid. The solution was evaporated and the resulting solid redissolved in water and evaporated to crystallization. The crystals were filtered and excess water removed by drying over conc sulfuric acid followed by drying over partially dehydrated $La(NO_3)_3 \cdot 6H_2O$ in a desiccator.
The alcohol was stock material obtained from a commercial supplier.

ESTIMATED ERROR:

See COMMENTS above.

REFERENCES:

1. Van Tassel, J.H.; Wendlandt, W.W. *J. Am. Chem. Soc.* 1959, *81*, 813.

COMPONENTS:

(1) Lanthanum nitrate; $La(NO_3)_3$; [10099-59-9]

(2) 2-Propene-1-ol (allyl alcohol; C_3H_6O; [107-18-6]

(3) Water; H_2O; [7732-18-5]

ORIGINAL MEASUREMENTS:

Stewart, D.F.; Wendlandt, W.W.
J. Phys. Chem. 1959, *63*, 1330-1.

VARIABLES:

One temperature: 25.0°C

PREPARED BY:

Mark Salomon

EXPERIMENTAL VALUES:

Original data:

$La(NO_3)_3\cdot 6H_2O$/mass %	ΔH_{sln}/kcal mol^{-1}
47.01	9.85
46.89	9.79

Compiler's conversions for anhydrous salt:

average solubility of $La(NO_3)_3$ = 35.23 mass % (1.674 mol kg^{-1})

COMMENTS AND/OR ADDITIONAL DATA:

The solid phase was not analysed and the nature of the solvate formed in the equilibrated solution is unknown. The water content of the equilibrated solution saturated with $La(NO_3)_3$ was calculated by the compiler as 11.72 mass % H_2O.

Based on the duplicate analysis, an average precision of ± 0.06 mass % is indicated. However, due to the unknown purity of the solvent and uncertainties associated with the experimental technique (e.g., temperature at which centrifuging was carried out), the compiler estimates an accuracy of no better than ± 1 % for the solubility.

AUXILIARY INFORMATION

METHOD/APPARATUS/PROCEDURE:

Isothermal method used. Excess lanthanum nitrate hexahydrate and 10 ml of $H_2C{=}CHCH_2OH$ were placed in 25 x 160 mm screw-cap vials, sealed with paraffin, and allowed to stand at room temperature (25-35°C) for several hours with periodic shaking. The vial was then placed in a water bath at 25 ± 0.2°C and mechanically rotated end-over-end at 30 r.p.m. Preliminary experiments indicated that equilibrium was established in 4 days, but the solutions were equilibrated for at least 7 days.

The saturated solution was centrifuged and analysed for lanthanum gravimetrically by precipitating as the oxalate and ignition to the oxide.

Heats of solution were measured calorimetrically as described in (1). The solvent used for this determination was dried and fractionated.

SOURCE AND PURITY OF MATERIALS:

$La(NO_3)_3\cdot 6H_2O$ was prepared from 99.99 % La_2O_3 and aqueous nitric acid. The solution was evaporated and the resulting solid redissolved in water and evaporated to crystallization. The crystals were filtered and excess water removed by drying over conc sulfuric acid followed by drying over partially dehydrated $La(NO_3)_3\cdot 6H_2O$ in a desiccator.

The alcohol was stock material obtained from a commercial supplier.

ESTIMATED ERROR:

See COMMENTS above.

REFERENCES:

1. Van Tassel, J.H.; Wendlandt, W.W. *J. Am. Chem. Soc.* 1959, *81*, 813.

COMPONENTS:

(1) Lanthanum nitrate; $La(NO_3)_3$; [10099-59-9]

(2) 1,2,3-Propanetriol (glycerol); $C_3H_8O_3$; [56-81-5]

(3) Water; H_2O; [7732-18-5]

ORIGINAL MEASUREMENTS:

Stewart, D.F.; Wendlandt, W.W.
J. Phys. Chem. 1959, *63*, 1330-1.

VARIABLES:

One temperature: 25.0°C

PREPARED BY:

Mark Salomon

EXPERIMENTAL VALUES:

Original data:

$La(NO_3)_3 \cdot 6H_2O$/mass %
81.66
81.45

Compiler's conversions for anhydrous salt:

average solubility for $La(NO_3)_3$ = 61.20 mass % (4.854 mol kg^{-1})

COMMENTS AND/OR ADDITIONAL DATA:

The solid phase was not analysed and the nature of the solvate formed in the equilibrated solution is unknown. The water content of the equilibrated solution saturated with $La(NO_3)_3$ was calculated by the compiler as 20.36 mass % H_2O.

Based on the duplicate analysis, an average precision of ± 0.10 mass % is indicated. However, due to the unknown purity of the solvent and uncertainties associated with the experimental technique (e.g., temperature at which centrifuging was carried out), the compiler estimates an accuracy of no better than ± 1 % for the solubility.

AUXILIARY INFORMATION

METHOD/APPARATUS/PROCEDURE:

Isothermal method used. Excess lanthanum nitrate hexahydrate and 10 ml of $CH_2OHCH_2OHCH_2OH$ were placed in 25 x 160 mm screw-cap vials, sealed with paraffin, and allowed to stand at room temperature (25-35°C) for several hours with periodic shaking. The vial was then placed in a water bath at 25 ± 0.2°C and mechanically rotated end-over-end at 30 r.p.m. Preliminary experiments indicated that equilibrium was established in 4 days, but the solutions were equilibrated for at least 7 days.

The saturated solution was centrifuged and analysed for lanthanum gravimetrically by precipitating as the oxalate and ignition to the oxide.

SOURCE AND PURITY OF MATERIALS:

$La(NO_3)_3 \cdot 6H_2O$ was prepared from 99.99 % La_2O_3 and aqueous nitric acid. The solution was evaporated and the resulting solid redissolved in water and evaporated to crystallization. The crystals were filtered and excess water removed by drying over conc sulfuric acid followed by drying over partially dehydrated $La(NO_3)_3 \cdot 6H_2O$ in a desiccator.
Glycerol was stock material obtained from a commercial supplier.

ESTIMATED ERROR:

See COMMENTS above.

REFERENCES:

COMPONENTS:

(1) Lanthanum nitrate; $La(NO_3)_3$; [10099-59-9]
(2) 1-Butanol; $C_4H_{10}O$; [71-36-3]
(3) Water; H_2O; [7732-18-5]

ORIGINAL MEASUREMENTS:

Stewart, D.F.; Wendlandt, W.W.
J. Phys. Chem. 1959, *63*, 1330-1.

VARIABLES:

One temperature: 25.0°C

PREPARED BY:

Mark Salomon

EXPERIMENTAL VALUES:

Original data:

$La(NO_3)_3 \cdot 6H_2O$/mass %	ΔH_{sln}/kcal mol^{-1}
28.74	8.46
28.75	8.34

Compiler's conversions for anhydrous salt:

average solubility of $La(NO_3)_3$ = 21.57 mass % (0.846 mol kg^{-1})

COMMENTS AND/OR ADDITIONAL DATA:

The solid phase was not analysed and the nature of the solvate formed in the equilibrated solution is unknown. The water content of the equilibrated solution saturated with $La(NO_3)_3$ was calculated by the compiler as 7.18 mass % H_2O.

Based on the duplicate analysis, an average percision of ± 0.01 mass % is indicated. However, due to the unknown purity of the solvent and uncertainties associated with the experimental technique (e.g., temperature at which centrifuging was carried out), the compiler estimates an accuracy of no better than ± 1 % for the solubility.

AUXILIARY INFORMATION

METHOD/APPARATUS/PROCEDURE:

Isothermal method used. Excess lanthanum nitrate hexahydrate and 10 ml of $CH_3CH_2CH_2CH_2OH$ were placed in 25 x 160 mm screw-cap vials, sealed with paraffin, and allowed to stand at room temperature (25-35°C) for several hours with periodic shaking. The vial was then placed in a water bath at 25 ± 0.2°C and mechanically rotated end-over-end at 30 r.p.m. Preliminary experiments indicated that equilibrium was established in 4 days, but the solutions were equilibrated for at least 7 days.

The saturated solution was centrifuged and analysed for lanthanum gravimetrically by precipitating as the oxalate and ignition to the oxide.

Heats of solution were measured calorimetrically as described in (1). The solvent used for this determination was dried and fractionated.

SOURCE AND PURITY OF MATERIALS:

$La(NO_3)_3 \cdot 6H_2O$ was prepared from 99.99 % La_2O_3 and aqueous nitric acid. The solution was evaporated and the resulting solid redissolved in water and evaporated to crystallization. The crystals were filtered and excess water removed by drying over conc sulfuric acid followed by drying over partially dehydrated $La(NO_3)_3 \cdot 6H_2O$ in a desiccator.
The alcohol was stock material obtained from a commercial supplier.

ESTIMATED ERROR:

See COMMENTS above.

REFERENCES:

1. Van Tassel, J.H.; Wendlandt, W.W. *J. Am. Chem. Soc.* 1959, *81*, 813.

COMPONENTS:

(1) Lanthanum nitrate; $La(NO_3)_3$; [10099-59-9]

(2) 2-Butanol (sec-butyl alcohol); $C_4H_{10}O$; [78-92-2]

(3) Water; H_2O; [7732-18-5]

ORIGINAL MEASUREMENTS:

Stewart, D.F.; Wendlandt, W.W.
J. Phys. Chem. 1959, *63*, 1330-1.

VARIABLES:

One temperature: 25.0°C

PREPARED BY:

Mark Salomon

EXPERIMENTAL VALUES:

Original data: $La(NO_3)_3 \cdot 6H_2O$/mass %

13.84

13.65

Compiler's conversions for anhydrous salt:

average solubility of $La(NO_3)_3$ = 10.32 mass % (0.354 mol kg^{-1})

COMMENTS AND/OR ADDITIONAL DATA:

The solid phase was not analysed and the nature of the solvate formed in the equilibrated solution is unknown. The water content of the equilibrated solution saturated with $La(NO_3)_3$ was calculated by the compiler as 3.43 mass % H_2O.

Based on the duplicate analysis, an average precision of ± 0.1 mass % is indicated. However, due to the unknown purity of the solvent and uncertainties associated with the experimental technique (e.g., temperature at which centrifuging was carried out), the compiler estimates an accuracy of no better than ± 1 % for the solubility.

AUXILIARY INFORMATION

METHOD/APPARATUS/PROCEDURE:

Isothermal method used. Excess lanthanum nitrate hexahydrate and 10 ml of $CH_3CH_2CHOHCH_3$ were placed in 25 x 160 mm screw-cap vials, sealed with paraffin, and allowed to stand at room temperature (25-35°C) for several hours with periodic shaking. The vial was then placed in a water bath at 25 ± 0.2°C and mechanically rotated end-over-end at 30 r.p.m. Preliminary experiments indicated that equilibrium was established in 4 days, but the solutions were equilibrated for at least 7 days.

The saturated solution was centrifuged and analysed for lanthanum gravimetrically by precipitating as the oxalate and ignition to the oxide.

SOURCE AND PURITY OF MATERIALS:

$La(NO_3)_3 \cdot 6H_2O$ was prepared from 99.99 % La_2O_3 and aqueous nitric acid. The solution was evaporated and the resulting solid redissolved in water and evaporated to crystallization. The crystals were filtered and excess water removed by drying over conc sulfuric acid followed by drying over partially dehydrated $La(NO_3)_3 \cdot 6H_2O$ in a desiccator.

The alcohol was stock material obtained from a commercial supplier.

ESTIMATED ERROR:

See COMMENTS above.

REFERENCES:

COMPONENTS:

(1) Lanthanum nitrate; $La(NO_3)_3$; [10099-59-9]

(2) 2-Methyl-1-propanol (isobutyl alcohol); $C_4H_{10}O$; [78-83-1]

(3) Water; H_2O; [7732-18-5]

ORIGINAL MEASUREMENTS:

Stewart, D.F.; Wendlandt, W.W.
J. Phys. Chem. 1959, *63*, 1330-1.

VARIABLES:

One temperature: 25.0°C

PREPARED BY:

Mark Salomon

EXPERIMENTAL VALUES:

Original data:

$La(NO_3)_3 \cdot 6H_2O$/mass %
14.98
15.15

Compiler's conversions for anhydrous salt:

average solubility of $La(NO_3)_3$ = 11.30 mass % (0.392 mol kg^{-1})

COMMENTS AND/OR ADDITIONAL DATA:

The solid phase was not analysed and the nature of the solvate formed in the equilibrated solution is unknown. The water content of the equilibrated solution saturated with $La(NO_3)_3$ was calculated by the compiler as 3.76 mass % H_2O.

Based on the duplicate analysis, an average precision of ± 0.08 mass % is indicated. However, due to the unknown purity of the solvent and uncertainties associated with the experimental technique (e.g., temperature at which centrifuging was carried out), the compiler estimates an accuracy of no better than ± 1 % for the solubility.

AUXILIARY INFORMATION

METHOD/APPARATUS/PROCEDURE:

Isothermal method used. Excess lanthanum nitrate hexahydrate and 10 ml of isobutyl alcohol were placed in 25 x 160 mm screw-cap vials, sealed with paraffin, and allowed to stand at room temperature (25-35°C) for several hours with periodic shaking. The vial was then placed in a water bath at 25 ± 0.02°C and mechanically rotated end-over-end at 30 r.p.m. Preliminary experiments indicated that equilibrium was established in 4 days, but the solutions were equilibrated for at least 7 days.

The saturated solution was centrifuged and analysed for lanthanum gravimetrically by precipitating as the oxalate and ignition to the oxide.

SOURCE AND PURITY OF MATERIALS:

$La(NO_3)_3 \cdot 6H_2O$ was prepared from 99.99 % La_2O_3 and aqueous nitric acid. The solution was evaporated and the resulting solid redissolved in water and evaporated to crystallization. The crystals were filtered and excess water removed by drying over conc sulfuric acid followed by drying over partially dehydrated $La(NO_3)_3 \cdot 6H_2O$ in a desiccator.

The alcohol was stock material obtained from a commercial supplier.

ESTIMATED ERROR:

See COMMENTS above.

REFERENCES:

COMPONENTS:

(1) Lanthanum nitrate; $La(NO_3)_3$; [10099-59-9]

(2) 2-Methyl-2-propanol (t-butyl alcohol); $C_4H_{10}O$; [75-65-0]

(3) Water; H_2O; [7732-18-5]

ORIGINAL MEASUREMENTS:

Stewart, D.F.; Wendlandt, W.W.
J. Phys. Chem. 1959, *63*, 1330-1.

VARIABLES:

One temperature: 25.0°C

PREPARED BY:

Mark Salomon

EXPERIMENTAL VALUES:

Original data:

$La(NO_3)_3 \cdot 6H_2O$/mass %
21.23
21.63

Compiler's conversions for anhydrous salt:

average solubility of $La(NO_3)_3$ = 16.08 mass % (0.590 mol kg^{-1})

COMMENTS AND/OR ADDITIONAL DATA:

The solid phase was not analysed and the nature of the solvate formed in the equilibrated solution is unknown. The water content of the equilibrated solution saturated with $La(NO_3)_3$ was calculated by the compiler as 5.35 mass % H_2O.

Based on the duplicate analysis, an average precision of ± 0.2 mass % is indicated. However, due to the unknown purity of the solvent and uncertainties associated with the experimental technique (e.g., temperature at which centrifuging was carried out), the compiler estimates an accuracy of no better than ± 1 % for the solubility.

AUXILIARY INFORMATION

METHOD/APPARATUS/PROCEDURE:

Isothermal method used. Excess lanthanum nitrate hexahydrate and 10 ml of $(CH_3)_3COH$ were placed in 25 x 160 mm screw-cap vials, sealed with paraffin, and allowed to stand at room temperature (25-35°C) for several hours with periodic shaking. The vial was then placed in a water bath at 25 ± 0.2°C and mechanically rotated end-over-end at 30 r.p.m. Preliminary experiments indicated that equilibrium was established in 4 days, but the solutions were equilibrated for at least 7 days.

The saturated solution was centrifuged and analysed for lanthanum gravimetrically by precipitating as the oxalate and ignition to the oxide.

SOURCE AND PURITY OF MATERIALS:

$La(NO_3)_3 \cdot 6H_2O$ was prepared from 99.99 % La_2O_3 and aqueous nitric acid. The solution was evaporated and the resulting solid redissolved in water and evaporated to crystallization. The crystals were filtered and excess water removed by drying over conc sulfuric acid followed by drying over partially dehydrated $La(NO_3)_3 \cdot 6H_2O$ in a desiccator.
The alcohol was stock material obtained from a commercial supplier.

ESTIMATED ERROR:

See COMMENTS above.

REFERENCES:

COMPONENTS:

(1) Lanthanum nitrate; $La(NO_3)_3$; [10099-59-9]

(2) 1-Pentanol (amyl alcohol); $C_5H_{12}O$; [71-41-0]

(3) Water; H_2O; [7732-18-5]

ORIGINAL MEASUREMENTS:

Stewart, D.F.; Wendlandt, W.W.
J. Phys. Chem. 1959, *63*, 1330-1.

VARIABLES:

One temperature: 25.0°C

PREPARED BY:

Mark Salomon

EXPERIMENTAL VALUES:

Original data:

$La(NO_3)_3 \cdot 6H_2O$/mass %
13.52
13.53

Compiler's conversions for anhydrous salt:

average solubility of $La(NO_3)_3$ = 10.15 mass % (0.348 mol kg^{-1})

COMMENTS AND /OR ADDITIONAL DATA:

The solid phase was not analysed and the nature of the solvate formed in the equilibrated solution is unknown. The water content of the equilibrated solution saturated with $La(NO_3)_3$ was calculated by the compiler as 3.38 mass % H_2O.

Based on the duplicate analysis, an average precision of ± 0.01 mass % is indicated. However, due to the unknown purity of the solvent and uncertainties associated with the experimental technique (e.g., temperature at which centrifuging was carried out), the compiler estimates an accuracy of no better than ± 1 % for the solubility.

AUXILIARY INFORMATION

METHOD/APPARATUS/PROCEDURE:

Isothermal method used. Excess lanthanum nitrate hexahydrate and 10 ml of alcohol $CH_3(CH_2)_3CH_2OH$ were placed in 25 x 160 mm screw-cap vials, sealed with paraffin, and allowed to stand at room temperature (25-35°C) for several hours with periodic shaking. The vial was then placed in a water bath at 25 ± 0.2°C and mechanically rotated end-over-end at 30 r.p.m. Preliminary experiments indicated that equilibrium was established in 4 days, but the solutions were equilibrated for at least 7 days.

The saturated solution was centrifuged and analysed for lanthanum gravimetrically by precipitating as the oxalate and ignition to the oxide.

SOURCE AND PURITY OF MATERIALS:

$La(NO_3)_3 \cdot 6H_2O$ was prepared from 99.99 % La_2O_3 and aqueous nitric acid. The solution was evaporated and the resulting solid redissolved in water and evaporated to crystallization. The crystals were filtered and excess water removed by drying over conc sulfuric acid followed by drying over partially dehydrated $La(NO_3)_3 \cdot 6H_2O$ in a desiccator.

The alcohol was stock material obtained from a commercial supplier.

ESTIMATED ERROR:

See COMMENTS above.

REFERENCES:

COMPONENTS:

(1) Lanthanum nitrate; $La(NO_3)_3$; [10099-59-9]

(2) 3-Pentanol; $C_5H_{12}O$; [584-02-1]

(3) Water; H_2O; [7732-18-5]

ORIGINAL MEASUREMENTS:

Stewart, D.F.; Wendlandt, W.W.
J. Phys. Chem. 1959, *63*, 1330-1.

VARIABLES:

One temperature: 25.0°C

PREPARED BY:

Mark Salomon

EXPERIMENTAL VALUES:

Original data:

$La(NO_3)_3 \cdot 6H_2O$/mass %
5.68
5.78

Compiler's conversions for anhydrous salt:

average solubility of $La(NO_3)_3$ = 4.30 mass % (0.138 mol kg^{-1})

COMMENTS AND/OR ADDITIONAL DATA:

The solid phase was not analysed and the nature of the solvate formed in the equilibrated solution is unknown. The water content of the equilibrated solution saturated with $La(NO_3)_3$ was calculated by the compiler as 1.43 mass % H_2O.

Based on the duplicate analysis, an average precision of ± 0.05 mass % is indicated. However, due to the unknown purity of the solvent and uncertainties associated with the experimental technique (e.g., temperature at which centrifuging was carried out), the compiler estimates an accuracy of no better than ± 1 % for the solubility.

AUXILIARY INFORMATION

METHOD/APPARATUS/PROCEDURE:

Isothermal method used. Excess lanthanum nitrate hexahydrate and 10 ml of $(C_2H_5)_2CHOH$ were placed in 25 x 160 mm screw-cap vials, sealed with paraffin, and allowed to stand at room temperature (25-35°C) for several hours with periodic shaking. The vial was then placed in a water bath at 25 ± 0.2°C and mechanically rotated end-over-end at 30 r.p.m. Preliminary experiments indicated that equilibrium was established in 4 days, but the solutions were equilibrated for at least 7 days.

The saturated solution was centrifuged and analysed for lanthanum gravimetrically by precipitating as the oxalate and ignition to the oxide.

SOURCE AND PURITY OF MATERIALS:

$La(NO_3)_3 \cdot 6H_2O$ was prepared from 99.99 % La_2O_3 and aqueous nitric acid. The solution was evaporated and the resulting solid redissolved in water and evaporated to crystallization. The crystals were filtered and excess water removed by drying over conc sulfuric acid followed by drying over partially dehydrated $La(NO_3)_3 \cdot 6H_2O$ in a desiccator.

The alcohol was stock material obtained from a commercial supplier.

ESTIMATED ERROR:

See COMMENTS above.

REFERENCES:

COMPONENTS:

(1) Lanthanum nitrate; $La(NO_3)_3$; [10099-59-9]

(2) 3-Methyl-1-butanol (isoamyl alchol); $C_5H_{12}O$; [123-51-3]

(3) Water; H_2O; [7732-18-5]

ORIGINAL MEASUREMENTS:

Stewart, D.F.; Wendlandt, W.W.
J. Phys. Chem. 1959, *63*, 1330-1.

VARIABLES:

One temperature: 25.0°C

PREPARED BY:

Mark Salomon

EXPERIMENTAL VALUES:

Original data:

$La(NO_3)_3 \cdot 6H_2O$/mass %
11.62
11.89

Compiler's conversions for anhydrous salt:

average solubility of $La(NO_3)_3$ = 8.821 mass % (0.298 mol kg^{-1})

COMMENTS AND/OR ADDITIONAL DATA:

The solid phase was not analysed and the nature of the solvate formed in the equilibrated solution is unknown. The water content of the equilibrated solution saturated with $La(NO_3)_3$ was calculated by the compiler as 2.93 mass % H_2O.

Based on the duplicate analysis, an average precision of ± 0.14 mass % is indicated. However, due to the unknown purity of the solvent and uncertainties associated with the experimental technique (e.g., temperature at which centrifuging was carried out), the compiler estimates an accuracy of no better than ± 1 % for the solubility.

AUXILIARY INFORMATION

METHOD/APPARATUS/PROCEDURE:

Isothermal method used. Excess lanthanum nitrate hexahydrate and 10 ml of $(CH_3)_2CHCH_2CH_2OH$ were placed in 25 x 160 screw-cap vials, sealed with paraffin, and allowed to stand at room temperature (25-35°C) for several hours with periodic shaking. The vial was then placed in a water bath at 25 ± 0.2°C and mechanically rotated end-over-end at 30 r.p.m. Preliminary experiments indicated that equilibrium was established in 4 days, but the solutions were equilibrated for a least 7 days.

The saturated solution was centrifuged and analysed for lanthanum gravimetrically by precipitating as the oxalate and ignition to the oxide.

SOURCE AND PURITY OF MATERIALS:

$La(NO_3)_3 \cdot 6H_2O$ was prepared from 99.99 % La_2O_3 and aqueous nitric acid. The solution was evaporated and the resulting solid redissolved in water and evaporated to crystallization. The crystals were filtered and excess water removed by drying over conc sulfuric acid followed by drying over partially dehydrated $La(NO_3)_3 \cdot 6H_2O$ in a desiccator.

The alcohol was stock material obtained from a commercial supplier.

ESTIMATED ERROR:

See COMMENTS above.

REFERENCES:

COMPONENTS:

(1) Lanthanum nitrate; $La(NO_3)_3$; [10099-59-9]

(2) 2-Methyl-2-butanol (t-amyl alcohol); $C_5H_{12}O$; [75-85-4]

(3) Water; H_2O; [7732-18-5]

ORIGINAL MEASUREMENTS:

Stewart, D.F.; Wendlandt, W.W.
J. Phys. Chem. 1959, *63*, 1330-1.

VARIABLES:

One temperature: 25.0°C

PREPARED BY:

Mark Salomon

EXPERIMENTAL VALUES:

Original data:

$La(NO_3)_3 \cdot 6H_2O$/mass %
8.98
9.22

Compiler's conversions for anhydrous salt:

average solubility of $La(NO_3)_3$ = 6.83 mass % (0.23 mol kg^{-1})

COMMENTS AND/OR ADDITIONAL DATA:

The solid phase was not analysed and the nature of the solvate formed in the equilibrated solution is unknown. The water content of the equilibrated solution saturated with $La(NO_3)_3$ was calculated by the compiler as 2.27 mass % H_2O.

Based on the duplicate analysis, an average precision of ± 0.12 % is indicated. However, due to the unknown purity of the solvent and uncertainties associated with the experimental technique (e.g., temperature at which centrifuging was carried out), the compiler estimates an accuracy of no better than ± 1 % for the solubility.

AUXILIARY INFORMATION

METHOD/APPARATUS/PROCEDURE:

Isothermal method used. Excess lanthanum nitrate hexahydrate and 10 ml of $C_2H_5C(CH_3)_2OH$ were placed in 25 x 160 mm screw-cap vials, sealed with paraffin, and allowed to stand at room temperature (25-35°C) for several hours with periodic shaking. The vial was then placed in a water bath at 25 ± 0.2°C and mechanically rotated end-over-end at 30 r.p.m. Preliminary experiments indicated that equilibrium was established in 4 days, but the solutions were equilibrated for at least 7 days.

The saturated solution was centrifuged and analysed for lanthanum gravimetrically by precipitating as the oxalate and ignition to the oxide.

SOURCE AND PURITY OF MATERIALS:

$La(NO_3)_3 \cdot 6H_2O$ was prepared from 99.99 % La_2O_3 and aqueous nitric acid. The solution was evaporated and the resulting solid redissolved in water and evaporated to crystallization. The crystals were filtered and excess water removed by drying over conc sulfuric acid followed by drying over partially dehydrated $La(NO_3)_3 \cdot 6H_2O$ in a desiccator.
The alcohol was stock material obtained from a commercial supplier.

ESTIMATED ERROR:

See COMMENTS above.

REFERENCES:

COMPONENTS:

(1) Lanthanum nitrate; $La(NO_3)_3$; [10099-59-9]

(2) 1-Hexanol; $C_6H_{14}O$; [111-27-3]

(3) Water ; H_2O ; [7732-18-5]

ORIGINAL MEASUREMENTS:

Templeton, C.C.; Daly, L.K. *J. Am. Chem. Soc.* 1951, *73*, 3989-91.

VARIABLES:

Solvent composition at 25°C

PREPARED BY:

T. Mioduski, S. Siekierski and M. Salomon

EXPERIMENTAL VALUES:

The $La(NO_3)_3$ - $CH_3(CH_2)_5OH$ - H_2O system at 25.00°C

	solubility of $La(NO_3)_3$[a]	
aqueous phase[b,c,d]	mass %	mol kg^{-1}
	60.6	4.73
	60.1	4.64
alcohol phase[c,d]		
4.80 mass % water	7.85	0.262
4.78 mass % water	7.95	0.266

a. Molalities calculated by the compilers.
b. Alcohol content of aqueous phase not specified (analysis not carried out).
c. Solid phase said to be the hexahydrate $La(NO_3)_3 \cdot 6H_2O$, but no analyses were carried out (see comments below).
d. For the binary 1-hexanol - H_2O system at 20°C, the solubility of alcohol in water is 7.4 mass %, and the solubility of water in the alcohol is 0.71 mass % (3).

AUXILIARY INFORMATION

METHOD/APPARATUS/PROCEDURE:

The isothermal method was used. All samples were brought to equilibrium at 25 ± 0.05°C. Lanthanum was determined by ignition to the oxide. In the alcohol phase water was determined by Karl Fischer titration and the alcohol detmd by difference. An attempt to determine the water content of the aqueous phase by Karl Fischer titration failed due to precipitation (compilers presume the ppt is lanthanum hydroxide).

SOURCE AND PURITY OF MATERIALS:

C.p. grade $La(NO_3)_3 \cdot 6H_2O$ (Eimer and Amend) was used as received. Practical grade n-hexanol (Eastman Kodak) was fractionated twice, the middle fraction being retained after each distillation.

No other information given.

COMMENTS AND/OR ADDITIONAL DATA:

The authors state that in all cases the liquid-liquid region ends in an invariant liquid-liquid-solid state in which the solid is the hexahydrate, and cite Seidell's third edition (1) in support of this conclusion. However the compilers are not aware of any studies on the solubility of lanthanum nitrate in n-hexanol or aqueous n-hexanol prior to the present work. Templeton (2) did publish a paper on the distribution of $La(NO_3)_3 \cdot 6H_2O$ between water and n-hexanol (unsaturated solutions).

ESTIMATED ERROR:

Soly: precision about ± 1 % (compilers).

Temp: control of thermostat was ± 0.05 K.

REFERENCES:

1. Seidell, A. *Solubilities of Inorganic and Metal Organic Compounds*. D. Van Nostrand. New York. 3rd edition. 1940.
2. Templeton, C.C. *J. Am. Chem. Soc.* 1949, *71*, 2187.
3. Addison, C. *J. Chem. Soc.* 1945, 98.

COMPONENTS:	ORIGINAL MEASUREMENTS:
(1) Lanthanum nitrate; $La(NO_3)_3$; [10099-59-9] (2) 1-Hexanol (n-hexyl alcohol); $C_6H_{14}O$; [111-27-3] (3) Water; H_2O; [7732-18-5]	Stewart, D.F.; Wendlandt, W.W. *J. Phys. Chem.* 1959, *63*, 1330-1.
VARIABLES:	PREPARED BY:
One temperature: 25.0°C	Mark Salomon

EXPERIMENTAL VALUES:

Original data:	$La(NO_3)_3 \cdot 6H_2O$/mass %	ΔH_{sln}/kcal mol^{-1}
	16.01	10.4
	15.88	9.24

Compiler's conversions for anhydrous salt:

average solubility of $La(NO_3)_3$ = 11.96 mass % (0.418 mol kg^{-1})

COMMENTS AND/OR ADDITIONAL DATA:

The solid phase was not analysed and the nature of the solvate formed in the equilibrated solution is unknown. The water content of the equilibrated solution saturated with $La(NO_3)_3$ was calculated by the compiler as 3.98 mass % H_2O. Note that the solubility of water in 1-hexanol at 20°C is reported to be 0.71 mass % (2). The authors do not report the appearance of an aqueous phase.

Based on the duplicate analysis, an average precision of ± 0.07 mass % is indicated. However, due to the unknown purity of the solvent and uncertainties associated with the experimental technique (e.g., temperature at which centrifuging was carried out), the compiler estimates an accuracy of no better than ± 1 % for the solubility.

AUXILIARY INFORMATION

METHOD/APPARATUS/PROCEDURE:

Isothermal method used. Excess lanthanum nitrate hexahydrate and 10 ml of $CH_3(CH_2)_4CH_2OH$ were placed in a 25 x 160 mm screw-cap vial, sealed with paraffin, and allowed to stand at room temperature (25-35°C) for several hours with periodic shaking. The vial was then placed in a water bath at 25 ± 0.2°C and mechanically rotated end-over-end at 30 r.p.m. Preliminary experiments indicated that equilibrium was established in 4 days, but the solutions were equilibrated for at least 7 days.

The saturated solution was centrifuged and analysed for lanthanum gravimetrically by precipitating as the oxalate and ignition to the oxide.

Heats of solution were measured calorimetrically as described in (1). The solvent used for this determination was dried and fractionated.

SOURCE AND PURITY OF MATERIALS:

$La(NO_3)_3 \cdot 6H_2O$ was prepared from 99.99 % La_2O_3 and aqueous nitric acid. The solution was evaporated and the resulting solid redissolved in water and evaporated to crystallization. The crystals were filtered and excess water removed by drying over conc sulfuric acid followed by drying over partially dehydrated $La(NO_3)_3 \cdot 6H_2O$ in a desiccator.

The alcohol was stock material obtained from a commercial supplier.

ESTIMATED ERROR:

See COMMENTS above.

REFERENCES:

1. Van Tassel, J.H.; Wendlandt, W.W. *J. Am. Chem. Soc.* 1959, *81*, 813.

2. Addison, C.C. *J. Chem. Soc.* 1945, 98.

COMPONENTS:

(1) Lanthanum nitrate; $La(NO_3)_3$; [10099-59-9]

(2) Cyclohexanol; $C_6H_{12}O$; [108-93-0]

(3) Water; H_2O; [7732-18-5]

ORIGINAL MEASUREMENTS:

Stewart, D.F.; Wendlandt, W.W.
J. Phys. Chem. 1959, *63*, 1330-1.

VARIABLES:

One temperature: 25.0°C

PREPARED BY:

Mark Salomon

EXPERIMENTAL VALUES:

Original data:

$La(NO_3)_3 \cdot 6H_2O$/mass %
17.00
16.97

Compiler's conversions for anhydrous salt:

average solubility of $La(NO_3)_3$ = 12.75 mass % (0.450 mol kg^{-1})

COMMENTS AND/OR ADDITIONAL DATA:

The solid phase was not analysed and the nature of the solvate formed in the equilibrated solution is unknown. The water content of the equilibrated solution saturated with $La(NO_3)_3$ was calculated by the compiler as 4.24 mass % H_2O.

Based on the duplicate analysis, an average precision of ± 0.02 mass % is indicated. However, due to the unknown purity of the solvent and uncertainties associated with the experimental technique (e.g., temperature at which centrifuging was carried out), the compiler estimates an accuracy of no better than ± 1 % for the solubility.

AUXILIARY INFORMATION

METHOD/APPARATUS/PROCEDURE:

Isothermal method used. Excess lanthanum nitrate hexahydrate and 10 ml of $C_6H_{11}OH$ were placed in 25 x 160 mm screw-cap vials, sealed with paraffin, and allowed to stand at room temperature (25-35°C) for several hours with periodic shaking. The vial was then placed in a water bath at 25 ± 0.2°C and mechanically rotated end-over-end at 30 r.p.m. Preliminary experiments indicated that equilibrium was established in 4 days, but the solutions were equilibrated for at least 7 days.

The saturated solution was centrifuged and analysed for lanthanum gravimetrically by precipitating as the oxalate and ignition to the oxide.

SOURCE AND PURITY OF MATERIALS:

$La(NO_3)_3 \cdot 6H_2O$ was prepared from 99.99 % La_2O_3 and aqueous nitric acid. The solution was evaporated and the resulting solid redissolved in water and evaporated to crystallization. The crystals were filtered and excess water removed by drying over conc sulfuric acid followed by drying over partially dehydrated $La(NO_3)_3 \cdot 6H_2O$ in a desiccator.
The alcohol was stock material obtained from a commercial supplier.

ESTIMATED ERROR:

See COMMENTS above.

REFERENCES:

COMPONENTS:

(1) Lanthanum nitrate; $La(NO_3)_3$; [10099-59-9]

(2) Benzyl alcohol; C_7H_8O; [100-51-6]

(3) Water; H_2O; [7732-18-5]

ORIGINAL MEASUREMENTS:

Stewart, D.F.; Wendlandt, W.W.
J. Phys. Chem. 1959, *63*, 1330-1.

VARIABLES:

One temperature: 25.0°C

PREPARED BY:

Mark Salomon

EXPERIMENTAL VALUES:

Original data:

$La(NO_3)_3 \cdot 6H_2O$/mass %
10.96
11.01

Compiler's conversions for anhydrous salt:

average solubility of $La(NO_3)_3$ = 8.243 mass % (0.276 mol kg^{-1})

COMMENTS AND/OR ADDITIONAL DATA:

The solid phase was not analysed and the nature of the solvate formed in the equilibrated solution is unknown. The water content of the equilibrated solution saturated with $La(NO_3)_3$ was calculated by the compiler as 2.74 mass % H_2O.

Based on the duplicate analysis, an average precision of ± 0.02 mass % is indicated. However, due to the unknown purity of the solvent and uncertainties associated with the experimental technique (e.g., temperature at which centrifuging was carried out), the compiler estimates an accuracy of no better than ± 1 % for the solubility.

AUXILIARY INFORMATION

METHOD/APPARATUS/PROCEDURE:

Isothermal method used. Excess lanthanum nitrate hexahydrate and 10 ml of $C_6H_5CH_2OH$ were placed in 25 x 160 mm screw-cap vials, sealed with paraffin, and allowed to stand at room temperature (25-35°C) for several hours with periodic shaking. The vial was then placed in a water bath at 25 ± 0.2°C and mechanically rotated end-over-end at 30 r.p.m. Preliminary experiments indicated that equilibrium was established in 4 days, but the solutions were equilibrated for at least 7 days.

The saturated solution was centrifuged and analysed for lanthanum gravimetrically by precipitating as the oxalate and ignition to the oxide.

SOURCE AND PURITY OF MATERIALS:

$La(NO_3)_3 \cdot 6H_2O$ was prepared from 99.99 % La_2O_3 and aqueous nitric acid. The solution was evaporated and the resulting solid redissolved in water and evaporated to crystallization. The crystals were filtered and excess water removed by drying over conc sulfuric acid followed by drying over partially dehydrated $La(NO_3)_3 \cdot 6H_2O$ in a desiccator.
The alcohol was stock material obtained from a commercial supplier.

ESTIMATED ERROR:

See COMMENTS above.

REFERENCES:

COMPONENTS:

(1) Lanthanum nitrate; $La(NO_3)_3$; [10099-59-9]

(2) Diethyl ether; $C_4H_{10}O$; [60-29-7]

ORIGINAL MEASUREMENTS:

Wells, R.C. *J. Wash. Acad. Sci.* 1930, *20*, 146-8.

VARIABLES:

Room temperature (about 20°C)

PREPARED BY:

T. Mioduski, S. Siekierski, M. Salomon

EXPERIMENTAL VALUES:

Experiment 1. This experiment involves the hydrated lanthanum nitrate as the initial solid, and which the compilers assume to be the hexahydrate.

Authors report the solubility as 0.0002 g La_2O_3 in 10 ml ether.

This is equivalent to a $La(NO_3)_3$ soly of 1.2×10^{-4} mol dm^{-3} (compilers).

Experiment 2. This experiment involves lanthanum nitrate dehydrated as described in the METHOD/APPARATUS/PROCEDURE box below.

Authors report the solubility as 0.0001 g La_2O_3 in 10 ml ether.

This is equivalent to a $La(NO_3)_3$ soly of 6.1×10^{-5} mol dm^{-3} (compilers).

AUXILIARY INFORMATION

METHOD/APPARATUS/PROCEDURE:

The isothermal method was used. The soly of lanthanum nitrate was determined in two experiments in which the nature of the initial solid phase differs.

Experiment 1. A few grams of lanthanum nitrate (presumably the hexahydrate, compilers) was added to about 20 ml of ether in small stoppered flasks. The flasks were periodically agitated and permitted to stand at about 20°C overnight. A 10 ml sample was removed, filtered, the solvent evaporated and the salt ignited to the oxide and weighed.

Experiment 2. The remaining salt in the flask was freed from ether, dissovled in water and a few drops of HNO_3 added. The solution was evaporated to dryness and heated to 150°C. The solubility in ether was then determined again with this "dehydrated" salt.

SOURCE AND PURITY OF MATERIALS:

Nothing specified.

ESTIMATED ERROR:

Soly: precision probably around ± 10 % (compilers).

Temp: precision probably ± 4 K (compilers).

REFERENCES:

COMPONENTS:

(1) Lanthanum nitrate; $La(NO_3)_3$; [10099-59-9]

(2) Diethyl ether; $C_4H_{10}O$; [60-29-7]

ORIGINAL MEASUREMENTS:

Hopkins, B.S.; Quill, L.L. *Proc. Natl. Acad. Sci. U.S.A.* 1933, *19*, 64-8.

VARIABLES:

One temperature: 25°C

PREPARED BY:

T. Mioduski, S. Siekierski, M. Salomon

EXPERIMENTAL VALUES:

The solubility of $La(NO_3)_3$ in diethyl ether at 25°C was given in the form of a small diagram of solubility vs atomic number Z for Z = 57-64. In the absence of numerical data, the compilers interpolated the solubility from the published diagram. The result is:

$$\text{soly of } La(NO_3)_3 = 0.6 \text{ g dm}^{-3} \quad (0.0018 \text{ mol dm}^{-3}).$$

COMMENTS AND/OR ADDITIONAL DATA:

The name Philip Kalischer appears on the diagram published in the source paper. The compilers suspected that Mr. Kalischer was an MSc student of Prof. Hopkins and thus contacted Ms. Susanne Redalje, the Assistant Chemistry Librarian at the University of Illinois at Urbana-Champaign. Ms. Redalje searched the University records for references to a thesis or any publication by Mr. Kalischer. The records show that Mr. Kalischer attended classes for the Fall, Spring, and Summer semesters of 1930-1931. There is no indication that Mr. Kalischer had finished his studies or submitted a thesis, and it is therefore apparent that the original experimental data are lost. The compilers are most grateful to Ms. Redalje for all her help in searching the University records and providing important information on numerous other lanthanide systems.

AUXILIARY INFORMATION

METHOD/APPARATUS/PROCEDURE:

No information is available, but based on similar work by Hardy (1) being carried out at the University of Illinois at the time, it is likely that the isothermal method was employed. The solubility data for neodymium and praseodymium nitrates in several ethers from Hardy's MSc Thesis are compiled elsewhere in this volume, and the compilations contain detailed information on the experimental techniques which the compilers assume were similar to those used by Mr. Kalischer.

SOURCE AND PURITY OF MATERIALS:

No information available.

ESTIMATED ERROR:

No information available.

REFERENCES:

1. Hardy, Z.M. *Masters Thesis.* The University of Illinois. Urbana, Il. 1932.

COMPONENTS:

(1) Lanthanum nitrate; $La(NO_3)_3$; [10099-59-9]

(2) 2-Hydroxy ethyl ether (2,2'-oxydiethanol, diethylene glycol); $C_4H_{10}O_3$; [111-46-6]

(3) Water; H_2O; [7732-18-5]

ORIGINAL MEASUREMENTS:

Stewart, D.F.; Wendlandt, W.W.
J. Phys. Chem. 1959, *63*, 1330-1.

VARIABLES:

One temperature: 25.0°C

PREPARED BY:

Mark Salomon

EXPERIMENTAL VALUES:

Original data:

$La(NO_3)_3 \cdot 6H_2O$/mass %
75.09
74.69

Compiler's conversions for anhydrous salt:

average solubility of $La(NO_3)_3$ = 56.20 mass % (3.948 mol kg^{-1})

COMMENTS AND/OR ADDITIONAL DATA:

The solid phase was not analysed and the nature of the solvate formed in the equilibrated solution is unknown. The water content of the equilibrated solution saturated with $La(NO_3)_3$ was calculated by the compiler as 18.69 mass % H_2O.

Based on the duplicate analysis, an average precision of ± 0.2 mass % is indicated. However, due to the unknown purity of the solvent and uncertainties associated with the experimental technique (e.g., temperature at which centrifuging was carried out), the compiler estimates an accuracy of no better than ± 1 % for the solubility.

AUXILIARY INFORMATION

METHOD/APPARATUS/PROCEDURE:

Isothermal method used. Excess lanthanum nitrate hexahydrate and 10 ml of solvent $(HOCH_2CH_2)_2O$ were placed in 25 x 160 mm screw-cap vials, sealed with paraffin, and allowed to stand at room temperature (25-35°C) for several hours with periodic shaking. The vial was then placed in a water bath at 25 ± 0.2°C and mechanically rotated end-over-end at 30 r.p.m. Preliminary experiments indicated that equilibrium was established in 4 days, but the solutions were equilibrated for at least 7 days.

The saturated solution was centrifuged and analysed for lanthanum gravimetrically by precipitating as the oxalate and ignition to the oxide.

SOURCE AND PURITY OF MATERIALS:

$La(NO_3)_3 \cdot 6H_2O$ was prepared from 99.99 % La_2O_3 and aqueous nitric acid. The solution was evaporated and the resulting solid redissolved in water and evaporated to crystallization. The crystals were filtered and excess water removed by drying over conc sulfuric acid followed by drying over partially dehydrated $La(NO_3)_3 \cdot 6H_2O$ in a desiccator.

The alcohol was stock material obtained from a commercial supplier.

ESTIMATED ERROR:

See COMMENTS above.

REFERENCES:

COMPONENTS:

(1) Lanthanum nitrate; $La(NO_3)_3$; [10099-59-9]

(2) 1,4-Dioxane (p-dioxane); $C_4H_8O_2$; [123-91-1]

ORIGINAL MEASUREMENTS:

Hopkins, B.S.; Quill, L.L. *Proc. Natl. Acad. Sci. U.S.A.* 1933, *19*, 64-8.

VARIABLES:

One temperature: 25°C

PREPARED BY:

T. Mioduski, S. Siekierski, M. Salomon

EXPERIMENTAL VALUES:

The solubility of $La(NO_3)_3$ in p-dioxane at 25°C was given in the form of a small diagram of solubility vs atomic number Z for Z = 57-64. In the absence of numerical data, the compilers interpolated the solubility from the published diagram. The result is:

$$\text{soly of } La(NO_3)_3 = 3.8 \text{ g dm}^{-3} \quad (0.012 \text{ mol dm}^{-3})$$

COMMENTS AND/OR ADDITIONAL DATA:

It appears that the original experimental work was done by a Mr. P. Kalischer who was a student at the University of Illinois at Urbana-Champaign. Attempts to locate the original experimental data have failed, and it thus appears that these data are lost (see the COMMENTS in the compilation for the $La(NO_3)_3$ - diethyl ether system).

AUXILIARY INFORMATION

METHOD/APPARATUS/PROCEDURE:

No information is available, but based on similar work by Hardy (1) being carried out at the University of Illinois at the time, it is likely that the isothermal method was employed. The solubility data for neodymium and praseodymium nitrates in several ethers from Hardy's MSc Thesis are compiled elsewhere in this volume, and the compilations contain detailed information on the experimental techniquies which the compilers assume were similar to those used by Mr. Kalischer.

SOURCE AND PURITY OF MATERIALS:

No information available.

ESTIMATED ERROR:

No information available.

REFERENCES:

1. Hardy, Z.M. *Masters Thesis.* The University of Illinois. Urbana, IL. 1932.

COMPONENTS:

(1) Lanthanum nitrate; $La(NO_3)_3$; [10099-59-9]

(2) 1,4-Dioxane (p-dioxane); $C_4H_8O_2$; [123-91-1]

(3) Water; H_2O; [7732-18-5]

ORIGINAL MEASUREMENTS:

Stewart, D.F.; Wendlandt, W.W.
J. Phys. Chem. 1959, *63*, 1330-1.

VARIABLES:

One temperature: 25.0°C

PREPARED BY:

Mark Salomon

EXPERIMENTAL VALUES:

Original data:	$La(NO_3)_3\cdot6H_2O$/mass %	ΔH_{sln}/kcal mol^{-1}
	73.97	1.66
	74.13	1.56

Compiler's conversions for anhydrous salt:

average solubility of $La(NO_3)_3$ = 55.57 mass % (3.849 mol kg^{-1})

COMMENTS AND/OR ADDITIONAL DATA:

The solid phase was not analysed and the nature of the solvate formed in the equilibrated solution is unknown. The water content of the equilibrated solution saturated with $La(NO_3)_3$ was calculated by the compiler as 18.48 mass % H_2O.

Based on the duplicate analysis, an average precision of ± 0.08 mass % is indicated. However, due to the unknown purity of the solvent and uncertainties associated with the experimental technique (e.g., temperature at which centrifuging was carried out), the compiler estimates an accuracy of no better than ± 1 % for the solubility.

AUXILIARY INFORMATION

METHOD/APPARATUS/PROCEDURE:

Isothermal method used. Excess lanthanum nitrate hexahydrate and 10 ml of p-dioxane were placed in 25 x 160 mm screw-cap vials, sealed with pariffin, and allowed to stand at room temperature (25-35°C) for several hours with periodic shaking. The vial was then placed in a water bath at 25 ± 0.2°C and mechanically rotated end-over-end at 30 r.p.m. Preliminary experiments indicated that equilibrium was established in 4 days, but the solutions were equilibrated for at least 7 days.

The saturated solution was centrifuged and analysed for lanthanum gravimetrically by precipitating as the oxalate and ignition to the oxide.

Heats of solution were measured calorimetrically as described in (1). The solvent used for this determination was dried and fractionated.

SOURCE AND PURITY OF MATERIALS:

$La(NO_3)_3\cdot6H_2O$ was prepared from 99.99 % La_2O_3 and aqueous nitric acid. The solution was evaporated and the resulting solid redissolved in water and evaporated to crystallization. The crystals were filtered and excess water removed by drying over conc sulfuric acid followed by drying over partially dehydrated $La(NO_3)_3\cdot6H_2O$ in a desiccator.

Dioxane was stock material obtained from a commercial supplier.

ESTIMATED ERROR:

See COMMENTS above.

REFERENCES:

1. Van Tassel, J.H.; Wendlandt, W.W. *J. Am. Chem. Soc.* 1959, *81*, 813.

COMPONENTS:

(1) Lanthanum nitrate; $La(NO_3)_3$; [10099-59-9]

(2) Acetone (2-propanone); C_3H_6O; [67-64-1]

(3) Water; H_2O; [7732-18-5]

ORIGINAL MEASUREMENTS:

Stewart, D.F.; Wendlandt, W.W.
J. Phys. Chem. 1959, *63*, 1330-1.

VARIABLES:

One temperature: 25.0°C

PREPARED BY:

Mark Salomon

EXPERIMENTAL VALUES:

Original data:

$La(NO_3)_3 \cdot 6H_2O$/mass %	ΔH_{sln}/kcal mol^{-1}
76.64	5.00
76.43	4.89

Compiler's conversions for anhydrous salt:

average solubility of $La(NO_3)_3$ = 57.43 mass % (4.152 mol kg^{-1})

COMMENTS AND/OR ADDITIONAL DATA:

The solid phase was not analysed and the nature of the solvate found in the equilibrated solution is unknown. The water content of the equilibrated solution solution with $La(NO_3)_3$ was calculated by the compiler as 19.11 mass % H_2O.

Based on the duplicate analysis, an average precision of ± 0.11 mass % is indicated. However, due to the unknown purity of the solvent and uncertainties associated with the experimental technique (e.g., temperature at which centrifuging was carried out), the compiler estimates an accuracy of no better than ± 1 % for the solubility.

AUXILIARY INFORMATION

METHOD/APPARATUS/PROCEDURE:

Isothermal method used. Excess lanthanum nitrate hexahydrate and 10 ml of CH_3COCH_3 were placed in a 25 x 160 mm screw-cap vial, sealed with paraffin, and allowed to stand at room temperature (25-35°C) for several hours with periodic shaking. The vial was then placed in a water bath at 25 ± 0.2°C and mechanically rotated end-over-end at 30 r.p.m. Preliminary experiments indicated that equilibrium was established in 4 days, but the solutions were equilibrated for at least 7 days.

The saturated solution was centrifuged and analysed for lanthanum gravimetrically by precipitating as the oxalate and ignition to the oxide.

Heats of solution were measured calorimetrically as described in (1). The solvent used for this determination was dried and fractionated.

SOURCE AND PURITY OF MATERIALS:

$La(NO_3)_3 \cdot 6H_2O$ was prepared from 99.99 % La_2O_3 and aqueous nitric acid. The solution was evaporated and the resulting solid redissolved in water and evaporated to crystallization. The crystals were filtered and excess water removed by drying over conc sulfuric acid followed by drying over partially dehydrated $La(NO_3)_3 \cdot 6H_2O$ in a desiccator.

Acetone was stock material obtained from a commercial supplier.

ESTIMATED ERROR:

See COMMENTS above.

REFERENCES:

1. Van Tassel, J.H.; Wendlandt, W.W. *J. Am. Chem. Soc.* 1959, *81*, 813.

COMPONENTS:

(1) Lanthanum nitrate; $La(NO_3)_3$; [10099-59-9]
(2) Cyclohexanone; $C_6H_{10}O$; [108-94-1]
(3) Water; H_2O; [7732-18-5]

ORIGINAL MEASUREMENTS:

Stewart, D.F.; Wendlandt, W.W.
J. Phys. Chem. 1959, *63*, 1330-1.

VARIABLES:

One temperature: 25.0°C

PREPARED BY:

Mark Salomon

EXPERIMENTAL VALUES:

Original data:

$La(NO_3)_3 \cdot 6H_2O$/mass %	ΔH_{sln}/kcal mol^{-1}
42.84	12.4
43.24	16.4

Compiler's conversions for anhydrous salt:

average solubility of $La(NO_3)_3$ = 32.30 mass % (1.468 mol kg^{-1})

COMMENTS AND/OR ADDITIONAL DATA:

The solid phase was not analysed and the nature of the solvate formed in the equilibrated solution is unknown. The water content of the equilibrated solution saturated with $La(NO_3)_3$ was calculated by the compiler as 10.74 mass % H_2O. Note that the solubility of H_2O in cyclohexanone at 25°C is reported to be 8.0 mass % (2). The authors do not report the appearance of an aqueous phase.

Based on the duplicate analysis, an average precision of ± 0.2 mass % is indicated. However, due to the unknown purity of the solvent and uncertainties associated with the experimental technique (e.g., temperature at which centrifuging was carried out), the compiler estimates an accuracy of no better than ± 1 % for the solubility.

AUXILIARY INFORMATION

METHOD/APPARATUS/PROCEDURE:

Isothermal method used. Excess lanthanum nitrate hexahydrate and 10 ml of $C_6H_{10}(=O)$ were placed in 25 x 160 mm screw-cap vials, sealed with paraffin, and allowed to stand at room temperature (25-35°C) for several hours with periodic shaking. The vial was then placed in a water bath at 25 ± 0.2°C and mechanically rotated end-over-end at 30 r.p.m. Preliminary experiments indicated that equilibrium was established in 4 days, but the solutions were equilibrated for at least 7 days.

The saturated solution was centrifuged and analysed for lanthanum gravimetrically by precipitating as the oxalate and ignition to the oxide.

Heats of solution were measured calorimetrically as described in (1). The solvent used for this determination was dried and fractionated.

SOURCE AND PURITY OF MATERIALS:

$La(NO_3)_3 \cdot 6H_2O$ was prepared from 99.99 % La_2O_3 and aqueous nitric acid. The solution was evaporated and the resulting solid redissolved in water and evaporated to crystallization. The crystals were filtered and excess water removed by drying over conc sulfuric acid followed by drying over partially dehydrated $La(NO_3)_3 \cdot 6H_2O$ in a desiccator.

Cyclohexanone was stock material obtained from a commercial supplier.

ESTIMATED ERROR:

See COMMENTS above.

REFERENCES:

1. Van Tassel, J.H.; Wendlandt, W.W. *J. Am. Chem. Soc.* 1959, *81*, 813.
2. Doolittle, A.K. *The Technology of Solvents and Plasticizers*. Wiley. N.Y. 1954.

COMPONENTS:	ORIGINAL MEASUREMENTS:
(1) Lanthanum nitrate; $La(NO_3)_3$; [10099-59-9] (2) Ethyl formate; $C_3H_6O_2$; [109-94-4] (3) Water; H_2O; [7732-18-5]	Stewart, D.F.; Wendlandt, W.W. *J. Phys. Chem.* 1959, *63*, 1330-1.
VARIABLES:	PREPARED BY:
One temperature: 25.0°C	Mark Salomon

EXPERIMENTAL VALUES:

Original data:	$La(NO_3)_3 \cdot 6H_2O$/mass %	ΔH_{sln}/kcal mol^{-1}
	50.04	12.9
	50.01	12.0

Compiler's conversions for anhydrous salt:

average solubility of $La(NO_3)_3$ = 37.54 mass % (1.850 mol kg^{-1})

COMMENTS AND/OR ADDITIONAL DATA:

The solid phase was not analysed and the nature of the solvate formed in the equilibrated solution is unknown. The water content of the equilibrated solution saturated with $La(NO_3)_3$ was calculated by the compiler as 12.49 mass % H_2O.

Based on the duplicate analysis, an average precision of ± 0.02 mass % is indicated. However, due to the unknown purity of the solvent and uncertainties associated with the experimental technique (e.g., temperature at which centrifuging was carried out), the compiler estimates an accuracy of no better than ± 1 % for the solubility.

AUXILIARY INFORMATION

METHOD/APPARATUS/PROCEDURE:

Isothermal method used. Excess lanthanum nitrate hexahydrate and 10 ml of $HCOOC_2H_5$ were placed in 25 x 160 mm screw-cap vials, sealed with paraffin, and allowed to stand at room temperature (25-35°C) for several hours with periodic shaking. The vial was then placed in a water bath at 25 ± 0.2°C and mechanically rotated end-over-end at 30 r.p.m. Preliminary experiments indicated that equilibrium was established in 4 days, but the solutions were equilibrated for at least 7 days.

The saturated solution was centrifuged and analysed for lanthanum gravimetrically by precipitating as the oxalate and ignition to the oxide.

Heats of solution were measured calorimetrically as described in (1). The solvent used for this determination was dried and fractionated.

SOURCE AND PURITY OF MATERIALS:

$La(NO_3)_3 \cdot 6H_2O$ was prepared from 99.99 % La_2O_3 and aqueous nitric acid. The solution was evaporated and the resulting solid redissolved in water and evaporated to crystallization. The crystals were filtered and excess water removed by drying over conc sulfuric acid followed by drying over partially dehydrated $La(NO_3)_3 \cdot 6H_2O$ in a desiccator.

Ethyl formate was stock material obtained from a commercial supplier.

ESTIMATED ERROR:

See COMMENTS above.

REFERENCES:

1. Van Tassel, J.H.; Wendlandt, W.W. *J. Am. Chem. Soc.* 1959, *81*, 813.

COMPONENTS:

(1) Lanthanum nitrate; $La(NO_3)_3$; [10099-59-9]

(2) Methyl acetate; $C_3H_6O_2$; [79-20-9]

(3) Water; H_2O; [7732-18-5]

ORIGINAL MEASUREMENTS:

Stewart, D.F.; Wendlandt, W.W.
J. Phys. Chem. 1959, *63*, 1330-1.

VARIABLES:

One temperature: 25.0°C

PREPARED BY:

Mark Salomon

EXPERIMENTAL VALUES:

Original data:

$La(NO_3)_3 \cdot 6H_2O$/mass %	ΔH_{sln}/kcal mol^{-1}
62.88	6.03
62.85	5.93

Compiler's conversions for anhydrous salt:

average solubility of $La(NO_3)_3$ = 47.17 mass % (2.748 mol kg^{-1})

COMMENTS AND/OR ADDITIONAL DATA:

The solid phase was not analysed and the nature of the solvate formed in the equilibrated solution is unknown. The water content of the equilibrated solution saturated with $La(NO_3)_3$ was calculated by the compiler as 15.69 mass % H_2O. Note that the solubility of water in methyl acetate at 20°C is reported to be 8 mass % (2). The authors do not report the appearance of an aqueous phase.

Based on the duplicate analysis, an average precision of ± 0.02 mass % is indicated. However, due to the unknown purity of the solvent and uncertainties associated with the experimental technique (e.g., temperature at which centrifuging was carried out), the compiler estimates an accuracy of no better than ± 1 % for the solubility.

AUXILIARY INFORMATION

METHOD/APPARATUS/PROCEDURE:

Isothermal method used. Excess lanthanum nitrate hexahydrate and 10 ml of CH_3COOCH_3 were placed in 25 x 160 mm screw-cap vials, sealed with paraffin, and allowed to stand at room temperature (25-35°C) for several hours with periodic shaking. The vial was then placed in a water bath at 25 ± 0.2°C and mechanically rotated end-over-end at 30 r.p.m. Preliminary experiments indicated that equilibrium was established in 4 days, but the solutions were equilibrated for at least 7 days.

The saturated solution was centrifuged and analysed for lanthanum gravimetrically by precipitating as the oxalate and ignition to the oxide.

Heats of solution were measured calorimetrically as described in (1). The solvent used for this determination was dried and fractionated.

SOURCE AND PURITY OF MATERIALS:

$La(NO_3)_3 \cdot 6H_2O$ was prepared from 99.99 % La_2O_3 and aqueous nitric acid. The solution was evaporated and the resulting solid redissolved in water and evaporated to crystallization. The crystals were filtered and excess water removed by drying over conc sulfuric acid followed by drying over partially dehydrated $La(NO_3)_3 \cdot 6H_2O$ in a desiccator.

Methyl acetate was stock material obtained from a commercial supplier.

ESTIMATED ERROR:

See COMMENTS above.

REFERENCES:

1. Van Tassel, J.H.; Wendlandt, W.W. *J. Am. Chem. Soc.* 1959, *81*, 813.

2. Führer, H. *Ber. Deut. Chem. Gesell.* 1924, *57*, 510.

COMPONENTS:

(1) Lanthanum nitrate; $La(NO_3)_3$; [10099-59-9]

(2) Ethyl acetate; $C_4H_8O_2$; [141-78-6]

(3) Water; H_2O; [7732-18-5]

ORIGINAL MEASUREMENTS:

Stewart, D.F.; Wendlandt, W.W.
J. Phys. Chem. 1959, *63*, 1330-1.

VARIABLES:

One temperature: 25.0°C

PREPARED BY:

Mark Salomon

EXPERIMENTAL VALUES:

Original data:

$La(NO_3)_3 \cdot 6H_2O$/mass %
1.73
1.89

Compiler's conversions for anhydrous salt:

average solubility of $La(NO_3)_3$ = 1.36 mass % (0.042 mol kg^{-1})

COMMENTS AND/OR ADDITIONAL DATA:

The solid phase was not analysed and the nature of the solvate formed in the equilibrated solution is unknown. The water content of the equilibrated solution saturated with $La(NO_3)_3$ was calculated by the compiler as 0.45 mass % H_2O.

Based on the duplicate analysis, an average precision of ± 0.09 mass % is indicated. However, due to the unknown purity of the solvent and uncertainties associated with the experimental technique (e.g., temperature at which centrifuging was carried out), the compiler estimates an accuracy of no better than ± 1 % for the solubility.

AUXILIARY INFORMATION

METHOD/APPARATUS/PROCEDURE:

Isothermal method used. Excess lanthanum nitrate hexahydrate and 10 ml of $CH_3COOC_2H_5$ were placed in 25 x 160 mm screw-cap vials, sealed with paraffin, and allowed to stand at room temperature (25-35°C) for several hours with periodic shaking. The vial was then placed in a water bath at 25 ± 0.2°C and mechanically rotated end-over-end at 30 r.p.m. Preliminary experiments indicated that equilibrium was established in 4 days, but the solutions were equilibrated for at least 7 days.

The saturated solution was centrifuged and analysed for lanthanum gravimetrically by precipitating as the oxalate and ignition of the oxide.

SOURCE AND PURITY OF MATERIALS:

$La(NO_3)_3 \cdot 6H_2O$ was prepared from 99.99 % La_2O_3 and aqueous nitric acid. The solution was evaporated and the resulting solid redissolved in water and evaporated to crystallization. The crystals were filtered and excess water removed by drying over conc sulfuric acid followed by drying over partially dehydrated $La(NO_3)_3 \cdot 6H_2O$ in a desiccator.

Ethyl acetate was stock material obtained from a commercial supplier.

ESTIMATED ERROR:

See COMMENTS above.

REFERENCES:

COMPONENTS:

(1) Lanthanum nitrate; $La(NO_3)_3$; [10099-59-9]

(2) Acetonitrile; C_2H_3N; [75-05-8]

(3) Water; H_2O; [7732-18-5]

ORIGINAL MEASUREMENTS:

Stewart, D.F.; Wendlandt, W.W.
J. Phys. Chem. 1959, *63*, 1330-1.

VARIABLES:

One temperature: 25.0°C

PREPARED BY:

Mark Salomon

EXPERIMENTAL VALUES:

Original data: $La(NO_3)_3 \cdot 6H_2O$/mass %

72.45

72.37

Compiler's conversions for anhydrous salt:

average solubility of $La(NO_3)_3$ = 54.33 mass % (3.662 mol kg^{-1})

COMMENTS AND/OR ADDITIONAL DATA:

The solid phase was not analysed and the nature of the solvate formed in the equilibrated solution is unknown. The water content of the equilibrated solution saturated with $La(NO_3)_3$ was calculated by the compiler as 18.08 mass % H_2O.

Based on the duplicate analysis, an average precision of ± 0.04 mass % is indicated. However, due to the unknown purity of the solvent and uncertainties associated with the experimental technique (e.g., temperature at which centrifuging was carried out), the compiler estimates an accuracy of no better than ± 1% for the solubility.

AUXILIARY INFORMATION

METHOD/APPARATUS/PROCEDURE:

Isothermal method used. Excess lanthanum nitrate hexahydrate and 10 ml of CH_3CN were placed in 25 x 160 mm screw-cap vials, sealed with parafin, and allowed to stand at room temperature (25-35°C) for several hours with periodic shaking. The vial was then placed in a water bath at 25 ± 0.2°C and mechanically rotated end-over-end at 30 r.p.m. Preliminary experiments indicated that equilibrium was established in 4 days, but the solutions were equilibrated for at least 7 days.

The saturated solution was centrifuged and analysed for lanthanum gravimetrically by precipitating as the oxalate and ignition to the oxide.

SOURCE AND PURITY OF MATERIALS:

$La(NO_3)_3 \cdot 6H_2O$ was prepared from 99.99 % La_2O_3 and aqueous nitric acid. The solution was evaporated and the resulting solid redissolved in water and evaporated to crystallization. The crystals were filtered and excess water removed by drying over conc sulfuric acid followed by drying over partially dehydrated $La(NO_3)_3 \cdot 6H_2O$ in a desiccator.

Acetonitrile was stock material obtained from a commercial supplier.

ESTIMATED ERROR:

See COMMENTS above.

REFERENCES:

COMPONENTS:

(1) Lanthanum nitrate; $La(NO_3)_3$; [10099-59-9]

(2) Ethylenediamine; $C_2H_8N_2$; [107-15-3]

(3) Water ; H_2O ; [7732-18-5]

ORIGINAL MEASUREMENTS:

Moeller, T.; Cullen, G.W. *J. Inorg. Nucl. Chem.* 1959, *10*, 148-52.

VARIABLES:

Water content of $H_2N{\cdot}CH_2CH_2{\cdot}NH_2$ at 30°C

PREPARED BY:

Mark Salomon

EXPERIMENTAL VALUES:

H_2O content	solubility of $La(NO_3)_3$	
mass %	g dm^{-3}	10^2mol dm^{-3} [a]
0.00[b]	4.66	1.43
0.191	15.41	4.743
0.287	21.10	6.494
0.383	23.63	7.273
0.479	20.15	6.202
0.574	15.04	4.629
0.670	5.84	1.80

a. Conversions to mol dm^{-3} units made by compiler.

b. The anhydrous nitrate is markedly dependent upon trace quantities of water. The smallest solubility found was for freshly distilled solvent which was 3.81 g dm^{-3} (0.0117 mol dm^{-3}). The largest solubility found was 5.88 g dm^{-3} (0.0181 mol dm^{-3}): presumably this result is for the anhydrous solvent which was stored in the dry box for an unspecified period of time. The authors state that the previous result (3) of 24.5 g kg^{-1} (0.0754 mol kg^{-1}) is probably in error due to the presence of considerable moisture. For this reason the data in reference (3) for the $La(NO_3)_3$-$H_2N{\cdot}CH_2CH_2{\cdot}NH_2$ system was not compiled.

COMMENTS AND/OR ADDITIONAL DATA:

The anhydrous solution is colorless and the heat of solution is small. Traces of water evolve considerable heat upon dissolution, and the solution is pale yellow. The heat evolved upon dissolution is said to be sufficient to promote oxidation of the solvent by the nitrate ion.

AUXILIARY INFORMATION

METHOD/APPARATUS/PROCEDURE:

The isothermal method was used. A dry box with a nitrogen atmosphere was used for manipulation of all anhydrous materials. Mixtures of solvent and solute were sealed in 25 x 200 mm test tubes and thermostated at 30.00 ± 0.025°C for four days. The tubes were then centrifuged for 30 minutes, returned to the thermostat for 30 minutes, and removed to the dry box for analysis. In the dry box the supernatant was separated from the solid by pipetting. Analysis was carried out by addition of excess water to an aliquot, removal of the hydrous hydroxide, and ignition to the oxide which was weighed. Temperature of the dry box was not specified.

SOURCE AND PURITY OF MATERIALS:

Ethylenediamine was purified and dried as in (1). Freshly distilled product had an electrolytic conductivity of 4.6 x 10^{-7} S cm^{-1} at 30°C which increased during storage in the dry box to 1.36 x 10^{-6} S cm^{-1} due to absorption of water. Freshly ignited 99.9+ % lanthanum oxide was converted to the anhydrous nitrate by treatment with nitrogen (IV) oxide (2).

ESTIMATED ERROR:

Soly: accuracy probably around ± 1-2 % (compiler).

Temp: accuracy about ± 0.5 K (compiler).

REFERENCES:

1. Putnam, G.L.; Kobe, K.A. *Trans. Electrochem. Soc.* 1938, *74*, 609.
2. Moeller, T.; Aftandilian, V.D. *J. Am. Chem. Soc.* 1954, *76*, 5249.
3. Moeller, T.; Zimmerman, P.A. *J. Am. Chem. Soc.* 1953, *75*, 3950.

COMPONENTS:

(1) Lanthanum nitrate; $La(NO_3)_3$; [10099-59-9]

(2) Morpholine; C_4H_9NO; [110-91-8]

ORIGINAL MEASUREMENTS:

Moeller, T.; Zimmerman, P.A. *J. Am. Chem. Soc.* 1953, *75*, 3950-3.

VARIABLES:

One temperature: 30°C

PREPARED BY:

T. Mioduski and S. Siekierski

EXPERIMENTAL VALUES:

The solubility of lanthanum nitrate in morpholine at 30°C was reported to be

3.74 g/100 g solvent

The compilers have converted this value to molality:

0.115 mol kg^{-1}

AUXILIARY INFORMATION

METHOD/APPARATUS/PROCEDURE:

The isothermal method was used. Reaction mixtures were sealed in 25 x 200 mm test tubes and thermostated for one week at 30 ± 0.05°C with frequent agitation. The density of the supernatant liquid was determined pycnometrically, and the lanthanum content determined by precipitating the hydrous hydroxide from an aliquot by adding excess water, and igniting to the oxide and weighing.

All anhydrous substances were handled in a dry box through which a current of nitrogen was passed. The nitrogen was freed of CO_2 and moisture by passage through concentrated sulfuric acid, soda lime, and Drierite. All solutions were prepared in the dry box, and were sealed before being removed.

SOURCE AND PURITY OF MATERIALS:

Lanthanum oxide was converted to the nitrate by high temp reaction with NH_4NO_3. Unreacted NH_4NO_3 was removed by heating in N_2 and then in vacuo. The oxide, obtained from University stocks, contained traces of other rare earth metals. Morpholine was purified as in (1), and had a density of 0.9863 g/ml at 27°C and an electrolytic conductivity of 3.368×10^{-8} S cm^{-1}. The solvent was stored under nitrogen in glass stoppered flasks sealed with wax.

ESTIMATED ERROR:

Soly: precision probably around ± 1 % (compilers).

Temp: thermostat control was ± 0.05 K, but uncertainty during handling and analysis is closer to ± 0.5 K (compilers).

REFERENCES:

1. Dermer, V.H.; Dermer, O.C. *J. Am. Chem. Soc.* 1937, *59*, 1148.

COMPONENTS:

(1) Lanthanum nitrate; $La(NO_3)_3$; [10099-59-9]

(2) 2-Methyl aminobenzene (*o*-toluidine); C_7H_9N ; [95-53-4]

(3) Water ; H_2O ; [7732-18-5]

ORIGINAL MEASUREMENTS:

Stewart, D.F.; Wendlandt, W.W.

J. Phys. Chem. 1959, *63*, 1330-1.

VARIABLES:

One temperature: 25.0°C

PREPARED BY:

Mark Salomon

EXPERIMENTAL VALUES:

Original data:

Duplicate analyses gave identical results of 16.20 mass % $La(NO_3)_3 \cdot 6H_2O$

Compiler's conversions for anhydrous salt:

solubility of $La(NO_3)_3$ = 12.16 mass % (0.426 mol kg^{-1})

COMMENTS AND/OR ADDITIONAL DATA:

The solid phase was not analysed, and the type of solvate formed in the equilibrated solution is unkown.

Since the authors do not report the appearance of three phases in the equilibrated solution, it must be assumed that the water of hydration from the hexahydrate has dissolved in the organic solvent. Based on the solubility of 16.20 mass % for the hexahydrate, the solvent should contain 4.04 mass% water. Note that at 20°C the solubility of water in *o*-toluidine is 2.44 mass % (1).

Based on the results of the duplicate analyses, excellent precision is indicated. However due to the unkown purity of the solvent and uncertainties associated with the experimental technique (e.g. temperature at which centrifuging was carried out), the compiler estimates an accuracy of no better than ± 1 % for the solubility.

AUXILIARY INFORMATION

METHOD/APPARATUS/PROCEDURE:

Isothermal method used. Excess lanthanum nitrate hexahydrate and 10 ml of $CH_3C_6H_4NH_2$ were placed in a 25 x 160 mm screw-cap vial, sealed with paraffin, and allowed to stand at room temperature (25-35°C) for several hours with periodic shaking. The vial was then placed in a water bath at 25 ± 0.2°C and mechanically rotated end-over-end at 30 r.p.m. Preliminary experiments indicated that equilibrium was established in 4 days, but the solution was equilibrated for at least 7 days.

The saturated solution was centrifuged and analysed for lanthanum gravimetrically by precipitation as the oxalate and ignition to the oxide.

SOURCE AND PURITY OF MATERIALS:

$La(NO_3)_3 \cdot 6H_2O$ was prepared from 99.99 % La_2O_3 and aqueous nitric acid. The solution was evaporated and the remainging solid redissolved in water and evaporated to crystallization. The crystals were filtered and excess water removed by drying over conc sulfuric acid followed by drying over partially dehydrated $La(NO_3)_3 \cdot 6H_2O$ in a desiccator.

o-Toluidine was stock material obtained from a commercial supplier.

ESTIMATED ERROR:

See COMMENTS above.

REFERENCES:

1. Seidell, A. *Solubilities of Organic Compounds*. Vol. II. D. Van Nostrand. New York. 1941.

COMPONENTS:

(1) Lanthanum nitrate; $La(NO_3)_3$; [10099-59-9]

(2) Aniline (aminobenzene, phenylamine); C_6H_7N; [62-53-3]

(3) Water; H_2O; [7732-18-5]

ORIGINAL MEASUREMENTS:

Stewart, D.F.; Wendlandt, W.W.

J. Phys. Chem. 1959, *63*, 1330-1.

VARIABLES:

One temperature: 25.0°C

PREPARED BY:

Mark Salomon

EXPERIMENTAL VALUES:

Original data:

$La(NO_3)_3\cdot6H_2O$/mass %
15.90
16.05

Compiler's conversions for anhydrous salt:

average solubility of $La(NO_3)_3$ = 11.99 mass % (0.419 mol kg^{-1})

COMMENTS AND/OR ADDITIONAL DATA:

The solid phase was not analysed and the nature of the solvate formed in the equilibrated solution is unknown. The water content of the equilibrated solution saturated with $La(NO_3)_3$ was calculated by the compiler as 3.99 mass % H_2O.

Based on the duplicate analysis, an average precision of the ± 0.08 mass % is indicated. However, due to the unknown purity of the solvent and uncertainties associated with the experimental technique (e.g., temperature at which centrifuging was carried out), the compiler estimates an accuracy of no better than ± 1 % for the solubility.

AUXILIARY INFORMATION

METHOD/APPARATUS/PROCEDURE:

Isothermal method used. Excess lanthanum nitrate hexahydrate and 10 ml of $C_6H_5NH_2$ were placed in 25 x 160 mm screw-cap vial, sealed with paraffin, and allowed to stand at room temperature (25-35°C) for several hours with periodic shaking. The vial was then placed in a water bath at 25 ± 0.2°C and mechanically rotated end-over-end at 30 r.p.m. Preliminary experiments indicated that equilibrium was established in 4 days, but the solutions were equilibrated for at least 7 days.

The saturated solution was centrifuged and analysed for lanthanum gravimetrically by precipitating as the oxalate and ignition to the oxide.

SOURCE AND PURITY OF MATERIALS:

$La(NO_3)_3\cdot6H_2O$ was prepared from 99.99 % La_2O_3 and aqueous nitric acid. The solution was evaporated and the resulting solid redissolved in water and evaporated to crystallization. The crystals were filtered and excess water removed by drying over conc sulfuric acid followed by drying over partially dehydrated $La(NO_3)_3\cdot6H_2O$ in a desiccator.

Aniline was stock material obtained from a commercial supplier.

ESTIMATED ERROR:

See COMMENTS above.

REFERENCES:

COMPONENTS:

(1) Lanthanum nitrate; $La(NO_3)_3$; [10099-59-9]

(2) Tri-n-butylphosphate; $C_{12}H_{27}O_4P$; [126-73-8]

ORIGINAL MEASUREMENTS:

Wendlandt, W.W.; Bryant, J.M.
J. Phys. Chem. 1956, *60*, 1145-6.

VARIABLES:

Room temperature: 25-27°C

PREPARED BY:

Mark Salomon

EXPERIMENTAL VALUES:

Two analyses of a single equilibrated saturated solution was reported. The compiler assumes that the temperature was 26 ± 1°C.

Solubility of $La(NO_3)_3$ = 28.5 mass %

Solubility of $La(NO_3)_3$ = 28.3 mass %

The compiler has calculated the following.

Average solubility and average error = 28.4 ± 0.1 mass %

Converting to molality,

average solubility = 1.22 ± 0.01 mol kg^{-1}

COMMENTS AND/OR ADDITIONAL DATA:

An average precision of around ± 0.5 % is indicated for the chemical analyses. However the overall experimental precision is probably no better than 1-2 % due to uncertainties in the experimental temperature, the completeness of separation of the organic and aqueous phases in the equilibrated system, and the completeness of extraction of the salt from the organic phase by water.

The authors do not specify the water content of the equilibrated tri-butylphosphate phase.

AUXILIARY INFORMATION

METHOD/APPARATUS/PROCEDURE:

Isothermal method used. About 25 g of salt and 20 ml solvent were placed in a screw-cap bottle and mechanically agitated at room temperature (25-27°C) for 48-72 hours. At this time three phases were present in the bottle: the solid hydrated salt, an aqueous phase, and an organic phase. The organic phase was separated, centrifuged, and analysed for lanthanum as follows.

Duplicate 1-4 g samples of the organic phase were placed in separatory funnels containing 25 ml benzene and 50 ml water. After two minutes the aqueous phase was removed and 50 ml water added and the equilibration repeated. Two such extractions were sufficient to remove the salt from the organic phase. The lanthanum content in the aqueous phase was analysed by standard procedures (1).

SOURCE AND PURITY OF MATERIALS:

Reagent grade $La(NO_3)_3 \cdot 6H_2O$ was used as received.

Eastman Kodak tri-n-butylphosphate was used as received.

ESTIMATED ERROR:

See above.

REFERENCES:

1. Scott, W.W. *Standard Methods of Chemical Analysis*. Furnam, N.H., ed. D. Van Nostrand, Inc. New York. 5th ed, 1946.

COMPONENTS:	EVALUATOR:
(1) $La(NO_3)_3$ double salts (2) Water ; H_2O ; [7732-18-5]	Mark Salomon U.S. Army Electronics Technology and Devices Laboratory Fort Monmouth, NJ, USA November 1982

CRITICAL EVALUATION:

$La(NO_3)_3$ DOUBLE SALTS WITH INORGANIC NITRATES

The existence of a number of double nitrates has been established in multicomponent phase equilibria studies as discussed previously in the critical evaluation for multi-component systems with $La(NO_3)_3$. There are only three publications (1-3) which report the solubility of double nitrates in pure water, or in aqueous HNO_3 solutions (2). Detailed critical evaluation of these works is not possible at this time, but some discussion of the data is required.

Double salts involving alkali metal nitrates have been discussed previously, but it is interesting to note that in the $La(NO_3)_3$-$RbNO_3$-H_2O system (4), the stable solid phase for the double nitrate was reported to be the trihydrate, but according to Wyrouboff (5) the stable solid phase appears to be the tetrahydrate.

$2RbNO_3.La(NO_3)_3.4H_2O$ [65907-05-3] : $2RbNO_3.La(NO_3)_3.3H_2O$ [61391-42-2]

The tetrahydrate melts congruently at 359 K (5), and while thermal analyses data are given for the trihydrate in (4), it is not clearly stated if the trihydrate melts congruently or what the temperature of melting is.

Lanthanum ammonium nitrate was studied at 288 K by Holmberg (1), but the absense of descriptions on experimental methods makes the estimation of precision difficult to determine. The author did not specify the nature of the solid phase, but it is probably the tetrahydrate as reported in an earlier compilation of the paper by Urazov and Shevtsova (6).

Double salts of alkaline earth nitrates and transition metal nitrates tend to form stable tetracosahydrates as noted in the $La(NO_3)_3$-H_2O critical evaluation and subsequent compilations. The only system studied as a function of temperature is that for $3Mg(NO_3)_2.2La(NO_3)_3$ reported by Friend and Wheat (3), and the accuracy of these results is very difficult to estimate because other studies by Friend on binary $Ln(NO_3)_3$-H_2O systems (Ln = La, Pr, Nd) all contain a large negative systematic error. The data in (3) can be fitted successfully to the solubility equation (see eq. [1] in the $La(NO_3)_3$-H_2O critical evaluation) with the following result:

$$\ln(m/m_o) - 24M_2(m - m_o) = -41.09 + 1512/(T/K) + 6.304\ln(T/K) \quad [1]$$

For a reference molality m_o = 0.883 mol kg^{-1} at 320.0 K, the standard error of estimate for the molality is σ_m = 0.014 mol kg^{-1} which leads to an uncertainty of ± 0.02 mol kg^{-1} (i.e. ± 2%) at the 95% level of confidence (Student's t = 3.182). However the congruent melting point of the tetracosahydrate calculated from eq. [1] is 381.3 K compared to the experimental value of 386.7 K reported by Jantsch (2). Using data of precision of ± 1% or better, we have found that predicted congurent melting points are generally within ± 2 K of the observed values. On this basis the data of Friend and Wheat were assigned an overall precision of ± 3%.

$La(NO_3)_3$ DOUBLE SALTS WITH ORGANIC NITRATES

The only quantitative data available are from the papers by Gorshunova and Zhuravlev (7,8) which have been compiled previously. For the double salt with piperidine nitrate, $La(NO_3)_3.4(C_5H_{10}NH.HNO_3)$, the solubilities at 298.2 K and 323.2 K are, respectively, 4.25 and 8.15 mol kg^{-1} (7). The double salt does not form a hydrate. For the double salt with quinoline nitrate, $La(NO_3)_3.C_9H_7N.HNO_3$, the solubilities at 298.2 K and 323.2 K are, respectively, 1.62 and 3.12 mol kg^{-1} (8). The stable solid phase is the tetrahydrate.

REFERENCES

1. Holmberg, O. *Z. Anorg. Chem.* 1907, *53*, 83.
2. Jantsch, G. *Z. Anorg. Chem.* 1912, *76*, 303.
3. Friend, J.N.; Wheat, W.N. *J. Chem. Soc.* 1935, 356.
4. Molodkin, A.K.; Odinets, Z.K. Vargas Ponse, O. *Zh. Neorg. Khim.* 1976, *21*, 2590.
5. Wyrouboff, G. *Bull. Soc. Min.* 1901, *24*, 112.
6. Urazov, G.G.; Shevtsova, Z.N. *Zh. Neorg. Khim.* 1957, *2*, 655.
7. Gorshunova, V.P.; Zhuravlev, E.F. *Zh. Neorg. Khim.* 1972, *17*, 231.
8. Gorshunova, V.P.; Zhuravlev, E.F. *Zh. Neorg. Khim.* 1973, *18*, 1688.

COMPONENTS:

(1) Magnesium lanthanum nitrate; $3Mg(NO_3)_2 \cdot 2La(NO_3)_3$; [13826-42-1]

(2) Nitric acid; HNO_3; [7697-37-2]

(3) Water ; H_2O ; [7732-18-5]

ORIGINAL MEASUREMENTS:

Jantsch, C. *Z. Anorg. Chem.* 1912, *76*, 303-23.

VARIABLES:

One temperature: 16°C

PREPARED BY:

Mark Salomon

EXPERIMENTAL VALUES:

Soly of the double salt in HNO_3 sln of density d_4^{16} = 1.325 g cm^{-3}.

aliquot volume cm^{-3}	La_2O_3 g	Soly $3Mg(NO_3)_2 \cdot 2La(NO_3)_3$ [a] mol dm^{-3}
1.2324	0.0168	
1.2324	0.0168	0.0418

a. Average value calculated by the author.

ADDITIONAL DATA:

The melting point of the tetracosahydrate is 113.5°C, and the density at 0°C is 1.988 g cm^{-3}.

AUXILIARY INFORMATION

METHOD/APPARATUS/PROCEDURE:

Isothermal method used. The soly was studied in HNO_3 sln of density 1.325 g cm^{-3} at 16°C because the author did not have sufficient quantity of the rare earth to study the soly of the salt in pure water. Pulverized salt and HNO_3 sln were placed in glass-stoppered tubes and thermostated at 16°C for 24 h with periodic shaking. The solution was then allowed to settle for 2 h, and a pipet maintained at 16°C was used to withdraw aliquots for analysis. Two analyses were performed.

Solutions were analysed by adding 2-3 g NH_4Cl and 10% NH_3 sln followed by boiling to ppt the hydroxide. The ppt was filtered, dissolved in HNO_3, reprecipitated as the hydroxide, and ignited to the oxide. Mg in the filtrate was "determined by the usual method" (no details were given).

An attempt to determine the waters of hydration by dehydration was not successful because the temperature required (120°C or higher) resulted in decomposition of the salt with the formation of basic salts. Presumably the waters of hydration were found by difference.

SOURCE AND PURITY OF MATERIALS:

"Pure" lanthanum oxide was dissolved in dil HNO_3 and $Mg(NO_3)_2$ added to give a mole ratio of La/Mg = 2/3. The sln was evapd and a small crystal of $Bi_2Mg_3(NO_3)_{12}$ added, and the mixt cooled to ppt the tetracosahydrate. The double nitrate was recrystd before use.

ESTIMATED ERROR:

REFERENCES:

Soly: reproducibility about ± 1-5% (compilers).

Temp: nothing specified.

COMPONENTS:

(1) Magnesium lanthanum nitrate; $3Mg(NO_3)_2 \cdot 2La(NO_3)_3$; [13826-42-1]

(2) Water; H_2O; [7732-18-5]

ORIGINAL MEASUREMENTS:

Friend, J.N.; Wheat, W.N.
J. Chem. Soc. 1935, 356-9.

VARIABLES:

Temperature

PREPARED BY:

Mark Salomon

EXPERIMENTAL VALUES:

t/°C	mass ratio [a] La_2O_3/MgO	$3Mg(NO_3)_2 \cdot 2La(NO_3)_3 \cdot 24H_2O$ [b] mass %	$3Mg(NO_3)_2 \cdot 2La(NO_3)_3$ [c] mass %	mol kg^{-1}
18.6	2.68	62.19	44.58	0.735
31.6	2.61	63.96	45.85	0.773
46.8	2.68	68.57	49.16	0.883
50.8	2.60	70.17	50.30	0.925
61.4	2.63	73.05	52.37	1.004
74.8	2.59	77.43	55.51	1.140
113.5 [d]	——	100	71.69	2.313

a. Results of gravimetric analyses of the oxides. The theoretical ratio is 2.695 (compiler).

b. Authors' conversions. Since the individual mass % of each oxide was not given (only the mass ratios were given), the compiler cannot check these calculations using IUPAC recommended atomic weights.

c. Compiler's calculations.

d. Melting point of the tetracosahydrate (2): i.e., the system is infinitely miscible.

AUXILIARY INFORMATION

METHOD/APPARATUS/PROCEDURE:

The isothermal method was used as described in (1). In order to obtain reproducible results, equilibrium had to be approached from above. Saturated or near saturated solutions were prepared at 90-100°C and quickly cooled in a thermostat. The solutions were stirred for several hours which the authors claim is essential to remove supersaturation.

La determined gravimetrically by precipitation of the oxalate from 50 cc of saturated solution followed by ignition to the oxide. Total La + Mg determined by evaporating 50 cc of solution to dryness followed by ignition to the oxides. Mg found by difference.

SOURCE AND PURITY OF MATERIALS:

La_2O_3 was "free from other rare earths" with the exception of a trace of Pr_6O_{11} which gave slightly brown solutions. The double salt was prepared by dissolving stoichiometric amounts of La_2O_3 and MgO in dilute HNO_3 and crystallizing. The product was recrystallized from dilute HNO_3.

ESTIMATED ERROR:

Soly: precision ± 3 % (compiler).

Temp: accuracy probably ± 0.05 K as in (1) (compiler).

REFERENCES:

1. Friend, J.N. *J. Chem. Soc.* 1930, 1633.

2. Jantsch, G. *Z. Anorg. Chem.* 1912, *76*, 303.

COMPONENTS:

(1) Lanthanum manganese nitrate; $2La(NO_3)_3 \cdot 3Mn(NO_3)_2$; [53368-21-1]

(2) Nitric acid; HNO_3; [7697-37-2]

(3) Water ; H_2O ; [7732-18-5]

ORIGINAL MEASUREMENTS:

Jantsch, G. *Z. Anorg. Chem.* 1912, *76*, 303-23.

VARIABLES:

One temperature: 16°C

PREPARED BY:

Mark Salomon

EXPERIMENTAL VALUES:

Soly of the double salt in HNO_3 sln of density d_4^{16} = 1.325 g cm^{-3}

aliquot volume	$La_2O_3 + Mn_3O_4$	Soly $2La(NO_3)_3 \cdot 3Mn(NO_3)_2$	
cm^3	g	mol dm^{-3}	
1.2324	0.0822		
1.2324	0.0826	0.1192[a]	0.1206[b]

a. Author's calculation (average value).

b. Compiler's calculation (average value) based on 1977 IUPAC recommended atomic masses.

ADDITIONAL DATA:

The melting point of the tetracosahydrate is 87.2°C, and the density at 0°C is 2.080 g cm^{-3}.

AUXILIARY INFORMATION

METHOD/APPARATUS/PROCEDURE:

Isothermal method used. The soly was studied in HNO_3 sln of density 1.325 g cm^{-3} at 16°C because the author did not have sufficient quantity of the rare earth to study the soly of the salt in pure water. Pulverized salt and HNO_3 sln were placed in glass-stoppered tubes and thermostated at 16°C for 24 h with periodic shaking. The solution was then allowed to settle for 2 h, and a pipet maintained at 16°C was used to withdraw aliquots for analysis. Two analyses were performed.

Solutions were analysed by precipitating both La and Mn hydroxides by respective addition of NH_3 and H_2O_2. The ppt was ignited to give $La_2O_3 + Mn_3O_4$.

An attempt to determine the waters of hydration by dehydration was not successful because the temperature required (120°C or higher) resulted in decomposition of the salt with the formation of basic salts. Presumably the waters of hydration were found by difference.

SOURCE AND PURITY OF MATERIALS:

"Pure" lanthanum oxide was dissolved in dil HNO_3 and $Mn(NO_3)_2$ added to give a mole ratio of La/Mn = 2/3. The sln was evapd and a small crystal of $Bi_2Mg_3(NO_3)_{12}$ added, and the mixt cooled to ppt the tetracosahydrate. The double nitrate was recrystd before use.

ESTIMATED ERROR:

Soly: reproducibility about ± 1-5% (compiler).

Temp: nothing specified.

REFERENCES:

COMPONENTS:

(1) Lanthanum cobalt nitrate; $2La(NO_3)_3 \cdot 3Co(NO_3)_2$; [22465-27-6]

(2) Nitric acid; HNO_3; [7697-37-2]

(2) Water ; H_2O ; [7732-18-5]

ORIGINAL MEASUREMENTS:

Jantsch, G. *Z. Anorg. Chem.* 1912, *76*, 303-23.

VARIABLES:

One temperature: 16°C

PREPARED BY:

Mark Salomon

EXPERIMENTAL VALUES: Soly of the double salt in HNO_3 sln of density d_4^{16} = 1.325 g cm^{-3}.

aliquot volume	La_2O_3	soly $2La(NO_3)_3 \cdot 3Co(NO_3)_2$ [a]
cm^3	g	mol dm^{-3}
1.2324	0.0266	
1.2324	0.0272	0.669

a. Author's calculation of average solubility.

ADDITIONAL DATA:

The melting point of the tetracosahydrate is 101.8°C, and the density at 0°C is 2.131 g cm^{-3}.

AUXILIARY INFORMATION

METHOD/APPARATUS/PROCEDURE:

Isothermal method used. The soly was studied in HNO_3 sln of density 1.325 g cm^{-3} at 16°C because the author did not have sufficient quantity of the rare earth to study the soly of the salt in pure water. Pulverized salt and HNO_3 sln were placed in glass-stoppered tubes and thermostated at 16°C for 24 h with periodic shaking. The solution was then allowed to settle for 2 h, and a pipet maintained at 16°C was used to withdraw aliquots for analysis. Two analyses were performed.

Solutions were analyzed by adding 2-3 g NH_4Cl and 10% NH_3 sln followed by boiling to ppt the hydroxide. The ppt was filtered, dissolved in HNO_3, reprecipitated as the hydroxide, and ignited to the oxide. Co in the filtrate was "determined by the usual method" (no details were given).

An attempt to determine the waters of hydration by dehydration was not successful because the temperature required (120°C or higher) resulted in decomposition of the salt with the formation of basic salts. Presumably the waters of hydration were found by difference.

SOURCE AND PURITY OF MATERIALS:

"Pure" lanthanum oxide was dissolved in dil HNO_3 and $Co(NO_3)_2$ added to give a mole ratio of La/Co = 2/3. The sln was evapd and a small crystal of $Bi_2Mg_3(NO_3)_{12}$ added, and the mixt cooled to ppt the tetracosahydrate. The double nitrate was recrystd before use.

ESTIMATED ERROR:

Soly: reproducibility about ± 1-5% (compiler).

Temp: nothing specified.

REFERENCES:

COMPONENTS:

(1) Lanthanum nickel nitrate; $2La(NO_3)_3 \cdot 3Ni(NO_3)_2$; [25822-29-1]

(2) Nitric acid; HNO_3; [7697-37-2]

(3) Water ; H_2O ; [7732-18-5]

ORIGINAL MEASUREMENTS:

Jantsch, G. *Z. Anorg. Chem.* 1912, *76*, 303-23.

VARIABLES:

One temperature: 16°C

PREPARED BY:

Mark Salomon

EXPERIMENTAL VALUES:

Soly of the double salt in HNO_3 sln of density d_4^{16} = 1.325 g cm^{-3}.

aliquot volume	La_2O_3	soly $2La(NO_3)_2 \cdot 3Ni(NO_3)_2$ [a]
cm^3	g	mol dm^{-3}
1.2324	0.0197	
1.2324	0.0199	0.0492

a. Author's calculation (average value).

ADDITIONAL DATA:

The melting point of the tetracosahydrate is 110.5°C, and the density at 0°C is 2.146 g cm^{-3}.

AUXILIARY INFORMATION

METHOD/APPARATUS/PROCEDURE:

Isothermal method used. The soly was studied in HNO_3 sln of density 1.325 g cm^{-3} at 16°C because the author did not have sufficient quantity of the rare earth to study the soly of the salt in pure water. Pulverized salt and HNO_3 sln were placed in glass-stoppered tubes and thermostated at 16°C for 24 h with periodic shaking. The solution was then allowed to settle for 2 h, and a pipet maintained at 16°C was used to withdraw aliquots for analysis. Two analyses were performed.

Solutions were analysed by adding 2-3 g NH_4Cl and 10% NH_3 sln followed by boiling to ppt the hydroxide. The ppt was filtered, dissolved in HNO_3, reprecipitated as the hydroxide, and ignited to the oxide. Ni in the filtrate was "determined by the usual method" (no details were given).

An attempt to determine the waters of hydration by dehydration was not successful because the temperature required (120°C or higher) resulted in decomposition of the salt with the formation of basic salts. Presumably the waters of hydration were found by difference.

SOURCE AND PURITY OF MATERIALS:

"Pure" lanthanum oxide was dissolved in dil HNO_3 and $Ni(NO_3)_2$ added to give a mole ratio of La/Ni = 2/3. The sln was evapd and a small crystal of $Bi_2Mg_3(NO_3)_{12}$ added, and the mixt cooled to ppt the tetracosahydrate. The double nitrate was recrystd before use. The double salt was analysed gravimetrically for La_2O_3 and metallic Ni. 0.5117 g samples of the tetracosahydrate yielded 0.1031 g La_2O_3 (20.15 mass %), and 0.0544 g Ni (10.63 mass %). Theor values are 19.98 mass % La_2O_3 and 10.80 mass % Ni (compiler). NO_3 analysis by pptn with nitron gave 45.65 mass % compared to the theor value of 45.64 mass % NO_3 (compiler).

ESTIMATED ERROR:

Soly: reproducibility about ± 1-5% (compiler).

Temp: nothing specified.

COMPONENTS:

(1) Lanthanum zinc nitrate; $2La(NO_3)_3 \cdot 3Zn(NO_3)_2$; [32074-09-2]

(2) Nitric acid; HNO_3; [7697-37-2]

(3) Water ; H_2O ; [7732-18-5]

ORIGINAL MEASUREMENTS:

Jantsch, G. *Z. Anorg. Chem.* 1912, *76*, 303-23.

VARIABLES:

One temperature: 16°C

PREPARED BY:

Mark Salomon

EXPERIMENTAL VALUES:

Soly of the double salt in HNO_3 sln of density d_4^{16} = 1.325 g cm^{-3}

aliquot volume	La_2O_3	soly $2La(NO_3)_3 \cdot 3Zn(NO_3)_2$ [a]
cm^3	g	mol dm^{-3}
1.4638	0.0357	
1.4638	0.0360	0.0751

a. Author's calculation (average value).

ADDITIONAL DATA:

The melting point of the tetracosahydrate is 98.0°C, and the density at 0°C is 2.161 g cm^{-3}.

AUXILIARY INFORMATION

METHOD/APPARATUS/PROCEDURE:

Isothermal method used. The soly was studied in HNO_3 sln of density 1.325 g cm^{-3} at 16°C because the author did not have sufficient quantity of the rare earth to study the soly of the salt in pure water. Pulverized salt and HNO_3 sln were placed in glass-stoppered tubes and thermostated at 16°C for 24 h with periodic shaking. The solution was then allowed to settle for 2 h, and a pipet maintained at 16°C was used to withdraw aliquots for analysis. Two analyses were performed.

Solutions were analysed by adding 2-3 g NH_4Cl and 10% NH_3 sln followed by boiling to ppt the hydroxide. The ppt was filtered, dissolved in HNO_3, reprecipitated as the hydroxide, and ignited to the oxide. Zn in the filtrate was "determined by the usual method" (no details were given).

An attempt to determine the waters of hydration by dehydration was not successful because the temperature required (120°C or higher) resulted in decomposition of the salt with the formation of basic salts. Presumably the waters of hydration were found by difference.

SOURCE AND PURITY OF MATERIALS:

"Pure" lanthanum oxide was dissolved in dil HNO_3 and $Zn(NO_3)_2$ added to give a mole ratio of La/Zn = 2/3. The sln was evapd and a small crystal of $Bi_2Mg_3(NO_3)_{12}$ added, and the mixt cooled to ppt the tetracosahydrate. The double nitrate was recrystd before use.

ESTIMATED ERROR:

Soly: reproducibility about ± 1-5% (compiler).

Temp: nothing specified.

REFERENCES:

COMPONENTS:

(1) Lanthanum ammonium nitrate; $La(NO_3)_3 \cdot 2NH_4NO_3$; [13566-21-7]

(2) Water ; H_2O ; [7732-18-5]

ORIGINAL MEASUREMENTS:

Holmberg, O. *Z. Anorg. Chem.* 1907, *53*, 83-134.

VARIABLES:

One temperature: 15°C

PREPARED BY:

Mark Salomon

EXPERIMENTAL VALUES:

Solubility of $La(NO_3)_3 \cdot 2NH_4NO_3$ at 15°C[a,b]

	g/100 g H_2O	mol kg^{-1}
	182	3.75
	180	3.71
	181	3.73
	183	3.77
	181	3.73
mean values	181.4	3.74

a. Molalities calculated by the compiler.
b. Nature of the solid phase not specified.

AUXILIARY INFORMATION

METHOD/APPARATUS/PROCEDURE:

No information given.

SOURCE AND PURITY OF MATERIALS:

Lanthanum "material" (presumably the oxide, compiler) was obtained from the author's colleague Prof. P.T. Cleve. The "material" was ignited and the atomic mass of La determined by comparing the mass of oxide to the mass of sulfate. An atomic mass % of 138.9 g /g atom was found. No other information given.

ESTIMATED ERROR:

Nothing specified.

REFERENCES:

COMPONENTS:	EVALUATOR:
(1) Cerium nitrate; $Ce(NO_3)_3$; [10108-73-3] (2) Water ; H_2O ; [7732-18-5]	S. Siekierski, T. Mioduski Institute for Nuclear Research Warsaw, Poland and M. Salomon U.S. Army ET & DL Ft. Monmouth, NJ, USA December 1982

CRITICAL EVALUATION:

THE BINARY SYSTEM

INTRODUCTION

Data for the solubility of $Ce(NO_3)_3$ in water have been reported in 26 publications (1-26). Many of these studies deal with multicomponent systems, and the solubility in the binary system is given as one point on a phase diagram. It appears that for many of the studies involving ternary systems, the solubility in the binary system was measured only once by a given research group, and that the same value was reported in subsequent publications (7,8,13,14,16-23) and (24,25). Thus in critically evaluating the solubility data for the binary system, we have considered the data on the basis of 13 independent studies. In all of the solubility studies given in the compilations, the only hydrates identified as equilibrium solid phases are the hexa-, penta-, and tetrahydrates which exist over the temperature range of 240 K to about 353 K. Above 353 K Mironov and Popov (3) report the stable solid phase to be a "lower hydrate", and below we offer evidence suggesting that this "lower hydrate" is probably the monohydrate. In studies not involving solubility determinations, the only mention of a trihydrate was made by Jolin (27) and Lange (28) who reported the conversion of the hexahydrate to the trihydrate by heating. On the other hand Löwenstein (29) measured the vapor pressure of $Ce(NO_3)_3.nH_2O$ crystals and found evidence for the hexa-, tetra-, and monohydrates. The evaluators could not find any references in the literature to a dihydrate, and we have concluded that the solid phases in equilibrium with saturated solutions over the temperature range of 240 K to 410 K are:

$Ce(NO_3)_3.6H_2O$	[10294-41-4]	$Ce(NO_3)_3.4H_2O$	[15878-76-9]
$Ce(NO_3)_3.5H_2O$	[15878-75-8]	$Ce(NO_3)_3.H_2O$	[81201-28-7]

EVALUATION PROCEDURE

The data in the compilations were examined and either rejected immediately because of large obvious errors, or were analysed by a weighted least squares method. It should be noted that in all but one case, only experimental solubility data were used in the least squares analyses: i.e. smoothed or extrapolated data were not used. The one exception refers to the solubility at 298.2 K in the tetrahydrate system which is an extrapolated value (see below and ref. 4). The data were fitted to a general solubility equation based on the treatments in (30,31) and in the INTRODUCTION to this volume:

$$Y = \ln(m/m_o) - nM_2(m - m_o) = a + b/(T/K) + c\ \ln(T/K) \qquad [1]$$

In eq. [1] m is the molality at temperature T, m_o is an arbitrarily selected reference molality (usually the 298.2 K value), n is the hydrate number of the solid, M_2 is the molar mass of the solvent, and a, b, c are constants from which enthalpies and heat capacitites of solution, ΔH_{solv} and ΔC_p, can be estimated (see INTRODUCTION). In fitting the solubility data to eq. [1], weight factors of 0, 1, 2, 3 were assigned to each published value depending upon the precision of the experimental value. In this procedure, if the calculated error between the observed and calculated molalities, Δm, was larger than twice the standard error of estimate for m, σ_m, the data point was either rejected or its weight factor decreased. The fitting of the data was repeated until all Δm values were equal to or less than $\pm 2\sigma_m$.

The experimental methods used to determine the solubilities are the isothermal method (1,2,5,6,10-12,15,24,25), the synthetic method (3, 4 (?)), and the method of isothermal sections (actually an isothermal method) in which refractometric analysis was employed to determine concentrations (7-9, 13, 14, 16-23, 26). With very few exceptions the precision of the data was not specified by the authors, and the compilers and evaluators had to estimate these quantities. In general the synthetic method should yield very precise solubility values due to the accuracy one can achieve in the quantitative make-up (i.e. the "synthesis") of the composition of the binary system. The major source of error in this method is due to the precision given to the temperatures of crystallization. Mironov and Popov (3) measured these temperatures visually, and report the solubilities to three significant figures: thus at best we assign an overall precision to their data of around ± 0.5%. Temperature control for isothermal studies can, generally, be very precise, and one major source of imprecision is attributed to chemical analyses of saturated solutions. In the study of Brunisholz et al. (2) the temperature control was

COMPONENTS:	EVALUATOR:
(1) Cerium nitrate; $Ce(NO_3)_3$; [10108-73-3] (2) Water ; H_2O ; [7732-18-5]	S. Siekierski, T. Mioduski Institute for Nuclear Research Warsaw, Poland and M. Salomon U.S. Army ET & DL Ft. Monmouth, NJ, USA December 1982

CRITICAL EVALUATION:

at least ± 0.05 K and probably represents an accuracy, and analysis of the cerium content of saturated solutions was accomplished by complexometric titration: the compilers have assigned an overall precision of ± 0.2% (or possibly better) to these values. In Angelov's isothermal studies (5) the compilers estimated an experimental precision of ± 0.3% mainly due to the fact that the results are reported to four significant figures. However because of the wide divergence of Angelov's results from those of other investigators (similar to the situation found for Angelov's data for other lanthanide nitrate solubilities), the evaluators have concluded that there is an obvious systematic error in all of Angelov's studies, and his data have therefore been rejected. In the method of isothermal sections reported in (7-9, 14, 16-23, 26), temperature control was generally ± 0.1 - 0.2 K, and refractive indices were reported to five significant figures. However solubilities were reported to three significant figures, and the evaluators estimate that the overall precision for these studies is around ± 1%. While the data of Khisaeva et al. (26) appear to be precise to ± 1% at best, the accuracy is very poor and the publication was rejected. This paper (26) reports solubilities in the cerium nitrate-quinoline nitrate-water ternary system at 298 K and 323 K, and an obvious large negative systematic error is present for all data given by the authors: e.g. at 298 K and 323 K the solubility of $Ce(NO_3)_3$ in pure water (hexahydrate solid phase) is given as 56.0 mass % (3.90 mol kg^{-1}) and 62.0 mass % (5.00 mol kg^{-1}), respectively. The recommended solubilities (see discussion below and Table 4) at 298.2 K and 323.2 K are 5.232 mol kg^{-1} and 7.698 mol kg^{-1}, respectively (at 323.2 K the hexahydrate system is metastable).

SOLUBILITY OF $Ce(NO_3)_3$ IN THE HEXAHYDRATE SYSTEM

Data for the solubility of $Ce(NO_3)_3$ in the hexahydrate system are given in Table 1 along with the initial and final weight factors used in the least squares fitting of the data to the smoothing equation. The data from references (5, 26) were not included

Table 1. Solubility of $Ce(NO_3)_3$ in water for the hexahydrate system.

T/K	solubility mol kg^{-1}	ref	weight initial/final	T/K	mol kg^{-1}	ref	weight initial/final
239.7	3.57	3	2/2	298.2	5.161	10	0/0
263.9	3.87	3	2/0	298.2	5.11	14	0/0
273.15	4.238	2	3/3	298.2	5.230	11	1/2
273.2	4.20	9	1/2	298.2	5.243	12	1/2
277.5	4.23	3	2/0	298.2	5.22	18-23	1/2
278.1	4.39	3	2/2	302.1	5.38	3	2/0
283.15	4.567	2	3/3	303.2	5.22	7,8,13,14,16,17	0/0
283.2	4.60	7,8,13,16,17	2/1	303.2	5.472	24,25	2/1
288.2	4.70	14	2/1	308.15	5.852	2	3/3
291.1	4.78	3	2/0	308.2	5.33	14	0/0
293.15	5.007	2	3/3	310.8	6.14	3	2/1
293.2	4.90	7,8,13,16,17	1/0	323.15	7.331	2	3/0
293.2	5.11	9	1/0	323.2	8.673	1	0/0
296.5	5.24	3	2/0	323.2	8.29	18,21,23	0/0
298.15	5.383	1	1/0	323.2	8.42	19,22	0/0
298.2	5.31	6	2/0				

in Table 1 because, as discussed above, these data contain a large systematic error of unknown origin (note that of all the studies reported in 1-26, Angelov is the only investigator to use the Knorre method for analysis of saturated solutions; cerium was analysed as Ce^{4+}, and details on experimental techniques were not given). The results of the fitting of the data in Table 1 to the smoothing eq. [1] are given in Table 3 and Figure 1. Some of the data reported by Mironov and Popov (3) fall either above or below the smoothed polytherm (see Fig 1), and we attribute this to a non-systematic error probably associated with the visual recording of the temperatures of crystallization. Other data which deviate from the smoothed polytherm thus indicating errors which are systematically high (1) or low (8, 9, 14, 15, 17, 18) are probably due to incorrect identification of the solid phase (1), or to general experimental imprecision. Quill and Robey (1) stated that the salt they used was dried in a desiccator over 55% H_2SO_4, and it is possible that lower hydrates were formed.

COMPONENTS:	EVALUATOR:
(1) Cerium nitrate; $Ce(NO_3)_3$; [10108-73-3] (2) Water ; H_2O ; [7732-18-5]	S. Siekierski, T. Mioduski Institute for Nuclear Research Warsaw, Poland and M. Salomon U.S. Army ET & DL Ft. Monmouth, NJ, USA December 1982

CRITICAL EVALUATION:

The value of the congruent melting point predicted from eq. [1] is 326.4 K which is in good agreement with the experimental values of 324.6 (32) and 325 K (33). Wendlandt and Sewell's value of 369 K for the congruent melting point of the hexahydrate (34) is obviously in error.

SOLUBILITY OF $Ce(NO_3)_3$ in the $Ce(NO_3)_3nH_2O - H_2O$ SYSTEMS: $n \leq 5$

The solubility of $Ce(NO_3)_3$ in the penta- and tetrahydrate systems has been reported in (3-6), and for an unidentified "lower hydrate" (i.e. $n < 4$) in ref (3). The data are summarized in Table 2 where again the results of Angelov (5) were omitted because of the existence of a large negative systematic error in this study.

Table 2. Solubility of $Ce(NO_3)_3$ in $Ce(NO_3)_3.nH_2O - H_2O$ systems for $n \leq 5$.

T/K	solubility mol kg^{-1}	ref	solid phase
298.2	6.11	4 , 6	pentahydrate
311.9	6.20	3	"
321.7	6.64	3	"
322.9	6.67	3	"
325.1	6.79	3	"
298.2	7.51[a]	4	tetrahydrate
326.2	7.15	3	"
330.9	7.81	3	"
335.4	8.37	3	"
340.5	8.73	3	"
347.3	9.87	3	"
353.2	11.68	3	lower hydrate (probably the monohydrate)
367.2	13.69	3	"
381.2	17.51	3	"
392.7	22.70	3	"

a. Extrapolated value.

The data for the penta- and tetrahydrate systems were fitted to the smoothing eq. [1], and the results are given in Table 3. The evaluators could not locate references for the congruent melting points of these solids in order to compare them with the predicted melting points given in Table 3 (assuming these solids melt congruently). In analysing the solubility data for the "lower hydrate" results of Mironov and Popov (3), it was found that the assumption that the solid phase is the monohydrate gives the most consistent results. Analyses of these data using eq. [1] results in the best fit (i.e. lowest σ values) when n = 1. The assumption of a dihydrate solid phase would appear to be improbable due to the fact that this solid has not been found to exist in any solubility or thermal study as discussed above. If a trihydrate solid phase exists, its maximum solubility (at the congruent melting point) would be 18.503 mol kg^{-1}, but the experimental data point at 22.70 mol kg^{-1} which cannot correspond to the trihydrate lies on the polytherm for the monohydrate system with a residual error a $\Delta m = m_{obs} - m_{calc} = -0.048$ mol kg^{-1}. Therefore at the present time, we identify Mironov and Popov's data for "lower hydrate" with the monohydrate.

COMPONENTS:

(1) Cerium nitrate; $Ce(NO_3)_3$; [10108-73-3]

(2) Water ; H_2O ; [7732-18-5]

EVALUATOR:
S. Siekierski, T. Mioduski
Institute for Nuclear Research
Warsaw, Poland
and
M. Salomon
U.S. Army ET & DL
Ft. Monmouth, NJ, USA
December 1982

CRITICAL EVALUATION:

Table 3. Derived parameters for the smoothing equation [1].

parameter	hexahydrate	pentahydrate	tetrahydrate	monohydrate
a	-31.349	-120.410	-176.06	-267.071
b	1058.2	5418.2	8155.4	12965
c	4.8950	17.9377	26.092	39.2431
σ_a	0.002	0.002	0.01	0.004
σ_b	0.7	0.6	4.4	1.6
σ_c	0.0004	0.0003	0.002	0.0007
σ_Y	0.002	0.002	0.01	0.004
σ_m	0.03	0.03	0.3	0.1
ΔH_{sln}/kJ mol^{-1}	-35.0	-179.6	-270.4	-429.9
ΔC_p/J K^{-1} mol^{-1}	162.8	596.6	867.7	1305
congruent melting point/K	326.4	344.6	353.0	410.2
concn at congr melting point/mol kg^{-1}	9.251	11.102	13.877	55.508

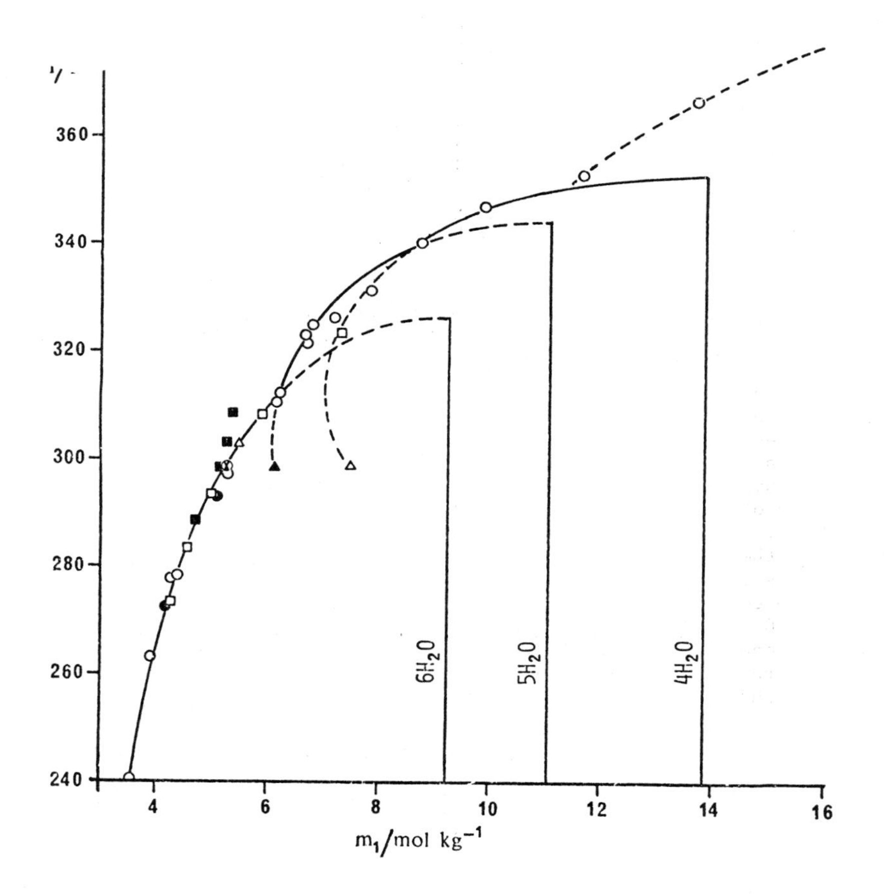

Figure 1. Phase diagram for the $Ce(NO_3)_3$-H_2O system. Solid lines (stable systems) and dashed lines (metastable systems) based on results from smoothing equation. Experimental points are: ○ ref 3; □ ref 2; ● ref 9; ■ ref 14; ⊗ refs 11, 12, 18-23; △ refs 24, 25; ▲ refs 4, 6.

COMPONENTS:	EVALUATOR:
(1) Cerium nitrate; $Ce(NO_3)_3$; [10108-73-3] (2) Water ; H_2O ; [7732-18-5]	S. Siekierski, T. Mioduski Institute for Nuclear Research Warsaw, Poland and M. Salomon U.S. Army ET & DL Ft. Monmouth, NJ, USA December 1982

CRITICAL EVALUATION:

Table 4. Recommended and Tentative solubilities.[a]

T/K	hexahydrate[b]	pentahydrate[c]	tetrahydrate[c]	monohydrate[c]
243.2	3.612			
253.2	3.768			
263.2	3.973			
273.2	4.233			
278.2	4.388			
283.2	4.562			
288.2	4.757			
293.2	4.979			
298.2	5.232	6.11[d]	7.49[d]	
303.2	5.526	6.09[d]	7.22[d]	
308.2	5.874	6.14[d]	7.08[d]	
312.4[e]	6.23	6.23		
313.2	6.300[d]	6.24	7.04[d]	
318.2	6.854[d]	6.43	7.10[d]	
320.3[f]	7.15[d]		7.15[d]	
323.2	7.698[d]	6.69	7.25[d]	
328.2		7.05	7.51[d]	
333.2		7.55	7.88[d]	
338.2		8.29	8.41[d]	
339.9[g]		8.64	8.64	
343.2		9.68[d]	9.17	
348.2			10.35	11.20[d]
350.9[h]			11.44	11.44
353.2				11.67
363.2				13.02
373.2				15.07
383.2				18.16

a. $Ce(NO_3)_3$ solubilities in mol kg^{-1}. b. *Recommended* values. c. *Tentative* values.

d. Metastable equilibria. e. Hexahydrate → pentahydrate transition temperature.

f. Hexahydrate → tetrahydrate transition temperature.

g. Pentahydrate → tetrahydrate temperature.

h. Tetrahydrate → monohydrate transition temperature.

RECOMMENDED AND TENTATIVE SOLUBILITIES

Table 4 gives *recommended* solubility data in the hexahydrate system at selected temperatures as calculated from the smoothing equation. At the 95% level of confidence and a Student's t = 2.060, the uncertainty in the calculated values is ± 0.010 mol kg^{-1}. Combining this uncertainty with the weighted average experimental precision of ± 0.25% results in an overall uncertainty of ± 0.016 mol kg^{-1} in these *recommended* values.

The solubility data in the penta-, tetra-, and monohydrate systems are designated as *tentative* since they are based on the studies of one group of researchers (3,4). The smoothed *tentative* solubility data calculated from the smoothing equation are given in Table 4. For the pentahydrate system, the uncertainty in the smoothed values is ± 0.06 mol kg^{-1} (95% confidence level, Student's t = 4.303), and considering an experimental precision of ± 0.5%, the overall uncertainty in these smoothed values is ± 0.071 mol kg^{-1}.

Similarly the uncertainty in the smoothed solubility data in the tetrahydrate system is ± 0.04 mol kg^{-1} at the 95% level of confidence (Student's t = 3.182), and including the experimental precision of ± 0.5% leads to an overall uncertainty of ± 0.05 mol kg^{-1} in the *tentative* values calculated from the smoothing equation.

COMPONENTS:	EVALUATOR:
(1) Cerium nitrate; $Ce(NO_3)_3$; [10108-73-3] (2) Water ; H_2O ; [7732-18-5]	S. Siekierski, T. Mioduski Institute for Nuclear Research Warsaw, Poland and M. Salomon U.S. Army ET & DL Ft. Monmouth, NJ, USA December 1982

CRITICAL EVALUATION:

For the *tentative* solubility data in the monohydrate system, we estimate a total uncertainty of ± 0.66 mol kg^{-1} (imprecision of fitting of data to eq. [1] is ± 0.65 mol kg^{-1} at the 95% level of confidence, and the estimated experimental precision is ± 0.08 mol kg^{-1}).

The phase diagram for the $Ce(NO_3)_3 \cdot nH_2O-H_2O$ system is given in Fig. 1 where the solid lines represent stable systems, and the dashed lines represent metastable systems. All lines were calculated from the smoothing equation. The polytherm for the monohydrate system is represented by a dashed line although it is a stable system above 350.9 K because we are not able to present quantitative proof of the existence of this phase. We do however conclude that the evidence strongly supports the present assignment of the monohydrate to this portion of the phase diagram. The transition temperatures given in Table 4 were determined graphically from the phase diagram.

MULTICOMPONENT SYSTEMS

THE TERNARY $Ce(NO_3)_3-HNO_3-H_2O$ SYSTEM

Of the two studies reported in the literature (5,6), comparisons are not possible because the latter study reports the solubilities as a function of HNO_3 concentration at one temperature (298.2 K) while the former study reports solubilities as a function of HNO_3 concentration and temperature, but not at 298.2 K. In addition, the fact that Angelov's data (5) for the binary system was shown above to contain a large systematic error, the results of this author for solubilities in HNO_3 solutions must be designated as *doubtful*. The data of Popov and Mironov (6) at 298.2 K are designated as *tentative* solubility data. In HNO_3 solutions at 298.2 K Popov and Mironov (6) report the following smoothing equations:

solid phase	smoothing equation
hexahydrate	$\log m_1 = \log 5.32 - 0.868 \log y_1$
pentahydrate	$\log m_1 = \log 6.11 - 0.718 \log y_1$
tetrahydrate	$\log m_1 = \log 7.51 - 0.622 \log y_1$

In these equations, m_1 is the solubility of $Ce(NO_3)_3$ in mol kg^{-1}, and y_1 is the solute mole fraction of $Ce(NO_3)_3$ which is defined as: mol fraction of $Ce(NO_3)_3$ + mol fraction HNO_3 = 1. The only solid phases reported in (5) and (6) are the hexa-, penta-, and tetrahydrates for HNO_3 concentrations up to about 60 mass %. No complex salts are formed.

SYSTEMS WITH ALKALI METAL AND AMMONIUM NITRATES

The dominant feature in these studies is the existence of hydrated double salts:

$2KNO_3 \cdot Ce(NO_3)_3 \cdot 2H_2O$	(8)
$2RbNO_3 \cdot Ce(NO_3)_3 \cdot 4H_2O$	(8)
$2CsNO_3 \cdot Ce(NO_3)_3 \cdot 4H_2O$	(8)
$Ce(NO_3)_3 \cdot 2NH_4NO_3 \cdot 4H_2O$	(12)

No double salts were reported for ternary systems with $LiNO_3$ (7) or $NaNO_3$ (8).

COMPONENTS:	EVALUATOR:
(1) Cerium nitrate; $Ce(NO_3)_3$; [10108-73-3] (2) Water ; H_2O ; [7732-18-5]	S. Siekierski, T. Mioduski Institute for Nuclear Research Warsaw, Poland and M. Salomon U.S. Army ET & DL Ft. Monmouth, NJ, USA December 1982

CRITICAL EVALUATION:

SYSTEMS WITH ALKALINE EARTH AND TRANSITION METAL NITRATES

The phase diagram for the $Be(NO_3)_2$-$Ce(NO_3)_3$-H_2O system is of the simple eutonic type (9), but other systems are generally characterized by the existence of a tetracosahydrate.

$3Mg(NO_3)_2 \cdot Ce(NO_3)_3 \cdot 24H_2O$	[13550-46-4]	(10-12)
$Ca(NO_3)_2 \cdot Ce(NO_3)_3 \cdot 4H_2O$		(13)
$2Ce(NO_3)_3 \cdot 3Mn(NO_3)_2 \cdot 24H_2O$		(16)
$2Ce(NO_3)_3 \cdot 3Ni(NO_3)_2 \cdot 24H_2O$	[20340-68-3]	(11)
$2Ce(NO_3)_3 \cdot 3Cu(NO_3)_2 \cdot 24H_2O$	[29990-16-7]	(10,17)

OTHER SYSTEMS

The remaining studies which cannot be critically evaluated involve hydrazine nitrates (18) and organic solutes with $Ce(NO_3)_3$ in water (19-26). The reader is referred directly to the compilations for experimental details and results. The paper by Khisaeva et al. (26) on the $Ce(NO_3)_3$-$C_9H_7N.HNO_3$-H_2O system was rejected, and a compilation has not been prepared.

REFERENCES

1. Quill. L.L.; Robey, R.F. *J. Am. Chem. Soc.* 1937, *59*, 2591.
2. Brunisholz, G.; Quinche, J.P.; Kalo, A.M. *Helv. Chim. Acta* 1964, *47*, 14.
3. Mironov, K.E.; Popov, A.P. *Rev. Roum. Chim.* 1966, *11*, 1373.
4. Popov, A.P.; Mironov, K.E. *Zh. Neorg. Khim.* 1971, *16*, 464.
5. Angelov, I.I. *Tr. Vses. Nauch-Issled, Inst. Khim. Reakt. Osobo Chist, Khim. Veshchesty.* 1958, *No. 39*, 18.
6. Popov, A.P.; Mironov, K.E. *Izv. Sib. Otd. Akad. Nauk. SSSR. Ser. Khim. Nauk.* 1967, *12*, 109.
7. Bogdanovskaya, R.L.; Sheveleva, A.D. *Zh. Neorg. Khim.* 1965, *10*, 1713.
8. Zhuravlev, E.F.; Sheveleva, A.D.; Bogdanovskaya, R.L.; Kudryashov, S.F.; Shchurov, V.A. *Zh. Neorg. Khim.* 1963, *8*, 1955.
9. Kudryasov, S.F.; Frolova, S.I.; Ushkova, A.V. *Zh. Neorg. Khim.* 1967, *12*, 2494.
10. Yakimov, M.A.; Gizhavina, E.I. *Zh. Neorg. Khim.* 1971, *16*, 507.
11. Mitina, N.K.; Shevchuk, V.G. *Zh. Neorg. Khim.* 1977, *22*, 1376.
12. Shevchuk, V.G.; Mitina, N.K. *Zh. Neorg. Khim.* 1976, *21*, 1943.
13. Bogdanovskaya, R.L.; Kudryashov, S.F. *Uch. Zap. Permsk. Gos. Univ. im. A.M. Gor'kogo* 1966, *141*, 27.
14. Bogdanovskaya, R.L. *Uch. Zap. Permsk. Gos. Univ. im. A.M. Gor'kogo* 1966, *141*, 32.
15. Onishchenko, M.K.; Shevchuk, V.G. *Zh. Neorg. Khim.* 1977, *22*, 2879.
16. Shchurov, V.A.; Mochalov, K.I.; Volkov, A.A. *Uch. Zap. Permsk. Gos. Univ. im A.M. Gor'kogo* 1966, *141*, 18.
17. Sheveleva, A.D.; Shchurov, V.A. *Uch. Zap. Permsk. Gos. Univ. im. A.M. Gor'kogo* 1968, *178*, 85.
18. Mininkov, N.E.; Zhuravlev, E.F. *Zh. Neorg. Khim.* 1970, *15*, 205.
19. Zhuravlev, E.F.; Mininkov, N.E. *Zh. Neorg. Khim.* 1972, *17*, 1736.
20. Mininkov, N.E.; Zhuravlev, E.F. *Zh. Neorg. Khim.* 1976, *21*, 242.
21. Mininkov, N.E.; Zhuravlev, E.F. *Zh. Neorg. Khim.* 1969, *14*, 2565.
22. Mininkov, N.E.; Zhuravlev, E.F. *Zh. Neorg. Khim.* 1973, *18*, 3091.
23. Mininkov, N.E.; Zhuravlev, E.F. *Zh. Neorg. Khim.* 1974, *19*, 1656.
24. Zholalieva, Z.M.; Sulaimankulov, K.S.; Ismailov, M. *Zh. Neorg. Khim,* 1975, *20*, 2243.
25. Zholalieva, Z.M.; Sulaimankulov, K.S.; Ismailov, M.; Abykeev, K. *Zh. Neorg. Khim.* 1976, *21*, 2583.
26. Khisaeva, D.A.; Zhuravlev, E.F.; Semenova, E.B. *Zh. Neorg. Khim.* 1982, *27*, 2105.
27. Jolin, S. *Bull. Soc. Chim.* 1874, *21*, 533.
28. Lange, T. *J. Prakt. Chem.* 1861, *82*, 129.
29. Lowenstein, E. *Z. Anorg. Chem.* 1909, *63*, 69.
30. Williamson, A.T. *Trans. Faraday, Soc.* 1944, *40*, 421.
31. Counioux, J.-J.; Tenu, R. *J. Chim. Phys.* 1981, *78*, 816 and 823.
32. Quill, L.L.; Robey, R.F.; Seifter, S. *Ind. Eng. Chem. Anal. Ed.* 1937, *9*, 389.
33. Claudel, B.; Trambouze, Y.; Veron, J. *Bull. Soc. Chim. Fr.* 1963, 409.
34. Wendlandt, W.W.; Sewell, R.G. *Texas J. Sci,* 1961, *13*, 231.

COMPONENTS:

(1) Cerium nitrate; $Ce(NO_3)_3$; [10108-73-3]

(2) Water; H_2O; [7732-18-5]

ORIGINAL MEASUREMENTS:

Quill, L.L.; Robey, R.F.
J. Am. Chem. Soc. 1937, *59*, 2591-5

VARIABLES:

Two temperatures

PREPARED BY:

T. Mioduski, S. Siekierski, M. Salomon

EXPERIMENTAL VALUES:

At 25°C, soly = 63.71 mass %, density = 1.88 kg m^{-3}

At 50°C, soly = 73.88 mass %, density = 2.04 kg m^{-3}

Compilers' Conversions

25°C, soly = 5.383 mol kg^{-1} or 3.67 mol dm^{-3}

50°C, soly = 8.673 mol kg^{-1} or 4.62 mol dm^{-3}

The solid phase at both temperatures is the hexahydrate.

AUXILIARY INFORMATION

METHOD/APPARATUS/PROCEDURE:

Isothermal method. The salt and water were placed in pyrex tubes, heated to induce supersaturation, and thermostated. The pyrex tubes were sealed after a small crystal of $Bi(NO_3)_3 \cdot 5H_2O$ was added to "seed" crystallization. The sealed tubes were shaken in the thermostat for at least 8 hours (equilibrium was reached in 4 hours). Authors state that approach to equilibrium from undersaturation gave identical results within experimental error. All data reported above are the results obtained by approach from supersaturation.

A "filtering pipet" maintained at a temp slightly higher than the thermostat temp was used to withdraw samples for analyses. Weighed samples of liquid and solid phases were analysed. Ce was pptd as the oxalate, filtered, washed with hot dilute oxalic acid, and ignited to the oxide (CeO_2).

Ce(IV) detd in the presence of Ce(III) by adding excess standard Fe(II) sln to the sample in H_2SO_4 sln, and titrating the excess Fe(II) with standard ceric ammonium sulfate. The solid phase analysed as the hexahydrate (calcd anhydr salt 75.10%, found 75.10, 75.00%).

SOURCE AND PURITY OF MATERIALS:

HNO_3 prepd from c.p. grade by distillation in an all pyrex still and retaining the middle fraction. For very high HNO_3 concs, reagent grade fuming HNO_3 used as received.

"Pure" cerium nitrate (G.F. Smith) was pptd three times (the basic bromate method). The ceric basic bromate was washed, mixed with HNO_3, and dissolved by addn of 3% H_2O_2 sln. Cerium was then pptd as the oxalate, washed and digested in HNO_3 on a steam bath. The sln was evap to crystallation and the salt dried over 55% sulfuric acid in a desiccator. Distilled water was used which had an electrolytic conductivity of 2 x 10^{-6} S cm^{-1}.

ESTIMATED ERROR:

Soly: precision about ± 0.5 % (compilers).

Temp: at 25°C, accuracy was ± 0.03 K.
at 50°C, accuracy was ± 0.1 K.

COMPONENTS:

(1) Cerium (III) nitrate; $Ce(NO_3)_3$; [10108-73-3]

(2) Water; H_2O; [7732-18-5]

ORIGINAL MEASUREMENTS:

Brunisholz, G.; Quinche, J.P.; Kalo, A.M. *Helv. Chim. Acta* 1964, *47*, 14-27.

VARIABLES:

Temperature

PREPARED BY:

T. Mioduski and S. Siekierski

EXPERIMENTAL VALUES:

Solubility of $Ce(NO_3)_3$ in water [a]

t/°C	mass %	mol kg^{-1}
0	58.02	4.238
10	59.83	4.567
20	62.02	5.007
35	65.62	5.852
50	70.51	7.331

a. Molalities calculated by compilers.

Authors report the solid phase to be $Ce(NO_3)_3 \cdot 6H_2O$

AUXILIARY INFORMATION

METHOD/APPARATUS/PROCEDURE:

The isothermal method was used. Cerium determined by complexometric titration using xylenol orange indicator in the presence of a small quantity of urotropine buffer (10 mg of ascorbic acid was added to avoid oxidation of Ce(III)). Water was determined by difference.

SOURCE AND PURITY OF MATERIALS:

Cerium nitrate was prepared from the oxide of purity better than 99.7% (obtained by the ion exchange chromatography).

ESTIMATED ERROR:

Soly: precision ± 0.1% (compilers).
Temp: precision probably ± 0.05K (compilers).

REFERENCES:

COMPONENTS:

(1) Cerium nitrate; $Ce(NO_3)_3$; [10108-73-3]

(2) Water; H_2O; [7732-18-5]

ORIGINAL MEASUREMENTS:

Mironov, K.E.; Popov, A.P. *Rev. Roum. Chim.* 1966, *11*, 1373-81.

VARIABLES:

Temperature

PREPARED BY:

T. Mioduski and S. Siekierski

EXPERIMENTAL VALUES:

Solubility of $Ce(NO_3)_3$ in water [a,b]

t/°C	mass %	mol kg^{-1}	solid phase [c]	t/°C	mass %	mol kg^{-1}	solid phase [c]
- 0.5	16.5	0.606	A	38.7	66.9	6.20	C
- 2.4	23.1	0.921	A	48.5	68.4	6.64	C
- 3.0	27.5	1.16	A	49.7	68.5	6.67	C
-10.0	36.5	1.76	A	51.9	68.9	6.79	C
-21.4	46.4	2.65	A				
-28.2	51.8	3.30	A	53.0	70.0	7.15	D
				57.7	71.8	7.81	D
-33.5	53.8	3.57	A + B	62.2	73.2	8.37	D
				67.3	74.0	8.73	D
- 9.3	55.8	3.87	B	74.1	76.3	9.87	D
4.3	58.0	4.23	B				
4.9	58.9	4.39	B	80.0	79.2	11.68	E
17.9	60.9	4.78	B	94.0	81.7	13.69	E
23.3	63.1	5.24	B	108.0	85.1	17.51	E
28.9	63.7	5.38	B	119.5	88.1	22.70	E
37.6	66.7	6.14	B				

a. Molalities calculated by the compilers.

b. Interpolated value at 25°C is 5.27 mol kg^{-1} (compilers).

c. Solid phases: A = ice; B = $Ce(NO_3)_3 \cdot 6H_2O$; C = $Ce(NO_3)_3 \cdot 5H_2O$; D = $Ce(NO_3)_3 \cdot 4H_2O$; E = lower hydrate

AUXILIARY INFORMATION

METHOD/APPARATUS/PROCEDURE:

The synthetic method was used.
The temperature of crystallization was determined visually.

No other details given.

SOURCE AND PURITY OF MATERIALS:

$Ce(NO_3)_3 \cdot 6H_2O$ prepared by dissolving 99.8 % pure CeO_2 in HNO_3 followed by crystallization.

Doubly distilled water was used.

ESTIMATED ERROR:

Soly: precision about ± 0.2 to 0.5 % (compilers).

Temp: precision about ± 0.2 K (compilers).

REFERENCES:

COMPONENTS:

(1) Cerium nitrate; $Ce(NO_3)_3$; [10108-73-3]

(2) Water; H_2O; [7732-18-5]

ORIGINAL MEASUREMENTS:

Popov, A.P.; Mironov, K.E. *Zh. Neorg. Khim.* 1971, *16*, 464-6; *Russ, J. Inorg. Chem. Engl. Transl.* 1971, *16*, 244-6.

VARIABLES:

One temperature: 25°C

PREPARED BY:

T. Mioduski, S. Siekierski, M. Salomon

EXPERIMENTAL VALUES:

The solubility of metastable $Ce(NO_3)_3 \cdot 5H_2O$ in water at 25°C was reported as:

66.6 mass %

The corresponding molality value was calculated by the compilers as:

6.11 mol kg^{-1}

COMMENTS AND/OR ADDITIONAL DATA:

It is not clearly stated whether the solubility of the pentahydrate was actually measured in the present work or whether the authors are reporting the value obtained from earlier work (1). In reference (1) the solubility of the pentahydrate at 25°C was given as 66.6 mass % (see the compilation of the data from this paper).

By extrapolation of the solubility branch for $Ce(NO_3)_3 \cdot 4H_2O$ from data in the ternary system $Ce(NO_3)_3$-HNO_3-H_2O, the authors calculated the solubility of the metastable tetrahydrate in pure water as 71.0 mass % (7.51 mol kg^{-1}). See the critical evaluation for additional details.

AUXILIARY INFORMATION

METHOD/APPARATUS/PROCEDURE:

Experimental method not specified. Metastable (at 25°C) $Ce(NO_3)_3 \cdot 5H_2O$ could have been obtained by slow cooling of a saturated $Ce(NO_3)_3$ solution from 40°-50°C to 25°C (1).

SOURCE AND PURITY OF MATERIALS:

Nothing specified.

ESTIMATED ERROR:

Soly: Not specified, but precision probably ± 0.5 mass % units (compilers).

Temp: Nothing Specified.

REFERENCES:

1. Popov, A.P.; Mironov, K.E. *Izv. Sib. Otd. Adad. Nauk. SSSR, Ser. Khim. Nauk.* 1967, *12*, 109-11.

COMPONENTS:

(1) Cerium nitrate; $Ce(NO_3)_3$; [10108-73-3]

(2) Nitric acid; HNO_3; [7697-37-2]

(3) Water; H_2O; [7732-18-5]

ORIGINAL MEASUREMENTS:

Angelov, I.I. *Tr., Vses. Nauch-Issled. Inst. Khim. Reakt. Osobo Chist. Khim. Veshchestv.* 1958, *No. 39*, 18.

VARIABLES:

Composition and temperature

PREPARED BY:

T. Mioduski and S. Siekierski

EXPERIMENTAL VALUES:

The $Ce(NO_3)_3$ - HNO_3 - H_2O system [a]

t/°C	$Ce(NO_3)_3$ mass %	$Ce(NO_3)_3$ mol kg^{-1}	HNO_3 mass %	HNO_3 mol kg^{-1}	nature of the solid phase
0	59.15	4.440	——	——	$Ce(NO_3)_3 \cdot 6H_2O$
	53.51	3.880	4.20	1.576	"
	52.22	3.756	5.15	1.917	"
	48.71	3.484	8.42	3.117	"
	43.41	3.073	13.30	4.875	"
	34.78	2.507	22.68	8.461	"
	26.30	1.867	30.50	11.20	"
	24.30	1.774	33.70	12.73	"
	20.42	1.602	40.50	16.45	"
	20.67	1.858	45.22	21.04	$Ce(NO_3)_3 \cdot 6H_2O + Ce(NO_3)_2 \cdot 4H_2O$
	20.11	1.841	46.40	21.99	$Ce(NO_3)_3 \cdot 4H_2O$
	17.88	1.763	51.03	26.05	"
	16.11	1.735	55.42	30.89	"
	13.37	1.600	61.00	37.77	"
5	60.02	4.603	——	——	$Ce(NO_3)_3 \cdot 6H_2O$
10	61.05	4.806	——	——	$Ce(NO_3)_3 \cdot 6H_2O$

continued........

AUXILIARY INFORMATION

METHOD/APPARATUS/PROCEDURE:

The isothermal method was used. Solutions were equilibrated for 3-4 hours. Cerium was determined as C^{4+} by the Knorre method.

No other details given.

SOURCE AND PURITY OF MATERIALS:

Nothing specified.

ESTIMATED ERROR:

Soly: precision about ± 0.3% (compilers).
Temp: precision ± 0.1 K.

REFERENCES:

COMPONENTS:	ORIGINAL MEASUREMENTS:
(1) Cerium nitrate; $Ce(NO_3)_3$; [10108-73-3] (2) Nitric acid; HNO_3; [7697-37-2] (3) Water; H_2O; [7732-18-5]	Angelov, I.I. *Tr., Vses. Nauch-Issled. Inst. Khim. Reakt. Osobo Chist. Khim. Veshchestv.* 1958, *No. 39*, 18.

EXPERIMENTAL VALUES: continued........

The $Ce(NO_3)_3 - HNO_3 - H_2O$ system [a]

t/°C	$Ce(NO_3)_3$ mass %	$Ce(NO_3)_3$ mol kg^{-1}	HNO_3 mass %	HNO_3 mol kg^{-1}	nature of the solid phase
15	62.12	5.028	———	———	$Ce(NO_3)_3 \cdot 6H_2O$
	58.04	4.618	3.42	1.408	"
	49.85	3.893	10.89	4.402	"
	47.00	3.686	13.90	5.642	"
	41.63	3.284	19.50	7.961	"
	39.45	3.090	21.40	8.675	"
	31.90	2.574	30.10	12.57	"
	29.10	2.499	35.20	15.65	"
	27.50	2.643	40.60	20.20	$Ce(NO_3)_3 \cdot 6H_2O + Ce(NO_3)_3 \cdot 4H_2O$
	22.60	2.159	45.30	22.40	$Ce(NO_3)_3 \cdot 4H_2O$
	20.20	1.830	45.95	21.54	"
	19.10	1.968	51.14	27.27	"
	17.10	1.855	54.63	30.67	"
	15.30	2.022	61.50	42.07	"
	14.80	2.085	63.44	46.27	"
30	65.46	5.811	———	———	$Ce(NO_3)_3 \cdot 6H_2O$
	59.32	5.314	6.45	2.990	"
	53.80	4.591	10.27	4.536	"
	49.21	4.255	15.33	6.861	"
	40.70	3.828	26.70	13.00	"
	34.23	3.242	33.40	16.37	"
	34.70	3.570	35.50	18.91	$Ce(NO_3)_3 \cdot 6H_2O + Ce(NO_3)_3 \cdot 4H_2O$
	32.17	3.229	37.28	19.37	$Ce(NO_3)_3 \cdot 4H_2O$
	28.87	3.022	41.84	22.67	"
	26.40	2.875	45.44	25.61	"
	21.85	2.355	59.70	27.72	"
	20.60	2.301	51.95	30.03	"
	21.25	2.548	53.18	33.01	"
	18.80	2.597	59.00	42.18	"
	19.29	3.079	61.50	50.81	"
45	68.34	6.619	———	———	$Ce(NO_3)_3 \cdot 6H_2O$
	66.20	6.712	3.56	1.868	"
	63.95	6.371	5.27	2.717	"
	59.83	6.150	10.34	5.501	"
	58.37	6.219	12.85	7.086	$Ce(NO_3)_3 \cdot 6H_2O + Ce(NO_3)_3 \cdot 4H_2O$
	43.98	4.565	26.48	14.23	$Ce(NO_3)_3 \cdot 4H_2O$
	31.50	3.672	42.20	25.46	"
	26.16	3.242	49.10	31.50	"
	24.40	3.597	54.80	41.81	"
	24.40	3.817	56.00	45.34	"
50	68.85	6.777	———	———	$Ce(NO_3)_3 \cdot 6H_2O$
52	69.35	6.938	———	———	$Ce(NO_3)_3 \cdot 6H_2O$

continued.......

COMPONENTS:

(1) Cerium nitrate; $Ce(NO_3)_3$; [10108-73-3]

(2) Nitric acid; HNO_3; [7697-37-2]

(3) Water; H_2O; [7732-18-5]

ORIGINAL MEASUREMENTS:

Angelov, I.I. *Tr., Vses. Nauch-Issled. Inst. Khim. Reakt. Osobo Chist. Khim. Veshchestv.* 1958, *No. 39*, 18.

EXPERIMENTAL VALUES: continued.......

The $Ce(NO_3)_3$ - HNO_3 - H_2O system [a]

t/°C	$Ce(NO_3)_3$ mass %	mol kg^{-1}	HNO_3 mass %	mol kg^{-1}	nature of the solid phase
53	69.50	6.987	——	——	$Ce(NO_3)_3 \cdot 6H_2O$
54	69.12	6.863	——	——	$Ce(NO_3)_3 \cdot 4H_2O$
55	69.50	6.987	——	——	$Ce(NO_3)_3 \cdot 4H_2O$
56	70.87	7.460	——	——	$Ce(NO_3)_3 \cdot 4H_2O$
60	73.80	8.637	——	——	$Ce(NO_3)_3 \cdot 4H_2O$
	70.86	8.302	2.97	1.801	"
	65.08	7.522	8.39	5.019	"
	39.40	4.891	35.90	23.07	"
	35.93	4.309	38.50	23.89	"
	34.81	4.330	40.54	26.10	"
	25.80	3.629	52.40	38.15	"
80	77.77	10.727	——	——	$Ce(NO_3)_3 \cdot 4H_2O$

a. Molalities calculated by M. Salomon.

COMPONENTS:

(1) Cerium nitrate; $Ce(NO_3)_3$; [10108-73-3]

(2) Nitric acid; HNO_3; [7697-37-2]

(3) Water; H_2O; [7732-18-5]

ORIGINAL MEASUREMENTS:

Popov, A.P.; Mironov, K.E. *Izv. Sib. Otd. Akad. Nauk. SSSR, Ser. Khim. Nauk.* 1967, *12*, 109-11.

VARIABLES:

Composition at 25°C

PREPARED BY:

T. Mioduski and S. Siekierski

EXPERIMENTAL VALUES:

The $Ce(NO_3)_3$ - HNO_3 - H_2O system at 25°C

Composition of saturated solutions [a]

$Ce(NO_3)_3$ mass %	$Ce(NO_3)_3$ mol kg^{-1}	HNO_3 mass %	HNO_3 mol kg^{-1}	nature of the solid phase
63.4	5.31	——	——	$Ce(NO_3)_3 \cdot 6H_2O$
58.1	4.88	5.4	2.35	"
45.8	3.99	19.0	8.57	"
41.6	3.76	24.5	11.47	"
40.5	3.87	27.4	13.55	$Ce(NO_3)_3 \cdot 6H_2O + Ce(NO_3)_3 \cdot 5H_2O$
39.8	3.92	29.1	14.85	$Ce(NO_3)_3 \cdot 5H_2O$
38.0	3.81	31.4	16.28	$Ce(NO_3)_3 \cdot 4H_2O$
34.9	3.47	34.3	17.67	"
33.2	3.25	35.5	18.00	"
26.4	2.49	41.1	20.07	"
23.5	3.68	57.1	46.47	"
METASTABLE EQUILIBRIUM				
66.6	6.11	——	——	$Ce(NO_3)_3 \cdot 5H_2O$
59.0	5.67	9.1	4.53	"
47.4	4.47	20.1	9.81	"
38.1	3.93	32.2	17.21	"

a. Molalities calculated by M. Salomon.

AUXILIARY INFORMATION

METHOD/APPARATUS/PROCEDURE:

The isothermal method was used. Equilibrium was approach from above and below, and 4-5 hours were required to reach equilibrium. The composition of the solid phases were determined by Schreinemakers' method, and by refractometric measurements.

SOURCE AND PURITY OF MATERIALS:

Nothing specified.

ESTIMATED ERROR:

Soly: precision about 0.2 to 0.3 % (compilers).

Temp: precision probably ± 0.2K (compilers).

REFERENCES:

COMPONENTS:	ORIGINAL MEASUREMENTS:
(1) Lithium nitrate; $LiNO_3$; [7790-69-4] (2) Cerium nitrate; $Ce(NO_3)_3$; [10108-73-3] (3) Water; H_2O; [7732-18-5]	Bogdanovskaya, R.L.; Sheveleva, A.D. *Zh. Neorg. Khim.* 1965, *10*, 1713-5; *Russ. J. Inorg. Chem. Engl. Transl.* 1965, *10*,
VARIABLES:	PREPARED BY:
Composition and temperature	T. Mioduski and S. Siekierski

EXPERIMENTAL VALUES:

Composition of saturated solutions at 10°C [a]

$Ce(NO_3)_3$		$Li(NO_3)_3$		
mass %	mol kg^{-1}	mass %	mol kg^{-1}	nature of the solid phase
——	——	37.5	8.70	$LiNO_3 \cdot 3H_2O$
3.6	0.18	35.6	8.49	"
14.5	0.83	31.7	8.55	"
21.4	1.31	28.5	8.25	"
31.5	2.14	23.4	7.53	"
39.2	3.06	21.5	7.94	"
48.0	4.46	19.0	8.35	$LiNO_3 \cdot 3H_2O + Ce(NO_3)_3 \cdot 6H_2O$
49.0	4.35	16.5	6.94	$Ce(NO_3)_3 \cdot 6H_2O$
50.0	4.21	13.6	5.42	"
53.7	4.36	8.5	3.26	"
58.0	4.48	2.3	0.84	"
60.0	4.60	——	——	"

a. Molalities calculated by M. Salomon.

continued.......

AUXILIARY INFORMATION

METHOD/APPARATUS/PROCEDURE:

The method of isothermal sections was used with refractometric analysis (1). Homogeneous and heterogeneous mixtures of known compositions were equilibrated until the refractive index remained constant: this required 8 hours at 30°C, 10 hours at 20°C, and 12-14 hours at 10°C. The composition of the saturated solutions and the corresponding solid phases were found as inflection or "break" points on a plot of composition against refractive index. All refractive indices were measured at 30°C.

The phase diagrams are of the simple eutonic type.

SOURCE AND PURITY OF MATERIALS:

C.p. grade $Ce(NO_3)_3 \cdot 6H_2O$ was recrystallized before use: analysis for water yielded a value of 25.9 mass %.

C.p. grade $LiNO_3 \cdot 3H_2O$ was recrystallized before use: analysis for water yielded a value of 43.1%. The anhydrous salt was prepared by drying at 110°C followed by "prolonged" storage over $CaCl_2$.

ESTIMATED ERROR:

Nothing specified (see critical evaluation).

REFERENCES:

1. Zhuravlev, E.F.; Sheveleva, A.D. *Zh. Neorg. Khim.* 1960, *5*, 2630.

COMPONENTS:

(1) Lithium nitrate; $LiNO_3$; [7790-69-4]

(2) Cerium nitrate; $Ce(NO_3)_3$; [10108-73-3]

(3) Water; H_2O; [7732-18-5]

ORIGINAL MEASUREMENTS:

Bogdanovskaya, R.L.; Sheveleva, A.D.
Zh. Neorg. Khim. 1965, *10*, 1713-5;
Russ. J. Inorg. Chem. Engl. Transl. 1965, *10*,

EXPERIMENTAL VALUES: continued......

Composition of saturated solutions at 20°C [a]

$Ce(NO_3)_3$		$Li(NO_3)_3$		
mass %	mol kg^{-1}	mass %	mol kg^{-1}	nature of the solid phase
——	——	41.5	10.29	$LiNO_3 \cdot 3H_2O$
2.7	0.15	40.9	10.52	"
12.7	0.74	34.5	9.48	"
20.7	1.31	31.0	9.31	"
27.5	1.94	29.0	9.67	"
34.0	2.64	26.5	9.73	"
38.5	3.11	23.5	8.97	"
43.0	3.88	23.0	9.81	"
46.0	4.34	21.5	9.60	"
47.0	4.43	20.5	9.15	$LiNO_3 \cdot 3H_2O + Ce(NO_3)_3 \cdot 6H_2O$
48.5	4.31	17.0	7.15	$Ce(NO_3)_3 \cdot 6H_2O$
52.2	4.50	12.2	4.97	"
57.5	4.97	7.0	2.86	"
60.0	4.87	2.2	0.84	"
61.5	4.90	——	——	"

Composition of saturated solutions at 30°C

$Ce(NO_3)_3$		$Li(NO_3)_3$		
mass %	mol kg^{-1}	mass %	mol kg^{-1}	nature of the solid phase
——	——	58.0	20.03	$LiNO_3 \cdot 3H_2O$
16.1	1.32	46.5	18.03	"
32.2	3.49	39.5	20.24	"
37.7	4.31	35.5	19.21	"
44.5	5.81	32.0	19.75	$CeNO_3 \cdot 3H_2O + Ce(NO_3)_3 \cdot 6H_2O$
45.8	5.69	29.5	17.32	$Ce(NO_3)_3 \cdot 6H_2O$
51.0	5.21	19.0	9.19	"
54.1	4.78	11.2	4.68	"
58.2	5.10	6.8	2.82	"
62.3	5.28	1.5	0.60	"
63.0	5.22	——	——	"

a. Molalities calculated by M. Salomon.

COMPONENTS:

(1) Sodium nitrate; $NaNO_3$; [7631-99-4]

(2) Cerium nitrate; $Ce(NO_3)_3$; [10108-73-3]

(3) Water; H_2O; [7732-18-5]

ORIGINAL MEASUREMENTS:

Zhuravlev, E.F.; Sheveleva, A.D.; Bogdanovskaya, R.L.; Kudryashov, S.F.; Shchurov, V.A. *Zh. Neorg. Khim.* 1963, *8*, 1955-63; *Russ. J. Inorg. Chem. Engl. Transl.* 1963, *8*, 1017-21.

VARIABLES:

Composition at 10°C, 20°C, and 30°C

PREPARED BY:

T. Mioduski and S. Siekierski

EXPERIMENTAL VALUES: Composition of saturated solutions [a]

t/°C	$Ce(NO_3)_3$ mass %	$Ce(NO_3)_3$ mol kg^{-1}	$NaNO_3$ mass %	$NaNO_3$ mol kg^{-1}	nature of the solid phase
10	——	——	44.5	9.43	$NaNO_3$
	13.6	0.77	32.2	6.99	"
	35.0	2.32	18.7	4.75	"
	51.3	4.06	10.0	3.04	"
	56.3	4.77	7.5	2.44	$NaNO_3 + Ce(NO_3)_3 \cdot 6H_2O$
	57.7	4.54	3.3	1.00	$Ce(NO_3)_3 \cdot 6H_2O$
	60.0	4.60	——	——	"
20	——	——	46.8	10.35	$NaNO_3$
	12.9	0.77	35.4	8.06	"
	34.3	2.31	20.2	5.22	"
	50.3	4.06	11.7	3.62	"
	59.0	5.32	7.0	2.42	$NaNO_3 + Ce(NO_3)_3 \cdot 6H_2O$
	60.0	5.11	4.0	1.31	$Ce(NO_3)_3 \cdot 6H_2O$
	61.5	4.90	——	——	"
30	——	——	49.0	11.30	$NaNO_3$
	12.4	0.77	38.0	9.01	"
	33.2	2.31	22.7	6.06	"
	49.5	4.07	13.2	4.16	"
	62.3	6.10	6.4	2.41	$NaNO_3 + Ce(NO_3)_3 \cdot 6H_2O$
	62.5	5.60	3.3	1.14	$Ce(NO_3)_3 \cdot 6H_2O$
	63.0	5.22	——	——	"

a. Molalities calculated by M. Salomon.

AUXILIARY INFORMATION

METHOD/APPARATUS/PROCEDURE:

The method of isothermal sections was used with refractometric analysis (1). Homogeneous and heterogeneous mixtures of known compositions were equilibrated until the refractive index remained constant. The composition of the saturated solutions and the corresponding solid phases were found as inflection or "break" points on a plot of refractive index against composition. All refractive indices were measured at 30°C.

The phase diagrams are of the simple eutonic type.

SOURCE AND PURITY OF MATERIALS:

"Pure" and c.p. grade salts were used.

ESTIMATED ERROR:

Nothing specified (see critical evaluation).

REFERENCES:

1. Zhuravlev, E.F.; Sheveleva, A.D. *Zh. Neorg. Khim.* 1960, *5*, 2630.

COMPONENTS:

(1) Potassium nitrate; KNO_3; [7757-79-1]

(2) Cerium nitrate; $Ce(NO_3)_3$; [10108-73-3]

(3) Water; H_2O; [7732-18-5]

ORIGINAL MEASUREMENTS:

Zhuravlev, E.F.; Sheveleva, A.D.; Bogdanovskaya, R.L.; Kudryashov, S.P.; Shchurov, V.A. *Zh. Neorg. Khim.* 1963, *8*, 1955-63; *Russ. J. Inorg. Chem. Engl. Transl.* 1963, *8*, 1017-21.

VARIABLES:

Composition at 10°C, 20°C, and 30°C

PREPARED BY:

T. Mioduski and S. Siekierski

EXPERIMENTAL VALUES:

Composition of saturated solutions at 10°C [a]

$Ce(NO_3)_3$		KNO_3		nature of the solid phase
mass %	mol kg^{-1}	mass %	mol kg^{-1}	
——	——	17.7	2.13	KNO_3
17.8	0.77	11.0	1.53	"
37.6	2.31	12.5	2.48	"
43.0	3.07	14.0	3.22	"
46.5	3.82	16.2	4.30	$KNO_3 + 2KNO_3\cdot Ce(NO_3)_3\cdot 2H_2O$
47.0	3.85	15.6	4.13	$2KNO_3\cdot Ce(NO_3)_3\cdot 2H_2O$
49.8	4.04	12.4	3.24	"
55.0	4.68	9.0	2.47	"
58.0	5.11	7.2	2.05	$2KNO_3\cdot Ce(NO_3)_3\cdot 2H_2O + Ce(NO_3)_3\cdot 6H_2O$
59.0	4.93	4.3	1.16	$Ce(NO_3)_3\cdot 6H_2O$
60.0	4.60	——	——	"

a. Molalities calculated by M. Salomon.

continued......

AUXILIARY INFORMATION

METHOD/APPARATUS/PROCEDURE:

The method of isothermal sections was used with refractometric analyses (1). Heterogeneous and homogeneous mixtures of known composition were equilibrated until their refractive indices remained constant. The composition of the saturated solutions and the corresponding solid phases were found as inflection or "break" points on a plot of composition against refractive index. All refractive indices were measured at 30°C.

SOURCE AND PURITY OF MATERIALS:

"Pure" and c.p. grade salts were used.

No other information given.

ESTIMATED ERROR:

Nothing specified (see critical evaluation).

REFERENCES:

1. Zhuravlev, E.F.; Sheveleva, A.D. *Zh. Neorg. Khim.* 1960, *5*, 2630.

COMPONENTS:

(1) Potassium nitrate; KNO_3; [7757-79-1]

(2) Cerium nitrate; $Ce(NO_3)_3$; [10108-73-3]

(3) Water; H_2O; [7732-18-5]

ORIGINAL MEASUREMENTS:

Zhuravlev, E.F.; Sheveleva, A.D.; Bogdanovskaya, R.L.; Kudryashov, S.P.; Shchurov, V.A. *Zh. Neorg. Khim.* 1963, *8*, 1955-63; *Russ. J. Inorg. Chem. Engl. Transl.* 1963, *8*, 1017-21.

EXPERIMENTAL VALUES: continued.......

Composition of saturated solutions at 20°C [a]

$Ce(NO_3)_3$		KNO_3		
mass %	mol kg^{-1}	mass %	mol kg^{-1}	nature of the solid phase
——	——	24.1	3.14	KNO_3
16.4	0.77	18.0	2.71	"
36.1	2.31	16.0	3.30	"
41.0	3.07	18.0	4.34	"
45.5	3.80	17.8	4.80	$KNO_3 + 2KNO_3 \cdot Ce(NO_3)_3 \cdot 2H_2O$
47.5	4.05	16.5	4.53	$2KNO_3 \cdot Ce(NO_3)_3 \cdot 2H_2O$
50.4	4.18	12.6	3.37	"
55.5	4.86	9.5	2.68	"
60.0	5.64	7.4	2.25	$2KNO_3 \cdot Ce(NO_3)_3 \cdot 2H_2O + Ce(NO_3)_3 \cdot 6H_2O$
60.5	5.21	3.9	1.08	$Ce(NO_3)_3 \cdot 6H_2O$
61.6	4.90	——	——	"

Composition of saturated solutions at 30°C

$Ce(NO_3)_3$		KNO_3		
mass %	mol kg^{-1}	mass %	mol kg^{-1}	nature of the solid phase
——	——	31.5	4.55	KNO_3
15.4	0.77	23.0	3.69	"
34.4	2.31	20.0	4.34	"
40.0	3.07	20.0	4.95	"
45.0	4.03	20.8	6.02	$KNO_3 + 2KNO_3 \cdot Ce(NO_3)_3 \cdot 2H_2O$
48.0	4.27	17.5	5.02	$2KNO_3 \cdot Ce(NO_3)_3 \cdot 2H_2O$
51.8	4.53	13.1	3.69	"
56.0	5.13	10.5	3.10	"
62.0	6.34	8.0	2.64	$2KNO_3 \cdot Ce(NO_3)_3 \cdot 2H_2O + Ce(NO_3)_3 \cdot 6H_2O$
62.5	5.62	3.4	0.99	$Ce(NO_3)_3 \cdot 6H_2O$
63.0	5.22	——	——	"

a. Molalities calculated by M. Salomon.

COMMENTS AND/OR ADDITIONAL DATA:

The solubility isotherms consist of three crystallization branches representing $Ce(NO_3)_3 \cdot 6H_2O$, $2KNO_3 \cdot Ce(NO_3)_3 \cdot 2H_2O$, and KNO_3. The composition of saturated solutions at the eutonic points were determined by graphical extrapolation. The double nitrate is incongruently soluble.

COMPONENTS:

(1) Rubidium nitrate; $RbNO_3$; [13126-12-0]

(2) Cerium nitrate; $Ce(NO_3)_3$; [10108-73-3]

(3) Water; H_2O; [7732-18-5]

ORIGINAL MEASUREMENTS:

Zhuravlev, E.F.; Sheveleva, A.D.; Bogdanovskaya, R.L.; Kudryashov, S.P.; Shchurov, V.A. *Zh. Neorg. Khim.* 1963, *8*, 1955-63; *Russ. J. Inorg. Chem. Engl. Transl.* 1963, *8*, 1017-21.

VARIABLES:

Composition at 10°C, 20°C, and 30°C

PREPARED BY:

T. Mioduski and S. Siekierski

EXPERIMENTAL VALUES:

Composition of saturated solutions at 10°C [a]

$Ce(NO_3)_3$ mass %	$Ce(NO_3)_3$ mol kg^{-1}	$RbNO_3$ mass %	$RbNO_3$ mol kg^{-1}	nature of the solid phase
———	———	25.5	2.32	$RbNO_3$
16.0	0.77	20.2	2.15	"
23.0	1.38	26.0	3.46	"
35.5	3.05	28.8	5.47	"
36.5	3.29	29.5	5.88	$RbNO_3 + 2RbNO_3\cdot Ce(NO_3)_3\cdot 4H_2O$
37.5	3.29	27.5	5.33	$2RbNO_3\cdot Ce(NO_3)_3\cdot 4H_2O$
43.8	3.56	18.5	3.33	"
55.5	4.86	9.5	1.84	"
58.2	5.13	7.0	1.36	$2RbNO_3\cdot Ce(NO_3)_3\cdot 4H_2O + Ce(NO_3)_3\cdot 6H_2O$
59.0	4.82	3.5	0.63	$Ce(NO_3)_3\cdot 6H_2O$
60.0	4.60	———	———	"

a. Molalities calculated by M. Salomon.

continued......

AUXILIARY INFORMATION

METHOD/APPARATUS/PROCEDURE:

The method of isothermal sections was used with refractometric analyses (1). Heterogeneous and homogeneous mixtures of known composition were equilibrated until their refractive indices remained constant. The composition of the saturated solutions and the corresponding solid phase were found as inflection of "break" points on a plot of composition against refractive index. All refractive indices were measured at 30°C.

SOURCE AND PURITY OF MATERIALS:

"Pure" and c.p. grade salts were used.

No other information given.

ESTIMATED ERROR:

Nothing specified (see critical evaluation).

REFERENCES:

1. Zhuravlev, E.F.; Sheveleva, A.D. *Zh. Neorg. Khim.* 1960, 5, 2630.

COMPONENTS:

(1) Rubidium nitrate; $RbNO_3$; [13126-12-0]

(2) Cerium nitrate; $Ce(NO_3)_3$; [10108-73-3]

(3) Water; H_2O; [7732-18-5]

ORIGINAL MEASUREMENTS:

Zhuravlev, E.F.; Sheveleva, A.D.; Bogdanovskaya, R.L.; Kudryashov, S.P.; Shchurov, V.A. *Zh. Neorg. Khim.* 1963, *8*, 1955-63; *Russ. J. Inorg. Chem. Engl. Transl.* 1963, *8*, 1017-21.

EXPERIMENTAL VALUES: continued......

Composition of saturated solutions at 20°C [a]

$Ce(NO_3)_3$ mass %	$Ce(NO_3)_3$ mol kg^{-1}	$RbNO_3$ mass %	$RbNO_3$ mol kg^{-1}	nature of the solid phase
——	——	35.0	3.65	$RbNO_3$
14.0	0.77	30.0	3.63	"
29.3	2.30	31.7	5.51	"
33.6	3.10	33.2	6.78	"
35.6	3.64	34.4	7.78	$RbNO_3 + 2RbNO_3 \cdot Ce(NO_3)_3 \cdot 4H_2O$
38.5	3.68	29.5	6.25	$2RbNO_3 \cdot Ce(NO_3)_3 \cdot 4H_2O$
45.1	4.12	21.3	4.30	"
54.0	5.02	13.0	2.67	"
59.9	5.63	7.5	1.56	$2RbNO_3 \cdot Ce(NO_3)_3 \cdot 4H_2O + Ce(NO_3)_3 \cdot 6H_2O$
60.3	5.52	6.2	1.25	$Ce(NO_3)_3 \cdot 6H_2O$
61.2	5.27	3.2	0.61	"
61.5	4.90	——	——	"
Composition of saturated solutions at 30°C				
——	——	42.3	4.97	$RbNO_3$
11.8	0.76	40.4	5.73	"
26.5	2.31	38.4	7.42	"
30.5	3.07	39.0	8.67	"
34.5	4.07	39.5	10.30	$RbNO_3 + 2RbNO_3 \cdot Ce(NO_3)_3 \cdot 4H_2O$
39.5	4.15	31.3	7.27	$2RbNO_3 \cdot Ce(NO_3)_3 \cdot 4H_2O$
45.2	4.54	24.3	5.40	"
53.0	5.36	16.7	3.74	"
62.7	6.68	8.5	2.00	$2RbNO_3 \cdot Ce(NO_3)_3 \cdot 4H_2O + Ce(NO_3)_3 \cdot 6H_2O$
62.8	6.06	5.4	1.15	$Ce(NO_3)_3 \cdot 6H_2O$

a. Molalities calculated by M. Salomon.

COMMENTS AND/OR ADDITIONAL DATA:

The solubility isotherms consist of three crystallization branches representing $Ce(NO_3)_3 \cdot 6H_2O$, $2RbNO_3 \cdot Ce(NO_3)_3 \cdot 4H_2O$, and $RbNO_3$. The double salt is incongruently soluble at 10°C, and is congruently soluble at 20°C and 30°C.

COMPONENTS:

(1) Cesium nitrate; $CsNO_3$; [7789-18-6]

(2) Cerium nitrate; $Ce(NO_3)_3$; [10108-73-3]

(3) Water; H_2O; [7732-18-5]

ORIGINAL MEASUREMENTS:

Zhuravlev, E.F.; Sheveleva, A.D.; Bogdanovskaya, R.L.; Kudryashov, S.F.; Shchurov, V.A. *Zh. Neorg. Khim.* 1963, *8*, 1955-63; *Russ. J. Inorg. Chem. Engl. Transl.* 1963, *8*, 1017-21.

VARIABLES:

Composition at 10°C, 20°C, and 30°C

PREPARED BY:

T. Mioduski and S. Siekierski

EXPERIMENTAL VALUES:

Composition of saturated solutions at 10°C [a]

$Ce(NO_3)_3$ mass %	$Ce(NO_3)_3$ mol kg^{-1}	$CsNO_3$ mass %	$CsNO_3$ mol kg^{-1}	nature of the solid phase
——	——	11.8	0.69	$CsNO_3$
18.6	0.78	8.7	0.61	"
34.5	2.07	14.5	1.46	"
40.5	3.08	19.2	2.44	"
44.4	4.08	22.2	3.41	"
45.8	4.62	23.8	4.02	$CsNO_3 + 2CsNO_3 \cdot Ce(NO_3)_3 \cdot 4H_2O$
51.5	5.26	18.5	3.16	$2CsNO_3 \cdot Ce(NO_3)_3 \cdot 4H_2O$
55.0	5.78	15.8	2.78	$2CsNO_3 \cdot Ce(NO_3)_3 \cdot 4H_2O + Ce(NO_3)_3 \cdot 6H_2O$
55.4	5.59	14.2	2.40	$Ce(NO_3)_3 \cdot 6H_2O$
57.7	5.11	7.7	1.14	"
60.0	4.60	——	——	"

a. Molalities calculated by M. Salomon.

continued.......

AUXILIARY INFORMATION

METHOD/APPARATUS/PROCEDURE:

The method of isothermal sections was used with refractometric analyses (1). Heterogeneous and homogeneous mixtures of known composition were equilibrated until their refractive indices remained constant. The composition of the saturated solutions and the corresponding solid phases were found as inflection or "break" points on a plot of composition against refractive index. All refractive indices were measured at 30°C.

The double nitrate, $2CsNO_3 \cdot Ce(NO_3)_3 \cdot 4H_2O$, is congruently soluble.

SOURCE AND PURITY OF MATERIALS:

"Pure" and c.p. grade salts were used.

No other information given.

ESTIMATED ERROR:

Nothing specified (see critical evaluation).

REFERENCES:

1. Zhuravlev, E.F.; Sheveleva, A.D. *Zh. Neorg. Khim.* 1960, *5*, 2630.

COMPONENTS:

(1) Cesium nitrate; $CsNO_3$; [7789-18-6]

(2) Cerium nitrate; $Ce(NO_3)_3$; [10108-73-3]

(3) Water; H_2O; [7732-18-5]

ORIGINAL MEASUREMENTS:

Zhuravlev, E.F.; Sheveleva, A.D.; Bogdanovskaya, R.L.; Kudryashov, S.F.; Shchurov, V.A. *Zh. Neorg. Khim.* 1963, *8*, 1955-63; *Russ. J. Inorg. Chem. Engl. Transl.* 1963, *8*, 1017-21.

EXPERIMENTAL VALUES: continued.......

Composition of saturated solutions at 20°C [a]

$Ce(NO_3)_3$		$CsNO_3$		
mass %	mol kg^{-1}	mass %	mol kg^{-1}	nature of the solid phase
——	——	17.5	1.09	$CsNO_3$
17.7	0.79	13.3	0.99	"
33.0	2.07	18.1	1.90	"
39.0	3.08	22.2	2.94	"
42.7	4.08	25.2	4.03	"
44.8	4.87	27.0	4.91	$CsNO_3 + 2CsNO_3 \cdot Ce(NO_3)_3 \cdot 4H_2O$
45.5	5.00	26.6	4.89	$2CsNO_3 \cdot Ce(NO_3)_3 \cdot 4H_2O$
50.3	5.53	21.8	4.01	"
54.3	6.24	19.0	3.65	"
58.5	6.82	15.2	2.97	$2CsNO_3 \cdot Ce(NO_3)_3 \cdot 4H_2O + Ce(NO_3)_3 \cdot 6H_2O$
59.2	6.24	11.7	2.06	$Ce(NO_3)_3 \cdot 6H_2O$
60.4	5.63	6.7	1.04	"
61.5	4.90	——	——	"

Composition of saturated solutions at 30°C [a]

$Ce(NO_3)_3$		$CsNO_3$		
mass %	mol kg^{-1}	mass %	mol kg^{-1}	nature of the solid phase
——	——	25.5	1.76	$CsNO_3$
16.3	0.77	18.8	1.49	"
31.7	2.07	21.3	2.33	"
37.8	3.07	24.5	3.33	"
41.3	4.03	27.3	4.46	"
43.7	5.19	30.5	6.07	$CsNO_3 + 2CsNO_3 \cdot Ce(NO_3)_3 \cdot 4H_2O$
44.9	5.27	29.0	5.70	$2CsNO_3 \cdot Ce(NO_3)_3 \cdot 4H_2O$
49.7	5.91	24.5	4.87	"
53.5	6.48	21.2	4.30	"
62.3	8.06	14.0	3.03	$2CsNO_3 \cdot Ce(NO_3)_3 \cdot 4H_2O + Ce(NO_3)_3 \cdot 6H_2O$
62.8	6.90	9.3	1.71	$Ce(NO_3)_3 \cdot 6H_2O$
63.0	6.09	5.3	0.86	"
63.0	5.22	——	——	"

a. Molalities calculated by M. Salomon.

COMPONENTS:

(1) Beryllium nitrate; $Be(NO_3)_2$; [13597-99-4]
(2) Cerium nitrate; $Ce(NO_3)_3$; [10108-73-3]
(3) Water; H_2O; [7732-18-5]

ORIGINAL MEASUREMENTS:

Kudryashov, S.F.; Frolova, S.I.; Ushkova, A.V. *Zh. Neorg. Khim.* 1967, *12*, 2494-6; *Russ. J. Inorg. Chem. Engl. Transl.* 1967, *12*, 1316-7.

VARIABLES:

Composition at 0°C and 20°C

PREPARED BY:

T. Mioduski and S. Siekierski

EXPERIMENTAL VALUES:

Composition of saturated solutions at 0°C [a]

$Ce(NO_3)_3$ mass %	$Ce(NO_3)_3$ mol kg^{-1}	$Be(NO_3)_2$ mass %	$Be(NO_3)_2$ mol kg^{-1}	n_D^{40}	nature of the solid phase
——	——	51.50	7.983	1.4200	$Be(NO_3)_2 \cdot 4H_2O$
5.24	0.342	47.80	7.652	1.4243	"
13.58	0.955	42.80	7.376	1.4313	"
24.80	1.960	36.40	7.053	1.4425	"
26.40 [b]	2.103	35.10	6.854	1.4450	$Be(NO_3)_2 \cdot 4H_2O + Ce(NO_3)_3 \cdot 6H_2O$
26.66	1.982	32.10	5.851	1.4370	$Ce(NO_3)_3 \cdot 6H_2O$
27.30	1.856	27.61	4.603	1.4295	"
28.58	1.882	24.85	4.011	1.4260	"
36.20	2.359	16.75	2.676	1.4255	"
50.25	3.610	7.07	1.245	1.4360	"
57.80	4.200	——	——	1.4450	"

a. Molalities calculated by M. Salomon.

b. Eutonic point found by extrapolation.

continued.......

AUXILIARY INFORMATION

METHOD/APPARATUS/PROCEDURE:

The solubilities of the pure salts and their mixtures was studied by the method of isothermal sections (1). Solutions were prepared along eight sections of the phase diagram, and concentrations were determined by measuring the refractive indices and comparing with refractive indices of standard solutions. Solutions were equilibrated until their refractive indices remained constant. All refractive indices were measured at 40°C.

The solid solid phase was identified microscopically which is facilitated by the fact that $Be(NO_3)_2 \cdot 4H_2O$ crystals float to the surface of the saturated solutions and $Ce(NO_3)_3 \cdot 6H_2O$ crystals sink to the bottom.

SOURCE AND PURITY OF MATERIALS:

A.R. grade nitrates were recrystallized from water to give $Ce(NO_3)_3 \cdot 6H_2O$ and $Be(NO_3)_2 \cdot 4H_2O$: analyses for water gave 24.88 and 35.14 mass %, respectively.

No other information given.

ESTIMATED ERROR:

Soly: precision about ± 1 % (compilers).

Temp: precision probably ± 0.2 K (compilers).

REFERENCES:

1. Zhuravlev, E.F.; Sheveleva, A.D. *Zh. Neorg. Khim.* 1960, *5*, 2630.

COMPONENTS:	ORIGINAL MEASUREMENTS:
(1) Beryllium nitrate; $Be(NO_3)_2$; [13597-99-4] (2) Cerium nitrate; $Ce(NO_3)_3$; [10108-73-3] (3) Water; H_2O; [7732-18-5]	Kudryashov, S.F.; Frolova, S.I.; Ushkova, A.V. *Zh. Neorg. Khim.* 1967, *12*, 2494-6; *Russ. J. Inorg. Chem. Engl. Transl.* 1967, *12*, 1316-7.

EXPERIMENTAL VALUES: continued.......

Composition of saturated solutions at 20°C [a]

$Ce(NO_3)_3$ mass %	$Ce(NO_3)_3$ mol kg^{-1}	$Be(NO_3)_2$ mass %	$Be(NO_3)_2$ mol kg^{-1}	n_D^{40}	nature of the solid phase
——	——	53.70	8.719	1.4260	$Be(NO_3)_2 \cdot 4H_2O$
4.31	0.295	50.90	8.543	1.4292	"
10.90	0.800	47.30	8.507	1.4350	"
25.80	2.197	38.20	7.977	1.4490	"
33.77	3.196	33.83	7.849	1.4595	"
36.50 [b]	3.669	33.00	8.134	1.4630	$Be(NO_3)_2 \cdot 4H_2O + Ce(NO_3)_3 \cdot 6H_2O$
36.80	2.985	25.40	5.051	1.4480	$Ce(NO_3)_3 \cdot 6H_2O$
41.00	3.143	19.00	3.571	1.4420	"
41.00	3.086	18.26	3.369	1.4420	"
55.00	4.268	5.49	1.045	1.4500	"
62.50	5.110	——	——	1.4570	"

a. Molalities calculated by

b. Eutonic point found by extrapolation.

COMMENTS AND/OR ADDITIONAL DATA:

The phase diagram at 0°C and 20°C are of the simple eutonic type. The solubility isotherms consist of two branches corresponding to the crystallization of $Ce(NO_3)_3 \cdot 6H_2O$ and $Be(NO_3)_2 \cdot 4H_2O$. The cerium nitrate composition at the eutonic point increases with temperature.

COMPONENTS:

(1) Magnesium nitrate; $Mg(NO_3)_2$; [10377-60-3]

(2) Cerium nitrate; $Ce(NO_3)_3$; [10108-73-3]

(3) Water; H_2O; [7732-18-5]

ORIGINAL MEASUREMENTS:

Yakimov, M.A.; Gizhavina, E.I. *Zh. Neorg. Khim.* 1971, *16*, 507-9; *Russ. J. Inorg. Chem. Engl. Transl.* 1971, *16*, 268-9.

VARIABLES:

Composition at 25°C

PREPARED BY:

T. Mioduski and S. Siekierski

EXPERIMENTAL VALUES:

Composition of saturated solutions[a]

$Ce(NO_3)_3$ mass %	$Ce(NO_3)_3$ mol kg^{-1}	$Mg(NO_3)_2$ mass %	$Mg(NO_3)_2$ mol kg^{-1}	nature of the solid phase
62.73[b]	5.161	——	——	$Ce(NO_3)_3 \cdot 6H_2O$
61.81	5.082	0.90	0.163	$Ce(NO_3)_3 \cdot 6H_2O + Mg_3Ce_2(NO_3)_{12} \cdot 24H_2O$
48.51	3.381	7.50	1.149	$Mg_3Ce_2(NO_3)_{12} \cdot 24H_2O$
41.82	2.724	11.10	1.590	"
27.06	1.565	19.92	2.533	"
18.06	0.986	25.80	3.098	"
7.50	0.388	33.24	3.782	"
2.02	0.107	40.12	4.675	"
0.56[c]	0.030	41.87	4.903	$Mg_3Ce_2(NO_3)_{12} \cdot 24H_2O + Mg(NO_3)_2 \cdot 6H_2O$
——	——	41.80	4.842	$Mg(NO_3)_2 \cdot 6H_2O$

a. Molalities calculated by M. Salomon.

b. Vapor pressure of this solution is 13.05 mm Hg (1).

c. 0.55 mass % in the English Translation.

AUXILIARY INFORMATION

METHOD/APPARATUS/PROCEDURE:

The isothermal method was used. The composition of the saturated solutions was determined by chemical analysis, and the composition of the solid phases was determined by Schreinemakers' method.

Cerium was separated from magnesium by double precipitation with ammonium buffer ($NH_4Cl + NH_4OH$) followed by heating the cerium hydroxide to give CeO_2. The magnesium in the filtrate was determined by trilonometric titration.

SOURCE AND PURITY OF MATERIALS:

Nothing specified.

ESTIMATED ERROR:

Soly: precision about ± 0.5% (compilers).

Temp: precision probably ± 0.2K (compilers).

REFERENCES:

(1) Yakimov, M.A.; Gizhavina, E.J. *Zh. Neorg. Khim.* 1971, *16*, 1756.

COMPONENTS:

(1) Magnesium nitrate; $Mg(NO_3)_2$; [10377-60-3]

(2) Cerium nitrate; $Ce(NO_3)_3$; [10108-73-3]

(3) Nickel nitrate; $Ni(NO_3)_2$; [13138-45-9]

(4) Water; H_2O; [7732-18-5]

ORIGINAL MEASUREMENTS:

Mitina, N.K.; Shevchuk, V.G. *Zh. Neorg. Khim.* 1977, *22*, 1376-81; *Russ. J. Inorg. Chem. Engl. Transl.* 1977, *22*, 749-51.

VARIABLES:

Composition at 25°C

PREPARED BY:

T. Mioduski and S. Siekierski

EXPERIMENTAL VALUES:

Composition of saturated solutions [a]

$Ce(NO_3)_3$		$Mg(NO_3)_2$		$Ni(NO_3)_2$		
mass %	mol kg^{-1}	mass %	mol kg^{-1}	mass %	mol kg^{-1}	solid phase [b]
5.90	0.323	38.14	4.595	———	———	A + B
4.64	0.379	53.59	9.630	4.25	0.620	"
5.36	0.396	41.34	6.713	11.78	1.553	"
5.86	0.434	35.86	5.841	16.89	2.233	A + B + C
———	———	35.81	4.121	5.60	0.523	A + C
4.13	0.383	54.91	11.205	7.92	1.312	"
5.92	0.612	43.06	9.781	21.34	3.395	B + C
5.83	0.443	25.10	4.192	28.70	3.891	"
5.36	0.424	12.52	2.176	43.32	6.111	B + C + D
———	———	19.60	2.351	24.20	2.357	C + D
2.05	0.199	27.91	5.955	38.44	6.658	"
3.60	0.275	20.44	3.431	35.80	4.879	"
———	———	3.90	0.529	46.40	5.110	D + E
1.81	0.147	4.38	0.785	56.17	8.167	"

continued......

AUXILIARY INFORMATION

METHOD/APPARATUS/PROCEDURE:

The isothermal method was used. The solns were thermostated, and equilibrium was reached in 15-20 days with continuous stirring. For those solns in which cerium was absent, the sum of magnesium and nickel nitrates was determined by titration with Trilon; when cerium was present, it was precipitated with conc NH_4OH and an excess of NH_4Cl prior to the complexometric titration. The total nitrates was determined by Devarda's method. Nickel in the presence of magnesium was determined by titration with Trilon after addition of NH_4F. Nickel in the presence of cerium and water were determined by the difference. The compilers assume that the solid phases were determined by Schreinemakers' method.

SOURCE AND PURITY OF MATERIALS:

No information given.

ESTIMATED ERROR:

Nothing specified (see critical evaluation).

REFERENCES:

COMPONENTS:

(1) Magnesium nitrate; $Mg(NO_3)_2$; [10377-60-3]

(2) Cerium nitrate; $Ce(NO_3)_3$; [10108-73-3]

(3) Nickel nitrate; $Ni(NO_3)_2$; [13138-45-9]

(4) Water; H_2O; [7732-18-5]

ORIGINAL MEASUREMENTS:

Mitina, N.K.; Shevchuk, V.G. *Zh. Neorg. Khim.* 1977, *22*, 1376-81, *Russ. J. Inorg. Chem. Engl. Transl.* 1977, *22*, 749-51.

EXPERIMENTAL VALUES: continued......

Composition of saturated solutions [a]

$Ce(NO_3)_3$		$Mg(NO_3)_2$		$Ni(NO_3)_2$		
mass %	mol kg^{-1}	mass %	mol kg^{-1}	mass %	mol kg^{-1}	solid phase [b]
5.76	0.447	7.00	1.194	42.72	6.609	B + D + E
8.99	0.617	0.33	0.050	46.03	5.642	B + E
17.46	1.495	4.74	0.893	42.00	6.421	"
4.60	0.304	———	———	49.04	5.789	E + F
8.16	0.640	2.87	0.495	49.85	6.974	"
10.51	0.829	2.37	0.411	48.27	6.800	"
15.85	1.314	3.01	0.548	44.14	6.529	"
20.66	1.701	3.11	0.563	38.98	5.727	B + E + F
28.77	3.151	3.19	0.768	40.04	7.826	B + F
34.06	3.616	3.12	0.728	33.94	6.432	"
41.35	4.351	2.56	0.592	26.65	5.062	"
50.66	5.580	3.18	0.770	18.32	3.602	"
58.20	5.343	———	———	8.40	1.376	F + G
62.73	7.336	1.28	0.329	9.77	2.039	"
58.54	7.365	3.86	1.068	13.23	2.971	B + F + G
61.80	8.385	5.24	1.563	10.36	2.509	B + G
71.07	10.852	5.45	1.830	3.40	0.927	"
58.63	4.790	3.84	0.690	———	———	"

a. Molalities calculated by M. Salomon.

b. Solid phases:

A = $Mg(NO_3)_2 \cdot 6H_2O$

B = $3Mg(NO_3)_2 \cdot 2Ce(NO_3)_3 \cdot 24H_2O$

C = solid solutions based on $Mg(NO_3)_2$ (hydrated)

D = solid solutions based on $Ni(NO_3)_2$ (hydrated)

E = $Ni(NO_3)_2 \cdot 6H_2O$

F = $2Ce(NO_3)_3 \cdot 3Ni(NO_3)_2 \cdot 24H_2O$

G = $Ce(NO_3)_3 \cdot 6H_2O$

COMMENTS AND/OR ADDITIONAL DATA:

The phase diagram for the quaternary $Ce(NO_3)_3$ - $Mg(NO_3)_2$ - $Ni(NO_3)_2$ - H_2O system has seven crystallization fields for each of the phases A - G listed above. The relative size of each field is:

A = 4.5%, B = 71.1%, C = 8.1%, D = 5.2%, E = 3.9%, F = 5.6% and G = 1.6%.

COMPONENTS:

(1) Magnesium nitrate; $Mg(NO_3)_2$; [10377-60-3]

(2) Cerium nitrate; $Ce(NO_3)_3$; [10108-73-3]

(3) Ammonium nitrate; NH_4NO_3; [6484-52-2]

(4) Water; H_2O; [7732-18-5]

ORIGINAL MEASUREMENTS:

Shevchuk, V.G.; Mitina, N.K. *Zh. Neorg. Khim.* 1976, *21*, 1943-7, *Russ. J. Inorg. Chem. Engl. Transl.* 1976, *21*, 1067-70.

VARIABLES:

Computation at 25°C

PREPARED BY:

T. Mioduski and S. Siekierski

EXPERIMENTAL VALUES:

Composition of saturated solutions at 25°C [a]

$Ce(NO_3)_3$		NH_4NO_3		$Mg(NO_3)_2$		
mass %	mol kg^{-1}	mass %	mol kg^{-1}	mass %	mol kg^{-1}	solid phase [b]
5.90	0.323	———	———	38.14	4.595	A + E
4.94	0.456	3.07	1.15	58.78	11.93	"
———	———	5.61	1.29	40.23	5.008	A + B
4.43	0.337	9.77	3.03	45.47	7.601	A + B + E
———	———	11.90	2.969	38.02	5.119	B + C
4.30	0.333	15.73	4.958	40.33	6.859	B + C + E
4.18	0.319	18.30	5.694	37.37	6.275	"
5.02	0.400	26.41	8.566	30.05	5.260	"
4.26	0.353	30.52	10.31	28.23	5.145	"
———	———	29.43	9.595	32.25	5.674	C + D
0.41	0.026	20.65	5.424	31.38	4.448	"
0.88	0.076	31.80	11.19	31.81	6.040	"
4.71	0.418	33.01	11.94	27.74	5.415	C + D + E
6.60	0.619	33.42	12.78	27.30	5.632	"

continued.......

AUXILIARY INFORMATION

METHOD/APPARATUS/PROCEDURE:

The isothermal method was used. Equilibrium was reached in 15 days. Cerium was determined by oxidation to Ce^{4+} with ammonium persulfate followed by addition of standard iron (II) sulfate, and the excess titrated with potassium dichromate. Total nitrogen was determined by the method of Devarda (no reference given), and magnesium was determined by difference.

SOURCE AND PURITY OF MATERIALS:

"Chemically pure" salts were recrystallized from their aqueous solutions.

ESTIMATED ERROR:

Nothing specified (see critical evaluation).

REFERENCES:

COMPONENTS:

(1) Magnesium nitrate; $Mg(NO_3)_2$; [10377-60-3]

(2) Cerium nitrate; $Ce(NO_3)_3$; [10108-73-3]

(3) Ammonium nitrate; NH_4NO_3; [6484-52-2]

(4) Water; H_2O; [7732-18-5]

ORIGINAL MEASUREMENTS:

Shevchuk, V.G.; Mitina, N.K. *Zh. Neorg. Khim.* 1976, *21*, 1943-7; *Russ. J. Inorg. Chem. Engl. Transl.* 1976, *21*, 1067-70.

EXPERIMENTAL VALUES: continued.......

Composition of saturated solutions at 25°C [a]

$Ce(NO_3)_3$		NH_4NO_3		$Mg(NO_3)_2$		
mass %	mol kg^{-1}	mass %	mol kg^{-1}	mass %	mol kg^{-1}	solid phase [b]
14.31	1.343	30.38	11.62	22.64	4.672	D + E + F
29.27	2.812	38.81	15.19	——	——	"
25.29	2.034	26.61	8.721	9.98	1.765	"
37.03	3.179	21.32	7.457	5.93	1.119	"
47.65	4.934	17.82	7.519	4.92	1.120	"
41.99	4.083	21.79	8.634	4.69	1.002	"
49.94	5.081	15.43	6.396	4.49	1.004	G + E + F
59.50	5.866	9.40	3.776	——	——	G + F
60.52	5.938	4.92	1.967	3.31	0.714	G + E
58.63	4.790	——	——	3.84	0.690	"

a. Molalities calculated by M. Salomon.

b. A = $Mg(NO_3)_2\cdot 6H_2O$

B = $5Mg(NO_3)_2\cdot 2NH_4NO_3\cdot 30H_2O$

C = $5Mg(NO_3)_2\cdot NH_4NO_3\cdot 30H_2O$

D = NH_4NO_3

E = $3Mg(NO_3)_2\cdot 2Ce(NO_3)_3\cdot 24H_2O$

F = $Ce(NO_3)_3\cdot 2NH_4NO_3\cdot 4H_2O$

G = $Ce(NO_3)_3\cdot 6H_2O$

COMMENTS AND/OR ADDITIONAL DATA:

The system has seven crystallisation fields:

the initial salts NH_4NO_3, $Mg(NO_3)_2\cdot 6H_2O$, and $Ce(NO_3)_3\cdot 6H_2O$, the incongruently soluble double salts $5Mg(NO_3)_2\cdot 2NH_4NO_3\cdot 30H_2O$, and $5Mg(NO_3)_2\cdot NH_4NO_3\cdot 30H_2O$, and the congruently soluble magnesium and and ammonium cerium (III) nitrates $3Mg(NO_3)_2\cdot 2Ce(NO_3)_3\cdot 24H_2O$ and $Ce(NO_3)_3\cdot 2NH_4NO_3\cdot 4H_2O$.

COMPONENTS:

(1) Calcium nitrate; $Ca(NO_3)_2$; [10124-37-5]
(2) Cerium nitrate; $Ce(NO_3)_3$; [10108-73-3]
(3) Water; H_2O; [7732-18-5]

ORIGINAL MEASUREMENTS:

Bogdanovskaya, R.L.; Kudryashov, S.F. *Uch. Zap. Permsk. Gos. Univ. im. A.M. Gor'kogo* 1966, *141*, 27.

VARIABLES:

Composition and temperature

PREPARED BY:

T. Mioduski and S. Siekierski

EXPERIMENTAL VALUES:

Composition of saturated solutions [a]

t/°C	$Ce(NO_3)_3$ mass %	$Ce(NO_3)_3$ mol kg^{-1}	$Ca(NO_3)_2$ mass %	$Ca(NO_3)_2$ mol kg^{-1}	nature of the solid phase
10	——	——	53.6	7.04	$Ca(NO_3)_2\cdot 4H_2O$
	6.6	0.43	46.6	6.07	"
	11.1	0.77	44.6	6.14	"
	19.1	1.38	38.5	5.53	"
	27.9	2.33	35.4	5.88	"
	29.4	2.53	35.0	5.99	$Ca(NO_3)_2\cdot 4H_2O + Ca(NO_3)_2\cdot Ce(NO_3)_3\cdot 4H_2O$
	29.7	2.51	34.0	5.71	$Ca(NO_3)_2\cdot Ce(NO_3)_3\cdot 4H_2O$
	35.9	2.86	25.6	4.05	"
	43.8	3.53	18.2	2.92	"
	48.9	3.77	11.5	1.77	"
	57.0	4.56	4.7	0.75	"
	58.3	4.81	4.5	0.74	"
	58.6	4.86	4.4	0.72	$Ce(NO_3)_3\cdot 6H_2O$
	60.0	4.60	——	——	"

a. Molalities calculated by M. Salomon.

continued........

AUXILIARY INFORMATION

METHOD/APPARATUS/PROCEDURE:

The method of isothermal sections was used with refractometric analyses. Heterogeneous and homogeneous mixtures of known composition were equilibrated until their refractive indices remained constant. No other information given.

Additional details on the method of isothermal sections can be found in (1), and elsewhere in this volume (compilers).

SOURCE AND PURITY OF MATERIALS:

Nothing specified.

ESTIMATED ERROR:

Soly: precision about ± 1 % (compilers).
Temp: precision ± 0.1 K.

REFERENCES:

1. Zhuravlev, E.F.; Sheveleva, E.D. *Zh. Neorg. Khim.* 1960, *5*, 2030.

COMPONENTS:

(1) Calcium nitrate; $Ca(NO_3)_2$; [10124-37-5]

(2) Cerium nitrate; $Ce(NO_3)_3$; [10108-73-3]

(3) Water; H_2O; [7732-18-5]

ORIGINAL MEASUREMENTS:

Bogdanovskaya, R.L.; Kudryashov, S.F. *Uch. Zap. Permsk. Gos. Univ. im. A.M. Gor'kogo* 1966, *141*, 27.

EXPERIMENTAL VALUES: continued......

Composition of saturated solutions [a]

t/°C	$Ce(NO_3)_3$ mass %	$Ce(NO_3)_3$ mol kg^{-1}	$Ca(NO_3)_2$ mass %	$Ca(NO_3)_2$ mol kg^{-1}	nature of the solid phase
20	——	——	56.4	7.88	$Ca(NO_3)_2 \cdot 4H_2O$
	5.5	0.38	50.4	6.96	"
	10.8	0.77	46.0	6.49	"
	17.4	1.29	41.3	6.09	"
	25.9	2.19	37.8	6.35	"
	26.9	2.32	37.5	6.42	"
	31.2	2.87	35.5	6.50	$Ca(NO_3)_2 \cdot 4H_2O + Ca(NO_3)_3 \cdot Ce(NO_3)_3 \cdot 4H_2O$
	31.1	2.72	33.8	5.87	$Ca(NO_3)_2 \cdot Ce(NO_3)_3 \cdot 4H_2O$
	38.2	3.24	25.6	4.31	"
	44.4	3.68	18.6	3.06	"
	49.6	4.06	12.9	2.10	"
	57.9	4.82	5.3	0.88	"
	59.5	5.11	4.8	0.82	"
	60.5	5.26	4.2	0.73	$Ce(NO_3)_3 \cdot 6H_2O$
	61.5	4.90	——	——	"
30	——	——	60.4	9.30	$Ca(NO_3)_2 \cdot 4H_2O$
	4.2	0.31	54.9	8.18	"
	10.2	0.77	49.0	7.32	"
	13.5	1.06	47.6	7.46	"
	24.5	2.09	39.6	6.72	"
	27.0	2.40	38.5	6.80	"
	31.8	3.08	36.5	7.02	"
	32.8	3.22	36.0	7.03	$Ca(NO_3)_2 \cdot 4H_2O + Ca(NO_3)_2 \cdot Ce(NO_3)_3 \cdot 4H_2O$
	35.7	3.53	33.3	6.55	$Ca(NO_3)_2 \cdot Ce(NO_3)_3 \cdot 4H_2O$
	40.5	3.74	26.3	4.83	"
	45.4	3.95	19.4	3.36	"
	50.4	4.38	14.3	2.47	"
	57.1	4.84	6.7	1.13	"
	60.3	5.41	5.5	0.98	$Ca(NO_3)_2 \cdot Ce(NO_3)_3 \cdot 4H_2O + Ce(NO_3)_3 \cdot 6H_2O$
	60.8	5.27	3.8	0.65	$Ce(NO_3)_3 \cdot 6H_2O$
	63.0	5.22	——	——	"

a. Molalities calculated by M. Salomon.

COMPONENTS:

(1) Strontium nitrate; $Sr(NO_3)_2$; [10042-76-9]
(2) Cerium nitrate; $Ce(NO_3)_3$; [10108-73-3]
(3) Water; H_2O; [7732-18-5]

ORIGINAL MEASUREMENTS:

Bogdanovskaya, R.L. *Uch. Zap. Permsk. Gos. Univ. im. A.M. Gor'kogo.* 1966, *141*, 32.

VARIABLES:

Composition and temperature

PREPARED BY:

T. Mioduski and S. Siekierski

EXPERIMENTAL VALUES:

Composition of saturated solutions [a]

t/°C	$Ce(NO_3)_3$ mass %	$Ce(NO_3)_3$ mol kg^{-1}	$Sr(NO_3)_2$ mass %	$Sr(NO_3)_2$ mol kg^{-1}	nature of the solid phase
15	——	——	38.5	2.96	$Sr(NO_3)_2\cdot 4H_2O$
	6.9	0.34	31.5	2.42	"
	13.7	0.70	26.2	2.06	"
	22.9	1.22	19.5	1.60	"
	27.1	1.46	15.9	1.32	"
	30.0	1.65	14.2	1.20	"
	35.4	2.04	11.5	1.02	"
	37.4	2.17	9.7	0.87	"
	42.0	2.60	8.5	0.81	$Sr(NO_3)_2\cdot 4H_2O + Sr(NO_3)_2$
	47.4	3.07	5.3	0.53	$Sr(NO_3)_2$
	58.3	4.41	1.2	0.14	"
	60.0	4.69	0.8	0.096	$Sr(NO_3)_2 + Ce(NO_3)_3\cdot 6H_2O$
	60.5	4.70	——	——	$Ce(NO_3)_3\cdot 6H_2O$

a. Molalities calculated by M. Salomon.

continued.......

AUXILIARY INFORMATION

METHOD/APPARATUS/PROCEDURE:

The method of isothermal sections was used with refractometric analyses. Heterogeneous and homogeneous mixtures of known composition were equilibrated until their refractive indices remained constant. No other information given. Additional details on the method of isothermal sections can be found in (1), and elsewhere in this volume (compilers).

SOURCE AND PURITY OF MATERIALS:

Nothing specified.

ESTIMATED ERROR:

Soly: precision about ± 1 % (compilers).
Temp: precision ± 0.1 K.

REFERENCES:

1. Zhuravlev, E.F.; Sheveleva, A.D. *Zh. Neorg. Khim.* 1960, *5*, 2630.

COMPONENTS:

(1) Strontium nitrate; $Sr(NO_3)_2$; [10042-76-9]

(2) Cerium nitrate; $Ce(NO_3)_3$; [10108-73-3]

(3) Water; H_2O; [7732-18-5]

ORIGINAL MEASUREMENTS:

Bogdanovskaya, R.L. *Uch. Zap. Permsk. Gos. Univ. im. A.M. Gor'kogo.* 1966, *141*, 32.

EXPERIMENTAL VALUES: continued.......

Composition of saturated solutions [a]

t/°C	$Ce(NO_3)_3$ mass %	$Ce(NO_3)_3$ mol kg^{-1}	$Sr(NO_3)_2$ mass %	$Sr(NO_3)_2$ mol kg^{-1}	nature of the solid phase
25	——	——	44.0	3.71	$Sr(NO_3)_2 \cdot 4H_2O$
	6.3	0.34	37.5	3.15	"
	12.5	0.70	32.6	2.81	"
	21.8	1.22	23.5	2.03	"
	24.3	1.38	21.9	1.92	"
	26.8	1.56	20.5	1.84	"
	30.5	1.85	19.0	1.78	$Sr(NO_3)_2 \cdot 4H_2O + Sr(NO_3)_2$
	33.8	2.04	15.5	1.44	$Sr(NO_3)_2$
	45.9	2.94	6.2	0.61	"
	58.1	4.41	1.5	0.18	"
	61.5	5.06	1.2	0.15	$Sr(NO_3)_2 + Ce(NO_3)_3 \cdot 6H_2O$
	62.5	5.11	——	——	$Ce(NO_3)_3 \cdot 6H_2O$
35	——	——	47.5	4.28	$Sr(NO_3)_2$
	5.8	0.34	42.0	3.80	"
	11.9	0.69	35.5	3.19	"
	20.9	1.22	26.5	2.38	"
	33.4	2.04	16.5	1.56	"
	46.6	3.06	6.7	0.68	"
	57.4	4.41	2.7	0.32	"
	63.0	5.52	2.0	0.27	"
	63.5	5.33	——	——	$Ce(NO_3)_3 \cdot 6H_2O$

a. Molalities calculated by M. Salomon.

COMPONENTS:

(1) Barium nitrate; $Ba(NO_3)_2$; [10022-31-8]
(2) Cerium nitrate; $Ce(NO_3)_3$; [10108-73-3]
(3) Water; H_2O; [7732-18-5]

ORIGINAL MEASUREMENTS:

Bogdanovskaya, R.L. *Uch. Zap. Permsk. Gos. Univ.* 1966, *141*, 32.

VARIABLES:

Composition at 30°C

PREPARED BY:

T. Mioduski and S. Siekierski

EXPERIMENTAL VALUES:

The $Ce(NO_3)_3$ - $Ba(NO_3)_2$ - H_2O System

Composition of saturated solutions [a]

$Ce(NO_3)_3$		$Ba(NO_3)_2$		
mass %	mol kg^{-1}	mass %	mol kg^{-1}	nature of the solid phase
———	———	10.5	0.449	$Ba(NO_3)_2$
5.8	0.205	7.3	0.32	"
11.4	0.416	4.7	0.21	"
18.9	0.737	2.5	0.12	"
25.9	1.10	2.2	0.12	"
32.6	1.51	1.2	0.069	"
37.6	1.87	0.8	0.050	"
49.8	3.07	0.4	0.031	"
63.0	5.22	———	———	$Ce(NO_3)_3 \cdot 6H_2O$

a. Molalities calculated by M. Salomon.

AUXILIARY INFORMATION

METHOD/APPARATUS/PROCEDURE:

The solubility was studied by the method of isothermal sections using refractometric analysis (1). Homogeneous and heterogeneous solutions of known composition were prepared and equilibrated until their refractive indices were constant. The composition of the saturated solutions and of the corresponding solid phases were found as inflection or break points on the composition-refractive index diagrams. In addition, the composition of saturated and solid phases were determined by chemical analyses. No other information given.

SOURCE AND PURITY OF MATERIALS:

Nothing specified.

ESTIMATED ERROR:

Nothing specified.

REFERENCES:

Zhuravlev, E.F.; Sheveleva, A.D. *Zh. Neorg. Khim.* 1960, 5, 2630.

COMPONENTS:

(1) Cerium nitrate; $Ce(NO_3)_3$; [10108-73-3]

(2) Cerium chloride; $CeCl_3$; [7790-86-5]

(3) Water; H_2O; [7732-18-5]

ORIGINAL MEASUREMENTS:

Onishchenko, M.K.; Shevchuk, V.G.
Zh. Neorg. Khim. 1977, *22*, 2879-81;
Russ. J. Inorg. Chem. Engl. Transl. 1977, *22*, 1565-6.

VARIABLES:

Composition at 25°C

PREPARED BY:

T. Mioduski, S. Siekierski, M. Salomon

EXPERIMENTAL VALUES: Composition of saturated solutions [a]

$Ce(NO_3)_3$ mass %	$Ce(NO_3)_3$ mole kg^{-1}	$CeCl_3$ mass %	$CeCl_3$ mole kg^{-1}	n_D	density kg m^{-3}	nature of the solid phase
———	———	48.63	3.841	1.4690	1.6379	$CeCl_3 \cdot 7H_2O$ (A)
3.28	0.203	47.08	3.848	1.4720	1.6476	"
9.72	0.626	42.67	3.636	1.4765	1.6793	"
11.70	0.775	42.02	3.684	1.4790	1.7025	"
15.43	1.059	39.88	3.620	1.4820	———	"
19.92	1.424	37.18	3.516	1.4850	———	"
23.31	1.724	35.23	3.447	1.4872	1.7310	"
28.04	2.198	32.84	3.406	1.4912	———	"
32.23	2.643	30.38	3.296	1.4968	1.7523	"
36.73	3.184	27.90	3.200	1.4980	1.8234	"
38.71	3.471	27.09	3.214	1.4991	2.1814	"
39.72	3.653	26.94	3.278	1.5000	2.5762	"
44.85	4.330	23.39	2.988	1.5030	2.8953	A + B
46.89	4.037	17.50	1.994	1.4850	2.5678	$Ce(NO_3)_3 \cdot 6H_2O$ (B)
49.23	4.150	14.40	1.606	1.4800	2.1713	"
52.07	4.272	10.56	1.146	1.4732	1.9516	"
55.06	4.544	7.79	0.851	1.4698	1.8117	"
58.32	4.764	4.14	0.447	1.4650	———	"
59.93	4.949	2.94	0.321	1.4618	1.7897	"
62.87	5.192	———	———	1.4600	1.7627	"

a. Molalities calculated by compilers.

AUXILIARY INFORMATION

METHOD/APPARATUS/PROCEDURE:

The isothermal method was used. The mixtures reached equilibrium after 7-30 days at 25°C. Both the saturated solutions and solid residues were analysed: cerium was determined by complexometric titration with Trilon, and chloride was determined by the Mohr method. The compilers assume that the compositions of the solid phases were determined by Schreinemakers' method.

The compilers also assume that the refractive indices were measured at 25°C: i.e., the n_D^{25} values. The source paper also lists viscosities and electrolytic conductivities for the saturated solutions.

SOURCE AND PURITY OF MATERIALS:

Both $Ce(NO_3)_3 \cdot 6H_2O$ and $CeCl_3 \cdot 7H_2O$ were c.p. grade salts and recrystallized twice prior to use.

No other information given.

ESTIMATED ERROR:

Nothing specified.

REFERENCES:

COMPONENTS:

(1) Cerium nitrate; $Ce(NO_3)_3$; [10108-73-3]

(2) Manganese nitrate; $Mn(NO_3)_2$; [10377-66-9]

(3) Water ; H_2O ; [7732-18-5]

ORIGINAL MEASUREMENTS:

Shurov, V.A.; Mochalov, K.I.; Volkov, A.A. *Uch. Zap. Permsk. Gos. Univ.* 1966, *141*, 18.

VARIABLES:

Composition at 10°C, 20°C, 30°C

PREPARED BY:

T. Mioduski and S. Siekierski

EXPERIMENTAL VALUES: The $Ce(NO_3)_3 - Mn(NO_3)_2 - H_2O$ system [a]

t/°C	$Ce(NO_3)_3$ mass %	$Ce(NO_3)_3$ mol kg^{-1}	$Mn(NO_3)_2$ mass %	$Mn(NO_3)_2$ mol kg^{-1}	nature of the solid phase
10	60.0	4.60	——	——	$Ce(NO_3)_3 \cdot 6H_2O$
	56.6	4.30	3.0	0.41	"
	55.5	4.19	3.9	0.54	$Ce(NO_3)_3 \cdot 6H_2O + Ce_2Mn_3(NO_3)_{12} \cdot 24H_2O$
	48.9	3.49	8.1	1.05	$Ce_2Mn_3(NO_3)_{12} \cdot 24H_2O$
	35.7	2.31	16.9	1.99	"
	27.9	1.74	22.9	2.60	"
	17.5	1.04	30.8	3.33	"
	3.9	0.23	43.9	4.70	"
	——	——	54.2	6.61	$Mn(NO_3)_2 \cdot 6H_2O$
20	61.5	4.90	——	——	$Ce(NO_3)_3 \cdot 6H_2O$
	58.9	4.69	2.6	0.38	"
	57.3	4.58	4.3	0.63	$Ce(NO_3)_3 \cdot 6H_2O + Ce_2Mn_3(NO_3)_{12} \cdot 24H_2O$
	49.9	3.70	8.7	1.17	$Ce_2Mn_3(NO_3)_{12} \cdot 24H_2O$
	36.1	2.42	18.2	2.23	"
	28.9	1.87	23.7	2.79	"
	19.5	1.21	31.0	3.50	"
	6.1	0.37	43.2	4.76	"
	0	——	58.5	7.88	$Mn(NO_3)_2 \cdot 6H_2O$

a. Molalities calculated by M. Salomon.

continued.......

AUXILIARY INFORMATION

METHOD/APPARATUS/PROCEDURE:

The method of isothermal sections was used with refractometric analyses of both the liquid and solid phases. Solutions were equilibrated until the refractive indices were constant.

No other information given.

SOURCE AND PURITY OF MATERIALS:

Nothing specified.

ESTIMATED ERROR:

Soly: precision about 1 % (compilers).

Temp: precision ± 0.1 K.

REFERENCES:

COMPONENTS:

(1) Cerium nitrate; $Ce(NO_3)_3$; [10108-73-3]

(2) Manganese nitrate; $Mn(NO_3)_2$; [10377-66-9]

(3) Water; H_2O; [7732-18-5]

ORIGINAL MEASUREMENTS:

Shurov, V.A.; Mochalov, K.I.; Volkov, A.A.
Uch. Zap. Permsk. Gos. Univ. 1966, *141*, 18.

EXPERIMENTAL VALUES: continued.......

The $Ce(NO_3)_3 - Mn(NO_3)_2 - H_2O$ system [a]

t/°C	$Ce(NO_3)_3$ mass %	$Ce(NO_3)_3$ mol kg^{-1}	$Mn(NO_3)_2$ mass %	$Mn(NO_3)_2$ mol kg^{-1}	nature of the solid phase
30	63	5.22	——	——	$Ce(NO_3)_3 \cdot 6H_2O$
	61.2	5.13	2.2	0.34	"
	59.3	5.04	4.6	0.71	$Ce(NO_3)_3 \cdot 6H_2O + Ce_2Mn_3(NO_3)_{12} \cdot 24H_2O$
	51.4	4.03	9.5	1.36	$Ce_2Mn_3(NO_3)_{12} \cdot 24H_2O$
	36.5	2.56	19.8	2.53	"
	29.9	2.01	24.5	3.00	"
	21.6	1.40	31.2	3.69	"
	9.1	0.58	42.4	4.89	"
	3.6	0.24	50.6	6.17	"
	——	——	67.4	11.55	$Mn(NO_3)_2 \cdot 6H_2O$

a. Molalities calculated by M. Salomon.

COMPONENTS:

(1) Cerium nitrate; $Ce(NO_3)_3$; [10108-73-3]

(2) Nickel nitrate; $Ni(NO_3)_2$; [13138-45-9]

(3) Water; H_2O; [&&32-18-5]

ORIGINAL MEASUREMENTS:

Mitina, N.K.; Shevchuk, V.G. *Zh. Neorg. Khim.* 1977, *22*, 1376-81; *Russ. J. Inorg. Chem. Engl. Transl.* 1977, *22*, 749-51.

VARIABLES:

Composition at 25°C

PREPARED BY:

T. Mioduski, S. Siekierski, M. Salomon

EXPERIMENTAL VALUES:

Composition of saturated solutions [a]

$Ce(NO_3)_3$ mass %	$Ce(NO_3)_3$ mol kg^{-1}	$Ni(NO_3)_2$ mass %	$Ni(NO_3)_2$ mol kg^{-1}	density kg m^{-3}	nature of the solid phase
———	———	49.93	5.458	1.5342	$Ni(NO_3)_2\cdot 6H_2O$
0.63	0.039	49.61	5.457		"
1.97	0.123	48.96	5.461	1.5374	"
3.26	0.209	48.82	5.576		"
4.60	0.304	49.04	5.789	1.5417	$Ni(NO_3)_2\cdot 6H_2O + Ce_2Ni_3(NO_3)_{12}\cdot 24H_2O$
9.31	0.602	43.28	4.996		$Ce_2Ni_3(NO_3)_{12}\cdot 24H_2O$
10.60	0.674	41.16	4.670	1.5493	"
15.42	0.940	34.26	3.726		"
17.80	1.069	31.13	3.336	1.5681	"
23.38	1.401	25.44	2.720		"
27.60	1.684	22.15	2.412	1.6440	"
30.44	1.894	20.27	2.251		"
34.98	2.289	18.16	2.121		"
42.07	2.979	14.63	1.849		"
51.39	4.200	11.09	1.618		"
58.20	5.343	8.40	1.376	1.9026	$Ce_2Ni_3(NO_3)_{12}\cdot 24H_2O + Ce(NO_3)_3\cdot 6H_2O$
59.67	5.141	4.74	0.729		$Ce(NO_3)_3\cdot 6H_2O$
62.35	5.372	2.06	0.317	1.9003	"
63.04	5.230	———	———	1.8981	"

a. Molalities calculated by M. Salomon

AUXILIARY INFORMATION

METHOD/APPARATUS/PROCEDURE:

The isothermal method was used. The slns were thermostated and continually stirred. Equilibrium was reached in 15-20 days. Ce^{3+} was precipitated as the hydroxide by addition of conc NH_4OH and an excess of NH_4Cl: Nickel was determined by titration with Trilon in the presence of NH_4F. The total nitrates were determined by Devarda's method.

The compilers assume that the compositions of the solid phases were determined by Schreinemakers' method.

The source paper also gives viscosities and electrolytic conductivities for those satd slns for which the densities were measured.

SOURCE AND PURITY OF MATERIALS:

No information given.

ESTIMATED ERROR:

Soly: precision probably around ± 0.3 to 0.5 % (compilers).

Temp: precision is probably ± 0.2 K (compilers).

COMPONENTS:

(1) Cerium nitrate; $Ce(NO_3)_3$; [10108-73-3]
(2) Copper nitrate; $Cu(NO_3)_2$; [3251-23-8]
(3) Water; H_2O; [7732-18-5]

ORIGINAL MEASUREMENTS:

Sheveleva, A.D.; Shchurov, V.A.
Uch. Zap. Permsk. Gos. Univ. 1968, *178*, 85-91.

VARIABLES:

Components at 10°C, 20°C, 30°C

PREPARED BY:

T. Mioduski and S. Siekierski

EXPERIMENTAL VALUES:

Composition of the saturated solutions at 10°C [a]

$Ce(NO_3)_3$ mass %	$Ce(NO_3)_3$ mol kg^{-1}	$Cu(NO_3)_2$ mass %	$Cu(NO_3)_2$ mol kg^{-1}	nature of the solid phase
60.0	4.60	——	——	$Ce(NO_3)_3 \cdot 6H_2O$
56.3	4.26	3.2	0.42	"
52.7	3.98	6.7	0.88	"
44.9	3.64	17.3	2.44	"
38.2	2.92	21.7	2.89	"
37.1	2.84	22.9	3.05	$Ce(NO_3)_3 \cdot 6H_2O + 2Ce(NO_3)_3 \cdot 3Cu(NO_3)_2 \cdot 24H_2O$
33.0	2.48	26.2	3.42	$2Ce(NO_3)_3 \cdot 3Cu(NO_3)_2 \cdot 24H_2O$
25.2	1.81	32.1	4.01	"
11.4	0.82	45.9	5.73	"
8.2	0.61	50.7	6.58	$2Ce(NO_3)_3 \cdot 3Cu(NO_3)_2 \cdot 24H_2O + Cu(NO_3)_2 \cdot 6H_2O$
7.0	0.50	49.9	6.17	$Ce(NO_3)_2 \cdot 6H_2O$
2.6	0.16	48.6	5.31	"
——	——	50.0	5.33	"

a. Molalities calculated by M. Salomon.

continued......

AUXILIARY INFORMATION

METHOD/APPARATUS/PROCEDURE:

The method of isothermal sections was used with refractometric analyses (1). Heterogeneous and homogeneous mixtures of known composition were equilibrated until their refractive indices remained constant. The composition of the saturated solutions and the corresponding solid phases were found as inflection or "break" points on a plot of composition against refractive index. All refractive indices were measured at 50°C.

SOURCE AND PURITY OF MATERIALS:

C.p. grade $Ce(NO_3)_3 \cdot 6H_2O$ recrystallized prior to use.

C.p. grade $Cu(NO_3)_2 \cdot 3H_2O$ recrystallized prior to use.

The water contents of each salt were analysed and stated to be almost theoretical.

ESTIMATED ERROR:

Nothing specified (see critical evaluation).

REFERENCES:

1. Zhuravlev, E.F.; Sheveleva, A.D. *Zh. Neorg. Khim.* 1960, *5*, 2630.

COMPONENTS:	ORIGINAL MEASUREMENTS:
(1) Cerium nitrate; $Ce(NO_3)_3$; [10108-73-3] (2) Copper nitrate; $Cu(NO_3)_2$; [3251-23-8] (3) Water; H_2O; [7732-18-5]	Sheveleva, A.D.; Shchurov, V.A. *Uch. Zap. Permsk. Gos. Univ.* 1968, *178*, 85-91.

EXPERIMENTAL VALUES: continued......

Composition of saturated solutions at 20°C [a]

$Ce(NO_3)_3$ mass %	$Ce(NO_3)_3$ mol kg^{-1}	$Cu(NO_3)_2$ mass %	$Cu(NO_3)_2$ mol kg^{-1}	nature of the solid phase
61.5	4.90	——	——	$Ce(NO_3)_3 \cdot 6H_2O$
50.4	4.11	12.0	1.70	"
42.3	3.35	19.0	2.62	"
39.1	3.14	22.7	3.17	"
35.8	2.91	26.5	3.75	$Ce(NO_3)_3 \cdot 6H_2O + 2Ce(NO_3)_3 \cdot 3Cu(NO_3)_2 \cdot 24H_2O$
32.4	2.57	29.0	4.01	$2Ce(NO_3)_3 \cdot 3Cu(NO_3)_2 \cdot 24H_2O$
27.6	2.11	32.2	4.27	"
18.8	1.39	39.8	5.13	"
14.8	1.11	44.5	5.83	"
10.5	0.79	48.9	6.42	"
7.5	0.59	53.5	7.31	$2Ce(NO_3)_3 \cdot 3Cu(NO_3)_2 \cdot 24H_2O + Cu(NO_3)_2 \cdot 3H_2O$
6.1	0.47	54.1	7.25	$Cu(NO_3)_2 \cdot 3H_2O$
4.7	0.35	54.4	7.09	$Cu(NO_3)_2 \cdot 3H_2O + Cu(NO_3)_2 \cdot 6H_2O$
2.3	0.17	55.0	6.87	$Cu(NO_3)_2 \cdot 6H_2O$
——	——	55.2	6.57	"

Composition of saturated solutions at 30°C

$Ce(NO_3)_3$ mass %	$Ce(NO_3)_3$ mol kg^{-1}	$Cu(NO_3)_2$ mass %	$Cu(NO_3)_2$ mol kg^{-1}	nature of the solid phase
63.0	5.22	——	——	$Ce(NO_3)_3 \cdot 6H_2O$
58.0	4.87	5.5	0.80	"
53.8	4.60	10.3	1.53	"
43.8	3.82	21.0	3.18	"
41.6	3.69	23.8	3.67	$Ce(NO_3)_3 \cdot 6H_2O + 2Ce(NO_3)_3 \cdot 3Cu(NO_3)_2 \cdot 24H_2O$
38.1	3.31	26.6	4.02	$2Ce(NO_3)_3 \cdot 3Cu(NO_3)_2 \cdot 24H_2O$
29.1	2.41	33.8	4.86	"
19.9	1.61	42.2	5.94	"
15.5	1.27	47.2	6.75	"
12.5	1.05	51.0	7.45	$2Ce(NO_3)_3 \cdot 3Cu(NO_3)_2 \cdot 24H_2O + Cu(NO_3)_3 \cdot 3H_2O$
8.8	0.71	53.2	7.46	$Cu(NO_3)_3 \cdot 3H_2O$
5.9	0.47	55.2	7.57	"
2.0	0.16	59.6	8.28	"
——	——	62.0	8.70	"

a. Molalities calculated by M. Salomon.

COMPONENTS:

(1) Cerium nitrate; $Ce(NO_3)_3$; [10108-73-3]
(2) Copper nitrate; $Cu(NO_3)_2$; [3261-23-8]
(3) Water; H_2O; [7732-18-5]

ORIGINAL MEASUREMENTS:

Yakimov, M.A.; Gizhavina, E.I. *Zh. Neorg. Khim.* 1971, *16*, 507-9; *Russ. J. Inorg. Chem. Engl. Transl.* 1971, *16*, 268-9.

VARIABLES:

Composition at 25°C

PREPARED BY:

T. Mioduski and S. Siekierski

EXPERIMENTAL VALUES:

Composition of saturated solutions [a]

$Ce(NO_3)_3$ mass %	$Ce(NO_3)_3$ mol kg^{-1}	$Cu(NO_3)_2$ mass %	$Cu(NO_3)_2$ mol kg^{-1}	nature of the solid phase
62.7	5.15	——	——	$Ce(NO_3)_3 \cdot 6H_2O$
56.3	4.59	6.1	0.86	"
52.1	4.29	10.7	1.53	"
51.3	4.18	11.1	1.57	"
50.1	4.15	12.9	1.86	"
45.3	3.76	17.8	2.57	"
36.6	3.10	27.2	4.01	$Ce(NO_3)_3 \cdot 6H_2O + Ce_2Cu_3(NO_3)_{12} \cdot 24H_2O$
32.4	2.66	30.3	4.33	$Ce_2Cu_3(NO_3)_{12} \cdot 24H_2O$
28.3	2.28	33.7	4.73	"
25.2	2.01	36.3	5.03	"
19.1	1.48	41.4	5.59	"
16.1	1.26	44.8	6.11	"
15.2	1.19	45.7	6.23	"
14.4	1.13	46.5	6.34	"
13.0	1.04	48.6	6.75	"
10.3	0.85	52.6	7.56	$Ce_2Cu_3(NO_3)_{12} \cdot 24H_2O + Cu(NO_3)_2 \cdot 3H_2O$
8.0	0.64	53.4	7.38	$Cu(NO_3)_2 \cdot 3H_2O$
3.6	0.28	56.4	7.52	"
——	——	59.4	7.80	"

a. Molalities calculated by M. Salomon.

AUXILIARY INFORMATION

METHOD/APPARATUS/PROCEDURE:

The isothermal method was used. The composition of the saturated solutions was determined by chemical analysis, and the composition of the solid phases was determined by Schreinemakers' method.
In analysing the saturated solutions, copper was first determined gravimetrically by precipitation of the sulfide with sodium thiosulfate. After destroying the thiosulfate in the filtrate, cerium was determined gravimetrically by precipitation of the oxalate followed by ignition to the oxide.

SOURCE AND PURITY OF MATERIALS:

Nothing specified.

ESTIMATED ERROR:

Soly: precision about ± 0.5 % (compilers).

Temp: precision probably ± 0.2 K (compilers).

REFERENCES:

COMPONENTS:

(1) Cerium nitrate; $Ce(NO_3)_3$; [10108-73-3]
(2) Cadmium nitrate; $Cd(NO_3)_2$; [10325-94-7]
(3) Water; H_2O; [7732-18-5]

ORIGINAL MEASUREMENTS:

Shchurov, V.A.; Mochalov, K.I.; Volkov, A.A. *Uch. Zap. Permsk. Gos. Univ. Im. A. M. Gor'kogo.* 1966, *141*, 18.

VARIABLES:

Composition at 10°C, 20°C, and 30°C

PREPARED BY:

T. Mioduski and S. Siekierski

EXPERIMENTAL VALUES:

Composition of saturated solutions at 10°C [a]

$Ce(NO_3)_3$ mass %	$Ce(NO_3)_3$ mol kg^{-1}	$Cd(NO_3)_2$ mass %	$Cd(NO_3)_2$ mol kg^{-1}	nature of the solid phase
——	——	57.5	5.72	$Cd(NO_3)_2\cdot 4H_2O$
11.7	0.87	47.3	4.88	"
28.7	2.28	32.7	3.58	"
33.6	2.72	28.5	3.18	"
36.2	2.98	26.6	3.02	$Cd(NO_3)_2\cdot 4H_2O + Ce(NO_3)_3\cdot 6H_2O$
38.3	3.16	24.5	2.79	$Ce(NO_3)_3\cdot 6H_2O$
40.5	3.26	21.4	2.38	"
44.7	3.56	16.8	1.85	"
45.8	3.64	15.6	1.71	"
55.1	4.27	5.3	0.57	"
60.0	4.60	——	——	"

a. Molalities calculated by M. Salomon.

continued......

AUXILIARY INFORMATION

METHOD/APPARATUS/PROCEDURE:

The method of isothermal sections was used with refractometric analyses (1). Heterogeneous and homogeneous mixtures of known composition were equilibrated until their refractive indices remained constant. The composition of the saturated solutions and the corresponding solid phases were found as inflection or "break" points on a plot of composition against refractive index. All refractive indices were measured at 30°C.

SOURCE AND PURITY OF MATERIALS:

Nothing specified.

ESTIMATED ERROR:

Nothing specified (see critical evaluation).

REFERENCES:

1. Zhuravlev, E.F.; Sheveleva, A.D. *Zh. Neorg. Khim.* 1960, *5*, 2630.

COMPONENTS:	ORIGINAL MEASUREMENTS:
(1) Cerium nitrate; $Ce(NO_3)_3$; [10108-73-3] (2) Cadmium nitrate; $Cd(NO_3)_2$; [10325-94-7] (3) Water; H_2O; [7732-18-5]	Shchurov, V.A.; Mochalov, K.I.; Volkov, A.A. *Uch. Zap. Permsk. Gos. Univ. Im. A. M. Gor'kogo.* 1966, *141*, 18.

EXPERIMENTAL VALUES: continued......

Composition of saturated solutions at 20°C [a]

$Ce(NO_3)_3$		$Cd(NO_3)_2$		
mass %	mol kg^{-1}	mass %	mol kg^{-1}	nature of the solid phase
——	——	60.0	6.34	$Cd(NO_3)_2·4H_2O$
10.0	0.79	51.1	5.56	"
25.2	2.09	37.9	4.34	"
34.1	2.96	30.6	3.67	"
37.6	3.36	28.1	3.47	$Cd(NO_3)_2·4H_2O$ + $Ce(NO_3)_3·6H_2O$
41.5	3.57	22.9	2.72	$Cd(NO_3)_3·6H_2O$
43.2	3.73	21.3	2.54	"
45.9	3.89	17.9	2.09	"
50.1	4.20	13.3	1.54	"
57.5	4.66	4.7	0.53	"
61.5	4.90	——	——	"

Composition of saturated solutions at 30°C

$Ce(NO_3)_3$		$Cd(NO_3)_2$		
mass %	mol kg^{-1}	mass %	mol kg^{-1}	nature of the solid phase
——	——	63.0	7.20	$Cd(NO_3)_2·4H_2O$
8.4	0.71	55.2	6.41	"
21.8	1.92	43.3	5.25	"
34.7	4.08	39.2	6.35	"
39.4	3.84	29.1	3.91	$Cd(NO_3)_2·4H_2O$ + $Ce(NO_3)_3·6H_2O$
42.7	4.02	24.7	3.20	$Cd(NO_3)_3·6H_2O$
43.7	4.33	19.2	2.42	"
48.1	4.35	18.0	4.78	"
54.1	4.78	11.2	1.37	"
59.7	5.06	4.1	0.48	"
63.0	5.22	——	——	"

a. Molalities calculated by M. Salomon.

COMPONENTS:

(1) Cerium nitrate; $Ce(NO_3)_3$; [10108-73-3]

(2) Hydrazine mononitrate; $N_2H_4 \cdot HNO_3$; [13464-97-6]

(3) Water; H_2O; [7732-18-5]

ORIGINAL MEASUREMENTS:

Mininkov, N.E.; Zhuravlev, E.F. *Zh. Neorg. Khim.* 1970, *15*, 205-10; *Russ. J. Inorg. Chem. Engl. Transl.* 1970, *15*, 105-7.

VARIABLES:

Composition at 25°C and 50°C

PREPARED BY:

T. Mioduski and S. Siekierski

EXPERIMENTAL VALUES:

Composition of saturated solutions at 25°C [a]

$Ce(NO_3)_3$ mass %	$Ce(NO_3)_3$ mol kg^{-1}	$N_2H_4 \cdot HNO_3$ mass %	$N_2H_4 \cdot HNO_3$ mol kg^{-1}	n_D^{50}	nature of the solid phase
———	———	76.0	33.31	1.4680	$N_2H_4 \cdot HNO_3$
15.0	2.04	62.5	29.22	1.4765	"
22.0	2.93	55.0	25.16	1.4815	"
31.0	4.53	48.0	24.05	1.4870	"
45.0	7.26	36.0	19.93	1.4950	"
47.0	7.58	34.0	18.83	1.4960	$N_2H_4 \cdot HNO_3 + Ce(NO_3)_3 \cdot 2N_2H_4 \cdot HNO_3$
50.0	7.30	29.0	14.53	1.4900	$Ce(NO_3)_3 \cdot 2N_2H_4 \cdot HNO_3$
54.5	7.60	23.5	11.24	1.4905	"
57.0	8.74	23.0	12.10	1.4915	"
58.0	7.90	19.5	9.12	1.4940	$Ce(NO_3)_3 \cdot 2N_2H_4 \cdot HNO_3 + Ce(NO_3)_3 \cdot 6H_2O$
58.5	6.90	15.5	6.27	1.4870	$Ce(NO_3)_3 \cdot 6H_2O$
60.0	5.84	8.5	2.84	1.4760	"
61.5	5.47	4.0	1.22	1.4685	"
63.0	5.22	———	———	1.4620	"

a. Molalities calculated by M. Salomon.

continued........

AUXILIARY INFORMATION

METHOD/APPARATUS/PROCEDURE:

The method of isothermal sections was used with refractometric analyses (1). Heterogeneous and homogeneous mixtures of known composition were equilibrated until their refractive indices remained constant. The composition of the saturated solutions and the corresponding solid phases were found as inflection or "break" points on a plot of composition against refractive index. All refractive indices were measured at 50°C.

For the study of the 50°C isotherm, the initial cerium nitrate was partially dehydrated in a desiccator over 97 % sulfuric acid until it contained 80.5 mass % $Ce(NO_3)_3$.

SOURCE AND PURITY OF MATERIALS:

"Pure" grade $Ce(NO_3)_3 \cdot 6H_2O$ used. Analysis for Ce(III) yield 75.0 mass %.

"Pure" grade hydrazine mononitrate was recrystallized.

Doubly distilled water was used.

ESTIMATED ERROR:

Soly: precision ± 1% (compilers).
Temp: precision probably ± 0.2 K (compilers).

REFERENCES:

1. Zhuravlev, E.F.; Sheveleva, A.D. *Zh. Neorg. Khim.* 1960, *5*, 2630.

COMPONENTS:

(1) Cerium nitrate; $Ce(NO_3)_3$; [10108-73-3]

(2) Hydrazine mononitrate; $N_2H_4 \cdot HNO_3$; [13464-97-6]

(3) Water; H_2O; [7732-18-5]

ORIGINAL MEASUREMENTS:

Mininkov, N.E.; Zhuravlev, E.F. *Zh. Neorg. Khim.* 1970, *15*, 205-10; *Russ. J. Inorg. Chem. Engl. Transl.* 1970, *15*, 105-7.

EXPERIMENTAL VALUES: continued........

Composition of saturated solutions at 50°C [a]

$Ce(NO_3)_3$		$N_2H_4 \cdot HNO_3$			
mass %	mol kg^{-1}	mass %	mol kg^{-1}	n_D^{50}	nature of the solid phase
——	——	91.0	106.4	1.4985	$N_2H_4 \cdot HNO_3$
6.0	2.04	85.0	99.35	1.5020	"
9.0	2.90	81.5	90.25	1.5040	"
13.5	4.36	77.0	85.27	1.5060	"
21.5	7.32	69.5	81.24	1.5100	"
26.5	8.55	64.0	70.87	1.5130	"
37.0	15.13	55.5	77.85	1.5185	$N_2H_4 \cdot HNO_3 + Ce(NO_3)_3 \cdot 2N_2H_4 \cdot HNO_3$
40.0	11.15	49.0	46.86	1.5140	$Ce(NO_3)_3 \cdot 2N_2H_4 \cdot HNO_3$
45.0	9.86	41.0	30.81	1.5090	"
52.0	9.11	30.5	18.33	1.5045	"
57.0	8.96	23.5	12.68	1.5050	"
62.0	9.51	18.0	9.47	1.5063	"
65.5	10.04	14.5	7.63	1.5090	$Ce(NO_3)_3 \cdot 2N_2H_4 \cdot HNO_3 + Ce(NO_3)_3 \cdot 6H_2O$
67.0	8.93	10.0	4.57	1.5005	$Ce(NO_3)_3 \cdot 6H_2O$
69.5	8.36	5.0	2.06	1.4940	"
73.0	8.29	——	——	1.4900	"

a. Molalities calculated by M. Salomon.

COMPONENTS:

(1) Cerium nitrate; $Ce(NO_3)_3$; [10108-73-3]

(2) Ammonium nitrate; NH_4NO_3; [6484-52-2]

(3) Water; H_2O; [7732-18-5]

ORIGINAL MEASUREMENTS:

Shevchuk, V.G.; Mitina, N.K. *Zh. Neorg. Khim.* 1976, *21*, 1943-7; *Russ. J. Inorg. Chem. Engl. Transl.* 1976, *21*, 1067-70.

VARIABLES:

Composition at 25°C

PREPARED BY:

T. Mioduski and S. Siekierski

EXPERIMENTAL VALUES:

Composition of saturated solutions [a]

$Ce(NO_3)_3$		NH_4NO_3		density	
mass %	mol kg^{-1}	mass %	mol kg^{-1}	kg m^{-3}	nature of the solid phase
——	——	67.43	25.86	1.3059	NH_4NO_3
0.82		? [b]			"
3.43	0.301	61.64	22.05		"
8.00	0.745	59.06	22.40		"
11.42	1.038	54.85	20.32		"
13.16	1.193	53.01	19.58		"
15.26	1.348	50.04	18.02		"
18.15	1.687	48.86	18.50	1.4054	"
20.25	1.901	47.09	18.01		"
24.16	2.251	42.93	16.30	1.4351	"
26.22	2.562	42.40	16.88		"

a. Molalities calculated by M. Salomon.

b. Data point missing from table in original publication.

continued......

AUXILIARY INFORMATION

METHOD/APPARATUS/PROCEDURE:

The isothermal method was used. Equilibrium was reached after 15 days. Ammonia was determined by the Kjeldahl method, and cerium was determined by titration with Trilon B.

In the source paper, viscosities and electrolytic conductivities are given for the same saturated solutions for which the densities are reported (see data table above).

SOURCE AND PURITY OF MATERIALS:

"Chemically pure" salts were recrystallized twice from their aqueous solutions.

ESTIMATED ERROR:

Nothing specified (see critical evaluation).

REFERENCES:

COMPONENTS:

(1) Cerium nitrate; $Ce(NO_3)_3$; [10108-73-3]

(2) Ammonium nitrate; NH_4NO_3; [6484-52-2]

(3) Water; H_2O; [7732-18-5]

ORIGINAL MEASUREMENTS:

Shevchuk, V.G.; Mitina, N.K. *Zh. Neorg. Khim.* 1976, *21*, 1943-7, *Russ. J. Inorg. Chem. Engl. Transl.* 1976, *21*, 1067-70.

EXPERIMENTAL VALUES: continued......

Composition of saturated solutions [a]

$Ce(NO_3)_3$ mass %	$Ce(NO_3)_3$ mol kg^{-1}	NH_4NO_3 mass %	NH_4NO_3 mol kg^{-1}	density kg m^{-3}	nature of the solid phase
29.27	2.812	38.81	15.19	1.5135	$NH_4NO_3 + Ce(NO_3)_3 \cdot 2NH_4NO_3 \cdot 4H_2O$
40.80	4.112	28.78	11.82	1.4821	$Ce(NO_3)_3 \cdot 2NH_4NO_3 \cdot 4H_2O$
54.75	5.739	16.00	6.834	1.7139	"
57.22	5.549	11.16	4.409	1.8874	"
59.50	5.866	9.40	3.776	2.0003	$Ce(NO_3)_3 \cdot 2NH_4NO_3 \cdot 4H_2O + Ce(NO_3)_3 \cdot 6H_2O$
60.52	5.544	6.01	2.243	1.9615	$Ce(NO_3)_3 \cdot 6H_2O$
61.07	5.499	4.88	1.791		"
61.18	5.325	3.59	1.273		"
62.45	5.265	1.18	0.405	1.9188	"
63.10	5.243	———	———	1.9038	"

a. Molalities calculated by M. Salomon.

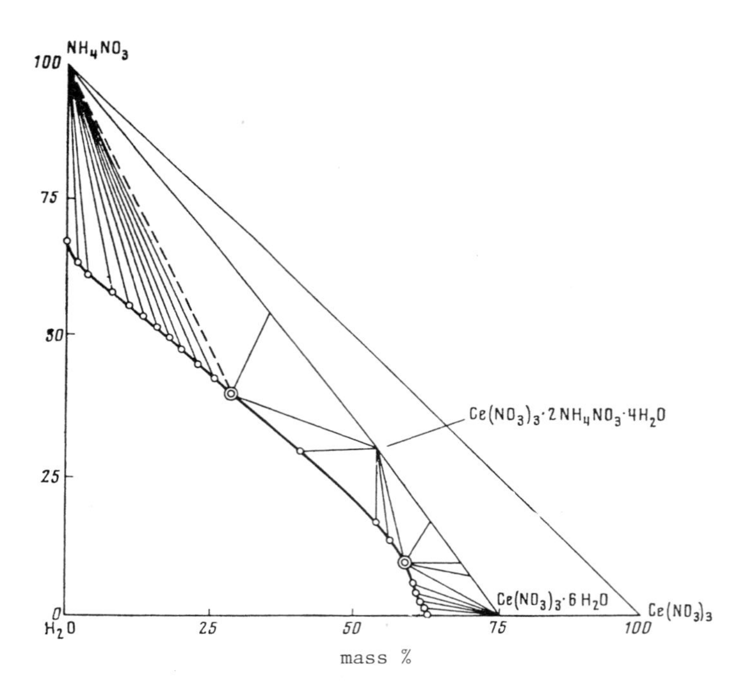

Solubility isotherm of the $Ce(NO_3)_3$ - NH_4NO_3 - H_2O system at 25°C

COMPONENTS:

(1) Cerium nitrate; $Ce(NO_3)_3$; [10108-73-3]

(2) Methanamine nitrate; $CH_6N_2O_3$; [22113-87-7]

(3) Water; H_2O; [7732-18-5]

ORIGINAL MEASUREMENTS:

Zhuravlev, E.F.; Mininkov, N.E. *Zh. Neorg. Khim.* 1972, *17*, 1736-40; *Russ. J. Inorg. Chem. Engl. Transl.* 1972, *17*, 899-901.

VARIABLES:

Composition at 25°C and 50°C

PREPARED BY:

T. Mioduski, S. Siekierski, M. Salomon

EXPERIMENTAL VALUES:

Composition of saturated solutions at 25°C [a]

$Ce(NO_3)_3$ mass %	$Ce(NO_3)_3$ mol kg^{-1}	$CH_3NH_3NO_3$ mass %	$CH_3NH_3NO_3$ mol kg^{-1}	n_D^{50}	nature of the solid phase
——	——	82.0	48.42	1.4280	$CH_3NH_3NO_3$
4.5	0.81	78.5	49.08	1.4317	"
11.0	2.11	73.0	48.50	1.4380	"
20.0	4.54	66.5	52.36	1.4492	"
29.5	8.61	60.0	60.74	1.4615	"
31.5	9.66	58.5	62.18	1.4625	$CH_3NH_3NO_3 + Ce(NO_3)_3\cdot 3CH_3NH_3NO_3$
35.5	7.51	50.0	36.65	1.4600	$Ce(NO_3)_3\cdot 3CH_3NH_3NO_3$
38.5	7.15	45.0	28.99	1.4593	"
43.5	6.84	37.0	20.17	1.4585	"
48.5	6.20	27.5	12.18	1.4607	"
55.5	6.19	17.0	6.57	1.4695	"
56.5	6.19	15.5	5.88	1.4730	$Ce(NO_3)_3\cdot 3CH_3NH_3NO_3 + Ce(NO_3)_3\cdot 6H_2O$
58.5	5.88	11.0	3.83	1.4700	$Ce(NO_3)_3\cdot 6H_2O$
63.0	5.22	——	——	1.4620	"

a. Molalities calculated by the compilers.

continued.......

AUXILIARY INFORMATION

METHOD/APPARATUS/PROCEDURE:

The method of isothermal sections was used with refractometric analyses (1,2). Heterogeneous and homogeneous mixtures of known composition were equilibrated until their refractive indices remained constant. The composition of the saturated solutions and the corresponding solid phases were found as inflection or "break" points on a plot of composition against refractive index. All refractive indices were measured at 50°C.

The double salt was isolated and identified by IR spectroscopy. Authors state the composition of this double salt satifies the empirical formula $Ce(NO_3)_3\cdot 3CH_3NH_3NO_3$. Details on the analysis of the double salt were not given.

SOURCE AND PURITY OF MATERIALS:

"Analytical reagent" grade cerium nitrate hexahydrate was used. Ce/H_2O ratio was "approximately 1:6."

Source and purity of $CH_3NH_3NO_3$ was not specified.

Doubly distilled water was used.

ESTIMATED ERROR:

Soly: precision around ± 1% (compilers).

Temp: precision probably ± 0.2 K (compilers).

REFERENCES:

1. Zhuravlev, E.F.; Sheveleva, A.D. *Zh. Neorg. Khim.* 1960, *5*, 2630.
2. Mininkov, N.E.; Zhuravlev, E.F. *Zh. Neorg. Khim.* 1969, *14*, 2565.

COMPONENTS:

(1) Cerium nitrate; $Ce(NO_3)_3$; [10108-73-3]

(2) Methanamine nitrate; $CH_6N_2O_3$; [22113-87-7]

(3) Water; H_2O; [7732-18-5]

ORIGINAL MEASUREMENTS:

Zhuravlev, E.F.; Mininkov, N.E. *Zh. Neorg. Khim.* 1972, *17*, 1736-40; *Russ. J. Inorg. Chem. Engl. Transl.* 1972, *17*, 899-901.

EXPERIMENTAL VALUES: continued.......

Composition of saturated solutions at 50°C [a]

$Ce(NO_3)_3$		$CH_3NH_3NO_3$			
mass %	mol kg^{-1}	mass %	mol kg^{-1}	n_D^{50}	nature of the solid phase
———	———	89.5	90.61	1.4400	$CH_3NH_3NO_3$
3.0	0.95	87.3	95.67	1.4413	"
7.0	2.38	84.0	99.21	1.4450	"
13.5	5.17	78.5	104.3	1.4502	"
21.5	9.42	71.5	108.6	1.4590	"
32.5	27.68	63.9 [b]	188.7	1.4750	$CH_3NH_3NO_3 + Ce(NO_3)_3 \cdot 3CH_3NH_3NO_3$
37.0	10.80	52.5	53.15	1.4730	$Ce(NO_3)_3 \cdot 3CH_3NH_3NO_3$
41.0	9.31	45.5	35.83	1.4717	"
44.0	7.50	38.0	22.44	1.4715	"
49.0	7.15	30.0	15.19	1.4733	"
57.0	7.28	19.0	8.42	1.4810	"
65.5	9.80	14.0	7.26	1.5000	$Ce(NO_3)_3 \cdot 3CH_3NH_3NO_3 + Ce(NO_3)_3 \cdot 6H_2O$
71.0	8.22	2.5	1.00	1.4920	$Ce(NO_3)_3 \cdot 6H_2O$
73.3	8.42	———	———	1.4900	"

a. Molalities calculated by the compilers.

b. English translation gives 63.0 for this data point.

COMMENTS AND/OR ADDITIONAL DATA:

The double salt is congruently soluble. The crystallization field of this double salt decreases appreciably upon reduction of the temperature to 25°C, and the authors suggest that at lower temperatures the system will be eutonic.

COMPONENTS:

(1) Cerium nitrate; $Ce(NO_3)_3$; [10108-73-3]

(2) Dimethylamine nitrate; $C_2H_8N_2O_3$; [30781-73-8]

(3) Water; H_2O; [7732-18-5]

ORIGINAL MEASUREMENTS:

Mininkov, N.E.; Zhuravlev, E.F. *Zh. Neorg. Khim.* 1976, *21*, 242-6; *Russ. J. Inorg. Chem. Engl. Transl.* 1976, *21*, 131-3.

VARIABLES:

Composition at 25°C

PREPARED BY:

T. Mioduski and S. Siekierski

EXPERIMENTAL VALUES: Composition of saturated solutions [a]

$Ce(NO_3)_3$		$(CH_3)_2NH\cdot HNO_3$			
mass %	mol kg^{-1}	mass %	mol kg^{-1}	n_D^{50}	nature of the solid phase [b]
——	——	91.0	93.5	1.4310	$(CH_3)_2NH\cdot HNO_3$
2.0	0.68	89.0	91.5	1.4320	"
5.0	1.92	87.0	100.6	1.4350	"
8.0	3.50	85.0	112.3	1.4400	"
11.5	5.42	82.0	116.7	1.4450	$(CH_3)_2NH\cdot HNO_3 + Ce(NO_3)_3\cdot 5(CH_3)_2NH\cdot HNO_3$
22.0	5.00	64.5	44.2	1.4470	$Ce(NO_3)_3\cdot 5(CH_3)_2NH\cdot HNO_3$
29.0	5.39	54.5	30.6	1.4500	"
35.0	5.80	46.5	23.3	1.4550	"
42.5	6.52	37.5	17.3	1.4600	"
60.0	9.94	21.5	10.8	1.4810	$Ce(NO_3)_3\cdot 5(CH_3)_2NH\cdot HNO_3 + Ce(NO_3)_3\cdot 6H_2O$
60.0	8.56	18.5	7.96	1.4790	$Ce(NO_3)_3\cdot 6H_2O$
60.0	6.94	13.5	4.71	1.4735	"
60.0	6.13	10.0	3.08	1.4700	"
61.0	5.94	7.5	2.20	1.4680	"
61.5	5.55	4.5	1.22	1.4660	"
63.0	5.22	——	——	1.4620	"

a. Molalities calculated by M. Salomon.

b. The solubility isotherm for the ternary system consists of three branches intersecting at two invariant points. The double salt is congruently soluble.

AUXILIARY INFORMATION

METHOD/APPARATUS/PROCEDURE:

The method of isothermal sections was used with refractometric analyses (1). Heterogeneous and homogeneous mixtures of known composition were equilibrated until thier refractive indices remained constant. The composition of the saturated solutions and the corresponding solid phases were found as inflection or "break" points on a plot of composition against refractive index. All refractive indices were measured at 50°C.

The double salt was separated and studied by micro-optical, x-ray, IR, and thermal methods. The melting point of this double salt is 100°C. At 180°C it partially decomposes to give $CeONO_3$ (autooxidized by NO_3 groups).

SOURCE AND PURITY OF MATERIALS:

1. Analar grade hexahydrate presumably used as received.
2. Nothing specified.
3. Doubly distilled water.

ESTIMATED ERROR:

Nothing specified (see critical evaluation).

REFERENCES:

1. Zhuravlev, E.F.; Sheveleva, A.D. *Zh. Neorg. Khim.* 1960, *5*, 2630.

COMPONENTS:

(1) Cerium nitrate; $Ce(NO_3)_3$; [10108-73-3]

(2) Diethylamine nitrate; $C_4H_{12}N_2O_3$; [27096-30-6]

(3) Water; H_2O; [7732-18-5]

ORIGINAL MEASUREMENTS:

Mininkov, N.E.; Zhuravlev, E.F. *Zh. Neorg. Khim.* 1976, *21*, 242-6; *Russ. J. Inorg. Chem. Engl. Transl.* 1976, *21*, 131-3.

VARIABLES:

Composition at 25°C and 50°C

PREPARED BY:

T. Mioduski, S. Siekierski, M. Salomon

EXPERIMENTAL VALUES: Composition of saturated solutions at 25°C [a,b]

$Ce(NO_3)_3$ mass %	$Ce(NO_3)_3$ mol kg^{-1}	$(C_2H_5)_2NH \cdot HNO_3$ mass %	$(C_2H_5)_2NH \cdot HNO_3$ mol kg^{-1}	n_D^{50}	nature of the solid phase
——	——	81.5	32.4	1.4260	$(C_2H_5)_2NH \cdot HNO_3$
9.5	2.16	77.0	41.9	1.4365	"
12.5	3.19	75.5	46.2	1.4400	"
22.5	9.20	70.0	68.6	1.4565	"
26.5	14.77	68.0	90.8	1.4630	$(C_2H_5)_2NH \cdot HNO_3 + Ce(NO_3)_3 \cdot 5(C_2H_5)_2 \cdot NH \cdot HNO_3$
29.5	8.61	60.0	42.0	1.4605	$Ce(NO_3)_3 \cdot 5(C_2H_5)_2NH \cdot HNO_3$ [c]
33.0	7.50	53.5	29.1	1.4595	"
38.0	7.77	47.0	23.0	1.4605	"
41.7	7.84	42.0	18.9	1.4620	"
49.3	10.65	36.5	18.9	1.4710	$Ce(NO_3)_3 \cdot 5(C_2H_5)_2NH \cdot HNO_3 + Ce(NO_3)_3 \cdot 6H_2O$
52.5	7.85	27.0	9.67	1.4690	$Ce(NO_3)_3 \cdot 6H_2O$
55.5	6.19	17.0	4.54	1.4660	"
58.0	5.83	11.5	2.77	1.4650	"
63.0	5.22	——	——	1.4620	"

a. Molalities calculated by the compilers.

b. Data for 50°C were presented only on the phase diagram, and they were rejected (see COMMENTS below).

c. $Ce(NO_3)_3 \cdot 5(C_2H_5)_2NH \cdot HNO_3$ said to be congruently soluble which changes to incongruently soluble at lower temperatures: compilers assume this means at temperatures below 25°C.

AUXILIARY INFORMATION

METHOD/APPARATUS/PROCEDURE:

The method of isothermal sections was used with refractometric analyses (1). Heterogeneous and homogeneous mixtures of known composition were equilibrated until their refractive indices remained constant. The composition of the saturated solutions and the corresponding solid phases were found as inflection or "break" points on a plot of composition against refractive index. All refractive indices were measured at 50°C.

COMMENTS AND/OR ADDITIONAL DATA:

The numerical solubility data at 50°C were not given, but the authors did present portions of the solubility branches for $(C_2H_5)_2NH \cdot HNO_3$ and $Ce(NO_3)_3$ in a phase diagram. For the latter, only the mass % for the binary system could be interpolated from the diagram. The interpolated value of 73.3 mass % appears to be in considerable error (value is too high), and we therefore, reject all the data at 50°C.

SOURCE AND PURITY OF MATERIALS:

"AR" grade $Ce(NO_3)_3 \cdot 6H_2O$ was used.

Diethylamine nitrate, $(C_2H_5)_2NH \cdot HNO_3$, was probably prepared as described earlier (2): see the compilation of this paper for the $La(NO_3)_3 - (C_2H_5)_2NH \cdot HNO_3 - H_2O$ system for details.

Doubly distilled water was used.

ESTIMATED ERROR:

Nothing specified (see critical evaluation).

REFERENCES:

1. Zhuravlev, E.F.; Sheveleva, A.D. *Zh. Neorg. Khim.* 1960, *5*, 2630.

2. Gorshunova, V.P.; Zhuravlev, E.F. *Zh. Neorg. Khim.* 1970, *15*, 3355.

COMPONENTS:

(1) Cerium nitrate; $Ce(NO_3)_3$; [10108-73-3]

(2) Ethylenediamine dinitrate; $C_2H_{10}N_4O_6$; [20829-66-7]

(3) Water; H_2O; [7732-18-5]

ORIGINAL MEASUREMENTS:

Mininkov, N.E.; Zhuravlev, E.F.; *Zh. Neorg. Khim.* 1969, *14*, 2565-9; *Russ. J. Inorg. Chem. Engl. Transl.* 1969, *14*, 1348-50.

VARIABLES:

Composition at 25°C and 50°C

PREPARED BY:

T. Mioduski, S. Siekierski, M. Salomon

EXPERIMENTAL VALUES: The $Ce(NO_3)_3$ - $H_2NCH_2CH_2NH_2 \cdot 2HNO_3$ - H_2O system

Composition of saturated solutions at 25°C [a]

$Ce(NO_3)_3$		$en \cdot 2HNO_3$			
mass %	mol kg^{-1}	mass %	mol kg^{-1}	n_D^{50}	nature of the solid phase
——	——	51.0	5.59	1.4175	$en \cdot 2HNO_3$
5.5	0.34	45.5	4.99	1.4195	"
12.5	0.82	40.5	4.63	1.4235	"
19.5	1.34	36.0	4.35	1.4290	"
28.0	2.08	30.8	4.02	1.4380	"
36.0	2.90	26.0	3.68	1.4580	$en \cdot 2HNO_3 + Ce(NO_3)_3 \cdot en \cdot 2HNO_3 \cdot 6H_2O$
38.0	2.91	22.0	2.96	1.4550	$Ce(NO_3)_3 \cdot en \cdot 2HNO_3 \cdot 6H_2O$
41.0	3.07	18.0	2.36	1.4493	"
46.0	3.36	12.0	1.54	1.4455	"
51.0	3.75	7.3	0.94	1.4460	"
53.0	3.94	5.8	0.76	1.4490	"
58.0	4.62	3.5	0.49	1.4560	"
62.5	5.48	2.5	0.38	1.4640	$Ce(NO_3)_3 \cdot en \cdot 2HNO_3 \cdot 6H_2O + Ce(NO_3)_3 \cdot 6H_2O$
62.3	5.38	2.2	0.33	1.4638	$Ce(NO_3)_3 \cdot 6H_2O$
63.0	5.22	——	——	1.4620	"

a. Molalities calculated by M. Salomon.

continued........

AUXILIARY INFORMATION

METHOD/APPARATUS/PROCEDURE:

The method of isothermal sections was used with refractometric analyses (1). Heterogeneous and homogeneous mixtures of known composition were equilibrated until their refractive indices remained constant. The composition of the saturated solutions and the corresponding solid phases were found as inflection or "break" points on a plot of composition against refractive index. All refractive indices were measured at 50°C.

SOURCE AND PURITY OF MATERIALS:

A.R. grade $Ce(NO_3)_3 \cdot 6H_2O$ used as received. Analysis by ovalate method showed it to contain 75.0 mass % $Ce(NO_3)_3$.

Ethylenediamine (en) and "pure" grade HNO_3 mixed in stoichiometric amounts with cooling. The resulting crystals of $en \cdot 2HNO_3$ were vacuum filtered and dried in a desiccator over $CaCl_2$.

Doubly distilled water was used.

ESTIMATED ERROR:

Nothing specified (see critical evaluation).

REFERENCES:

1. Zhuravlev, E.F.; Sheveleva, A.D. *Zh. Neorg. Khim.* 1960, *5*, 2630.

COMPONENTS:	ORIGINAL MEASUREMENTS:
(1) Cerium nitrate; $Ce(NO_3)_3$; [10108-73-3] (2) Ethylenediamine dinitrate; $C_2H_{10}N_4O_6$; [20829-66-7] (3) Water; H_2O; [7732-18-5]	Mininkov, N.E.; Zhuravlev, E.F.; *Zh. Neorg. Khim.* 1969, *14*, 2565-9; *Russ. J. Inorg. Chem. Engl. Transl.* 1969, *14*, 1348-50.

EXPERIMENTAL VALUES: continued.......

Composition of saturated solutions at 50°C [a]

$Ce(NO_3)_3$		en·$2HNO_3$			
mass %	mol kg^{-1}	mass %	mol kg^{-1}	n_D^{50}	nature of the solid phase
——	——	67.0	6.23	1.4460	en·$2HNO_3$
4.0	0.35	60.7	9.24	1.4470	"
9.0	0.79	56.0	8.60	1.4510	"
14.5	1.29	51.0	7.94	1.4540	"
21.5	2.00	45.5	7.41	1.4610	"
29.5	3.02	40.5	7.25	1.4720	"
34.0	3.72	38.0	7.29	1.4795	"
38.0	4.48	36.0	7.44	1.4850	en·$2HNO_3$ + $Ce(NO_3)_3$·en·$2HNO_3$·$6H_2O$
40.5	4.36	31.0	5.84	1.4800	$Ce(NO_3)_3$·en·$2HNO_3$·$6H_2O$
48.0	4.53	19.5	3.22	1.4740	"
57.5	5.34	9.5	1.55	1.4730	"
69.5	7.35	1.5	0.28	1.4910	$Ce(NO_3)_3$·en·$2HNO_3$·$6H_2O$ + $Ce(NO_3)_3$·$6H_2O$
70.0	7.40	1.0	0.19	1.4860	$Ce(NO_3)_3$·$6H_2O$
73.0	8.29	——	——	1.4900	"

a. Molalities calculated by M. Salomon.

COMMENTS AND/OR ADDITIONAL DATA:

The solubility isotherms consist of three branches representing the initial components and $Ce(NO_3)_3$·en·$2HNO_3$·$6H_2O$. The lines of the monosaturated components intersect in two invariant points. Both points for the 25°C and 50°C isotherms are eutonic, but the double nitrate is incongruently soluble.

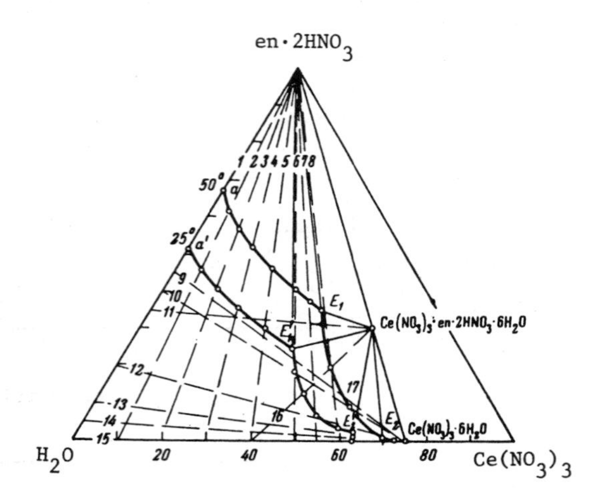

Solubility isotherms and direction of the sections (mass % units).

COMPONENTS:

(1) Cerium nitrate; $Ce(NO_3)_3$; [10108-73-3]

(2) Piperidine nitrate; $C_5H_{12}N_2O_3$; [6091-45-8]

(3) Water; H_2O; [7732-18-5]

ORIGINAL MEASUREMENTS:

Mininkov, N.E.; Zhuravlev, E.F. *Zh. Neorg. Khim.* 1973, *18*, 3091-5; *Russ. J. Inorg. Chem. Engl. Transl.* 1973, *18*, 1645-8.

VARIABLES:

Composition at 25°C and 50°C

PREPARED BY:

T. Mioduski, S. Siekierski, and M. Salomon

EXPERIMENTAL VALUES: Composition of saturated solutions at 25°C [a]

$Ce(NO_3)_3$		$C_5H_{11}N{\cdot}HNO_3$			
mass %	mol kg^{-1}	mass %	mol kg^{-1}	n_D^{50}	nature of the solid phase
——	——	86.0	41.46	1.4610	$C_5H_{11}N{\cdot}HNO_3$
3.5	0.86	84.0	45.36	1.4620	"
7.5	2.00	81.0	47.54	1.4635	"
10.0 [b]	2.92	79.5	51.10	1.4640	$C_5H_{11}N{\cdot}HNO_3$ + $Ce(NO_3)_3{\cdot}5C_5H_{11}N{\cdot}HNO_3$
12.0	2.54	73.5	34.21	1.4515	$Ce(NO_3)_3{\cdot}5C_5H_{11}N{\cdot}HNO_3$
15.5	2.21	63.0	19.78	1.4450	"
21.5	2.23	49.0	11.21	1.4410	"
30.0	2.97	39.0	8.49	1.4430	"
34.5	3.47	35.0	7.75	1.4475	"
37.5 [c]	3.96	33.5	7.80	1.4540	$Ce(NO_3)_3{\cdot}5C_5H_{11}N{\cdot}HNO_3$ + $Ce(NO_3)_3{\cdot}3C_5H_{11}N{\cdot}HNO_3{\cdot}6H_2O$
42.5	4.34	27.5	6.19	1.4550	$Ce(NO_3)_3{\cdot}3C_5H_{11}N{\cdot}HNO_3{\cdot}6H_2O$
49.5	5.06	20.5	4.61	1.4600	"
53.5	5.76	18.0	4.26	1.4640	"
58.65 [d]	6.58	14.0	3.45	1.4730	$Ce(NO_3)_3{\cdot}3C_5H_{11}N{\cdot}HNO_3{\cdot}6H_2O$ + $Ce(NO_3)_3{\cdot}6H_2O$
60.0	6.13	10.0	2.25	1.4680	$Ce(NO_3)_3{\cdot}6H_2O$
61.0	5.76	6.5	1.35	1.4650	"
63.0	5.22	——	——	1.4620	"

continued.......

AUXILIARY INFORMATION

METHOD/APPARATUS/PROCEDURE:

The method of isothermal sections was used with refractometric analyses (1). Heterogeneous and homogeneous mixtures of known composition were equilibrated until their refractive indices remained constant. The composition of the saturated solutions and the corresponding solid phases were found as inflection or "break" points on a plot of composition against refractive index. The compilers assume that all refractive indices were measured at 50°C.

X-ray diffraction studies on the initial salts and $Ce(NO_3)_3{\cdot}5C_5H_{11}N{\cdot}HNO_3$ are reported in the source paper. This latter double salt was also studied by IR spectroscopy, and by differential thermal analysis.

SOURCE AND PURITY OF MATERIALS:

"AR" grade cerium nitrate used as received.

$C_5H_{11}N{\cdot}HNO_3$ prepd by neutn of equiv amounts of "AR" grade piperidine and HNO_3 solutions. The neutd solution was evapd to crystallization and the crystals separated by vac filtration. The salt was washed with alcohol and dried at 50-70°C over KOH in vacuum.

Doubly distilled water was used.

ESTIMATED ERROR:

Nothing specified (see critical evaluation).

REFERENCES:

1. Zhuravlev, E.F.; Sheveleva, A.D. *Zh. Neorg. Khim.* 1960, *5*, 2630.

COMPONENTS:	ORIGINAL MEASUREMENTS:
(1) Cerium nitrate; $Ce(NO_3)_3$; [10108-73-3] (2) Piperidine nitrate; $C_5H_{12}N_2O_3$ [6091-45-8] (3) Water; H_2O; [7732-18-5]	Mininkov, N.E.; Zhuravlev, E.F. *Zh. Neorg. Khim.* 1973, *18*, 3091-5; *Russ. J. Inorg. Chem. Engl. Transl.* 1973, *18*, 1645-8.

EXPERIMENTAL VALUES: continued.......

Composition of saturated solutions at 50°C [a]

$Ce(NO_3)_3$		$C_5H_{11}N\cdot HNO_3$			
mass %	mol kg^{-1}	mass %	mol kg^{-1}	n_D^{50}	nature of the solid phase
——	——	89.0	54.61	1.4640	$C_5H_{11}N\cdot HNO_3$
2.5	0.73	87.0	55.92	1.4650	"
6.5	2.21	84.5	63.37	1.4665	"
13.0	6.64	81.0	91.12	1.4695	"
16.0 [b]	9.81	79.0	106.6	1.4710	$C_5H_{11}N\cdot HNO_3 + Ce(NO_3)_3\cdot 5C_5H_{11}N\cdot HNO_3$
17.5	7.15	75.0	67.49	1.4690	$Ce(NO_3)_3\cdot 5C_5H_{11}N\cdot HNO_3$
21.5	4.55	64.0	29.79	1.4635	"
25.0	4.26	57.0	21.37	1.4615	"
28.5	4.37	51.5	17.38	1.4630	"
34.0	4.74	44.0	13.50	1.4655	"
42.5 [c]	6.06	36.0	11.30	1.4725	$Ce(NO_3)_3\cdot 5C_5H_{11}N\cdot HNO_3$ + $Ce(NO_3)_3\cdot 3C_5H_{11}N\cdot HNO_3\cdot 6H_2O$
46.2	6.21	31.0	9.18	1.4750	$Ce(NO_3)_3\cdot 3C_5H_{11}N\cdot HNO_3\cdot 6H_2O$
52.0	7.25	26.0	7.98	1.4810	"
56.0	7.47	21.0	6.16	1.4830	"
59.0 [d]	8.61	20.0	6.43	1.4845	$Ce(NO_3)_3\cdot 3C_5H_{11}N\cdot HNO_3\cdot 6H_2O + Ce(NO_3)_3\cdot 6H_2O$
61.0	7.79	15.0	4.22	1.4815	$Ce(NO_3)_3\cdot 6H_2O$
64.5	7.76	10.0	2.65	1.4810	"
69.5	8.04	4.0	1.02	1.4850	"
73.3	8.42	——	——	1.4900	"

a. Molalities calculated by the compilers.

b. Eutonic points.

c. Peritonic points.

d. Eutonic points.

COMMENTS AND/OR ADDITIONAL DATA:

The solubility isotherms consist of four branches which intersect at three invariant points b, c, and d. The double nitrate $Ce(NO_3)_3\cdot 5C_5H_{11}N\cdot HNO_3$ is congruently soluble and can be isolated from solution.

The double nitrate $Ce(NO_3)_3\cdot 3C_5H_{11}N\cdot HNO_3\cdot 6H_2O$ is incongruently soluble, and the authors report great difficulties in isolating this compound.

COMPONENTS:

(1) Cerium nitrate; $Ce(NO_3)_3$; [10108-73-3]

(2) Pyridine nitrate; $C_5H_6N_2O_3$; [543-53-3]

(3) Water; H_2O; [7732-18-5]

ORIGINAL MEASUREMENTS:

Mininkov, N.E.; Zhuravlev, E.F. *Zh. Neorg. Khim.* 1974, *19*, 1656-9; *Russ. J. Inorg. Chem. Engl. Transl.* 1974, *19*, 901-4.

VARIABLES:

Composition at 25°C and 50°C

PREPARED BY:

T. Mioduski, S. Siekierski, M. Salomon

EXPERIMENTAL VALUES:

The $Ce(NO_3)_3$ - $C_5H_5N{\cdot}HNO_3$ - H_2O system at 25°C

Composition of saturated solutions [a]

$Ce(NO_3)_3$ mass %	$Ce(NO_3)_3$ mol kg^{-1}	$C_5H_5N{\cdot}HNO_3$ mass %	$C_5H_5N{\cdot}HNO_3$ mol kg^{-1}	n_D	nature of the solid phase	
——	——	76.5	22.91	1.4770	$C_5H_5N{\cdot}HNO_3$	(A)
6.00	0.84	72.0	23.03	1.4820	"	
13.0	1.99	67.0	23.57	1.4590	"	
21.0	3.79	62.0	25.66	1.4070	"	
26.5	5.42	58.5	27.44	1.5050	A + B	
27.0	5.17	57.0	25.07	1.5040	$Ce(NO_3)_3{\cdot}5C_5H_5N{\cdot}HNO_3$	(B)
69.0 [b]		52.0		1.4975	"	
33.0	4.13	42.5	12.21	1.4830	"	
34.5	3.92	38.5	10.03	1.4785	"	
38.0	4.02	33.0	8.01	1.4700	"	
39.5	4.11	31.0	7.39	1.4670	"	
43.0	4.00	24.0	5.12	1.4620	"	
49.0	4.42	17.0	3.52	1.4620	"	
61.5	5.99	7.0	1.56	1.4655	B + C	
62.0	5.51	3.5	0.71	1.4630	$Ce(NO_3)_3{\cdot}6H_2O$	(C)
63.0	5.22	——	——	1.4620	"	

continued........

AUXILIARY INFORMATION

METHOD/APPARATUS/PROCEDURE:

The method of isothermal sections was used with refractometric analyses (1). Heterogeneous and homogeneous mixtures of known composition were equilibrated until their refractive indices remained constant. The composition of the saturated solutions and the corresponding solid phases were found as inflection or "break" points on a plot of composition against refractive index. The compilers assume that all refractive indices were measured at 50°C.

The solid phases were analysed for cerium by the oxalate method, and for total nitrate by precipitation with nitron. IR, x-ray, and thermal studies are also reported.

SOURCE AND PURITY OF MATERIALS:

A.R. grade $Ce(NO_3)_3$ was presumably used as received.

Pyridine nitrate, $C_5H_5N{\cdot}HNO_3$, was prepared as in (2): see also the compilation of the $La(NO_3)_3$ - $C_5H_5N{\cdot}HNO_3$ - H_2O system (3).

Water was doubly distilled.

ESTIMATED ERROR:

Nothing specified (see critical evaluation).

REFERENCES:

1. Zhuravlev, E.F.; Sheveleva, A.D. *Zh. Neorg. Khim.* 1960, *5*, 2630.
2. Mininkov, N.E. *Candidate's Thesis.* Khabarovsk. 1971. p. 47.
3. Gorshunova, V.P.; Zhuravlev, E.F. *Zh. Neorg. Khim.* 1974, *19*, 249.

COMPONENTS:

(1) Cerium nitrate; $Ce(NO_3)_3$; [10108-73-3]

(2) Pyridine nitrate; $C_5H_6N_2O_3$; [543-53-3]

(3) Water; H_2O; [7732-18-5]

ORIGINAL MEASUREMENTS:

Mininkov, N.E.; Zhuravlev, E.F. *Zh. Neorg. Khim.* 1974, *19*, 1656-9; *Russ. J. Inorg. Chem. Engl. Transl.* 1974, *19*, 901-4.

EXPERIMENTAL VALUES: continued.......

The $Ce(NO_3)_3$ - $C_5H_5N{\cdot}HNO_3$ - H_2O system at 50°C

Composition of saturated solutions [a]

$Ce(NO_3)_3$ mass %	$Ce(NO_3)_3$ mol kg^{-1}	$C_5H_5N{\cdot}HNO_3$ mass %	$C_5H_5N{\cdot}HNO_3$ mol kg^{-1}	n_D	nature of the solid phase [c]
——	——	84.5	38.36	1.4960	A
3.5	0.77	82.5	41.47	1.4990	A
8.5	1.93	78.0	40.66	1.5060	A
16.0	4.67	73.5	49.26	1.5130	A
23.0	8.82	69.0	60.69	1.5225	A
28.0	12.26	65.0	65.34	1.5300	A + B
31.5	3.12	37.5	8.51	1.5210	B
36.5	6.99	47.5	20.89	1.5085	B
38.5	6.75	44.0	17.69	1.5050	B
40.0	6.29	40.5	14.61	1.5015	B
41.5	6.53	39.0	14.07	1.5000	B
42.5	6.36	37.0	12.70	1.4995	B
46.5	6.48	31.5	10.08	1.4990	B
48.0	9.60	30.0	7.50	1.4990	B
57.5	7.50	19.0	5.69	1.5020	B
62.7	8.82	15.5	5.00	1.5060	B + C
63.0	8.05	13.0	3.81	1.5025	C
64.0	7.01	8.0	2.01	1.4975	C
66.0	6.86	4.5	1.07	1.4940	C
68.0	6.84	1.5	0.35	1.4910	C
73.0	8.29	——	——	1.4900	C

a. Molalities calculated by the compilers.

b. Total mass % exceeds 100%. $Ce(NO_3)_3$ mass % of 69.0 appears in both the original publication and the English translation, and could be a misprint (29.0 mass % ?).

c. See previous page for A, B, C.

COMMENTS AND/OR ADDITIONAL DATA:

The solubility diagram consists of three lines corresponding to the crystallation of pyridine nitrate, cerium (III) nitrate hexahydrate, and the double nitrate $Ce(NO_3)_3{\cdot}5C_5H_5N{\cdot}HNO_3$. The three branches intersect at two points of invariant equilibrium: the A and B branches intersect at a peritonic point, and the B and C branches intersect at a eutonic point. The double nitrate is incongruently soluble, melts at 115°C, and decomposes slightly at 215°C due to autooxidation to $CeONO_3$: final decomposition at 240°C is violent.

COMPONENTS:

(1) Cerium nitrate; $Ce(NO_3)_3$; [10108-73-3]

(2) Aniline nitrate; $C_6H_8N_2O_3$; [542-15-4]

(3) Water; H_2O; [7732-18-5]

ORIGINAL MEASUREMENTS:

Mininkov, N.E.; Zhuravlev, E.F. *Zh. Neorg. Khim.* 1970, *15*, 205-10; *Russ. J. Inorg. Chem. Engl. Transl.* 1970, *15*, 105-7.

VARIABLES:

Composition at 25°C and 50°C

PREPARED BY:

T. Mioduski and S. Siekierski

EXPERIMENTAL VALUES:

	$Ce(NO_3)_3$		$C_6H_5NH_3NO_3$			
t/°C	mass %	mol kg^{-1}	mass %	mol kg^{-1}	n_D^{50}	nature of the solid phase
25	——	——	22.0	1.81	1.3750	$C_6H_5NH_3NO_3$
	12.5	0.51	13.0	1.12	1.3760	"
	24.5	1.14	9.5	0.92	1.3850	"
	38.0	2.06	5.5	0.62	1.4100	"
	48.0	3.07	4.0	0.53	1.4350	"
	59.0	4.64	2.0	0.33	1.4560	"
	62.0	5.25	1.8	0.32	1.4630	$C_6H_5NH_3NO_3 + Ce(NO_3)_3 \cdot 6H_2O$
	62.5	5.25	1.0	0.18	1.4625	$Ce(NO_3)_3 \cdot 6H_2O$
	63.0	5.22	——	——	1.4620	"
50	——	——	47.0	5.68	1.4220	$C_6H_5NH_3NO_3$
	9.0	0.51	36.5	4.29	1.4150	"
	19.5	1.12	27.0	3.23	1.4140	"
	33.0	2.02	17.0	2.18	1.4250	"
	45.0	3.00	9.0	1.25	1.4450	"
	52.5	3.88	6.0	0.93	1.4550	"
	58.0	4.62	3.5	0.58	1.4650	"
	64.5	5.82	1.5	0.28	1.4840	"
	71.2	7.63	0.2	0.045	1.4910	$C_6H_5NH_3NO_3 + Ce(NO_3)_3 \cdot 6H_2O$
	71.0	7.51	0.001	0.0002	1.4905	$Ce(NO_3)_3 \cdot 6H_2O$
	73.0	8.29	——	——	1.4900	"

AUXILIARY INFORMATION

METHOD/APPARATUS/PROCEDURE:

The method of isothermal sections was used with refractometric analyses (1). Heterogeneous and homogeneous mixtures of known composition were equilibrated until their refractive indices remained constant. The composition of the saturated solutions and the corresponding solid phases were found as inflection or "break" points on a plot of composition against refractive index. All refractive indices were measured at 50°C.

The phase diagram is of the simple eutonic type.

Molalities in the above table were calculated by M. Salomon.

SOURCE AND PURITY OF MATERIALS:

"Pure" grade $Ce(NO_3)_3 \cdot 6H_2O$ was used as received. Analysis for $Ce(NO_3)_3$ gave 75.0 mass %.
Source and purity of aniline nitrate not given, but the salt was probably recrystallized.
Doubly distilled water was used.

ESTIMATED ERROR:

Soly: precision about ± 1% (compilers).

Temp: precision probably ± 0.2% K (compilers).

REFERENCES:

1. Zhuravlev, E.F.; Sheveleva, A.D. *Zh. Neorg. Khim.* 1960, *5*, 2630.

COMPONENTS:

(1) Cerium nitrate; $Ce(NO_3)_3$; [10108-73-3]

(2) Urea; CH_4N_2O; [57-13-6]

(3) Water; H_2O; [7732-18-5]

ORIGINAL MEASUREMENTS:

Zholalieva, Z.M.; Sulaimankulov, K.S.; Ismailov, M. *Zh. Neorg. Khim.* 1975, *20*, 2243-5; *Russ. J. Inorg. Chem. Engl. Trans.* 1975, *20*, 1246-7.

VARIABLES:

Composition at 30°C

PREPARED BY:

T. Mioduski and S. Siekierski

EXPERIMENTAL VALUES:

Composition of saturated solutions [a]

$Ce(NO_3)_3$ mass %	$Ce(NO_3)_3$ mol kg^{-1}	$CO(NH_2)_2$ mass %	$CO(NH_2)_2$ mol kg^{-1}	nature of the solid phase
——	——	57.5	22.5	$CO(NH_2)_2$
3.91	0.303	56.52	23.78	"
8.57	0.761	56.90	27.44	"
16.51	1.874	56.48	34.82	"
21.54	3.035	56.70	43.39	"
31.22	8.493	57.51	84.97	"
35.73	15.95	57.40	139.1	"
35.55	8.051	50.91	62.61	$Ce(NO_3)_3 \cdot 8CO(NH_2)_2$
38.21	8.73	48.37	60.02	"
41.05	7.694	42.59	43.35	"
41.46	7.285	41.09	39.21	$Ce(NO_3)_3 \cdot 8CO(NH_2)_2 + Ce(NO_3)_3 \cdot 5CO(NH_2)_2$
41.54	7.258	40.91	38.81	$Ce(NO_3)_3 \cdot 5CO(NH_2)_2$
42.49	6.789	38.32	33.25	"
42.73	6.177	36.06	28.31	"
48.52	7.088	30.49	24.19	"

a. Molalities calculated by M. Salomon.

continued......

AUXILIARY INFORMATION

METHOD/APPARATUS/PROCEDURE:

The isothermal method was used. Equilibrium was reached after 8 hours. The urea content of the saturated solutions was determined by nitrogen analysis using the Kjeldahl ammonia method, and Ce(III) was determined by titration with Trilon B.

SOURCE AND PURITY OF MATERIALS:

No information given.

ESTIMATED ERROR:

Nothing specified (see critical evaluation).

REFERENCES:

COMPONENTS:	ORIGINAL MEASUREMENTS:
(1) Cerium nitrate; $Ce(NO_3)_3$; [10108-73-3] (2) Urea; CH_4N_2O; [57-13-6] (3) Water; H_2O; [7732-18-5]	Zholalieva, Z.M.; Sulaimankulov, K.S.; Ismailov, M. *Zh. Neorg. Khim.* 1975, *20*, 2243-5; *Russ. J. Inorg. Chem. Engl. Transl.* 1975, *20*, 1246-7.

EXPERIMENTAL VALUES: continued......

Composition of saturated solutions [a]

$Ce(NO_3)_3$		$CO(NO_2)_2$		nature of the solid phase
mass %	mol kg^{-1}	mass %	mol kg^{-1}	
51.61	9.190	31.17	30.14	$Ce(NO_3)_3 \cdot 5CO(NH_2)_2$ + $Ce(NO_3)_3 \cdot 2CO(NH_2)_2 \cdot 2H_2O$
51.61	8.923	30.91	29.13	$Ce(NO_3)_3 \cdot 2CO(NH_2)_2 \cdot 2H_2O$
53.06	7.627	25.61	19.99	"
54.29	7.863	24.54	19.30	"
59.34	7.786	17.29	12.32	"
63.46	7.183	9.45	5.809	"
65.15	7.152	6.92	4.13	$Ce(NO_3)_3 \cdot 6H_2O$
65.05	6.921	6.13	3.54	"
64.25	6.565	5.74	3.18	"
64.03	6.033	3.43	1.76	"
64.09	5.472	——	——	"

a. Molalities calculated by M. Salomon.

COMMENTS AND/OR ADDITIONAL DATA:

The phase diagram is given below. Two anhydrous and one hydrated double compounds are formed in the system: $Ce(NO_3)_3 \cdot 8CO(NH_2)_2$, $Ce(NO_3)_3 \cdot 5CO(NH_2)_2$, and $Ce(NO_3)_3 \cdot 2CO(NH_2)_2 \cdot 2H_2O$. All these compounds are congruently soluble in water. The first and last solubility branches belong to the crystallisation of the initial components.

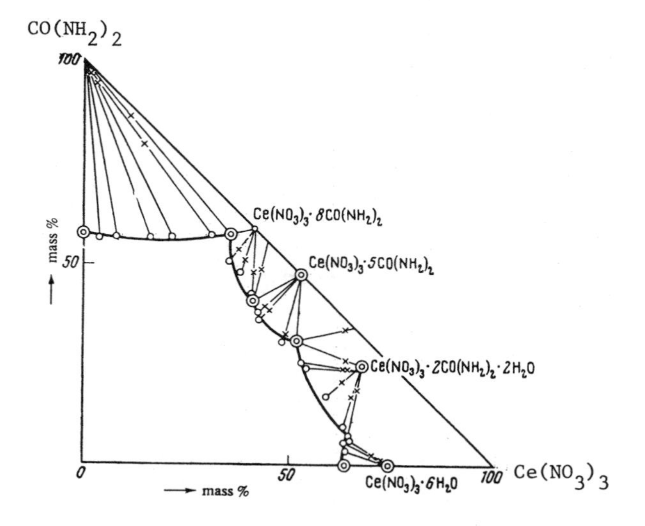

30°C solubility isotherm of the $Ce(NO_3)_3$ - $CO(NH_2)_2$ - H_2O system.

COMPONENTS:

(1) Cerium nitrate; $Ce(NO_3)_3$; [10108-73-3]

(2) Urea mononitrate; $CH_5N_3O_4$; [124-47-0]

(3) Water; H_2O; [7732-18-5]

ORIGINAL MEASUREMENTS:

Zhuravlev, E.F.; Mininkov, N.E. *Zh. Neorg. Khim.* 1972, *17*, 1736-40, *Russ. J. Inorg. Chem. Engl. Transl.* 1972, *17*, 899-901.

VARIABLES:

Composition at 25°C and 50°C

PREPARED BY:

T. Mioduski and S. Siekierski

EXPERIMENTAL VALUES:

	$Ce(NO_3)_3$		$CO(NH_2)_2 \cdot HNO_3$			
t/°C	mass %	mol kg^{-1}	mass %	mol kg^{-1}	n_D^{50}	nature of the solid phase
25	——	——	18.0	1.78	1.3570	$CO(NH_2)_2 \cdot HNO_3$
	8.5	0.34	14.0	1.47	1.3640	"
	18.0	0.77	10.5	1.19	1.3765	"
	27.5	1.30	7.5	0.94	1.3915	"
	37.5	2.02	5.5	0.78	1.4125	"
	48.0	3.05	3.7	0.62	1.4320	"
	59.0	4.64	2.0	0.42	1.4550	"
	61.5	5.12	1.7	0.38	1.4625	$CO(NH_2)_2 \cdot HNO_3 + Ce(NO_3)_3 \cdot 6H_2O$
	62.0	5.14	1.0	0.22	1.4622	$Ce(NO_3)_3 \cdot 6H_2O$
	63.0	5.22	——	——	1.4620	"
50	——	——	30.0	3.48	1.3740	$CO(NH_2)_2 \cdot HNO_3$
	7.5	0.34	25.0	3.01	1.3780	"
	16.0	0.77	20.0	2.54	1.3860	"
	25.5	1.31	15.0	2.05	1.3990	"
	36.0	2.04	10.0	1.50	1.4170	"
	46.5	3.10	7.5	1.32	1.4360	"
	57.5	4.52	3.5	0.73	1.4575	"
	69.5	7.13	0.6	0.16	1.4860	"
	73.3	8.42	——	——	1.4900	$Ce(NO_3)_3 \cdot 6H_2O$

AUXILIARY INFORMATION

METHOD/APPARATUS/PROCEDURE:

The method of isothermal sections was used with refractometric analyses (1,2). Heterogeneous and homogeneous mixtures of known composition were equilibrated until their refractive indices remained constant. The composition of the saturated solutions and the corresponding solid phases were found as inflection or "break" points on a plot of composition against refractive index. All refractive indices were measured at 50°C.

Molalities in the above table were calculated by M. Salomon.

SOURCE AND PURITY OF MATERIALS:

"Analytical reagent" grade cerium nitrate hexahydrate was used. Ce/H_2O ratio was "approximately 1:6."

Urea mononitrate, $CO(NH_2)_2 \cdot HNO_3$, was recrystallized.

Doubly distilled water was used.

ESTIMATED ERROR:

Soly: precision around ± 1% (compilers).

Temp: precision probably ± 0.2K (compilers).

REFERENCES:

1. Zhuravlev, E.F.; Sheveleva, A.D. *Zh. Neorg. Khim.* 1960, *5*, 2630.

2. Mininkov, N.E.; Zhuravlev, E.F. *Zh. Neorg. Khim.* 1969, *14*, 2565.

COMPONENTS:

(1) Cerium nitrate; $Ce(NO_3)_3$; [10108-73-3]
(2) Thiourea; CH_4N_2S; [62-56-6]
(3) Water; H_2O; [7732-18-5]

ORIGINAL MEASUREMENTS:

Zholalieva, Z.; Sulaimankulov, K.; Ismailov, M.; Abykeev, K. *Zh. Neorg. Khim.* 1976, *21*, 2583-5; *Russ. J. Inorg. Chem. Engl. Transl.* 1976, *21*, 1421-2.

VARIABLES:

Composition at 30°C

PREPARED BY:

T. Mioduski and S. Siekierski

EXPERIMENTAL VALUES:

Composition of saturated soluitons [a]

$Ce(NO_3)_3$ mass %	$Ce(NO_3)_3$ mol kg^{-1}	$CS(NH_2)_2$ mass %	$CS(NH_2)_2$ mol kg^{-1}	nature of the solid phase
———	———	18.00	2.884	$CS(NH_2)_2$
6.64	0.27	18.01	3.140	"
15.26	0.667	14.61	2.737	"
19.91	0.882	10.88	2.065	"
27.75	1.358	9.59	2.011	"
32.03	1.658	8.75	1.941	"
44.96	2.924	7.90	2.202	"
56.14	5.039	9.70	3.730	"
62.04	6.322	7.87	3.436	"
62.00	6.229	7.48	3.220	$CS(NH_2)_2 + Ce(NO_3)_3\cdot 6H_2O$
61.66	6.287	8.27	3.613	$Ce(NO_3)_3\cdot 6H_2O$
64.09	5.472	———	———	"

a. Molalities calculated by M. Salomon.

AUXILIARY INFORMATION

METHOD/APPARATUS/PROCEDURE:

Isothermal method used. No other information given.

SOURCE AND PURITY OF MATERIALS:

Nothing specified.

ESTIMATED ERROR:

Nothing specified (see critical evaluation).

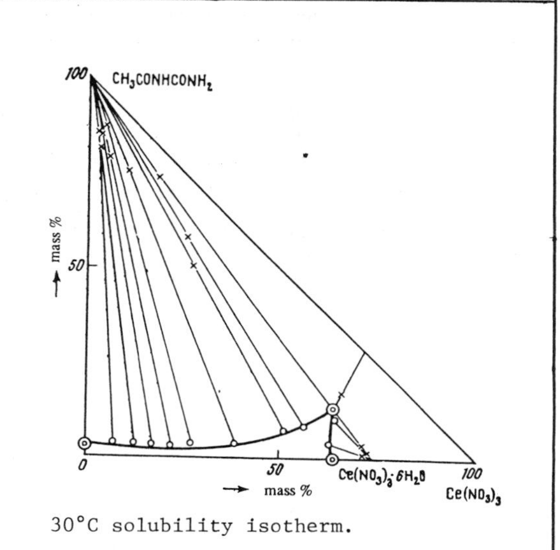

30°C solubility isotherm.

COMPONENTS:

(1) Cerium Nitrate; $Ce(NO_3)_3$; [10108-73-3]

(2) N-Acetylurea; $C_3H_6N_2O_2$; [591-07-1]

(3) Water; H_2O; [7732-18-5]

ORIGINAL MEASUREMENTS:

Zholalieva, Z.M.; Sulaimankulov, K.S.; Ismailov, M. *Zh. Neorg. Khim.* 1975, *20*, 2243-5; *Russ. J. Inorg. Chem. Engl. Transl.* 1975, *20*, 1246-7.

VARIABLES:

Composition at 30°C

PREPARED BY:

T. Mioduski and S. Siekierski

EXPERIMENTAL VALUES:

Composition of saturated solutions [a]

$Ce(NO_3)_3$ mass %	$Ce(NO_3)_3$ mol kg^{-1}	$CH_3CONHCONH_2$ mass %	$CH_3CONHCONH_2$ mol kg^{-1}	nature of the solid phase
———	———	3.23	0.327	$CH_3CONHCONH_2$
7.20	0.249	3.99	0.440	"
12.55	0.458	3.50	0.408	"
17.27	0.668	3.51	0.434	"
21.99	0.898	2.90	0.378	"
26.88	1.197	4.28	0.609	"
38.85	2.091	4.18	0.719	"
51.76	3.901	7.56	1.820	"
56.70	5.000	8.53	2.403	"
63.76	8.815	13.38	5.665	"
64.06	8.714	13.40	5.823	$CH_3CONHCONH_2 + Ce(NO_3)_3 \cdot 6H_2O$
63.81	8.608	13.46	5.800	$Ce(NO_3)_3 \cdot 6H_2O$
63.77	7.809	11.19	4.377	"
62.57	5.814	4.43	1.315	"
64.09	5.472	———	———	"

a. Molalities calculated by M. Salomon.

AUXILIARY INFORMATION

METHOD/APPARATUS/PROCEDURE:

The isothermal method was used. Equilibrium was reached after 8 hours. The saturated solutions were analysed for N-acetylurea by nitrogen determination using the Kjeldahl ammonia method, and Ce(III) was determined by titration with Trilon B.

COMMENTS AND/OR ADDITIONAL DATA:

The phase diagram is of the simple eutonic type.

SOURCE AND PURITY OF MATERIALS:

No information given.

ESTIMATED ERROR:

Nothing specified.

REFERENCES:

COMPONENTS:

(1) Cerium nitrate: $Ce(NO_3)_3$; [10108-73-3]

(2) Organic solvents

EVALUATOR:

Mark Salomon

U.S. Army Electronics Technology and Devices Laboratory
Fort Monmouth, NJ, USA

November, 1982

CRITICAL EVALUATION:

There are very few studies on the solubility of $Ce(NO_3)_3$ in organic solvents (1-4), and due to the absence of descriptions on purities of materials or to poorly defined experimental conditions, none of the data can be critically evaluated or designated as *tentative* solubility values.

The oldest publication by Vauquelin (1) simply states that 100 parts of alcohol (probably ethanol) dissolves 50 parts of $Ce(NO_3)_3$ (i.e. 33.3 mass % of salt, evaluator). Neither the temperature or the purity of the salt were given, and it is most probable that the initial salt was the hexahydrate and of low purity.

Wells (2) reported a solubility of 0.0081 mol dm^{-3} in diethyl ether at about 293 K, and Hopkins and Quill (3) reported graphical data from which the compilers estimated a solubility of 0.028 mol dm^{-3} at 298 K. Because it is improbable that the solubility of $Ce(NO_3)_3$ in diethyl ether would increase by a factor of three for a 5 K increase in temperature, both of the solubility data in (2) and (3) must be questioned.

Healy and McKay (4) carefully measured the solubility of $Ce(NO_3)_3 \cdot nH_2O$ (the evaluator assumes n = 6) in tri-n-butylphosphate. An aqueous phase formed during equilibration, and it is assumed that at equilibrium the solid phase is the hexahydrate in a three phase system. The equilibrated solid phase was not analysed.

REFERENCES

1. Vauquelin, L.N. *Ann. Chim. Phys.* 1801, *35*, 143.
2. Wells, R.C. *J. Wash. Acad. Sci.* 1930, *20*, 146.
3. Hopkins, B.S.; Quill, L.L. *Proc. Natl. Acad. Sci., USA* 1933, *19*, 64.
4. Healy, T.V.; McKay, H.A.C. *Trans. Faraday Soc.* 1956, *52*, 633.

COMPONENTS:	ORIGINAL MEASUREMENTS:
(1) Cerium nitrate; $Ce(NO_3)_3$; [10108-73-3] (2) Diethyl ether; $C_4H_{10}O$; [60-29-7]	Wells, R.C. *J. Wash. Acad. Sci.* 1930, *20*, 146-8.

VARIABLES:	PREPARED BY:
Room temperature (about 20°C)	T. Mioduski, S. Siekierski, M. Salomon

EXPERIMENTAL VALUES:

Experiment 1. This experiment involves hydrated cerium nitrate as the initial solid, and which the compilers assume to be the hexahydrate.

Authors report the solubility as 0.0010 g CeO_2 in 10 ml solution. This is equivalent to a $Ce(NO_3)_3$ soly of 5.8×10^{-4} mol dm^{-3} (compilers).

Experiment 2. The initial solid is dehydrated cerium nitrate.

Authors report the solubility as 0.0139 g CeO_2 in 10 ml solution. This is equivalent to a $Ce(NO_3)_3$ soly of 8.1×10^{-3} mol dm^{-3} (compilers).

AUXILIARY INFORMATION

METHOD/APPARATUS/PROCEDURE:

The isothermal method was used. The soly of cerium nitrate was determined in two experiments in which the nature of the initial solid phase differs.

Experiment 1. A few grams of cerium nitrate (presumably the hexahydrate, compilers) was added to about 20 ml of ether in small stoppered flasks. The flasks were periodically agitated and permitted to stand overnight at about 20°C. A 10 ml sample was removed, filtered, the solvent evaporated, and the salt ignited to the oxide and weighed.

Experiment 2. The remaining salt in the flask was freed from ether, dissolved in water and a few drops of HNO_3 added. The solution was evaporated to dryness, and heated to 150°C. The solubility in ether was then determined again with this "dehydrated" salt.

SOURCE AND PURITY OF MATERIALS:

Nothing specified.

ESTIMATED ERROR:

Soly: precision probably ± 10% at best (compilers).

Temp: precision probably ± 4K (compilers).

REFERENCES:

COMPONENTS:

(1) Cerium nitrate; $Ce(NO_3)_3$; [10108-73-3]

(2) Diethyl ether; $C_4H_{10}O$; [60-29-7]

ORIGINAL MEASUREMENTS:

Hopkins, B.S.; Quill, L.L. *Proc. Natl. Acad. Sci. U.S.A.* 1933, *19*, 64-8.

VARIABLES:

One temperature: 25°C

PREPARED BY:

T. Mioduski, S. Siekierski, M. Salomon

EXPERIMENTAL VALUES:

The solubility of $Ce(NO_3)_3$ in diethyl ether at 25°C was given in the form of a small diagram of solubility vs atomic number Z for Z = 57-64. In the absence of numerical data, the compilers interpolated the solubility from the published diagram. The result is:

$$\text{soly of } Ce(NO_3)_3 = 9.1 \text{ g dm}^{-3} \text{ (0.028 mol dm}^{-3}\text{)}$$

COMMENTS AND/OR ADDITIONAL DATA:

The name Philip Kalischer appears on the diagram published in the source paper. The compilers suspected that Mr. Kalischer was an MSc student of Prof. Hopkins and thus contracted Ms. Susanne Redalje, the Assistant Chemistry Librarian at the University of Illinois at Urbana-Champaign. Ms. Redalje searched the University records for reference to a thesis or any publication by Mr. Kalischer. The records show that Mr. Kalischer attended classes for the Fall, Spring, and Summer semesters of 1930-1931. There is no indication that Mr. Kalischer had finished his studies or submitted a thesis, and it is therefore apparent that the original experimental data are lost. The compilers are most grateful to Ms. Redalje for all her help in searching the University records and providing important information on numerous other lanthanide systems.

AUXILIARY INFORMATION

METHOD/APPARATUS/PROCEDURE:

No information is available, but based on similar work by Hardy (1) being carried out at the University of Illinois at the time, it is likely that the isothermal method was employed. The solubility data for neodymium and prascodymium nitrates in several ethers from Hardy's MSc Thesis are compiled elsewhere in this volume, and the compilations contain detailed information on the experimental techniques which the compilers assume were similar to those used by Mr. Kalischer.

SOURCE AND PURITY OF MATERIALS:

No information given.

ESTIMATED ERROR:

No information given.

REFERENCES:

1. Hardy, Z.M. *Masters Thesis.* The University of Illinois. Urbana, Il. 1932.

COMPONENTS:

(1) Cerium nitrate; $Ce(NO_3)_3$; [10108-73-3]

(2) 1,4-Dioxane (p-dioxane); $C_4H_8O_2$; [123-91-1]

ORIGINAL MEASUREMENTS:

Hopkins, B.S.; Quill, L.L. *Proc. Natl. Acad. Sci. U.S.A.* 1933, *19*, 64-8.

VARIABLES:

One temperature: 25°C

PREPARED BY:

T. Mioduski, S. Siekierski, M. Salomon

EXPERIMENTAL VALUES:

The solubility of $Ce(NO_3)_3$ in p-dioxane at 25°C was given the form of a small diagram of solubility vs atomic number Z for Z = 57-64. In the absence of numerical data, the compilers interpolated the solubility from the published diagram. The result is:

$$\text{soly of } Ce(NO_3)_3 = 4.8 \text{ g dm}^{-3} \ (0.015 \text{ mol dm}^{-3})$$

COMMENTS AND/OR ADDITIONAL DATA:

It appears that the original experimental work was done by a Mr. P. Kalischer who was a student at the University of Illinois at Urbana-Champaign. Attempts to locate the original experimental data have failed, and it thus appears that these data are lost (see the COMMENTS in the compilation for the $Ce(NO_3)_3$ - diethyl ether system).

AUXILIARY INFORMATION

METHOD/APPARATUS/PROCEDURE:

No information is available, but based on similar work by Hardy (1) being carried out at the University of Illinois at the time, it is likely that the isothermal method was employed. The solubility data for neodymium and praseodymium nitrates in several ethers from Hardy's MSc Thesis are compiled elsewhere in this volume, and the compilations contain detailed information on the experimental techniques which the compilers assume were similar to those used by Mr. Kalischer.

SOURCE AND PURITY OF MATERIALS:

No information available.

ESTIMATED ERROR:

No information available.

REFERENCES:

1. Hardy, Z.M. *Masters Thesis.* The University of Illinois. Urbana, IL. 1932.

COMPONENTS:

(1) Cerium nitrate; $Ce(NO_3)_3$; [10108-73-3]

(2) Tri-n-butylphosphate; $C_{12}H_{27}O_4P$; [126-73-8]

ORIGINAL MEASUREMENTS:

Healy, T.V.; McKay, H.A.C. *Trans. Faraday Soc.* 1956, *52*, 633-42.

VARIABLES:

One temperature: 25°C

PREPARED BY:

Mark Salomon

EXPERIMENTAL VALUES:

The solubility of $Ce(NO_3)_3$ in $[CH_3(CH_2)_3O]_3PO$ at 25°C was reported as

$$1.2 \text{ mol dm}^{-3}$$

COMMENTS AND/OR ADDITIONAL DATA:

Since the initial solid was probably the hexahydrate $Ce(NO_3)_3 \cdot 6H_2O$, it is probable that an aqueous phase formed during the equilibration. The authors state that the appearance of an aqueous phase does not interfere with the equilibrium between the solid and $[CH_3(CH_2)_3O]_3PO$ provided the aqueous phase is also in equilibrium.

AUXILIARY INFORMATION

METHOD/APPARATUS/PROCEDURE:

Isothermal method used. The solubility was determined by shaking the solid nitrate with $[CH_3(CH_2)_3O]_3PO$ for several days followed by analysis of the organic phase. Cerium was analysed volumetrically using ferrous sulfate and permanganate after Ce(III) was oxidized to Ce(IV) with sodium bismuthate.

SOURCE AND PURITY OF MATERIALS:

Cerium nitrate was an AR product which the compiler assumes is the hexahydrate. $[CH_3(CH_2)_3O]_3PO$ purified as in (1). It was boiled with dilute aqueous NaOH until all volatile impurities were distilled off. The remaining solvent was washed repeatedly with water and dried by warming in vacuum.

ESTIMATED ERROR:

Nothing specified.

REFERENCES:

1. Alcock, K.; Grimley, S.S.; Healy, T.V.; McKay, H.A.C. *Trans. Faraday Soc.* 1956, *52*, 39.

COMPONENTS:	EVALUATOR:
(1) Cerium double nitrates (2) Water ; H_2O ; [7732-18-5]	Mark Salomon U.S. Army Electronics Technology and Devices Laboratory Fort Monmouth, NJ, USA December 1982

CRITICAL EVALUATION:

Ce(III) DOUBLE NITRATES

Ce(III) nitrate forms a number of double salts with transition metal nitrates for which the stable solid phase is the tetracosahydrate (1-4). The cerium (III) ammonium nitrate double salt forms a stable tetrahydrate (3-5). The following stable solid phases have been found to exist in equilibrium with saturated solutions:

$3Mg(NO_3)_2.2Ce(NO_3)_3.24H_2O$	[13550-46-4]	(1,2)
$2Ce(NO_3)_3.3Mn(NO_3)_2.24H_2O$	[84682-55-3]	(2)
$2Ce(NO_3)_3.3Co(NO_3)_2.24H_2O$	[20394-14-3]	(2)
$2Ce(NO_3)_3.3Ni(NO_3)_2.24H_2O$	[20346-68-3]	(2)
$2Ce(NO_3)_3 3Zn(NO_3)_2.24H_2O$	[15276-92-3]	(2)
$Ce(NO_3)_3.2NH_4NO_3.4H_2O$	[13083-04-0]	(3,4)

Several of these salts have been discussed previously in the $Ce(NO_3)_3$-H_2O critical evaluation.

The solubility of $3Mg(NO_3)_2.2Ce(NO_3)_3$ has been studied as a function of temperature by Friend and Wheat (1), and at 289.2 K in concentrated nitric acid by Jantsch (2). The latter author also reported the congruent melting point of the tetracosahydrate to be 384.7 K. The data for this salt were fitted to the smoothing equation

$$Y = \ln(m/m_o) - nM_2(m - m_o) = a + b/(T/K) + c \ln(T/K) \qquad [1]$$

where all terms were defined previously in the $Ce(NO_3)_3$-H_2O critical evaluation (see eq. [1] in this critical evaluation). Friend and Wheat's solubility value at 297.8 K deviates significantly from the smoothed value obtained from eq. [1], and this data point is *rejected*. The remaining five data points over the temperature range of 284-330 K appear to be satisfactory. At the congruent melting point (384.7 K according to Jantsch) the concentration of the saturated solution is (theoretically) 2.313 mol kg^{-1}, and this value has been combined with the five acceptable solubility values of Friend and Wheat and fitted to eq. [1] by the method of least squares. The results of fitting these six solubility values to eq. [1] are given in Tables 1 and 2. The *tentative* solubility values given in Table 2 were calculated from eq. [1], and are associated with an uncertainty of ± 0.03 mol kg^{-1} at the 95% level of confidence (Student's t = 3.182).

The only other data which could be fitted to the smoothing eq. [1] are those of Wolff (3) for the double salt $Ce(NO_3)_3.2NH_4NO_3$. The results of fitting these data to the smoothing equation are given in Table 1, and the smoothed *tentative* values at selected temperatures are given in Table 2. The uncertainty in these calculated *tentative* values is ± 0.3 mol kg^{-1} at the 95% level of confidence (Student's t = 4.303). The congruent melting point predicted from eq. [1] is 344.4 K, and is not in very good agreement with the experimental value of 347 K (6): with solubility data of precision around ± 0.5% or better, we have found that congruent melting points calculated from eq. [1] are generally within ± 2 K of the experimental value. Angelov (4) reported the solubility of cerium (III) ammonium nitrate in two solutions of 10 and 30 mass % HNO_3 at 298.2 K, and these data cannot be compared with those of Jantsch or Friend and Wheat because of the differences in experimental conditions.

Ce(IV) DOUBLE NITRATES

The solubility of the double nitrate $Ce(NO_3)_4.2NH_4NO_3$ has been studied as a function of temperature by Wolff (3) and by Angelov and Poslavskaya (4). Both studies agree that the stable solid phase is the anhydrous double salt, but there is significant disagreement in the reported solubility values as a function of temperature. At this time it is not possible to give preference to either set of data.

COMPONENTS:	EVALUATOR:
(1) Cerium double nitrates (2) Water ; H_2O ; [7732-18-5]	Mark Salomon U.S. Army Electronics Technology and Devices Laboratory Fort Monmouth, NJ, USA December 1982

CRITICAL EVALUATION:

Table 1. Derived parameters for the smoothing equation [1].

parameter	$3Mg(NO_3)_2 \cdot 2Ce(NO_3)_3$ [a]	$Ce(NO_3)_3 \cdot 2NH_4NO_3$ [b]
a	-28.713	-85.78
b	961	3198
c	4.469	13.171
σ_a	0.004	0.01
σ_b	1.4	4.1
σ_c	0.001	0.002
σ_Y	0.004	0.01
σ_m	0.002	0.17
congruent melting point/K	348.7	344.4
concn at the congr melting pt/mol kg^{-1}	2.313	13.877

[a] Solid phase is the tetracosahydrate (see ref 1).

[b] Solid phase is the tetrahydrate (see ref 3).

Table 2. Tentative solubility data at selected temperatures.[a]

T/K	$3Mg(NO_3)_2 \cdot 2Ce(NO_3)_3$ [b]	$Ce(NO_3)_3 \cdot 2NH_4NO_3$ [c]
273.2	0.688	3.49
283.2	0.725	3.79
293.2	0.768	4.19
298.2	0.792	4.44
303.2	0.819	4.73
313.2	0.877	5.46
323.2	0.946	6.48
333.2	1.026	8.07
343.2	1.123	11.68
353.2	1.241	
363.2	1.393	
373.2	1.608	
383.2	2.034	

[a] All solubilities calculated from the smoothing equation; units are mol kg^{-1}.

[b] Solid phase is the tetracosahydrate.

[c] Solid phase is the tetrahydrate.

REFERENCES

1. Friend, J.N.; Wheat, W.N. *J. Chem. Soc.* 1935, 356.
2. Jantsch, G. *Z. Anorg. Chem.* 1912, *76*, 303.
3. Wolff, H. *Z. Anorg. Chem.* 1905, *45*, 89.
4. Angelov, I.I.; Poslavskaya, K.D. *Trudy. Vsesoyuz. Nauch. Issledovatel. Inst. Khim. Reaktivov,* 1958, *No. 22*, 26.
5. Shevchuk, V.G.; Mitina, N.K. *Zh. Neorg. Khim.* 1976, *21*, 1943 (this paper was compiled in the preceeding section).
6. Fock, A. *Zeit. Kryst.* 1894, *22*, 34.

COMPONENTS:

(1) Magnesium cerium nitrate; $3Mg(NO_3)_2 \cdot 2Ce(NO_3)_3$; [15276-91-2]

(2) Water; H_2O; [7732-18-5]

ORIGINAL MEASUREMENTS:

Friend, J.N.; Wheat, W.N.
J. Chem. Soc. 1935, 356-9.

VARIABLES:

Temperature

PREPARED BY:

Mark Salomon

EXPERIMENTAL VALUES:

t/°C	mass ratio [a] CeO_2/MgO	$3Mg(NO_3)_2 \cdot 2Ce(NO_3)_3 \cdot 24H_2O$ [b] mass %	$3Mg(NO_3)_2 \cdot 2Ce(NO_3)_3$ [c] mass %	$3Mg(NO_3)_2 \cdot 2Ce(NO_3)_3$ [c] mol kg^{-1}
11	2.84	61.98	44.46	0.730
16	2.78	62.89	45.11	0.749
24.6	2.87	65.63	47.08	0.811
31.8	——	66.20	47.49	0.824
42.2	——	69.26	49.68	0.900
57.0	2.77	72.78	52.21	0.996
111.5 [d]	——	100	64.07	2.313

a. Results of gravimetric analyses. Theoretical value is 2.847 (compiler).

b. Authors' conversions from mass % CeO_2 results (these mass % values were not given in the source paper).

c. Compiler's calculations.

d. Melting point of the tetracosahydrate (2).

AUXILIARY INFORMATION

METHOD/APPARATUS/PROCEDURE:

The isothermal method was used as described in (1). In order to obtain reproducible results, equilibrium had to be approached from above. Saturated or near saturated slns were prepared at 90 - 100°C and quickly cooled in a thermostat. The slns were stirred for several hours which the authors claim is essential to remove supersaturation.

Cerium was determined by precipitation of the oxalate from an aliquot and ignition to CeO_2. Mg was determined as MgO by difference in which a second aliquot of equal volume was evaporated to dryness followed by ignition to yield the mixed oxides of CeO_2 and MgO. This method was checked by precipitation of Mg as the phosphate, and the results agreed to around ± 1.3 % (compiler). The average error in the reported CeO_2MgO mass ratios (see above table) is also around ± 1.3 % leading the compiler to estimate a total precision of no better than ± 3 %.

SOURCE AND PURITY OF MATERIALS:

In the preparation of the tetracosahydrate $3Mg(NO_3)_2 \cdot 2Ce(NO_3)_3 \cdot 24H_2O$, Kahlbaum's cerous nitrate was used. No other information was given.

ESTIMATED ERROR:

Soly: precision ± 3 % at best (compiler).

Temp: accuracy probably ± 0.05K as in (1) (compiler).

REFERENCES:

1. Friend, J.N. *J. Chem. Soc.* 1930, 1633.

2. Jantsch, G.Z. *Anorg. Chem.* 1912, *76*, 303.

COMPONENTS:

(1) Magnesium cerium (III) nitrate; $3Mg(NO_3)_2 \cdot 2Ce(NO_3)_3$; [15276-91-2]

(2) Nitric acid; HNO_3; [7697-37-2]

(3) Water; H_2O; [7732-18-5]

ORIGINAL MEASUREMENTS:

Jantsch, G. *Z. Anorg. Chem.* 1912, *76*, 303-23.

VARIABLES:

One temperature: 16°C

PREPARED BY:

Mark Salomon

EXPERIMENTAL VALUES:

Solubility of the double salt in HNO_3 sln of density

$d_4^{16} = 1.325$ g cm^{-3}

aliquot volume	mass Ce_2O_3 [a]	soly $3Mg(NO_3)_2 \cdot 2Ce(NO_3)_3$ [b]
cm^3	g	mol dm^{-3}
1.2324	0.0155	
1.2324	0.0155	0.0382

a. Experimental quantity is mass CeO_2, but author converted the experimental quantity to mass Ce_2O_3. Neither the original mass of CeO_2 or the atomic masses used in the conversions were given.

b. Author's calculations for average solubility.

ADDITIONAL DATA:

The melting point of the tetracosahydrate was reported as 111.5°C, and the density at 0°C as 2.002 g cm^{-3}.

AUXILIARY INFORMATION

METHOD/APPARATUS/PROCEDURE:

Isothermal method used. The soly was studied in HNO_3 sln of density 1.325 g cm^{-3} at 16°C because the author did not have sufficient quantity of the rare earth to study the soly of the salt in pure water. Pulverized salt and HNO_3 sln were placed in glass-stoppered tubes and thermostated at 16°C for 24 h with periodic shaking. The solutions were then allowed to settle for 2 h, and a pipet maintained at 16°C was used to withdraw aliquots for analysis. Two analyses were performed.

Solutions were analysed by adding 2-3 g NH_4Cl and 10 % NH_3 sln followed by boiling to ppt the hydroxide. The ppt was filtered, dissolved in HNO_3, reprecipitated as the hydroxide, and ignited to CeO_2. Mg in the filtrate was "determined by the usual method" (no details were given).

An attempt to determine the waters of hydration by dehydration was not successful because the temperature required (120°C or higher) resulted in decomposition of the salt with the formation of basic salts. Presumably the waters of hydration were found by difference.

SOURCE AND PURITY OF MATERIALS:

"Pure" cerous oxide was dissolved in dil HNO_3 and $Mg(NO_3)_2$ added to give a mole ratio of Ce/Mg = 2/3. The solution was evaporated with additions of H_2O_2 to prevent cerric formation, and after adding a small crystal of $Bi_2Mg_3(NO_3)_{12}$ the sln was evaporated to ppt the double salt as the tetracosahydrate. The salt was recrystallized, but the author does not report any analysis of the purified double salt.

ESTIMATED ERROR:

Soly: reproducibility about ± 1-5 % (compiler).

Temp: nothing specified.

REFERENCES:

COMPONENTS:

(1) Cerium (III) manganese nitrate; $2Ce(NO_3)_3 \cdot 3Mn(NO_3)_2$; [84682-54-2]

(2) Nitric acid; HNO_3; [7697-37-2]

(3) Water ; H_2O ; [7732-18-5]

ORIGINAL MEASUREMENTS:

Jantsch, G. *Z. Anorg. Chem.* 1912, *76*, 303-23.

VARIABLES:

One temperature: 16°C

PREPARED BY:

Mark Salomon

EXPERIMENTAL VALUES:

Solubility of the double salt in HNO_3 sln of density d_4^{16} = 1.325 g cm^{-3}. (10 cm^3 sln contains 5.159 g HNO_3)

aliquot volume	mass Ce_2O_3 [a]	soly $2Ce(NO_3)_3 \cdot 3Mn(NO_3)_2$ [b]
cm^3	g	mol dm^{-3}
1.2324	0.0444	
1.2324	0.0449	0.1103

a. Experimental quantity is mass CeO_2, but the author converterted the experimental quantity to mass Ce_2O_3. Neither the original mass of CeO_2 or the atomic masses used in the conversions were given.

b. Author's calculation for average solubility.

ADDITIONAL DATA:

The melting point of the tetracosahydrate was reported as 83.7°C, and the density at 0°C as 2.102 g cm^{-3}.

AUXILIARY INFORMATION

METHOD/APPARATUS/PROCEDURE:

Isothermal method used. The soly was studied in HNO_3 sln of density 1.325 g cm^{-3} at 16°C because the author did not have sufficient quantity of the rare earth to study the soly of the salt in pure water. Pulverized salt and HNO_3 sln were placed in glass-stoppered tubes and thermostated at 16°C for 24 h with periodic shaking. The solution was then allowed to settle for 2 h, and a pipet maintained at 16°C was used to withdraw aliquots for analysis. Two analyses were performed.

Solutions were analysed by precipitating both Ce and Mn hydroxides by respective addition of NH_3 and H_2O_2. The ppt was ignited to give $CeO_2 + Mn_3O_4$.

An attempt to determine the waters of hydration by dehydration was not successful because the temperature required (120°C or higher) resulted in decomposition of the salt with the formation of basic salts. Presumably the waters of hydration were found by difference

SOURCE AND PURITY OF MATERIALS:

"Pure" cerous oxide was dissolved in dil HNO_3 and $Mn(NO_3)_2$ added to give a mole ratio of Ce/Mn = 2/3. The solution was evaporated with additions of H_2O_2 to prevent ceric formation, and after adding a small crystal of $Bi_2Mg_3(NO_3)_{12}$ the sln was evaporated to ppt the double salt as the tetracosahydrate. The salt was recrystallized, but the author does not report any analysis of the purified double salt.

ESTIMATED ERROR:

Soly: reproducibility about ± 1-5% (compiler).

Temp: nothing specified.

COMPONENTS:

(1) Cerium (III) cobalt nitrate; $2Ce(NO_3)_3 \cdot 3Co(NO_3)_2$; [84682-56-4]

(2) Nitric acid; HNO_3; [7697-37-2]

(3) Water ; H_2O ; [7732-18-5]

ORIGINAL MEASUREMENTS:

Jantsch, G. *Z. Anorg. Chem.* 1912, *76*, 303-23.

VARIABLES:

One temperature: 16°C

PREPARED BY:

Mark Salomon

EXPERIMENTAL VALUES:

Solubility of the double salt in HNO_3 sln of density d_4^{16} = 1.325 g cm^{-3}.

aliquot volume	mass Ce_2O_3 [a]	soly $2Ce(NO_3)_3 \cdot 3Co(NO_3)_2$ [b]
cm^3	g	mol dm^{-3}
1.2324	0.0254	
1.2324	0.0257	0.0632

a. Experimental quantity is mass CeO_2, but author converted the experimental quantity to mass Ce_2O_3. Neither the original mass of CeO_2 or the atomic masses used in the conversions were given.

b. Author's calculation for average solubility.

ADDITIONAL DATA:
The melting point of the tetracosahydrate was reported as 98.5°C, and the density at 0°C as 2.157 g cm^{-3}.

AUXILIARY INFORMATION

METHOD/APPARATUS/PROCEDURE:

Isothermal method used. The soly was studied in HNO_3 sln of density 1.325 g cm^{-3} at 16°C because the author did not have sufficient quantity of the rare earth to study the soly of the salt in pure water. Pulverized salt and HNO_3 sln were placed in glass-stoppered tubes and thermostated at 16°C for 24 h with periodic shaking. The solution was then allowed to settle for 2 h, and a pipet maintained at 16°C was used to withdraw aliquots for analysis. Two analyses were performed.

Solutions were analysed by adding 2-3 g NH_4Cl and 10% NH_3 sln followed by boiling to ppt the hydroxide. The ppt was filtered, dissolved in HNO_3, reprecipitated as the hydroxide, and ignited to CeO_2. Co in the filtrate was "determined by the usual method" (no details were given).

An attempt to determine the waters of hydration by dehydration was not successful because the temperature required (120°C or higher) resulted in decomposition of the salt with the formation of basic salts. Presumably the waters of hydration were found by difference.

SOURCE AND PURITY OF MATERIALS:

"Pure" cerous oxide was dissolved in dil HNO_3 and $Co(NO_3)_2$ added to give a mole ratio of Ce/Co = 2/3. The solution was evaporated with additions of H_2O_2 to prevent cerric formation, and after adding a small crystal of $Bi_2Mg_3(NO_3)_{12}$ the sln was evaporated to ppt the double salt as the tetracosahydrate. The salt was recrystallized, but the author does not report any analysis of the purified double salt.

ESTIMATED ERROR:

Soly: reproducibility about ± 1-5% (compiler).

Temp: nothing specified.

REFERENCES:

COMPONENTS:

(1) Cerium (III) nickel nitrate; $2Ce(NO_3)_3 \cdot 3Ni(NO_3)_2$; [84682-57-5]

(2) Nitric acid; HNO_3; [7697-37-2]

(3) Water; H_2O; [7732-18-5]

ORIGINAL MEASUREMENTS:

Jantsch, G. *Z. Anorg. Chem.* 1912, *76*, 303-23.

VARIABLES:

One temperature: 16°C

PREPARED BY:

Mark Salomon

EXPERIMENTAL VALUES:

Solubility of the double salt in HNO_3 sln of density

$d_4^{16} = 1.325$ g cm^{-3}

aliquot volume	mass Ce_2O_3 [a]	soly $2Ce(NO_3)_3 \cdot 3Ni(NO_3)_2$ [b]
cm^3	g	mol dm^{-3}
1.2324	0.0182	
1.2324	0.0191	0.0460

a. Experimental quantity is mass CeO_2, but author converted the experimental quantity to mass Ce_2O_3. Neither the original mass of CeO_2 or the atomic masses used in the conversions were given.

b. Author's calculations for average solubility.

ADDITIONAL DATA:

The melting point of the tetracosahydrate was reported as 108.5°C, and the density at 0°C as 2.173 g cm^{-3}.

AUXILIARY INFORMATION

METHOD/APPARATUS/PROCEDURE:

Isothermal method used. The soly was studied in HNO_3 sln of density 1.325 g cm^{-3} at 16°C because the author did not have sufficient quantity of the rare earth to study the soly of the salt in pure water. Pulverized salt and HNO_3 sln were placed in glass-stoppered tubes and thermostated at 16°C for 24 h with periodic shaking. The solutions were then allowed to settle for 2 h, and a pipet maintained at 16°C was used to withdraw aliquots for analysis. Two analyses were performed.

Solutions were analysed by adding 2-3 g NH_4Cl and 10 % NH_3 sln followed by boiling to ppt the hydroxide. The ppt was filtered, dissolved in HNO_3, reprecipitated as the hydroxide, and ignited to CeO_2. Ni in the filtrate was "determined by the usual method" (no details were given).

An attempt to determine the waters of hydration by dehydration was not successful because the temperature required (120°C or higher) resulted in decomposition of the salt with the formation of basic salts. Presumably the waters of hydration were found by difference.

SOURCE AND PURITY OF MATERIALS:

"Pure" cerous oxide was dissolved in dil HNO_3 and $Ni(NO_3)_2$ added to give a mole ratio of Ce/Ni = 2/3. The solution was evaporated with additions of H_2O_2 to prevent ceric formation, and after adding a small crystal of $Bi_2Mg_3(NO_3)_{12}$ the sln was evaporated to ppt the double salt as the tetracosahydrate. The salt was recrystallized, but the author does not report any analysis of the purified double salt.

ESTIMATED ERROR:

Soly: reproducibility about ± 1-5 % (compiler).

Temp: nothing specified.

REFERENCES:

COMPONENTS:

(1) Cerium (III) zinc nitrate; $2Ce(NO_3)_3 \cdot 3Zn(NO_3)_2$; [13773-54-1]

(2) Nitric acid; HNO_3; [7697-37-2]

(3) Water ; H_2O ; [7732-18-5]

ORIGINAL MEASUREMENTS:

Jantsch, G. *Z. Anorg. Chem.* 1912, *76*, 303-23.

VARIABLES:

One temperature: 16°C

PREPARED BY:

Mark Salomon

EXPERIMENTAL VALUES:

Solubility of the double salt in HNO_3 sln of density d_4^{16} = 1.325 g cm^{-3}

aliquot volume cm^3	mass Ce_2O_3 [a] g	soly $2Ce(NO_3)_3 \cdot 3Zn(NO_3)_2$ [b] mol dm^{-3}
1.4638	0.0326	
1.4638	0.0324	0.0675

a. Experimental quantity is mass CeO_2, but author converted the experimental quantity to mass Ce_2O_3. Neither the original mass of CeO_2 or the atomic masses used in the conversions were given.

b. Author's calculation for average solubility.

ADDITIONAL DATA:
The melting point of the tetracosahydrate was reported as 92.8°C, and the density at 0°C as 2.188 g cm^{-3}.

AUXILIARY INFORMATION

METHOD/APPARATUS/PROCEDURE:

Isothermal method used. The soly was studied in HNO_3 sln of density 1.325 g cm^{-3} at 16°C because the author did not have sufficient quantity of the rare earth to study the soly of the salt in pure water. Pulverized salt and HNO_3 sln were placed in glass-stoppered tubes and thermostated at 16°C for 24 h with periodic shaking. The solution was then allowed to settle for 2 h, and a pipet maintained at 16°C was used to withdraw aliquots for analysis. Two analyses were performed.

Solutions were analysed by adding 2-3 g NH_4Cl and 10% NH_3 sln followed by boiling to ppt the hydroxide. The ppt was filtered, dissolved in HNO_3, reprecipitated as the hydroxide, and ignited to CeO_2. Zn in the filtrate was "determined by the usual method" (no details were given).

An attempt to determine the waters of hydration by dehydration was not successful because the temperature required (120°C or higher) resulted in decomposition of the salt with the formation of basic salts. Presumably the waters of hydration were found by difference.

SOURCE AND PURITY OF MATERIALS:

"Pure" cerous oxide was dissolved in dil HNO_3 and $Zn(NO_3)_2$ added to give a mole ratio of Ce/Zn = 2/3. The solution was evaporated with additions of H_2O_2 to prevent ceric formation, and after adding a small crystal of $Bi_2Mg_3(NO_3)_{12}$ the sln was evaporated to ppt the double salt as the tetracosahydrate. The salt was recrystallized, but the author does not report any analysis of the purified double salt.

ESTIMATED ERROR:

Soly: reproducibility about ± 1-5% (compiler).

Temp: nothing specified.

REFERENCES:

COMPONENTS:

(1) Cerium (III) ammonium nitrate; $Ce(NO_3)_3 \cdot 2NH_4NO_3$; [15318-60-2]

(2) Water; H_2O; [7732-18-5]

ORIGINAL MEASUREMENTS:

Wolff, H. *Z. Anorg. Chem.* 1905, *45*, 89-115.

VARIABLES:

Temperature

PREPARED BY:

Mark Salomon

EXPERIMENTAL VALUES:

	Experimental [a]		Calculated $Ce(NO_3)_3 \cdot 2NH_4NO_3$ solubilities				
	NH_4	Ce	from NH_4	from Ce	mean values		
t/°C	mass %	mass %	mass % [b]	mass % [b]	mass % [c]	mass % [b]	mol kg^{-1} [b]
8.75	4.787	18.56	64.52	64.40	70.2	64.46	3.730
25.00	5.09	19.80	68.60	68.71	74.8	68.65	4.504
45.00	5.53	21.06	74.53	73.08	80.4	73.80	5.794
60.00	6.01	22.77	81.00	79.01	87.2	80.01	8.230
65.06	6.11	23.42	82.35	81.27	89.1	81.81	9.248

a. Mass % values based on total weight of the *tetrahydrate*.

b. Calculated by the compiler for the *anhydrous* salt based on 1977 IUPAC recommended atomic weights.

c. Author's original mean values for the *tetrahydrate*.

AUXILIARY INFORMATION

METHOD/APPARATUS/PROCEDURE:

Isothermal method used. Water and excess salt were placed in a cylinder and thermostated with stirring for 5 hours. Weighed aliquots were diluted to 250 cm^3 and analysed. Ce was determined by the Knorre method: Ce was oxidized in H_2SO_4 with ammonium persulfate, heated to decompose the excess persulfate, and titrated with H_2O_2. The end-point is identified by the color change of yellow Ce(IV) to colorless Ce(III). NH_4 was determined by adding an excess KOH to precipitate Ce and expel NH_3. The remaining KOH was back-titrated with standard acid.

SOURCE AND PURITY OF MATERIALS:

$Ce(NO_3)_3 \cdot 2NH_4NO_3 \cdot 4H_2O$ prepd by reduction of purified $Ce(NO_3)_4 \cdot 2NH_4NO_3$ with H_2O_2. The preparation of the Ce(IV) salt is described in the following compilation. A solution of the Ce(III) salt was slowly evaporated to crystallization, and the salt twice recrystallized. The salt was analysed for NH_4, and for Ce by the Knorre method. Results for the NH_4 are 6.40 and 6.66 mass %: results for Ce are 24.29 and 25.15 mass %. Theor mass % calcd by author are NH_4 6.57, and Ce 25.07. Using 1977 IUPAC recommended atomic weights, the compiler computes the following based upon the tetrahydrate molecular weight of 558.28 g mol^{-1}: NH_4 6.46 mass %, and Ce 25.10 mass %.

ESTIMATED ERROR:

Soly: nothing specified for individual NH_4 and Ce analyses. Mean error based on average soly is about ± 1% (compiler).

Temp: precision about ± 0.1 K (author).

COMPONENTS:

(1) Cerium (III) ammonium nitrate; $Ce(NO_3)_3 \cdot 2NH_4NO_3$; [15318-60-2]

(2) Nitric acid; HNO_3; [7697-37-2]

(3) Water ; H_2O ; [7732-18-5]

ORIGINAL MEASUREMENTS:

Angelov, I.I.; Poslavskaya, K.D.
Trudy. Vsesoyuz. Nauch. Issledovatel. Inst. Khim. Reaktivov, 1958, *No. 22*, 26-9.

VARIABLES:

Composition at 25°C

PREPARED BY:

Mark Salomon and Orest Popovych

EXPERIMENTAL VALUES:

Soly $Ce(NO_3)_3 \cdot 2NH_4NO_3$		Conc HNO_3	
mass %	mol kg^{-1}	mass %	mol kg^{-1}
63.5	4.59	10	5.99
42.5	2.96	30	17.31

a. Molalities calculated by the compilers.

AUXILIARY INFORMATION

METHOD/APPARATUS/PROCEDURE:

Nothing specified. Presumably the isothermal method was used.

SOURCE AND PURITY OF MATERIALS:

Initial solid is the tetrahydrate $Ce(NO_3)_3 \cdot 2NH_4NO_3 \cdot 4H_2O$. Nothing else specified.

ESTIMATED ERROR:

Soly: nothing specified.

Temp: precision probably ± 0.1 K (compilers).

REFERENCES:

COMPONENTS:

(1) Cerium (IV) ammonium nitrate; $Ce(NO_3)_4 \cdot 2NH_4NO_3$; [16774-21-3]

(2) Water; H_2O; [7732-18-5]

ORIGINAL MEASUREMENTS:

Wolff, H. *Z. Anorg. Chem.* 1905, *45*, 89-115.

VARIABLES:

Temperature

PREPARED BY:

Mark Salomon

EXPERIMENTAL VALUES:

	Experimental		Calculated $Ce(NO_3)_4 \cdot 2NH_4NO_3$ solubilities				
	NH_4	Ce	from NH_4	from Ce	mean values		
t/°C	mass %	mass %	mass % [a]	mass % [a]	mass % [b]	mass % [a]	mol kg^{-1} [a]
25.00	4.065	15.16	61.77	59.31	58.49	60.54	2.798
35.2	4.278	16.08	65.01	62.91	61.79	63.96	3.237
35.2	4.268	16.117	64.86	63.059	61.79	63.96	3.237
45.3	4.489	16.695	68.213	65.320	64.51	66.77	3.664
64.5	4.625	17.401 [c]	70.28	68.08	66.84	69.18	4.094
		15.034 [d]		———			
85.60	4.778	18.158 [c]	72.61	71.044	69.40	71.83	4.650
		15.79 [d]		———			
122	6.117	22.82 [c]	92.954	89.28	88.03	91.12	18.71
		16.22 [d]		———			

a. Calculated by the compiler using IUPAC recommended atomic weights.
b. Author's original calculated results.
c. Total Ce content obtained by adding persulfate prior to titration with H_2O_2.
d. Ce(IV) content obtained by titration of an aliquot of satd sln without pretreatment with persulfate (see below).

AUXILIARY INFORMATION

METHOD/APPARATUS/PROCEDURE:

Isothermal method used. Water and excess salt were placed in a cylinder and thermostated with stirring for 5 hours. Weighed aliquots were diluted to 250 cm^3 and analysed. Ce was detd by the Knorre method: Ce was oxidized in H_2SO_4 with ammonium persulfate, heated to decompose the excess persulfate, and titrated with H_2O_2. The end-point is identified by the color change of yellow Ce(IV) to colorless Ce(III). NH_4 was detd by adding an excess of KOH to ppt Ce and expel NH_3, and the remaining KOH was back-titrated with standard acid.

Above 60°C, there is partial reduction of Ce(IV) to Ce(III). Analysis of saturated solutions without pre-treatment with persulfate gives the Ce(IV) content (in mass % in the above table). Adding persulfate to the solution prior to analysis gives the total Ce content in the saturated solutions.

SOURCE AND PURITY OF MATERIALS:

2 kg of finely ground impure $CeO_2 \cdot 2H_2O$ was added to dil HNO_3 and the mixt stirred overnight: this removed the major impurities Ca, Fe, La, Yb, and didymium (didmium was assumed to be a rare earth but is now known to be a mixt of Pr and Nd, compiler). The remaining residue was dissolved in conc HNO_3, partially evaporated, and the basic nitrate pptd by addn of excess (boiled) water. The yellow ppt was filtered, washed, and redissolved in conc HNO_3. At this stage, about half of the original Ce content remained. Sufficient NH_4NO_3 was added (i.e. mole ratio $NH_4/Ce=2$), and the solution evaporated to crystallization at which point the solution was cooled in an ice bath. The ppt $Ce(NO_3)_4 \cdot 2NH_4NO_3$ was then recrystallized twice. Spectroscopic analysis of an aq sln of the salt showed it to be reasonably pure.

ESTIMATED ERROR:

Soly: nothing specified for individual NH_4 and Ce analysis. Mean error based on average soly is about ± 2% (compiler).

Temp: precision ± 0.1 K (author).

COMPONENTS:

(1) Cerium (IV) ammonium nitrate; $(NH_4)_2Ce(NO_3)_6$; [16774-21-3]

(2) Water; H_2O; [7732-18-5]

ORIGINAL MEASUREMENTS:

Angelov, I.I.; Poslavskaya, R.D. *Trudy. Vsesoyuz. Nauch. Issledovatel. Inst. Khim. Reaktivov*, 1958, *No.* 22, 26-9.

VARIABLES:

Temperature: range -18 to +50°C

PREPARED BY:

Mark Salomon and Orest Popovych

EXPERIMENTAL VALUES:

Solubility of $(NH_4)_2Ce(NO_3)_6$ as a function of temperature [a]

t/°C	mass %	mol kg^{-1}	solid phase	t/°C	mass %	mol kg^{-1}	solid phase
0 [b]	55.74	2.297	DS [c]	-17.95	52.96	2.053	ice
15 [b]	57.93	2.511	"				
25 [b]	59.51	2.681	"	-18.25 [d]	53.21	2.074	ice + DS
50 [b]	65.55	3.470	"				
				-10.53	54.19	2.157	DS
- 8.30	24.52	0.592	ice	- 3.76	55.20	2.247	"
-10.88	32.11	0.863	"	3.27	56.10	2.331	"
-12.25	36.57	1.051	"	10.40	57.11	2.428	"
-13.10	38.50	1.142	"	16.10	57.98	2.517	"
-14.80	43.70	1.416	"	22.00	58.83	2.606	"
-16.40	48.18	1.696	"	26.40	59.71	2.703	"
-17.25	50.60	1.868	"	31.30	60.45	2.788	"

a. Molalities calculated by the compilers.

b. Solubilities determined by the isothermal method. All other solubilities determined by the synthetic method.

c. DS = double salt = $(NH_4)_2Ce(NO_3)_6$.

d. Extrapolated data for the eutectic point.

AUXILIARY INFORMATION

METHOD/APPARATUS/PROCEDURE:

Data at 0.0°, 15.0°, 25.0°, and 50.0°C determined by the isothermal method. All other solubilities determined by the synthetic method. No other details given.

Ce(IV) was determined by reduction to Ce(III) with H_2O_2 in acidic medium, and then back-titrated with potassium permanganate solution.

SOURCE AND PURITY OF MATERIALS:

Authors state that the starting double salt, $(NH_4)_2Ce(NO_3)_6$, had a purity of 99.97 mass %.

No other information given.

ESTIMATED ERROR:

Soly: nothing specified.

Temp: for isothermal measurements, precision was ± 0.1 K. No other information given.

REFERENCES:

COMPONENTS:

(1) Cerium (IV) ammonium nitrate; $(NH_4)_2Ce(NO_3)_6$; [16774-21-3]

(2) Nitric acid; HNO_3; [7697-37-2]

(3) Water; H_2O; [7732-18-5]

ORIGINAL MEASUREMENTS:

Angelov, I.I.; Poslavskaya, K.D. *Trudy. Vsesoyuz. Nauch. Issledovatel. Inst. Khim. Reaktivov*, 1958, *No. 22*, 26-9.

VARIABLES:

Composition at 0°, 25°, and 50°C

PREPARED BY:

Mark Salomon and Orest Popovych

EXPERIMENTAL VALUES:

Solubility of $(NH_4)_2Ce(NO_3)_6$ in nitric acid solutions [a,b]

t/°C	$(NH_4)_2Ce(NO_3)_6$ mass %	$(NH_4)_2Ce(NO_3)_6$ mol kg^{-1}	HNO_3 mass %	HNO_3 mol kg^{-1}
0.0	55.74	2.297	0	———
	40.93	1.374	4.75	1.388
	32.04	1.008	9.98	2.732
	21.70	0.669	19.16	5.141
	12.06	0.388	31.31	8.774
	5.86	0.207	42.52	13.07
	2.83	0.122	55.02	20.72
	1.75	0.105	67.75	35.25
	1.41	0.130	78.81	63.23
	0.90	0.16	88.95	139.1

a. Molalities calculated by the compilers.

b. In all cases, the solid phase is the anhydrous double salt.

continued.......

AUXILIARY INFORMATION

METHOD/APPARATUS/PROCEDURE:

Isothermal method used. Equilibrium was reached after 3 hours of constant stirring.

Ce(IV) determined by reduction to Ce(III) with H_2O_2 in acid medium, and back-titrated with potassium permanganate solution. Nitric acid could not be determined directly in the presence of Ce(IV) because of the decomposition of indicators in the presence of Ce(IV). Therefore, the acid was determined by reduction of the total NO_3^- to ammonia using Devarda's alloy in alkaline medium, and absorption of the ammonia in 0.05 mol dm^{-3} sulfuric acid. The amount of nitric acid was then obtained from the difference between the total ammonia and the ammonia corresponding to the double nitrate.

The composition of the solid phase was determined by Schreinemakers' method of residues and checked by direct chemical analysis of crystals quickly pressed between sheets of filter paper at low acid concentrations, or by microscopic analysis at higher acid concentrations.

SOURCE AND PURITY OF MATERIALS:

Authors state that the starting material, $(NH_4)_2Ce(NO_3)_6$, had a purity of 99.97 mass %.

No other information given.

ESTIMATED ERROR:

Soly: nothing specified.

Temp: precision ± 0.1 K.

REFERENCES:

COMPONENTS:

(1) Cerium (IV) ammonium nitrate; $(NH_4)_2Ce(NO_3)_6$; [16774-21-3]

(2) Nitric acid; HNO_3; [7697-37-2]

(3) Water; H_2O; [7732-18-5]

ORIGINAL MEASUREMENTS:

Angelov, I.I.; Poslavskaya, K.D.
Trudy. Vsesoyuz. Nauch. Issledovatel. Inst. Khim. Reaktivov, 1958, *No. 22*, 26-9.

EXPERIMENTAL VALUES: continued........

Solubility of $(NH_4)_2Ce(NO_3)_6$ in nitric acid solutions [a,b]

	$(NH_4)_2Ce(NO_3)_6$		HNO_3	
t/°C	mass %	mol kg^{-1}	mass %	mol kg^{-1}
25	59.51	2.681	0	———
	43.00	1.523	5.51	1.698
	35.20	1.171	9.98	2.889
	31.31	1.022	12.83	3.645
	24.12	0.776	19.16	5.361
	15.50	0.519	30.02	8.745
	6.69	0.274	48.81	17.41
	4.50	0.248	62.43	29.96
	3.13	0.239	73.02	48.59
	2.71	0.286	80.01	73.48
	2.60	0.561	88.95	167.1
50	65.55	3.470	0	———
	47.51	1.883	6.48	2.235
	40.49	1.491	9.98	3.198
	32.84	1.146	14.89	4.521
	22.51	0.773	24.39	7.289
	17.69	0.633	31.31	9.743
	12.11	0.487	42.52	14.87
	5.26	0.287	61.31	29.10
	4.30	0.500	80.01	80.93

a. Molalities calculated by the compilers.

b. In all cases, the solid phase is the anhydrous double salt.

COMPONENTS:

(1) Cerium (IV) ammonium nitrate; $(NH_4)_2Ce(NO_3)_6$; [16774-21-3]

(2) Tri-n-butylphosphate; $C_{12}H_{27}O_4P$; [126-73-8]

ORIGINAL MEASUREMENTS:

Healy, T.V.; McKay, H.A.C. *Trans. Faraday Soc.* 1956, *52*, 633-42.

VARIABLES:

One temperature: 25°C

PREPARED BY:

Mark Salomon

EXPERIMENTAL VALUES:

The solubility of $(NH_4)_2Ce(NO_3)_6$ in $[CH_3(CH_2)_3O]_3PO$ at 25°C was given as

1.65 mol dm^{-3}

AUXILIARY INFORMATION

METHOD/APPARATUS/PROCEDURE:

Isothermal method used. The solubility was determined by shaking the solid nitrate with $[CH_3(CH_2)_3O]_3PO$ for several days followed by analysis of the organic phase. Cerium was analysed volumetrically using ferrous sulfate and permanganate.

SOURCE AND PURITY OF MATERIALS:

A.R. grade $(NH_4)_2Ce(NO_3)_3$ was used. The compiler assumes that the salt was anhydrous. $[CH_3(CH_2)_3O]_3PO$ was purified as in (1). It was boiled with dilute aqueous NaOH until all volatile impurities were distilled off. The remaining solvent was washed repeatedly with water and dried by warming in vacuum.

ESTIMATED ERROR:

Nothing specified.

REFERENCES:

1. Alcock, K.; Grimley, S.S.; Healy, T.V.; McKay, H.A.C. *Trans. Faraday Soc.* 1956, *52*, 39.

COMPONENTS:	EVALUATOR:
(1) Praseodymium nitrate; $Pr(NO_3)_3$; [10361-80-5] (2) Water ; H_2O ; [7732-18-5]	S. Siekierski, T. Mioduski Institute for Nuclear Research Warsaw, Poland and M. Salomon U.S. Army ET & DL Ft. Monmouth, NJ, USA December 1982

CRITICAL EVALUATION:

THE BINARY SYSTEM

INTRODUCTION

The solubility of $Pr(NO_3)_3$ in water has been reported in 17 publications (1-17), and in spite of these numerous reports between the years 1935 (1) and 1979 (11), there exists large differences in the reported solubilities at given temperatures. The large discrepancies between the reported solubilities can probably be attributed in large part to incorrect identification of the solid phases in equilibrium with saturated solutions. In the binary system, the solid phases identified are:

$Pr(NO_3)_3.6H_2O$	[15878-77-0]	$Pr(NO_3)_3.4H_2O$	[15878-79-2]
$Pr(NO_3)_3.5H_2O$	[14483-17-1]	$Pr(NO_3)_3.3H_2O$	[81201-33-4]

In solutions of high nitric acid content (between 63-93 mass % HNO_3), the monohydrate, $Pr(NO_3)_3.H_2O$ [81201-31-2], is stable, and above 93 mass % HNO_3 the anhydrous salt is the stable solid phase (6). There have been no reports suggesting that a dihydrate solid phase may be stable or metastable.

In some studies saturated solutions were analysed by "wet" chemical methods such as complexometric titration using ethylenediamine tetraacetic acid (EDTA, Trilon) as in (2, 4-7, 9, 11). Other methods reported include refractometric analysis (10, 12-17), gravimetric sulfate analysis (4,5) and gravimetric analysis of Pr_6O_{11} (1,8). In those isothermal studies which employed refractometric analyses of saturated solutions (10, 12-17), the initial salt $Pr(NO_3)_3.nH_2O$ (and presumably the equilibrated solid phases) were analysed gravimetrically by the oxalate method; i.e. Pr was precipitated as the oxalate and ignited to the oxide. A possible source of error exists in the ignition method due to the fact (18, 19) that praseodymium oxide can form a number of non-stoichiometric oxides ranging from $PrO_{1.5}$ to $PrO_{2.0}$. Under normal atmospheric conditions the oxide Pr_6O_{11} will form (18, 19), and in all studies reported in (1, 8, 10, 12-17) it was assumed that Pr_6O_{11} was the oxide obtained. The solubility data in the hexahydrate system reported by Friend (1) are obviously much lower than any of the data in (2-17), and we conclude that Friend's data contain a large negative systematic error such as that which would occur if the oxide produced upon ignition were of higher oxygen content. Friend did not mention any problems with the analyses in (1), but in a preceeding paper (23) he stated that in the presence of MgO about 80% of the praseodymium oxide obtained upon ignition was PrO_2.

EVALUATION PROCEDURE

To evaluate the solubility data as a function of temperature, the data were fitted by least squares to the solubility equation based on the treatments in (20, 21), and in the INTRODUCTION to this volume:

$$Y = \ell n(m/m_o) - nM_2(m - m_o) = a + b/(T/K) + c\ \ell n(T/K) \quad [1]$$

In this smoothing equation m is the molality at temperature T, m_o an arbitrarily selected reference molality (usually the 298.2 K value), n the hydration number of the solid, M_2 is the molar mass of the solvent, and a, b, c are constants from which enthalpies and heat capacities of solution, ΔH_{sln} and ΔC_p, can be estimated (see the INTRODUCTION).

Since most authors did not report experimental errors, the compilers attempted to provide this information when possible. As discussed in the $Ce(NO_3)_3$-H_2O critical evaluation, the data of highest precision are those from the isothermal studies of Spedding et al.(4,5) who reported a precision of ± 0.1% or better, Brunisholz et al.(2,9) in which the compilers estimated a precision of around ± 0.2%, and the results from the synthetic method reported by Mironov and Popov (6,7) which are probably associated with a precision of around ± 0.5% at best. Based on the fact that four significant figures are given for the solubility data in (11), a precision of at least ± 0.5% must be assumed, but the accuracy appears to be poor: e.g. the solubility of $Pr(NO_3)_3$ in the hexahydrate system at 303.2 K is given as 62.12 mass % (5.016 mol kg^{-1}) which is considerably lower than the *tentative* value of 5.298 mol kg^{-1} at 303.15 K (see discussion below and Table 5). The solubility data obtained by refractometric measurements are, as discussed in previous critical evaluations, assigned an overall experimental precision of ± 1% at best.

COMPONENTS:

(1) Praseodymium nitrate; $Pr(NO_3)_3$; [10361-80-5]

(2) Water ; H_2O ; [7732-18-5]

EVALUATOR:
S. Siekierski, T. Mioduski
Institute for Nuclear Research
Warsaw, Poland
and
M. Salomon
U.S. Army ET & DL
Ft. Monmouth, NJ, USA
December 1982

CRITICAL EVALUATION:

SOLUBILITY OF $Pr(NO_3)_3$ IN THE HEXAHYDRATE SYSTEM

The solubility data for $Pr(NO_3)_3$ solutions in equilibrium with solid $Pr(NO_3)_3 \cdot 6H_2O$ at various temperatures (2-5, 7, 9-17) are summarized in Table 1. The data of Friend (1) were rejected (see above discussion), and were not included in this table.

Table 1. Solubility of $Pr(NO_3)_3$ in the $Pr(NO_3)_3 \cdot 6H_2O$ - H_2O system

T/K	m_1/mol kg^{-1}	ref	T/K	m_1/mol kg^{-1}	ref
265.1	3.94	3	298.15	4.990	5
273.15	4.132	2	298.15	5.0166	4
281.5	4.37	3	298.2	4.97	7
283.15	4.455	2	303.2	5.016	11
293.15	4.827	9	305.1	5.25	3
293.15	4.833	2	307.5	5.34	3
293.2	4.61	15, 16	308.15	5.646	2
293.2	4.63	14	313.2	5.68	15
293.2	4.69	12	313.2	5.71	10, 13, 17
293.2	4.70	13	313.2	5.78	17
293.2	4.78	3	313.2	5.83	12, 14, 16
294.5	4.78	3	323.15	6.937	2

Evaluating the data in Table 1 proved difficult due to a significant difference in the results reported by Mironov and Popov (3) and by Brunisholz et al. (2, 9). In other critical evaluations we have found that the results of Mironov and Popov, Brunisholz et al., and Spedding et al. are generally of high precision and in very good agreement with one another. However this is not the case for the results in the $Pr(NO_3)_3$-H_2O binary system as seen in the above table and in Figure 1. The data of Mironov and Popov (3) above 283 K are systematically lower than those of Brunisholz et al. (2, 9), and assuming all results are based upon a common solid phase (i.e. the hexahydrate), it is obvious that one or both sets of data are incorrect. To determine which data are the most accurate, the results in (2, 9) and (3, 7) were separately fitted to the smoothing eq. [1]. The results of fitting these data to eq. [1] are shown in Table 2 below. Both sets of data are seen to result in similar standard

Table 2. Comparison of results of fitting the data in (2, 9) and (3, 7) to eq. [1].[a]

ref	N	m.p./K	m_1(calcd)/mol kg^{-1} for 298.2 K	σ_m/mol kg^{-1}	total uncertainty[b] mol kg^{-1}
3, 7	6	339.6	4.94	0.022	±0.028
2, 9	6	331.1	5.068	0.017	±0.022

[a] N = number of experimental points, σ_m is the standard error of estimate, and m.p. is the predicted (calculated) congruent melting point.

[b] 95% level of confidence, Student's t = 3.182.

errors of estimate and total estimated uncertainties, but there are several reasons why the data in (2, 9) are selected as being more accurate. Considering the data from (2, 9), the congruent melting point calculated from the smoothing equation is 331.1 K which is very close to the average experimental value of 331.1 K. The experimental congruent melting point of the hexahydrate was reported as 329.2 K by Friend (1), and as 333 ± 1 K by Wendlandt and Sewell (22). In addition, combining the data from (2, 9) with those of Spedding et al. (4, 5) gives a much smaller σ_m than that obtained by combining the data of (3, 7) with those of Spedding et al. The results of fitting the combined data from (2, 4, 5, 9) to the smoothing equation are given in Table 4, and the polytherm for the hexahydrate system given in Fig. 1 was drawn from these smoothed data. These smoothed data are designated as *tentative* values, and are listed in Table 5. All other data are rejected.

It is interesting to note that three solubility values at 324.5 K, 326.3 K, and 329.2 K which Mironov and Popov (3) assign to the pentahydrate system fall on the hexahydrate polytherm as seen in Fig. 1. If these three data points are included with the hexahydrate

COMPONENTS:	EVALUATOR:
(1) Praseodymium nitrate; $Pr(NO_3)_3$; [10361-80-5] (2) Water ; H_2O ; [7732-18-5]	S. Siekierski, T. Mioduski Institute for Nuclear Research Warsaw, Poland and M. Salomon U.S. Army ET & DL Ft. Monmouth, NJ, USA December 1982

CRITICAL EVALUATION:

data from (2, 4, 5, 9) and fitted to the smoothing eq. [1], the results are very similar to those given in Tables 4 and 5 for the *tentative* data in the hexahydrate system. It is therefore concluded that these three data points belong to the hexahydrate system, and that Mironov and Popov have incorrectly assigned them to the pentahydrate system.

SOLUBILITY OF $Pr(NO_3)_3$ IN $Pr(NO_3)_3.nH_2O$ SYSTEMS: $n \leq 5$

The solubility for lower hydrate systems (i.e. $n \leq 5$) are summarized in Table 3. The

Table 3.

T/K	m_1/mol kg^{-1}	ref	value of n in $Pr(NO_3)_3.nH_2O$ solid phase: authors's assignment	evaluators' assignment
298.2	5.32	6,7	5	
315.7	5.91	3	5	
324.5	7.10	3	5	6
326.3	7.53	3	5	6
329.2	7.90	3	5+4	6
337.9	7.94	3	4	5
349.6	9.74	3	4	5
350.7	11.37	3	4	
366.0	16.06	3	3	
400.2	17.33	3	3	

data in Table 3 are the results of one research group (3, 6, 7), and assignments of solid phases do not appear to be correct in all instances. Above it was concluded that the three data points at 324.5 K, 326.3 K, and 329.2 K should be included in the hexahydrate system since these data points fall (well within the precision of the fit to eq.[1]) on the smoothed hexahydrate polytherm. The assignment in (6,7) of the 298.2 K solubility of 5.32 mol kg^{-1} to the pentahydrate system is reasonable since the pentahydrate system is metastable at 298 K, and the solubility of $Pr(NO_3)_3$ is therefore expected to be greater than in the stable hexahydrate system at this temperature. The assignment to the pentahydrate system of the solubility of 5.91 mol kg^{-1} at 315.7 K also appears reasonable since the hexahydrate system is not metastable at this temperature, and the calculated (eq. [1]) solubility of 6.15 mol kg^{-1} in the hexahydrate system at 315.7 K is thermodynamically consistent with this assignment.

The two data points at 337.9 K and 349.6 K also may be incorrectly assigned to a tetrahydrate system since they appear to lie on the pentahydrate polytherm: i.e. by assuming that the solubilities in Table 3 at 298.2 K, 315.7 K, 337.9 K and 349.6 K all belong to the pentahydrate system, these data can be fitted to eq. [1] with reasonable results for all calculated parameters. The results of fitting these four data points to eq. [1] are given in Tables 4 and 5, and the pentahydrate polytherm in Fig. 1 was drawn from the smoothed values given in Table 5. In the absence of additional experimental data (solubility data and congruent melting points), it is not possible to estimate the accuracy in the solubility data which the evaluators propose for the pentahydrate system.

TENTATIVE VALUES FOR THE SOLUBILITIES

For solubilities in the hexahydrate system, the data of Brunisholz et al. (2,9) and Spedding et al. (4,5) were, for reasons discussed above, selected as the most accurate data. The data were fitted to eq. [1] and the resulting parameters for the smoothing equation are given in Table 4. The smoothed solubility data calculated from eq. [1] are designated as *tentative* values, and which are given in Table 5. In Table 4, σ_a, σ_b, σ_c are the standard deviations for the constants a, b, c, and σ_Y and σ_m are the standard errors of estimate for the quantity Y in eq. [1] and the solubility, respectively. From the latter quantity, we calculate the total uncertainty in the *tentative* solubility values to be ± 0.03 mol kg^{-1} at the 95% level of confidence (Student's t = 2.571).

The solubilities in the pentahydrate system are also designated as *tentative* data as calculated from eq. [1]. However the evaluators would like to emphasize some caution in accepting the assignment of the pentahydrate solid phase to all the data assigned to this system. We conclude that additional experimental data are required for definitive identification of the pentahydrate polytherm, and for the determination of more accurate solubility data.

COMPONENTS:

(1) Praseodymium nitrate; $Pr(NO_3)_3$; [10361-80-5]

(2) Water ; H_2O ; [7732-18-5]

EVALUATOR:
S. Siekierski, T. Mioduski
Institute for Nuclear Research
Warsaw, Poland
and
M. Salomon
U.S. Army ET & DL
Ft. Monmouth, NJ, USA
December 1982

CRITICAL EVALUATION:

Table 4. Derived parameters for the smoothing equation [1].

parameter	hexahydrate	pentahydrate
a	-30.158	-43.62
b	1014.7	1708
c	4.696	6.641
σ_a	0.003	0.01
σ_b	0.9	4
σ_c	0.001	0.002
σ_Y	0.003	0.01
σ_m	0.036	0.58
ΔH_{sln}/kJ mol^{-1}	-33.6	-56.6
ΔC_p/J K^{-1} mol^{-1}	156.2	221
congruent melting point/K	330.8	350.2
concn at the congr melting pt/mol kg^{-1}	9.251	11.102

Table 5. Tentative solubilities for $Pr(NO_3)_3$ calculated from eq. [1] (all values in units of mol kg^{-1}).

T/K	hexahydrate	pentahydrate
273.15	4.138	
278.15	4.280	
283.15	4.439	
288.15	4.616	
293.15	4.816	
298.15	5.041	5.30[a]
303.15	5.298	5.46[a]
308.15	5.596	5.65[a]
310.60[b]	5.757	5.757
313.15	5.949[a]	5.87
318.15	6.383[a]	6.13
323.15	6.951[a]	6.43
328.2		6.78
333.2		7.20
338.2		7.72
343.2		8.43
348.2		9.60

[a] Metastable solubilities.

[b] Temperature of the hexahydrate ⟶ pentahydrate transition (determined graphically by the evaluators).

COMPONENTS:

(1) Praseodymium nitrate; $Pr(NO_3)_3$; [10361-80-5]

(2) Water ; H_2O ; [7732-18-5]

EVALUATOR:
S. Siekierski, T. Mioduski
Institute for Nuclear Research
Warsaw, Poland
and
M. Salomon
U.S. Army ET & DL
Ft. Monmouth, NJ, USA
December 1982

CRITICAL EVALUATION:

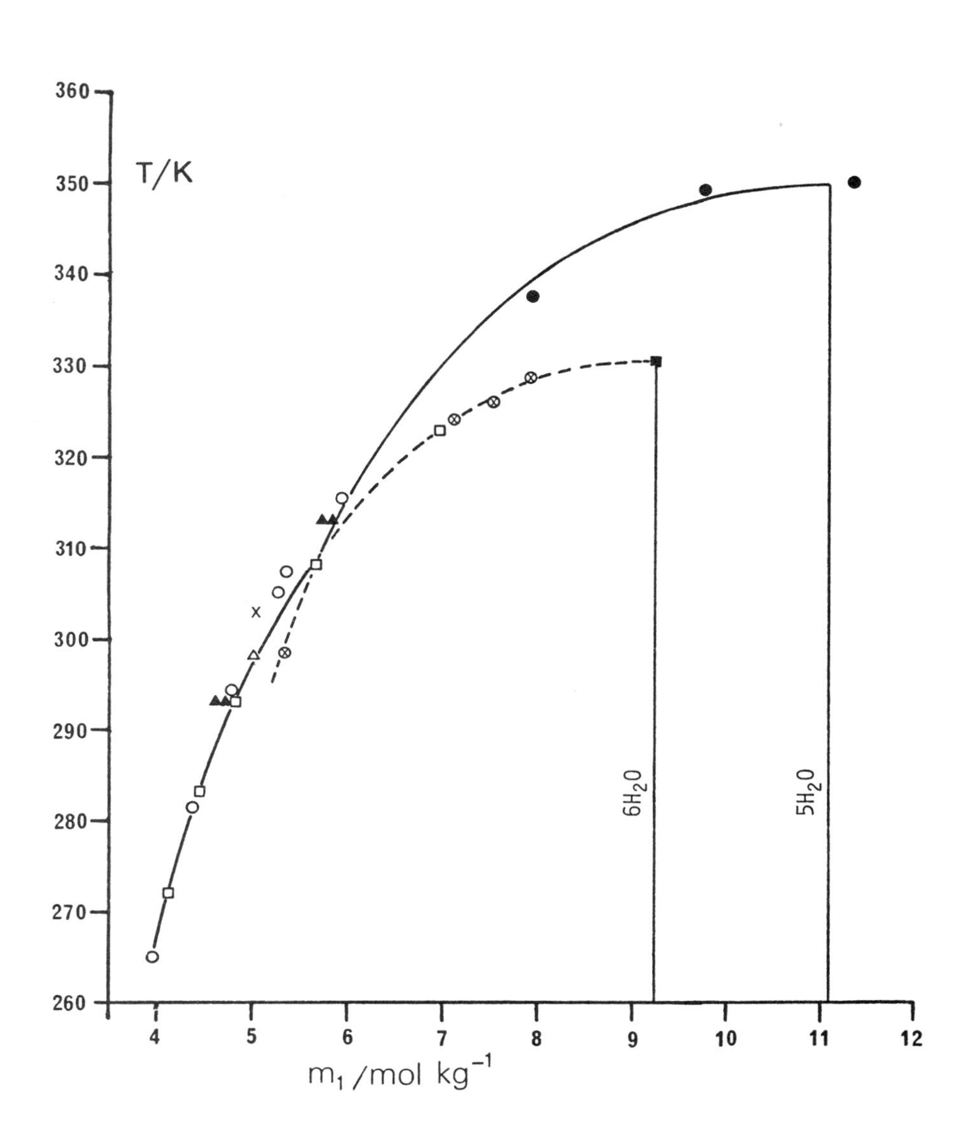

Figure 1. Phase diagram for the binary $Pr(NO_3)_3$ - H_2O system.

○ Mironov and Popov hexahydrate data (3).
⊗ Mironov and Popov pentahydrate data (3, 6, 7).
● Mironov and Popov tetrahydrate data (3).
□ Brunisholz et al. hexahydrate data (2, 9).
△ Spedding et al. hexahydrate data (4, 5).
▲ Zhuravlev et al. hexahydrate data (10, 12-17).
× Hexahydrate datum from ref (11).
■ Hexahydrate melting point: average value from (1 and 22).

COMPONENTS:	EVALUATOR:
(1) Praseodymium nitrate; $Pr(NO_3)_3$; [10361-80-5] (2) Water ; H_2O ; [7732-18-5]	S. Siekierski, T. Mioduski Institute for Nuclear Research Warsaw, Poland and M. Salomon U.S. Army ET & DL Ft. Monmouth, NJ, USA December 1982

CRITICAL EVALUATION:

REFERENCES

1. Friend, J.N. *J. Chem. Soc.* 1935, 1430.
2. Brunisholz, G.; Quinche, J.P.; Kalo, A.M. *Helv. Chim. Acta* 1964, *47*, 14.
3. Mironiv, K.E.; Popov, A.P. *Rev. Roum. Chim.* 1966, *11*, 1373.
4. Rard, J.A.; Spedding, F.H. *J. Phys. Chem.* 1975, *79*, 257.
5. Spedding, F.H.; Derer, J.L.; Mohs, M.A.; Rard, J.A. *J. Chem. Eng. Data* 1976, *21*, 474.
6. Mironov, K.E.; Snitsyna, E.D.; Popov, A.P. *Izv. Sibir. Otd. Akad. Nauk SSSR, Ser. Khim. Nauk.* 1964, *11*, 48.
7. Popov, A.P.; Mironov, K.E. *Zh. Neorg. Khim.* 1971, *16*, 464.
8. Perel'man, F.M.; Zvorykin, A. Ya.; Demina, G.A. *Zh. Neorg. Khim.* 1962, *7*, 641.
9. Brunisholz, G.; Klipfel, R. *Rev. Chem. Miner.* 1970, *7*, 349.
10. Zhuravlev, E.F.; Kuznetsova, L.S. *Zh. Neorg. Khim.* 1975, *20*, 1401.
11. Khudaibergenova, N.; Sulaimankulov, K.S. *Zh. Neorg. Khim.* 1979, *24*, 2005.
12. Zhuravlev, E.F.; Kuznetsova, L.S. *Zh. Neorg. Khim.* 1978, *23*, 1688.
13. Kuznetsova, L.S.; Zhuravlev, E.F. *Zh. Neorg. Khim.* 1977, *22*, 820.
14. Kuznetsova, L.S.; Zhuravlev, E.F. *V. sb. Fazovye Ravnovesiyo.* 1975, 57.
15. Zhuravlev, E.F.; Kuznetsova, L.S. *Zh. Neorg. Khim.* 1975, *20*, 1672.
16. Zhuravlev, E.F.; Kuznetsova, L.S.; Gabdulkhakova, A.Z. *Zh. Neorg. Khim.* 1977, *22*, 574.
17. Kuznetsova, L.S.; Zhuravlev, E.F. *Zh. Neorg. Khim.* 1977, *22*, 515.
18. Ferguson, R.E.; Guth, E.D.; Eyring, L. *J. Am. Chem. Soc.* 1954, *76*, 3890.
19. Hyde, B.G.; Bevan D.J.M.; Eyring, L. *Phil Trans. Roy. Soc. (London)* 1966, *A259*, 583.
20. Williamson, A.T. *Trans. Faraday Soc.* 1944, *40*, 421.
21. Counioux, J.-J.; Tenu, R. *J. Chim. Phys.* 1981, *78*, 816 and 823.
22. Wendlandt, W.W.; Sewell, R.G. *Texas J. Sci*, 1961, *13*, 231.
23. Friend, J.N.; Wheat, W.N. *J. Chem. Soc.* 1935, 356 (see the critical evaluation and compilation under Praseodymium double salts).

COMPONENTS:

(1) Praseodymium nitrate; $Pr(NO_3)_3$; [10361-80-5]

(2) Water; H_2O; [7732-18-5]

ORIGINAL MEASUREMENTS:

Friend, J.N. *J. Chem. Soc.* 1935, 1430-2.

VARIABLES:

Temperature

PREPARED BY:

T. Mioduski and S. Siekierski

EXPERIMENTAL VALUES:

Solubility of $Pr(NO_3)_3$ [a]

t/°C	mass %	mol kg^{-1}	solid phase
15.8	59.32	4.460	$Pr(NO_3)_3 \cdot 6H_2O$
22.0	60.18	4.623	"
30.4	61.94	4.978	"
43.0	65.00	5.681	"
56.0 [b]	75.15	9.250	"

a. Molalities calculated by the compilers.

b. Melting point.

AUXILIARY INFORMATION

METHOD/APPARATUS/PROCEDURE:

The isothermal method was used as described in (1). Aliquots of saturated solutions were diluted to 250 c.c. and Pr precipitated as the oxalate. The oxalate was ignited to Pr_6O_{11} in a Pt dish and weighed. The author reports the solubility in mass % of the anhydrous salt, and the original data of mass % of oxide was not given.

SOURCE AND PURITY OF MATERIALS:

Praseodymium nitrate was prepared by dissolving Pr_6O_{11} in dilute HNO_3, and seeding the concentrated solutions with $Bi(NO_3)_3 \cdot 5H_2O$. The crystals melted at approximately 56°C. (Compilers note: this is the approximate temperature of the congruent melting point).

ESTIMATED ERROR:

Soly: precision ± 1 % at best (compilers).

Temp: accuracy probably ± 0.5 to 0.1 K as in (1) (compilers).

REFERENCES:

1. Friend, J.N. *J. Chem. Soc.* 1930, 1633.

COMPONENTS:

(1) Praseodymium nitrate; $Pr(NO_3)_3$; [10361-80-5]

(2) Water; H_2O; [7732-18-5

ORIGINAL MEASUREMENTS:

Brunisholz, G.; Quinche, J.P.; Kalo, A.M. *Helv. Chim. Acta* 1964, *47*, 14-27.

VARIABLES:

Temperature

PREPARED BY:

T. Mioduski and S. Siekierski

EXPERIMENTAL VALUES:

Solubility of $Pr(NO_3)_3$ [a]

t/°C	mass %	mol kg^{-1}	solid phase
0	57.46	4.132	$Pr(NO_3)_3 \cdot 6H_2O$
10	59.29	4.455	"
20	61.24	4.833	"
35	64.86	5.646	"
50	69.40	6.937	"

a. Molalities calculated by the compilers.

AUXILIARY INFORMATION

METHOD/APPARATUS/PROCEDURE:

The isothermal method was used. Praseodymium determined by complexometric titration using Xylenol Orange indicator and a small quantity of urotropine buffer.

SOURCE AND PURITY OF MATERIALS:

Praseodymium nitrate was prepared from the oxide of purity better than 99.7 % (purified by ion exchange).

ESTIMATED ERROR:

Soly: precision ± 0.2 % (compilers).

Temp: precision probably better than ± 0.1 K (compilers).

REFERENCES:

COMPONENTS:

(1) Praseodymium nitrate; $Pr(NO_3)_3$; [10361-80-5]

(2) Water; H_2O; [7732-19-5]

ORIGINAL MEASUREMENTS:

Mironov, K.E.; Popov, A.P. *Rev. Roum. Chim.* 1966, *11*, 1373-81.

VARIABLES:

Temperature

PREPARED BY:

T. Mioduski and S. Siekierski

EXPERIMENTAL VALUES:

Solubility as a function of temperature [a]

t/°C	mass %	mol kg^{-1}	solid phase [b]	t/°C	mass %	mol kg^{-1}	solid phase [b]
- 0.2	8.8	0.30	A	42.5	65.9	5.91	C
- 0.7	15.3	0.55	A	51.3	69.9	7.10	C
- 2.1	22.8	0.90	A	53.1	71.1	7.53	C
- 5.4	30.0	1.31	A				
- 9.7	36.2	1.74	A	56.1	72.1	7.90	C+D
-12.3	39.5	2.00	A				
-15.9	43.1	2.32	A	64.7	72.2	7.94	D
-22.9	49.8	3.03	A	76.4	76.1	9.74	D
				77.5	78.8	11.37	D
-29.7	53.6	3.53	A+B				
				92.8	84.0	16.06	E
- 8.1	56.3	3.94	B	127.0	85.0	17.33	E
8.3	58.8	4.37	B				
21.3	61.0	4.78	B				
31.9	63.2	5.25	B				
34.3	63.6	5.34	B				

a. Molalities calculated by compilers.

b. A = ice, B = $Pr(NO_3)_3 \cdot 6H_2O$, C = $Pr(NO_3)_3 \cdot 5H_2O$, D = $Pr(NO_3)_3 \cdot 4H_2O$, E = $Pr(NO_3)_3 \cdot 3H_2O$

AUXILIARY INFORMATION

METHOD/APPARATUS/PROCEDURE:

The synthetic method was used. The temperatures of crystallization were determined visually. No other information given.

SOURCE AND PURITY OF MATERIALS:

$Pr(NO_3)_3 \cdot 6H_2O$ prepared by dissolving Pr_6O_4 of 99.5 % purity in HNO_3 followed by crystallization.

Doubly distilled water was used.

ESTIMATED ERROR:

Soly: nothing specified.

Temp: not specified.

REFERENCES:

COMPONENTS:

(1) Praseodymium nitrate; $Pr(NO_3)_3$; [10361-80-5]

(2) Water; H_2O; [7732-18-5]

ORIGINAL MEASUREMENTS:

Rard, J.A.; Spedding, F.H. *J. Phys. Chem.* 1975, *79*, 257-62.

VARIABLES:

One temperature: 25.0°C

PREPARED BY:

T. Mioduski, S. Siekierski, M. Salomon

EXPERIMENTAL VALUES:

The solubility of $Pr(NO_3)_3$ in water at 25°C was reported to be

$$m_1 = 5.0166 \text{ mol kg}^{-1}$$

COMMENTS AND/OR ADDITIONAL DATA:

Supplementary data available in the microfilm edition to *J. Phys. Chem.* 1975, *79* have enabled the compilers to provide the following additional data:

The density of the saturated solution was calculated by the compilers from the smoothing equation and found to be 1.8687 kg m^{-3} at 25.00°C. Using this density, the concentration of the saturated solution in volume units is

$$c_1 = 3.5510 \text{ mol dm}^{-3}$$

The electrolytic conductivity of the saturated solution corrected for the electrolytic conductivity of water was given as 0.015705 S cm^{-1}. The molar conductivity of the saturated solution is

$$\Lambda(\tfrac{1}{3}Pr(NO_3)_3) = 1.474 \text{ S cm}^2 \text{ mol}^{-1}$$

AUXILIARY INFORMATION

METHOD/APPARATUS/PROCEDURE:

The isothermal method was used. Experimental details on how equilibrium was ascertained not provided. Satd soln was analysed by both EDTA and sulfate methods, and resulting concn is reliable to ± 0.1 % or better. When sulfate analyses were performed, the $Pr(NO_3)_3$ samples were decomposed with HCl followed by evapn to dryness before the H_2SO_4 additions were made. This treatment eliminated the possibility of nitrate ion copptn. The solid phase is $Pr(NO_3)_3 \cdot 6H_2O$.

SOURCE AND PURITY OF MATERIALS:

1. The oxide was furnished by the Ames Laboratory Rare Earth Separation Group. No other details provided, but presumably the oxide was 99.85 % pure or better as used in other papers by the authors.
2. Presumably conductivity water was used.

ESTIMATED ERROR:

Soly: the reported result is reliable to ± 0.1 % or better.

Temp: the oil bath temp. was controlled to 24.99 ± 0.01°C.

REFERENCES:

COMPONENTS:

(1) Praseodymium nitrate; $Pr(NO_3)_3$; [10361-80-5]

(2) Water; H_2O; [7732-18-5]

ORIGINAL MEASUREMENTS:

Spedding, F.H.; Derer, J.L.; Mohs, M.A.; Rard, J.A. *J. Chem. Eng. Data* 1976, *21*, 474-88.

VARIABLES:

One temperature: 25.00°C

PREPARED BY:

T. Mioduski and S. Siekierski

EXPERIMENTAL VALUES:

The solubility of $Pr(NO_3)_3$ in water at 25°C was reported to be

$$m_1 = 4.990 \text{ mol kg}^{-1}$$

AUXILIARY INFORMATION

METHOD/APPARATUS/PROCEDURE:

The isothermal method was used. The satd solns (no details on how equilibrium was ascertained) were analysed by EDTA and gravimetric sulfate methods, and resulting concn is reliable to ± 0.1 % or better in terms of molality. The nitrate samples were decomposed by heating with HCl before the sulfate pptns were performed to avoid nitrate ion copptn. The solid phase is $Pr(NO_3)_3 \cdot 6H_2O$ (Hydrated crystals were grown from satd soln at 25°C and hydrate compns were detd by EDTA analyses. After drying, the hydrates were found to be within ± 0.16 water molecules of the indicated hydrate).

SOURCE AND PURITY OF MATERIALS:

1. The stoichiometric nitrate was obtained from the ion exchange purified oxide and HNO_3 (1). The purity of the oxide was 99.85 mass % or better with adjacent lanthanides, Ca, Fe, Si being the principal impurities.
2. Conductivity water; electrolytic conductivity less than 1×10^{-6} S cm^{-1}.

ESTIMATED ERROR:

Soly: The reported value is reliable to ± 0.1 % or better in terms of molality.

Temp: precision probably ± 0.02 K (compilers).

REFERENCES:

1. Spedding, F.H.; Pikal, M.J.; Ayers, B.O. *J. Phys. Chem.* 1966, *70*, 2440.

COMPONENTS:

(1) Praseodymium nitrate; $Pr(NO_3)_3$; [10361-80-5]

(2) Nitric acid; HNO_3; [7697-37-2]

(3) Water; H_2O; [7732-18-5]

ORIGINAL MEASUREMENTS:

Mironov, K.E.; Snitsyna, E.D.; Popov, A.P. *Izv. Sibir. Otd. Akad. Nauk SSSR, Ser. Khim. Nauk*, 1964, *11*, 48-53.

VARIABLES:

Composition

PREPARED BY:

T. Mioduski and S. Siekierski

EXPERIMENTAL VALUES:

Composition of saturated solutions at 25°C [a]

$Pr(NO_3)_3$		HNO_3			$Pr(NO_3)_3$		HNO_3		
mass %	mol kg	mass %	mol kg	solid phase [b]	mass %	mol kg^{-1}	mass %	mol kg^{-1}	solid phase [b]
63.5	5.32	——	——	A	21.4	3.62	60.5	53.05	B
60.3	5.07	3.3	1.44	A	29.6	6.00	55.3	58.12	B
50.8	4.26	12.7	5.52	A	29.3	5.82	55.3	56.99	B
43.6	3.66	20.0	8.72	A					
33.8	3.41	35.9	18.80	A	22.1	4.76	63.7	71.19	C
34.0	3.50	36.3	19.40	A	13.6	3.10	73.0	86.45	C
35.0	3.86	37.3	21.37	A	6.9	1.84	81.6	112.6	C
					0.9	0.44	92.8	233.8	C
36.1	4.00	36.7	21.41	B					
28.5	3.33	45.3	27.44	B	0.6	0.32	93.6	256.1	D
24.7	2.96	49.8	30.99	B	0.2	0.18	96.4	450.0	D
20.7	3.09	58.8	45.52	B	0.1	3.06	99.8	1584.	D

a. Molalities calculated by M. Salomon.

b. A = $Pr(NO_3)_3 \cdot 5H_2O$
B = $Pr(NO_3)_3 \cdot 3H_2O$
C = $Pr(NO_3)_3 \cdot H_2O$
D = $Pr(NO_3)_3$

AUXILIARY INFORMATION

METHOD/APPARATUS/PROCEDURE:

The isothermal method was used. Solutions were equilibrated for 4-6 h, and both the liquid and solid phases analysed.

Praseodymium determined by titration with Trilon B with Eriochrome Black indicator in the presence of ascorbic acid.

Method of determination of HNO_3 not specified, but probably the acid was titrated with standard base. In several cases, water was determined by the Karl Fischer method.

SOURCE AND PURITY OF MATERIALS:

$Pr(NO_3)_3 \cdot 5H_2O$ prepd by dissolving 99.5 % oxide in HNO_3. Analysis for Pr was 78.3 mass %, and using the Karl Fischer method, H_2O was found to be 21.6 mass %.

100 % HNO_3 prepd by the Brauer method.

Doubly distilled water was used.

ESTIMATED ERROR:

Soly: precision ± 0.3 to 0.5 % (compilers).

Temp: precision ± 0.1 K.

REFERENCES:

COMPONENTS:

(1) Praseodymium nitrate; $Pr(NO_3)_3$; [10361-80-5]

(2) Nitric acid; HNO_3; [7697-37-2]

(3) Water; H_2O; [7732-18-5]

ORIGINAL MEASUREMENTS:

Popov, A.P.; Mironov, K.E. *Zh. Neorg. Khim.* 1971, *16*, 464-6; *Russ. J. Inorg. Chem. Engl. Transl.* 1971, *16*, 244-6.

VARIABLES:

Composition at 25°C

PREPARED BY:

T. Mioduski, S. Siekierski, M. Salomon

EXPERIMENTAL VALUES:

Composition of saturated solutions [a]

$Pr(NO_3)_3$		HNO_3		
mass %	mol kg^{-1}	mass %	mol kg^{-1}	nature of the solid phase
61.9	4.97	——	——	$Pr(NO_3)_3 \cdot 6H_2O$
56.1	4.46	5.4	2.23	"
52.4	4.15	9.0	3.70	"
44.7	3.56	16.9	6.98	"
36.3	3.03	27.1	11.75	"
33.3	3.11	33.9	16.40	"
35.1 [b]	3.56	34.8	18.35	"
63.5 [b,c]	5.32	——	——	$Pr(NO_3)_3 \cdot 5H_2O$

a. Molalities calculated by the compilers.

b. Metastable.

c. Data point probably taken from reference (1), compilers.

AUXILIARY INFORMATION

METHOD/APPARATUS/PROCEDURE:

Not stated, but probably the isothermal method was used as described in (1). Reference (1) has been compiled on the previous page.

SOURCE AND PURITY OF MATERIALS:

Nothing specified, but probably similar to (1).

ESTIMATED ERROR:

Soly: precision about ± 0.5 mass % (compilers).

Temp: precision probably ± 0.1 K (compilers).

REFERENCES:

1. Mironov, K.E.; Sinitsyna, E.D.; Popov, A.P. *Izv. Sibir. Otd. Akad. Nauk SSSR, Ser. Khim. Nauk*, 1964, *11*, 48.

COMPONENTS:

(1) Rubidium nitrate; $RbNO_3$; [13126-12-0]

(2) Praseodymium nitrate; $Pr(NO_3)_3$; [10361-80-5]

(3) Nitric acid; HNO_3; [7697-37-2]

(4) Water; H_2O; [7732-18-5]

ORIGINAL MEASUREMENTS:

Perel'man, F.M.; Zvorykin, A.Ya.; Demina, G.A. *Zh. Neorg. Khim.* 1962, *7*, 641-4; *Russ. J. Inorg. Chem. Engl. Transl.* 1962 *7*, 325-6.

VARIABLES:

Composition at 0°C

PREPARED BY:

T. Mioduski, S. Siekierski, M. Salomon

EXPERIMENTAL VALUES: Composition of saturated solutions [a]

$Pr(NO_3)_3$		$RbNO_3$		HNO_3		
mass %	mol kg^{-1}	mass %	mol kg^{-1}	mass %	mol kg^{-1}	nature of the solid phase
34.31	3.293	———	———	33.82	16.841	$Pr(NO_3)_3$ (A)
35.27	3.446	3.02	0.654	30.40	15.409	"
36.96	3.655	4.76	1.044	27.35	14.033	"
38.48	3.654	3.67	0.773	25.64	12.633	A + B
36.93	3.367	3.62	0.732	25.90	12.251	$4Pr(NO_3)_3 \cdot 5RbNO_3$ (B)
35.10	3.526	7.81	1.739	26.64	13.884	"
33.19	3.272	9.00	1.967	26.78	13.696	"
31.37	3.180	11.71	2.632	26.75	14.071	"
29.93	2.857	11.28	2.387	26.75	13.250	"
29.03	3.062	17.33	4.052	24.64	13.484	"
29.48	3.587	19.20	5.179	26.18	16.526	"
14.69	2.390	31.37	11.315	35.14	29.663	$5Pr(NO_3)_3 \cdot 7RbNO_3$ (C)
12.36	1.950	32.55	11.383	35.70	29.219	"
10.09	2.025	38.32	17.050	36.35	37.852	C + D
9.37	2.142	41.04	20.799	36.21	42.948	$2Pr(NO_3)_3 \cdot 5RbNO_3$ (D)
6.64	1.579	48.60	25.626	31.90	39.366	"
5.73	1.443	50.38	28.117	31.74	41.457	"
6.21	0.835	44.23	13.189	26.82	18.717	$RbNO_3$
2.60	0.357	44.89	13.650	30.21	21.499	"
———	———	45.96	13.324	30.65	20.796	"

a. Molalities calculated by the compilers.

AUXILIARY INFORMATION

METHOD/APPARATUS/PROCEDURE:

The isothermal method was used. Equilibrium was reached after 2-3 days. Both the liquid and solid phases were analysed. Praseodymium was determined by precipitation as the hydroxide using ammonia followed by ignition to Pr_6O_{11}. Rubidium was determined gravimetrically as the perchlorate, and nitric acid was determined by acidimetric titration.

The compositions of the solid phases were determined by Schreinemakers' method of residues.

SOURCE AND PURITY OF MATERIALS:

C.p. grade Pr_6O_{11} and Rb_2CO_3 dissolved in aq HNO_3, and the slns evapd to crystallization. No further purifications were carried out. The Pr salt was analysed gravimetrically and found to contain 40.96 mass % Pr_6O_{11} confirming the hexahydrate composition. Gravimetric analysis of $RbNO_3$ gave 62.66 mass % Rb_2O confirming the composition $RbNO_3$.

ESTIMATED ERROR:

Soly: nothing specified.

Temp: precision ± 0.1 K.

REFERENCES:

COMPONENTS:

(1) Praseodymium nitrate; $Pr(NO_3)_3$; [10361-80-5]

(2) Neodymium nitrate; $Nd(NO_3)_3$; [10045-95-1]

(3) Water; H_2O; [7732-18-5]

ORIGINAL MEASUREMENTS:

Brunisholz, G.; Quinche, J.P.; Kalo, A.M. *Helv. Chim. Acta* 1964, *47*, 14-27.

VARIABLES:

Composition at 20°C

PREPARED BY:

T. Mioduski, S. Sierkierski, and M. Salomon

EXPERIMENTAL VALUES:

Composition of saturated solutions [a]

Nd mol %	moles H_2O per 100 moles Pr + Nd	moles Pr + Nd per kg H_2O [b]	solubility [b] /mol kg^{-1} $Pr(NO_3)_3$	$Nd(NO_3)_3$
0.0	1149	4.831	4.831	———
9.9	1155	4.806	4.330	0.476
19.2	1152	4.818	3.893	0.925
34.9	1184	4.688	3.052	1.636
48.1	1187	4.676	2.433	2.249
63.8	1219	4.554	1.648	2.905
69.7	1221	4.546	1.377	3.169
76.9	1222	4.542	1.049	3.493
92.0	1254	4.426	0.354	4.072
100.0	1259	4.409	———	4.409

a. The solid phase is a continuous series of solid solutions of $Pr_{1-x}Nd_x(NO_3)_3 \cdot 6H_2O$.

b. Calculated by the compilers.

AUXILIARY INFORMATION

METHOD/APPARATUS/PROCEDURE:

The isothermal method was used. The solid(s) were placed in glass-stoppered vials into which a glass dumb-bell shaped pestle was placed. These vials were placed in a second larger tube and sealed with a rubber stopper. The tubes were placed in a thermostat and rotated end-over-end so that the pestle pulverized the solid two times through each 360° rotation. The solutions were equilibrated for 2-3 weeks, and underwent 500,000 pulverizations per week. The total Pr + Nd was determined by complexometric titration with Xylenol Orange indicator in the presence of urotropine buffer. The ratio of rare earths was determined chromatographically using a morin (pentahydroxyflavone) cation resin. Water was determined by difference.

The compositions of the solid phases were determined by Schreinemakers' method of residues, and by X-ray diffraction.

SOURCE AND PURITY OF MATERIALS:

The rare earth nitrates were prepared from the oxides of purity of at least 99.7 % (obtained by ion exchange chromatography). No other information given.

ESTIMATED ERROR:

Soly: precision ± 0.2 % (compilers).

Temp: precision probably better than ± 0.05 K (compilers).

REFERENCES:

COMPONENTS:

(1) Praseodymium nitrate; $Pr(NO_3)_3$; [10361-80-5]

(2) Zinc nitrate; $Zn(NO_3)_2$; [7779-88-6]

(3) Water; H_2O; [7732-18-5]

ORIGINAL MEASUREMENTS:

Brunisholz, G.; Klipfel, K. *Rev. Chim. Miner.* 1970, *7*, 349-58.

VARIABLES:

Composition at 20°C

PREPARED BY:

T. Mioduski, S. Siekierski, M. Salomon

EXPERIMENTAL VALUES:

Composition of saturated solutions

Pr mol %	moles H_2O per 100 moles Pr + Zn	moles Pr + Zn per kg H_2O [a]	solubility [a] /mol kg^{-1} $Pr(NO_3)_3$	$Zn(NO_3)_2$	solid phase [b]
0.0	892	6.223	0.0	6.223	A
0.37	873	6.358	0.024	6.335	A + B
1.10	1100	5.046	0.056	4.991	B
11.53	1231	4.509	0.520	3.989	B
43.76	1269	4.374	1.914	2.460	B
74.80	1219	4.554	3.406	1.148	B
88.75	1100	5.046	4.479	0.568	B
89.19	1095	5.069	4.521	0.548	B + C
89.25	1095	5.069	4.524	0.545	B + C
89.98	1100	5.046	4.541	0.506	C
94.15	1125	4.934	4.645	0.289	C
100.0	1150	4.827	4.827	0.0	C

a. Calculated by the compilers.

b. A = $Zn(NO_3)_2 \cdot 6H_2O$

B = $2Pr(NO_3)_3 \cdot 3Zn(NO_3)_2 \cdot 24H_2O$

C = $Pr(NO_3)_3 \cdot 6H_2O$

AUXILIARY INFORMATION

METHOD/APPARATUS/PROCEDURE:

The isothermal method was used. The solids were placed in glass-stoppered vials into which a glass dumb-bell shaped pestle was placed. These vials were placed in a second larger tube and sealed with a rubber stopper. The tubes were placed in a thermostat and rotated end-over-end so that the pestle pulverized the solid two times through each 360° rotation. The solutions were equilibrated for 2 to 4 weeks.

For the saturated solutions Pr was determined by titration with Trilon using Xylenol Orange indicator and urotropine buffer, Zn was determined by precipitation with thioacetimide as the sulfide (using thioacetamide), or by ion exchange chromatography.

The compositions of the solid phases were determined by Schreinemakers' method of residues, and by X-ray analysis.

SOURCE AND PURITY OF MATERIALS:

The nitrates were obtained by dissolving the oxide in nitric acid followed by crystallization. The source and purities of the oxides was not specified. The hydrated nitrates were dried in a desiccator over KOH under vacuum.

ESTIMATED ERROR:

Soly: precision ± 0.2 % (compilers).

Temp: precision probably at least ± 0.05 K (compilers).

REFERENCES:

COMPONENTS:

(1) Praseodymium nitrate; $Pr(NO_3)_3$; [10361-80-5]

(2) Hydrazine dinitrate; $N_2H_4 \cdot 2HNO_3$; [13464-98-7]

(3) Water ; H_2O ; [7732-18-5]

ORIGINAL MEASUREMENTS:

Zhuravlev, E.F.; Kuznetsova, L.S. *Zh. Neorg. Khim.* 1975, *20*, 1401-5; *Russ. J. Inorg. Chem. Engl. Transl.* 1975, *20*, 788-90.

VARIABLES:

Composition at 20°C and 40°C

PREPARED BY:

T. Mioduski, S Siekierski, and M. Salomon

EXPERIMENTAL VALUES:

Composition of saturated solutions at 20°C[a]

$Pr(NO_3)_3$ mass %	$Pr(NO_3)_3$ mol kg^{-1}	$N_2H_4 \cdot 2HNO_3$ mass %	$N_2H_4 \cdot 2HNO_3$ mol kg^{-1}	n_D	nature of the solid phase
0	———	70.1	14.83	1.4670	$N_2H_4 \cdot 2HNO_3$
1.6	0.16	67.0	13.50	1.4660	"
5.6	0.54	62.5	12.39	1.4615	$Pr(NO_3)_3 \cdot 3[N_2H_4 \cdot 2HNO_3]$
9.0	0.80	56.8	10.51	1.4580	"
15.7	1.30	47.5	8.17	1.4530	"
26.5	2.03	33.5	5.30	1.4470	"
29.0	2.22	31.0	4.90	1.4465	"
38.0	2.89	21.8	3.43	1.4450	"
41.5	3.13	18.0	2.81	1.4472	"
50.7	3.92	9.7	1.55	1.4530	"
56.7	4.52	5.2	0.86	1.4570	$Pr(NO_3)_3 \cdot 6H_2O$
58.7	4.57	2.0	0.32	1.4550	"
100.0[b]		0	———	1.4540	"

a. Molalities calculated by the compilers.
b. This incorrect entry appears in both the original publication and the English Translation.

continued.........

AUXILIARY INFORMATION

METHOD/APPARATUS/PROCEDURE:

The method of isothermal sections with refractometric analyses was used as described in (1). No other information was given.

COMMENTS AND/OR ADDITIONAL DATA:

The double salt, $Pr(NO_3)_3 \cdot 3[N_2H_4 \cdot 2HNO_3]$, is congruently soluble and was isolated. Chemical analysis for nitrate by precipitation with nitron and for praseodymiun by the oxalate method gave: found (mass %) Pr 17.52, NO_3 68.93; calcd (mass %) Pr 17.59, NO_3 69.73.

SOURCE AND PURITY OF MATERIALS:

"Pure grade praseodymiun nitrate was recrystallized from dilute nitric acid. Analysis for water gave 24.9 mass %.

Hydrazine dinitrate, $N_2H_4 \cdot 2HNO_3$, was prepared by neutralization of "pure" grade hydrazine with nitric acid as described in (2): note reference (2) has been compiled previously (see lanthanum nitrates).

Doubly distilled water was used.

ESTIMATED ERROR:

Soly: based on the method, precision is about ± 1 % (compilers).

Temp: precision probably ± 0.2 K (compilers).

REFERENCES:

1. Zhuravlev, E.F.; Sheveleva, A.D. *Zh. Neorg. Khim.* 1960, *5*, 2630.

2. Gorshunova, V.P.; Zhuravlev, E.F. *Zh. Neorg. Khim.* 1971, *16*, 1700.

COMPONENTS:

(1) Praseodymium nitrate; $Pr(NO_3)_3$; [10361-80-5]

(2) Hydrazine dinitrate; $N_2H_4 \cdot 2HNO_3$; [13464-98-7]

(3) Water ; H_2O ; [7732-18-5]

ORIGINAL MEASUREMENTS:

Zhuravlev, E.F.; Kuznetsova, L.S. *Zh. Neorg. Khim.* 1975, *20*, 1401-5; *Russ. J. Inorg. Chem. Engl. Transl.* 1975, *20*, 788-90.

EXPERIMENTAL VALUES: continued ....

Composition of saturated solutions at 40°C[a]

$Pr(NO_3)_3$		$H_2H_4 \cdot 2HNO_3$			
mass %	mol kg^{-1}	mass %	mol kg^{-1}	n_D	nature of the solid phase
0	———	73.8	17.82	1.4750	$N_2H_4 \cdot 2HNO_3$
1.6	0.17	70.0	15.59	1.4720	$N_2H_4 \cdot 2HNO_3$ + $Pr(NO_3)_3 \cdot 3[N_2H_4 \cdot 2HNO_3]$
3.2	0.33	67.5	14.57	1.4705	$Pr(NO_3)_3 \cdot 3[N_2H_4 \cdot 2HNO_3]$
5.5	0.52	62.2	12.18	1.4630	"
9.0	0.86	58.9	11.61	1.4615	"
13.0	1.16	52.6	9.67	1.4575	"
17.5	1.47	46.0	7.97	1.4550	"
25.0	1.99	36.5	6.00	1.4519	"
33.2	2.63	28.2	4.62	1.4510	"
36.0	2.88	25.8	4.27	1.4502	"
49.0	4.02	13.7	2.32	1.4570	"
57.5	4.91	6.7	1.18	1.4655	"
62.5	5.62	3.5	0.65	1.4700	$Pr(NO_3)_3 \cdot 6H_2O$
64.25	5.70	1.25	0.23	1.4690	"
65.1	5.71	0	———	1.4678	"

a. Molalities calculated by the compilers.

COMPONENTS:

(1) Praseodymium nitrate; $Pr(NO_3)_3$; [10361-80-5]

(2) Guanidine mononitrate; $CH_6N_4O_3$; [506-93-4]

(3) Water; H_2O; [7732-18-5]

ORIGINAL MEASUREMENTS:

Kuznetsova, L.S.; Zhuravlev, E.F. *Zh. Neorg. Khim.* 1977, *22*, 820-2; *Russ. J. Inorg. Chem. Engl. Transl.* 1977, *22*, 454-6.

VARIABLES:

Composition at 20°C and 40°C

PREPARED BY:

T. Mioduski and S. Siekierski

EXPERIMENTAL VALUES:

Composition of saturated solutions at 20°C [a]

$Pr(NO_3)_3$ mass %	$Pr(NO_3)_3$ mol kg^{-1}	$CH_5N_3 \cdot HNO_3$ mass %	$CH_5N_3 \cdot HNO_3$ mol kg^{-1}	solid phase [b]
0	——	12.7	1.19	A
13.8	0.54	7.8	0.81	A
28.6	1.31	4.4	0.54	A
43.8	2.49	2.5	0.38	A
54.0	3.70	1.3	0.24	A
60.8	4.86	0.9	0.19	A + B
60.7	4.81	0.7	0.15	B
60.6	4.70	0	——	B

Composition of saturated solutions at 40°C [a]

$Pr(NO_3)_3$ mass %	$Pr(NO_3)_3$ mol kg^{-1}	$CH_5N_3 \cdot HNO_3$ mass %	$CH_5N_3 \cdot HNO_3$ mol kg^{-1}	solid phase [b]
0	——	22.6	2.39	A
12.6	0.53	14.7	1.66	A
27.3	1.31	8.8	1.13	A
42.5	2.48	5.0	0.78	A
53.5	3.80	3.4	0.65	A
65.5	5.98	1.0	0.24	A + B
65.4	5.92	0.8	0.19	B
65.3	5.82	0.4	0.10	B
65.1	5.71	0	——	B

a. Molalities calculated by M. Salomon.

b. A = $CH_5N_3 \cdot HNO_3$, B = $Pr(NO_3)_3 \cdot 6H_2O$

AUXILIARY INFORMATION

METHOD/APPARATUS/PROCEDURE:

The method of isothermal sections was used with refractometric analyses (1). Heterogeneous and homogeneous mixtures of known composition were equilibrated until their refractive indices remained constant.

Equilibrium for homogeneous compositions was reached in 24 hours, and for heterogeneous compositions, equilibrium was reached in 2-3 days.

SOURCE AND PURITY OF MATERIALS:

"Pure" grade praseodymium nitrate was recrystallized before use. The salt was analysed for Pr by the oxalate method and for NO_3^- by pptn with nitron. Water, found by difference, was 25.3 mass % (theor: 24.83 mass %).

"Pure" grade quanidine mononitrate was recrystallized and dried to constant weight over $CaCl_2$. The composition of the salt was verified by nitrate analysis by pptn with nitron.

ESTIMATED ERROR:

Soly: based on the method, precision is about ± 1 % (compilers).
Temp: precision probably 0.1-0.2 K (compilers)

REFERENCES:

1. Zhuravlev, E.F.; Sheveleva, A.D. *Zh. Neorg. Khim.* 1960, *5*, 2630.

COMPONENTS:

(1) Praseodymium nitrate; $Pr(NO_3)_3$; [10361-80-5]

(2) Diethylamine nitrate; $C_4H_{12}N_2O_3$; [27096-30-6]

(3) Water; H_2O; [7732-18-5]

ORIGINAL MEASUREMENTS:

Zhuravlev, E.F.; Kuznetsova, L.S. *Zh. Neorg. Khim.* 1975, *20*, 1672-5; *Russ. J. Inorg. Chem. Engl. Transl.* 1975, *20*, 937-9.

VARIABLES:

Composition and temperature

PREPARED BY:

T. Mioduski and S. Siekierski

EXPERIMENTAL VALUES:

Composition of saturated solutions at 20°C [a]

$Pr(NO_3)_3$ mass %	$Pr(NO_3)_3$ mol kg^{-1}	$(C_2H_5)_2NH\cdot HNO_3$ mass %	$(C_2H_5)_2NH\cdot HNO_3$ mol kg^{-1}	solid phase [b]
0	——	79.7	28.84	A
3.2	0.52	77.8	30.07	A
7.1	1.27	75.8	32.56	A
11.6	2.38	73.5	36.23	A
15.0	3.67	72.5	42.60	A
21.7	8.51	70.5	66.39	A
51.5	9.27	31.5	13.61	B
54.0	6.61	21.0	6.17	B
55.1	5.69	15.3	3.80	B
57.0	5.20	9.5	2.08	B
58.7	4.88	4.5	0.90	B
60.1	4.61	0	——	B

Composition of saturated solutions at 40°C [a]

$Pr(NO_3)_3$ mass %	$Pr(NO_3)_3$ mol kg^{-1}	$(C_2H_5)_2NH\cdot HNO_3$ mass %	$(C_2H_5)_2NH\cdot HNO_3$ mol kg^{-1}	solid phase [b]
0	——	87.2	50.04	A
2.0	0.51	86.0	52.63	A
4.4	1.27	85.0	58.90	A
7.2	2.42	83.7	67.56	A
9.5	3.63	82.5	75.74	A
15.2	8.02	79.0	100.0	A
59.5	9.33	21.0	7.91	B
61.5	7.52	13.5	3.97	B
62.5	6.95	10.0	2.67	B
63.5	6.37	6.0	1.44	B
64.0	5.88	2.7	0.60	B
65.0	5.68	0	——	B

a. Molalities calculated by M. Salomon.

b. A = $(C_2H_5)_2NH\cdot HNO_3$, B = $Pr(NO_3)_3\cdot 6H_2O$

AUXILIARY INFORMATION

METHOD/APPARATUS/PROCEDURE:

The solubilities were studied by the method of isothermal sections as described in (1). No other information given.

SOURCE AND PURITY OF MATERIALS:

"Pure grade praseodymium nitrate was recrystallized.

Diethylamine nitrate was prepd by neutralization of "pure" grade $(C_2H_5)_2NH$ with HNO_3 followed by evapn to crystn. The product was dried in air, and then in a desiccator over $CaCl_2$. The salt was analysed, but details and results not given.

Doubly distilled water was used.

ESTIMATED ERROR:

Soly: based on the method, precision is around ± 1 % (compilers).
Temp: precision probably ± 0.2 K (compilers).

REFERENCES:

1. Zhuravlev, E.F.; Sheveleva, A.D. *Zh. Neorg. Khim.* 1960, 5, 2630.

COMPONENTS:

(1) Praseodymium nitrate; $Pr(NO_3)_3$; [10361-80-5]

(2) Triethylamine nitrate; $C_6H_{16}N_2O_3$; [27096-31-7]

(3) Water; H_2O; [7732-18-5]

ORIGINAL MEASUREMENTS:

Zhuravlev, E.F.; Kuznetsova, L.S. *Zh. Neorg. Khim.* 1975, *20*, 1672-5; *Russ. J. Inorg. Chem. Engl. Transl.* 1975, *20*, 937-9.

VARIABLES:

Composition and temperature

PREPARED BY:

T. Mioduski and S. Siekierski

EXPERIMENTAL VALUES:

Composition of saturated solutions at 20°C

$Pr(NO_3)_3$ mass %	$Pr(NO_3)_3$ mol kg^{-1}	$(C_2H_5)_3N \cdot HNO_3$ mass %	$(C_2H_5)_3N \cdot HNO_3$ mol kg^{-1}	solid phase [b]
0	——	90.0	54.81	A
1.6	0.52	89.0	57.66	A
3.9	1.47	88.0	66.16	A
6.2	2.60	86.5	72.16	A
7.9	3.72	85.6	80.20	A + B
10.0	2.71	78.7	42.41	B
11.6	2.75	75.5	35.64	B
16.0	2.97	67.5	24.91	B
21.0	3.38	60.0	19.23	B
36.0	9.18	52.0	26.39	B
49.5	9.18	34.0	12.55	C
53.0	6.48	22.0	5.36	C
55.0	5.65	15.2	3.11	C
56.7	5.13	9.5	1.71	C
58.7	4.83	4.1	0.67	C
60.1	4.61	0	——	C

Composition of saturated solutions at 40°C

$Pr(NO_3)_3$ mass %	$Pr(NO_3)_3$ mol kg^{-1}	$(C_2H_5)_3N \cdot HNO_3$ mass %	$(C_2H_5)_3N \cdot HNO_3$ mol kg^{-1}	solid phase [b]
0	——	91.5	65.56	A
1.5	0.61	91.0	73.89	A
3.2	1.40	89.8	78.12	A
5.1	2.44	88.5	84.21	A
8.0	4.89	87.0	106.0	A + B
14.7	7.75	79.5	83.47	B
15.7	5.22	75.1	49.71	B
20.5	6.15	69.0	40.80	B
29.3	8.46	60.1	34.53	B
30.2	8.97	59.5	35.18	B
56.2	9.24	25.2	8.25	C
59.7	7.22	15.0	3.61	C
61.3	6.65	10.5	2.27	C
62.6	6.10	6.0	1.16	C
64.2	5.93	2.7	0.50	C
65.1	5.71	0	——	C

a. Molalities calculated by M. Salomon.

b. A = $(C_2H_5)_3N \cdot HNO_3$, B = $Pr(NO_3)_3 \cdot 4(C_2H_5)_3N \cdot HNO_3$, C = $Pr(NO_3)_3 \cdot 6H_2O$

AUXILIARY INFORMATION

METHOD/APPARATUS/PROCEDURE:

The solubilities were studied by the method of isothermal sections as described in (1). No other information given.

COMMENTS AND/OR ADDITIONAL DATA:

The double nitrate, $Pr(NO_3)_3 \cdot 4(C_2H_5)_3N \cdot HNO_3$, is congruently soluble and was thus isolated for further analyses. Chemical analysis gave a composition of Pr = 14.7 mass % and NO_3^- = 43.7 mass %. Based on the formula for the double salt, the calculated composition is Pr = 14.2 mass % and NO_3^- = 44.1 mass %. The results of X-ray diffraction studies on the double salt are given in the source paper.

SOURCE AND PURITY OF MATERIALS:

"Pure" grade praseodymium nitrate was recrystallized.

Triethylamine nitrate was prepd by neutralization of "pure" grade $(C_2H_5)_3N$ with HNO_3 followed by evapn to crystn. The product was dried in air, and then in a desiccator over $CaCl_2$. The salt was analysed, but details and results not given.

Doubly distilled water was used.

ESTIMATED ERROR:

Soly: based on the method, precision is about ± 1 % (compilers).
Temp: precision probably ± 0.2 K (compilers).

REFERENCES:

1. Zhuravlev, E.F.; Sheveleva, A.D. *Zh. Neorg. Khim.* 1960, *5*, 2630.

COMPONENTS:

(1) Praseodymium nitrate; $Pr(NO_3)_3$; [10361-80-5]

(2) Ethylenediamine dinitrate; $C_2H_{10}N_4O_6$; [20829-66-7]

(3) Water; H_2O; [7732-18-5]

ORIGINAL MEASUREMENTS:

Zhuravlev, E.F.; Kuznetsova, L.S. *Zh. Neorg. Khim.* 1975, *20*, 1401-5; *Russ. J. Inorg. Chem. Engl. Transl.* 1975, *20*, 788-90.

VARIABLES:

Composition at 20°C and 40°C

PREPARED BY:

T. Mioduski, S. Siekierski and M. Salomon

EXPERIMENTAL VALUES:

Composition of saturated solutions at 20°C [a]

$Pr(NO_3)_3$ mass %	$Pr(NO_3)_3$ mol kg^{-1}	$En\cdot 2HNO_3$ [b] mass %	$En\cdot 2HNO_3$ mol kg^{-1}	n_D	nature of the solid phase
0	——	45.5	4.49	1.4080	$En\cdot 2HNO_3$
9.0	0.53	39.0	4.03	1.4120	"
20.8	1.31	30.5	3.36	1.4212	"
34.0	2.48	24.0	3.07	1.4390	"
42.0	3.29	19.0	2.62	1.4465	$Pr(NO_3)_3\cdot En\cdot HNO_3\cdot 4H_2O$
45.2	3.50	15.3	2.08	1.4460	"
49.0	3.79	11.5	1.56	1.4480	"
53.5	4.34	8.8	1.25	1.4514	"
60.0	4.73	1.2	0.17	1.4570	$Pr(NO_3)_3\cdot 6H_2O$
60.5	5.18	3.8	0.57	1.4645	"
100.0 [c]	?	0	——	1.4540	"

a. Molalities calculated by the compilers.

b. En = $H_2NCH_2CH_2NH_2$

c. This incorrect entry appears in both the original publication and the English translation.

continued........

AUXILIARY INFORMATION

METHOD/APPARATUS/PROCEDURE:

The method of isothermal sections with refractometric analyses was used as described in (1). No other information was given.

COMMENTS AND/OR ADDITIONAL DATA:

The double salt, $Pr(NO_3)_3\cdot En2HNO_3\cdot 4H_2O$, is congruently soluble at 40°C, and at 20°C it has "a tendency" to become incongruently soluble. The double salt was isolated and analysed for nitrate by precipitation with nitron, and for praseodymium by the oxalate method. Water was determined by the Karl Fischer method: found (mass %) Pr 24.00, NO_3 52.41, H_2O 13.15; calcd (mass %) Pr 24.11, NO_3 52.99, H_2O 12.30.

SOURCE AND PURITY OF MATERIALS:

"Pure" grade praseodymium nitrate was recrystallized from dilute nitric acid. Analysis for water gave 24.9 mass %.

Ethylenediamine dinitrate, $H_2NCH_2CH_2NH_2\cdot 2HNO_3$, was prepared by neutralization of "pure" grade ethylenediamine with nitric acid as described in (2): note reference (2) has been compiled previously (see lanthanum nitrates).

Doubly distilled water was used.

ESTIMATED ERROR:

Soly: based on the method, precision is around ± 1 % (compilers).
Temp: precision probably ± 0.2 K (compilers).

REFERENCES:

1. Zhuravlev, E.F.; Sheveleva, A.D. *Zh. Neorg. Khim.* 1960, *5*, 2630.

2. Zhuravlev, E.F.; Gorshunova, V.P. *Zh. Neorg. Khim.* 1970, *15*, 195.

COMPONENTS:

(1) Praseodymium nitrate; $Pr(NO_3)_3$; [10361-80-5]

(2) Ethylenediamine dinitrate; $C_2H_{10}N_4O_6$; [20829-66-7]

(3) Water; H_2O; [7732-18-5]

ORIGINAL MEASUREMENTS:

Zhuravlev, E.F.; Kuznetsova, L.S. *Zh. Neorg. Khim.* 1975, *20*, 1401-5; *Russ. J. Inorg. Chem. Engl. Transl.* 1975, *20*, 788-90.

EXPERIMENTAL VALUES: continued.......

Composition of saturated solutions at 40°C [a]

$Pr(NO_3)_3$		En·$2HNO_3$ [b]			
mass %	mol kg^{-1}	mass %	mol kg^{-1}	n_D	nature of the solid phase
0	——	59.9	8.03	1.4330	En·$2HNO_3$
7.0	0.54	53.5	7.28	1.4390	"
15.8	1.28	46.4	6.60	1.4442	"
28.0	2.52	38.0	6.00	1.4560	"
36.8	3.67	32.5	5.69	1.4645	"
44.5	4.58	25.8	4.67	1.4665	$Pr(NO_3)_3$·En·$2HNO_3$·$4H_2O$
47.0	4.92	23.8	4.38	1.4660	"
50.5	4.72	16.8	2.76	1.4650	"
54.1	5.00	12.8	2.07	1.4665	"
58.5	5.47	8.8	1.45	1.4700	"
66.0	6.41	2.5	0.43	1.4721	$Pr(NO_3)_3$·$6H_2O$
65.8	6.10	1.2	0.20	1.4703	"
65.1	5.71	——	——	1.4678	"

a. Molalities calculated by the compilers.

b. En = $H_2NCH_2CH_2NH_2$.

COMPONENTS:

(1) Praseodymium nitrate; $Pr(NO_3)_3$; [10361-80-5]

(2) Hexamethyleneimine nitrate; (homopiperidine nitrate, perhydropyridinium nitrate); $C_6H_{14}N_2O_3$

(3) Water; H_2O; [7732-18-5]

ORIGINAL MEASUREMENTS:

Zhuravlev, E.F.; Kuznetsova, L.S.; Gabdulkhakova, A.Z. *Zh. Neorg. Khim.* 1977, *22*, 574-7; *Russ. J. Inorg. Chem. Engl. Transl.* 1977, *22*, 319-21.

VARIABLES:

Composition at 20°C and 40°C

PREPARED BY:

T. Mioduski and S. Siekierski

EXPERIMENTAL VALUES:

Composition of saturated solutions at 20°C [a]

$Pr(NO_3)_3$ mass %	$Pr(NO_3)_3$ mol kg^{-1}	$C_6H_{13}N\cdot HNO_3$ mass %	$C_6H_{13}N\cdot HNO_3$ mol kg^{-1}	nature of the solid phase
0	——	88.1	45.05	$C_6H_{13}N\cdot HNO_3$
1.7	0.43	86.3	44.34	"
4.5	1.21	84.1	45.48	"
7.5	2.34	82.7	52.03	"
12.6	4.88	79.5	62.05	$C_6H_{13}N\cdot HNO_3 + Pr(NO_3)_3\cdot 4C_6H_{13}N\cdot HNO_3$
14.4	4.95	76.7	53.14	$Pr(NO_3)_3\cdot 4C_6H_{13}N\cdot HNO_3$
20.6	4.57	65.6	29.31	"
27.4	4.53	54.1	18.03	"
33.7	4.96	45.5	13.49	"
37.3	5.23	40.9	11.57	"
41.2	5.65	36.5	10.09	"
45.4	6.34	32.7	9.21	"
48.1	6.88	30.5	8.79	$Pr(NO_3)_3\cdot 4C_6H_{13}N\cdot HNO_3 + Pr(NO_3)_3\cdot 6H_2O$
49.2	6.32	27.0	6.99	$Pr(NO_3)_3\cdot 6H_2O$
53.2	5.46	17.0	3.52	"
56.2	5.01	9.5	1.71	"
58.3	4.76	4.2	0.69	"
60.1	4.61	0	——	"

a. Molalities calculated by M. Salomon.

continued.........

AUXILIARY INFORMATION

METHOD/APPARATUS/PROCEDURE:

The method of isothermal sections was used with refractometric analysis (1). Refractive indices were not given in the source paper. The attainment of equilibrium was verified by repeated sampling and analysis of the liquid phase. Equilibrium was reached in 3 to 5 days.

SOURCE AND PURITY OF MATERIALS:

"Pure" grade praseodymium nitrate was recrystallized prior to use. Analysis for water gave 24.85 mass % corresponding to the hexahydrate.

$C_6H_{13}N\cdot HNO_3$ was prepared by neutralization of "pure" grade $C_6H_{13}N$ and dilute (1:5) nitric acid as described in (2). The purity of the salt was checked by analysis for NO_3^- by precipitation with nitron. Results were not given, but authors state that the composition of the salt was confirmed.

ESTIMATED ERROR:

Soly: based on the method, precision is around ± 1 % (compilers).

Temp: precision from ± 0.1-0.2 K (compilers).

REFERENCES:

1. Zhuravlev, E.F.; Sheveleva, A.D. *Zh. Neorg. Khim.* 1960, *5*, 2630.
2. Zhuravlev, E.F.; Kuznetsova, L.S. *Zh. Neorg. Khim.* 1975, *20*, 1401 and 1672.

COMPONENTS:

(1) Praseodymium nitrate; $Pr(NO_3)_3$; [10361-80-5]

(2) Hexamethyleneimine nitrate (hemopiperidine nitrate; perhydropyridinium nitrate); $C_6H_{14}N_2O_3$

(3) Water; H_2O; [7732-18-5]

ORIGINAL MEASUREMENTS:

Zhuravlev, E.F.; Kuznetsova, L.S.; Gabdulkhakova, A.Z. *Zh. Neorg. Khim.* 1977, *22*, 574-7; *Russ. J. Inorg. Chem. Engl. Transl.* 1977, *22*, 319-21.

EXPERIMENTAL VALUES: continued.......

Composition of saturated solutions at 40°C [a]

$Pr(NO_3)_3$		$C_6H_{13}N{\cdot}HNO_3$		
mass %	mol kg^{-1}	mass %	mol kg^{-1}	nature of the solid phase
0	———	94.5	105.9	$C_6H_{13}N{\cdot}HNO_3$
0.8	0.43	93.5	101.1	"
2.1	1.17	92.4	103.6	"
3.7	2.26	91.3	112.6	"
9.4	7.99	87.0	149.0	"
12.0	14.12	85.4	202.5	$C_6H_{13}N{\cdot}HNO_3 + Pr(NO_3)_3{\cdot}4_6H_{13}N{\cdot}HNO_3$
14.0	8.08	80.7	93.88	$Pr(NO_3)_3{\cdot}4C_6H_{13}N{\cdot}HNO_3$
14.7	7.37	79.2	80.05	"
21.6	6.67	68.5	42.66	"
29.1	7.24	58.6	29.37	"
36.5	8.59	50.5	23.95	"
56.7	8.99	24.0	7.67	$Pr(NO_3)_3{\cdot}6H_2O$
59.8	7.56	16.0	4.08	"
61.8	6.78	10.3	2.28	"
63.2	6.42	6.7	1.37	"
64.4	5.97	2.6	0.49	"
65.6	5.83	0	———	"

a. Molalities calculated by M. Salomon.

COMMENTS AND/OR ADDITIONAL DATA:

The composition of the double salt $Pr(NO_3)_3{\cdot}4C_6H_{13}N{\cdot}HNO_3$ was determined graphically. The compound is congruently soluble and was isolated for analyses. Pr was determined by the oxalate method, and nitrate ion by precipitation with nitron: found (mass %) Pr 14.35, NO_3 44.2; calcd (mass %) Pr 14.45, NO_3 44.5. Derivatograms (temperature vs time curves) are presented in the source paper. Above 155°C, the double salt undergoes thermal decomposition.

COMPONENTS:

(1) Praseodymium nitrate; $Pr(NO_3)_3$; [10361-80-5]

(2) Cyclohexylamine nitrate; $C_6H_{14}N_2O_3$; [6941-45-3]

(3) Water; H_2O; [7732-18-5]

ORIGINAL MEASUREMENTS:

Zhuravlev, E.F.; Kuznetsova, L.S. *Zh. Neorg. Khim.* 1978, *23*, 1688-91; *Russ. J. Inorg. Chem. Engl. Transl.* 1978, *23*, 929-31.

VARIABLES:

Composition at 20°C and 40°C

PREPARED BY:

T. Mioduski and S. Siekierski

EXPERIMENTAL VALUES:

Composition of saturated solutions at 20°C [a]

$Pr(NO_3)_3$ mass %	$Pr(NO_3)_3$ mol kg^{-1}	$C_6H_{11}NH_2 \cdot HNO_3$ mass %	$C_6H_{11}NH_2 \cdot HNO_3$ mol kg^{-1}	nature of the solid phase
0	——	54.0	7.24	$C_6H_{11}NH_2 \cdot HNO_3$
4.0	0.25	47.7	6.09	"
14.6	0.89	35.0	4.28	"
18.6	1.12	30.8	3.75	"
21.5	1.31	28.2	3.46	"
28.8	1.85	23.5	3.04	"
41.5	3.04	16.7	2.46	"
52.0	5.43	18.7	3.94	$C_6H_{11}NH_2 \cdot HNO_3 + Pr(NO_3)_3 \cdot 6H_2O$
54.0	5.15	13.9	2.67	$Pr(NO_3)_3 \cdot 6H_2O$
54.7	5.02	12.0	2.22	"
57.0	4.84	7.0	1.20	"
58.7	4.71	3.2	0.52	"
60.5	4.69	0.0	——	"

a. Molalities calculated by M. Salomon.

continued........

AUXILIARY INFORMATION

METHOD/APPARATUS/PROCEDURE:

The method of isothermal sections was used as described in (1). Equilibrium was established in 4-5 days. No other information given.

SOURCE AND PURITY OF MATERIALS:

$Pr(NO_3)_3$ was prepared by dissolving the oxide in c.p. grade nitric acid followed by recrystallization on a water bath. Gravimetric anal of the hexahydrate by the oxalate method gave a Pr content of 32.9 mass % (theory: 32.4 mass %).

Cyclohexylamine nitrate, $C_6H_{11}NH_2 \cdot HNO_3$, was prepared by neutralization of cyclohexylamine with nitric acid. The stoichiometry of the salt was verified gravimetrically by precipitation of the nitrate ion with nitron.

Doubly distilled water was used.

ESTIMATED ERROR:

Soly: based on the method, precision is around ± 1 % (compilers).

Temp: precision probably ± 0.2 K (compilers).

REFERENCES:

1. Zhuravlev, E.F.; Sheveleva, A.D. *Zh. Neorg. Khim.* 1960, *5*, 2630.

COMPONENTS:	ORIGINAL MEASUREMENTS:
(1) Praseodymium nitrate; $Pr(NO_3)_3$; [10361-80-5] (2) Cyclohexylamine nitrate; $C_6H_{14}N_2O_3$; [6941-45-3] (3) Water; H_2O; [7732-18-5]	Zhuravlev, E.F.; Kuznetsova, L.S. *Zh. Neorg. Khim.* 1978, *23*, 1688-91; *Russ. J. Inorg. Chem. Engl. Transl.* 1978, *23*, 929-31.

EXPERIMENTAL VALUES: continued.........

Composition of saturated solutions at 40°C [a]

$Pr(NO_3)_3$		$C_6H_{11}NH_2 \cdot HNO_3$		
mass %	mol kg^{-1}	mass %	mol kg^{-1}	nature of the solid phase
0.0	——	75.0	18.50	$C_6H_{11}NH_2 \cdot HNO_3$
2.0	0.24	72.5	17.53	"
7.0	0.87	68.5	17.24	"
13.7	1.80	63.0	16.67	"
20.8	3.15	59.0	18.01	"
31.8	6.00	52.0	19.79	"
37.0	8.08	49.0	21.58	"
38.5	9.06	48.5	23.00	"
46.5	9.18	38.0	15.12	$Pr(NO_3)_3 \cdot 6H_2O$
59.2	6.27	11.9	2.54	"
61.3	6.11	8.0	1.61	"
63.2	6.04	4.6	0.88	"
64.3	5.84	2.0	0.37	"
65.6	5.83	0.0	——	"

a. Molalities calculated by M. Salomon.

COMPONENTS:

(1) Praseodymium nitrate; $Pr(NO_3)_3$; [10361-80-5]

(2) Piperidine nitrate; $C_5H_{12}N_2O_3$; [6091-45-8]

(3) Water; H_2O; [7732-18-5]

ORIGINAL MEASUREMENTS:

Kuznetsova, L.A.; Zhuravlev, E.F. *V. sb. Fazovye Ravnovesiya* 1975, 57-63.

VARIABLES:

Composition at 20°C and 40°C

PREPARED BY:

T. Mioduski and S. Siekierski

EXPERIMENTAL VALUES:

Composition of saturated solutions at 20°C [a]

$Pr(NO_3)_3$ mass %	$Pr(NO_3)_3$ mol kg^{-1}	Pip·HNO_3 [b] mass %	Pip·HNO_3 [b] mol kg^{-1}	solid phase [c]
0	——	89.2	55.74	A
1.8	0.50	87.2	53.50	A
2.9	0.81	86.2	53.38	A + B
4.5	0.89	80.0	34.83	B
9.2	1.13	66.0	17.96	B
12.0	1.32	60.2	14.62	B
19.7	1.82	47.2	9.62	B
27.5	2.45	38.1	7.48	B
35.5	3.11	29.6	5.72	B
42.4	3.76	23.1	4.52	B
55.5	5.62	14.3	3.20	B + C
56.1	5.30	11.5	2.40	C
57.8	5.01	6.9	1.32	C
59.1	4.81	3.3	0.59	C
60.2	4.63	0	——	C

Composition of saturated solutions at 40°C [a]

$Pr(NO_3)_3$ mass %	$Pr(NO_3)_3$ mol kg^{-1}	Pip·HNO_3 [b] mass %	Pip·HNO_3 [b] mol kg^{-1}	solid phase [c]
0	——	91.5	72.65	A
1.7	0.59	89.5	68.64	A
2.7	0.84	87.5	60.26	A
4.2	1.32	86.1	59.91	A
4.7	1.45	85.4	58.22	A + B
13.0	1.82	65.1	20.06	B
18.1	2.22	57.0	15.45	B
26.5	3.08	47.2	12.11	B
33.0	3.79	40.4	10.25	B
49.9	6.18	25.4	6.94	B
59.7	7.91	17.2	5.03	B + C
60.8	7.53	14.5	3.96	C
62.2	6.84	10.0	2.43	C
63.0	6.47	7.2	1.63	C
64.0	6.16	4.2	0.89	C
64.8	5.97	2.0	0.41	C
65.6	5.83	0	——	C

a. Molalities calculated by M. Salomon.
b. Pip = piperdine = $(CH_2)_5NH$
c. A = Pip·HNO_3 , B = $Pr(NO_3)_3 \cdot 4Pip \cdot HNO_3$, C = $Pr(NO_3)_3 \cdot 6H_2O$

AUXILIARY INFORMATION

METHOD/APPARATUS/PROCEDURE:

The method of isothermal sections was used with refractometric analyses (1). Heterogeneous and homogeneous mixtures of known composition were equilibrated until their refractive indices remained constant. The composition of the saturated solutions and the corresponding solid phases were found as inflection or "break" points on a plot of composition against refractive index.

COMMENTS AND/OR ADDITIONAL DATA:

The double salt was isolated and analysed for Pr by the oxalate method and for NO_3 by precipitation with nitron. The results confirmed the composition of the double salt.

SOURCE AND PURITY OF MATERIALS:

"Pure" grade $Pr(NO_3)_3 \cdot 6H_2O$ was recrystd before use. Analysis for Pr by the oxalate method gave 75.2 mass %.

Piperidine nitrate, $(CH_2)_5NH \cdot HNO_3$, was prepd by neutralization of "pure" grade piperidine with c.p. grade nitric acid followed by crystallization. Nitrate analysis by precipitation with nitron confirmed the composition of the salt.

Doubly distilled water was used.

ESTIMATED ERROR:

Soly: based on the method, precision is about ± 1 % (compilers).
Temp: precision probably 0.1-0.2K (compilers).

REFERENCES:

1. Zhuravlev, E.F.; Sheveleva, A.D. *Zh. Neorg. Khim.* 1960, *5*, 2630.

COMPONENTS:

(1) Praseodymium nitrate; $Pr(NO_3)_3$; [10361-80-5]

(2) Pyridine nitrate; $C_5H_6N_2O_3$; [543-53-3]

(3) Water; H_2O; [7732-18-5]

ORIGINAL MEASUREMENTS:

Kuznetsova, L.S.; Zhuravlev, E.F. *V. sb. Fazovye Ravnovesiya* 1975, 57-63.

VARIABLES:

Composition at 20°C and 40°C

PREPARED BY:

T. Mioduski and S. Siekierski

EXPERIMENTAL VALUES:

Composition of saturated solutions at 20°C [a]

$Pr(NO_3)_3$ mass %	$Pr(NO_3)_3$ mol kg^{-1}	$C_5H_5N{\cdot}HNO_3$ mass %	$C_5H_5N{\cdot}HNO_3$ mol kg^{-1}	nature of the solid phase [b]
0	——	72.1	18.18	$C_5H_5N{\cdot}HNO_3$
4.5	0.53	69.6	18.91	"
10.0	1.34	67.2	20.74	"
16.0	2.48	64.3	22.97	"
16.1	2.50	64.2	22.93	$C_5H_5N{\cdot}HNO_3 + Pr(NO_3)_3{\cdot}5py{\cdot}HNO_3$
19.3	2.85	60.0	20.40	$Pr(NO_3)_3{\cdot}5py{\cdot}HNO_3$
22.7	3.34	56.5	19.11	"
26.7	4.02	53.0	18.37	"
31.0	5.02	50.1	18.65	"
33.4	5.71	48.7	19.14	$Pr(NO_3)_3{\cdot}5C_5H_5N{\cdot}HNO_3 + Pr(NO_3)_3{\cdot}2py{\cdot}HNO_3{\cdot}2H_2O$
34.0	4.95	45.0	15.08	$Pr(NO_3)_3{\cdot}2py{\cdot}HNO_3{\cdot}2H_2O$
36.2	4.56	39.6	11.51	"
38.5	4.35	34.4	8.93	"
44.3	4.43	25.1	5.77	"
45.5	4.53	23.8	5.46	"
47.7	4.60	20.6	4.57	"
51.6	5.04	17.1	3.84	"
56.0	5.62	13.5	3.11	$Pr(NO_3)_3{\cdot}2py{\cdot}HNO_3{\cdot}2H_2O + Pr(NO_3)_3{\cdot}6H_2O$

continued.......

AUXILIARY INFORMATION

METHOD/APPARATUS/PROCEDURE:

The method of isothermal sections was used with refractometric analyses (1). Heterogeneous and homogeneous mixtures of known composition were equilibrated until their refractive indices remained constant. The composition of the saturated solutions and the corresponding solid phases were found as inflection or "break" points on a plot of composition against refractive index.

COMMENTS AND/OR ADDITIONAL DATA:

The double salts were isolated and analysed for Pr by the oxalate method and for NO_3 by precipitation with nitron. The results confirmed the composition of the double salts.

SOURCE AND PURITY OF MATERIALS:

"Pure" grade $Pr(NO_3)_3{\cdot}6H_2O$ was recrystd before use. Analysis for Pr by the oxalate method gave 75.2 mass %.

Pyridine nitrate, $C_5H_5N{\cdot}HNO_3$, was prepared by neutralization of "pure" grade pyridine with c.p. grade nitric acid followed by crystallization. Nitrate analysis by precipitation with nitron confirmed the composition of the salt.

Doubly distilled water was used.

ESTIMATED ERROR:

Soly: based on the method, precision is around ± 1 % (compilers).

Temp: precision probably 0.1-0.2 K (compilers)

REFERENCES:

1. Zhuravlev, E.F.; Sheveleva, A.D. *Zh. Neorg. Khim.* 1960, 5, 2630.

COMPONENTS:

(1) Praseodymium nitrate; $Pr(NO_3)_3$; [10361-80-5]

(2) Pyridine nitrate; $C_5H_6N_2O_3$; [543-53-3]

(3) Water; H_2O; [7732-18-5]

ORIGINAL MEASUREMENTS:

Kuznetsova, L.S.; Zhuravlev, E.F. *V. sb. Fazovye Ravnovesiya* 1975, 57-63.

EXPERIMENTAL VALUES: continued........

Composition of saturated solutions at 20°C [a]

$Pr(NO_3)_3$ mass %	$Pr(NO_3)_3$ mol kg^{-1}	$C_5H_5N\cdot HNO_3$ mass %	$C_5H_5N\cdot HNO_3$ mol kg^{-1}	nature of the solid phase [b]
57.2	5.45	10.7	2.35	$Pr(NO_3)_3\cdot 6H_2O$
58.2	5.04	6.5	1.30	"
59.3	4.82	3.1	0.58	"
60.1	4.61	——	——	"

Composition of saturated solutions at 40°C

$Pr(NO_3)_3$ mass %	$Pr(NO_3)_3$ mol kg^{-1}	$C_5H_5N\cdot HNO_3$ mass %	$C_5H_5N\cdot HNO_3$ mol kg^{-1}	nature of the solid phase [b]
0	——	80.7	29.42	$C_5H_5N\cdot HNO_3$
3.2	0.55	78.9	31.02	"
7.0	1.33	76.9	33.61	"
11.6	2.53	74.4	37.39	"
15.7	3.97	72.2	41.99	$C_5H_5N\cdot HNO_3$ + $Pr(NO_3)_3\cdot 5py\cdot HNO_3$
21.0	4.76	65.5	34.14	$Pr(NO_3)_3\cdot 5py\cdot HNO_3$
24.6	5.26	61.1	30.06	"
28.7	6.18	57.1	28.29	"
33.3	7.84	53.7	29.07	"
37.1	9.54	51.0	30.16	$Pr(NO_3)_3\cdot 5py\cdot HNO_3$ + $Pr(NO_3)_3\cdot 2py\cdot HNO_3\cdot 2H_2O$
37.8	9.10	49.5	27.43	$Pr(NO_3)_3\cdot 2py\cdot HNO_3\cdot 2H_2O$
40.6	8.17	44.2	20.46	"
43.7	7.68	38.9	15.73	"
51.0	7.72	28.8	10.03	"
54.9	8.00	24.1	8.08	"
59.3	8.56	19.5	6.47	"
60.0	8.74	19.0	6.37	$Pr(NO_3)_3\cdot 2py\cdot HNO_3\cdot 2H_2O$ + $Pr(NO_3)_3\cdot 6H_2O$
62.8	6.99	9.7	2.48	$Pr(NO_3)_3\cdot 6H_2O$
63.9	6.72	7.0	1.69	"
64.0	6.20	4.4	0.98	"
64.6	5.92	2.0	0.42	"
65.6	5.83	——	——	"

a. Molalities calculated by M. Salomon.

b. py = pyridine, C_5H_5N.

COMPONENTS:

(1) Praseodymium nitrate; $Pr(NO_3)_3$; [10361-80-5]

(2) Aniline nitrate; $C_6H_8N_2O_3$; [542-15-4]

(3) Water; H_2O; [7732-18-5]

ORIGINAL MEASUREMENTS:

Kuznetsova, L.S.; Zhuravlev, E.F. *Zh. Neorg. Khim.* 1977, *22*, 820-2; *Russ. J. Inorg. Chem. Engl. Transl.* 1977, *22*, 454-6.

VARIABLES:

Composition at 20°C and 40°C

PREPARED BY:

T. Mioduski and S. Siekierski

EXPERIMENTAL VALUES:

Composition of saturated solutions at 20°C [a]

$Pr(NO_3)_3$ mass %	$Pr(NO_3)_3$ mol kg^{-1}	$C_6H_5NH_2\cdot HNO_3$ mass %	$C_6H_5NH_2\cdot HNO_3$ mol kg^{-1}	solid phase [b]
0	——	20.7	1.67	A
13.7	0.54	9.0	0.75	A
17.2	0.70	7.6	0.65	A
28.7	1.29	3.5	0.33	A
44.0	2.50	2.1	0.25	A
53.9	3.73	1.9	0.28	A
59.5	4.64	1.3	0.21	A + B
59.6	4.63	1.0	0.16	B
60.6	4.70	0	——	B

Composition of saturated solutions at 40°C [a]

$Pr(NO_3)_3$ mass %	$Pr(NO_3)_3$ mol kg^{-1}	$C_6H_5NH_2\cdot HNO_3$ mass %	$C_6H_5NH_2\cdot HNO_3$ mol kg^{-1}	solid phase [b]
0	——	35.8	3.57	A
11.5	0.54	23.3	2.29	A
25.8	1.29	13.2	1.39	A
41.5	2.49	7.5	0.94	A
52.5	3.73	4.5	0.67	A
63.0	5.60	2.6	0.48	A + B
63.8	5.64	1.6	0.30	B
64.8	5.75	0.7	0.13	B
65.1	5.71	0	——	B

a. Molalities calculated by M. Salomon.

b. A = $C_6H_5NH_2\cdot HNO_3$, B = $Pr(NO_3)_3\cdot 6H_2O$

AUXILIARY INFORMATION

METHOD/APPARATUS/PROCEDURE:

The method of isothermal sections was used with refractometric analyses (1). Heterogeneous and homogeneous mixtures of known composition were equilibrated until their refractive indices remained constant.

Equilibrium for homogeneous compositions was reached in 24 hours, and for heterogeneous compositions, equilibrium was reached in 2-3 days.

SOURCE AND PURITY OF MATERIALS:

"Pure" grade praseodymium nitrate was recrystallized before use. The salt was analysed for Pr by the oxalate method and for NO_3^- by pptn with nitron. Water, found by difference, was 25.3 mass % (theor: 24.83 mass %).

Aniline nitrate prepd by neutralization of "pure" grade aniline with dil HNO_3, evapn to crystallization, and drying over $CaCl_2$. The composition of the salt was verified by nitrate analysis by pptn with nitron.

ESTIMATED ERROR:

Soly: based on the method, precision is about ± 1 % (compilers).

Temp: precision probably 0.1-0.2 K (compilers).

REFERENCES:

1. Zhuravlev, E.F.; Sheveleva, A.D. *Zh. Neorg. Khim.* 1960, *5*, 2630.

COMPONENTS:

(1) Praseodymium nitrate; $Pr(NO_3)_3$; [10361-80-3]

(2) Quinoline nitrate; $C_9H_8N_2O_3$; [21640-15-3]

(3) Water; H_2O; [7732-18-5]

ORIGINAL MEASUREMENTS:

Kuznetsova, L.S.; Zhuravlev, E.F. *Zh. Neorg. Khim.* 1977, 22, 515-9; *Russ. J. Inorg. Chem. Engl. Transl.* 1977, 22, 282-4.

VARIABLES:

Composition at 20°C and 40°C

PREPARED BY:

T. Mioduski and S. Siekierski

EXPERIMENTAL VALUES:

Composition of saturated solutions at 20°C [a]

$Pr(NO_3)_3$ mass %	$Pr(NO_3)_3$ mol kg^{-1}	$C_9H_7N\cdot HNO_3$ mass %	$C_9H_7N\cdot HNO_3$ mol kg^{-1}	nature of the solid phase
0	——	69.0	11.58	$C_9H_7N\cdot HNO_3$
3.2	0.32	66.5	11.42	"
6.0	0.64	65.2	11.78	$C_9H_7N\cdot HNO_3 + Pr(NO_3)_3\cdot 3C_9H_7N\cdot HNO_3\cdot H_2O$
8.2	0.77	59.3	9.49	$Pr(NO_3)_3\cdot 3C_9H_7N\cdot HNO_3\cdot H_2O$
15.7	1.07	39.6	4.61	"
20.3	1.23	29.1	2.99	"
23.0	1.31	23.2	2.24	"
24.1	1.33	20.4	1.91	"
29.0	1.50	12.0	1.06	"
35.5	1.82	4.8	0.42	"
38.5	2.03	3.5	0.31	"
60.4	4.79	1.0	0.13	$Pr(NO_3)_3\cdot 3C_9H_7N\cdot HNO_3\cdot H_2O + Pr(NO_3)_3\cdot 6H_2O$
60.5	4.78	0.8	0.11	$Pr(NO_3)_3\cdot 6H_2O$
60.7	4.72	0	——	"

a. Molalities calculated by M. Salomon.

continued.......

AUXILIARY INFORMATION

METHOD/APPARATUS/PROCEDURE:

The solubilities were studied by the method of isothermal sections with refractometric analyses (1). The refractive indices were not given in the source paper. Mixtures of known composition were thermostated, and equilibrium assertained by constancy in the refractive indices. Equilibrium in the heterogeneous systems was reached in 3-4 days.

SOURCE AND PURITY OF MATERIALS:

"Pure" grade $Pr(NO_3)_3\cdot 6H_2O$ was recrystallized from dil aq acid. Analysis for Pr and NO_3 confirmed the hexahydrate composition.

Quinoline nitrate was prepared by neutralization of "pure" grade quinoline with dil (1:5) c.p. grade HNO_3. The sln was evaporated to pptn. The ppt was filtered, recrystallized, and dried to constant mass. Chemical analysis of nitrate confirmed the composition of the salt.

Doubly distilled water was used.

ESTIMATED ERROR:

Soly: based on the method, precision is around ± 1 % (compilers).

Temp: precision about ± 0.1-0.2 K (compilers)

REFERENCES:

1. Zhuravlev, E.F.; Sheveleva, A.D. *Zh. Neorg. Khim.* 1960, 5, 2630.

COMPONENTS:

(1) Praseodymium nitrate; $Pr(NO_3)_3$; [10361-80-3]

(2) Quinoline nitrate; $C_9H_8N_2O_3$; [21640-15-3]

(3) Water; H_2O; [7732-18-5]

ORIGINAL MEASUREMENTS:

Kuznetsova, L.S.; Zhuravlev, E.F. *Zh. Neorg. Khim.* 1977, *22*, 515-9; *Russ. J. Inorg. Chem. Engl. Transl.* 1977, *22*, 282-4.

EXPERIMENTAL VALUES: continued.......

Composition of saturated solutions at 40°C [a]

$Ce(NO_3)_3$		$C_9H_7N \cdot HNO_3$		
mass %	mol kg^{-1}	mass %	mol kg^{-1}	nature of the solid phase
0	——	77.7	18.13	$C_9H_7N \cdot HNO_3$
2.3	0.32	76.0	18.22	"
5.1	0.78	74.9	19.49	"
6.9	1.08	73.6	19.64	$C_9H_7N \cdot HNO_3$ + $Pr(NO_3)_3 \cdot 3C_9H_7N \cdot HNO_3 \cdot H_2O$
12.5	1.28	57.6	10.02	$Pr(NO_3)_3 \cdot 3C_9H_7N \cdot HNO_3 \cdot H_2O$
22.6	1.63	35.0	4.30	"
27.2	1.76	25.5	2.81	"
31.5	1.89	17.5	1.79	"
35.0	2.04	12.6	1.25	"
37.8	2.22	10.0	1.00	"
52.5	3.61	3.0	0.35	"
65.1	5.96	1.5	0.23	$Pr(NO_3)_3 \cdot 3C_9H_7N \cdot HNO_3 \cdot H_2O$ + $Pr(NO_3)_3 \cdot 6H_2O$
65.2	5.83	0.6	0.09	$Pr(NO_3)_3 \cdot 6H_2O$
65.4	5.78	0	——	"

a. Molalities calculated by M. Salomon.

COMMENTS AND/OR ADDITIONAL DATA:

The composition of the double salt was determined graphically. It is congruently soluble. The double salt was isolated and analysed for Pr by the oxalate method, and for NO_3 by precipitation with nitron. The results in mass % units are:

found Pr 15.27, NO_3 40.00

calcd Pr 15.32, NO_3 40.38

The double salt was also studied by differential thermal analysis, and the results are discussed in the source paper.

COMPONENTS:

(1) Praseodymium nitrate; $Pr(NO_3)_3$; [10361-80-5]

(2) 8-Methylquinoline nitrate; $C_{10}H_{10}N_2O_3$; [60491-92-1]

(3) Water; H_2O; [7732-18-5]

ORIGINAL MEASUREMENTS:

Kuznetsova, L.S.; Zhuravlev, E.F. *Zh. Neorg. Khim.* 1977, *22*, 515-9; *Russ. J. Inorg. Chem. Engl. Transl.* 1977, *22*, 282-4.

VARIABLES:

Composition at 20°C and 40°C

PREPARED BY:

T. Mioduski and S. Siekierski

EXPERIMENTAL VALUES:

Composition of saturated solutions at 20°C [a]

$Pr(NO_3)_3$ mass %	$Pr(NO_3)_3$ mol kg^{-1}	$C_{10}H_9N \cdot HNO_3$ mass %	$C_{10}H_9N \cdot HNO_3$ mol kg^{-1}	nature of the solid phase
0	——	49.4	4.73	$C_{10}H_9N \cdot HNO_3$
5.9	0.34	41.0	3.74	"
13.0	0.76	35.0	3.26	"
16.5	1.00	33.0	3.17	"
20.7	1.32	31.5	3.20	"
28.0	2.00	29.2	3.31	$C_{10}H_9N \cdot HNO_3 + Pr(NO_3)_3 \cdot 2C_{10}H_9N \cdot HNO_3 \cdot 2H_2O$
28.7	1.94	26.0	2.78	$Pr(NO_3)_3 \cdot 2C_{10}H_9N \cdot HNO_3 \cdot 2H_2O$
29.4	1.95	24.5	2.58	"
30.1	1.92	22.0	2.23	"
32.0	1.93	17.2	1.64	"
38.2	2.16	7.7	0.69	"
44.9	2.59	2.0	0.18	"
50.0	3.10	0.6	0.06	"
60.7	4.72	0	——	$Pr(NO_3)_3 \cdot 6H_2O$

a. Molalities calculated by M. Salomon.

continued..........

AUXILIARY INFORMATION

METHOD/APPARATUS/PROCEDURE:

The solubilities were studied by the method of isothermal sections with refractometric analyses (1). The refractive indices were not given in the source paper. Mixtures of known composition were thermostated, and equilibrium assertained by constancy in the refractive indices. Equilibrium in the heterogeneous systems was reached in 3-4 days.

SOURCE AND PURITY OF MATERIALS:

"Pure" grade $Pr(NO_3)_3 \cdot 6H_2O$ was recrystallized from dil aq acid. Analysis for Pr and NO_3 confirmed the hexahydrate composition.

8-Methylquinoline nitrate was prepd by neutralization of "pure" grade 8-methylquinoline with dil (1:5) c.p. grade HNO_3. The sln was evaporated to pptn. The ppt was filtered, recrystallized and dried to constant mass. Chemical analysis of nitrate confirmed the composition of the salt.

Doubly distilled water was used.

ESTIMATED ERROR:

Soly: based on the method, precision is around ± 1 % (compilers).
Temp: precision about ± 0.1-0.2 K (compilers).

REFERENCES:

1. Zhuravlev, E.F.; Sheveleva, A.D. *Zh. Neorg. Khim.* 1960, *5*, 2630.

COMPONENTS:

(1) Praseodymium nitrate; $Pr(NO_3)_3$; [10361-80-5]

(2) 8-Methylquinoline nitrate; $C_{10}H_{10}N_2O_3$; [60491-92-1]

(3) Water; H_2O; [7732-18-5]

ORIGINAL MEASUREMENTS:

Kuznetsova, L.S.; Zhuravlev, E.F. *Zh. Neorg. Khim.* 1977, *22*, 515-9; *Russ. J. Inorg. Chem. Engl. Transl.* 1977, *22*, 282-4.

EXPERIMENTAL VALUES: continued........

Composition of saturated solutions at 40°C

$Pr(NO_3)_3$		$C_{10}H_9N \cdot HNO_3$		
mass %	mol kg^{-1}	mass %	mol kg^{-1}	nature of the solid phase
0	——	66.5	9.63	$C_{10}H_9N \cdot HNO_3$
3.7	0.34	62.7	9.05	"
8.0	0.77	60.4	9.27	"
12.6	1.34	58.7	9.92	"
16.8	2.04	58.0	11.16	"
23.2	3.70	57.6	14.55	$C_{10}H_9N \cdot HNO_3 + Pr(NO_3)_3 \cdot 2C_{10}H_9N \cdot HNO_3 \cdot 2H_2O$
25.5	3.29	50.8	10.39	$Pr(NO_3)_3 \cdot 2C_{10}H_9N \cdot HNO_3 \cdot 2H_2O$
29.8	3.01	39.9	6.39	"
33.3	2.74	29.5	3.85	"
35.0	2.70	25.3	3.09	"
37.0	2.63	20.0	2.26	"
40.3	2.69	13.8	1.46	"
47.7	3.12	5.6	0.58	"
54.6	3.81	1.6	0.18	"
61.0	4.85	0.5	0.06	"
65.0	5.75	0.4	0.06	$Pr(NO_3)_3 \cdot 2C_{10}H_9N \cdot HNO_3 \cdot 2H_2O + Pr(NO_3)_3 \cdot 6H_2O$
65.1	5.71	0	——	$Pr(NO_3)_3 \cdot 6H_2O$

a. Molalities calculated by M. Salomon.

COMMENTS AND/OR ADDITIONAL DATA:

The composition of the double salt was determined graphically. It is congruently soluble at 40°C and incongruently soluble at 20°C. The double salt was isolated and analysed for Pr by the oxalate method, and for NO_3 by precipitation with nitron. The results in mass % units are:

found Pr 18.20, NO_3 39.40

calcd Pr 18.20, NO_3 40.10

The double salt was also studied by differential thermal analysis, and the results are discussed in the source paper.

COMPONENTS:

(1) Praseodymium nitrate; $Pr(NO_3)_3$; [10361-80-5]

(2) Urea; CH_4N_2O; [57-13-6]

(3) Water; H_2O; [7732-18-5]

ORIGINAL MEASUREMENTS:

Khudaibergenova, N.; Sulaimankulov, K.S. *Zh. Neorg. Khim.* 1979, *24*, 2005-8; *Russ. J. Inorg. Chem. Engl. Transl.* 1979, *24*, 1112-4.

VARIABLES:

Composition at 30°C

PREPARED BY:

T. Mioduski and S. Siekierski

EXPERIMENTAL VALUES:

$Pr(NO_3)_3$		$CO(NH_2)_2$		
mass %	mol kg^{-1}	mass %	mol kg^{-1}	nature of the solid phase
62.12	5.016	——	——	$Pr(NO_3)_3 \cdot 6H_2O$
64.57	6.551	5.28	2.916	$Pr(NO_3)_3 \cdot 6H_2O + Pr(NO_3)_3 \cdot CO(NH_2)_2$
61.02	5.846	7.05	3.677	$Pr(NO_3)_3 \cdot 2CO(NH_2)_2$
58.80	5.958	11.01	6.073	"
55.37	6.504	18.59	11.89	"
53.77	7.099	23.06	16.57	"
52.31	7.569	26.55	20.91	"
52.49	8.876	29.42	27.08	"
62.74	325.3	36.67	1035.	"
50.88	13.80	37.84	55.86	$Pr(NO_3)_3 \cdot 4CO(NH_2)_2$
47.46	14.44	42.49	70.40	"
46.32	23.00	47.52	128.5	"
41.61	11.86	47.66	73.96	$Pr(NO_3)_3 \cdot 6CO(NH_2)_2$
41.48	13.97	49.44	90.66	"
38.85	12.03	51.27	86.41	"
38.73	23.41	56.21	185.0	$Pr(NO_3)_3 \cdot 6CO(NH_2)_2 + CO(NH_2)_2$
35.98	15.20	56.78	130.6	$CO(NH_2)_2$
28.75	5.962	56.50	63.78	"
14.75	1.556	56.25	32.30	"
——	——	57.50	22.53	"

AUXILIARY INFORMATION

METHOD/APPARATUS/PROCEDURE:

The isothermal method was used. Equilibrium was reached after 7-8 h. The urea nitrogen was determined by the Kjeldahl method, and Pr determined by complexometric titration using Xylenol Orange indicator.

The liquid phase was separated from the solid using a Schott No. 3 filter. The compilers assume that the solid phases were found by Schreinemakers' method.

The authors reported the solubilities in mass % units: conversions to molality were made by M. Salomon. All three double salts are congruently soluble in water.

SOURCE AND PURITY OF MATERIALS:

Nothing specified.

ESTIMATED ERROR:

Soly: nothing specified.

Temp: precision probably ± 0.1-0.2 K (compilers).

REFERENCES:

COMPONENTS:

(1) Praseodymium nitrate; $Pr(NO_3)_3$; [10361-80-5]

(2) Urea mononitrate; $CH_5N_3O_4$; [124-47-0]

(3) Water; H_2O; [7732-18-5]

ORIGINAL MEASUREMENTS:

Zhuravlev, E.F.; Kuznetsova, L.S. *Zh. Neorg. Khim.* 1978, *23*, 1688-91; *Russ. J. Inorg. Chem. Engl. Transl.* 1978, *23*, 929-31.

VARIABLES:

Composition at 20°C and 40°C

PREPARED BY:

T. Mioduski and S. Siekierski

EXPERIMENTAL VALUES:

	$Pr(NO_3)_3$		$CO(NH_2)_2 \cdot HNO_3$		
t/°C	mass %	mol kg^{-1}	mass %	mol kg^{-1}	nature of the solid phase
20	0.0	——	16.7	1.63	$CO(NH_2)_2 \cdot HNO_3$
	9.0	0.35	12.8	1.33	"
	18.3	0.77	9.3	1.04	"
	28.2	1.28	4.5	0.54	"
	38.0	2.06	5.5	0.79	"
	48.4	3.08	3.5	0.59	"
	60.1	4.88	2.2	0.47	$CO(NH_2)_2 \cdot HNO_3 + Pr(NO_3)_3 \cdot 6H_2O$
	60.4	4.80	1.1	0.23	$Pr(NO_3)_3 \cdot 6H_2O$
	60.5	4.69	0.0	——	"
40	0.0	——	25.5	2.78	$CO(NH_2)_2 \cdot HNO_3$
	8.0	0.34	20.1	2.27	"
	17.0	0.77	15.5	1.87	"
	26.5	1.32	12.0	1.59	"
	36.8	2.06	8.5	1.26	"
	47.0	3.09	6.5	1.14	"
	65.2	6.08	2.0	0.50	$CO(NH_2)_2 \cdot HNO_3 + Pr(NO_3)_3 \cdot 6H_2O$
	65.4	6.08	1.7	0.42	$Pr(NO_3)_3 \cdot 6H_2O$
	65.5	5.96	0.9	0.22	"
	65.6	5.83	0.0	——	"

Molalities calculated by M. Salomon

AUXILIARY INFORMATION

METHOD/APPARATUS/PROCEDURE:

The method of isothermal sections was used (1). No other information given. The compilers assume that the compositions of saturated solutions were determined by refractrometric analysis.

The phase diagram is of the simple eutonic type.

SOURCE AND PURITY OF MATERIALS:

$Pr(NO_3)_3 \cdot 6H_2O$ prepd by dissolving the oxide in c.p. grade HNO_3 followed by recrystn on a water bath. The hydrate was analysed by the oxalate method and found to contain 32.9 mass % Pr (theory: 32.4 mass %).

$CO(NH_2)_2 \cdot HNO_3$ prepd by neutralization of urea with nitric acid. The stoichiometry of the salt was checked by gravimetric nitrate analysis using nitron.

Doubly distilled water was used.

ESTIMATED ERROR:

Soly: based on the method, precision ia about ± 1 % (compilers).

Temp: precision about ± 0.1-0.2 K (compilers).

REFERENCES:

1. Zhuravlev, E.F.; Sheveleva, A.D. *Zh. Neorg. Khim.* 1960, *5*, 2630.

COMPONENTS:	EVALUATOR:
(1) Praseodymium nitrate; $Pr(NO_3)_3$; [10361-80-5] (2) Organic solvents	Mark Salomon U.S. Army Electronics Technology and Devices Laboratory Fort Monmouth, NJ, USA December 1982

CRITICAL EVALUATION:

The existing publications on the solubility of praseodymium nitrate in organic solvents were all published prior to 1935 (1-3). Most of the results are probably very imprecise, and detailed discussions on these results are therefore not waranted. The reader is referred directly to the compilations for additional information.

The only multiple study of a given system is that for $Pr(NO_3)_3$ in diethyl ether (2,3). The only interesting comparison that can made is that both publications agree that at 293 K the anhydrous salt has a very low solubility. Wells (3) states that the solubility of the anhydrous salt in ether at 293 K is zero, and from the diagram given by Hopkins and Quill (2), it appears as if the solubility is given as zero.

REFERENCES

1. Hardy, Z.M. *Masters Thesis*. The University of Illinois. Urbana, IL. 1931 (graphical data from this thesis were also published in reference 2 below).

2. Hopkins, B.S.; Quill, L.L. *Proc. Natl. Acad. Sci. U.S.A.* 1933, *19*, 64.

3. Wells, R.C. *J. Wash. Acad. Sci.* 1930, *20*, 146.

COMPONENTS:

(1) Praseodymium nitrate; $Pr(NO_3)_3$; [10361-80-5]

(2) Neodymium nitrate; $Nd(NO_3)_3$; [10045-95-1]

(3) 2-Methoxyethanol (methyl cellosolve); $C_3H_8O_2$; [109-86-4]

ORIGINAL MEASUREMENTS:

Hardy, Z.M.; *Masters Thesis*. The University of Illinois. Urbana, IL. 1932.[1]

VARIABLES:

Temperature

PREPARED BY:

Mark Salomon

EXPERIMENTAL VALUES:

Composition of initial solid is 83.4 mass % $Nd(NO_3)$ - 16.6 mass % $Pr(NO_3)_3$

t/°C	volume of aliquot c.c.	total mass of oxide g	soly fraction[a] mass % Nd	Pr	solubility[b] 10^2 mol kg^{-1} $Nd(NO_3)_3$	$Pr(NO_3)_3$
- 8	25	0.1310	79.4	19.1	2.47	0.59
9	25	0.2423	82.0	26.0	4.72	1.48
20	25	0.2383	92.3	10.0	5.23	0.56
26	50	0.4398	90.0	13.0	4.71	0.67
30	25	0.1322	83.2	15.9	2.62	0.49
37	25	0.2375	75.0	29.0	4.24	1.62
47	25	0.1286	85.5	9.7	2.61	0.29

a. Total mass % should equal 100.

b. Calculated by compiler. For conversions to mol kg^{-1} for $Pr(NO_3)_3$, the compiler assumed the oxide to be Pr_6O_{11}.

AUXILIARY INFORMATION

METHOD/APPARATUS/PROCEDURE:

Isothermal method used. One g of mixed solid and about 200 cc solvent in 250 c.c. bottle and sealed with a cork covered with tin foil. Rubber tubing was placed over the cork and neck of the bottle, and the open end sealed with a rubber stopper. The bottles were agitated in a thermostat for 24 h after which the cork was removed and the mixture permitted to settle for 3 h. Calibrated pipets were used to withdraw aliquots. Pr and Nd were pptd as the oxalate and ignited to the oxide at 600°C, and weighed. The mixed oxides were dissolved in nitric acid and analysed for Pr and Nd with a Hilger spectroscope by comparison to standard solutions of known $Pr(NO_3)_3$ and $Nd(NO_3)_3$ mass %.

SOURCE AND PURITY OF MATERIALS:

The Nd + Pr mixed nitrate was prepd by addn of HNO_3 to the oxides, evapn of the excess acid, and heating. The mixture was heated at 100°C for 6 d, 140°C for 2 d, and 160°C for 2 d.

The ether, $CH_3OCH_2CH_2OH$, was distilled from anhydrous $CaCl_2$.

C.p. grade oxalic acid and HNO_3 were used.

ESTIMATED ERROR:

Soly: nothing specified, but probably very large (compiler).

Temp: accuracy about ± 0.1 K (compiler).

REFERENCES:

1. Graphical data from Hardy's thesis were reported by Hopkins, B.S.; Quill, L.L. *Proc. Natl. Acad. Sci. U.S.A.* 1933, *19*, 64.

COMPONENTS:

(1) Praseodymium nitrate; $Pr(NO_3)_3$; [10361-80-5]

(2) Neodymium nitrate; $Nd(NO_3)_3$; [10045-95-1]

(3) 2-Ethoxyethanol (ethyl cellosolve, ethylene glycol ethyl ether); $C_4H_{10}O_2$; [110-80-5]

ORIGINAL MEASUREMENTS:

Hardy, Z. M. *Masters Thesis*. The University of Illinois. Urbana, IL. 1932.[1]

VARIABLES:

Temperature

PREPARED BY:

Mark Salomon

EXPERIMENTAL VALUES:

Composition of initial solid is 83.4 mass % $Nd(NO_3)$ - 16.6 mass % $Pr(NO_3)_3$

	volume of aliquot	total mass of oxide	soly fraction[a] mass %		solubility[b] 10^2 mol kg^{-1}	
t/°C	c.c.	g	Nd	Pr	$Nd(NO_3)_3$	$Pr(NO_3)_3$
- 8	25	0.1098	72.9	27.3	1.90	0.70
30	25	0.1106	76.9	22.0	2.02	0.57
47	25	0.1109	85.5	10.0	2.25	0.26

a. Total mass % should equal 100.

b. Calculated by the compiler. For conversions to mol kg^{-1} for $Pr(NO_3)_3$, the compiler assumed the oxide to be Pr_6O_{11}.

AUXILIARY INFORMATION

METHOD/APPARATUS/PROCEDURE:

Isothermal method used. One g of mixed solid and about 200 cc solvent placed in 250 c.c. bottle and sealed with a cork covered with tin foil. Rubber tubing was placed over the cork and neck of the bottle, and the open end sealed with a rubber stopper. The bottles were agitated in a thermostat for 24 h after which the cork was removed and the mixture permitted to settle for 3 h. Calibrated pipets were used to withdraw aliquots. Pr and Nd were pptd as the oxalate and ignited to the oxide at 600°C, and weighed. The mixed oxides were dissolved in nitric acid and analysed for Pr and Nd with a Hilger spectroscope by comparison to standard solutions of known $Pr(NO_3)_3$ and $Nd(NO_3)_3$ mass %.

SOURCE AND PURITY OF MATERIALS:

The Nd + Pr mixed nitrate was prepd by addn of HNO_3 to the oxides, evapn of excess acid, and heating. The mixture was heated at 100°C for 6 d, 140°C for 2 d, and 160°C for 2 d.

The ether, $C_2H_5OCH_2CH_2OH$, was distilled from anhydrous $CaCl_2$.

C.p. grade oxalic acid and HNO_3 were used.

ESTIMATED ERROR:

Soly: nothing specified, but probably very large (compiler).

Temp: accuracy about ± 0.1 K (compiler).

REFERENCES:

1. Graphical data from Hardy's thesis were reported by Hopkins, B.S.; Quill, L.L. *Proc. Natl. Acad. Sci. U.S.A.* 1933, *19*, 64.

COMPONENTS:

(1) Praseodymium nitrate; $Pr(NO_3)_3$; [10361-80-5]

(2) Diethyl ether; $C_4H_{10}O$; [60-29-7]

ORIGINAL MEASUREMENTS:

Wells, R.C. *J. Wash. Acad. Sci.* 1930, *20*, 146-8.

VARIABLES:

Room temperature (about 20°C)

PREPARED BY:

T. Mioduski, S. Siekierski, M. Salomon

EXPERIMENTAL VALUES:

Experiment 1. This experiment involves the hydrated lanthanum nitrate as the initial solid, and which the compilers assume to be the hexahydrate.

Authors report the solubility as 0.0004 g oxide in 10 ml ether.

Assuming the oxide to be Pr_6O_{11}, the solubility in volume units is

2.3×10^{-4} mol dm^{-3} (compilers).

Experiment 2. This experiment involves praseodymium nitrate dehydrated as described in the METHOD/APPARATUS/PROCEDURE box below.

Author reports the anhydrous nitrate to be insoluble in ether.

AUXILIARY INFORMATION

METHOD/APPARATUS/PROCEDURE:

The isothermal method was used. The soly of praseodymium nitrate was determined in two experiments in which the nature of the initial solid phase differs.

Experiment 1. A few grams of praseodymium nitrate (presumably the hexahydrate, compilers) was added to about 20 ml of ether in small stoppered flasks. The flasks were periodically agitated and permitted to stand at about 20°C overnight. A 10 ml sample was removed, filtered, the solvent evaporated and the salt ignited to the oxide and weighed.

Experiment 2. The remaining salt in the flask was freed from ether, dissolved in water and a few drops of HNO_3 added. The solution was evaporated to dryness and heated to 150°C. The solubility in ether was determined again with this "dehydrated" salt.

SOURCE AND PURITY OF MATERIALS:

Nothing specified.

ESTIMATED ERROR:

Soly: precision probably around ± 10 % (compilers).

Temp: precision probably ± 4 K (compilers).

REFERENCES:

COMPONENTS:

(1) Praseodymium nitrate; $Pr(NO_3)_3$; [10361-80-5]

(2) Neodymium nitrate; $Nd(NO_3)_3$; [10045-95-1]

(3) Diethyl ether; $C_4H_{10}O$; [60-29-7]

ORIGINAL MEASUREMENTS:

Hardy, Z.M. *Masters Thesis*. The University of Illinois. Urbana, IL. 1932.[1]

VARIABLES:

Temperature

PREPARED BY:

Mark Salomon

EXPERIMENTAL VALUES:

Composition of initial solid is 83.4 mass % $Nd(NO_3)_3$ - 16.6 mass % $Pr(NO_3)_3$

t/°C	volume of aliquot c.c.	total mass of oxide g	soly fraction[a] mass % Nd	Pr	solubility[b] 10^3 mol kg^{-1} $Nd(NO_3)_3$	$Pr(NO_3)_3$
		Experiment No. 1				
- 8	100	0.0426	76.3	23.5	1.93	0.59
9	100	0.0434	97.0	4.0	2.50	0.10
20	100	0.0647	95.8	5.0	3.68	0.19
26	150	0.1224	89.0	12.0	4.32	0.58
30	100	0.0611	84.5	18.5	3.07	0.66
34	150	0.0704	89.0	14.0	2.48	0.39
		Experiment No. 2				
- 8	100	0.0392	86.7	12.0	2.02	0.28
20	100	0.0614	81.3	9.7	2.97	0.35
26	100	0.1781	89.8	13.0	9.51	1.36
30	100	0.0692	86.7	14.5	3.57	0.59

a. Total mass % should equal 100.

b. Calculated by the compiler. For conversions to mol kg^{-1} for $Pr(NO_3)_3$, the compiler assumed the oxide to be Pr_6O_{11}.

AUXILIARY INFORMATION

METHOD/APPARATUS/PROCEDURE:

Two isothermal experiments were reported. Experiment 1. One g of mixed solid and about 200 cc solvent placed in 250 c.c. bottle and sealed with a cork covered with tin foil. Rubber tubing was placed over the cork and neck of the bottle, and the open end sealed with a rubber stopper. The bottles were agitated in a thermostat for 24 h after which the cork was removed and the mixture permitted to settle for 3 h. Calibrated pipets were used to withdraw aliquots. Pr and Nd were pptd as the oxalate and ignited to the oxide at 600°C, and weighed. The mixed oxides were dissolved in nitric acid and analysed for Pr and Nd with a Hilger spectroscope by comparison to standard solutions of known $Pr(NO_3)_3$ and $Nd(NO_3)_3$ mass %.

Experiment 2. Saturated solutions were prepd as described above. The saturated solutions were then drawn off and placed in new bottles with 1 g of the mixed nitrate solids. The solutions were then equilibrated and analysed as described above.

SOURCE AND PURITY OF MATERIALS:

The Nd + Pr mixed nitrate was prepd by addn of HNO_3 to the oxides, evapn of excess acid, and heating. The mixture was heated at 100°C for 6 d, 140°C for 2 d, and 160°C for 2 d.

Diethyl ether was treated with a satd aqueous $CaCl_2$ sln for 1 h followed by drying over P_2O_5 for 2 h. Several c.c.'s of ethyl magnesium bromide were added and allowed to react. The ether was then distilled and stored over P_2O_5.

C.p. grade oxalic acid and HNO_3 were used.

ESTIMATED ERROR:

Soly: nothing specified, but probably very large (compiler).

Temp: accuracy about ± 0.1 K (compiler).

REFERENCES:

1. Graphical data from Hardy's thesis were reported by Hopkins, B.S.; Quill, L.L. *Proc. Natl. Acad. Sci. U.S.A.* 1933, *19*, 64.

COMPONENTS:

(1) Praseodymium nitrate; $Pr(NO_3)_3$; [10361-80-5]

(2) Diethyl ether; $C_4H_{10}O$; [60-29-7]

ORIGINAL MEASUREMENTS:

Hopkins, B.S.; Quill, L.L. *Proc. Natl. Acad. Sci. U.S.A.* 1933, *19*, 64-8.

VARIABLES:

Temperature

PREPARED BY:

T. Mioduski, S. Siekierski, M. Salomon

EXPERIMENTAL VALUES:

The solubility of $Pr(NO_3)_3$ in diethyl ether as a function of temperature was given in the form of a small diagram. In the absence of numerical data, the compilers interpolated the solubilities from the published diagram. The results are:

t/°C	Solubility 10^3 g dm^{-3}	10^5 mol dm^{-3} [a]
20	0	——
22	0	——
24	2.8	0.9
27.5	6.0	1.8
28.4	7.2	2.2
29.5	7.8	2.4
30	8.3	2.5

a. Calculated by the compilers.

COMMENTS AND/OR ADDITIONAL DATA:

The name Philip Kalischer appears on the diagram published in the source paper. The compilers suspected that Mr. Kalischer was an MSc student of Prof. Hopkins and thus contacted Ms. Susanne Redalje, the Assistant Chemistry Librarian at the University of Illinois at Urbana-Champaign. Ms. Redalje searched the University records for references to a thesis or any publication by Mr. Kalischer. The records show that Mr. Kalischer attended classes for the Fall, Spring, and Summer semesters of 1930-1931. There is no indication that Mr. Kalischer had finished his studies or submitted a thesis, and it is therefore apparent that the original experimental data are lost. The compilers are most grateful to Ms. Redalje for all her help in searching the University records and providing important information on numerous other lanthanide systems.

AUXILIARY INFORMATION

METHOD APPARATUS/PROCEDURE:

No information is available, but based on similar work by Hardy (1) being carried out at the University of Illinois at the time, it is likely that the isothermal method was employed. The solubility data for neodymium and praseodymium nitrates in several ethers from Hardy's MSc Thesis are compiled elsewhere in this volume, and the compilations contain detailed information on the experimental techniques which the compilers assume were similar to those used by Mr. Kalischer.

SOURCE AND PURITY OF MATERIALS:

No information available.

ESTIMATED ERROR:

No information available.

REFERENCES:

1. Hardy, Z.M. *Masters Thesis*. The University of Illinois. Urbana, IL. 1932.

COMPONENTS:

(1) Praseodymium nitrate; $Pr(NO_3)_3$; [10361-80-5]

(2) Neodymium nitrate; $Nd(NO_3)_3$; [10045-95-1]

(3) Diisopropyl ether; (isopropyl ether); $C_6H_{14}O$; [108-20-3]

ORIGINAL MEASUREMENTS:

Hardy, Z.M. *Masters Thesis*. The University of Illinois. Urbana, IL. 1932.[1]

VARIABLES:

Temperature

PREPARED BY:

Mark Salomon

EXPERIMENTAL VALUES:

Composition of initial solid is 83.4 mass % $Nd(NO_3)_3$ - 16.6 mass % $Pr(NO_3)_3$

t/°C	volume of aliquot c.c.	total mass of oxide g	soly fraction[a] mass % Nd	Pr	solubility[b] 10^3 mol kg^{-1} $Nd(NO_3)_3$	$Pr(NO_3)_3$
- 7	30	0.0116	85.47	14.9	1.96	0.34
9	50	0.0524	95.0	4.0	5.92	0.25
20	100	0.0730	89.0	10.0	3.86	0.43
26	100	0.0561	90.0	10.0	3.00	0.33
30	30	0.0144	80.0	17.3	2.28	0.49
37	100	0.0322	90.0	8.8	1.72	0.17
47	25	0.0095	89.0	not detected	2.01	(0.25) [c]

a. Total mass % should equal 100.

b. Calculated by the compiler. For conversions to mol kg^{-1} for $Pr(NO_3)_3$, the compiler assumed the oxide to be Pr_6O_{11}.

c. Since Pr was not detected by the spectroscopic technique, the compiler assumed its mass % to equal 0.11 (i.e.; 100-mass % Nd).

AUXILIARY INFORMATION

METHOD/APPARATUS/PROCEDURE:

The isothermal method was used. One g of mixed solid and about 200 cc solvent placed in 250 c.c. bottle and sealed with cork covered with tin foil. Rubber tubing was placed over the cork and neck of the bottle, and the open end sealed with a rubber stopper. The bottles were agitated in a thermostat for 24 h after which the cork was removed and the mixture permitted to settle for 3 h. Calibrated pipets were used to withdraw aliquots. Pr and Nd were pptd as the oxalate and ignited to the oxide at 600°C, and weighed. The mixed oxides were dissolved in nitric acid and analysed for Pr and Nd with a Hilger spectroscope by comparison to standard solutions of known $Pr(NO_3)_3$ and $Nd(NO_3)_3$ mass %.

SOURCE AND PURITY OF MATERIALS:

The Nd + Pr mixed nitrate was prepd by addn of HNO_3 to the oxides, evapn of excess acid, and heating. The mixture was heated at 100°C for 6 d, 140°C for 2 d, and 160°C for 2 d.

Two sources. (1) Isopropyl ether prepd by distn from ethyl magnesium bromide; (2) prepd from the alcohol and H_2SO_4 followed by neutn with aq NaOH, drying with anhydr $CaCl_2$, and distn.

C.p. grade oxalic acid and HNO_3 were used.

ESTIMATED ERROR:

Soly: nothing specified, but probably very large (compiler).

Temp: accuracy about ± 0.1 K (compiler).

REFERENCES:

1. Graphical data from Hardy's thesis were reported by Hopkins, B.S.; Quill, L.L. *Proc. Natl. Acad. Sci. U.S.A.* 1933, *19*, 64.

COMPONENTS:

(1) Praseodymium nitrate; $Pr(NO_3)_3$; [10361-80-5]

(2) 1,4-Dioxane (p-dioxane); $C_4H_8O_2$; [123-91-1]

ORIGINAL MEASUREMENTS:

Hopkins, B.S.; Quill, L.L. *Proc. Natl. Acad. Sci. U.S.A.* 1933, *19*, 64-8.

VARIABLES:

One temperature: 25°C

PREPARED BY:

T. Mioduski, S. Siekierski, M. Salomon

EXPERIMENTAL VALUES:

The solubility of $Pr(NO_3)_3$ in p-dioxane at 25°C was given in the form of a small diagram of solubility vs atomic number Z for Z = 57-64. In the absence of numerical data, the compilers interpolated the solubility from the published diagram. The result is:

$$\text{soly of } Pr(NO_3)_3 = 3.2 \text{ g dm}^{-3} \text{ (0.0098 mol dm}^{-3})$$

COMMENTS AND/OR ADDITIONAL DATA:

It appears that the original experimental work was done by a Mr. P. Kalischer who was a student at the University of Illinois at Urbana-Champaign. Attempts to locate the original experimental data have failed, and it thus appears that these data are lost (see COMMENTS in the compilation for the $Pr(NO_3)_3$-diethyl ether system).

AUXILIARY INFORMATION

METHOD/APPARATUS/PROCEDURE:

No information is abailable, but based on similar work by Hardy (1) being carried out at the University of Illinois at the time, it is likely that the isothermal method was employed. The solubility data for neodymium and praseodymium nitrates in several ethers from Hardy's MSc Thesis are compiled elsewhere in this volume, and the compilations contain detailed information on the experimental techniques which the compilers assume were similar to those used by Mr. Kalischer.

SOURCE AND PURITY OF MATERIALS:

No information available.

ESTIMATED ERROR:

No information available.

REFERENCES:

1. Hardy, Z.M. *Masters Thesis*. The University of Illinois. Urbana, IL. 1932.

COMPONENTS:

(1) Praseodymium double nitrates

(2) Water ; H_2O ; [7732-18-5]

EVALUATOR:

Mark Salomon

U.S. Army Electronics Technology and Devices Laboratory
Fort Monmouth, NJ, USA

December 1982

CRITICAL EVALUATION:

$Pr(NO_3)_3$ DOUBLE SALTS WITH INORGANIC NITRATES

INTRODUCTION

Studies on the direct determination of the solubilities of $Pr(NO_3)_3$ double nitrates involving inorganic nitrates are relatively few (1-3). The double nitrates reported in (1-3) are all characterized by the formation of the following tetracosahydrate solid phases:

$3Mg(NO_3)_2 \cdot Pr(NO_3)_3 \cdot 24H_2O$	[19478-66-1]	(1-3)
$2Pr(NO_3)_3 \cdot 3Mn(NO_3)_2 \cdot 24H_2O$	[34216-91-6]	(1,2)
$2Pr(NO_3)_3 \cdot 3Co(NO_3)_2 \cdot 24H_2O$	[34342-98-8]	(1,2)
$2Pr(NO_3)_3 \cdot 3Ni(NO_3)_2 \cdot 24H_2O$	[36153-28-3]	(1,2)
$2Pr(NO_3)_3 \cdot 3Cu(NO_3)_2 \cdot 24H_2O$	[84682-61-1]	(2)
$2Pr(NO_3)_3 \cdot 3Zn(NO_3)_2 \cdot 24H_2O$	[28876-81-5]	(1,2)

EVALUATION PROCEDURE

Where possible, the solubility data were fitted by least squares to the smoothing equation

$$Y = \ln(m/m_o) - nM_2(m - m_o) = a + b/(T/K) + c \ln(T/K) \quad [1]$$

All terms in eq. [1] have been previously defined (see eq. [1] in the $Pr(NO_3)_3$-H_2O critical evaluation). Due to the absense from the literature of sufficient publications to provide bases for critical comparisons of most data, and due to the large experimental errors associated with existing data, a detailed statistical treatment of the data is not possible. A simplified method of estimating the accuracy of the data was therefore adopted. The solubility data were fitted to eq. [1] and the value of the congruent melting point calculated. If the congruent melting point calculated from eq. [1] is in agreement with the experimental value of the melting point (within experimental and calculated errors), we consider this as strong support for designating the solubility data as either *tentative* or *recommended*. In most cases considered below, the least squares fitted data are fairly precise (i.e. standard errors of estimate, σ_m, are generally small), and the accuracy in the smoothed solubility data is governed mainly by the experimental errors.

Jantsch (1) reported the solubilities of a number of double nitrates in concentrated nitric acid solution of density = 1.325 kg m^{-3} at 289 K. The solubilities were determined in this concentrated HNO_3 solution because the author did not have sufficient quantity of the double nitrates to determine solubilities in pure water. One of the most useful results reported by Jantsch are the congruent melting points of the hydrated salts. Jantsch determined the solubilities by a gravimetric method by precipitation of Pr as the hydroxide followed by ignition to the oxide. He assumed that the resulting oxide was Pr_4O_7 but reported the oxide content in terms of mass Pr_2O_3. Since the oxide obtained upon ignition of $Pr(OH)_3$ was probably Pr_6O_{11}, the compiler had to recalculate the mass of Pr_4O_7 based upon Jantsch's data for mass Pr_2O_3, and an error of unknown magnitude arises due to the unknown differences in atomic masses used by Jantsch and by the compiler. The compiler's calculated mass of Pr_4O_7 actually corresponds to the experimental mass of oxide which is Pr_6O_{11}, and the compiler's calculations of solubilities in the compilations on Jantsch's studies were based upon this treatment. The total correction to the solubilities reported by Jantsch amounts to about + 3% which is close to the estimated error of around 1-5% in Jantsch's results.

COMPONENTS:	EVALUATOR:
(1) Praseodymium double nitrates (2) Water ; H_2O ; [7732-18-5]	Mark Salomon U.S. Army Electronics Technology and Devices Laboratory Fort Monmouth, NJ, USA December 1982

CRITICAL EVALUATION:

The data of Prandtl and Ducrue (2) were assigned a precision of around ± 0.4% at best based upon the highest reproducibilities achieved by these authors. Considering the unknown error in temperature (not reported by the authors) and average reproducibility of the analyses, the total uncertainty in Prandtl and Ducrue's results is probably around ± 1%.

Friend and Wheat's results must be carefully reviewed because of our previous experience with Friend's data which contain systematic errors. In the study on magnesium praseodymium nitrate (3), the authors experienced problems with the gravimetric analysis of praseodymium oxide: they state that 80% of the oxide was completely oxidized to PrO_2, and that this was taken into account in computing the solubilities from the experimental mass of oxide. The authors do not provide sufficient information which can be used to recalculate and thus check these data. A precision of around ± 3% was estimated for these results.

Magnesium praseodymium nitrate. Solubility data for this double salt has been reported in the three publications (1-3). The data of Prandtl and Ducrue (2) and Friend and Wheat (3) were fitted to the smoothing equation, and the results are given in Table 1. The precision of the fit is very good: $\sigma_m = 0.01$ and all residual errors, $m_{obs} - m_{calcd}$, are within $\pm \sigma_m$. The predicted congruent melting point is 383.0 K which is in very good agreement with the observed value (1) of 384.4 K. These smoothed data are therefore designated as *recommended* values, and the smoothed data at selected temperatures are given in Table 2. The total uncertainty in the smoothed (*recommended*) data is around ± 3%.

Double nitrates with Mn, Co, Ni, and Zn nitrates. For each double salt, the only comparable data available are the solubilities as a function of temperature reported by Prandtl and Ducrue (2), and the congruent melting points of the tetracosahydrates reported by Jantsch (1). All the solubility data were fitted to eq. [1], and the results are given in Table 1. For all cases except $2Pr(NO_3)_3 \cdot 3Ni(NO_3)_2$, there is good agreement between the predicted melting points for the tetracosahydrates and the experimental melting points. The evaluator regards this agreement as sufficient justification to designate the smoothed solubilities for all double salts (except the double salt with $Ni(NO_3)_2$) as *tentative* data. The smoothed (*tentative*) solubility data at selected temperatures are given in Table 2.

The failure of the solubility data for praseodymium nickel nitrate to predict the observed melting point of the tetracosahydrate when fitted to the smoothing equation suggests a large error in the solubility data rather than in the experimental melting point. The values of the constants a, b,c (see Table 1) appear to be trivial, and the positive value for the constant "a" suggests an unlikely positive value for the heat of solution. Because only four data points were used in the least squares fitting to eq. [1], it is highly probable that one imprecise datum would invalidate the least squares treatment. Thus while some or most of the data for $2Pr(NO_3)_3 \cdot 3Ni(NO_3)_2$ may be acceptable (i.e. accurate to within about ± 3-4%), the uncertainty as to which of these data may be highly inaccurate leads the evaluator to the conclusion that none of the data can be assigned the *tentative* designation.

$Pr(NO_3)_3$ DOUBLE NITRATES WITH ORGANIC NITRATES

A number of praseodymium double nitrates with organic nitrates have been identified in the preceeding sections. While some of these double salts are congruently soluble, there are no studies available dealing with the direct determination of the solubilities of these salts. For details on the specific salts which form stable solid phases, the reader is referred to the section on the compilations of ternary aqueous $Pr(NO_3)_3$ systems.

REFERENCES

1. Jantsch, G. *Z. Anorg. Chem.* 1912, *76*, 303.
2. Prandtl, W.; Ducrue, H. *Z. Anorg. Chem.* 1926, *150*, 105.
3. Friend, J.N.; Wheat, W.N. *J. Chem. Soc.* 1935, 356.

COMPONENTS:

(1) Praseodymium double nitrates

(2) Water ; H_2O ; [7732-18-5]

EVALUATOR:

Mark Salomon

U.S. Army Electronics Technology and Devices Laboratory
Fort Monmouth, NJ, USA

December 1982

CRITICAL EVALUATION:

Table 1. Smoothing equation parameters for $2Pr(NO_3)_3 \cdot 3M(NO_3)_2$ solubilities.

parameter	M = Mg	Mn	Co	Ni	Zn
a	-32.665	-52.869	-37.264	0.245	-36.696
b	1142	2061.5	1392	-388	1253
c	5.043	8.0623	5.778	0.180	5.6980
σ_m	0.01	0.002	0.01	0.01	0.007
tetracosahydrate melting point/K					
calcd (eq. [1])	383.0	350.8	372.2	405.3	362.1
obsd (ref 1)	384.4	354.2	370.2	381.2	364.7

Table 2. Recommended and tentative solubility data at selected temperatures calculated from equation [1].

	solubility of $2Pr(NO_3)_3 \cdot 3M(NO_3)_2$/mol kg^{-1} [a]			
T/K	M = Mg[b]	Mn[c]	Co[c]	Zn[c]
273.2	0.693	0.880	0.771	0.719
283.2	0.727	0.915	0.805	0.766
293.2	0.769	0.965	0.847	0.823
298.2	0.792	0.996	0.872	0.855
303.2	0.818	1.032	0.899	0.891
313.2	0.876	1.122	0.963	0.973
323.2	0.944	1.242	1.040	1.075
333.2	1.026	1.410	1.137	1.203
343.2	1.125	1.678	1.258	1.374
353.2	1.248		1.420	1.631
363.2	1.409		1.663	
373.2	1.644			

[a] In all cases the solid phase is the tetracosahydrate.

[b] *Recommended* solubility data.

[c] *Tentative* solubility data.

COMPONENTS:

(1) Magnesium praseodymium nitrate; $3Mg(NO_3)_2 \cdot Pr(NO_3)_3$; [32074-07-0]

(2) Nitric acid; HNO_3; [7697-37-2]

(3) Water ; H_2O ; [7732-18-5]

ORIGINAL MEASUREMENTS:

Jantsch, G. *Z. Anorg. Chem.* 1912, *76*, 303-23.

VARIABLES:

One temperature: 16°C

PREPARED BY:

Mark Salomon

EXPERIMENTAL VALUES: Soly of the double salt in HNO_3 sln of density d_4^{16} = 1.325 g cm^{-3}.

aliquot volume	Pr_2O_3 [a]	Pr_6O_{11} [b]	soly of $3Mg(NO_3)_2 \cdot 2Pr(NO_3)_3$ [c]
cm^3	g	g	mol dm^{-3}
1.4638	0.0239	0.0247	0.0502
1.4638	0.0246	0.0254	

a. Original values reported by the author. The author states that the oxide produced upon ignition of the pptd hydroxide is Pr_4O_7, and he evidently converted the mass of oxide (from Pr_4O_7) to the equiv mass of Pr_2O_3. However, in the opinion of the compiler, the oxide produced upon ignition of the hydroxide at "normal" atmospheric conditions is Pr_6O_{11}, and the author's conversion to equiv mass of Pr_2O_3 is therefore in error.

b. Assuming that the oxide produced upon ignition of $Pr(OH)_3$ is Pr_6O_{11}, the reported mass of Pr_2O_3 based upon the author's assumption that the original oxide was Pr_4O_7 is incorrect. To obtain the correct mass of Pr_6O_{11} produced upon ignition of $Pr(OH)_3$, the author's value of mass Pr_2O_3 must be multiplied by the factor 1.0323. Thus the values of mass Pr_6O_{11} in the table were calculated by the compiler by multiplying the author's values for mass Pr_2O_3 by the appropriate factor.

c. Average value calculated by the compiler based upon the corrected mass of oxide (i.e. based on the mass Pr_6O_{11}).

ADDITIONAL DATA:

The melting point of the tetracosahydrate is 111.2°C, and the density of the double salt at 0°C is 2.0195 g cm^{-3}.

AUXILIARY INFORMATION

METHOD/APPARATUS/PROCEDURE:

Isothermal method used. The soly was studied in HNO_3 sln of density 1.325 g cm^{-3} at 16°C because the author did not have sufficient quantity of the rare earth to study the soly of the salt in pure water. Pulverized salt and HNO_3 sln were placed in glass-stoppered tubes and thermostated at 16°C for 24 h with periodic shaking. The solution was then allowed to settle for 2 h, and a pipet maintained at 16°C was used to withdraw aliquots for analysis. Two analyses were performed.

Solutions were analysed by adding 2-3 g NH_4Cl and 10% NH_3 sln followed by boiling to ppt the hydroxide. The ppt was filtered, dissolved in HNO_3, reprecipitated as the hydroxide, and ignited to the oxide. Mg in the filtrate was "determined by the usual method" (no details were given).

An attempt to determine the waters of hydration by dehydration was not successful because the temperature required (120°C or higher) resulted in decomposition of the salt with the formation of basic salts. Presumably the waters of hydration were found by difference.

SOURCE AND PURITY OF MATERIALS:

"Pure" praseodymium oxide was dissolved in dil HNO_3 and $Mg(NO_3)_2$ added to give a mole ratio of Pr/Mg = 2/3. The sln was evapd and a small crystal of $Bi_2Mg_3(NO_3)_{12}$ added, and the mixt cooled to ppt the tetracosahydrate. The double nitrate was recrystd before use. The double salt was analysed gravimetrically for praseodymium oxide. A 0.3668 g sample of the tetracosahydrate yielded 0.0816 g oxide (i.e. 22.25 mass %). This is in fair agreement with the author's theor value of 22.03 mass % based upon the assumption the oxide is Pr_4O_7. However since the oxide was probably Pr_6O_{11}, the compiler calculated the theor oxide content as 22.24 mass % which is in much better agreement with the experimental value. Analysis for NO gave 23.44 mass %: theor value is 23.52 mass %.

ESTIMATED ERROR:

Soly: reproducibility about ± 1-5% (compiler).

Temp: nothing specified

COMPONENTS:

(1) Magnesium praseodymium nitrate ; $3Mg(NO_3)_2 \cdot 2Nd(NO_3)_3$; [32074-07-0]

(2) Water ; H_2O ; [7732-18-5]

ORIGINAL MEASUREMENTS:

Prandtl, W.; Ducrue, H.

Z. Anorg. Chem. 1926, *150*, 105-16.

VARIABLES:

Temperature

PREPARED BY:

Mark Salomon

EXPERIMENTAL VALUES:

solubility

			oxides		double salt			
	mole ratio[a]	density	Pr_2O_3	MgO	hydrate[b]	hydrate[c]	anhydrous salt[d]	
t/°C	MgO/Pr_2O_3	kg m^{-3}	mass %	mass %	mass %	mass %	mass %	mol kg^{-1}
15	2.99[e]	1.49	13.55	4.97	63.0	62.92	45.15	0.749
30	2.99[e]	1.52	14.29	5.24	66.4	66.34	47.61	0.827
50	2.97	1.55	15.27	5.55	71.0	70.58	50.65	0.934
70	3.01[f]	1.61	16.65	6.14	77.5	77.52	55.63	1.141

a. Experimental value: theoretical value = 3.00.
b. Authors' values apparently based on mass % Pr_2O_3. The hydrate which is the equilibrium solid phase is the tetracosahydrate $2Pr(NO_3)_3 \cdot 3Mg(NO_3)_2 \cdot 24H_2O$.
c. Compiler's calculations based on average from mass % Pr_2O_3 and MgO.
d. Compiler's calculations based on results from c above.
e. Compiler computes 3.00.
f. Compiler computes 3.02.

AUXILIARY INFORMATION

METHOD/APPARATUS/PROCEDURE:

Isothermal method used. Pulverized double salt (hydrate) and conductivity water were placed in two 50 cc flasks and agitated for 1 day in a thermostat. The slns were then permitted to settle and aliquots of approx 4 cc removed with pipets maintained at the same temp as the satd slns. The aliquots were placed in graduated flasks and weighed, and then diluted with 50 cc of water for analysis. The results for densities and mass % of oxides are the mean of two determinations. The mass % of the tetracosahydrate was apparently calculated by the authors from the mass % Pr_2O_3: i.e. the mass % MgO was not considered.

Both metals were determined gravimetrically.

Pr was precipitated as the oxalate, filtered and ignited to the oxide. Mg in the filtrate was precipitated as $MgNH_4PO_4$, and presumably ignited to the pyrophosphate $Mg_2P_2O_7$.

SOURCE AND PURITY OF MATERIALS:

Pr_2O_3 prepared by W. Prandtl was analysed by X-ray spectroscopy and found to be "very pure," particularly with respect to lanthanum. It was dissolved in nitric acid and the required amount of commercial "pure" grade $Mg(NO_3)_2$ added. The solution was evaporated to crystallization, and the double salt recrystallized several times from conductivity water. The salt was dried over $CaCl_2$ in a desiccator to give the tetracosahyrdate. Results of the analysis of the double salt are:
Pr_2O_3 found 20.83 %, calcd 21.54 mass %.
MgO found 7.68 %, calcd 7.90 mass %.
Conductivity water was used.

ESTIMATED ERROR:

Soly: precision ± 0.4 % at best (compiler).

Temp: not specified

COMPONENTS:

(1) Magnesium praseodymium nitrate ; $3Mg(NO_3)_2 \cdot 2Pr(NO_3)_3$; [32074-07-0]

(2) Water; H_2O; [7732-18-5]

ORIGINAL MEASUREMENTS:

Friend, J.N.; Wheat, W.N. *J. Chem. Soc.* 1935, 356-9.

VARIABLES:

Temperature

PREPARED BY:

Mark Salomon

EXPERIMENTAL VALUES:

t/°C	$3Mg(NO_3)_2 \cdot 2Pr(NO_3)_3 \cdot 24H_2O$ mass %	$3Mg(NO_3)_2 \cdot 2Pr(NO_3)_3$ [b] mass %	mol kg^{-1}
17.8	63.1	45.3	0.753
37.4	67.74	48.61	0.861
61.8	74.25	53.28	1.038
74.6	78.39	56.25	1.170
111.2 [a]	100	71.76	2.313

a. Melting point of the tetracosahydrate (2).

b. Compiler's calculations.

AUXILIARY INFORMATION

METHOD/APPARATUS/PROCEDURE:

The isothermal method was used as described in (1). In order to obtain reproducible results, equilibrium had to be approached from above. Satd or near satd slns were prepd at 90°C to 100°C and quickly cooled in a thermostat. The slns were stirred for several h, which the authors claim is essential to remove supersatn.

Pr was detd by pptn as the oxalate followed by ignition in a Pt crucible to Pr_6O_{11}. This result was not given in the source paper, but was used to calc the mass % of the tetracosahydrate given in the above table. MgO content of each sln was calcd by method described below.

Three std slns of Pr and Mg nitrates were evaporated to dryness and ignited to the oxides. The resulting weight of Pr_6O_{11} was 1.30% higher than expected suggesting that 80% of Pr_6O_{11} was completely oxidized to PrO_2. In calculating the mass % MgO content in the satd slns by difference, this correction was applied. However, the authors do not report the mass % values for either MgO or Pr_6O_{11} for satd slns. From examples of results of typical Pr_6O_{11} analyses, the compiler ests average error of ± 4 % in the solubility.

SOURCE AND PURITY OF MATERIALS:

The tetracosahydrate, $3Mg(NO_3)_2 \cdot 2Pr(NO_3)_3 \cdot 24H_2O$, was prepared by dissolving stoichiometric quantities of Pr_6O_{11} and MgO in dilute nitric acid and crystallizing. The oxide Pr_6O_{11} was a "pure" grade commerical product which was dissolved in nitric acid, precipitated as the oxalate, and ignited to the oxide. The source and purity of MgO was not specified.

The source and purity of water was not specified.

ESTIMATED ERROR:

Soly: precision ± 3 % at best (compiler).

Temp: accuracy probably ± 0.05 K as in (1) (compiler).

REFERENCES:

1. Friend, J.N. *J. Chem. Soc.* 1930, 1633.

2. Jantsch, G. *Z. Anorg. Chem.* 1912, *76*, 303.

COMPONENTS:

(1) Praseodymium manganese nitrate; $2Pr(NO_3)_3 \cdot 3Mn(NO_3)_2$; [84682-58-6]

(2) Nitric acid; HNO_3; [7697-37-2]

(3) Water ; H_2O ; [7732-18-5]

ORIGINAL MEASUREMENTS:

Jantsch, G. *Z. Anorg. Chem.* 1912, *76*, 303-23.

VARIABLES:

One temperature: 16°C

PREPARED BY:

Mark Salomon

EXPERIMENTAL VALUES:

Soly of the double salt in HNO_3 sln of density d_4^{16} = 1.325 g cm^{-3}.

aliquot volume	Pr_2O_3 [a]	Pr_6O_{11} [b]	soly of $2Pr(NO_3)_3 \cdot 3Mn(NO_3)_2$ [c]
cm^3	g	g	mol dm^{-3}
1.4538	0.0699	0.0722	
1.4638	0.0691	0.0713	0.1329

a. Original values reported by the author. The author states that the oxide produced upon ignition of the pptd hydroxide is Pr_4O_7, and he evidently converted the mass of oxide (from Pr_4O_7) to the equiv mass of Pr_2O_3. However, in the opinion of the compiler, the oxide produced upon ignition of the hydroxide at "normal" atmospheric conditions is Pr_6O_{11}, and the author's conversion to equiv mass of Pr_2O_3 is therefore in error.

b. Assuming that the oxide produced upon ignition of $Pr(OH)_3$ is Pr_6O_{11}, the reported mass of Pr_2O_3 based upon the author's assumption that the original oxide was Pr_4O_7 is incorrect. To obtain the correct mass of Pr_6O_{11} produced upon ignition of $Pr(OH)_3$, the author's value of mass Pr_2O_3 must be multiplied by the factor 1.0323. Thus the values of mass Pr_6O_{11} in the table were calculated by the compiler by multiplying the author's values for mass Pr_2O_3 by the appropriate factor.

c. Average value calculated by the compiler based upon the corrected mass of oxide (i.e. based on the mass Pr_6O_{11}).

ADDITIONAL DATA:

The melting point of the tetracosahydrate is 81.0°C, and the density of the double salt at 0°C is 2.109 g cm^{-3}.

AUXILIARY INFORMATION

METHOD/APPARATUS/PROCEDURE:

Isothermal method used. The soly was studied in HNO_3 sln of density 1.325 g cm^{-3} at 16°C because the author did not have sufficient quantity of the rare earth to study the soly of the salt in pure water. Pulverized salt and HNO_3 sln were placed in glass-stoppered tubes and thermostated at 16°C for 24 h with periodic shaking. The solution was then allowed to settle for 2 h, and a pipet maintained at 16°C was used to withdraw aliquots for analysis. Two analyses were performed.

Solutions were analysed by precipitating both Pr and Mn hydroxides by respective addition of NH_3 and H_2O_2. The ppt was ignited to give Pr_6O_{11} + Mn_3O_4 (note the author states the combined oxide to be Pr_4O_7 + Mn_3O_4).

An attempt to determine the waters of hydration by dehydration was not successful because the temperature required (120°C or higher) resulted in decomposition of the salt with the formation of basic salts. Presumably the waters of hydration were found by difference.

SOURCE AND PURITY OF MATERIALS:

"Pure" praseodymiun oxide was dissolved in dil HNO_3 and $MN(NO_3)_2$ added to give a mole ratio of Pr/Mn = 2/3. The sln was evapd and a small crystal of $Bi_2Mg_3(NO_3)_{12}$ added, and the mixt cooled to ppt the tetracosahydrate. The double nitrate was recrystd before use. The double salt was analysed gravimetrically for combined Pr + Mn oxides. A 0.4640 g sample of the tetracosahydrate yielded 0.1613 g oxide (i.e. 34.76 mass %). This is in fair agreement with the theor value of 34.91 mass % based upon the assumption the oxide in Pr_4O_7 + Mn_3O_4. However, since the oxide was probably Pr_6O_{11} + Mn_3O_4, the compiler calculated the theor oxide content as 35.08 mass % which is also in fair agreement with the experimental value.

ESTIMATED ERROR:

Soly: reproducibility about ± 1-5% (compiler).

Temp: nothing specified.

COMPONENTS:

(1) Praseodymium manganese nitrate; $2Pr(NO_3)_3 \cdot 3Mn(NO_3)_2$; [84682-58-6]

(2) Water ; H_2O ; [7732-18-5]

ORIGINAL MEASUREMENTS:

Prandtl, W.; Ducrue, H.

Z. Anorg. Chem. 1926, *150*, 105-16.

VARIABLES:

Temperature

PREPARED BY:

Mark Salomon

EXPERIMENTAL VALUES:

			solubility					
			oxides		double salt			
	mole ratio[a]	density	Pr_2O_3	MnO	hydrate[b]	hydrate[c]	anhydrous salt[d]	
t/°C	MnO/Pr_2O_3	kg m^{-3}	mass %	mass %	mass %	mass %	mass %	mol kg^{-1}
15	3.00[e]	1.63	14.60	9.44	71.8	71.92	52.76	0.938
30	2.94	1.68	15.42	9.75	75.9	75.12	55.11	1.031
45	2.98	1.72	16.24	10.40	79.9	79.62	58.41	1.179
60	2.98[f]	1.79	17.43	11.15	85.8	85.41	62.66	1.409

a. Experimental value: theoretical value is 3.00.
b. Authors' values apparently based on mass % Pr_2O_3. The hydrate which is also the equilibrium solid phase is the tetracosahydrate $2Pr(NO_3)_3 \cdot Mn(NO_3)_2 \cdot 24H_2O$.
c. Compiler's calculations based on average from mass % Pr_2O_3 and MnO.
d. Compiler's calculations based on results from c above.
e. Compiler computes 3.01.
f. Compiler computes 2.97.

AUXILIARY INFORMATION

METHOD/APPARATUS/PROCEDURE:

Isothermal method used. Pulverized double salt (hydrate) and conductivity water were placed in two 50 cc flasks and agitated for 1 day in a thermostat. The slns were then permitted to settle and aliquots of approx 4 cc removed with pipets maintained at the same temp as the satd slns. The aliquots were placed in graduated flasks and weighed, and then diluted with 50 cc of water for analysis. The results for densities and mass % of oxides are the mean of two determinations. The mass % of the tetracosahydrate was apparently calculated by the authors from the mass % Pr_2O_3: i.e. the mass % MnO was not considered.

Both metals were determined gravimetrically.

Diluted aliquots of the satd solutions were heated and Dr pptd as the oxalate, ignited, dissolved in dilute HNO_3, and pptd two more times as the oxalate. From the combined filtrates, Mn was pptd as $MnNH_4PO_4$, and presumably ignited to the pyrophosphate $Mn_2P_2O_7$.

SOURCE AND PURITY OF MATERIALS:

Pr_2O_3 prepared by W. Prandtl was analysed by X-ray spectroscopy and found to be "very pure," particularly with respect to lanthanum. It was dissolved in nitric acid and the required amount of commercial "pure" grade $Mn(NO_3)_2$ added. The solution was evaporated to crystallization, and the double salt recrystallized several times from conductivity water. The salt was dried over $CaCl_2$ in a desiccator to give the tetracosahydrate. Results of the analysis of the double salt are:

Pr_2O_3 found 19.55, 19.68 %, calcd 20.32 mass %.

MnO found 12.70, 12.60 %, calcd 13.11 mass %.

Conductivity water was used.

ESTIMATED ERROR:

Soly: precision ± 0.3 to ± 1 % (compiler).

Temp: not specified.

COMPONENTS:

(1) Praseodymium cobalt nitrate; $2Pr(NO_3)_3 \cdot 3Co(NO_3)_2$; [76637-28-0]

(2) Nitric acid; HNO_3; [7697-37-2]

(3) Water ; H_2O ; [7732-18-5]

ORIGINAL MEASUREMENTS:

Jantsch, G. *Z. Anorg. Chem.* 1912, *76*, 303-23.

VARIABLES:

One temperature: 16°C

PREPARED BY:

Mark Salomon

EXPERIMENTAL VALUES: Soly of the double salt in HNO_3 sln of density d_4^{16} = 1.325 g cm^{-3}.

aliquot volume	Pr_2O_3 [a]	Pr_6O_{11} [b]	soly of $2Pr(NO_3)_3 \cdot 3Co(NO_3)_2$ [c]
cm^3	g	g	mol dm^{-3}
1.4638	0.0380	0.0392	
1.4638	0.0386	0.0399	0.0725

a. Original values reported by the author. The author states that the oxide produced upon ignition of the pptd hydroxide is Pr_4O_7, and he evidently converted the mass of oxide (from Pr_4O_7) to the equiv mass of Pr_2O_3. However, in the opinion of the compiler, the oxide produced upon ignition of the hydroxide at "normal" atmospheric conditions is Pr_6O_{11}, and the author's conversion to equiv mass of Pr_2O_3 is therefore in error.

b. Assuming that the oxide produced upon ignition of $Pr(OH)_3$ is Pr_6O_{11}, the reported mass of Pr_2O_3 based upon the author's assumption that the original oxide was Pr_4O_7 is incorrect. To obtain the correct mass of Pr_6O_{11} produced upon ignition of $Pr(OH)_3$, the author's value of mass Pr_2O_3 must be multiplied by the factor 1.0323. Thus the values of mass Pr_6O_{11} in the table were calculated by the compiler by multiplying the author's values for mass Pr_2O_3 by the appropriate factor.

c. Average value calculated by the compiler based upon the corrected mass of oxide (i.e. based on the mass Pr_6O_{11}).

ADDITIONAL DATA:

The melting point of the tetracosahydrate is 97.0°C, and the density of the double salt at 0°C is 2.176 g cm^{-3}.

AUXILIARY INFORMATION

METHOD/APPARATUS/PROCEDURE:

Isothermal method used. The soly was studied in HNO_3 sln of density 1.325 g cm^{-3} at 16°C because the author did not have sufficient quantity of the rare earth to study the soly of the salt in pure water. Pulverized salt and HNO_3 sln were placed in glass-stoppered tubes and thermostated at 16°C for 24 h with periodic shaking. The solution was then allowed to settle for 2 h, and a pipet maintained at 16°C was used to withdraw aliquots for analysis. Two analyses were performed.

Solutions were analysed by adding 2-3 g NH_4Cl and 10% NH_3 sln followed by boiling to ppt the hydroxide. The ppt was filtered, dissolved in HNO_3, reprecipitated as the hydroxide, and ignited to the oxide. Co in the filtrate was "determined by the usual method" (no details were given).

An attempt to determine the waters of hydration by dehydration was not successful because the temperature required (120°C or higher) resulted in decomposition of the salt with the formation of basic salts. Presumably the waters of hydration were found by difference.

SOURCE AND PURITY OF MATERIALS:

"Pure" praseodymium oxide was dissolved in dil HNO_3 and $Co(NO_3)_2$ added to give a mole ratio of Pr/Co = 2/3. The sln was evapd and a small crystal of $Bi_2Mg_3(NO_3)_{12}$ added, and the mixt cooled to ppt the tetracosahydrate. The double nitrate was recrystd before use. The double salt was analysed gravimetrically for praseodymium oxide. A 0.3844 g sample of the tetracosahydrate yielded 0.0795 g oxide (i.e. 20.68 mass %). This is in good agreement with theor value of 20.66 mass % based upon the assumption the oxide is Pr_4O_7. However since the oxide was probably Pr_6O_{11}, the compiler calculated the theor oxide content as 20.82 mass % which is in poorer agreement with the experimental value. Analysis for metallic cobalt (details not given) resulted in 10.92 mass % Co: theor value is 10.82 mass % (compiler).

ESTIMATED ERROR:

Soly: reproducibility about ± 1-5% (compiler).

Temp: nothing specified.

COMPONENTS:

(1) Praseodymium cobalt nitrate; $2Pr(NO_3)_3 \cdot 3Co(NO_3)_2$; [76637-28-0]

(2) Water ; H_2O ; [7732-18-5]

ORIGINAL MEASUREMENTS:

Prandtl, W.; Ducrue, H.

Z. Anorg. Chem. 1926, *150*, 105-16.

VARIABLES:

Temperature

PREPARED BY:

Mark Salomon

EXPERIMENTAL VALUES:

			solubility					
			oxides		double salt			
	mole ratio[a]	density	Pr_2O_3	CoO	hydrate[b]	hydrate[c]	anhydrous salt[d]	
$t/^\circ C$	CoO/Pr_2O_3	$kg\ m^{-3}$	mass %	mass %	mass %	mass %	mass %	$mol\ kg^{-1}$
15	3.00	1.62	13.64	9.30	67.6	67.63	49.75	0.823
30	3.00[e]	1.65	14.28	9.75	70.8	70.85	52.12	0.905
45	2.97[f]	1.69	14.98	10.13	74.3	73.97	54.41	0.992
60	2.99	1.72	15.89	10.79	78.8	78.63	57.84	1.140

a. Experimental value: theoretical value = 3.00.
b. Authors' values apparently based on mass % Pr_2O_3. The hydrate which is also the equilibrium solid phase is the tetracosahydrate $2Pr(NO_3)_3 \cdot 3Co(NO_3)_2 \cdot 24H_2O$.
c. Compiler's calculations based on average from mass % Pr_2O_3 and CoO.
d. Compiler's calculations based on results from c above.
e. Compiler computes 3.01.
f. Compiler computes 2.98.

AUXILIARY INFORMATION

METHOD/APPARATUS/PROCEDURE:

Isothermal method used. Pulverized double salt (hydrate) and conductivity water were placed in two 50 cc flasks and agitated for 1 day in a thermostat. The slns were then permitted to settle and aliquots of approx 4 cc removed with pipets maintained at the same temp as the satd slns. The aliquots were placed in graduated flasks and weighed, and then diluted with 50 cc of water for analysis. The results for densities and mass % of oxides are the mean of two determinations. The mass % of the tetracosahydrate was apparently calculated by the authors from the mass % Pr_2O_3: i.e. the mass % CoO was not considered.

Both metals were determined gravimetrically.

Pr was precipitated as the hydroxide by addn of NH_4Cl/NH_4OH solution. The hydroxide was dissolved in dil HNO_3, pptd as the oxalate and ignited to the oxide.
The filtrates from the hydroxide and oxalate separations were combined, and nitric and oxalic acids eliminated by addn of sulfuric acid and heating. Co was determined by electrolytic deposition.

SOURCE AND PURITY OF MATERIALS:

Pr_2O_3 prepared by W. Prandtl was analyzed by X-ray spectroscopy and found to be "very pure," particularly with respect to lanthanum. It was dissolved in nitric acid and the required amount of commercial "pure" grade $Co(NO_3)_2$ added. The solution was evaporated to crystallization, and the double salt recrystallized several times from conductivity water. The salt was dried over $CaCl_2$ in a desiccator to give the tetracosahydrate. Results of the analysis of the double salt are:
Pr_2O_3 found 20.13, 20.08 %, calcd 20.17 mass %.
CoO found 13.85, 13.93 %, calcd 13.75 mass %.
Conductivity water was used.

ESTIMATED ERROR:

Soly: precision ± 0.4 % at best (compiler).

Temp: not specified

COMPONENTS:

(1) Praseodymium nickel nitrate; $2Pr(NO_3)_3 \cdot 3Ni(NO_3)_2$; [84682-59-7]

(2) Nitric acid; HNO_3; [7697-37-2]

(3) Water ; H_2O ; [7732-18-5]

ORIGINAL MEASUREMENTS:

Jantsch, G. *Z. Anorg. Chem.* 1912, *76*, 303-23.

VARIABLES:

One temperature: 16°C

PREPARED BY:

Mark Salomon

EXPERIMENTAL VALUES:

Soly of the double salt in HNO_3 sln of density d_4^{16} = 1.325 g cm^{-3}.

aliquot volume	Pr_2O_3 [a]	Pr_6O_{11} [b]	soly of $2Pr(NO_3)_3 \cdot 3Ni(NO_3)_2$ [c]
cm^3	g	g	mol dm^{-3}
1.4638	0.0270	0.0279	
1.4638	0.0278	0.0287	0.0519

a. Original values reported by the author. The author states that the oxide produced upon ignition of the pptd hydroxide is Pr_4O_7, and he evidently converted the mass of oxide (from Pr_4O_7) to the equiv mass of Pr_2O_3. However, in the opinion of the compiler, the oxide produced upon ignition of the hydroxide at "normal" atmospheric conditions is Pr_6O_{11}, and the author's conversion to equiv mass of Pr_2O_3 is therefore in error.

b. Assuming that the oxide produced upon ignition of $Pr(OH)_3$ is Pr_6O_{11}, the reported mass of Pr_2O_3 based upon the author's assumption that the original oxide was Pr_4O_7 is incorrect. To obtain the correct mass of Pr_6O_{11} produced upon ignition of $Pr(OH)_3$, the author's value of mass Pr_2O_3 must be multiplied by the factor 1.0323. Thus the values of mass Pr_6O_{11} in the table were calculated by the compiler by multiplying the author's values for mass Pr_2O_3 by the appropriate factor.

c. Average value calculated by the compiler based upon the corrected mass of oxide (i.e. based on the mass Pr_6O_{11}).

ADDITIONAL DATA:

The melting point of the tetracosahydrate is 108.0°C, and the density of the double salt at 0°C is 2.195 g cm^{-3}.

AUXILIARY INFORMATION

METHOD/APPARATUS/PROCEDURE:

Isothermal method used. The soly was studied in HNO_3 sln of density 1.325 g cm^{-3} at 16°C because the author did not have sufficient quantity of the rare earth to study the soly of the salt in pure water. Pulverized salt and HNO_3 sln were placed in glass-stoppered tubes and thermostated at 16°C for 24 h with periodic shaking. The solution was then allowed to settle for 2 h, and a pipet maintained at 16°C was used to withdraw aliquots for analysis. Two analyses were performed.

Solutions were analysed by adding 2-3 g NH_4Cl and 10% NH_3 sln followed by boiling to ppt the hydroxide. The ppt was filtered, dissolved in HNO_3, reprecipitated as the hydroxide, and ignited to the oxide. Ni in the filtrate was "determined by the usual method" (no details were given).

An attempt to determine the waters of hydration by dehydration was not successful because the temperature required (120°C or higher) resulted in decomposition of the salt with the formation of basic salts. Presumably the waters of hydration were found by difference.

SOURCE AND PURITY OF MATERIALS:

"Pure" praseodymium oxide was dissolved in dil HNO_3 and $Ni(NO_3)_2$ added to give a mole ratio of Pr/Ni = 2/3. The sln was evapd and a small crystal of $Bi_2Mg_3(NO_3)_{12}$ added, and the mixt cooled to ppt the tetracosahydrate. The double nitrate was recrystd before use. The double salt was analysed gravimetrically for praseodymium oxide. A 0.4196 g sample of the tetracosahydrate yielded 0.0876 g oxide (i.e. 20.88 mass %). This is in fair agreement with the theor value of 20.67 mass % based upon the assumption the oxide is Pr_4O_7. However since the oxide was probably Pr_6O_{11}, the compiler calculated the theor oxide content as 20.83 mass % which is in much better agreement with the experimental value. Ni analysis (no details given) gave 10.77 mass %, and for NO_3 by pptn with nitron gave 45.43 mass %. Theor values are 10.77 mass % and 45.53 mass %, resp (compiler).

ESTIMATED ERROR:

Soly: reproducibility about ± 1-5% (compiler).

Temp: nothing specified

COMPONENTS:

(1) Pradeodymium nickel nitrate; $2Pr(NO_3)_3 \cdot 3Ni(NO_3)_2$; [84682-59-7]

(2) Water ; H_2O ; [7732-18-5]

ORIGINAL MEASUREMENTS:

Prandtl, W.; Ducrue, H.

Z. Anorg. Chem. 1926, *150*, 105-16.

VARIABLES:

Temperature

PREPARED BY:

Mark Salomon

EXPERIMENTAL VALUES:

solubility

			oxides		double salt			
	mole ratio[a]	density	Pr_2O_3	NiO	hydrate[b]	hydrate[c]	anhydrous salt[d]	
t/°C	NiO/Pr_2O_3	kg m^{-3}	mass %	mass %	mass %	mass %	mass %	mol kg^{-1}
15	3.00	1.60	12.99	8.83	64.4	64.38	47.35	0.748
30	3.00	1.63	13.83	9.41	68.5	68.58	50.43	0.847
45	3.01	1.66	14.50	9.87	71.8	71.91	52.89	0.934
60	3.03	1.70	15.27	10.47	75.7	76.01	55.90	1.055

a. Experimental value: theoretical value = 3.00.
b. Author's values apparently based on mass % Pr_2O_3. The hydrate which is the equilibrium solid phase is the tetracosahydrate $2Pr(NO_3)_3 \cdot 3Ni(NO_3)_2 \cdot 24H_2O$.
c. Compiler's calculations based on average from mass % Pr_2O_3 and NiO.
d. Compiler's calculations based on results from c above.

AUXILIARY INFORMATION

METHOD/APPARATUS/PROCEDURE:

Isothermal method used. Pulverized double salt (hydrate) and conductivity water were placed in two 50 cc flasks and agitated for 1 day in a thermostat. The slns were then permitted to settle and aliquots of approx 4 cc removed with pipets maintained at the same temp as the satd slns. The aliquots were placed in graduated flasks and weighed, and then diluted with 50 cc of water for analysis. The results for densities and mass % of oxides are the mean of two determinations. The mass % of the tetracosahydrate was apparently calcd by the authors from the mass % Pr_2O_3: i.e. the mass % NiO was not considered.
Both metals were determined gravimetrically. Pr was pptd as the hydroxide by addn of NH_4Cl/NH_4OH solution. The hydroxide was dissolved in dil HNO_3, pptd as the oxalate and ignited to the oxide.
The filtrates from the hydroxide and oxalate separations were combined, and nitric and oxalic acids eliminated by addn of sulfuric acid and heating. Ni was determined by electrolytic deposition.

SOURCE AND PURITY OF MATERIALS:

Pr_2O_3 prepared by W. Prandtl was analyzed by X-ray spectroscopy and found to be "very pure," particulary with respect to lanthaum. It was dissolved in nitric acid and the required amount of commercial "pure" grade $Ni(NO_3)_2$ added. The solution was evaporated to crystallization, and the double salt recrystallized several times from conductivity water. The salt was dried over $CaCl_2$ in a desiccator to give the tetracosahydrate. Results of the analyses of the double salt are:
Pr_2O_3 found 20.18, 20.18 %, calcd 20.18 mass %.
NiO found 13.72, 13.80 %, calcd 13.71 mass %.
Conductivity water was used.

ESTIMATED ERROR:

Soly: precision ± 0.3 % at best (compiler).

Temp: not specified

COMPONENTS:

(1) Praseodymium copper nitrate; $2Pr(NO_3)_3 \cdot 3Cu(NO_3)_2$; [84682-60-0]

(2) Water; H_2O; [7732-18-5]

ORIGINAL MEASUREMENTS:

Prandtl, W.; Ducrue, H. *Z. Anorg. Chem.* 1926, *150*, 105-16.

VARIABLES:

One temperature: 15°C

PREPARED BY:

Mark Salomon

EXPERIMENTAL VALUES:

mole ratio [a]	density	solubility	
		Pr_2O_3	CuO
CuO/Pr_2O_3	kg m^{-3}	mass %	mass %
3.98	1.82	13.84	13.28

a. Experimental value: theoretical value is 3.00. Apparently this large difference is not due to experimental error, but rather to the probable decomposition of the double nitrate in aqueous solution. The solid phase is almost copper-free, and is light green in color, and largely $Pr(NO_3)_3$.

COMMENTS AND/OR ADDITIONAL DATA:

It is fairly certain that the authors have reported equilibrated solubility data, but the nature of the solid phase is undefined. Since the solid phase could be a mixture of Pr and Cu nitrates and solid solutions, and also some double salt, it is not possible to compute the solubility in terms of a specific salt or double salt. Because of these problems, the authors did not attempt to determine the solubilities at other temperatures.

AUXILIARY INFORMATION

METHOD/APPARATUS/PROCEDURE:

Isothermal method used. Pulverized double salt (hydrate) and conductivity water were placed in two 50 cc flasks and agitated for 1 day in a thermostat. The slns were then permitted to settle and aliquots of approx 4 cc removed with pipets maintained at the same temp as the satd slns. The aliquots were placed in graduated flasks and weighed, and then diluted with 50 cc of water for analysis. The results for densities and mass % of oxides are the mean of two determinations.

Both metals were determined gravimetrically. Cu was first deposited electrolytically, and Pr then determined by the oxalate method.

SOURCE AND PURITY OF MATERIALS:

Pr_2O_3 prepared by W. Prandtl was analysed by X-ray spectroscopy and found to be "very pure," particularly with respect to lanthanum. The computed amounts of Pr_2O_3 and CuO were dissolved in the required amount of HNO_3, and then the solution was concentrated by evaporation with the addition of HNO_3 to prevent the pptn of basic salts. The cooled concentrated solution crystallized only after the addn of a small seed of $2Pr(NO_3)_3 \cdot 3Mg(NO_3)_2 \cdot 24H_2O$; the tetracosahydrate crystallized as green hexagonal plates. Analysis of the double salt yielded:

Pr_2O_3 found 19.80, 19.80 %, calcd 20.00 mass%

CuO found 14.32, 14.44 %, calcd 14.48 mass %.

The melting point of the tetracosahydrate was reported as 42.5°C. Conductivity water was used.

ESTIMATED ERROR:

Soly: unknown.

Temp: not specified.

COMPONENTS:

(1) Praseodymium zinc nitrate; $2Pr(NO_3)_3 \cdot 3Zn(NO_3)_2$; [84682-62-2]

(2) Nitric acid; HNO_3; [7697-37-2]

(3) Water ; H_2O ; [7732-18-5]

ORIGINAL MEASUREMENTS:

Jantsch, G. *Z. Anorg. Chem.* 1912, *76*, 303-23.

VARIABLES:

One temperature: 16°C

PREPARED BY:

Mark Salomon

EXPERIMENTAL VALUES: Soly of the double salt in HNO_3 sln of density d_4^{16} = 1.325 g cm^{-3}.

aliquot volume	Pr_2O_3 [a]	Pr_6O_{11} [b]	soly of $2Pr(NO_3)_3 \cdot 3Zn(NO_3)_2$ [c]
cm^3	g	g	mol dm^{-3}
1.4638	0.0428	0.442	
1.4638	0.0428	0.442	0.0797

a. Original values reported by the author. The author states that the oxide produced upon ignition of the pptd hydroxide is Pr_4O_7, and he evidently converted the mass of oxide (from Pr_4O_7) to the equiv mass of Pr_2O_3. However, in the opinion of the compiler, the oxide produced upon ignition of the hydroxide at "normal" atmospheric conditions is Pr_6O_{11}, and the author's conversion to equiv mass of Pr_2O_3 is therefore in error.

b. Assuming that the oxide produced upon ignition of $Pr(OH)_3$ is Pr_6O_{11}, the reported mass of Pr_2O_3 based upon the author's assumption that the original oxide was Pr_4O_7 is incorrect. To obtain the correct mass of Pr_6O_{11} produced upon ignition of $Pr(OH)_3$, the author's value of mass Pr_2O_3 must be multiplied by the factor 1.0323. Thus the values of mass Pr_6O_{11} in the table were calculated by the compiler by multiplying the author's values for mass Pr_2O_3 by the appropriate factor.

c. Average value calculated by the compiler based upon the corrected mass of oxide (i.e. based on the mass Pr_6O_{11}).

ADDITIONAL DATA:

The melting point of the tetracosahydrate is 91.5°C, and the density of the double salt at 0°C is 2.203 g cm^{-3}.

AUXILIARY INFORMATION

METHOD/APPARATUS/PROCEDURE:

Isothermal method used. The soly was studied in HNO_3 sln of density 1.325 g cm^{-3} at 16°C because the author did not have sufficient quantity of the rare earth to study the soly of the salt in pure water. Pulverized salt and HNO_3 sln were placed in glass-stoppered tubes and thermostated at 16°C for 24 h with periodic shaking. The solution was then allowed to settle for 2 h, and a pipet maintained at 16°C was used to withdraw aliquots for analysis. Two analyses were performed.

Solutions were analysed by adding 2-3 g NH_4Cl and 10% NH_3 sln followed by boiling to ppt the hydroxide. The ppt was filtered, dissolved in HNO_3, reprecipitated as the hydroxide, and ignited to the oxide. Zn in the filtrate was "determined by the usual method" (no details were given).

An attempt to determine the waters of hydration by dehydration was not successful because the temperature required (120°C or higher) resulted in decomposition of the salt with the formation of basic salts. Presumably the waters of hydration were found by difference.

SOURCE AND PURITY OF MATERIALS:

"Pure" praseodymium oxide was dissolved in dil HNO_3 and $Zn(NO_3)_2$ added to give a mole ratio of Pr/Zn = 2/3. The sln was evapd and a small crystal of $Bi_2Mg_3(NO_3)_{12}$ added, and the mixt cooled to ppt the tetracosahydrate. The double nitrate was recrystd before use. The double salt was analysed gravimetrically for praseodymium oxide. A 0.2236 g sample of the tetracosahydrate yielded 0.0465 g oxide (i.e. 20.80 mass %). This is in fair agreement with the theor value of 20.42 mass % based upon the assumption the oxide is Pr_4O_7. However since the oxide was probably Pr_6O_{11}, the compiler calculated the theor oxide content as 20.58 mass % which is in better agreement with the experimental value. Analysis for NO gave 21.50 mass %: theor value is 21.77 mass %.

ESTIMATED ERROR:

Soly: reproducibility about ± 1-5% (compiler).

Temp: nothing specified.

COMPONENTS:

(1) Praseodymiun zinc nitrate; $2Pr(NO_3)_3 \cdot 3Zn(NO_3)_2$; [84682-62-2]

(2) Water ; H_2O ; [7732-18-5]

ORIGINAL MEASUREMENTS:

Prandtl, W.; Ducrue, H.

Z. Anorg. Chem. 1926, *150*, 105-16.

VARIABLES:

Temperature

PREPARED BY:

Mark Salomon

EXPERIMENTAL VALUES:

			solubility					
			oxides		double salt			
	mole ratio[a]	density	Pr_2O_3	ZnO	hydrate[b]	hydrate[c]	anhydrous salt[d]	
t/°C	ZnO/Pr_2O_3	kg m^{-3}	mass %	mass %	mass %	mass %	mass %	mol kg^{-1}
15	3.03	1.63	13.22	9.87	66.3	66.60	49.19	0.792
30	3.01	1.67	14.08	10.45	70.6	70.72	52.24	0.895
45	2.97	1.71	15.02	11.01	75.4	74.98	55.38	1.016
60	3.04	1.76	15.97	11.98	80.1	80.65	59.57	1.206

a. Experimental value: theoretical value = 3.00.

b. Authors' values apparently based on mass % Pr_2O_3. The hydrate which is also the equilibrium solid phase is the tetracosahydrate $2Pr(NO_3)_3 \cdot 3Zn(NO_3)_2 \cdot 24H_2O$.

c. Compiler's calculations based on average from mass % Pr_2O_3 and ZnO.

d. Compiler's calculations based on results from c above.

AUXILIARY INFORMATION

METHOD/APPARATUS/PROCEDURE:

Isothermal method used. Pulverized double salt (hydrate) and conductivity water were placed in two 50 cc flasks and agitated for 1 day in a thermostat. The slns were then permitted to settle and aliquots of approx 4 cc removed with pipets maintained at the same temp as the satd slns. The aliquots were placed in graduated flasks and weighed, and then diluted with 50 cc of water for analysis. The results for densities and mass % of oxides are the mean of two determinations. The mass % of the tetracosahydrate was apparently calculated by the authors from the mass % Pr_2O_3: i.e. the mass % ZnO was not considered.

Both metals were determined gravimetrically.

Pr was ppt as the hydroxide with NH_4Cl/NH_4OH solution, filtered, redissolved in dil HNO_3, pptd as the oxalate and ignited to the oxide. From the combined filtrates, Zn was pptd as $ZnNH_4PO_4$ and presumably ignited to the pyrophosphate $Zn_2P_2O_7$.

SOURCE AND PURITY OF MATERIALS:

Pr_2O_3 prepared by W. Prandtl was analysed by X-ray spectroscopy and found to be "very pure," particularly with respect to lanthanum. It was dissolved in nitric acid and the required amount of commercial "pure" grade $Zn(NO_3)_2$ added. The solution was evaporated to crystallization, and the double salt recrystallized several times from conductivity water. The salt was dried over $CaCl_2$ in a desiccator to give the tetracosahydrate. Results of the analysis of the double salt are:

Pr_2O_3 found 19.41, 19.43 %, calcd 19.93 mass %.

ZnO found 14.25, 14.49 %, calcd 14.76 mass %.
Conductivity water was used.

ESTIMATED ERROR:

Soly: precision around ± 1% (compiler).

Temp: not specified.

COMPONENTS:	EVALUATOR:
(1) Neodymium nitrate; $Nd(NO_3)_3$; [10045-95-1] (2) Water ; H_2O ; [7732-18-5]	S. Siekierski and T. Mioduski Institute for Nuclear Research Warsaw, Poland and M. Salomon U.S. Army ET & DL Ft. Monmouth, NJ, USA December 1982

CRITICAL EVALUATION:

THE BINARY SYSTEM

INTRODUCTION

The solubility of $Nd(NO_3)_3$ in water has been given in eighteen publications (1-18), but numerical data in (9) were not given, and the paper was rejected. For those studies dealing with ternary systems (13-17), it appears that the solubility of $Nd(NO_3)_3$ at 293.2 K and 313.2 K in pure water was measured only once, and the same value reported in all five publications. Thus of these eighteen publications, it appears that only fourteen report independently determined solubilities in the binary system.

Only one study (3) reports solubilities determined by the synthetic method, and all other studies employ isothermal methods. The only papers which report experimental uncertainties are those of Spedding et al. (4-6), and for the remaining investigations the compilers and evaluators estimated the experimental uncertainties when possible.

Two solid phases have been identified in the binary system depending upon temperature:

$Nd(NO_3)_3 \cdot 6H_2O$ [16454-60-7] ; $Nd(NO_3)_3 \cdot 5H_2O$ [14517-29-4]

Popov and Mironov report (3) that for temperatures equal to or greater than 358.8 K, the stable solid phase is a lower hydrate, but its composition could not be determined. In ternary systems of high nitric acid content (see below), both the tetrahydrate, $Nd(NO_3)_3 \cdot 4H_2O$ [26635-06-3], and monohydrate, $Nd(NO_3)_3 \cdot H_2O$ [68028-01-3], have been identified (7,8). No reports could be found suggesting the possible existence of tri-hydrate or dihydrate solid phases, and we therefore conclude that the unidentifiable solid phase reported by Popov and Mironov is either the tetrahydrate or the monohydrate.

EVALUATION PROCEDURE

The data in the compilations were examined and either rejected immediately because of large obvious errors, or were analysed by a weighted least squares method. In all cases only experimental solubility data were used in the least squares analyses: i.e. smoothed or extrapolated data were not used. The data were fitted to the general solubility equation based on the treatments in (20, 21) and in the INTRODUCTION to this volume:

$$Y = \ln(m/m_0) - nM_2(m - m_0) = a + b/(T/K) + c\ln(T/K) \qquad [1]$$

In eq. [1] m is the molality at temperature T, m_0 is an arbitrarily selected reference molality (usually the 298.2 K value), n is the hydration number of the solid, M_2 is the molar mass of the solvent, and a, b, c are constants from which the enthalpies and heat capacities of solution, ΔH_{sln} and ΔC_p, can be estimated (see INTRODUCTION). In fitting the solubility data to eq. [1], weight factors of 0, 1, 2, 3 were assigned to each published value depending upon the experimental precision. In this procedure, if the difference between the observed and calculated solubilities, Δm (also called the residual error), for a given data point was larger than twice the standard error of estimate, σ_m, the data point was either rejected or its weight factor decreased for the next calculation. The fitting of the data in this manner was repeated until all Δm values were equal to or less than $2\sigma_m$.

The data of Friend (1) were rejected immediately because they show a systematic negative deviation of at least 3% from most other data. Friend also reports the existence of α- and β-forms of the solid hexahydrate, and a temperature of about 295 K for the α→β transition. No other authors have reported α- or β-forms of the hexahydrate.

As discussed in the $Ce(NO_3)_3-H_2O$ critical evaluation, the data of highest precision are those of Spedding et al. (4, 5, 6) which are precise to within ± 0.1%, and those of Brunisholz et al. (2, 11) and other authors (7, 8, 10, 18) which are probably precise to within ± 0.2% to 0.5%. All the solubility data in (2, 4-8, 10-12, 18) were determined by the isothermal method using direct chemical analyses to determine concentrations of saturated solutions. The results of Popov and Mironov (3) based on the synthetic method are probably precise to ± 0.5% at best. For those isothermal studies in which refractometric analyses were employed (9, 13-17), an experimental precision of ± 1% at best was estimated by the compilers.

COMPONENTS:	EVALUATOR:
(1) Neodymium nitrate; $Nd(NO_3)_3$; [10045-95-1] (2) Water ; H_2O ; [7732-18-5]	S. Siekierski and T. Mioduski Institute for Nuclear Research Warsaw, Poland and M. Salomon U.S. Army ET & DL Ft. Monmouth, NJ, USA December 1982

CRITICAL EVALUATION:

SOLUBILITY OF $Nd(NO_3)_3$ IN THE HEXAHYDRATE SYSTEM

The solubility data for $Nd(NO_3)_3$ as a function of temperature in the hexahydrate system are summarized in Table 1. As discussed above, the data of Friend were rejected and were not included in Table 1. The initial and final weight factors for all other data are included in Table 1. Table 2 gives the results of the best fit to the smoothing equation, and Table 3 gives smoothed (i.e. calculated from eq. [1]) solubility data at selected temperatures. These smoothed solubilities in Table 3 are designated as *recommended* values because of the excellent agreement of results from a number of independent studies as also indicated by the weight factors in Table 1. From the standard error of estimate $\sigma_m = 0.01_5$ (see Table 2), we obtain an uncertainty of ± 0.005 mol kg^{-1} in the smoothed data (95% level of confidence, Student's t = 2.03). Combining this uncertainty with the average experimental precision of ± 0.3% yields a total uncertainty of ± 0.015 mol kg^{-1} in the smoothed (*recommended*) solubilities.

The value of the congruent melting point calculated from eq. [1] is 341.9 K which is close to the experimental values of 337.3 K (22), 340.7 K (1), and 343.2 ± 1 K (23).

In the determination of the *recommended* solubilities in the hexahydrate system, the solubilities above 277.5 K reported by Popov and Mironov (3) were rejected in the final analysis (see Table 1). These authors state that the hexahydrate is the stable solid phase over the temperature range 244-295 K, and above 295 K the stable solid phase is stated to be the pentahydrate. It is interesting to note that the $\alpha \rightarrow \beta$ phase transition temperature reported by Friend (1) is 295 K, and if Friend's α - and β -phases can be identified with the hexa- and pentahydrate solid phases (similar to the $La(NO_3)_3$-H_2O system as discussed in the critical evaluation), then this would lend support to Popov and Mironov's assignment of stabilities of solid phases. At the present time we cannot ignore this possibility which would suggest that the pentahydrate and hexahydrate polytherms lie very close to each other and cross at 295 K. However in order to confirm this possibility, we believe that additional studies are required. The present recommended solubilities for the hexahydrate system are independent of the temperature for the hexahydrate $\rightarrow$ pentahydrate transition: i.e. since data for both stable and metastable systems are simultaneously fitted to eq. [1], the recommended values are correct providing we have correctly identified the solid phase. For those studies which report direct chemical analyses of the equilibrated solid phases (1, 6, and probably 13-17), the results in all cases show that the hexahydrate is the solid phase from 291.4 K to 339.4 K. At 298.15 K Spedding et al.(6) analysed the "dried" hydrate grown from saturated solutions, and the results were within ± 0.016 water molecules of the hexahydrate. In two other studies by Mironov and his co-workers (8, 24), the hexahydrate is stated to be the stable solid phase at 298.2 K which is in contradiction to the conclusion in (3): note that in ref (8) the solubilities were determined by the isothermal method whereas in (3) the solubilities were determined by the synthetic method. Ref (24) does not report original solubility data for the $Nd(NO_3)_3$-H_2O system. In (24) the authors calculate a solubility of 4.57 mol kg^{-1} for the hexahydrate system at 298.2 K using a smoothing equation based upon previously published data for the $Nd(NO_3)_3$-HNO_3-H_2O system (see eq. [2] below): a solubility of 4.64 mol kg^{-1} at 298.2 K for the pentahydrate system is also reported in (24), and which is stated to be based upon the experimental data in (3). However the results in (3) for the pentahydrate system (see the compilation for this reference) when fitted to the smoothing equation [1] gives a value of 4.40 mol kg^{-1} at 298.2 K, not 4.64 mol kg^{-1}. We conclude that in addition to these inconsistencies in (3, 8, 24), the results of Popov and Mironov (3) above 295 K contain a negative systematic error.

SOLUBILITY OF $Nd(NO_3)_3$ IN THE $Nd(NO_3)_3 \cdot nH_2O$ - H_2O SYSTEM: $n \leq 5$

Popov and Mironov (3) are the only researchers to report solubility data for lower hydrate (binary) systems. In the pentahydrate system, the solubilities at 306.9 K and 316.5 K are lower than the *recommended* solubilities in the hexahydrate system suggesting that the pentahydrate system is the stable system at these temperatures, but at and above 318.6 K, Popov and Mironov's solubilities for the pentahydrate system are greater than the solubilities in the hexahydrate system. Since this cannot be correct, all results at and above 318.6 K must be rejected. However the data points below 318.6 K assigned to the pentahydrate system may indeed lie on the pentahydrate polytherm, and in which case we

COMPONENTS:	EVALUATOR:
(1) Neodymium nitrate; $Nd(NO_3)_3$; [10045-95-1] (2) Water ; H_2O ; [7732-18-5]	S. Siekierski and T. Mioduski Institute for Nuclear Research Warsaw, Poland and M. Salomon U.S. Army ET & DL Ft. Monmouth, NJ, USA December 1982

CRITICAL EVALUATION:

Table 1. Solubility of $Nd(NO_3)_3$ in the $Nd(NO_3)_3.6H_2O-H_2O$ system.

T/K	solubility mol kg^{-1}	ref	weight initial/final	T/K	solubility mol kg^{-1}	ref	weight initial/final
244.5	3.22	3	2/2	298.2	4.595	8	3/1
255.9	3.43	3	2/2	298.2	4.630	10	1/2
269.5	3.72	3	2/2	298.2	4.595	12	3/1
273.15	3.809	2	3/3	303.2	4.872	18	2/0
277.5	3.95	3	2/2	308.15	5.073	2	3/3
283.15	4.117	2	3/3	312.1	5.45	3	2/0
293.15	4.410	2	3/3	313.2	4.94	13-17	0/0
293.15	4.405	11	3/3	313.2	5.00	16	0/0
293.2	4.36	13-17	0/0	317.9[a]	5.96	3	2/0
295.3	4.58	3	2/0	323.15	5.954	2	3/3
298.15	4.582	6	3/1	323.2	5.946	7	2/2
298.15	4.6184	4	3/3	328.2[a]	7.49	3	2/0
298.15	4.6184	5	3/3	333.9[a]	8.80	3	2/0

[a]Metastable according to Popov and Mironov (3).

Table 2. Derived parameters for the smoothing equation.[a]

parameter	hexahydrate system
a	-21.749
b	643.4
c	3.4384
σ_a	0.002
σ_b	0.5
σ_c	0.0003
σ_Y	0.002
σ_m	0.01_5
ΔH_{sln}/k J mol^{-1}	-21.3
ΔC_p/J K^{-1} mol^{-1}	114.4
congruent melting point/K	341.85
concn at the congruent m.p./mol kg^{-1}	9.251

[a] σ_a, σ_b, σ_c are standard deviations for the derived parameters a, b, c and σ_Y and σ_m are the standard errors of estimate for Y in eq. [1] and the molality, respectively.

Table 3. Recommended solubilities of $Nd(NO_3)_3$ in the $Nd(NO_3)_3.6H_2O-H_2O$ system.[a]

T/K	solubility mol kg^{-1}	T/K	solubility mol kg^{-1}
243.15	3.200	303.15	4.822
253.15	3.374	308.15	5.052
263.15	3.578	313.15	5.312
273.15	3.816	318.15	5.607
278.15	3.950	323.15	5.950
283.15	4.095	328.15	6.363
288.15	4.253	333.15	6.889
293.15	4.425	338.15	7.656
298.15	4.614	341.82[b]	9.251

[a]All solubilities calculated from the smoothing equation.

[b]Congruent melting point calculated from the smoothing equation.

COMPONENTS:

(1) Neodymium nitrate; $Nd(NO_3)_3$; [10045-95-1]

(2) Water ; H_2O ; [7732-18-5]

EVALUATOR:

S. Siekierski and T. Mioduski
Institute for Nuclear Research
Warsaw, Poland
and
M. Salomon
U.S. Army ET & DL
Ft. Monmouth, NJ, USA
December 1982

CRITICAL EVALUATION:

would have to conclude that the pentahydrate polytherm lies very close to the hexahydrate polytherm, and that the polytherms probably cross at about 295 ± 5 K. It is entirely possible that while the hexahydrate is metastable above 295 K, it does not easily convert to the more stable pentahydrate even at temperatures as high as 330 K. Clearly, additional experimental data are required to completely define the phase relationships.

For the two solubilities involving unidentified lower hydrates (3), the value of 12.83 mol kg^{-1} at 358.8 K probably lies on the tetrahydrate polytherm, and the value of 16.76 mol kg^{-1} at 395.0 K probably lies on the monohydrate polytherm. Based on solubility data in the ternary $Nd(NO_3)_3$-HNO_3-H_2O system, Popov and Mironov (24) calculated the solubility in the tetrahydrate system at 298.2 K by extrapolation to zero HNO_3 concentration: the result is 9.33 mol kg^{-1} (see below and eq. [3]).

MULTICOMPONENT SYSTEMS

TERNARY SYSTEMS WITH NITRIC ACID

The three studies (7, 8, 19) which report the solubility of $Nd(NO_3)_3$ in aqueous HNO_3 are in basic agreement with each other. The 298.2 K isotherms are given in Fig. 1, showing the regions of stability for the various hydrated solid phases: the hexahydrate is stable up to 45 mass % HNO_3, the tetrahydrate is stable up to around 64-65 mass % HNO_3, and the monohydrate is stable up to about 94 mass % HNO_3. Above 94 mass % HNO_3, the solid phase is an unidentifiable hydrate according to Mironov and Sinitsyna (8). Popov and Mironov (24) give smoothing equations for the hexahydrate and tetrahydrate isotherms at 298.2 K which we presume are based on the results of Mironov and Sinitsyna. The smoothing equations are based upon Kirginestsev's equation (25) and are

hexahydrate isotherm: $\log m_1 = \log(4.57) - 0.649 \log(y_1)$ [2]

tetrahydrate isotherm: $\log m_1 = \log(9.33) - 0.735 \log(y_1)$ [3]

In eq. [2] and [3], m_1 is the molality of $Nd(NO_3)_3$, and y_1 is the solute mole fraction of the salt defined as $y_1 + y_2 = 1$ where y_2 is the solute mole fraction of HNO_3. There is no compound formation in this ternary system either at 298.2 K (7, 8) or 323.2 K (7).

SYSTEMS WITH TWO SATURATING COMPONENTS

(a) Mutual solubilities of $Nd(NO_3)_3$ with inorganic salts. Since only one publication exists for each multicomponent system involving the mutual solubilities of neodymium nitrate with inorganic salts, critical evaluations of the data are not possible, and the reader is referred directly to the compilations for further information. Data for the $La(NO_3)_3$-$Nd(NO_3)_3$-H_2O (2) and $La(NO_3)_3$-$Nd(NO_3)_3$-TBP-H_2O (26) systems have been compiled in the chapter on $La(NO_3)_3$ solubilities: note TBP = tributylphosphate, $[CH_3(CH_2)_3O]PO$. Mutual solubilities in the $Pr(NO_3)_3$-$Nd(NO_3)_3$-H_2O system (2) were compiled in the previous chapter.

One paper by Zhuravlev and Boeva (9) on the $Nd(NO_3)_3$-$N_2H_4.2HNO_3$-H_2O ternary system was rejected and not compiled because numerical solubility data are not given, and the published phase diagram does not contain grid marks denoting the concentration scales. The double salt $Nd(NO_3)_3.3[N_2H_4.2HNO_3]$ is stable at 293.2 K and at 313.2 K.

(b) Mutual solubilities of $Nd(NO_3)_3$ with organic compounds. As in (a) above, these data cannot be critically evaluated because only one publication exists for each system given in the compilations. Some phase diagrams not given in the compilations are given in Figures 2-8 below. The phase diagram in Figs. 4-7 include lines indicating the directions of each of the sections studied (i.e. the method of isothermal sections). One paper (9) on the neodynium nitrate-ethylenediamine dinitrate-water ternary system was rejected and not compiled becuase numerical solubility data were not given, and the published phase diagram does not include grid marks indicating the concentration scales. No double salts are formed in this system, and the phase diagrams at 293.2 K and 313.2 K are of the simple eutonic type. All concentrations in Figs 2-8 are mass %.

COMPONENTS:	EVALUATOR:
(1) Neodymium nitrate; $Nd(NO_3)_3$; [10045-95-1]	S. Siekierski and T. Mioduski
	Institute for Nuclear Research
	Warsaw, Poland
	and
	M. Salomon
(2) Water ; H_2O ; [7732-18-5]	U.S. Army ET & DL
	Ft. Monmouth, NJ, USA
	December 1982

CRITICAL EVALUATION:

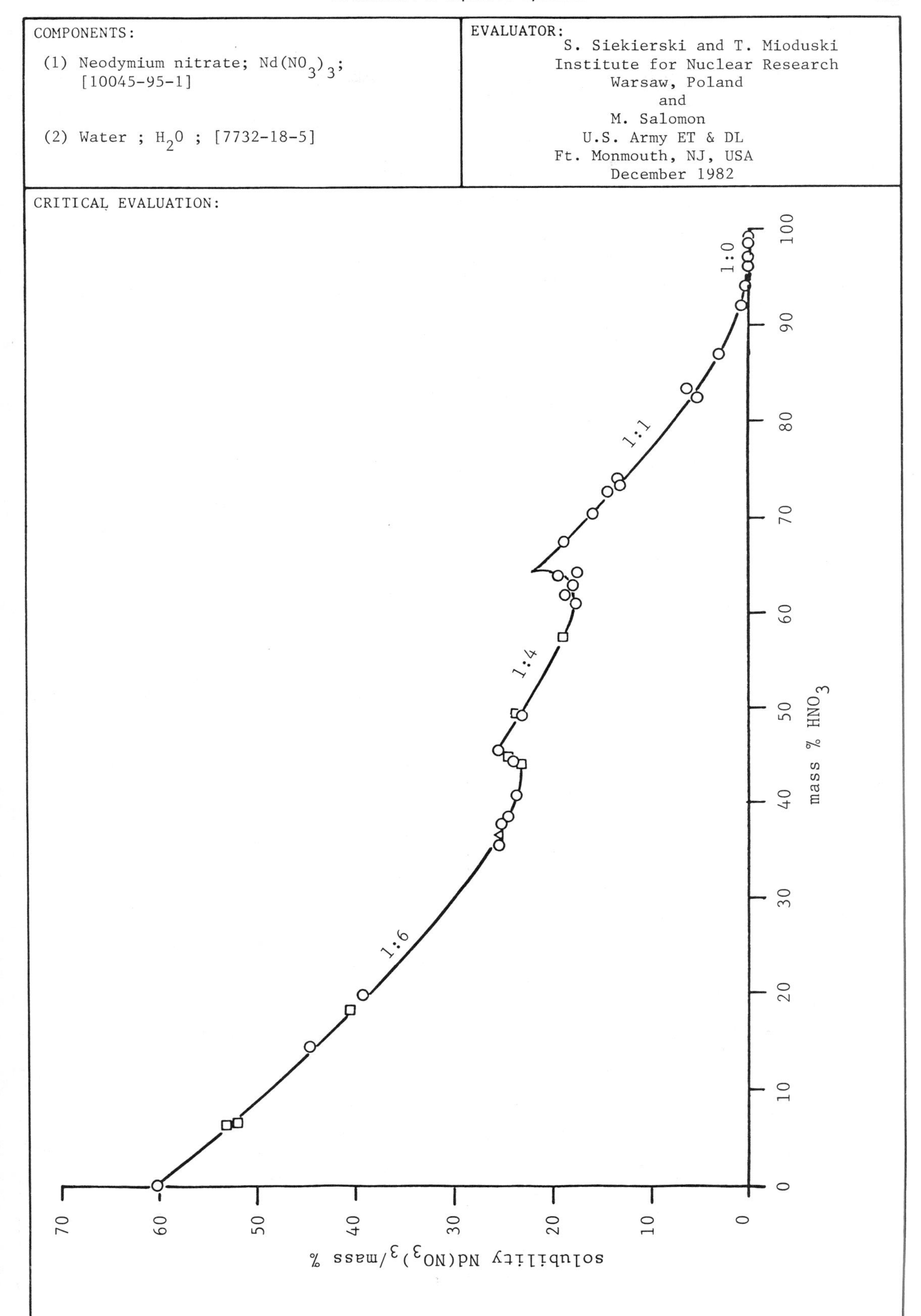

Fig. 1. 298.2 K isotherms in the $Nd(NO_3)_3$-HNO_3-H_2O system. Curves were hand-drawn by the evaluators, and the experimental points are: □ ref. (7), ○ ref. (8), △ ref. (19).

COMPONENTS:	EVALUATOR:
(1) Neodymium nitrate; $Nd(NO_3)_3$; [10045-95-1] (2) Water ; H_2O ; [7732-18-5]	S. Siekierski and T. Mioduski Institute for Nuclear Research Warsaw, Poland and M. Salomon U.S. Army ET & DL Ft. Monmouth, NJ, USA December 1982

CRITICAL EVALUATION:

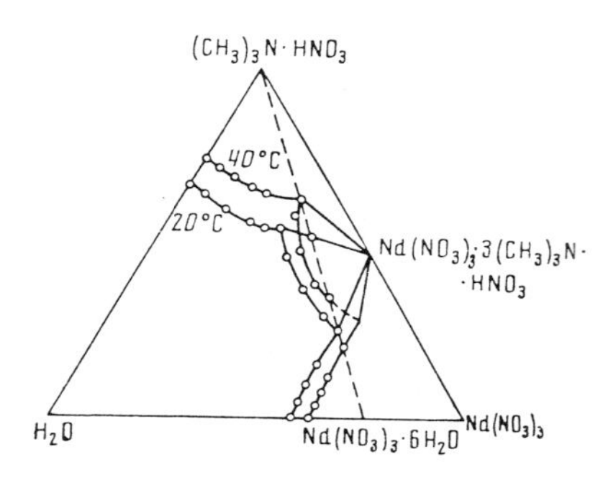

Fig. 2. Phase diagram for the $Nd(NO_3)_3$-$(CH_3)_3N.HNO_3$-H_2O system (14).

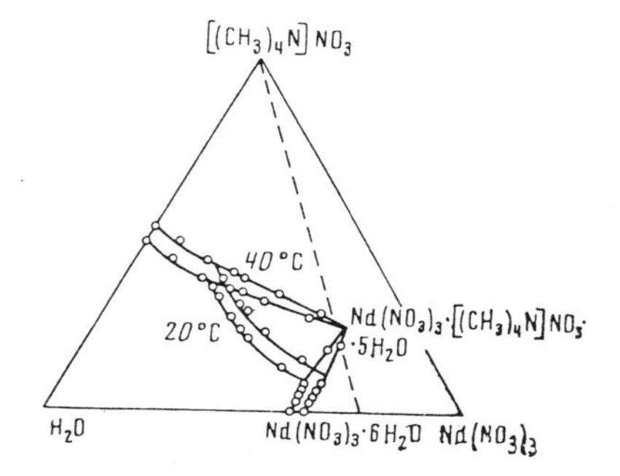

Fig. 3. Phase diagram for the $Nd(NO_3)_3$-$(CH_3)_4N.NO_3$-H_2O system (14).

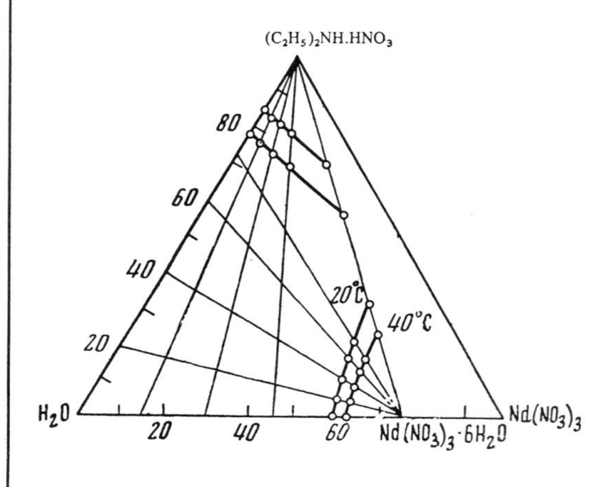

Fig. 4. Phase diagram for the $Nd(NO_3)_3$-$(C_2H_5)_2NH.HNO_3$-H_2O system (15).

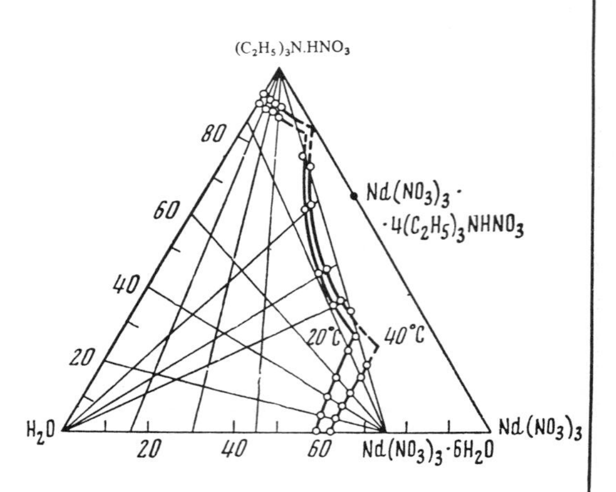

Fig. 5. Phase diagram for the $Nd(NO_3)_3$-$(C_2H_5)_3N{\cdot}HNO_3$-H_2O system (15).

COMPONENTS:	EVALUATOR:
(1) Neodymium nitrate; $Nd(NO_3)_3$; [10045-95-1] (2) Water ; H_2O ; [7732-18-5]	S. Siekierski and T. Mioduski Institute for Nuclear Research Warsaw, Poland and M. Salomon U.S. Army ET & DL Ft. Monmouth, NJ, USA December 1982

CRITICAL EVALUATION:

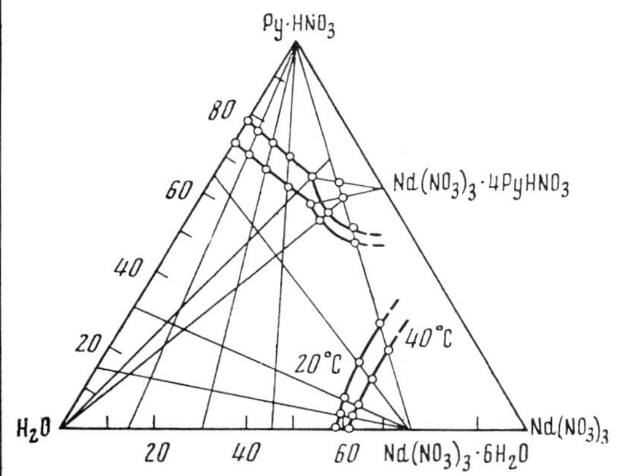

Fig. 6. Phase diagram for the $Nd(NO_3)_3$-$C_5H_{10}NH.HNO_3$-H_2O system (15).

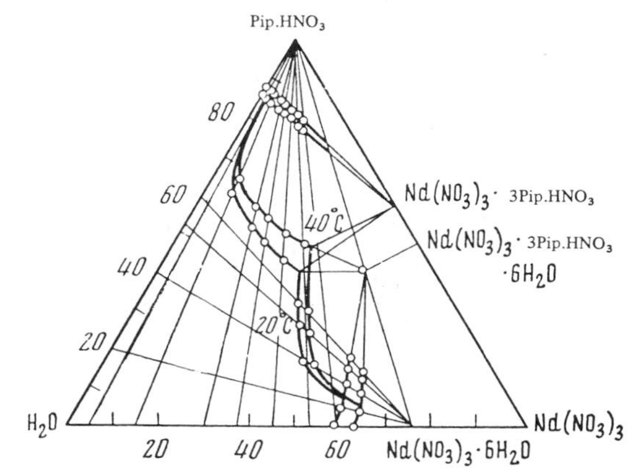

Fig. 7. Phase diagram for the $Nd(NO_3)_3$-$C_5H_5N.HNO_3$-H_2O system (15).

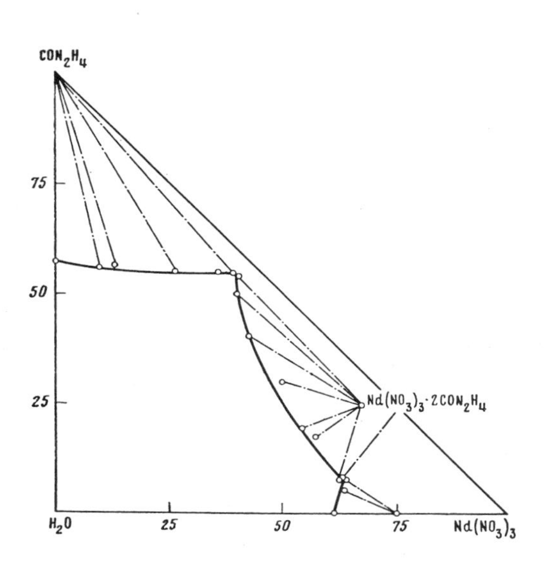

Fig. 8. 303.2 K isotherm of the $Nd(NO_3)_3$-$CO(NH_2)_2$-H_2O system (18).

COMPONENTS:	EVALUATOR:
(1) Neodymium nitrate; $Nd(NO_3)_3$; [10045-95-1] (2) Water ; H_2O ; [7732-18-5]	S. Siekierski and T. Mioduski Institute for Nuclear Research Warsaw, Poland and M. Salomon U.S. Army ET & DL Ft. Monmouth, NJ, USA December 1982

CRITICAL EVALUATION:

REFERENCES

1. Friend, J.N. *J. Chem. Soc.* 1935, 1430.
2. Brunisholz, G.; Quinche, J.P.; Kalo, A.M. *Helv. Chim. Acta* 1964, *47*, 14.
3. Popov, A.P.; Mironov, K.E. *Rev. Roum. Chim.* 1968, *13*, 765.
4. Spedding, F.H.; Shiers, L.E.; Rard, J.A. *J. Chem. Eng. Data* 1975, *20*, 88.
5. Rard, J.A.; Spedding, F.H.; *J. Phys. Chem.* 1975, *79*, 257.
6. Spedding, F.H.; Derer, J.L.; Mohs, M.A.; Rard, J.A. *J. Chem. Eng. Data* 1976, *21*, 474.
7. Quill, L.L.; Robey, R.F. *J. Am. Chem. Soc.* 1937, *59*, 2591.
8. Mironov, K.E.; Sinitsyna, E.D. *Izv. Sib. Otd. Akad. Nauk SSSR, Ser. Khim. Nauk* 1963, *11*, 3.
9. Zhuravlev, E.F.; Boeva, M.K. *Vses. Sb. Fazovye Ravnovesiya* 1975, *No. 2*, 45.
10. James, C.; Robinson, J.E. *J. Am. Chem. Soc.* 1913, *35*, 754.
11. Brunisholz, G.; Klipfel, K. *Rev. Chem. Miner.* 1970, *7*, 349.
12. Kuznetsova, G.P.; Yakimova, Z.P.; Yastrebova, L.F.; Stepin, B.D. *Zh. Neorg. Khim.* 1981, *26*, 3161.
13. Boeva, M.K.; Zhuravlev, E.F. *Zh. Neorg. Khim.* 1977, *22*, 1977.
14. Boeva, M.K.; Zhuravlev, E.F. *Zh. Neorg. Khim.* 1977, *22*, 1112.
15. Zhuravlev, E.F.; Boeva, M.K. *Zh. Neorg. Khim.* 1974, *19*, 3369.
16. Boeva, M.K.; Zhuravlev, E.F. *Zh. Neorg. Khim.* 1977, *22*, 263.
17. Boeva, M.K.; Ishmuratov, G.Y. *Issled Mnogokomponent. Sistem s Razl. Vzaimodeistviem Komponentov, Izd. Saratov Univ.* 1977, *No. 2*, 95.
18. Khudaibergenova, N.; Sulaimankulov, K. *Zh. Neorg. Khim.* 1981, *26*, 1107.
19. Perel'man, F.M.; Zvorykin, A.Ya.; Demina, G.A. *Zh. Neorg. Khim.* 1963, *8*, 1753.
20. Williamson, A.T. *Trans. Faraday Soc.* 1944, *40*, 421.
21. Counioux, J.-J.; Tenu, R. *J. Chim. Phys.* 1981, *78*, 816 and 823.
22. Quill, L.L.; Robey, R.F.; Seifter, S. *Ind. Eng. Chem. Anal. Ed.* 1937, *9*, 389.
23. Wendlandt, W.W.; Sewell, R.G. *Texas J. Sci.* 1961, *13*, 231.
24. Popov, A.P.; Mironov, K.E. *Zh. Neorg. Khim.* 1971, *16*, 464.
25. Kirgintsev, A.N. *Izv. Akad. Nauk SSSR. Ser. Khim. Nauk* 1965, *No. 8*, 1591.
26. Kolesnikov, A.A.; Korotkevich, I.B.; Bui Van Tuan; Stepin, B.D. *Zh. Neorg. Khim.* 1978, *23*, 2833.

COMPONENTS:

(1) Neodymium nitrate; $Nd(NO_3)_3$; [10045-95-1]

(2) Water ; H_2O ; [7732-18-5]

ORIGINAL MEASUREMENTS:

Friend, J.N. *J. Chem. Soc.* 1935, 1430-2.

VARIABLES:

Temperature: range 0°C to 67.5°C

PREPARED BY:

T. Mioduski, S. Siekierski and M. Salomon

EXPERIMENTAL VALUES:

Composition of saturated solutions[a]

t/°C	mass %	mol kg^{-1}	nature of the solid phase[b]
0.0[c]	55.97	3.711	α-$Nd(NO_3)_3 \cdot 6H_2O$
13.2	57.37	4.075	"
18.2[c]	58.03	4.187	"
23.0[c]	59.59	4.465	"
26.2[c]	60.69	4.675	"
27.2	58.17	4.211	β-$Nd(NO_3)_3 \cdot 6H_2O$
29.4	59.18	4.390	"
37.2	60.95	4.726	"
42.4	61.91	4.922	"
50.0	64.86	5.589	"
57.2[c]	67.00	6.148	"
66.2[c]	73.13	8.241	"
67.5[d]	75.34	9.251	"

a. Molalities calculated by the compilers.
b. Transition temperature for the α and β phases is approx 22°C.
c. Solid phase analysed and found to be the hexahydrate.
d. Melting point solubility calcd by the compilers.

AUXILIARY INFORMATION

METHOD/APPARATUS/PROCEDURE:

The isothermal method used as described earlier (1). Aliquots of around 50 cc were taken from 250 cc "saturation bottles" and Nd precipitated as the oxalate. Nd was determined gravimetrically by ignition of the oxalate to the oxide. As a check on the method, several samples of the saturated solutions were evaporated to dryness and directly ignited to the oxide. These results were always higher by 1.5 to 2 %, and the solubilities were therefore calculated from the results obtained by the oxalate method.

COMMENTS AND/OR ADDITIONAL DATA:

This paper is the only study that reports the existence of α and β phases for the hexahydrate. This is discussed further in the critical evaluation.

SOURCE AND PURITY OF MATERIALS:

Neodymium nitrate was prepared by dissolution of the oxide in dilute HNO_3. The filtered solution was concentrated on a water-bath and was seeded on the first occasion with the lanthanum salt (presumably the salt was then recrystallized, compilers). The melting point of the hexahydrate was 67.5°C.

The source and purity of water was not specified.

ESTIMATED ERROR:

Soly: precision around ± 1-2 % (compilers).

Temp: accuracy probably ± 0.05 K or better as in (1) (compilers).

REFERENCES:

1. Friend, J.N. *J. Chem. Soc.* 1930, 1633.

COMPONENTS:

(1) Neodymium nitrate; $Nd(NO_3)_3$; [10045-95-1]

(2) Water; H_2O; [7732-18-5]

ORIGINAL MEASUREMENTS:

Brunisholz, G.; Quinche, J.P.; Kalo, A.M. *Helv. Chim. Acta* 1964, *47*, 14-27.

VARIABLES:

Temperature

PREPARED BY:

T. Mioduski and S. Siekierski

EXPERIMENTAL VALUES:

Solubility of $Nd(NO_3)_3$ as a function of temperature [a]

t/°C	mass %	mol kg^{-1}	solid phase
0	55.71	3.809	$Nd(NO_3)_3 \cdot 6H_2O$
10	57.62	4.117	"
20	59.29	4.410	"
35	62.62	5.073	"
50	66.29	5.954	"

a. Molalities calculated by the compilers.

AUXILIARY INFORMATION

METHOD/APPARATUS/PROCEDURE:

The isothermal method was used. Neodymium was determined by complexometric titration using Xylenol orange indicator in the presence of a small quantity of urotropine buffer.

SOURCE AND PURITY OF MATERIALS:

Nd_2O_3 was purified by the ion exchange method, and had a purity of better than 99.7 %. The nitrate was prepared by dissolving the oxide in nitric acid followed by crystallization: details not given.

ESTIMATED ERROR:

Soly: precision around ± 0.2 % (compilers).

Temp: precision ± 0.05 K or better (compilers).

REFERENCES:

COMPONENTS:

(1) Neodymium nitrate; $Nd(NO_3)_3$; [10045-95-1]

(2) Water ; H_2O ; [7732-18-5]

ORIGINAL MEASUREMENTS:

Popov, A.P.; Mironov, K.E. *Rev. Roum. Chim.* 1968, *13*, 765-73.

VARIABLES:

Temperature: range -29°C to 122°C

PREPARED BY:

T. Mioduski and S. Siekierski

EXPERIMENTAL VALUES:

Solubility of $Nd(NO_3)_3$ as a function of temperature[a]

t/°C	mass %	mol kg^{-1}	solid phase
- 1.5	16.1	0.58	ice
- 2.3	20.7	0.79	"
- 6.3	32.8	1.48	"
-12.9	38.8	1.92	"
-21.0	45.2	2.50	"
-24.6	48.4	2.84	"
-28.7	51.5	3.22	$Nd(NO_3)_3\cdot6H_2O$
-17.3	53.1	3.43	"
- 3.7	55.1	3.72	"
4.3	56.6	3.95	"
22.1	60.2	4.58	"
38.9	64.3	5.45	$Nd(NO_3)_3\cdot6H_2O$ (metastable equil)
43.9	66.3	5.96	"
55.0[b]	71.2	7.49	"
60.7[b]	74.4	8.80	"
33.7	61.3	4.80	$Nd(NO_3)_3\cdot5H_2O$
43.3	63.5	5.27	"
45.4	64.7	5.55	"
50.8	66.2	5.93	"
55.6	68.6	6.62	"
63.0	72.7	8.06	"
63.0	72.6	8.02	"
68.9	76.4	9.80	"
85.6	80.9	12.83	lower hydrate
121.8	84.7	16.76	"

a. Molalities calculated by the compilers.
b. Isothermal method.

AUXILIARY INFORMATION

METHOD/APPARATUS/PROCEDURE:

For most of the determinations the synthetic method was used. The temperatures of crystallization were determined visually. Two data points were determined by the isothermal method.

No other information given.

SOURCE AND PURITY OF MATERIALS:

$Nd(NO_3)_3\cdot6H_2O$ was prepared by dissolving Nd_2O_3 (purity of 99.5 %) in nitric acid followed by crystallization.

Doubly distilled water was used.

ESTIMATED ERROR:

Soly: for data based on the synthetic method, reproducibility appears to be around ± 0.5 %. Isothermal measurements also appear to be precise to ± 0.5 % (compilers).

Temp: Authors state temperature control was ± 0.02 K.

COMPONENTS:

(1) Neodymium nitrate; $Nd(NO_3)_3$; [10045-95-1]

(2) Water; H_2O; [7732-18-5]

ORIGINAL MEASUREMENTS:

1. Spedding, F.H.; Shiers, L.E.; Rard, J.A. *J. Chem. Eng. Data* 1975, *20*, 88-93.
2. Rard, J.A.; Spedding, F.H. *J. Phys. Chem.* 1975, *79*, 257-62.
3. Spedding, F.H.; Derer, J.L.; Mohs, M.A.; Rard, J.A. *J. Chem. Eng. Data* 1976, *21*, 474-88.

VARIABLES:

One temperature: 25.00°C

PREPARED BY:

T. Mioduski, S. Siekierski, and M. Salomon

EXPERIMENTAL VALUES:

The solubility of $Nd(NO_3)_3$ in water at 25.00°C has been reported by Spedding and co-workers in three publications. Source paper [3] reports the solubility to be 4.582 mol kg^{-1}, but the preferred value is given in source papers [1] and [2] as 4.6184 mol kg^{-1}.

COMMENTS AND/OR ADDITIONAL DATA:

Source paper [1] reports the relative viscosity, η_R, of a saturated solution to be 20.440. Taking the viscosity of water at 25°C to equal 0.008903 poise, the viscosity of a saturated $Nd(NO_3)_3$ solution at 25°C is 0.18198 poise (compilers' calculation).

Supplementary data available in the microfilm edition to *J. Phys. Chem.* 1975, *79* have enabled compilers to provide the following additional data.

The density of the saturated solutions was calculated by the compilers from the smoothing equation, and at 25°C the value is 1.8400 kg m^{-3}. Using this density, the solubility in volume units is (based on the preferred value of 4.6184 mol kg^{-1})

$$c_{satd} = 3.365 \text{ mol dm}^{-3}$$

Source paper [2] reports the electrolyte conductivity of the saturated solution to be (corrected for the electrolytic conductivity of the solvent) κ = 0.021295 S cm^{-1}.

The molar conductivity of the saturated solution is calculated from 1000 $\kappa/3c_{satd}$ and is

$$\Lambda\ (\tfrac{1}{3}Nd(NO_3)_3) = 2.109 \text{ S cm}^2 \text{ mol}^{-1}$$

AUXILIARY INFORMATION

METHOD/APPARATUS/PROCEDURE:

Isothermal method used. Solutions were prepared as described in (1) and (3). The concentration of the saturated solution was determined by both EDTA (1) and sulfate (2) methods which is said to be reliable to 0.1% or better. In the sulfate analysis, the salt was first decomposed with HCl followed by evaporation to dryness before sulfuric acid additions were made. This eliminated the possibility of nitrate ion coprecipitation.

SOURCE AND PURITY OF MATERIALS:

$Nd(NO_3)_3 \cdot 6H_2O$ was prepd by addn of HNO_3 to the oxide. The oxide was purified by an ion exchange method and the upper limit for the impurities Ca, Fe, Si and adjacent rare earths was given as 0.15 %.

In source paper [3] the salt was analysed for water of hydration and found to be within ± 0.016 water molecules of the hexahydrate.

ESTIMATED ERROR:

Soly: duplicate analyses agreed to at least ± 0.1 %.

Temp: not specified, but probably accurate to at least ± 0.01 K as in (3) (comp)

REFERENCES:

1. Spedding, F.H.; Cullen, P.F.; Habenschuss, A. *J. Phys. Chem.* 1974, *78*, 1106.
2. Spedding, F.H.; Pikal, M.J.; Ayers, B.O. *J. Phys. Chem.* 1966, *70*, 2440.
3. Spedding, F.H.; et.al. *J. Chem. Eng. Data* 1975, *20*, 72.

COMPONENTS:

(1) Neodymium nitrate; $Nd(NO_3)_3$; [10045-95-1]

(2) Nitric acid; HNO_3; [7697-37-2]

(3) Water ; H_2O ; [7732-18-5]

ORIGINAL MEASUREMENTS:

Quill, L.L.; Robey, R.F. *J. Am. Chem. Soc.* 1937, *59*, 2591-5.

VARIABLES:

Composition at 25°C and 50°C

PREPARED BY:

T. Mioduski, S. Siekierski, and M. Salomon

EXPERIMENTAL VALUES:

Solubility of $Nd(NO_3)_3$ in nitric adic solutions[a]

	$Nd(NO_3)_3$		HNO_3		density	
t/°C	mass %	mol kg^{-1}	mass %	mol kg^{-1}	kg m^{-3}	nature of the solid phase
25	53.31	3.986	6.20	2.43	1.741	$Nd(NO_3)_3 \cdot 6H_2O$
	52.15	3.806	6.36	2.43		"
	40.79	3.018	18.28	7.09		"
	23.64	2.209	43.95	21.52		"
	24.08	2.330	44.63	22.64		"
	23.47	2.600	49.20	28.57	1.595	$Nd(NO_3)_3 \cdot 4H_2O$
	19.11	2.387	56.95	37.09	1.572	"
50	66.26	5.946	0	———	1.963	$Nd(NO_3)_3 \cdot 6H_2O$
	64.62	5.886	2.14	1.02	1.948	"
	60.34	5.581	6.92	3.35	1.901	"
	57.74	5.489	10.41	5.19	1.885	"
	59.47	6.296	11.93	6.62	1.997	$Nd(NO_3)_3 \cdot 6H_2O + Nd(NO_3)_3 \cdot 4H_2O$
	51.24	5.946	22.67	13.79		$Nd(NO_3)_3 \cdot 4H_2O$
	47.36	5.556	26.83	16.50	1.819	"
	32.49	4.694	46.55	35.25		"
	34.97	5.411	45.46	36.86		"

a. Molalities calculated by M. Salomon

AUXILIARY INFORMATION

METHOD/APPARATUS/PROCEDURE:

Isothermal method . Appropriate quantities of $Nd(NO_3)_3$, HNO_3 and H_2O were placed in Pyrex tubes, heated to induce supersaturation, and thermostated. The tubes were sealed after a small crystal of $Bi(NO_3)_3 \cdot 5H_2O$ was added to "seed" crystallization. The sealed tubes were shaken in the thermostat for at least 8 hours (equilibrium was reached in 4 hours). Authors state that approach to equilibrium from undersaturation gave identical results within experimental error. All data reported in the above table are the results obtained by approach from supersaturation.

A "filtering pipet" maintained at a temperature slightly higher than the thermostat temperature was used to withdraw samples for analyses. Weighed samples of liquid and solid phases were analysed for HNO_3 by titrn with 0.1 mol dm^{-3} NaOH using methyl red indicator. Nd was pptd as the oxalate, filtered, washed with hot dilute oxalic acid, and ignited to the oxide.

SOURCE AND PURITY OF MATERIALS:

C.p. grade HNO_3 was distilled in an all-Pyrex still, and the middle fraction retained for use. For very high HNO_3 concentrations, reagent grade fuming HNO_3 was used as received. $Nd(NO_3)_3 \cdot 6H_2O$ prepared by dissolving the oxide in pure HNO_3, recrystallizing twice, and drying over 55% sulfuric acid in a desiccator. No trace of any other rare earth was found in the oxide by arc emission spectroscopy.

Distilled water was used which had a conductivitity of 2 x 10^{-6} S cm^{-1}.

ESTIMATED ERROR:

Soly: precision probably ± 0.2 % (compilers).
Temp: at 25°C accuracy was ± 0.03 K.
at 50°C accuracy was ± 0.1 K.

REFERENCES:

COMPONENTS:

(1) Neodymium nitrate; $Nd(NO_3)_3$; [10045-95-1]

(2) Nitric acid; HNO_3; [7697-37-2]

(3) Water ; H_2O ; [7732-18-5]

ORIGINAL MEASUREMENTS:

Mironov, K.E.; Sinitsyna, E.D. *Izv. Sib. Otd. Akad. Nauk SSSR, Ser. Khim. Nauk,* 1963, *11*, 3-7.

VARIABLES:

Concentration of HNO_3 at 25.0°C

PREPARED BY:

T. Mioduski and S. Siekierski

EXPERIMENTAL VALUES:

$Nd(NO_3)_3$ mass %	$Nd(NO_3)_3$ mol kg^{-1}	HNO_3 mass %	HNO_3 mol kg^{-1}	solid phase
60.28	4.595	0	——	A
44.8	3.34	14.6	5.71	A
39.3	2.92	19.9	7.74	A
25.3	1.96	35.6	14.5	A
25.2	2.05	37.6	16.0	A
25.2	2.06	37.8	16.2	A
24.5	2.00	38.4	16.4	A
23.8	2.04	40.8	18.3	A
23.9	2.25	44.0	21.8	A
25.5	2.65	45.4	24.8	A
23.8	2.56	48.0	27.0	B
17.7	2.59	61.6	47.2	B
18.9	2.97	61.8	50.8	B
18.0	2.85	62.9	52.3	B
19.7	3.66	64.0	62.3	B
17.5	2.91	64.3	56.1	B
18.9	4.18	67.4	78.1	C
18.8	4.16	67.5	78.2	C
16.0	3.51	70.2	80.7	C
14.5	3.43	72.7	90.1	C
13.3	2.98	73.2	86.0	C
13.5	3.27	74.0	93.9	C
5.5	1.39	82.5	109.1	C
6.2	1.82	83.5	128.7	C
3.2	0.99	87.0	140.9	C
1.0	0.43	92.0	208.6	C
1.5	0.73	92.3	236.3	C
0.5	0.28	94.0	271.2	C
0.3	0.25	96.0	411.8	D
0.3	0.28	96.4	463.6	D
0.3	0.76	98.5	1302	D
0.1	0.30	98.9	1570	D

a. Molalities calculated by M. Salomon.

b. Solid phases are:

A = $Nd(NO_3)_3 \cdot 6H_2O$, B = $Nd(NO_3)_3 \cdot 4H_2O$

C = $Nd(NO_3)_3 \cdot H_2O$, D = $Nd(NO_3)_3 \cdot nH_2O$

AUXILIARY INFORMATION

METHOD/APPARATUS/PROCEDURE:

The isothermal method was used. Solutions were equilibrated for 2-3 hours, and both the saturated solutions and solid phases were analysed. Neodymium was determined either gravimetrically or by complexometric titration, and nitric acid was determined by titration with standard 0.1 mol dm^{-3} NaOH solution.

SOURCE AND PURITY OF MATERIALS:

$Nd(NO_3)_3 \cdot 6H_2O$ was prepared as in (1) by dissolving the oxide in HNO_3. Impurities in Nd_2O_3 were: CeO_2 and Sm_2O_3 less than 0.1 %, and La_2O_3 and Pr_6O_{11} less than 0.3 %.

Nitric acid was concentrated to 100 % acid by the Brauer method, and was slightly yellow.

Doubly distilled water was used.

ESTIMATED ERROR:

Soly: precision ± 0.2-0.5 % (compilers).

Temp: precision ± 0.1 K.

REFERENCES:

1. Quill, L.L.; Robey, R.F. *J. Am. Chem. Soc.* 1937, *59*, 2591.

COMPONENTS:	ORIGINAL MEASUREMENTS:
(1) Rubidium nitrate; $RbNO_3$; [13126-12-0] (2) Neodymium nitrate; $Nd(NO_3)_3$; [10045-95-1] (3) Nitric acid; HNO_3; [7697-37-2] (4) Water; H_2O; [7732-18-5]	Perel'man, F.M.; Zvorykin, A. Ya.; Demina, G.A. *Zh. Neorg. Khim.* 1963, *8*, 1753-5; *Russ. J. Inorg. Chem. Engl. Transl.* 1963, *8*, 909-11.
VARIABLES:	PREPARED BY:
Composition at 25°C	T. Mioduski and S. Siekierski

EXPERIMENTAL VALUES:

$Nd(NO_3)_3$		$RbNO_3$		HNO_3			
mass %	mol kg^{-1}	mass %	mol kg^{-1}	mass %	mol kg^{-1}	nature of the solid phase	
25.67	2.07	———	———	36.7	15.48	$Nd(NO_3)_3 \cdot 6H_2O$	
29.57	2.74	6.5	1.35	31.2	15.13	"	
34.33	3.67	8.81	2.11	28.5	15.95	"	
31.33	3.42	11.41	2.79	29.55	16.92	"	
34.88	4.01	14.36	3.69	24.40	14.69	"	
22.49	3.36	20.84	6.98	36.42	28.54	$4Nd(NO_3)_3 \cdot 5RbNO_3$	
18.26	2.93	29.09	10.47	33.81	28.48	"	
18.19	3.70	34.61	17.75	32.30	34.40	"	
13.57	2.83	36.76	17.16	35.14	38.38	$2Nd(NO_3)_3 \cdot 7.5RbNO_3$	(A)
15.58	3.17	38.24	17.44	31.31	33.42	"	
13.28	3.18	39.19	21.12	35.0	44.15	"	
11.84	1.74	39.57	13.03	28.0	21.58	"	
12.0	2.64	41.37	20.42	32.89	37.99	"	
8.20	1.68	43.88	20.13	33.14	35.58	"	
8.25	1.90	45.35	23.35	33.23	40.04	A + B	
5.17	0.70	44.52	13.42	27.82	19.63	$RbNO_3$	(B)
———	———	44.02	12.24	31.60	20.57	"	
———	———	44.71	13.06	32.08	21.93	"	

Molalities calculated by M. Salomon.

AUXILIARY INFORMATION

METHOD/APPARATUS/PROCEDURE:

The isothermal method was used. The equilibration period varied from 2 to 3 days. Both the liquid and solid phases were analysed for Nd and Rb. Neodymium was determined by precipitation with ammonia and ignition to the oxide, and rubidium was determined as the perchlorate. The total nitric acid content in solution was determined by acidimetric titration.

The compositions of the solid phases were determined by Schreinemakers' method of residues.

SOURCE AND PURITY OF MATERIALS:

$Nd(NO_3)_3 \cdot 6H_2O$ prepared by dissolving c.p. grade oxide in nitric acid followed by crystallization.

$RbNO_3$ prepared by dissolving c.p. grade rubidium carbonate in nitric acid followed by crystallization.

ESTIMATED ERROR:

Soly: nothing specified.

Temp: precision ± 0.1 K.

REFERENCES:

COMPONENTS:

(1) Neodymium nitrate; $Nd(NO_3)_3$; [10045-95-1]

(2) Neodymium oxalate; $Nd_2(C_2O_4)_3$; [1186-50-1]

(3) Water; H_2O; [7732-18-5]

ORIGINAL MEASUREMENTS:

James, C.; Robinson, J.E. *J. Am. Chem. Soc.* 1913, *35*, 754-9.

VARIABLES:

Composition at 25°C

PREPARED BY:

T. Mioduski, S. Siekierski, and M. Salomon

EXPERIMENTAL VALUES:

Composition of saturated solutions at 25°C [a]

$Nd(NO_3)_3$ mass %	$Nd(NO_3)_3$ mol kg^{-1}	$Nd_2(C_2O_4)_3$ mass %	$Nd_2(C_2O_4)_3$ mol kg^{-1}
6.46	0.210	0.18	0.003
12.23	0.425	0.54	0.011
17.78	0.661	0.76	0.017
22.67	0.898	0.85	0.020
24.43	1.160	0.96	0.024
31.36	1.410	1.28	0.034
35.26	1.685	1.38	0.039
38.70	1.965	1.66	0.050
42.13	2.278	1.88	0.061
44.82	2.550	1.96	0.067
47.64	2.868	2.07	0.074
50.52	3.259	2.54	0.098
52.82	3.611	2.89	0.118
54.67	3.926	3.17	0.136
56.48	4.140	2.21	0.097
59.70	4.624	1.21	0.056
59.67	4.633	1.33	0.062
59.68	4.648	1.44	0.067
59.75	4.605	0.96	0.044
60.46	4.630	——	——

a. Molalities calculated by the compilers. Nature of the solid phases not completely identified (see discussions below).

COMMENTS AND/OR ADDITIONAL DATA:

The authors state that it was not possible to analyse the solid phases involving $Nd(NO_3)_3$ due to the lack of sufficient data points. The compilers assume that these solid phases include $Nd(NO_3)_3 \cdot 6H_2O$.

AUXILIARY INFORMATION

METHOD/APPARATUS/PROCEDURE:

The isothermal method was used. Mixtures were sealed in glass-stoppered bottles with paraffin and rotated in a thermostat at 25°C for 12 weeks. The authors assumed equilibrium had been attained after this period.

An aliquot was removed for analysis with a pipet and weighed. Nd was precipitated as the oxalate by addition of a precisely known quantity of excess oxalic acid. The ppt was filtered, washed, and ignited to the oxide. The excess oxalic acid in the filtrate was titrated with standard $KMnO_4$ solution after the addition of 10 % H_2SO_4. The amount, in grams, of oxalate was calculated from the weight of Nd_2O_3. To this weight was added the number of grams of oxalate from the $KMnO_4$ titration giving the total oxalate composition of the saturated solution + the initial addition of the known quantity of oxalic acid: the amount of $Nd_2(C_2O_4)_3$ originally present in the satd solution is thus the difference of this total oxalate and the amount added to precipitate Nd. $Nd(NO_3)_3$ was obtained by difference. In slns of lower NO_3 content, the solid phase was analysed by the method of residues. Construction of the tie-lines

METHOD/APPARATUS/PROCEDURE: (continued)

joined at a point corresponding to $Nd_2(C_2O_4)_3 \cdot 11H_2O$.

SOURCE AND PURITY OF MATERIALS:

Neodymium oxalate was precipitated from a $2Nd(NO_3)_3 \cdot 3Mg(NO_3)_2$ solution, ignited to the oxide, dissolved in HCl, and reprecipitated as the oxalate. The oxalate used for soly determinations was dried at room temp. The nitrate was prepared by dissolving the oxide in slight excess of HNO_3, and evaporating to crystallization.

ESTIMATED ERROR:

Soly: precision around ± 0.3 % (compilers).

Temp: precision probably ± 0.2 K (compilers).

COMPONENTS:

(1) Neodymium nitrate; $Nd(NO_3)_3$; [10045-95-1]

(2) Samarium nitrate; $Sm(NO_3)_3$; [10361-83-8]

(3) Water ; H_2O ; [7732-18-5]

ORIGINAL MEASUREMENTS:

Brunisholz, G.; Quinche, J.P.; Kalo, A.M.
Helv. Chim. Acta 1964, *47*, 14-27.

VARIABLES:

Composition at 20°C

PREPARED BY:

T. Mioduski, S. Siekierski, and M. Salomon

EXPERIMENTAL VALUES:

mol % Sm of total Nd + Sm	moles H_2O per 100 moles Nd + Sm	$Nd(NO_3)_3$ [a] mol kg^{-1}	$Sm(NO_3)_3$ [a] mol kg^{-1}	solid phase
0.0	1259	4.409	——	$Nd(NO_3)_3 \cdot 6H_2O$
10.0	1254	3.984	0.443	solid solution[b]
19.4	1271	3.520	0.847	"
47.3	1298	2.254	2.023	"
76.6	1326	0.980	3.207	"
88.5	1334	0.479	3.683	"
100.0	1350	——	4.112	$Sm(NO_3)_3 \cdot 6H_2O$

a. Molalities calculated by the compilers.
b. Continuous series of solid solutions $(Nd, Sm)_3(NO_3)_3 \cdot 6H_2O$

AUXILIARY INFORMATION

METHOD/APPARATUS/PROCEDURE:

The isothermal method was used. The solids were placed in glass-stoppered vials into which a glass dumb-bell shaped pestle was placed. These vials were placed in a second larger tube and sealed with a rubber stopper. The tubes were placed in a thermostat and rotated end-over-end so that the pestle pulverized the solid two times through each 360° rotation. The solutions were equilibrated for 3 to 5 weeks undergoing about 500,000 pulverizations per week.

For the saturated solutions Ln was determined by titration with Trilon using Xylenol Orange indicater and urotropine buffer , and by ion exchange chromatography.

The compositions of the solid phases were determined by Schreinemakers' method of residues, and by X-ray analysis.

SOURCE AND PURITY OF MATERIALS:

The nitrates were obtained by dissolving the oxide in nitric acid followed by crystallization. The source and purities of the oxides were not specified. The hydrated nitrates were dried in a desiccator over KOH under vacuum.

ESTIMATED ERROR:

Soly: precision ± 0.2 % (compilers).

Temp: precision ± 0.05 K or better (compilers).

REFERENCES:

COMPONENTS:

(1) Neodymiun nitrate; $Nd(NO_3)_3$; [10045-95-1]

(2) Zinc nitrate; $Zn(NO_3)_2$; [7779-88-6]

(3) Water ; H_2O ; [7732-18-5]

ORIGINAL MEASUREMENTS:

Brunisholz, G. Klipfel, K. *Rev. Chem. Miner.* 1970, *7*, 349-58.

VARIABLES:

Composition at 20°C

PREPARED BY:

T. Mioduski and S. Siekierski

EXPERIMENTAL VALUES:

Composition of saturated solutions at 20.000°C[a]

mol % Nd of total Nd + Zn	moles H_2O per 100 moles Nd + Zn	$Nd(NO_3)_3$ mol kg^{-1}	$Zn(NO_3)_2$ mol kg^{-1}	solid phase[b]
0.0	892	——	6.223	A
0.27	860	0.017	6.437	A+B
0.82	961	0.047	5.729	B
8.49	1141	0.413	4.452	B
22.50	1200	1.041	3.585	B
39.57	1215	1.808	2.761	B
69.70	1176	3.290	1.430	B
75.20	1155	3.614	1.192	B+C
75.22	1153	3.621	1.193	B+C
80.11	1184	3.756	0.932	C
88.31	1218	4.025	0.533	C
92.50	1235	4.158	0.337	C
100.0	1260	4.405	——	C

a. Molalities calculated by M. Salomon.

b. Solid phases: A = $Zn(NO_3)_2 \cdot 6H_2O$
B = $2Nd(NO_3)_3 \cdot 3Zn(NO_3)_2 \cdot 24H_2O$
C = $Nd(NO_3)_3 \cdot 6H_2O$

AUXILIARY INFORMATION

METHOD/APPARATUS/PROCEDURE:

The isothermal method was used. The solids were placed in glass-stoppered vials into which a glass dumb-bell shaped pestle was placed. These vials were placed in a second larger tube and sealed with a rubber stopper. The tubes were placed in a thermostat and rotated end-over-end so that the pestle pulverized the solid two times through each 360° rotation. The solutions were equilibrated for 2 to 4 weeks.

For the saturated solutions Nd was determined by titration with Trilon uning Xylenol Orange indicater and urotropine buffer. Zn was determined by precipitation as the sulfide (using thioacetamide), or by ion exchange chromatography.

The compositions of the solid phases were determined by Schreinemakers' method of residues, and by X-ray analysis.

SOURCE AND PURITY OF MATERIALS:

The nitrates were obtained by dissolving the oxide in nitric acid followed by crystallization. The source and purities of the oxides were not specified. The hydrated nitrates were dried in a desiccator over KOH under vacuum.

ESTIMATED ERROR:

Soly: precision ± 0.2 % (compilers).

Temp: precision ± 0.005 K.

REFERENCES:

COMPONENTS:

(1) Neodymium nitrate; $Nd(NO_3)_3$; [10045-95-1]

(2) Aluminum nitrate; $Al(NO_3)_3$; [13473-90-0]

(3) Water; H_2O; [7732-18-5]

ORIGINAL MEASUREMENTS:

Kuznetsova, G.P.; Yakimova, Z.P.; Yastrebova, L.F.; Stepin, B.D. *Zh. Neorg. Khim.* 1981, *26*, 3161-4; *Russ. J. Inorg. Chem. Engl. Transl.* 1981, *26*, 1692-3.

VARIABLES:

Composition at 25°C

PREPARED BY:

Mark Salomon

EXPERIMENTAL VALUES:

Composition of saturated solutions [a]

$Nd(NO_3)_3$		$Al(NO_3)_3$		
mass %	mol kg^{-1}	mass %	mol kg^{-1}	nature of the solid phase
60.28	4.595	———	———	$Nd(NO_3)_3 \cdot 6H_2O$
48.04	3.333	8.32	0.895	$Nd(NO_3)_3 \cdot 6H_2O + Al(NO_3)_3 \cdot 9H_2O$
———	———	40.73	3.226	$Al(NO_3)_3 \cdot 9H_2O$

a. Molalities calculated by the compilers.

AUXILIARY INFORMATION

METHOD/APPARATUS/PROCEDURE:

The isothermal method was used. Neodymium was determined complexometrically using sulfosalicylate as a masking-agent for aluminum. Aluminum was determined by back-titration of excess EDTA with standard $ZnSO_4$ solution using dithizone indicator (1).

SOURCE AND PURITY OF MATERIALS:

Neodymium nitrate was prepared by dissolving high purity oxide in nitric acid. The nitric acid was c.p. grade material which was distilled.

A.R. grade $Al(NO_3)_3 \cdot 9H_2O$ was recrystallized before use.

ESTIMATED ERROR:

Soly: precision probably ± 0.2 % (compilers).

Temp: nothing specified.

REFERENCES:

1. Grosskreutz, W.; Schultze, D.; Wilke, K.T. *Z. Anal. Chem.* 1967, *232*, 278.

COMPONENTS:

(1) Neodymium nitrate; $Nd(NO_3)_3$; [10045-95-1]

(2) Hydrazine mononitrate; $N_2H_4 \cdot HNO_3$; [13464-97-6]

(3) Water; H_2O; [7732-18-5]

ORIGINAL MEASUREMENTS:

Boeva, M.K.; Zhuravlev, E.F. *Zh. Neorg. Khim.* 1977, *22*, 1977-81; *Russ. J. Inorg. Chem. Engl. Transl.* 1977, *22*, 1073-5.

VARIABLES:

Composition at 20°C and 40°C

PREPARED BY:

T. Mioduski and S. Siekierski

EXPERIMENTAL VALUES:

Composition of saturated solutions at 20°C [a]

$Nd(NO_3)_3$ mass %	$Nd(NO_3)_3$ mol kg^{-1}	$N_2H_4 \cdot HNO_3$ mass %	$N_2H_4 \cdot HNO_3$ mol kg^{-1}	nature of the solid phase
0	——	71.0	25.76	$N_2H_4 \cdot HNO_3$
5.5	0.61	67.0	25.63	"
12.0	1.30	60.0	22.54	"
14.0	1.51	58.0	21.79	"
21.0	2.45	53.0	21.44	"
28.0	4.24	52.0	27.35	"
29.0	3.99	49.0	23.43	"
36.0	6.81	48.0	31.56	$N_2H_4 \cdot HNO_3 + Nd(NO_3)_3 \cdot 3[N_2H_4 \cdot HNO_3]$
37.0	6.59	46.0	28.47	$Nd(NO_3)_3 \cdot 3[N_2H_4 \cdot HNO_3]$
40.0	7.12	43.0	26.61	$Nd(NO_3)_3 \cdot 3[N_2H_4 \cdot HNO_3] + Nd(NO_3)_3$
44.5	6.42	34.5	17.28	$Nd(NO_3)_3$
49.5	6.66	28.0	13.09	$Nd(NO_3)_3 \cdot 6H_2O$
51.0	6.71	26.0	11.89	"
52.0	6.56	24.0	10.52	"
53.5	4.63	11.5	3.46	"
59.0	4.36	0	———	"

continued.........

AUXILIARY INFORMATION

METHOD/APPARATUS/PROCEDURE:

The method of isothermal sections was used with refractometric analyses (1). Heterogeneous and homogeneous mixtures of known composition were equilibrated until their refractive indices remained constant. The composition of the saturated solutions and the corresponding solid phases were found as inflection or "break" points on a plot of composition against refractive index.

SOURCE AND PURITY OF MATERIALS:

A.R. grade $Nd(NO_3)_3 \cdot 6H_2O$, c.p. grade nitric acid, A.R. grade hydrazine, and doubly distilled water used. No other details given.

The neodymium salt was probably used as received. Hydrazine mononitrate probably prepared as in (2): this paper has been compiled elsewhere in this volume (see chapter on lanthanum nitrates).

ESTIMATED ERROR:

Soly: precision about ± 1 % (compilers).

Temp: precision probably ± 0.2 K (compilers).

REFERENCES:

1. Zhuravlev, E.F.; Sheveleva, A.D. *Zh. Neorg. Khim.* 1960, *5*, 2630.

2. Gorshunova, V.P.; Zhuravlev, E.F. *Zh. Neorg. Khim.* 1971, *16*, 1700.

COMPONENTS:

(1) Neodymium nitrate; $Nd(NO_3)_3$; [10045-95-1]

(2) Hydrazine mononitrate; $N_2H_4\cdot HNO_3$; [13464-97-6]

(3) Water; H_2O; [7732-18-5]

ORIGINAL MEASUREMENTS:

Boeva, M.K.; Zhuravlev, E.F. *Zh. Neorg. Khim.* 1977, 22, 1977-81; *Russ. J. Inorg. Chem. Engl. Transl.* 1977, 22, 1073-5.

EXPERIMENTAL VALUES: continued........

Composition of saturated solutions at 40°C [a]

$Nd(NO_3)_3$ mass %	$Nd(NO_3)_3$ mol kg^{-1}	$N_2H_4\cdot HNO_3$ mass %	$N_2H_4\cdot HNO_3$ mol kg^{-1}	nature of the solid phase
0	——	85.0	59.61	$N_2H_4\cdot HNO_3$
3.0	0.61	82.0	57.51	"
6.0	1.21	79.0	55.40	"
11.0	2.22	74.0	51.90	"
14.5	2.37	67.0	38.10	"
19.0	2.50	58.0	26.53	"
28.0	4.71	54.0	31.56	"
34.0	6.06	49.0	30.22	$N_2H_4\cdot HNO_3$ + $Nd(NO_3)_3\cdot 3[N_2H_4\cdot HNO_3]$
36.0	6.81	48.0	31.56	$Nd(NO_3)_3\cdot 3[N_2H_4\cdot HNO_3]$
39.0	6.95	44.0	27.23	"
43.0	7.23	39.0	22.79	"
60.0	10.09	22.0	12.86	$Nd(NO_3)_3\cdot 6H_2O$
64.0	8.43	13.0	5.95	"
61.5	6.11	8.0	2.76	"
62.0	5.36	3.0	0.90	"
62.0	4.94	0	——	"

a. Molalities calculated by M. Salomon.

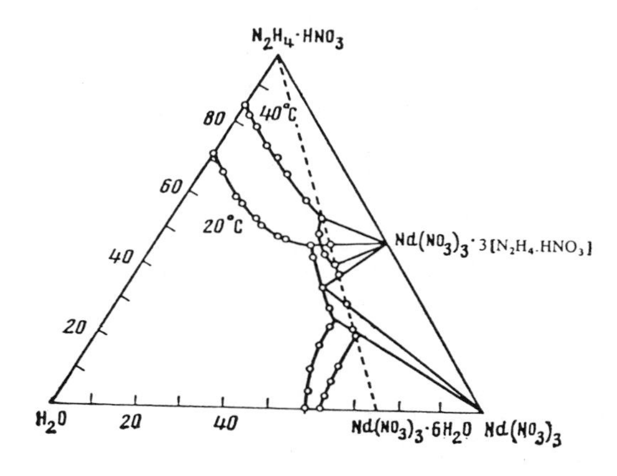

COMPONENTS:

(1) Neodymium nitrate; $Nd(NO_3)_3$; [10045-92-1]

(2) Trimethylamine nitrate; $C_3H_{10}N_2O_3$; [25238-43-1]

(3) Water ; H_2O ; [7732-18-5]

ORIGINAL MEASUREMENTS:

Boeva, M.K.; Zhuravlev, E.F. *Zh. Neorg. Khim.* 1977, *22*, 1112-5, *Russ. J. Inorg. Chem. Engl. Transl.* 1977, *22*, 610-2.

VARIABLES:

Composition at 20°C and 40°C

PREPARED BY:

T. Mioduski and S. Siekierski

EXPERIMENTAL VALUES:

20°C Isotherm[a]					40°C Isotherm[a]			
$Nd(NO_3)_3$		$(CH_3)_3N \cdot HNO_3$			$Nd(NO_3)_3$		$(CH_3)_3N \cdot HNO_3$	
mass %	mol kg^{-1}	mass %	mol kg^{-1}	solid phase[b]	mass %	mol kg^{-1}	mass %	mol kg^{-1}
0	——	66.5	16.25	A	0	——	73.5	22.71
5.8	0.56	63.0	16.53	A	4.5	0.57	71.5	24.39
12.5	1.38	60.0	17.87	A	9.6	1.36	69.0	26.40
20.5	2.59	55.5	18.94	A	16.0	2.62	65.5	28.99
25.0	3.52	53.5	20.38	A	20.0	3.67	63.5	31.51
28.0	4.76	54.2	24.93	A+B	28.5	9.59	62.5	56.86
34.8	5.30	45.3	18.64	B	29.8	7.10	57.5	37.07
42.5	6.28	37.0	14.78	B	36.8	7.33	48.0	25.86
51.5	7.92	28.8	11.97	B	44.0	7.61	38.5	18.01
				B	50.0	9.18	33.5	16.62
57.0	9.33	24.5	10.84	B+C	60.5	9.64	20.5	8.83
56.8	6.14	15.2	4.45	C	61.2	6.71	11.2	3.32
57.7	4.68	5.0	1.10	C	61.8	6.06	7.3	1.93
59.0	4.36	0	——	C	62.0	5.44	3.5	0.83
				C	62.0	4.94	0	——

a. Molalities calculated by M. Salomon.

b. Solid phases: A = $(CH_3)_3N \cdot HNO_3$, B = $Nd(NO_3)_3 \cdot 3(CH_3)_3N \cdot HNO_3$

C = $Nd(NO_3)_3 \cdot 6H_2O$

AUXILIARY INFORMATION

METHOD/APPARATUS/PROCEDURE:

The method of isothermal sections was used with refractometric analyses (1). Heterogeneous and homogeneous mixtures of known composition were equilibrated until their refractive indices remained constant. The time required for equilibrium was not specified. The composition of the saturated solutions and the corresponding solid phases were found as inflection or "break" points on a plot of composition against refractive index.

COMMENTS AND/OR ADDITIONAL DATA:

The hydrated double salt occupies a very large crystallization field on the phase diagram. The compound was isolated and subjected to differential thermal analysis studies. The melting point of the hydrate was 25°C, and the anhydrous double salt melted at 250°C. Decomposition occurred in the range of 325-360°C.

SOURCE AND PURITY OF MATERIALS:

All materials were prepared as described previously (2,3). No other information given.

ESTIMATED ERROR:

Soly: precision about ± 1 % (compilers).

Temp: precision probably ± 0.2 K (compilers).

REFERENCES:

1. Zhuravlev, E.F.; Sheveleva, A.D. *Zh. Neorg. Khim.* 1960, *5*, 2630.
2. Zhuravlev, E.F.; Boeva, M.K. *Zh. Neorg. Khim.* 1974, *19*, 3369.
3. Boeva, M.K.; Ishmuratov, G. Yu. in *Tez. Dokl Vses. Konf. po R.E.E. Izd. Saratov. Gos. Univ.* 1975, 76.

COMPONENTS:

(1) Neodymium nitrate; $Nd(NO_3)_3$; [10045-95-1]

(2) Tetramethylammonium nitrate; $C_4H_{12}N_2O_3$; [1941-24-8]

(3) Water; H_2O; [7732-18-5]

ORIGINAL MEASUREMENTS:

Boeva, M.K.; Zhuravlev, E.F. *Zh. Neorg. Khim.* 1977, 22, 1112-5; *Russ. J. Inorg. Chem. Engl. Transl.* 1977, 22, 610-2.

VARIABLES:

Composition at 20°C and 40°C

PREPARED BY:

T. Mioduski and S. Siekierski

EXPERIMENTAL VALUES:

20°C Isotherm[a]					40°C Isotherm[a]			
$Nd(NO_3)_3$		$(CH_3)_4N{\cdot}NO_3$			$Nd(NO_3)_3$		$(CH_3)_4N{\cdot}NO_3$	
mass %	mol kg^{-1}	mass %	mol kg^{-1}	solid phase[b]	mass %	mol kg^{-1}	mass %	mol kg^{-1}
0	——	48.2	7.68	A	0	——	52.5	9.13
8.5	0.57	46.0	8.35	A	7.6	0.52	48.5	9.12
19.0	1.31	37.0	6.94	A	17.0	1.30	43.5	9.09
24.3	1.86	36.2	7.57	A+B	19.0	1.48	42.0	8.89
22.0	1.57	35.5	6.90	B	23.0	1.79	38.0	8.04
24.5	1.73	32.5	6.24	B	26.5	2.08	35.0	7.51
30.5	2.20	27.5	5.41	B	30.5	2.43	31.5	6.84
34.5	2.55	24.5	4.93	B	32.5	2.59	29.5	6.41
38.5	2.90	21.3	4.37	B	41.0	3.40	22.5	5.09
50.5	4.25	13.5	3.10	B	53.5	5.14	15.0	3.93
57.2	5.28	10.0	2.52	B+C	61.4	6.74	11.0	3.29
57.5	5.12	8.5	2.06	C	61.8	6.16	7.8	2.12
58.0	5.09	7.5	1.79	C	61.9	5.93	6.5	1.70
58.0	4.88	6.0	1.38	C	62.0	5.52	4.0	0.97
52.2[c]	3.57	3.5	0.65	C	62.2	5.29	2.2	0.51
59.0	4.36	0	——	C	62.0	4.94	0	——

a. Molalities calculated by M. Salomom.

b. Solid phases: A = $(CH_3)_4N{\cdot}NO_3$, B = $Nd(NO_3)_3{\cdot}(CH_3)_4N{\cdot}NO_3{\cdot}5H_2O$

C = $Nd(NO_3)_3{\cdot}6H_2O$

c. Compilers would like to point out the possibility of a typographical error for this mass %.

AUXILIARY INFORMATION

METHOD/APPARATUS/PROCEDURE

The method of isothermal sections were used with refractometric analyses (1). Heterogeneous and homogeneous mixtures of known composition were equilibrated until their refractive indices remained constant. The time required for equilibrium was not specified. The composition of the saturated solutions and the corresponding solid phases were found as inflection or "break" points on a plot of composition against refractive index.

COMMENTS AND/OR ADDITIONAL DATA:

The double nitrate is congruently soluble in the temperature range studied, and it occupies a well developed crystn field on the phase diagram. The compound was isolated and its composition confirmed by chemical analysis (no details given). Differential thermal analysis studies gave a melting point of 40°C for this compound which is the same value obtained by the capillary method. The DTA curve also showed extensive decomposition at 205-205°C.

SOURCE AND PURITY OF MATERIALS

All materials were prepared as described previously (2,3). No other information given.

ESTIMATED ERROR:

Soly: precision about ± 1 % (compilers).

Temp: precision probably ± 0.2 K (compilers).

REFERENCES:

1. Zhuravlev, E.F.; Sheveleva, A.D. *Zh. Neorg. Khim.* 1960, 5, 2630.
2. Zhuravlev, E.F.; Boeva, M.K. *Zh. Neorg. Khim.* 1974, 19, 3369.
3. Boeva, M.K.; Ishmuratov, G. Yu. in *Tez. Dokl. Vses. Konf. po R.E.E. Izd. Saratov. Gos. Univ.* 1975, 76.

COMPONENTS:

(1) Neodymium nitrate; $Nd(NO_3)_3$; [10045-95-1]

(2) Diethylamine nitrate; $C_4H_{12}N_2O_3$; [27096-30-6]

(3) Water ; H_2O ; [7732-18-5]

ORIGINAL MEASUREMENTS:

Zhuravlev, E.F.; Boeva, M.K. *Zh. Neorg. Khim.* 1974, *19*, 3369-73; *Russ. J. Inorg. Chem. Eng. Transl.* 1974, *19*, 1846-9.

VARIABLES:

Composition at 20°C and 40°C

PREPARED BY:

T. Mioduski and S. Siekierski

EXPERIMENTAL VALUES:

20°C Isotherm[a]				40°C Isotherm[a]				
$Nd(NO_3)_3$		$(C_2H_5)_2NH\cdot HNO_3$		$Nd(NO_3)_3$		$(C_2H_5)_2NH\cdot HNO_3$		
mass %	mol kg^{-1}	mass %	mol kg^{-1}	mass %	mol kg^{-1}	mass %	mol kg^{-1}	solid phase[b]
0.0	——	78.5	26.82	0.0	——	85.0	41.62	A
5.0	0.76	75.2	27.90	4.8	1.12	82.2	46.44	A
9.0	1.50	72.8	29.38	7.0	1.77	81.0	49.58	A
14.5	2.66	69.0	30.71	11.5	3.32	78.0	54.56	A
33.7	9.28	55.3	36.92	23.5	10.95	70.0	79.10	A
51.5	10.33	33.4	16.25	58.5	9.84	23.5	9.59	B
54.0	6.65	21.4	6.39	59.8	7.33	15.5	4.61	B
55.0	5.84	16.5	4.25	60.2	6.75	12.8	3.48	B
56.6	5.21	10.5	2.34	61.0	6.16	9.0	2.20	B
58.0	4.79	5.3	1.06	61.5	5.54	4.9	1.07	B
59.0	4.36	0.0	——	62.0	4.94	0.0	——	B

a. Molalities calculated by M. Salomon.
b. Solid phases: A = $(C_2H_5)_2NH\cdot HNO_3$, B = $Nd(NO_3)_3\cdot 6H_2O$

The system is of the simple eutonic type.

AUXILIARY INFORMATION

METHOD/APPARATUS/PROCEDURE:

The method of isothermal sections was used with refractometric analyses (1). Heterogeneous and homogeneous mixtures of known composition were equilibrated until their refractive indices remained constant (3-5 days). The composition of the saturated solutions and the corresponding solid phases were found as inflection or "break" points on a plot of composition against refractive index.

SOURCE AND PURITY OF MATERIALS:

$Nd(NO_3)_3\cdot 6H_2O$ prepd by dissolving "pure" grade Nd_2O_3 in dil (1:4) c.p. grade HNO_3 and crystn. Analysis for $Nd(NO_3)_3$ gave 75.36 mass % (theor value is 75.34 mass %, compilers).
$(C_2H_5)_2NH\cdot HNO_3$ was obtained by dissolving "pure" grade $(C_2H_5)_2NH$ in an equivalent quantity of c.p. grade HNO_3. The solutions were mixed in small quantities to avoid extreme heating. The nitrate was dried in a desiccator over anhydrous $CaCl_2$.
Doubly distilled water was used.

ESTIMATED ERROR:

Nothing specified.

REFERENCES:

1. Zhuravlev, E.F.; Sheveleva, A.D. *Zh. Neorg. Khim.* 1960, *5*, 2630.

COMPONENTS:	ORIGINAL MEASUREMENTS:
(1) Neodymium nitrate; $Nd(NO_3)_3$; [10045-95-1] (2) Triethylamine nitrate; $C_6H_{16}N_2O_3$; [27096-31-7] (3) Water ; H_2O ; [7732-18-5]	Zhuravlev, E.F.; Boeva, M.K. *Zh. Neorg. Khim.* 1974, *19*, 3369-73; *Russ. J. Inorg. Chem. Engl. Transl.* 1974, *19*, 1846-9.
VARIABLES:	**PREPARED BY:**
Composition at 20°C and 40°C	T. Mioduski and S. Siekierski

EXPERIMENTAL VALUES:

20°C Isotherm[a]				40°C Isotherm[a]				
$Nd(NO_3)_3$		$(C_2H_5)_3N\cdot HNO_3$		$Nd(NO_3)_3$		$(C_2H_5)_3N\cdot HNO_3$		
mass %	mol kg^{-1}	mass %	mol kg^{-1}	mass %	mol kg^{-1}	mass %	mol kg^{-1}	solid phase[b]
0	—	90.0	54.81	0	—	92.0	70.03	A
3.2	1.33	89.5	74.66	2.9	1.44	91.0	90.85	A
5.0	2.52	89.0	90.33	4.1	2.30	90.5	102.1	A
6.8	3.96	88.0	103.1	5.6	3.85	90.0	124.6	A
18.8	9.99	75.5	80.66	21.3	10.40	72.5	71.21	B
26.5	6.42	61.0	29.72	27.0	7.43	62.0	34.32	B
38.5	6.66	44.0	15.31	39.0	7.29	44.8	16.84	B
46.5	7.61	35.0	11.52	47.0	8.00	35.2	12.04	B
				40.8	4.75	33.2	7.78	B
55.0	9.41	27.3	9.39					B+C
55.8	5.99	16.0	3.46	62.0	8.53	16.0	4.43	C
56.3	5.13	10.5	1.93	62.0	7.99	14.5	3.76	C
58.0	4.75	5.0	0.82	62.0	6.70	10.0	2.17	C
59.0	4.36	0	—	62.0	6.02	6.8	1.33	C
				62.0	5.44	3.5	0.62	C
				62.0	4.94	0	—	C

a. Molalities calculated by M. Salomon

b. Solid phases: A = $(C_2H_5)_3N\cdot HNO_3$, B = $Nd(NO_3)_3\cdot 4(C_2H_5)_3N\cdot HNO_3$

C = $Nd(NO_3)_3\cdot 6H_2O$

AUXILIARY INFORMATION

METHOD/APPARATUS/PROCEDURE:

The method of isothermal sections was used with refractometric analyses (1). Heterogeneous and homogeneous mixtures of known composition were equilibrated until their refractive indices remained constant (3-5 days). The composition of the saturated solutions and the corresponding solid phases were found as inflection or "break" points on a plot of composition against refractive index.

SOURCE AND PURITY OF MATERIALS:

$Nd(NO_3)_3\cdot 6H_2O$ prepd by dissolving "pure" grade Nd_2O_3 in dil (1:4) c.p. grade HNO_3 and crystn. Analysis for $Nd(NO_3)_3$ gave 75.36 mass % (theor value is 75.34 mass %, compilers).
$(C_2H_5)_3N\cdot HNO_3$ was obtained by dissolving "pure" grade $(C_2H_5)_3N$ in an equivalent quantity of c.p. grade HNO_3. The solutions were mixed in small quantities to avoid extreme heating. The nitrate was dried in a desiccator over anhydrous $CaCl_2$.
Doubly distilled water was used.

ESTIMATED ERROR:

Nothing specified.

REFERENCES:

1. Zhuravlev, E.F.; Sheveleva, A.D. *Zh. Neorg. Khim.* 1960, *5*, 2630.

COMPONENTS:

(1) Neodymium nitrate; $Nd(NO_3)_3$; [10045-95-1]

(2) 2-Butanamine nitrate; $C_4H_{12}N_2O_3$;

(3) Water; H_2O; [7732-18-5]

ORIGINAL MEASUREMENTS:

Boeva, M.K.; Zhuravlev, E.F. *Zh. Neorg. Khim.* 1977, *22*, 263-5; *Russ. J. Inorg. Chem. Engl. Transl.* 1977, *22*, 146-7.

VARIABLES:

Composition at 20°C and 40°C

PREPARED BY:

T. Mioduski and S. Siekierski

EXPERIMENTAL VALUES:

20°C Isotherm[a]

$Nd(NO_3)_3$ mass %	$Nd(NO_3)_3$ mol kg^{-1}	$C_4H_9NH_2\cdot HNO_3$ mass %	$C_4H_9NH_2\cdot HNO_3$ mol kg^{-1}	solid phase[b]
0	——	93.0	97.58	A
1.5	0.61	91.0	89.12	A
3.0	1.21	89.5	87.65	A
4.5	1.95	88.5	92.86	A
7.0	3.85	87.5	116.9	A
11.0	8.33	85.0	156.1	A
46.0	9.95	40.0	20.98	B
47.0	6.19	30.0	9.58	B
50.5	5.10	19.5	4.77	B
54.0	4.67	11.0	2.31	B
56.0	4.24	4.0	0.73	B
59.0	4.36	0	——	B

40°C Isotherm[a]

$Nd(NO_3)_3$ mass %	$Nd(NO_3)_3$ mol kg^{-1}	$C_4H_9NH_2\cdot HNO_3$ mass %	$C_4H_9NH_2\cdot HNO_3$ mol kg^{-1}
0	——	97.0	237.5
0.5	0.50	96.5	236.3
1.5	1.51	95.5	233.8
2.5	2.52	94.5	231.4
3.5	3.52	93.5	228.9
6.5	7.87	91.0	267.3
52.5	9.08	30.0	12.59
57.0	5.08	9.0	1.94
57.5	4.46	3.5	0.66
58.0	5.17	8.0	1.73
61.0	5.20	3.5	0.72
62.3	5.00	0	——

a. Molalities calculated by M. Salomon.
b. Solid phases: A = $C_4H_9NH_2\cdot HNO_3$, B = $Nd(NO_3)_3\cdot 6H_2O$

AUXILIARY INFORMATION

METHOD/APPARATUS/PROCEDURE:

The method of isothermal sections was used with refractometric analyses (1). Heterogeneous and homogeneous mixtures of known composition were equilibrated until their refractive indices remained constant. The composition of the saturated solutions and the corresponding solid phases were found as inflection or "break" points on a plot of composition against refractive index.

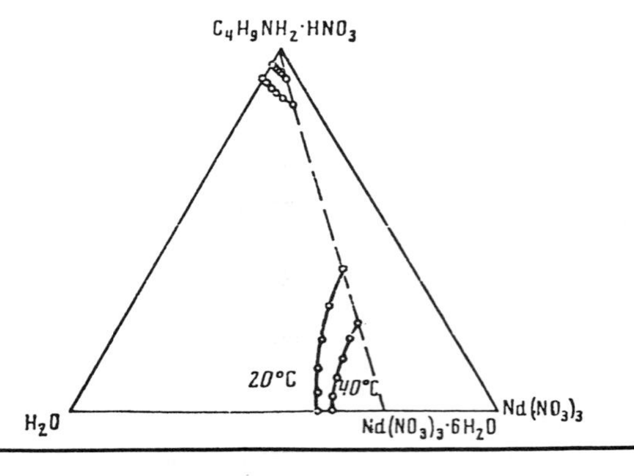

SOURCE AND PURITY OF MATERIALS:

No information given.

ESTIMATED ERROR:

Soly: precision around ± 1 % (compilers).

Temp: precision probably ± 0.2 K (compilers).

REFERENCES:

1. Zhuravlev, E.F.; Sheveleva, A.D. *Zh. Neorg.* 1960, *5*, 2630.

COMPONENTS:

(1) Neodymium nitrate; $Nd(NO_3)_3$; [10045-95-1]

(2) Hexamethylenediamine dinitrate; $C_6H_{18}N_4O_6$; [6143-53-9]

(3) Water ; H_2O ; [7732-18-5]

ORIGINAL MEASUREMENTS:

Boeva, M.K.; Ishmuratov, G. Y. *Issled Mnogokomponent. Sistem s Razl. Vzaimodeistviem Komponentov, Izd. Saratov Univ.* 1977, *No. 2*, 95-9.

VARIABLES:

Composition at 20°C and 40°C

PREPARED BY:

T. Mioduski and S. Siekierski

EXPERIMENTAL VALUES:

t/°C	$Nd(NO_3)_3$ mass %	$Nd(NO_3)_3$ mol kg^{-1}	$H_2N(CH_2)_6NH_2 \cdot 2HNO_3$ mass %	$H_2N(CH_2)_6NH_2 \cdot 2HNO_3$ mol kg^{-1}	nature of the solid phase	
20	0	——	78.0	14.64	$(CH_2)_6(NH_2)_2 \cdot 2HNO_3$	(A)
	4.0	0.55	74.0	13.89	"	
	8.5	1.32	72.0	15.24	"	
	16.0	2.31	63.0	12.38	$Nd(NO_3)_3 \cdot 4(CH_2)_6(NH_2)_2 \cdot 2HNO_3$	(B)
	17.5	2.52	61.5	12.09	"	
	19.5	2.88	60.0	12.08	"	
	25.5	4.06	55.5	12.06	"	
	35.5	7.17	49.5	13.62	B+D	
	40.0	5.27	37.0	6.64	$2Nd(NO_3)_3 \cdot (CH_2)_6(NH_2)_2 \cdot 2HNO_3 \cdot 8H_2O$	(C)
	42.5	4.68	30.0	4.50	"	
	46.0	4.49	23.0	3.06	"	
	50.5	4.43	15.0	1.79	"	
	55.5	4.60	8.0	0.90	C+D	
	57.5	4.52	4.0	0.43	$ND(NO_3)_3 \cdot 6H_2O$	(D)
	58.9	4.34	0	——	"	
40	0	——	84.0	23.03	A	
	3.0	0.57	81.0	20.90	A	
	6.0	1.17	78.5	20.91	A+B	
	15.0	2.67	68.0	16.51	A+B	
	19.5	3.81	65.0	17.31	A+B	

Molalities calculated by M. Salomon

AUXILIARY INFORMATION

METHOD/APPARATUS/PROCEDURE:

The method of isothermal sections was used with refractometric analyses (1). Heterogeneous and homogeneous mixtures of known composition were equilibrated until their refractive indices remained constant. The composition of the saturated solutions and the corresponding solid phases were found as inflection or "break" points on a plot of composition against refractive index.

COMMENTS AND/OR ADDITIONAL DATA:

The double salts are congruently soluble and were isolated for additional analysis. Nd was determined gravimetrically, and elemental analyses for C, N. and H were carried out (no details given). The results confirmed the composition of the double salts.

SOURCE AND PURITY OF MATERIALS:

All materials were reagent grade and were recrystallized twice. Their physical constants corresponded to the literature values. No other information given

ESTIMATED ERROR:

Soly: precision about ± 1 % (compilers).

Temp: precision probably ± 0.1 to 0.2 K (compilers).

REFERENCES:

1. Nikurashina, N.I.; Mertslin, R.V. *Metod. Sechenii, Saratov Univ.* 1969 (see also ref. 2, compilers).
2. Zhuravlev, E.F.; Sheveleva, A.D. *Zh. Neorg. Khim.* 1960, *5*, 2630

COMPONENTS:

(1) Neodymium nitrate; $Nd(NO_3)_3$; [10045-95-1]

(2) Cyclohexanamine nitrate; $C_6H_{14}N_2O_3$; [6941-45-3]

(3) Water ; H_2O ; [7732-18-5]

ORIGINAL MEASUREMENTS:

Boeva, M.K.; Zhuravlev, E.F. *Zh. Neorg. Khim.* 1977, *22*, 263-5; *Russ. J. Inorg. Chem. Engl. Transl.* 1977, *22*, 146-7.

VARIABLES:

Composition at 20°C and 40°C

PREPARED BY:

T. Mioduski and S. Siekierski

EXPERIMENTAL VALUES:

20°C Isotherm[a]					40°C Isotherm[a]			
$Nd(NO_3)_3$		$C_6H_{11}NH_2\cdot HNO_3$			$Nd(NO_3)_3$		$C_6H_{11}NH_2\cdot HNO_3$	
mass %	mol kg^{-1}	mass %	mol kg^{-1}	solid phase[b]	mass %	mol kg^{-1}	mass %	mol kg^{-1}
0	——	5.8[c]	0.45	A	0	——	71.8	18.70
9.0	0.53	40.0	5.76	A	5.5	0.61	67.0	17.89
21.5	1.30	28.5	4.19	A	12.0	1.91	69.0	26.67
37.0	2.51	18.4	3.03	A	18.5	2.38	58.0	18.13
46.0	3.48	14.0	2.57	A	24.0	3.63	56.0	20.57
				A	36.0	10.90	54.0	39.66
53.0	4.59	12.0	2.52	A+B				
56.5	4.44	5.0	0.95	B	53.0	10.84	32.2	15.98
59.0	4.36	0	——	B	58.0	5.85	12.0	2.94
				B	59.0	5.67	9.5	2.22
				B	61.0	5.28	4.0	0.84
				B	62.0	4.94	0	——

a. Molalities calculated by M. Salomon.
b. Solid phases: A = $C_6H_{11}NH_2\cdot HNO_3$, B = $Nd(NO_3)_3\cdot 6H_2O$

c. This appears to be a typographical error. If the correct figure is 58.0 mass %, then the solubility of $C_6H_{11}NH_2\cdot HNO_3$ at 20°C would be 10.14 mol kg^{-1}.

AUXILIARY INFORMATION

METHOD/APPARATUS/PROCEDURE:

The method of isothermal sections was used with refractometric analyses (1). Heterogeneous and homogeneous mixtures of known composition were equilibrated until their refractive indices remained constant. The composition of the saturated solutions and the corresponding solid phases were found as inflection or "break" points on a plot of composition against refractive index.

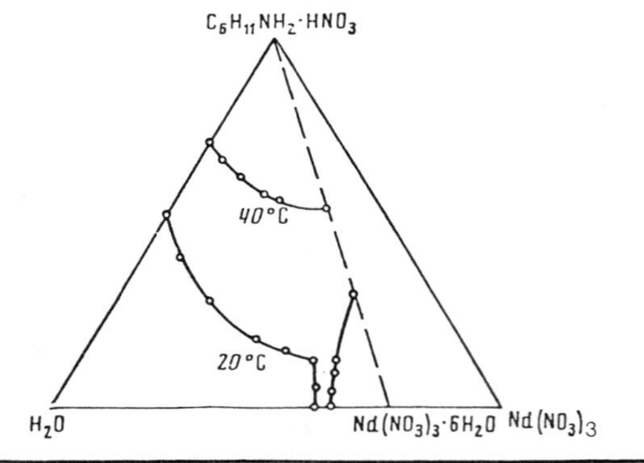

SOURCE AND PURITY OF MATERIALS:

No information given.

ESTIMATED ERROR:

Soly: precision around ± 1 % (compilers).

Temp: precision probably ± 0.2 K (compilers).

REFERENCES:

1. Zhuravlev, E.F.; Sheveleva, A.D. *Zh. Neorg. Khim.* 1960, *5*, 2360.

COMPONENTS:

(1) Neodymium nitrate; $Nd(NO_3)_3$; [10045-95-1]

(2) Piperidine nitrate; $C_5H_{12}N_2O_3$; [6091-45-8]

(3) Water; H_2O; [7732-]8-5]

ORIGINAL MEASUREMENTS:

Zhuravlev, E.F.; Boeva, M.K. *Zh. Neorg. Khim.* 1974, *19*, 3369-73; *Russ. J. Inorg. Chem. Engl. Transl.* 1974, *19*, 1846-9.

VARIABLES:

Composition at 20°C and 40°C

PREPARED BY:

T. Mioduski and S. Siekierski

EXPERIMENTAL VALUES:

20°C Isotherm[a]

$Nd(NO_3)_3$ mass %	$Nd(NO_3)_3$ mol kg^{-1}	$C_5H_{10}NH·HNO_3$ mass %	$C_5H_{10}NH·HNO_3$ mol kg^{-1}
0	——	85.8	40.78
1.0	0.22	85.0	40.98
5.5	0.55	64.0	14.16
14.3	1.28	52.0	10.41
19.5	1.79	47.5	9.71
25.5	2.45	43.0	9.21
34.6	3.18	32.5	6.67
28.3	1.88	26.0	3.84
43.2	3.29	17.0	2.88
54.0	4.30	8.0	1.42
57.3	4.58	4.8	0.85
59.0	4.36	0	——

40°C Isotherm

$Nd(NO_3)_3$ mass %	$Nd(NO_3)_3$ mol kg^{-1}	$C_5H_{10}NH·HNO_3$ mass %	$C_5H_{10}NH·HNO_3$ mol kg^{-1}	solid phase[b]
0	——	87.5	47.25	A
0.8	0.19	86.5	45.97	A
5.5	0.55	64.0	14.16	B
12.8	1.28	57.0	12.74	B
17.0	1.76	53.8	12.44	B
21.8	2.43	51.0	12.65	B
37.8	3.50	29.5	6.09	C
40.0	3.46	25.0	4.82	C
45.0	3.49	16.0	2.77	C
53.5	4.30	8.8	1.58	C
61.5	5.37	3.8	——	D
62.0	4.94	0	——	D

a. Molalities calculated by M. Salomon

b. Solid phases: A = $C_5H_{10}NH·HNO_3$, B = $Nd(NO_3)_3·3C_5H_{10}NH·HNO_3$

C = $Nd(NO_3)_3·2C_5H_{10}NH·HNO_3·6H_2O$, D = $Nd(NO_3)_3·6H_2O$

AUXILIARY INFORMATION

METHOD/APPARATUS/PROCEDURE:

The method of isothermal sections was used with refractometric analyses (1). Heterogeneous and homogeneous mixtures of known composition were equilibrated until their refractive indices remained constant (3-5 days). The composition of the saturated solutions and the corresponding solid phases were found as inflection or "break" points on a plot of composition against refractive index.

SOURCE AND PURITY OF MATERIALS:

$Nd(NO_3)_3·6H_2O$ prepd by dissolving "pure" grade Nd_2O_3 in dil (1:4) c.p. grade HNO_3 and crystn. Analysis for $Nd(NO_3)_3$ gave 75.36 mass % (theoretical is 75.34 mass %, compilers).

$C_5H_{10}NH·HNO_3$ was obtained by dissolving "pure" grade $C_5H_{10}NH$ in an equivalent quantity of c.p. grade HNO_3. The solutions were mixed in small quantities to avoid extreme heating. The nitrate was dried in a desiccator over anhydrous $CaCl_2$.

Doubly distilled water was used.

ESTIMATED ERROR:

Nothing specified.

REFERENCES:

1. Zhuravlev, E.F.; Sheveleva, A.D. *Zh. Neorg. Khim.* 1960, *5*, 2630.

COMPONENTS:

(1) Neodymium nitrate; $Nd(NO_3)_3$; [10045-95-1]

(2) Piperazine dinitrate; $C_4H_{12}N_4O_6$; [10308-78-8]

(3) Water ; H_2O ; [7732-18-5]

ORIGINAL MEASUREMENTS:

Boeva, M.K.; Ishmuratov, G. Yu. *Issled. Mnogokomponent, Sistem s Razl. Vzaimodeistviem Komponentov, Izd. Saratov Univ.* 1977, *No. 2*, 95-9.

VARIABLES:

Composition at 20°C and 40°C

PREPARED BY:

T. Mioduski and S. Siekierski

EXPERIMENTAL VALUES:

20°C Isotherm[a]

$Nd(NO_3)_3$ mass %	$Nd(NO_3)_3$ mol kg^{-1}	$C_4H_{10}N_2\cdot 2HNO_3$ mass %	$C_4H_{10}N_2\cdot 2HNO_3$ mol kg^{-1}	solid phase[b]
0	——	24.0	1.49	A
12.5	0.55	19.0	1.31	A
26.0	1.33	15.0	1.20	A
40.0	2.50	11.5	1.12	A
48.0	3.46	10.0	1.12	A
55.8	4.87	9.5	1.29	A+B
57.0	4.60	5.5	0.69	B
58.0	4.39	2.0	0.24	B
58.9	4.34	0	——	B

40°C Isotherm[a]

$Nd(NO_3)_3$ mass %	$Nd(NO_3)_3$ mol kg^{-1}	$C_4H_{10}N_2\cdot 2HNO_3$ mass %	$C_4H_{10}N_2\cdot 2HNO_3$ mol kg^{-1}
0	——	37.0	2.77
10.5	0.53	30.0	2.38
22.7	1.30	24.5	2.19
36.0	2.56	21.5	2.38
42.8	3.44	19.5	2.44
56.2	6.60	18.0	3.29
61.0	5.28	4.0	0.54
62.0	5.21	2.0	0.26
62.0	4.94	0	——

a. Molalities calcualted by M. Salomon.
b. Solid phase solid phase: A = $C_4H_{10}N_2\cdot 2HNO_3$, B = $Nd(NO_3)_3\cdot 6H_2O$

AUXILIARY INFORMATION

METHOD/APPARATUS/PROCEDURE:

The method of isothermal sections was used with refractometric analyses (1). Heterogeneous and homogeneous mixtures of known composition were equilibrated until their refractive indices remained constant. The composition of the saturated solutions and the corresponding solid phases were found as inflection or "break" points on a plot of composition against refractive index.

The phase diagram is of the simple eutonic type.

SOURCE AND PURITY OF MATERIALS:

All materials were reagent grade and were recrystallized twice. Their physical constants corresponded to the literature values. No other information given.

ESTIMATED ERROR:

Soly: precision about ± 1 % (compilers).

Temp: precision probably ± 0.1 to 0.2 K (compilers).

REFERENCES:

1. Nikurashins, N.I.; Mertslin, R.V. *Metod. Sechenii, Saratov Univ.* 1969 (see also ref. 2, compilers).
2. Zhuravlev, E.F.; Sheveleva, A.D. *Zh. Neorg. Khim.* 1960, *5*, 2630.

COMPONENTS:

(1) Neodymium nitrate; $Nd(NO_3)_3$; [10045-95-1]

(2) Pyridine nitrate; $C_5H_6N_2O_3$; [543-53-3]

(3) Water ; H_2O ; [7732-18-5]

ORIGINAL MEASUREMENTS:

Zhuravlev, E.F.; Boeva, M.K. *Zh. Neorg. Khim.* 1974, *19*, 3369-73; *Russ. J. Inorg. Chem. Engl. Transl.* 1974, *19*, 1846-9.

VARIABLES:

Composition at 20°C and 40°C

PREPARED BY:

T. Mioduski and S. Siekierski

EXPERIMENTAL VALUES:

20°C Isotherm[a]				40°C Isotherm[a]				
$Nd(NO_3)_3$		$C_5H_5N \cdot HNO_3$		$Nd(NO_3)_3$		$C_5H_5N \cdot HNO_3$		
mass %	mol kg^{-1}	mass %	mol kg^{-1}	mass %	mol kg^{-1}	mass %	mol kg^{-1}	solid phase[b]
0	——	78.0	24.95	0	——	79.0	26.47	A
5.0	0.62	70.5	20.25	4.0	0.61	76.0	26.74	A
11.0	1.48	66.5	20.80	8.8	1.51	73.5	29.22	A
25.0	6.06	62.5	35.18	14.5	2.91	70.4	32.81	A
25.6	4.64	57.7	24.31	21.3	4.89	65.5	34.92	A+B
28.7	4.75	53.0	20.38	30.0	6.26	55.5	26.93	B
39.0	9.84	49.0	28.73	37.0	10.00	51.8	32.54	B
54.0	9.08	28.0	10.95	59.8	9.19	20.5	7.32	C
55.3	5.94	16.5	4.12	60.5	6.91	13.0	3.45	C
56.8	4.89	8.0	1.60	61.0	5.68	6.5	1.41	C
58.0	4.72	4.8	0.91	61.6	5.34	3.5	0.71	C
59.0	4.36	0	——	62.0	4.94	0	——	C

a. Molalities calculated by M. Salomon

b. Solid phases: A = $C_5H_5N \cdot HNO_3$, B = $Nd(NO_3)_3 \cdot 4C_5H_5N \cdot HNO_3$

C = $Nd(NO_3)_3 \cdot 6H_2O$

AUXILIARY INFORMATION

METHOD/APPARATUS/PROCEDURE:

The method of isothermal sections was used with refractometric analyses (1). Heterogeneous and homogeneous mixtures of known composition were equilibrated until their refractive indices remained constant (3-5 days). The composition of the saturated solutions and the corresponding solid phases were found as inflection or "break" points on a plot of composition against refractive index.

SOURCE AND PURITY OF MATERIALS:

$Nd(NO_3)_3 \cdot 6H_2O$ prepd by dissolving "pure" grade Nd_2O_3 in dil (1:4) c.p. grade HNO_3 and crystn. Analysis for $Nd(NO_3)_3$ gave 75.36 mass % (theoretical is 75.34 mass %, compilers).
$C_5H_5N \cdot HNO_3$ was obtained by dissolving "pure" grade C_5H_5N in an equivalent quantity of c.p. grade HNO_3. The solutions were mixed in small quantities to avoid extreme heating. The nitrate was dried in a desiccator over anhydrous $CaCl_2$.
Doubly distilled water was used.

ESTIMATED ERROR:

Nothing specified

REFERENCES:

1. Zhuravlev, E.F.; Sheveleva, A.D. *Zh. Neorg. Khim.* 1960, *5*, 2630.

COMPONENTS:

(1) Neodymium nitrate; $Nd(NO_3)_3$; [10045-95-1]

(2) Quinoline nitrate; $C_9H_8N_2O_3$; [21640-15-3]

(3) Water; H_2O; [7732-18-5]

ORIGINAL MEASUREMENTS:

Boeva, M.K.; Zhuravlev, E.F. *Zh. Neorg. Khim.* 1977, *22*, 1977-81; *Russ. J. Inorg. Chem. Engl. Transl.* 1977, *22*, 1073-5.

VARIABLES:

Composition at 20°C and 40°C

PREPARED BY:

T. Mioduski and S. Siekierski

EXPERIMENTAL VALUES:

Composition of saturated solutions at 20°C [a]

$Nd(NO_3)_3$		$C_9H_7N\cdot HNO_3$		
mass %	mol kg^{-1}	mass %	mol kg^{-1}	nature of the solid phase
0	——	68.0	11.06	$C_9H_7N\cdot HNO_3$
5.0	0.50	65.0	11.27	"
11.0	1.28	63.0	12.61	$C_9H_7N\cdot HNO_3 + Nd(NO_3)_3\cdot 2C_9H_7N\cdot HNO_3\cdot 3H_2O$
15.0	1.68	58.0	11.18	$Nd(NO_3)_3\cdot 2C_9H_7N\cdot HNO_3\cdot 3H_2O$
17.0	1.91	56.0	10.79	"
21.0	1.82	44.0	6.54	"
32.0	3.03	36.0	5.85	"
37.0	2.24	13.0	1.35	"
41.0	2.34	6.0	0.59	"
55.0	4.06	4.0	0.51	"
57.0	4.31	3.0	0.39	"
58.0	4.39	2.0	0.26	$Nd(NO_3)_3\cdot 2C_9H_7N\cdot HNO_3\cdot 3H_2O + Nd(NO_3)_3\cdot 6H_2O$
59.0	4.36	0	——	$Nd(NO_3)_3\cdot 6H_2O$

continued.........

AUXILIARY INFORMATION

METHOD/APPARATUS/PROCEDURE:

The method of isothermal sections was used with refractometric analyses (1). Heterogeneous and homogeneous mixtures of known composition were equilibrated until their refractive indices remained constant. The composition of the saturated solutions and the corresponding solid phases were found as inflection or "break" points on a plot of composition against refractive index.

SOURCE AND PURITY OF MATERIALS:

A.R. grade $Nd(NO_3)_3\cdot 6H_2O$, c.p. grade nitric acid, A.R. grade quinoline, and doubly distilled water used. No other details given.

The neodymium salt was probably used as received. Quinoline nitrate, $C_9H_7N\cdot HNO_3$, was probably prepared by neutralization of quinoline with HNO_3.

ESTIMATED ERROR:

Soly: precision about ± 1 % (compilers).

Temp: precision probably ± 0.2 K (compilers).

REFERENCES:

1. Zhuravlev, E.F.; Sheveleva, A.D. *Zh. Neorg. Khim.* 1960, *5*, 2630.

COMPONENTS:	ORIGINAL MEASUREMENTS:
(1) Neodymium nitrate; $Nd(NO_3)_3$; [10045-95-1] (2) Quinoline nitrate; $C_9H_8N_2O_3$; [21640-15-3] (3) Water; H_2O; [7732-18-5]	Boeva, M.K.; Zhuravlev, E.F. *Zh. Neorg. Khim.* 1977, *22*, 1977-81; *Russ. J. Inorg. Chem. Engl. Transl.* 1977, *22*, 1073-5.

EXPERIMENTAL VALUES: continued.........

Composition of saturated solutions at 40°C [a]

$Nd(NO_3)_3$ mass %	$Nd(NO_3)_3$ mol kg^{-1}	$C_9H_7N \cdot HNO_3$ mass %	$C_9H_7N \cdot HNO_3$ mol kg^{-1}	nature of the solid phase
0	———	75.0	15.61	$C_9H_7N \cdot HNO_3$
4.0	0.50	72.0	15.61	"
7.0	0.90	69.5	15.39	"
10.5	1.48	68.0	16.46	$C_9H_7N \cdot HNO_3 + Nd(NO_3)_3 \cdot 2C_9H_7N \cdot HNO_3 \cdot 3H_2O$
12.0	1.77	67.5	17.13	$Nd(NO_3)_3 \cdot 2C_9H_7N \cdot HNO_3 \cdot 3H_2O$
14.0	2.12	66.0	17.17	"
18.0	1.47	45.0	6.33	"
28.5	2.36	35.0	4.99	"
37.0	2.73	22.0	2.79	"
44.0	2.96	11.0	1.27	"
58.0	4.75	5.0	0.70	"
59.5	4.80	3.0	0.42	"
60.0	4.78	2.0	0.27	$Nd(NO_3)_3 \cdot 6H_2O$
62.0	4.94	0	———	"

a. Molalities calculated by M. Salomon.

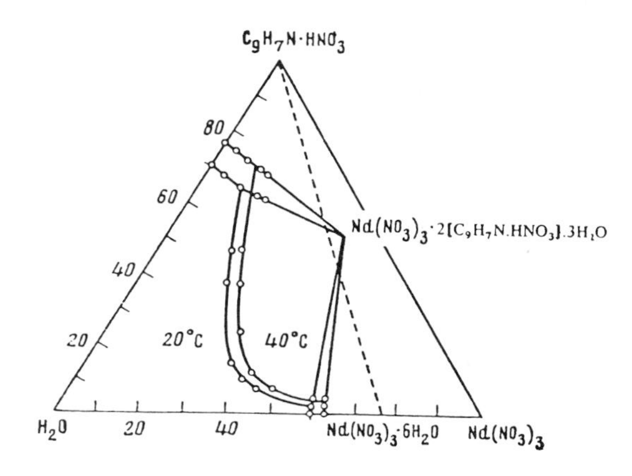

COMPONENTS:

(1) Neodymium nitrate; $Nd(NO_3)_3$; [10045-95-1]

(2) Urea; CH_4N_2O; [57-13-6]

(3) Water; H_2O; [7732-18-5]

ORIGINAL MEASUREMENTS:

Khudaibergenova, N.; Sulaimankulov, K.
Zh. Neorg. Khim. 1981, *26*, 1107-9;
Russ. J. Inorg. Chem. Engl. Transl. 1981, *26*, 599-600.

VARIABLES:

Composition at 30°C

PREPARED BY:

T. Mioduski and S. Siekierski

EXPERIMENTAL VALUES:

Composition of saturated solutions at 30°C [a]

$Nd(NO_3)_3$ mass %	$Nd(NO_3)_3$ mol kg^{-1}	$CO(NH_2)_2$ mass %	$CO(NH_2)_2$ mol kg^{-1}	nature of the solid phase
———	———	57.50	22.53	$CO(NH_2)_2$
9.56	0.824	55.30	26.20	"
12.77	1.272	56.84	31.14	"
26.22	4.289	55.27	49.72	"
35.61	11.29	54.84	95.62	"
39.26	19.95	54.78	153.0	"
39.76	19.29	54.00	144.1	$CO(NH_2)_2 + Nd(NO_3)_3 \cdot 2CO(NH_2)_2 \cdot 2H_2O$
40.17	21.15	54.08	156.6	$Nd(NO_3)_3 \cdot 2CO(NH_2)_2 \cdot 2H_2O$
39.99	11.76	49.71	80.36	"
42.37	7.525	40.58	39.63	"
50.41	7.776	29.96	25.41	"
54.40	6.268	19.32	12.24	"
57.74	7.124	17.72	12.02	"
63.09	6.480	7.43	4.197	"
64.11	7.002	8.17	4.908	$Nd(NO_3)_3 \cdot 2CO(NH_2)_2 \cdot 2H_2O + Nd(NO_3)_3 \cdot 6H_2O$
63.86	6.933	8.25	4.926	$Nd(NO_3)_3 \cdot 6H_2O$
63.03	6.053	5.44	2.873	"
61.67	4.872	———	———	"

a. Molalities calculated by M. Salomon.

AUXILIARY INFORMATION

METHOD/APPARATUS/PROCEDURE:

The isothermal method was used. Equilibrium was reached within 7-9 hours. The liquid phase was separated using a No. 3 Schott filter. Nitrogen was determined by the Kjeldahl method, and neodymium was determined by titration with Trilon B using Xylenol orange indicator.

SOURCE AND PURITY OF MATERIALS:

Nothing specified.

ESTIMATED ERROR:

Nothing specified.

REFERENCES:

COMPONENTS:	EVALUATOR:
(1) Neodymium nitrate; $Nd(NO_3)_3$; [10045-95-1] (2) Organic solvents	Mark Salomon U.S. Army Electronics Technology and Devices Laboratory Fort Monmouth, NJ, USA December 1982

CRITICAL EVALUATION:

The solubility of $Nd(NO_3)_3$ has been reported in only two organic solvents: diethyl ether (1,2) and dioxane (2). Solubilities of $Pr(NO_3)_3$ + $Nd(NO_3)_3$ mixtures were reported by Hardy (3) in 2-methoxyethanol and in 2-ethoxyethanol. The compilations for these data are included in the chapter on $Pr(NO_3)_3$ - Organic solvents.

Wells (1) and Hopkins and Quill (2) reported the solubility of neodymium nitrate in diethyl ether, but the results at 293 K are so divergent that one or both of the studies are in error. At the present time, all solubility data for $Nd(NO_3)_3$ in organic solvents must be classified as *doubtful*.

REFERENCES

1. Wells, R.C. *J. Wash. Acad. Sci.* 1930, *20*, 145.
2. Hopkins, B.S.; Quill, L.L. *Proc. Natl. Acad. Sci. U.S.A.* 1933, *19*, 64.
3. Hardy, Z.M. *Masters Thesis*. The University of Illinois. Urbana, IL. 1931.

COMPONENTS:

(1) Neodymium nitrate; $Nd(NO_3)_3$; [10045-95-1]

(2) Diethyl ether; $C_4H_{10}O$; [60-29-7]

ORIGINAL MEASUREMENTS:

Wells, R.C. *J. Wash. Acad. Sci.* 1930, *20*, 145-8.

VARIABLES:

Room temperature (about 20°C)

PREPARED BY:

T. Mioduski, S. Siekierski, M. Salomon

EXPERIMENTAL VALUES:

Experiment 1. This experiment involves the hydrated neodymium nitrate as the initial solid, and which the compilers assume to be the hexahydrate.

Authors report the solubility as 0.020 g Nd_2O_3 in 10 ml ether.

This is equivalent to a $Nd(NO_3)_3$ soly of 0.0012 mol dm^{-3} (compilers).

Experiment 2. This experiment involves neodymium nitrate dehydrated as described in the METHOD/APPARATUS/PROCEDURE box below.

Authors report the solubility as 0.267 g Nd_2O_3 in 10 ml ether.

This is equivalent to a $Nd(NO_3)_3$ soly of 0.159 mol dm^{-3} (compilers).

AUXILIARY INFORMATION

METHOD/APPARATUS/PROCEDURE:

The isothermal method was used. The soly of neodymium nitrate was determined in two experiments in which the nature of the initial solid phase differs.

Experiment 1. A few grams of neodymium nitrate (presumably the hexahydrate, compilers) was added to about 20 ml of ether in small stoppered flasks. The flasks were periodically agitated and permitted to stand at about 20°C overnight. A 10 ml sample was removed, filtered, the solvent evaporated and the salt ignited to the oxide and weighed.

Experiment 2. The remaining salt in the flask was freed from ether, dissolved in water and a few drops of HNO_3 added. The solution was evaporated to dryness and heated to 150°C. The solubility in ether was then determined again with this "dehydrated" salt.

SOURCE AND PURITY OF MATERIALS:

Nothing specified.

ESTIMATED ERROR:

Soly: precision probably around ± 10 % (compilers).

Temp: precision probably ± 4 K (compilers).

REFERENCES:

COMPONENTS:

(1) Neodymium nitrate; $Nd(NO_3)_3$; [10045-95-1]

(2) Diethyl ether; $C_4H_{10}O$; [60-29-7]

ORIGINAL MEASUREMENTS:

Hopkins, B.S.; Quill, L.L. *Proc. Natl. Acad. Sci. U.S.A.* 1933, *19*, 64-8.

VARIABLES:

Temperature: range 18.5°C to 29°C

PREPARED BY:

T. Mioduski, S. Siekierski, M. Salomon

EXPERIMENTAL VALUES:

The solubility of $Nd(NO_3)_3$ in diethyl ether was given in the form of a small diagram of solubility vs temperature. In the absence of numerical data, the compilers interpolated the solubility from the published diagram. The results are:

t/°C	g dm^{-3}	mol dm^{-3}
18.5	4.8	0.015
19	5.2	0.016
22.2	6.0	0.018
24.4	6.25	0.019
26	6.75	0.020
29.4	7.2	0.022

COMMENTS AND/OR ADDITIONAL DATA:

It appears that the original experimental work was done by a Mr. P. Kalischer who was a student at the University of Illinois at Urbana-Champaign. Attempts to locate the original experimental data have failed, and it thus appears that these data are lost (see COMMENTS in the compilation for the $Nd(NO_3)$-dioxane system).

AUXILIARY INFORMATION

METHOD/APPARATUS/PROCEDURE:

No information is available, but based on similar work by Hardy (1) being carried out at the University of Illinois at the time, it is likely that the isothermal method was employed. The solubility data for neodymium and praseodymium nitrates in several ethers from Hardy's MSc Thesis are compiled elsewhere in this volume, and the compilations contain detailed information on the experimental techniques which the compilers assume were similar to those used by Mr. Kalischer.

SOURCE AND PURITY OF MATERIALS:

No information available.

ESTIMATED ERROR:

No information available.

REFERENCES:

1. Hardy, Z.M. *Masters Thesis.* The University of Illinois. Urbana, IL. 1932.

COMPONENTS:

(1) Neodymium nitrate; $Nd(NO_3)_3$; [10045-95-1]

(2) Dioxane; $C_4H_8O_2$; [505-22-6]

ORIGINAL MEASUREMENTS:

Hopkins, B.S.; Quill, L.L. *Proc. Natl. Acad. Sci. U.S.A.* 1933, *19*, 64-8.

VARIABLES:

One temperature: 25°C

PREPARED BY:

T. Mioduski, S. Siekierski, M. Salomon

EXPERIMENTAL VALUES:

The solubility of $Nd(NO_3)_3$ in dioxane at 25°C was given in the form of a small diagram of solubility vs atomic number Z for Z = 57-64. In the absence of numerical data, the compilers interpolated the solubility from the published diagram. The result is:

$$\text{soly of } Nd(NO_3)_3 = 8.2 \text{ g dm}^{-3} \text{ (0.025 mol dm}^{-3}\text{)}$$

COMMENTS AND/OR ADDITIONAL DATA:

The name Philip Kalischer appears on the diagram published in the source paper. Also in related work by Hardy (1), reference was given to a Masters Thesis by P. Kalischer. The compilers thus assumed that Mr. Kalischer was an MSc student of Prof. Hopkins and thus contacted Ms. Susanne Redalje, the Assistant Chemistry Librarian at the University of Illinois at Urbana-Champaign. Ms. Redalje searched the University records for references to a thesis or any publication by Mr. Kalischer. The records show that Mr. Kalischer attended classes for the Fall, Spring, and Summer semesters of 1930-1931. There is no indication that Mr. Kalischer had finished his studies or submitted a thesis, and it is therefore apparent that the original experimental data are lost. The compilers are most grateful to Ms. Redalje for all her help in searching the University records and providing important information on numerous other lanthanide systems.

AUXILIARY INFORMATION

METHOD/APPARATUS/PROCEDURE:

No information is available, but based on similar work by Hardy (1) being carried out at the University of Illinois at the time, it is likely that the isothermal method was employed. The solubility data for neodymium and praseodymium nitrates in several ethers from Hardy's MSc Thesis are compiled elsewhere in this volume, and the compilations contain detailed information on the experimental techniques which the compilers assume were similar to those used by Mr. Kalischer.

SOURCE AND PURITY OF MATERIALS:

No information available.

ESTIMATED ERROR:

No information available.

REFERENCES:

1. Hardy, Z.M. *Masters Thesis.* The University of Illinois. Urbana, Il. 1932.

COMPONENTS:	EVALUATOR:
(1) Neodymium double nitrates (2) Water ; H_2O ; [7732-18-5]	Mark Salomon U.S. Army Electronics Technology and Devices Laboratory Fort Monmouth, NJ, USA December 1982

CRITICAL EVALUATION:

$Nd(NO_3)_3$ DOUBLE SALTS WITH INORGANIC NITRATES

INTRODUCTION

Studies on the direct determination of the solubilities of $Nd(NO_3)_3$ double nitrates involving inorganic nitrates are relatively few (1-3). The double nitrates reported in (1-3) are all characterized by the formation of the following tetracosahydrate solid phases:

$3Mg(NO_3)_2.2Nd(NO_3)_3.24H_2O$	[17203-49-5]	(1-3)
$2Nd(NO_3)_3.3Mn(NO_3)_2.24H_2O$	[84682-64-4]	(1,2)
$2Nd(NO_3)_3.3Co(NO_3)_2.24H_2O$	[84682-65-5]	(1,2)
$2Nd(NO_3)_3.3Ni(NO_3)_2.24H_2O$	[84682-67-7]	(1)
$2Nd(NO_3)_3.3Zn(NO_3)_2.24H_2O$	[28876-82-6]	(1,2)

EVALUATION PROCEDURE

Where possible, the solubility data were fitted by least squares to the smoothing equation

$$Y = \ln(m/m_o) - nM_2(m - m_o) = a + b/(T/K) + c \ln(T/K) \qquad [1]$$

All terms in eq. [1] have been previously defined (see eq. [1] in the $Nd(NO_3)_3$-H_2O critical evalution). Due to the absense from the literature of sufficient publications to provide bases for critical comparisons of most data, and due to the large experimental errors associated with existing data, a detailed statistical treatment of the data is not possible. A simplified method of estimating the accuracy of the data was therefore adopted. The solubility data were fitted to eq. [1] and the value of the congruent melting point calculated. If the congruent melting point calculated from eq. [1] is in agreement with the experimental value of the melting point (within experimental and calculated errors), we consider this as strong support for designating the solubilities as either *tentative* or *recommended* values. In most cases considered below, the least squares fitted data are fairly precise (i.e. standard errors of estimate, σ_m, are generally small), and the accuracy in the smoothed solubility data is determined by the experimental errors.

Jantsch (1) reported the solubilities of a number of double nitrates in concentrated nitric acid solution of density = 1.325 kg m^{-3} at 289 K. The solubilities were determined in this concentrated HNO_3 solution because the author did not have sufficient quantities of the double nitrates to determine solubilities in pure water. One of the most useful results reported by Jantsch are the congruent melting points of the hydrated salts. Jantsch determined the solubilities by a gravimetric method by precipitation of Nd as the hydroxide followed by ignition to the oxide Nd_2O_3.

The data of Prandtl and Ducrue (2) were assigned a precision of around ± 0.4% at best based upon the highest reproducibilities achieved by these authors. Considering the unknown error in temperature (not reported by the authors) and average reproducibility in the analyses, the total uncertainty in Prandtl and Ducrue's results is probably around ± 1%.

Friend and Wheat's results must be carefully reviewed because of our previous experience with Firend's data which contain systematic errors. In the study on magnesium neodymium nitrate (3), the authors present typical results for analyses of the solid tetracosa-hydrate $3Mg(NO_3)_2.2Nd(NO_3)_3.24H_2O$, and in all cases their results for $Nd(NO_3)_3$ were too low (from theoretical) by about 3-5%, their results for $Mg(NO_3)_2$ were too low by about 1-2%, and their results for waters of hydration (obtained by difference) were too high approaching $27H_2O$. The results of Friend and Wheat are slightly lower than those of Prandtl and Ducrue (2), and below we present calculations and discussions which suggest that the data of Friend and Wheat probably contain a small negative systematic error.

COMPONENTS:	EVALUATOR:
(1) Neodymium double nitrates (2) Water ; H_2O ; [7732-18-5[	M. Salomon U.S. Army Electronics Technology and Devices Laboratory Fort Monmouth, NJ, USA December 1982

CRITICAL EVALUATION:

Magnesium neodymium nitrate. The combined data of Friend and Wheat (3) and Prandtl and Ducrue (2) were fitted to the smoothing equation [1], and the results do not appear to be satisfactory. The standard error of estimate $\sigma_m = 0.021$ is double the value of $\sigma_m = 0.010$ obtained by fitting only those data of Prandtl and Ducrue to the smoothing equation. In addition, the fitting of all the results from (2, 3) to eq. [1] results in a predicted congruent melting point of 387.2 K which is considerably higher than the experimental value of 382.2 K (1). Considering these factors and the low results of the chemical analyses reported by Friend and Wheat, we conclude that these data from (3) contain a small negative systematic error, and that they should be rejected. The results of fitting the data of Prandtl and Ducrue (2) to the smoothing equation are given in Table 1, and the smoothed solubility data at selected temperatures calculated from eq. [1] are given in Table 2: these smoothed values are designated as *tentative* solubility data. Combining the precision of the least squares fit with the experimental error of around 0.4%, the overall uncertainty in the smoothed values is ± 0.06 mol kg^{-1} (95% level of confidence, Student's t = 12.706).

Double nitrates with Mn, Co, and Zn nitrates. For each double salt, the only comparable data available are the solubilities as a function of temperature reported by Prandtl and Ducrue (2), and the congruent melting points of the tetracosahydrates reported by Jantsch (1). All the solubility data were fitted to eq. [1], and the results are given in Table 1. For all cases except $2Nd(NO_3)_3.3Zn(NO_3)_2$, there is acceptable agreement between the predicted melting points for the tetracosahydrates and the experimental melting points. The evaluator regards this agreement as sufficient justification to designate the smoothed solubilities for all double salts (except the double salt with $Zn(NO_3)_2$) as *tentative* data. The smoothed (*tentative*) solubility data at selected temperatures are given in Table 2.

The failure of the solubility data for neodymium zinc nitrate to predict the observed melting point of the tetracosahydrate when fitted to the smoothing equation suggests a large error in the solubility data rather than in the experimental melting point. The values of the constants a, b, c (see Table 1) appear to be trivial, and the high value for the constant "a" and the negative value for "b" suggests an unlikely positive value for the heat of solution. Because only four data points were used in the least squares fitting to eq. [1], it is highly probable that one imprecise datum would invalidate the least squares treatment. Thus while some or most of the data for $2Nd(NO_3)_3.3Zn(NO_3)_2$ may be acceptable (i.e. accurate to within about ± 3-4%), the uncertainty as to which of these data may be highly inaccurate leads the evaluator to the conclusion that none of the data can be assigned the *tentative* designation.

OTHER $Nd(NO_3)_3$ DOUBLE NITRATES

A number of neodymium double nitrates with inorganic and organic nitrates have been identified in the preceeding sections. While some of these double salts are congruently soluble, there are no studies available dealing with the direct determination of the solubilities of these salts. For details on the specific salts which form stable solid phases, the reader is referred to the section on the compilations of ternary aqueous $Nd(NO_3)_3$ systems.

REFERENCES

1. Jantsch, G. *Z. Anorg. Chem.* 1912, *76*, 303.

2. Prandtl. W.; Ducrue, H. *Z. Anorg. Chem.* 1926, *150*, 105.

3. Friend, J.N.; Wheat, W.N. *J. Chem. Soc.* 1935, 356.

COMPONENTS:

(1) Neodymium double nitrates

(2) Water ; H_2O ; [7732-18-5]

EVALUATOR:
Mark Salomon
U.S. Army Electronics Technology and Devices Laboratory
Fort Monmouth, NJ, USA

December 1982

CRITICAL EVALUATION:

Table 1. Smoothing equation parameters for $2Nd(NO_3)_3 \cdot 3M(NO_3)_2$ solubilities.

parameters M =	Mg	Mn	Co	Zn
a	-15.108	-27.595	-19.109	-2.181
b	357	880.4	478.6	-267
c	2.437	4.3206	3.0677	0.5353
σ_m	0.010	0.008	0.003	0.01
tetracosahydrate melting point/K				
calcd (eq. [1])	385.3	347.2	366.5	378.2
obsd (ref 1)	382.2	350.2	368.7	361.7

Table 2. Tentative solubility data at selected temperatures calculated from the smoothing equation [1] (solid phase is the tetracosahydrate).

solubility of $2Nd(NO_3)_3 \cdot 3M(NO_3)_2$/mol kg^{-1}

T/K	M = Mg	Mn	Co
273.2	0.743	0.877	0.744
283.2	0.791	0.940	0.801
293.2	0.843	1.014	0.864
298.2	0.871	1.056	0.899
303.2	0.901	1.103	0.936
313.2	0.965	1.213	1.020
323.2	1.038	1.354	1.117
333.2	1.121	1.547	1.234
343.2	1.218	1.877	1.380
353.2	1.332		1.580
363.2	1.475		1.924
373.2	1.670		
383.2	2.027		

COMPONENTS:

(1) Magnesium neodymium nitrate; $3Mg(NO_3)_2 \cdot 2Nd(NO_3)_3$; [13568-66-6]

(2) Nitric acid; HNO_3; [7697-37-2[

(3) Water ; H_2O ; [7732-18-5]

ORIGINAL MEASUREMENTS:

Jantsch, G. *Z. Anorg. Chem.* 1912, *76*, 303-23.

VARIABLES:

One temperature: 16°C

PREPARED BY:

Mark Salomon

EXPERIMENTAL VALUES:

Soly of the double salt in HNO_3 sln of density d_4^{16} = 1.325 g cm^{-3}.

aliquot volume	Nd_2O_3	soly $3Mg(NO_3)_2 \cdot 2Nd(NO_3)_3$ [a]
cm^3	g	mol dm^{-3}
1.4638	0.0317	
1.4638	0.0309	0.0635

a. Author's calculation (average value).

ADDITIONAL VALUE:

The melting point of the tetracosahydrate is 109.0°C, and the density at 0°C is 2.023 g cm^{-3}.

AUXILIARY INFORMATION

METHOD/APPARATUS/PROCEDURE:

Isothermal method used. The soly was studied in HNO_3 sln of density 1.325 g cm^{-3} at 16°C because the author did not have sufficient quantity of the rare earth to study the soly of the salt in pure water. Pulverized salt and HNO_3 sln were placed in glass-stoppered tubes and thermostated at 16°C for 24 h with periodic shaking. The solution was then allowed to settle for 2 h, and a pipet maintained at 16°C was used to withdraw aliquots for analysis. Two analyses were performed.

Solutions were analysed by adding 2-3 g NH_4Cl and 10% NH_3 sln followed by boiling to ppt the hydroxide. The ppt was filtered, dissolved in HNO_3, reprecipitated as the hydroxide, and ignited to the oxide. Mg in the filtrate was "determined by the usual method" (no details were given).

An attempt to determine the waters of hydration by dehydration was not successful because the remperature required (120°C or higher) resulted in decomposition of the salt with the formation of basic salts. Presumably the waters of hydration were found by difference.

SOURCE AND PURITY OF MATERIALS:

"Pure" neodymium oxide was dissolved in dil HNO_3 and $Mg(NO_3)_2$ added to give a mole ratio of Nd/Mg = 2/3. The sln was evapd and a small crystal of $Bi_2Mg_3(NO_3)_{12}$ added, and the mixt cooled to ppt the tetracosahydrate. The double nitrate was recrystd before use.

ESTIMATED ERROR:

Soly: reproducibility about ± 1-5% (compiler).

Temp: nothing specified.

REFERENCES:

COMPONENTS:

(1) Magnesium neodymium nitrate ; $3Mg(NO_3)_2 \cdot 2Nd(NO_3)_3$; [13568-66-6]

(2) Water ; H_2O ' [7732-18-5]

ORIGINAL MEASUREMENTS:

Prandtl, W.; Ducrue, H.

Z. Anorg. Chem. 1926, *150*, 105-16.

VARIABLES:

Temperature

PREPARED BY:

Mark Salomon

EXPERIMENTAL VALUES:

			solubility					
			oxides		double salt			
	mole ratio[a]	density	Nd_2O_3	MgO	hydrate[b]	hydrate[c]	anhydrous salt[d]	
t/°C	MgO/Nd_2O_3	$kg\ m^{-3}$	mass %	mass %	mass %	mass %	mass %	$mol\ kg^{-1}$
15	2.99	1.52	14.45	5.17	66.0	65.89	47.37	0.814
30	3.01	1.55	15.20	5.49	69.5	69.64	50.06	0.907
50	2.96[e]	1.59	16.34	5.78	74.4	74.09	53.26	1.031
70	3.00[f]	1.64	17.47	6.29	79.8	79.92	57.45	1.221

a. Experimental value: theoretical value = 3.00.
b. Authors' values apparently based on mass % Nd_2O_3. The hydrate which is also the equilibrium solid phase is the tetracosahydrate $2Nd(NO_3)_3 \cdot 3Mg(NO_3)_2 \cdot 24H_2O$.
c. Compiler's calculations based on average from mass % Nd_2O_3 and MgO.
d. Compiler's calculations based on results from c above.
e. Compiler computes 2.95.
f. Compiler computes 3.01.

AUXILIARY INFORMATION

METHOD/APPARATUS/PROCEDURE:

Isothermal method used. Pulverized double salt (hydrate) and conductivity water were placed in two 50 cc flasks and agitated for 1 day in a thermostat. The slns then permitted to settle and aliquots of approx 4 cc removed with pipets maintained at the same temp as the satd slns. The aliquots were placed in graduated flasks and weighed, and then diluted with 50 cc of water for analysis. The results for densities and mass % of oxides are the mean of two determinations. The mass % of the tetracosahydrate was apparently calculated by the authors from the mass % Nd_2O_3: i.e. the mass % MgO was not considered.

Both metals were determined gravimetrically. Nd was precipitated as the oxalate, filtered, and ignited to the oxide. Mg in the filtrate was precipitated as $MgNH_4PO_4$, and presumably ignited to the pyrophosphate $Mg_2P_2O_7$.

SOURCE AND PURITY OF MATERIALS:

Nd_2O_3 prepared by W. Prandtl was analysed by X-ray spectroscopy and found to be "very pure." It was dissolved in nitric acid and the required amount of commercial "pure" grade $Mg(NO_3)_2$ added. The solution was evaporated to crystallization and the double salt recrystallized several times from conductivity water. The salt was dried over $CaCl_2$ in a desiccator to give the tetracosahydrate. Results of the analysis of the double salt are:

Nd_2O_3 found 22.03 %, calcd 21.89 mass %.

MgO found 7.94 %, calcd 7.86 mass %.

Conductivity water was used.

ESTIMATED ERROR:

Soly: based on the deviation of the mole ratio MgO/Nd_2O_3 and the theor value of 3.00, the compiler estimates a precision of ± 0.3 to ± 1 %.

Temp: not specified.

COMPONENTS:

(1) Magnesium neodymium nitrate ; $3Mg(NO_3)_2 \cdot 2Nd(NO_3)_3$; [13568-66-6]

(2) Water ; H_2O ; [7732-18-5]

ORIGINAL MEASUREMENTS:

Friend, J.A.N. *J. Chem. Soc.* 1930, 1903-8.

VARIABLES:

Temperature

PREPARED BY:

Mark Salomon

EXPERIMENTAL VALUES:

	$3Mg(NO_3)_2 \cdot 2Nd(NO_3)_3 \cdot 24H_2O$[a]	$3Mg(NO_3)_2 \cdot 2Nd(NO_3)_3$[b]	
t/°C	mass %	mass %	mol kg^{-1}
0.4	60.42	43.43	0.695
0.6	60.84	43.73	0.703
4.4	61.92	44.51	0.726
14.2	64.00	46.01	0.771
17.0	63.81	45.87	0.767
27.2	67.28	48.36	0.847
40.6	70.72	50.84	0.935
48.8	72.50	52.12	0.985
65.4	77.64	55.81	1.143
75.6	81.07	58.28	1.263
87.0	84.35	60.63	1.393
90.6	86.89	62.46	1.505

a. Author's original calculations. The primary data, mass % Nd_2O_3, were not reported.
b. Compiler's conversions.

AUXILIARY INFORMATION

METHOD/APPARATUS/PROCEDURE:

The isothermal method was used as described in (1). About 10-15 g aliquots of saturated sln were filtered through glass wool or sintered glass and diluted to 250 ml for analysis. An aliquot of this diluted sln was taken for detn of neodymium. Nd was pptd as the oxalate by addn of hot oxalic acid, and the ppt allowed to stand for 1 d. The ppt was then filtered, washed with hot dil oxalic acid sln, and ignited to the oxide. The Mg content in satd slns was not determined.
Several wet residues for unspecified slns were analysed for total Nd + Mg by direct ignition to the oxide. The oxide was dissolved in HNO_3 and Nd detd by the oxalate method and Mg obtained by difference. In all cases the Nd content was too low by 2 to 5 %, and the Nd_2O_3/MgO mass ratio too low by as much as 9 %. Analysis of dry residues again always gave Nd content too low by as much as 5 mass %, and water (obtained by difference) was always too high by as much as 10 mass %. Analysis of the solid phases at 0°C suggested the existence of a hydrate with more than 24 waters of crystallization.

SOURCE AND PURITY OF MATERIALS:

The double salt obtained in (1) by fractional crystallizations was recrystallized several times from dilute nitric acid. The salt was washed with distilled water, and air dried on filter paper. Author states that recrystallization from pure water results in a slight loss of the Nd salt.
Presumably distilled water was used for the solubility measurements.

ESTIMATED ERROR:

Soly: precision ± 3 % at best (compiler).

Temp: accuracy probably ± 0.05 K as in (1) (compiler).

REFERENCES:

Friend, J.A.N. *J. Chem. Soc.* 1930, 1633.

COMPONENTS:

(1) Magnesium neodymium nitrate ; $3Mg(NO_3)_2 \cdot 2Nd(NO_3)_3$; [13568-66-6]

(2) Nitric acid; HNO_3; [7697-37-2]

(3) Water ; H_2O ; [7732-18-5]

ORIGINAL MEASUREMENTS:

Friend, J.A.N. *J. Chem. Soc.* 1930, 1903-8.

VARIABLES:

HNO_3 concentration and temperature

PREPARED BY:

Mark Salomon

EXPERIMENTAL VALUES:

t/°C	HNO_3 [a] mol dm^{-3}	$3Mg(NO_3)_2 \cdot 2Nd(NO_3)_3 \cdot 24H_2O$ [b] mass %	$3Mg(NO_3)_2 \cdot 2Nd(NO_3)_3$ [c] mass %	density kg m^{-3}
0.6	1.0	57.21	41.13	
14.4	1.0	60.25	43.31	1.499
24.2	1.0	63.06	45.33	
77.2	1.0	78.26	56.26	
15.2	2.2	54.49	39.17	
50.2	2.2	63.14	45.39	
14.8	5.2	42.01	30.20	
24.8	5.2	45.84	32.95	
74.0	5.2	64.40	46.29	
14.8	11.25	5.64	4.05	1.357
49.0	11.25	30.98	22.27	
78.0	11.25	62.47	44.91	

a. These are *initial* HNO_3 concentrations. The acid concentration in the equilibrated solutions was not determined.

b. Author's original calculations. The primary data, mass % Nd_2O_3, were not reported.

c. Compiler's calculations. Conversions to molality cannot be made because the acid concentrations in the equilibrated solutions are not known.

AUXILIARY INFORMATION

METHOD/APPARATUS/PROCEDURE:

The isothermal method was used as described in (1). About 10-15 g aliquots of saturated sln were filtered through glass wool or sintered glass and diluted to 250 ml for analysis. As aliquot of this diluted sln was taken for detn of neodymium. Nd was pptd as the oxalate by addn of hot oxalic acid, and the ppt allowed to stand for 1 d. The ppt was then filtered, washed with hot dil oxalic acid sln, and ignited to the oxide. The Mg content in satd slns was not determined.
Several wet residues for unspecified slns were analysed for total Nd + Mg by direct ingition to the oxide. The oxide was dissolved in HNO_3 and Nd detd by the oxalate method and Mg obtained by difference. In all cases the Nd content was too low by 2 to 5%, and the Nd_2O_3:MgO mass ratio too low by as much as 9%.
Analysis of dry residues again always gave Nd content too low by as much as 5 mass %, and water (obtained by difference) was always too high by as much as 10 mass %.
Analysis of solid phases at 0°C suggested the existence of a hydrate with more than 24 waters of crystallization.

SOURCE AND PURITY OF MATERIALS:

The double salt obtained in (1) by fractional crystallizations was recrystallized several times from dilute nitric acid. The salt was washed with distilled water, and air dried on filter paper. Author states that recrystallization from pure water results in a slight loss of the Nd salt.
Presumably distilled water was used for the solubility measurements.

ESTIMATED ERROR:

Soly: precision ± 3 % at best (compiler).

Temp: accuracy probably ± 0.05 K as in (1) (compiler).

REFERENCES:

Friend, J.A.N. *J. Chem. Soc.* 1930, 1633.

COMPONENTS:	ORIGINAL MEASUREMENTS:
(1) Magnesium nitrate; $Mg(NO_3)_2$; [10377-60-3] (2) Magnesium neodymium nitrate ; $3Mg(NO_3)_2 \cdot 2Nd(NO_3)_3$; [13568-66-6] (3) Water ; H_2O ; [7732-18-5]	Friend, J.A.N. *J. Chem. Soc.* 1930, 1903-8.

VARIABLES:	PREPARED BY:
$Mg(NO_3)_2$ concentration	Mark Salomon

EXPERIMENTAL VALUES:

t/°C	$Mg(NO_3)_2$ [a] mol dm^{-3}	$3Mg(NO_3)_2 \cdot 2Nd(NO_3)_3 \cdot 24H_2O$ [c] mass %	$3Mg(NO_3)_2 \cdot 2Nd(NO_3)_3$ [d] mass %
24.8	0.35	59.23	42.58
24.4	0.885	50.31	36.17
24.2	satd[b]	2.92	2.10

a. These are the *initial* $Mg(NO_3)_2$ concentrations. The total Mg content in the equilibrated solutions was not determined.

b. Exact compositions of the liquid and solid phases not specified.

c. Author's original calculations. The primary data, mass % Nd_2O_3, were not reported.

d. Compiler's conversions. Conversions to molality could not be made since density data and total Mg concentrations were not reported.

AUXILIARY INFORMATION

METHOD/APPARATUS/PROCEDURE:

The isothermal method was used as described in (1). About 10-15 g aliquots of saturated sln were filtered through glass wool or sintered glass and diluted to 250 ml for analysis. An aliquot of this diluted sln was taken for detn of neodymium. Nd was pptd as the oxalate by addn of hot oxalic acid, and the ppt allowed to stand to 1 d. The ppt was then filtered, washed with hot dil oxalic acid sln, and ignited to the oxide. The Mg content in satd slns was not determined. Several wet residues for unspecified slns were analysed for total Nd + Mg by direct ingition to the oxide. The oxide was dissolved in HNO_3 and Nd detd by the oxalate method and Mg obtained by difference. In all cases the Nd content was too low by 2 to 5 %, and the Nd_2O_3/MgO mass ratio too low by as much as 9 %. Analysis of dry residues again always gave Nd content too low by as much as 5 mass %, and water (obtained by difference) was always too high by as much as 10 mass %. Analysis of solid phases at 0°C suggested the existence of a hydrate with more than 24 waters of crystallization.

SOURCE AND PURITY OF MATERIALS:

The double salt obtained in (1) by fractional crystallizations was recrystallized several times from dilute nitric acid. The salt was washed with distilled water, and air dried on filter paper. Author states that recrystallization from pure water results in a slight loss of the Nd salt. $Mg(NO_3)_2$ prepd by adding excess MgO to HNO_3. The solution was filtered and Mg content detd gravimetrically (no details given).
Presumably distilled water was used for the solubility measurements.

ESTIMATED ERROR:

Soly: precision ± 5 % at best (compiler).

Temp: accuracy probably ± 0.05 K as in (1) (compiler).

REFERENCES:

Friend, J.A.N. *J. Chem. Soc.* 1930, 1633.

COMPONENTS:

(1) Neodymium manganese nitrate; $2Nd(NO_3)_3 \cdot 3Mn(NO_3)_2$; [84682-63-3]

(2) Nitric acid; HNO_3; [7697-37-2]

(3) Water ; H_2O ; [7732-18-5]

ORIGINAL MEASUREMENTS:

Jantsch, G. *Z. Anorg. Chem.* 1912, *76*, 303-23.

VARIABLES:

One temperature: 16°C

PREPARED BY:

Mark Salomon

EXPERIMENTAL VALUES:

Soly of the double salt in HNO_3 sln of density d_4^{16} = 1.325 g cm^{-3}.

aliquot volume	Nd_2O_3 [a]	soly $2Nd(NO_3)_3 \cdot 3Mn(NO_3)_2$ [b]
cm^3	g	mol dm^{-3}
1.4638	0.0898	
1.4638	0.0892	0.1816

a. Saturated solutions were analysed for total oxides (Nd_2O_3 + Mn_3O_4), but the author does not report this experimental quantity.

b. Author's calculation (average value).

ADDITIONAL DATA:

The melting point of the tetracosahydrate is 77.0°C, and the density at 0°C is 2.114 g cm^{-3}.

AUXILIARY INFORMATION

METHOD/APPARATUS/PROCEDURE:

Isothermal method used. The soly was studied in HNO_3 sln of density 1.325 g cm^{-3} at 16°C because the author did not have sufficient quantity of the rare earth to study the soly of the salt in pure water. Pulverized salt and HNO_3 sln were placed in glass-stoppered tubes and thermostated at 16°C for 24 h with periodic shaking. The solution was then allowed to settle for 2 h, and a pipet maintained at 16°C was used to withdraw aliquots for analysis. Two analyses were performed.

Solutions were analysed by precipitating both Nd and Mn hydroxides by respective addition of NH_3 and H_2O_2. The ppt was ignited to give Nd_2O_3 + Mn_3O_4.

An attempt to determine the waters of hydration by dehydration was not successful because the temperature required (120°C or higher) resulted in decomposition of the salt with the formation of basic salts. Presumably the waters of hydration were found by difference.

SOURCE AND PURITY OF MATERIALS:

"Pure" neodymium oxide was dissolved in dil HNO_3 and $Mn(NO_3)_2$ added to give a mole ratio of Nd/Mn = 2/3. The sln was evapd and a small crystal of $Bi_2Mg_3(NO_3)_{12}$ added, and the mixt cooled to ppt the tetracowahydrate. The double nitrate was recrystd before use. The double salt was analysed gravimetrically for Nd_2O_3 + Mn_3O_4. A 0.4950 g sample of the tetracosahydrate yielded 0.1710 g oxides (i.e. 34.55 mass %O. Theor value is 34.69 mass % oxides (compiler).

ESTIMATED ERROR:

Soly: reproducibility about ± 1-5% (compiler).

Temp: nothing specified.

COMPONENTS:

(1) Neodymium magnanese nitrate; $2Nd(NO_3)_3 \cdot 3Mn(NO_3)_2$; [84682-63-3]

(2) Water ; H_2O ; [7732-18-5]

ORIGINAL MEASUREMENTS:

Prandtl, W.; Ducrue, H.

Z. Anorg. Chem. 1926, *150*, 105-16

VARIABLES:

Temperature

PREPARED BY:

Mark Salomon

EXPERIMENTAL VALUES:

			solubility					
			oxides		double salt			
	mole ratio[a]	density	Nd_2O_3	MnO	hydrate[b]	hydrate[c]	anhydrous salt[d]	
t/°C	MnO/Nd_2O_3	kg m^{-3}	mass %	mass %	mass %	mass %	mass %	mol kg^{-1}
15	2.96	1.66	15.24	9.50	73.8	73.28	53.84	0.974
30	3.01	1.70	15.99	10.15	77.4	77.59	57.00	1.107
45	2.97	1.75	17.05	10.68	82.6	82.18	60.38	1.273
60	2.97[e]	1.82	18.37	11.48	89.0	88.44	64.98	1.550

a. Experimental value; theoretical value = 3.00.

b. Authors' values apparently based on mass % Nd_2O_3. The hydrate which is also the equilibrium solid phase is the tetracosahydrate $2Nd(NO_3)_3 \cdot 3Mn(NO_3)_2 \cdot 24H_2O$.

c. Compiler's calculations based on average from mass % Nd_2O_3 and MnO.

d. Compiler's calculations based on results from c above.

e. Compiler computes 2.96.

AUXILIARY INFORMATION

METHOD/APPARATUS/PROCEDURE:

Isothermal method used. Pulverized double salt (hydrate) and conductivity water were placed in two 50 cc flasks and agitated for 1 day in a thermostat. The slns were then permitted to settle and aliquots of approx 4 cc removed with pipets maintained at the same temp as the satd slns. The aliquots were placed in graduated flasks and weighed, and then diluted with 50 cc of water for analysis. The results for densities and mass % of oxides are the mean of two determinations. The mass % of the tetracosahydrate was apparently calculated by the authors from the mass % Nd_2O_3: i.e. the mass % MnO was not considered.
Both metals were determined gravimetrically. Diluted aliquots were heated and Nd pptd as the oxalate, ignited, dissolved in dilute HNO_3, and pptd two more times as the oxalate. The oxalate was ignited to the oxide and weighed. From the combined filtrates, Mn was pptd as $MnNH_4PO_4$, and presumably ignited to the pyrophosphate $Mn_2P_2O_7$.

SOURCE AND PURITY OF MATERIALS:

Nd_2O_3 prepared by W. Prandtl was analysed by X-ray spectroscopy and found to be "very pure." It was dissolved in nitric acid and the required amount of commercial "pure" grade $Mn(NO_3)_2$ added. The solution was evaporated to crystallization, and the double salt recrystallized several times from conductivity water. The salt was dried over $CaCl_2$ in a desiccator to give the tetracosahydrate. Results of the analysis of the double salt are:

Nd_2O_3 found 20.58, 20.57 %, calcd 20.65 mass %.
MnO found 12.93, 12.64 %, calcd 13.06 mass %.
Conductivity water was used.

ESTIMATED ERROR:

Soly: precision ± 0.3 to ± 1 % (compiler).

Temp: not specified.

COMPONENTS:

(1) Neodymium cobalt nitrate; $2Nd(NO_3)_3 \cdot 3Co(NO_3)_2$; [84697-20-1]

(2) Nitric acid; HNO_3; [7697-37-2]

(3) Water ; H_2O ; [7732-18-5]

ORIGINAL MEASUREMENTS:

Jantsch, G. *Z. Anorg. Chem.* 1912, *76*, 303-23.

VARIABLES:

One temperature: 16°C

PREPARED BY:

Mark Salomon

EXPERIMENTAL VALUES:

Soly of the double salt in HNO_3 sln of density d_4^{16} = 1.325 g cm^{-3}

aliquot volume	Nd_2O_3	soly $2Nd(NO_3)_3 \cdot 3Co(NO_3)_2$ [a]
cm^2	g	mol dm^{-3}
1.4638	0.0452	
1.4638	0.0458	0.0923

a. Author's calculation (average value).

ADDITIONAL DATA:

The melting point of the tetracosahydrate is 95.5°C, and the density at 0°C is 2.195 g cm^{-3}.

AUXILIARY INFORMATION

METHOD/APPARATUS/PROCEDURE:

Isothermal method used. The soly was studied in HNO_3 sln of density 1.325 g cm^{-3} at 16°C because the author did not have sufficient quantity of the rare earth to study the soly of the salt in pure water. Pulverized salt and HNO_3 sln were placed in glass-stoppered tubes and thermostated at 16°C for 24 h with periodic shaking. The solution was then allowed to settle for 2 h, and a pipet maintained at 16°C was used to withdraw aliquots for analysis. Two analyses were performed.

Solutions were analysed by adding 2-3 g NH_4Cl and 10% NH_3 sln followed by boiling to ppt the hydroxide. The ppt was filtered, dissolved in HNO_3, reprecipitated as the hydroxide, and ignited to the oxide. Co in the filtrate was "determined by the usual method" (no details were given).

An attempt to determine the waters of hydration by dehydration was not successful because the temperature required (120°C or higher) resulted in decomposition of the salt with the formation of basic salts. Presumably the waters of hydration were found by difference.

SOURCE AND PURITY OF MATERIALS:

"Pure" neodymium oxide was dissolved in dil HNO_3 and $Co(NO_3)_2$ added to give a mole ratio of Nd/Co = 2/3. The sln was evapd and a small crystal of $Bi_2Mg_3(NO_3)_{12}$ added, and the mixt cooled to ppt the tetracosahydrate. The double nitrate was recrystd before use. The double salt was analysed gravimetrically for Nd_2O_3 and metallic cobalt. 0.5750 g samples of the tetracosahydrate yielded 0.1162 g Nd_2O_3 (i.e. 20.21 mass %) and 0.0622 g metallic cobalt (i.e. 10.82 mass %). Theoretical values calculated by the compiler are 20.50 mass % Nd_2O_3 and 10.77 mass % Co.

ESTIMATED ERROR:

Soly: reproducibility about ± 1-5% (compiler).

Temp: nothing specified.

COMPONENTS:

(1) Neodymium cobalt nitrate; $2Nd(NO_3)_3 \cdot 3Co(NO_3)_2$; [84697-20-1]

(2) Water ; H_2O ; [7732-18-5]

ORIGINAL MEASUREMENTS:

Prandtl, W.; Ducrue, H.

Z. Anorg. Chem. 1926, *150*, 105-16.

VARIABLES:

Temperature

PREPARED BY:

Mark Salomon

EXPERIMENTAL VALUES:

			solubility					
			oxides		double salt			
	mole ratio[a]	density	Nd_2O_3	CoO	hydrate[b]	hydrate[c]	anhydrous salt[d]	
t/°C	CoO/Nd_2O_3	kg m^{-3}	mass %	mass %	mass %	mass %	mass %	mol kg^{-1}
15	2.98	1.65	14.00	9.28	68.3	68.04	50.12	0.831
30	2.98	1.68	14.84	9.84	72.4	72.13	53.14	0.938
45	2.98	1.72	15.70	10.43	76.6	76.39	56.27	1.064
60	3.01	1.77	16.63	11.16	81.1	81.32	59.90	1.235

a. Experimental value: theoretical value = 3.00.

b. Authors' values apparently based on mass % Nd_2O_3. The hydrate which is also the equilibrium solid phase is the tetracosahydrate $2Nd(NO_3)_3 \cdot 3Co(NO_3)_2 \cdot 24H_2O$.

c. Compiler's calculations based on average from mass % Nd_2O_3 and CoO.

d. Compiler's calculations based on results from c above.

AUXILIARY INFORMATION

METHOD/APPARATUS/PROCEDURE:

Isothermal method used. Pulverized double salt (hydrate) and conductivity water were placed in two 50 cc flasks and agitated for 1 day in a thermostat. The slns were them permitted to settle and aliquots of approx 4 cc removed with pipets maintained at the same temp as the satd slns. The aliquots were placed in graduated flasks and weighed, and then diluted with 50 cc of water for analysis. The results for densities and mass % of oxides are the mean of two determinations. The mass % of the tetracosahydrate was apparently calculated by the authors from the mass % Nd_2O_3: i.e. the mass % CoO was not considered.

Both metals were determined gravimetrically.

Nd was pptd as the hydroxide by addition of NH_4Cl/NH_4OH solution. The hydroxide was dissolved in dil HNO_3, pptd as the oxalate, and ignited to the oxide.
The filtrates from the hydroxide and oxalate separations were combined, and nitric and oxalic acids eliminated by addn of sulfuric acid and heating. Co was determined by electrolytic deposition.

SOURCE AND PURITY OF MATERIALS:

Nd_2O_3 prepared by W. Prandtl was analysed by X-ray spectroscopy and found to be "very pure." It was dissolved in nitric acid and the required amount of commercial "pure" grade added. The solution was evaporated to crystallization, and the double salt recrystallized several times from conductivity water. The salt was dried over $CaCl_2$ in a desiccator to give the tetracosahydrate. Results of the analysis of the double salt are:

Nd_2O_3 found 20.51, 20.25 %, calcd 20.50 mass %.
CoO found 13.52, 13.59 %, calcd 13.70 mass %.
Conductivity data was used.

ESTIMATED ERROR:

Soly: precision ± 0.4 % at best (compiler).

Temp: not specified

COMPONENTS:

(1) Neodymium nickel nitrate; $2Nd(NO_3)_3 \cdot 3Ni(NO_3)_2$; [84682-66-6]

(2) Nitric acid; HNO_3; [7697-37-2]

(3) Water ; H_2O ; [7732-18-5]

ORIGINAL MEASUREMENTS:

Jantsch, G. *Z. Anorg. Chem.* 1912, *76*, 303-23.

VARIABLES:

One temperature: 16°C

PREPARED BY:

Mark Salomon

EXPERIMENTAL VALUES:

Soly of the double salt in HNO_3 sln of density d_4^{16} = 1.325 g cm^{-3}.

aliquot volume	Nd_2O_3	soly $2Nd(NO_3)_3 \cdot 3Ni(NO_3)_2$ [a]
cm^3	g	mol dm^{-3}
1.4638	0.0350	
1.4638	0.0350	0.0710

a. Author's calculation (average value).

ADDITIONAL DATA:

The melting point of the tetracosahydrate is 104.6°C, and the density at 0°C is 2.202 g cm^{-3}.

AUXILIARY INFORMATION

METHOD/APPARATUS/PROCEDURE:

Isothermal method used. The soly was studied in HNO_3 sln of density 1.325 g cm^{-3} at 16°C because the author did not have sufficient quantity of the rare earth to study the soly of the salt in pure water. Pulverized salt and HNO_3 sln were placed in glass-stoppered tubes and thermostated at 16°C for 24 h with periodic shaking. The solution was then allowed to settle for 2 h, and a pipet maintained at 16°C was used to withdraw aliquots for analysis. Two analyses were performed.

Solutions were analysed by adding 2-3 g NH_4Cl and 10% NH_3 sln followed by boiling to ppt the hydroxide. The ppt was filtered, dissolved in HNO_3, reprecipitated as the hydroxide, and ignited to the oxide. Ni in the filtrate was "determined by the usual method" (no details were given).

An attempt to determine the waters of hydration by dehydration was not successful because the temperature required (120°C or higher) resulted in decomposition of the salt with the formation of basic salts. Presumably the waters of hydration were found by difference.

SOURCE AND PURITY OF MATERIALS:

"Pure" neodymium oxide was dissolved in dil HNO_3 and $Ni(NO_3)_2$ added to give a mole ratio of Nd/Ni = 2/3. The sln was evapd and a small crystal of $Bi_2Mg_3(NO_3)_{12}$ added, and the mixt cooled to ppt the tetracosahydrate. The double nitrate was recrystd before use. The double salt was analysed gravimetrically for Nd_2O_3 and NiO. A 0.4625 g sample of the tetracosahydrate yielded 0.0952 g Nd_2O_3 (20.58 mass %). Theor value is 20.50 mass % Nd_2O_3 (compiler). Analysis for NiO gave 13.83 mass % compared to the theor value of 13.66 mass % (compiler).

ESTIMATED ERROR:

Soly: reproducibility about ± 1-5% (compiler).

Temp: nothing specified.

COMPONENTS:

(1) Neodymium zinc nitrate; $2Nd(NO_3)_3.3Zn(NO_3)_2$; [31176-55-3]

(2) Nitric acid; HNO_3; [7697-37-2]

(3) Water ; H_2O ; [7732-18-5]

ORIGINAL MEASUREMENTS:

Jantsch, G. *Z. Anorg. Chem.* 1912, *76*, 303-23.

VARIABLES:

One temperature: 16°C

PREPARED BY:

Mark Salomon

EXPERIMENTAL VALUES:

Soly of the double salt in HNO_3 sln of density d_4^{16} = 1.325 g cm^{-3}.

aliquot volume	Nd_2O_3	Soly $2Nd(NO_3)_3.3Zn(NO_3)_2$ [a]
cm^3	g	mol dm^{-3}
1.4638	0.0522	
1.4638	0.0528	0.1066

a. Author's calculation (average value).

ADDITIONAL DATA:

The melting point of the tetracosahydrate is 88.5°C, and the density at 0°C is 2.2215 g cm^{-3}.

AUXILIARY INFORMATION

METHOD/APPARATUS/PROCEDURE:

Isothermal method used. The soly was studied in HNO_3 sln of density 1.325 g cm^{-3} at 16°C because the author did not have sufficient quantity of the rare earth to study the soly of the salt in pure water. Pulverized salt and HNO_3 sln were placed in glass-stoppered tubes and thermostated at 16°C for 24 h with periodic shaking. The solution was then allowed to settle for 2 h, and a pipet maintained at 16°C was used to withdraw aliquots for analysis. Two analyses were performed.

Solutions were analysed by adding 2-3 g NH_4Cl and 10% NH_3 sln followed by boiling to ppt the hydroxide. The ppt was filtered, dissolved in HNO_3, reprecipitated as the hydroxide, and ignited to the oxide. Zn in the filtrate was "determined by the usual method" (no details were given).

An attempt to determine the waters of hydration by dehydration was not successful because the temperature required (120°C or higher) resulted in decomposition of the salt with the formation of basic salts. Presumably the waters of hydration were found by difference.

SOURCE AND PURITY OF MATERIALS:

"Pure" neodymium oxide was dissolved in dil HNO_3 and $Zn(NO_3)_2$ added to give a mole ratio of Nd/Zn = 2/3. The sln was evapd and a small crystal of $Bi_2Mg_3(NO_3)_{12}$ added, and the mixt cooled to ppt the tetracosahydrate. The double nitrate was recrystd before use. The double salt was analysed gravimetrically for Nd_2O_3. A 0.5650 g sample of the tetracosahydrate yielded 0.1150 g oxide (i.e. 20.35 mass %). The theoretical value is 20.26 mass %. Analysis for NO (details not given): found 21.73 mass %, theor 21.68 mass %.

ESTIMATED ERROR:

Soly: reproducibility about ± 1-5% (compiler).

Temp: nothing specified.

COMPONENTS:

(1) Neodymium zinc nitrate; $2Nd(NO_3)_3 \cdot 3Zn(NO_3)_2$; [31176-55-3]

(2) Water ; H_2O ; [7732-18-5]

ORIGINAL MEASUREMENTS:

Prandtl, W.; Ducrue, H.

Z. Anorg. Chem. 1926, *150*, 105-16.

VARIABLES:

Temperature

PREPARED BY:

Mark Salomon

EXPERIMENTAL VALUES:

			solubility					
			oxides		double salt			
	mole ratio[a]	density	Nd_2O_3	ZnO	hydrate[b]	hydrate[c]	anhydrous salt[d]	
t/°C	ZnO/Nd_2O_3	kg m^{-3}	mass %	mass %	mass %	mass %	mass %	mol kg^{-1}
15	3.00	1.65	14.00	10.15	69.1	69.09	51.10	0.851
30	3.00	1.69	14.88	10.79	73.5	73.44	54.32	0.968
50	3.01	1.75	15.86	11.53	78.8	78.37	57.97	1.123
70	2.97	1.81	17.15	12.33	84.6	84.28	62.34	1.347

a. Experimental value: theoretical value = 3.00.

b. Authors' values apparently based on mass % Nd_2O_3. The hydrate which is the equilibrium solid phase is the tetracosahydrate $2Nd(NO_3)_3 \cdot 3Zn(NO_3)_2 \cdot 24H_2O$.

c. Compiler's calculations based on average from mass % Nd_2O_3 and ZnO.

d. Compiler's calculations based on results from c above.

AUXILIARY INFORMATION

METHOD/APPARATUS/PROCEDURE:

Isothermal method used. Pulverized double salt (hydrate) and conductivity water were placed in two 50 cc flasks and agitated for 1 day in a thermostat. The slns were then permitted to settle and aliquots of approx 4 cc removed with pipets maintained at the same temp as the satd slns. The aliquots were placed in graduated flasks and weighed, and then diluted with 50 cc of water for analysis. The results for densities and mass % of oxides are the mean of two determinations. The mass % of the tetracosahydrate was apparently calculated by the authors from the mass % Nd_2O_3: i.e. the mass % ZnO was not considered.

Both metals were determined gravimetrically. Nd was pptd with NH_4Cl/NH_4OH solution, filtered, redissolved in dil HNO_3, pptd as the oxalate and ignited to the oxide. From the combined filtrates, Zn was pptd as $ZnNH_4PO_4$, and presumably ignited to the pyrophosphate $Zn_2P_2O_7$.

SOURCE AND PURITY OF MATERIALS:

Nd_2O_3 prepared by W. Prandtl was analysed by X-ray spectroscopy and found to be "very pure." It was dissolved in nitric acid and the required amount of commercial "pure" grade $Zn(NO_3)_2$ added. The solution was evaporated to crystallization, and the double salt recrystallized several times from conductivity water. The salt was dried over $CaCl_2$ in a desiccator to give the tetracosahydrate. Results of the analysis of the double salt are:

Nd_2O_3 found 20.33 %, calcd 20.26 mass %.

ZnO found 14.45 %, calcd 14.70 mass %.

Conductivity water was used.

ESTIMATED ERROR:

Soly: precision ± 0.5 % at best (compiler).

Temp: not specified.

COMPONENTS:

(1) Samarium nitrate; $Sm(NO_3)_3$; [10361-83-8]

(2) Water ; H_2O ; [7732-18-5]

EVALUATOR:

S. Siekierski, T. Mioduski
Institute for Nuclear Research
Warsaw, Poland
and
M. Salomon
U.S. Army ET & DL
Ft. Monmouth, NJ, USA
May 1982

CRITICAL EVALUATION:

THE BINARY SYSTEM

INTRODUCTION

Solubility data in the binary $Sm(NO_3)_3$-H_2O binary system have been reported in 12 publications (1-12). Only one study (2) reports solubilities over a wide temperature range using the synthetic method, and all other studies employ the isothermal method. A number of hydrates have been reported to be equilibrium with saturated solutions:

$Sm(NO_3)_3 \cdot 6H_2O$	[13759-83-6]	$Sm(NO_3)_3 \cdot 4H_2O$	[37131-73-0]
$Sm(NO_3)_3 \cdot 5H_2O$	[24581-35-9]	$Sm(NO_3)_3 \cdot 3H_2O$	[81201-38-9]

Only the hexahydrate and pentahydrate have been quantitatively identified in the binary system, but Popov and Mironov (2) state that a lower hydrate exists above 359.6 K. The tetrahydrate, trihydrate and anhydrous solid phases were identified in nitric acid solutions at 298.2 K and 323.2 K (8-10).

Analyses of the solubilities in the binary system as a function of temperature (see below) clearly shows that both the hexahydrate and pentahydrate exist over a wide range of temperatures. Although the pentahydrate is metastable below 332.0 K, it is surprisingly slow to convert to the stable hexahydrate as demonstrated by the fact that its solubility can be measured at temperatures as low as 236.5 K (2). Attempts to convert the metastable pentahydrate solid to the hexahydrate be seeding did not succeed, and the hexahydrate could only be produced by first heating the metastable saturated solution followed by slow cooling (9). This fact that the metastable pentahydrate is extremely stable below 332.0 K is believed to result in errors in the experimentally determined congruent melting point as discussed below.

EVALUATION PROCEDURE

The data in the compilations were examined and either rejected immediately because of large obvious errors, or were analysed by a weighted least squares method. It should be noted that only experimental solubility data were used in the least squares analyses: i.e. smoothed or extrapolated data were not used. The data were fitted to a general solubility equation based on the treatments in (13, 14) and in the INTRODUCTION to this volume:

$$Y = \ell n(m/m_o) - nM_2(m - m_o) = a + b/(T/K) + c\ \ell n(T/K) \quad [1]$$

In eq. [1] m is the molality at temperature T, m_o is an arbitrarily selected reference molality (usually the 298.2 K value), n is the hydration number of the solid, M_2 is the molar mass of the solvent, and a, b, c are constants from which enthalpies and heat capacities of solution, ΔH_{sln} and ΔC_p, can be estimated (see INTRODUCTION). In fitting the solubility data to eq. [1] , weights of 0, 1, 2 were assigned to each published value depending upon the precision of the experimental value. In this procedure, if the residual error between the observed and calculated molalities, Δm, was larger than twice the standard error of estimate for m, σ_m, the data point was either rejected or its weight factor decreased. The fitting of the data was repeated in this manner until all Δm values were equal to or less than $\pm 2\sigma_m$.

Since most authors did not report experimental errors (except in 3-7), the compilers attempted to provide this information when possible. As discussed in previous critical evaluations, the data of highest precision are those from the isothermal studies of Spedding et al. (3-7) who reported a precision of ± 0.1% or better, Brunisholz et al. (1, 11) in which the compilers estimated a precision of around ± 0.2%, and Popov and Mironov (9, 10) for which the evaluators estimate an experimental precision of ± 0.5%. For the remaining two isothermal studies in (8, 12), the experimental precision is probably around ± 0.2% (based on the number of significant figures reported by these authors), but the accuracy in the binary system in (8) is poor, and very poor in (12). The results reported by Popov and Mironov based on the synthetic method (2) are assigned an overall precision of ± 0.5% at best.

COMPONENTS:	EVALUATOR:
(1) Samarium nitrate; $Sm(NO_3)_3$; [10361-83-8] (2) Water ; H_2O ; [7732-18-5]	S. Siekierski, T. Mioduski Institute for Nuclear Research Warsaw, Poland and M. Salomon U.S. Army ET & DL Ft. Monmouth, NJ, USA May 1982

CRITICAL EVALUATION:

SOLUBILITY OF $Sm(NO_3)_3$ IN THE $Sm(NO_3)_3 \cdot 6H_2O-H_2O$ SYSTEM

The solubility data in the binary system in which the solid phase is the hexahydrate are summarized in Table 1. The results of fitting the data in Table 1 to the smoothing equation [1] are given in Tables 3 and 4, and in Fig. 1. Table 3 gives the derived parameters for eq. [1] , and Table 4 gives the smoothed solubility data calculated from eq. [1] , and which are designated *recommended* values. The hexahydrate polytherm in

Table 1. Solubility of $Sm(NO_3)_3$ in the $Sm(NO_3)_3 \cdot 6H_2O-H_2O$ system.

T/K	soly mol kg^{-1}	ref	weight initial/final	T/K	soly mol kg^{-1}	ref	weight initial/final
243.2	3.21	2	1/1	298.15	4.2800	6	2/2
268.4	3.50	2	1/1	298.15	4.2774	7	1/1
273.15	3.596	1	2/2	303.2	4.225	12	0/0
273.15	3.623	11	2/2	303.5	4.50	2	1/1
283.15	3.861	1	2/2	308.15	4.689	1	2/2
287.0	3.85	2	1/0	308.15	4.704	11	2/1
293.15	4.112	1	2/2	314.6	5.15	2	1/0
293.15	4.113	11	2/2	323.15	5.444	1	2/0
298.2	4.261	10	1/0	323.2	5.475	8	1/0
298.15	4.269	8	1/0	337.0	7.42	2	1/0
298.15	4.2811	3,4	2/2	344.4	8.92	2	0/0
298.15	4.284	5	1/2	346.2	9.52	2	0/0

Fig. 1 was drawn from the smoothed solubility data, and the experimental data points are included for comparisons.

The data in (8, 10, 12) were rejected, and four data points from (2) were also rejected. The rejected data from (2) show small but significant negative deviations from the smoothed polytherm, and the solubility of 9.52 mol kg^{-1} at 346.2 K exceeds the limit of 9.25 mol kg^{-1} for the congruently melting solid (this latter experimental solubility could be attributed to supersaturation). The congruent melting point for the hexahydrate as calculated from eq. [1] is 344.1 K which differs significantly from early experimental values of 351-352 K (15) and 352.7 K (16), but is within experimental and calculated error to the more recent value of 341.2 ± 1K (17). The high melting points reported in (15, 16) are probably due to the presence of pentahydrate solid phases (m.p. of the pentahydrate is 359.5 K as discussed below). Quill et al. (16) do report the appearance of cloudiness and a temperature arrest in the experimental cooling curve before the appearance of the major temperature arrest at 352.7 K.

SOLUBILITY OF $Sm(NO_3)_3$ IN THE $Sm(NO_3)_3 \cdot nH_2O-H_2O$ SYSTEM: $n \leq 5$

The solubility data for binary systems involving the pentahydrate and lower hydrates are given in Table 2. Because all the data are from one research group (Mironov and

Table 2. Solubility of $Sm(NO_3)_3$ in the $Sm(NO_3)_3 \cdot nH_2O-H_2O$ system: $n \leq 5$

T/K	solubility mol kg^{-1}	solid phase[a]	ref	weight initial/final
236.5	3.55	$Sm(NO_3)_3 \cdot 5H_2O$	2	1/1
298.2	4.677	"	9	1/1
327.0	6.04	"	2	1/1
339.0	6.65	"	2	1/1
350.1	8.20	"	2	1/1
356.8	9.31	"	2	1/1
356.0	9.84	"	2	1/0
361.6	11.82	"	2	0/0
359.6	14.83	"	2	0/0
298.2	12.27[b]	$Sm(NO_3)_3 \cdot 4H_2O$	10	
360.1	14.94	?	2	
408.2	18.73	?	2	

a. Solid phases reported by authors.
b. Extrapolated value: See eq. [4] below.

COMPONENTS:	EVALUATOR:
(1) Samarium nitrate; $Sm(NO_3)_3$; [10361-83-8] (2) Water ; H_2O ; [7732-18-5]	S. Siekierski, T. Mioduski Institute for Nuclear Research Warsaw, Poland and M. Salomon U.S. Army ET & DL Ft. Monmouth, NJ, USA May 1982

CRITICAL EVALUATION:

co-workers), equal weights of unity were given to each solubility value in the initial fitting of the data to eq. [1]. The two solubility values at 361.6 K and 359.6 K which Popov and Mironov (2) assign to the pentahydrate system were immediately rejected because the solubilities of 11.82 mol kg^{-1} and 14.83 mol kg^{-1} exceed the theoretical solubility limit of 11.10 mol kg^{-1} (i.e. at the congruent melting point where the salt dissolves in its waters of hydration). The results of the least squares fitting of the pentahydrate data to eq. [1] are given in Table 3, and the smoothed solubility data are given in Table 4. The predicted congruent melting point of the pentahydrate is 359.5 K at a solubility of 11.102 mol kg^{-1}. There are no experimental values with which to compare this calculated melting point.

The evaluators attempted to fit the remaining data points in Table 2 (including the rejected data points in the pentahydrate system) to the smoothing equation, but a significant fit assuming tetrahydrate or trihydrate solid phases could not be achieved. The solubility of 12.27 mol kg^{-1} at 298.2 K in the tetrahydrate system is an extrapolated value (10), and is based on data in the $Sm(NO_3)_3$-HNO_3-H_2O ternary system (see discussion below and eq. [4]).

RECOMMENDED AND TENTATIVE SOLUBILITY DATA

The phase diagram of the $Sm(NO_3)_3.nH_2O$-H_2O system is given in Fig. 1. It is interesting to note that the polytherms for the hexa- and pentahydrates remain nearly parallel up to the transition point of 332.0 K. There is, in general, excellent agreement between the data of Mironov et al., Brunisholz et al., and Spedding et al. for solubilities below the hexahydrate to pentahydrate transition temperature of 332.0 K. All data for the hexahydrate system fall within $\pm$ $2\sigma_m$ of the solubilities calculated from eq. [1], and the smoothed solubilities are designated as *recommended* values. These *recommended* values are given in Table 4 for selected temperatures. At the 95% level of confidence and a Student's t = 2.086, the uncertainty in the smoothed solubilities is $\pm$ 0.006 mol kg^{-1}. Combining this uncertainty with the average experimental precision of $\pm$ 0.2% results in an overall precision in the *recommended* solubilities of $\pm$ 0.011 mol kg^{-1}.

The smoothed solubility data in the pentahydrate system given in Table 4 are designated as *tentative* solubilities. The uncertainty in the calculated values is $\pm$ 0.19 mol kg^{-1} based on a Student's t = 3.182 at the 95% level of confidence, and combining this with the experimental precision of $\pm$ 0.5% results in an overall uncertainty of $\pm$ 0.2 mol kg^{-1} for the values given in Table 4.

MULTICOMPONENT SYSTEMS

The $Sm(NO_3)_3$ - HNO_3 - H_2O SYSTEM

The solubility of $Sm(NO_3)_3$ in aqueous nitric acid solutions has been reported by Quill and Robey (8), and by Mironov et al. (9, 10). The solubility isotherms at 298.2 K are given in Fig. 2, and the branches for all hydrates are marked on the figure. The hexahydrate and pentahydrate solubility branches are essentially parallel almost up to the eutonic point at about 53 mass % HNO_3. For the hexahydrate isotherm the combined data of Quill and Robey and Mironov et al. are in excellent agreement, and these data are designated as *recommended* values: only the value of m_1 = 16.2 mass % and m_2 = 42.6 mass % (10) is *rejected*. The remaining solubility data in the pentahydrate, tetrahydrate, trihydrate, and anhydrous salt systems are designated as *tentative* solubilities.

There is agreement in (8-10) that complexes between $Sm(NO_3)_3$ and HNO_3 do not exist. Popov and Mironov (10) give smoothing equations for the hexa-, penta-, and tetrahydrate isotherms based on Kirgintsev's equation (18), and the equations are (for 298.2 K):

isotherm	smoothing equation	
hexahydrate	$\log m_1 = \log 4.26 - 0.703 \log y_1$	[2]
pentahydrate	$\log m_1 = \log 4.68 - 0.606 \log y_1$	[3]
tetrahydrate	$\log m_1 = \log 12.27 - 0.523 \log y_1$	[4]

In eqs. [2]-[4], m_1 is the molality of $Sm(NO_3)_3$, and y_1 is the solute mole fraction of the salt defined as $y_1 + y_2 = 1$ where y_2 is the solute mole fraction of HNO_3.

COMPONENTS:

(1) Samarium nitrate; $Sm(NO_3)_3$; [10361-83-8]

(2) Water ; H_2O ; [7732-18-5]

EVALUATOR:
S. Siekierski, T. Mioduski
Institute for Nuclear Research
Warsaw, Poland
and
M. Salomon
U.S. Army ET & DL
Ft. Monmouth, NJ, USA
May 1982

CRITICAL EVALUATION:

Table 3. Derived parameters for the smoothing equation.[a]

parameter	hexahydrate system	pentahydrate system
a	-36.708	-31.169
b	1298.9	1075
c	5.6821	4.814
σ_a	0.002	0.007
σ_b	0.5	3
σ_c	0.0003	0.001
σ_Y	0.002	0.007
σ_m	0.013	0.1_5
ΔH_{sln}/kJ mol^{-1}	-43.0	-35.6
ΔC_p/J K^{-1} mol^{-1}	189.0	160.1
congruent melting point/K	344.1	359.5
concn at the congruent m.p./mol kg^{-1}	9.251	11.102

[a] The values σ_a, σ_b, σ_c are standard deviations, and σ_Y and σ_m are standard errors of estimate for the quantity Y in eq. [1] and the molality, respectively.

Table 4. Recommended and tentative solubility data at selected temperatures calculated from the smoothing equation.

T/K	hexahydrate system[a] soly/mol kg^{-1}	pentahydrate system[b] soly/mol kg^{-1}
243.15	3.208	3.60
263.15	3.432	3.84
273.15	3.609	4.03
278.15	3.715	4.13
283.15	3.835	4.25
288.15	3.969	4.38
293.15	4.119	4.52
298.15	4.287	4.68
303.15	4.475	4.85
308.15	4.688	5.04
313.15	4.929	5.25
318.15	5.207	5.48
323.15	5.533	5.74
328.15	5.923	6.03
332.00[c]	6.283	6.283
333.15	6.413	6.36
338.15	7.085	6.75
343.15	8.340	7.20
348.2		7.77
353.2		8.54
358.2		9.89

[a] *Recommended* solubility data.

[b] *Tentative* solubility data.

[c] Hexahydrate → pentahydrate transition temperature (determined graphically by the evaluators).

COMPONENTS:

(1) Samarium nitrate; $Sm(NO_3)_3$; [10361-83-8]

(2) Water ; H_2O ; [7732-18-5]

EVALUATOR:

S. Siekierski, T. Mioduski
Institute for Nuclear Research
Warsaw, Poland
and
M. Salomon
U.S. Army ET & DL
Ft. Monmouth, NJ, USA
May 1982

CRITICAL EVALUATION:

MULTICOMPONENT SYSTEMS WITH TWO SATURATING COMPONENTS

Brunisholz et al. (1) have studied the solubilities in the ternary $La(NO_3)_3$-$Sm(NO_3)_3$-H_2O system at 273.2, 293.2 , and 308.2 K. These authors report the existence of solid solutions of general composition $La_xSm_{1-x}(NO_3)_3 \cdot 6H_2O$, and the miscibility limits defined by x vary as a function of temperature. In a rejected paper on this ternary system, ((19), see the $La(NO_3)_3$ critical evaluation for details), the existence of solid solutions was not reported. Solubilities in the quaternary system $La(NO_3)_3$-$Sm(NO_3)_3$-tributylphosphate-water have been reported by Kolesnikov et al. (20), and the compilation of this paper can be found in the $La(NO_3)_3$ chapter.

The $Sm(NO_3)_3$-$Zn(NO_3)_2$-H_2O system has been studied by Brunisholz and Klipfel (11), and the dominant feature in this system is the formation of the tetracosahydrate

$$2Sm(NO_3)_3 \cdot 3Zn(NO_3)_2 \cdot 24H_2O \quad [28876\text{-}83\text{-}7]$$

Although the solubility data in the $Sm(NO_3)_3$-$CH_3CONHCONH_2$-H_2O system at 303.2 K (12) cannot be critically evaluated, it should be noted that the solubility of $Sm(NO_3)_3$ found for the binary system is in error by -22%.

REFERENCES

1. Brunisholz, G.; Quinche, J.P.; Kalo, A.M. *Helv. Chim. Acta* 1964, *47*, 14.
2. Popov, A.P.; Mironov, K.E. *Rev. Roum. Chim.* 1968, *13*, 765.
3. Spedding, F.H.; Shiers, L.E.; Rard, J.A. *J. Chem. Eng. Data* 1975, *20*,88.
4. Rard, J.A.; Spedding, F.H. *J. Phys. Chem.* 1975, *79*, 257.
5. Spedding, F.H.; Derer, J.L.; Mohs, M.A.; Rard, J.A. *J. Chem. Eng. Data* 1976, *21*, 474.
6. Spedding, F.H.; Shiers, L.E.; Brown, M.A.; Baker, J.L.; Guiterrez, L.; McDowell, L.S.; Habenschuss, A. *J. Phys. Chem.* 1975, *79*, 1087.
7. Rard, J.A.; Shiers, L.E.; Heiser, D.J.; Spedding, F.H. *J. Chem. Eng. Data* 1977, *22*, 337.
8. Quill, L.L.; Robey, R.F. *J. Am. Chem. Soc.* 1937, 2591.
9. Mironov, K.E.; Sinitsyna, E.D.; Popov, A.P. *Zh. Neorg. Khim.* 1966, *11*, 2361.
10. Popov, A.P. Mironov, K.E. *Zh. Neorg. Khim.* 1971, *16*, 464.
11. Brunisholz, G.; Klipfel, K. *Rev. Chim. Miner.* 1970, *7*, 349.
12. Usubalieva, U.; Sulaimankulov, K. *Zh. Neorg. Khim.* 1982, *27*, 1338.
13. Williamson, A.T. *Trans. Faraday Soc.* 1944, *40*, 421.
14. Counioux, J.-J.; Tenu, R. *J. Chim. Phys.* 1981, *78*, 816 and 823.
15. Demarcay, E. *Compt. Rend.* 1898, *126*, 1039; 1900, *130*, 1185; 1900, *131*, 343.
16. Quill, L.L.; Robey, R.F.; Seifter, S. *Ind. Eng. Chem. Anal. Ed.* 1937, *9*, 389.
17. Wendlandt, W.W.; Sewell, R.G. *Texas J. Sci.* 1961, *13*, 231.
18. Kirgintsev, A.N. *Izv. Akad. Nauk SSSR, Ser. Khim. Nauk* 1965, *No. 8*, 1591.
19. Petelina, V.S.; Guzhvina, O.V. *Issled. Mnogokomponent. Sistem s Razl. Vzaimodeistviem Komponentov (Saratov)* 1977, 86.
20. Kolesnikov, A.S.; Korotkevich, T.B.; Shakhaleeva, N.N.; Stepin, B.D. *Zh. Neorg. Khim.* 1978, *23*, 1395 (the compilation for this paper is included in the $La(NO_3)_3$ chapter).

COMPONENTS:

(1) Samarium nitrate; $Sm(NO_3)_3$; [10361-83-8]

(2) Water ; H_2O ; [7732-18-5]

EVALUATOR:
S. Siekierski, T. Mioduski
Institute for Nuclear Research
Warsaw, Poland
and
M. Salomon
U.S. Army ET & DL
Ft. Monmouth, NJ, USA
May 1982

CRITICAL EVALUATION:

Fig. 1. Solubility polytherms for the $Sm(NO_3)_3$-H_2O system.

Solid lines (stable systems) and dashed lines (metastable systems) calculated from smoothing equation. Experimental points are:

□ refs (1, 11)
○ ref (2)
△ refs (3-6)
● refs (2,9)
▲ ref (12)

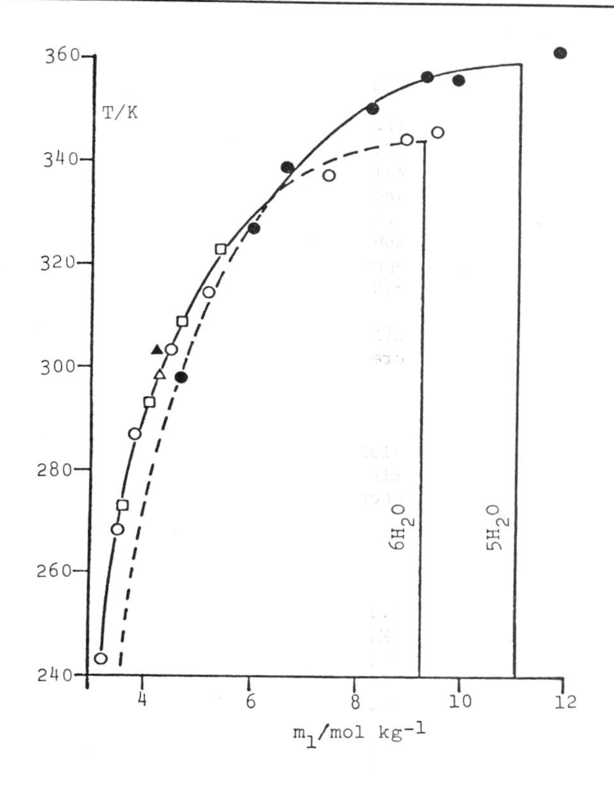

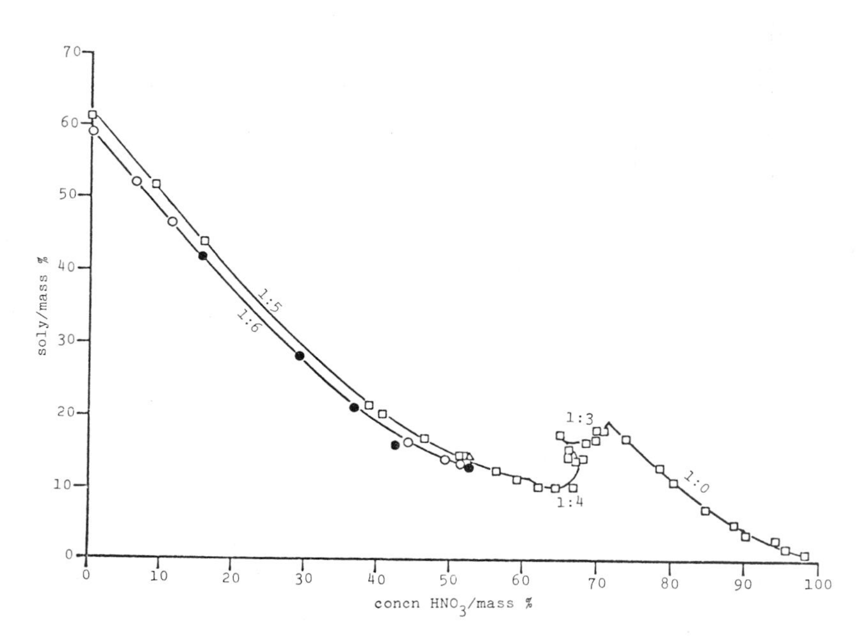

Fig. 2. Solubility isotherms at 298.2 K for the $Sm(NO_3)_3$ - HNO_3 - H_2O system. Lines hand-drawn by the evaluators. Experimental points are: ○ (8), ● (9,10), □ (9).

COMPONENTS:

(1) Samarium nitrate; $Sm(NO_3)_3$; [10361-83-8]

(2) Water; H_2O; [7732-18-5]

ORIGINAL MEASUREMENTS:

Brunisholz, G.; Quinche, J.P.; Kalo, A.M. *Helv. Chim. Acta.*, 1964, *47*, 14-27.

VARIABLES:

Temperature

PREPARED BY:

T. Mioduski and S. Siekierski

EXPERIMENTAL VALUES:

Solubility in the system $Sm(NO_3)_3$-H_2O [a]

t/°C	mass %	mol kg^{-1}
0	54.74	3.596
10	56.50	3.861
20	58.04	4.112
35	61.20	4.689
50	64.68	5.444

a. Molalities calculated by the compilers.

AUXILIARY INFORMATION

METHOD/APPARATUS/PROCEDURE:

The isothermal method was used. Sm was determined by complexometric titration using xylenol orange indicator in the presence of a small quantity of urotropine buffer. Water was determined by difference. The solid phase is $Sm(NO_3)_3 \cdot 6H_2O$.

SOURCE AND PURITY OF MATERIALS:

Samarium nitrate was prepared from Sm_2O_3 of purity better than 99.7 % (obtained by the ion exchange chromatography).

ESTIMATED ERROR:

Soly: precision about ± 0.2 % (compilers).

Temp: precision probably better than ± 0.05 K (compilers).

REFERENCES:

COMPONENTS:

(1) Samarium nitrate; $Sm(NO_3)_3$; [10361-83-8]

(2) Water; H_2O; [7732-18-5]

ORIGINAL MEASUREMENTS:

Popov, A.P.; Mironov, K.E. *Rev. Roum. Chim.* 1968, *13*, 765-73.

VARIABLES:

Temperature: range -30 to 135°C

PREPARED BY:

T. Mioduski and S. Siekierski

EXPERIMENTAL VALUES:

Solubility of $Sm(NO_3)_3$ as a function of temperature [a]

t/°C	mass %	mol kg^{-1}	solid phase
- 2.1	19.2	0.71	ice
- 3.9	25.1	1.00	"
- 8.5	33.7	1.51	"
-11.9	38.0	1.82	"
-15.9	42.5	2.20	"
-20.3	46.5	2.58	"
-26.5	49.9	2.96	"
-33.5	52.9	3.34	"
-30.0	51.9	3.21	$Sm(NO_3)_3 \cdot 6H_2O$
- 4.8	54.1	3.50	"
13.8	56.4	3.85	"
30.3	60.2	4.50	"
41.4	63.4	5.15	"
63.8	71.4	7.42	$Sm(NO_3)_3 \cdot 6H_2O$ [b]
71.2 [c]	75.0	8.92	"
73.0	76.2	9.52	"
-36.7	54.4	3.55	$Sm(NO_3)_3 \cdot 5H_2O$ [b]
53.8	67.0	6.04	$Sm(NO_3)_3 \cdot 5H_2O$
65.8	69.1	6.65	"
76.9	73.4	8.20	"
83.6	75.8	9.31	"
82.8	76.8	9.84	"
88.4	79.9	11.82	"
86.4	83.3	14.83	"
86.9	83.4	14.94	lower hydrate
135.0	86.3	18.73	"

a. Molalities calculated by M. Salomon.

b. Metastable equilibrium.

c. This solubility value was determined by the isothermal method.

AUXILIARY INFORMATION

METHOD/APPARATUS/PROCEDURE:

The synthetic method was used.

The temperatures of crystallization were determined visually.

SOURCE AND PURITY OF MATERIALS:

$Sm(NO_3)_3 \cdot 6H_2O$ was obtained by dissolving Sm_2O_3 of 99.5 % purity in HNO_3 and crystn. Water bidistd.

ESTIMATED ERROR:

Nothing specified.

REFERENCES:

COMPONENTS:

(1) Samarium nitrate; $Sm(NO_3)_3$; [10361-83-8]

(2) Water; H_2O; [7732-18-5]

ORIGINAL MEASUREMENTS:

1. Spedding, F.H.; Shiers, L.E.; Rard, J.A. *J. Chem. Eng. Data* 1975, *20*, 88-93.
2. Rard, J.A.; Spedding, F.H. *J. Phys. Chem.* 1975, *79*, 257-62.
3. Spedding, F.H.; Derer, J.L.; Mohs, M.A.; Rard, J.A. *J. Chem. Eng. Data* 1976, *21*, 474-88.

VARIABLES:

One temperature: 25.00°C

PREPARED BY:

T. Mioduski, S. Siekierski, and M. Salomon

EXPERIMENTAL VALUES:

The solubility of $Sm(NO_3)_3$ in water at 25.00°C has been reported by Spedding and co-workers in three publications. Source paper [3] reports the solubility to be 4.284 mol kg^{-1}, but the preferred value is given in source papers [1] and [2] as 4.2811 mol kg^{-1}.

COMMENTS AND/OR ADDITIONAL DATA:

Source paper [1] reports the relative viscosity, η_R, of a saturated solution to be 15.883. Taking the viscosity of water at 25°C to equal 0.008903 poise, the viscosity of a saturated $Sm(NO_3)_3$ solution at 25°C is 0.1410 poise (compilers calculation).

Supplementary data available in the microfilm edition to *J. Phys. Chem.* 1975, *79*, have enabled the compilers to provide the following additional data.

The density of the saturated solution was calculated by the compilers from the smoothing equation, and at 25°C the value is 1.8269 kg m^{-3}. Using this density, the solubility in volume units is (based on the preferred value of 4.2811 mol kg^{-1})

$$c_{satd} = 3.2054 \text{ mol dm}^{-3}$$

Source paper [2] reports the electrolyte conductivity of the saturated solution to be (corrected for the electrolytic conductivity of the solvent) $\kappa = 0.029531$ S cm^{-1}.

The molar conductivity of the saturated solution is calculated from $1000\ \kappa/3c_{satd}$ and is

$$\Lambda(\tfrac{1}{3}Sm(NO_3)_3) = 3.071 \text{ S cm}^2 \text{ mol}^{-1}$$

AUXILIARY INFORMATION

METHOD/APPARATUS/PROCEDURE:

Isothermal method used. Solutions were prepared as described in (1) and (3). The concentration of the saturated solution was determined by both EDTA (1) and sulfate (2) methods which is said to be reliable to 0.1 % or better. In the sulfate analysis, the salt was first decomposed with HCl followed by evaporation to dryness before sulfuric acid additions were made. This eliminated the possibility of nitrate ion coprecipitation.

SOURCE AND PURITY OF MATERIALS:

$Sm(NO_3)_3 \cdot 6H_2O$ was prepd by addn of HNO_3 to the oxide. The oxide was purified by an ion exchange method, and the upper limit for the impurities Ca, Fe, Si and adjacent rare earths was given as 0.15 %.

In source paper (3) the salt was analysed for water of hydration and found to be within ± 0.016 water molecules of the hexahydrate.

ESTIMATED ERROR:

Soly: duplicate analyses agreed to at least ± 0.1 %.

Temp: not specified, but probably accurate to at least ± 0.01 K as in (3) (compilers).

REFERENCES:

1. Spedding, F.H.; Cullen, P.F.; Habenschuss, A. *J. Phys. Chem.* 1974, *78*, 1106.
2. Spedding, F.H.; Pikal, M.J.; Ayers, B.O. *J. Phys. Chem.* 1966, *70*, 2440.
3. Spedding, F.H.; et. al. *J. Chem. Eng. Data* 1975, *20*, 72.

COMPONENTS:

(1) Samarium nitrate; $Sm(NO_3)_3$; [10361-83-8]

(2) Water; H_2O; [7732-18-5]

ORIGINAL MEASUREMENTS:

1. Spedding, F.H.; Shiers, L.E.; Brown, M.A.; Baker, J.L.; Guitierez, L.; McDowell, L.S.; Habenshuss, A. *J. Phys. Chem.* 1975, *79*, 1087-96:
2. Rard, J.A.; Shiers, L.E.; Heiser, D.J.; Spedding, F.H. *J. Chem. Eng. Data* 1977, *22*, 337-47.

VARIABLES:

One temperature: 25.00°C

PREPARED BY:

T. Mioduski, S. Siekierski, and M. Salomon

EXPERIMENTAL VALUES:

Source paper [1] reports a solubility of 4.2800 mol kg^{-1}

Source paper [2] reports a solubility of 4.2774 mol kg^{-1}

The solid phase in both studies is the hexahydrate $Sm(NO_3)_3 \cdot 6H_2O$

Source paper [1] reports a density of the saturated solution of 1.82668 kg m^3. Using this density value, the compilers have calculated the solubility in volume units (based on the value of 4.2800 mol kg^{-1} given for the saturated solution in source paper [1]:

$$c_{satd} = 3.2046 \text{ mol dm}^{-3}$$

AUXILIARY INFORMATION

METHOD/APPARATUS/PROCEDURE:

Isothermal method used. Solutions were prepared as described in (1) and (2). The concentration of the saturated solution was determined by both EDTA (1) and sulfate (2) methods which is said to be reliable to 0.1 % or better. In the sulfate analysis, the salt was first decomposed with HCl followed by evaporation to dryness before sulfuric acid additions were made. This eliminated the possibility of nitrate ion coprecipitation.

SOURCE AND PURITY OF MATERIALS:

$Sm(NO_3)_3 \cdot 6H_2O$ was prepd by addn of HNO_3 to the oxide. The oxide was purified by an ion exchange method, and the upper limit for the impurities Ca, Fe, Si and adjacent rare earths was given as 0.15 %.

ESTIMATED ERROR:

Soly: duplicate analyses agreed to at least ± 0.1 %.

Temp: not specified, but probably accurate to at least ± 0.01 K as in (3).

REFERENCES:

1. Spedding, F.H.; Cullen, P.F.; Habenschuss, A. *J. Phys. Chem.* 1974, *78*, 1106.
2. Spedding, F.H.; Pikal, M.J.; Ayers, B.O. *J. Phys. Chem.* 1966, *70*, 2440.
3. Spedding, F.H.; et. al. *J. Chem. Eng. Data* 1975, *20*, 72.

COMPONENTS:

(1) Samarium nitrate; $Sm(NO_3)_3$; [10361-83-8]
(2) Nitric acid; HNO_3; [7697-37-2]
(3) Water; H_2O; [7732-18-5]

ORIGINAL MEASUREMENTS:

Quill, L.L.; Robey, R.F. *J. Am. Chem. Soc.* 1937, *59*, 2591-5.

VARIABLES:

Composition at 25°C and 50°C

PREPARED BY:

T. Mioduski, S. Siekierski, and M. Salomon

EXPERIMENTAL VALUES:

Solubility of $Sm(NO_3)_3$ in nitric acid solutions [a]

	$Sm(NO_3)_3$		HNO_3		density	
t/°C	mass %	mol kg^{-1}	mass %	mol kg^{-1}	kg m^{-3}	nature of the solid phase
25	58.95	4.269	0	———	1.782	$Sm(NO_3)_3 \cdot 6H_2O$
	52.08	3.707	6.15	2.34		"
	46.52	3.282	11.34	4.27		"
	16.40	1.243	44.38	17.96		"
	14.20	1.128	48.40	20.54		"
	13.64	1.166	51.58	23.54		"
50	64.81	5.475	0	———	1.939	$Sm(NO_3)_3 \cdot 6H_2O$
	62.97	5.363	2.12	0.96	1.916	"
	57.57	4.904	7.53	3.42	1.859	"
	45.05	4.112	22.38	10.90		"
	33.20	3.247	36.40	19.00	1.648	"
	30.22	4.036	47.52	33.88	1.664	$Sm(NO_3)_3 \cdot 6H_2O + Sm(NO_3)_3 \cdot 4H_2O$
	27.89	3.201	46.21	28.31	1.638	$Sm(NO_3)_3 \cdot 4H_2O$?

a. Molalities calculated by the compilers.

AUXILIARY INFORMATION

METHOD/APPARATUS/PROCEDURE:

Isothermal method. Appropriate quantities of $Sm(NO_3)_3$ and aqueous HNO_3 were placed in Pyrex tubes, heated to induce supersaturation and thermostated. The Pyrex tubes were sealed after a small crystal of $Bi(NO_3)_3 \cdot 5H_2O$ was added to "seed" crystallization. The sealed tubes were shaken in the thermostat for at least 8 hours (equilibrium was reached in 4 hours). Authors state that approach to equilibrium from undersaturation gave indentical results within experimental error. All data reported in the above table are the results obtained by approach from supersaturation.

A "filtering pipet" maintained at a temperature slightly higher than the thermostat temperature was used to withdraw samples for analyses. Weighed samples of liquid and solid phases were analysed for HNO_3 by titrn with 0.1 mol dm^{-3} NaOH with methyl red indicator. Sm was pptd as the oxalate, filtered, washed with hot dilute oxalic acid, and ignited to the oxide. Composition of solid phases determined by Schreinemakers' method of residues.

SOURCE AND PURITY OF MATERIALS:

HNO_3 prepd from c.p. grade by distillation in an all Pyrex still and retaining the middle fraction. For very high HNO_3 concns, reagent grade fuming HNO_3 used as received. $Sm(NO_3)_3 \cdot 6H_2O$ prepared by dissolving the oxide in pure nitric acid, recrystallizing twice, and dried over 55 % sulfuric acid in a desiccator. No trace of any other rare earth was found in the oxide by arc emission spectroscopy.
Distilled water was used which had a conductivity of 2 x 10^{-6} S cm^{-1}.

ESTIMATED ERROR:

Soly: precision probably ± 0.2 % (compilers).

Temp: accuracy ± 0.03 K at 25°C
accuracy ± 0.1 K at 50°C

REFERENCES:

COMPONENTS:

(1) Samarium nitrate; $Sm(NO_3)_3$; [10361-83-8]

(2) Nitric acid; HNO_3; [7697-37-2]

(3) Water; H_2O; [7732-18-5]

ORIGINAL MEASUREMENTS:

Mironov, K.E.; Sinitsyna, E.D.; Popov, A.P. *Zh. Neorg. Khim.* 1966, *11*, 2361-5; *Russ. J. Inorg. Chem. Engl. Transl.* 1966, *11*, 1266-8.

VARIABLES:

Concentration of HNO_3 at 25°C

PREPARED BY:

T. Mioduski and S. Siekierski

EXPERIMENTAL VALUES:

Solubility of $Sm(NO_3)_3$ in HNO_3 solutions at 25°C [a]

$Sm(NO_3)_3$		HNO_3		
mass %	mol kg^{-1}	mass %	mol kg^{-1}	nature of the solid phase
61.14	4.677	0.0	———	$Sm(NO_3)_3 \cdot 5H_2O$
75.6[e]	10.70	3.4	2.57	"
51.9	3.94	8.9	3.60	"
43.6	3.19	15.8	6.18	"
21.6	1.62	38.8	15.55	"
20.3	1.55	40.7	16.56	"
16.9	1.38	46.6	20.26	"
14.5	1.27	51.5	24.04	"
14.5	1.29	52.1	24.75	"
14.5	1.29	52.2	24.88	$Sm(NO_3)_3 \cdot 5H_2O + Sm(NO_3)_3 \cdot 4H_2O$
14.6	1.32	52.5	25.32	$Sm(NO_3)_3 \cdot 4H_2O$
12.6	1.22	56.6	29.16	"
11.3	1.14	59.2	31.85	"
10.4	1.13	62.3	36.22	"
10.3	1.20	64.1	39.74	"
10.4	1.36	66.9	46.77	"
13.9	2.16	67.0	55.67	"
14.5	2.21	66.0	53.71	"
15.6	2.52	66.0	56.92	"
15.5	2.53	66.3	57.81	"

continued.........

AUXILIARY INFORMATION

METHOD/APPARATUS/PROCEDURE:

The isothermal method was used. Equilibrium was reached in 4-5 hours. Sampling techniques and methods of analysis for Sm, HNO_3, and H_2O were the same as in (1). The compositions of the solid phases were determined by Schreinemakers' method of residues using Cameron's computational method.

SOURCE AND PURITY OF MATERIALS:

Samarium nitrate was prepared from "Sm-2" grade oxide (USSR standard Sm-2) and nitric acid.

100% nitric acid was prepared by the Brauer method.

Doubly distilled water was used.

ESTIMATED ERROR:

Soly: precision about ± 0.5 % (compilers).

Temp: precision probably ± 0.1 K (compilers).

REFERENCES:

1. Mironov, K.E.; Sinitsyna, E.D.; Popov, A.P. *Izv. Sibir. Otd. Akad. Nauk Ser. Khim.* 1964, *No. 11*, 48.

2. Padova, J.I.; Soriano, J.R. *Israel J. Chem.* 1963, *1*, 310.

COMPONENTS:

(1) Samarium nitrate; $Sm(NO_3)_3$; [10361-83-8]

(2) Nitric acid; HNO_3; [7697-37-2]

(3) Water; H_2O; [7732-18-5]

ORIGINAL MEASUREMENTS:

Mironov, K.E.; Sinitsyna, E.D.; Popov, A.P. *Zh. Neorg. Khim.* 1966, *11*, 2361-5; *Russ. J. Inorg. Chem. Engl. Transl.* 1966, *11*, 1266-8.

EXPERIMENTAL VALUES: continued.......

Solubility of $Sm(NO_3)_3$ in HNO_3 solutions at 25°C [a]

$Sm(NO_3)_3$ mass %	$Sm(NO_3)_3$ mol kg^{-1}	HNO_3 mass %	HNO_3 mol kg^{-1}	nature of the solid phase
17.5	2.97	65.0	58.94	$Sm(NO_3)_3 \cdot 4H_2O + Sm(NO_3)_3 \cdot 3H_2O$
15.8	2.68	66.7	60.49	$Sm(NO_3)_3 \cdot 3H_2O$
14.5	2.48	68.1	62.11	"
16.5	3.27	68.5	72.47	"
16.9	3.75	69.7	82.55	"
18.3	4.53	69.7	92.18	"
19.5	5.52	70.0	105.8	$Sm(NO_3)_3 \cdot 3H_2O + Sm(NO_3)_3 \cdot H_2O$
18.5	5.24	71.0	107.3	$Sm(NO_3)_3$ [b]
17.1	5.65	73.9	130.3	"
13.2	4.73	78.5	150.1	"
11.4	3.99	80.1	149.5	"
7.5	2.86	84.7	172.3	"
5.4	2.72	88.7	238.6	"
3.8	1.91	90.3	242.9	"
3.2	3.66	94.2	575.0	"
2.0	2.48	95.6	632.1	"
1.6	47.57	98.3	15600	"
21.1	1.49	36.7	13.80	$Sm(NO_3)_3 \cdot 6H_2O$ [c]
13.4	1.22	54.0	26.29	$Sm(NO_3)_3 \cdot 6H_2O + Sm(NO_3)_3 \cdot 4H_2O$ [d]

a. Molalities calculated by M. Salomon.

b. Composition of this phase not well established. The solid contains some NO_2. The possibility of a compound between the anhyd salt and NO_2 was eliminated on the basis of the low solubility: extrapolation shows the solubility of the anhyd salt to be no more than 1.6 ± 0.2 mass % in an anhyd solution.

c. Hexahydrate phase obtained by heating followed by slow cooling.

d. Extrapolated based on data in (2).

e. Apparent typographical error. Probable values are m_1 = 57.6 mass % (4.39 mol kg^{-1}), and m_2 = 3.4 mass % (1.38 mol kg^{-1}). See the phase diagram in the critical evaluation.

COMMENTS AND/OR ADDITIONAL DATA:

An attempt to convert the pentahydrate into the hexahydrate by seeding did not succeed indicating a high stability of the metastable pentahydrate.

COMPONENTS:

(1) Samarium nitrate; $Sm(NO_3)_3$; [10361-83-8]

(2) Nitric acid; HNO_3; [7697-37-2]

(3) Water; H_2O; [7732-18-5]

ORIGINAL MEASUREMENTS:

Popov, A.P.; Mironov, K.E. *Zh. Neorg. Khim.* 1971, *16*, 464-6; *Russ. J. Inorg. Chem. Engl. Transl.* 1971, *16*, 244-6.

VARIABLES:

Concentration of HNO_3 at 25°C

PREPARED BY:

T. Mioduski and S. Siekierski

EXPERIMENTAL VALUES:

Solubility of $Sm(NO_3)_3$ in HNO_3 solutions at 25°C [a]

$Sm(NO_3)_3$		HNO_3		
mass %	mol kg^{-1}	mass %	mol kg^{-1}	nature of the solid phase
58.9	4.26	——	——	$Sm(NO_3)_3 \cdot 6H_2O$
42.0	2.94	15.5	5.79	"
28.5	2.00	29.2	10.96	"
21.1	1.49	36.7	13.80	"
16.2	1.17	42.6	16.41	"
13.2	1.15	52.7	24.53	"
80.5 [b]	12.27	——	——	$Sm(NO_3)_3 \cdot 4H_2O$

a. Molalities calculated by M. Salomon.

b. Calculated value based on fitting of Kirgintsev's equation (1) to data for the soly of the tetrahydrate in HNO_3 solutions (2): i.e., obtained by extrapolation to zero HNO_3 concentration.

AUXILIARY INFORMATION

METHOD/APPARATUS/PROCEDURE:

Nothing specified, but probably the isothermal method was used as in earlier work (2).

SOURCE AND PURITY OF MATERIALS:

No information given.

ESTIMATED ERROR:

Nothing specified.

REFERENCES:

1. Kirgintsev, A.N. *Izv. Adad. Nauk SSSR, Ser. Khim. Nauk* 1965, *No. 8*, 1591.

2. Mironov, K.E.; Sinitsyna, E.D.; Popov, A.P. *Zh. Neorg. Khim.* 1966, *11*, 2361.

COMPONENTS:

(1) Samarium nitrate; $Sm(NO_3)_3$; [10361-83-8]

(2) Zinc nitrate; $Zn(NO_3)_2$; [7779-88-6]

(3) Water; H_2O; [7732-18-5]

ORIGINAL MEASUREMENTS:

Brunisholz, G.; Klipfel, K. *Rev. Chim. Miner.* 1970, *7*, 349-58.

VARIABLES:

Composition at 0°C, 20°C, 35°C

PREPARED BY:

T. Mioduski and S. Siekierski

EXPERIMENTAL VALUES:

Composition of saturated solutions at 0°C [a]

mol % Sm of total Sm + Zn	moles H_2O per 100 moles Sm + Zn	$Sm(NO_3)_3$ mol kg^{-1}	$Zn(NO_3)_2$ mol kg^{-1}	nature of the solid phase
0.0	1110	———	5.001	$Zn(NO_3)_2 \cdot 6H_2O$ (A)
4.31	1198	0.200	4.434	"
9.90	1080	0.509	4.631	A + B
9.91	1085	0.507	4.609	"
9.93	1088	0.507	4.595	"
14.72	1107	0.738	4.276	$2Sm(NO_3)_3 \cdot 3Zn(NO_3)_2 \cdot 24H_2O$ (B)
20.15	1128	0.992	3.929	"
25.96	1135	1.270	3.621	"
26.06	1134	1.276	3.619	B + C
26.04	1132	1.277	3.627	"
45.37	1265	1.991	2.397	$Sm(NO_3)_3 \cdot 6H_2O$ (C)
71.54	1407	2.822	1.123	"
100.0	1532	3.623	———	"

continued.......

AUXILIARY INFORMATION

METHOD/APPARATUS/PROCEDURE:

The isothermal method was used. The solids were placed in glass-stoppered vials into which a glass dumb-bell shaped pestle was placed. These vials were placed in a second larger tube and sealed with a rubber stopper. The tubes were placed in a thermostat and rotated end-over-end so that the pestle pulverized the solid two times through each 360° rotation. The solutions were equilibrated for 2 to 4 weeks.

For the saturated solutions, Sm was determined by titration with Trilon using Xylenol Orange indicator and urotropine buffer. Zn was determined by precipitation as the sulfide (using thioacetamide), or by ion exchange chromatography.

The compositions of the solid phases were determined by Schreinemakers' method of residues, and by X-ray analysis.

SOURCE AND PURITY OF MATERIALS:

The nitrates were obtained by dissolving the oxide in nitric acid followed by crystallization. The source and purities of the oxides was not specified. The hydrated nitrates were dried in a desiccator over KOH under vacuum.

ESTIMATED ERROR:

Soly: precision ± 0.3-0.5 % (compilers).

Temp: precision ± 0.005 K.

REFERENCES:

COMPONENTS:

(1) Samarium nitrate; $Sm(NO_3)_3$; [10361-83-8]

(2) Zinc nitrate; $Zn(NO_3)_2$; [7779-88-6]

(3) Water; H_2O; [7732-18-5]

ORIGINAL MEASUREMENTS:

Brunisholz, G.; Klipfel, K. *Rev. Chim. Miner.* 1970, *7*, 349-58.

EXPERIMENTAL VALUES: continued........

Composition of saturated solutions at 20°C [a]

mol % Sm of total Sm + Zn	moles H_2O per 100 moles Sm + Zn	$Sm(NO_3)_3$ mol kg^{-1}	$Zn(NO_3)_2$ mol kg^{-1}	nature of the solid phase
0.0	892	———	6.223	A
2.67	871	0.170	6.203	A + B
2.79	876	0.177	6.160	"
11.45	956	0.665	5.141	B
19.12	1102	0.963	4.074	"
28.09	1015	1.536	3.933	"
32.96	1009	1.813	3.688	B + C
32.92	1006	1.816	3.701	"
49.08	1100	2.477	2.570	C
73.50	1245	3.277	1.182	"
100.0	1350	4.112	———	"

Composition of saturated solutions at 35°C [a]

mol % Sm of total Sm + Zn	moles H_2O per 100 moles Sm + Zn	$Sm(NO_3)_3$ mol kg^{-1}	$Zn(NO_3)_2$ mol kg^{-1}	nature of the solid phase
0.0	668	———	8.310	A
0.75	661	0.063	8.335	A + B
2.21	751	0.163	7.228	B
12.55	862	0.808	5.631	"
25.50	898	1.576	4.605	"
39.65	902	2.440	3.714	"
39.79	903	2.446	3.701	B + C
39.80	905	2.441	3.692	"
59.72	1054	3.145	2.121	C
85.50	1130	4.200	0.712	"
100.0	1180	4.704	———	"

a. Molalities calculated by M. Salomon.

COMPONENTS:

(1) Samarium nitrate; $Sm(NO_3)_3$; [10361-83-8]

(2) N-Acetylurea; $C_3H_6N_2O_2$; [591-07-1]

(3) Water ; H_2O ; [7732-18-5]

ORIGINAL MEASUREMENTS:

Usubalieva, U.; Sulaimankulov, K. *Zh. Neorg. Khim.* 1982, *27*, 1338-9; *Russ J. Inorg. Chem. Engl Transl.* 1982, *27*, 755-6.

VARIABLES:

Composition at 30°C

PREPARED BY:

T. Mioduski and S. Siekierski

EXPERIMENTAL VALUES:

Composition of saturated solutions at 30°C[a]

$Sm(NO_3)_3$ mass %	$Sm(NO_3)_3$ mol kg^{-1}	$CH_3CONHCONH_2$ mass %	$CH_3CONHCONH_2$ mol kg^{-1}	nature of the solid phase
0	——	3.23	0.327	$CH_3CONHCONH_2$
5.15	0.168	3.54	0.380	"
15.13	0.557	4.16	0.505	"
24.45	1.018	4.12	0.565	"
33.27	1.590	4.51	0.710	"
38.60	2.018	4.52	0.778	"
55.40	3.859	1.92	0.441	"
60.14	5.153	5.16	1.457	"
60.24	5.178	5.17	1.464	$CH_3CONHCONH_2 + Sm(NO_3)_3 \cdot 6H_2O$
61.35	5.310	4.30	1.226	$Sm(NO_3)_3 \cdot 6H_2O$
59.80	4.884	3.80	1.023	"
58.70	4.225	0	——	"

a. Molalities calculated by M. Salomon.

AUXILIARY INFORMATION

METHOD/APPARATUS/PROCEDURE:

The isothermal method was used. With constant agitation, equilibrium was reached within 6-9 hours. The liquid and solid phases were separated using a Schott No. 3 filter, and both phases were analysed. Nitrogen was determined by the Kjeldahl method, and Sm was determined by titration with Trilon using Xylenol Orange indicator. The compilers assume water was determined by difference.

No double compound was formed in this system.

SOURCE AND PURITY OF MATERIALS:

Nothing specified.

ESTIMATED ERROR:

Nothing specified, but the solubility of $Sm(NO_3)_3$ in water is too low by at least 5 % (see the critical evaluation).

REFERENCES:

COMPONENTS:

(1) Magnesium samarium nitrate; $3Mg(NO_3)_2 \cdot 2Sm(NO_3)_3$; [32074-08-1]

(2) Nitric acid; HNO_3; [7697-37-2]

(3) Water ; H_2O ; [7732-18-5]

ORIGINAL MEASUREMENTS:

Jantsch, G. *Z. Anorg. Chem.* 1912, *76*, 303-23.

VARIABLES:

One temperature: 16°C

PREPARED BY:

Mark Salomon

EXPERIMENTAL VALUES:

Soly of the double salt in HNO_3 sln of density d_4^{16} = 1.325 g cm^{-3}.

aliquot volume	Sm_2O_3	Soly $3Mg(NO_3)_2 \cdot 2Sm(NO_3)_3$ [a]
cm^3	g	mol dm^{-3}
1.4638	0.0810	
1.4638	0.0807	0.1583

a. Author's calculation (average value).

ADDITIONAL DATA:

The melting point of the tetracosahydrate is 96.2°C, and the density at 0°C is 2.088 g cm^{-3}.

AUXILIARY INFORMATION

METHOD/APPARATUS/PROCEDURE:

Isothermal method used. The soly was studied in HNO_3 sln of density 1.325 g cm^{-3} at 16°C because the author did not have sufficient quantity of the rare earth to study the soly of the salt in pure water. Pulverized salt and HNO_3 sln were placed in glass-stoppered tubes and thermostated at 16°C for 24 h with periodic shaking. The solution was then allowed to settle for 2 h, and a pipet maintained at 16°C was used to withdraw aliquots for analysis. Two analyses were performed.

Solutions were analysed by adding 2-3 g NH_4Cl and 10% NH_3 sln followed by boiling to ppt the hydroxide. The ppt was filtered, dissolved in HNO_3, reprecipitated as the hydroxide, and ignited to the oxide. Mg in the filtrate was "determined by the usual method" (no details were given).

An attempt to determine the waters of hydration by dehydration was not successful because the temperature required (120°C or higher) resulted in decomposition of the salt with the formation of basic salts. Presumably the waters of hydration were found by difference.

SOURCE AND PURITY OF MATERIALS:

"Pure" samarium oxide was dissolved in dil HNO_3 and $Mg(NO_3)_2$ added to give a mole ratio of Sm/Mg = 2/3. The sln was evapd and a small crystal of $Bi_2Mg_3(NO_3)_{12}$ added, and the mixt cooled to ppt the tetracosahydrate. The double nitrate was recrystd before use.

ESTIMATED ERROR:

Soly: reproducibility about ± 1-5% (compiler).

Temp: nothing specified.

REFERENCES:

COMPONENTS:

(1) Samarium manganese nitrate; $2Sm(NO_3)_3 \cdot 3Mn(NO_3)_2$; [84682-68-8]

(2) Nitric acid; HNO_3; [7697-37-2]

(3) Water ; H_2O ; [7732-18-5]

ORIGINAL MEASUREMENTS:

Jantsch, G. *Z. Anorg. Chem.* 1912, *76*, 303-23.

VARIABLES:

One temperature: 16°C

PREPARED BY:

Mark Salomon

EXPERIMENTAL VALUES:

Soly of the double salt in HNO_3 sln of density d_4^{16} = 1.325 g cm^{-3}.

aliquot volume	Sm_2O_3 [a]	soly $2Sm(NO_3)_3 \cdot 3Mn(NO_3)_2$ [b]
cc	g	mol dm^{-3}
1.4638	0.1553	
1.4638	0.1559	0.3047

a. Experimental quantity is total weight of Sm_2O_3 + Mn_3O_4 oxides. The author did not report the experimental quantity.

b. Calculated by the author (average value).

ADDITIONAL DATA:

The melting point of the tetracosahydrate is 70.2°C, and the density at 0°C is 2.188 g cm^{-3}.

AUXILIARY INFORMATION

METHOD/APPARATUS/PROCEDURE:

Isothermal method used. The soly was studied in HNO_3 sln of density 1.325 g cm^{-3} at 16°C because the author did not have sufficient quantity of the rare earth to study the soly of the salt in pure water. Pulverized salt and HNO_3 sln were placed in glass-stoppered tubes and thermostated at 16°C for 24 h with periodic shaking. The solution was then allowed to settle for 2 h, and a pipet maintained at 16°C was used to withdraw aliquots for analysis. Two analyses were performed.

Solutions were analysed by precipitating both Sm and Mn hydroxides by respective addition of NH_3 and H_2O_2. The ppt was ignited to give Sm_2O_3 + Mn_3O_4.

An attempt to determine the waters of hydration by dehydration was not successful because the temperature required (120°C or higher) resulted in decomposition of the salt with the formation of basic salts. Presumably the waters of hydration were found by difference.

SOURCE AND PURITY OF MATERIALS:

"Pure" samarium oxide was dissolved in dil HNO_3 and $Mn(NO_3)_2$ added to give a mole ratio of Sm/Mn = 2/3. The sln was evapd and a small crystal of $Bi_2Mg_3(NO_3)_{12}$ added, and the mixt cooled to ppt the tetracosahydrate. The double nitrate was recrystd before use. The double salt was analysed gravimetrically for total Sm_2O_3 + Mn_3O_4. A 0.5500 g sample of the tetracosahydrate yielded 0.1940 g oxides (i.e. 35.27 mass %). Theor for Sm_2O_3 + Mn_3O_4 is 35.17 mass % (compiler).

ESTIMATED ERROR:

Soly: reproducibility about ± 1-5% (compiler)

Temp: nothing specified

COMPONENTS:

(1) Samarium cobalt nitrate; $2Sm(NO_3)_3.3Co(NO_3)_2$; [84682-49-5]

(2) Nitric acid; HNO_3; [7697-37-2]

(3) Water ; H_2O ; [7732-18-5]

ORIGINAL MEASUREMENTS:

Jantsch, G. *Z. Anorg. Chem.* 1912, *76*, 303-23.

VARIABLES:

One temperature: 16°C

PREPARED BY:

Mark Salomon

EXPERIMENTAL VALUES:

Soly of the double salt in HNO_3 sln of density d_4^{16} = 1.325 g cm^{-3}.

aliquot volume	Sm_2O_3	Soly $2Sm(NO_3)_3.3Co(NO_3)_2$ [a]
cm^3	g	mol dm^{-3}
1.4638	0.1058	
1.4638	0.1058	0.2072

a. Average value calculated by the author.

ADDITIONAL DATA:

The melting point of the tetracosahydrate is 83.2°C, and the density at 0°C is 2.237 g cm^{-3}.

AUXILIARY INFORMATION

METHOD/APPARATUS/PROCEDURE:

Isothermal method used. The soly was studied in HNO_3 sln of density 1.325 g cm^{-3} at 16°C because the author did not have sufficient quantity of the rare earth to study the soly of the salt in pure water. Pulverized salt and HNO_3 sln were placed in glass-stoppered tubes and thermostated at 16°C for 24 h with periodic shaking. The solution was then allowed to settle for 2 h, and a pipet maintained at 16°C was used to withdraw aliquots for analysis. Two analyses were performed.

Solutions were analysed by adding 2-3 g NH_4Cl and 10% NH_3 sln followed by boiling to ppt the hydroxide. The ppt was filtered, dissolved in HNO_3, reprecipitated as the hydroxide, and ignited to the oxide. Co in the filtrate was "determined by the usual method" (no details were given).

An attempt to determine the waters of hydration by dehydration was not successful because the temperature required (120°C or higher) resulted in decomposition of the salt with the formation of basic salts. Presumably the waters of hydration were found by difference.

SOURCE AND PURITY OF MATERIALS:

"Pure" samarium oxide was dissolved in dil HNO_3 and $Co(NO_3)_2$ added to give a mole ratio of Sm/Co = 2/3. The sln was evapd and a small crystal of $Bi_2Mg_3(NO_3)_{12}$ added, and the mixt cooled to ppt the tetracosahydrate. The double nitrate was recrystd before use. The double salt was analysed gravimetrically for Sm_2O_3 and metallic Co. 0.2970 g samples of the tetracosahydrate yielded 0.0625 g oxide (21.04 mass %), and 0.0306 g Co (10.30 mass %). Theor values are 21.08 mass % Sm_2O_3 and 10.69 mass % Co (compiler).

ESTIMATED ERROR:

Soly: reproducibility about ± 1-5% (compiler).

Temp: nothing specified.

COMPONENTS:	ORIGINAL MEASUREMENTS:
(1) Samarium nickel nitrate; $2Sm(NO_3)_3 \cdot 3Ni(NO_3)_2$; [84682-71-3] (2) Nitric acid: HNO_3; [7697-37-2] (3) Water ; H_2O ; [7732-18-5]	Jantsch, G. *Z. Anorg. Chem.* 1912, *76*, 303-23.

VARIABLES:	PREPARED BY:
One temperature: 16°C	Mark Salomon

EXPERIMENTAL VALUES:

Soly of the double salt in HNO_3 sln of density d_4^{16} = 1.325 g cm^{-3}.

aliquot volume	Sm_2O_3	soly $2Sm(NO_3)_3 \cdot 3Ni(NO_3)_2$ [a]
cm^3	g	mol dm^{-3}
1.4638	0.0901	
1.4638	0.0897	0.1760

a. Average value calculated by the author.

ADDITIONAL DATA:

The melting point of the tetracosahydrate is 92.2°C and the density at 0°C is 2.272 g cm^{-3}.

AUXILIARY INFORMATION

METHOD/APPARATUS/PROCEDURE:

Isothermal method used. The soly was studied in HNO_3 sln of density 1.325 g cm^{-3} at 16°C because the author did not have sufficient quantity of the rare earth to study the soly of the salt in pure water. Pulverized salt and HNO_3 sln were placed in glass-stoppered tubes and thermostated at 16°C for 24 h with periodic shaking. The solution was then allowed to settle for 2 h, and a pipet maintained at 16°C was used to withdraw aliquots for analysis. Two analyses were performed.

Solutions were analysed by adding 2-3 g NH_4Cl and 10% NH_3 sln followed by boiling to ppt the hydroxide. The ppt was filtered, dissolved in HNO_3, reprecipitated as the hydroxide, and ignited to the oxide. Ni in the filtrate was "determined by the usual method" (no details were given).

An attempt to determine the waters of hydration by dehydration was not successful because the temperature required (120°C or higher) resulted in decomposition of the salt with the formation of basic salts. Presumably the waters of hydration were found by difference.

SOURCE AND PURITY OF MATERIALS:

"Pure" samarium oxide was dissolved in dil HNO_3 and $Ni(NO_3)_2$ added to give a mole ratio of Sm/Ni = 2/3. The sln was evapd and a small crystal of $Bi_2Mg_3(NO_3)_{12}$ added, and the mixt cooled to ppt the tetracosahydrate. The double nitrate was recrystd before use. The double salt was analysed gravimetrically for Sm_2O_3 and NiO. 0.2290 g samples of the tetracosahydrate yielded 0.0485 g Sm_2O_3 and 0.0312 g NiO. In units of mass % Sm_2O_3: found 21.18, theor 21.09. NiO: found 13.62, theor 13.56 (theor values calcd by compiler).

ESTIMATED ERROR:

Soly: reproducibility about ± 1-5% (compiler).

Temp: nothing specified

COMPONENTS:	ORIGINAL MEASUREMENTS:
(1) Samarium zinc nitrate; $2Sm(NO_3)_3.3Zn(NO_3)_2$; [84682-73-5] (2) Nitric acid; HNO_3; [7697-37-2] (3) Water ; H_2O ; [7732-18-5]	Jantsch, G. *Z. Anorg. Chem.* 1912, *76*, 303-23.
VARIABLES:	PREPARED BY:
One temperature: 16°C	Mark Salomon

EXPERIMENTAL VALUES:

Soly of the double salt in HNO_3 sln of density d_4^{16} = 1.325 g cm^{-3}.

aliquot volume	Sm_2O_3	Soly $2Sm(NO_3)_3.3Zn(NO_3)_2$ [a]
cm^3	g	mol dm^{-3}
1.4638	0.1553	
1.4638	0.1559	0.3047

a. Average value calculated by the author.

ADDITIONAL DATA:

The melting point of the tetracosahydrate is 76.5°C, and the density at 0°C is 2.293 g cm^{-3}.

AUXILIARY INFORMATION

METHOD/APPARATUS/PROCEDURE:

Isothermal method used. The soly was studied in HNO_3 sln of density 1.325 g cm^{-3} at 16°C because the author did not have sufficient quantity of the rare earth to study the soly of the salt in pure water. Pulverized salt and HNO_3 sln were placed in glass-stoppered tubes and thermostated at 16°C for 24 h with periodic shaking. The solution was then allowed to settle for 2 h, and a pipet maintained at 16°C was used to withdraw aliquots for analysis. Two analyses were performed.

Solutions were analysed by adding 2-3 g NH_4Cl and 10% NH_3 sln followed by boiling to ppt the hydroxide. The ppt was filtered, dissolved in HNO_3, reprecipitated as the hydroxide, and ignited to the oxide. Zn in the filtrate was "determined by the usual method" (no details were given).

An attempt to determine the waters of hydration by dehydration was not successful because the temperature required (120°C or higher) resulted in decomposition of the salt with the formation of basic salts. Presumably the waters of hydration were found by difference.

SOURCE AND PURITY OF MATERIALS:

"Pure" samarium oxide was dissolved in dil HNO_3 and $Zn(NO_3)_2$ added to give a mole ratio of Sm/Zn = 2/3. The sln was evapd and a small crystal of $Bi_2Mg_3(NO_3)_{12}$ added, and the mixt cooled to ppt the tetracosahydrate. The double nitrate was recrystd before use. The double salt was analysed gravimetrically for Sm_2O_3 and NO. A 0.5280 g sample of the tetracosahydrate yielded 0.1095 g Sm_2O_3 (20.74 mass %): Theor = 20.84 mass % (compiler). Analysis for NO (details not given) gave 21.22 mass %; Theor value is 21.52 mass % (compiler).

ESTIMATED ERROR:

Soly: reproducibility about ± 1-5% (compiler).

Temp: nothing specified.

COMPONENTS:

(1) Europium nitrate; $Eu(NO_3)_3$; [10138-01-9]

(2) Water ; H_2O ; [7732-18-5]

EVALUATOR:
S. Siekierski, T Mioduski
Institute for Nuclear Research
Warsaw, Poland
and
M. Salomon
U.S. Army ET & DL
Ft. Monmouth, NJ
January 1983

CRITICAL EVALUATION:

THE BINARY SYSTEM

INTRODUCTION

The direct determination of the solubility of europium nitrate in water has been reported in only two publications (1,2). Mironov et al. (1) studied the solubility as a function of temperature from 241.1 K to 362.4 K, and Khudaibergenova and Sulaimankulov (2) determined the solubility at 303.2 K. According to Mironov et al. (1), the following solid phases have been identified over the temperature range of 241 K to 372 K:

$Eu(NO_3)_3 \cdot 6H_2O$	[10031-53-5]	$Eu(NO_3)_3 \cdot 4H_2O$	[37131-74-1]
$Eu(NO_3)_3 \cdot 5H_2O$	[63026-01-7]		

Lower hydrates exist as stable solid phases in aqueous nitric acid solutions (1), but only the hexa-, penta-, and tetrahydrates have been identified in the binary system.

EVALUATION

An attempt was made to use the least squares method to fit the data of Mironov et al. to the general solubility equation

$$Y = \ln(m/m_o) - nM_2(m - m_o) = a + b/(T/K) + c\ln(T/K) \qquad [1]$$

All terms in eq. [1] have been previously defined (e.g. see eq. [1] in the preceeding critical evaluation for $Sm(NO_3)_3$-H_2O). We were unable to obtain a satisfactory fit to eq. [1] for any of the data in (1). For example, the solubility data in the hexahydrate system from (1) can be fitted to eq. [1] with the following results:

$$Y = -2.033 - 99.7/(T/K) + 0.418\ln(T/K) \qquad [2]$$

The standard error of estimate for the molality, σ_m, is 0.12, and the calculated uncertainty in these solubilities is ± 0.2 mol kg^{-1} (95% level of confidence, Student's t = 3.182). The constants in eq. [2] appear to be trival, and the calculated congruent melting point for the hexahydrate is 431.3 K which is considerably higher than the experimental melting points of 364.30 K (1) and 364 ± 1 K (3) for the tetrahydrate (note that Mironov et al. report that the hexahydrate and pentahydrate melt incongruently).

The problems encountered in determining the solubility of $Eu(NO_3)_3$ in water are probably due, in most part, to the existence mixed and metastable phases, and the difficulties in identifying these phases. For example, Mironov et al. state that at 372.4 K the solid phase is the tetrahydrate, and the solubility of $Eu(NO_3)_3$ at this temperature is 15.19 mol kg^{-1}. However as pointed out by the compilers, the solubility of the tetrahydrate at its congruent melting point is 13.88 mol kg^{-1}, and the solubility result of 15.19 mol kg^{-1} at 362.4 K probably belongs to a metastable lower hydrate system ($n \leqslant 3$). The solubility value of 5.023 mol kg^{-1} at 303.2 K reported in (2) appears to be too high by about 11%, and this value is rejected.

CONCLUSION

At the present time, it is not possible to assign either the recommended or tentative designations to any of the solubility data in the binary $Eu(NO_3)_3$-H_2O system. It is however interesting to note that it is still possible to estimate more reliable solubilities in the hexahydrate system by an interpolation procedure. Figure 1 shows the recommended solubilities for the lanthanide nitrates at 298.2 K as a function of atomic number. From this figure we estimate that at 298.2 K, the solubility of $Eu(NO_3)_3$ in the hexahydrate system is 4.24 mol kg^{-1}. Rard and Spedding (4) estimate a solubility of 4.33 ± 0.03 mol kg^{-1} at 298.2 K, presumably by this same interpolation method. Brunisholz et al. (5) have employed this interpolation method to estimate the solubility of $Eu(NO_3)_3$ in pure water at 273.2 K, 293.2 K, and 323.2 K. These interpolated values are summarized in Table 1, and probably represent the best estimates for the solubilities in the hexahydrate system. It is noted that the interpolated solubility at 298.2 K based on the smoothed curve (Fig. 1) is 4.24 mol kg^{-1} which is lower than the value of 4.33 mol kg^{-1} obtained by the method of linear extrapolation between Sm and Gd (4).

COMPONENTS:	EVALUATOR:
(1) Europium nitrate; $Eu(NO_3)_3$; [10138-01-9] (2) Water ; H_2O ; [7732-18-5]	S. Siekierski, T. Mioduski Institute for Nuclear Research Warsaw, Poland and M. Salomon U.S. Army ET & DL Ft. Monmouth, NJ January 1983

CRITICAL EVALUATION:

REFERENCES

1. Mironov, K.E.; Popov, A.P.; Vorob'eva, V.Ya.; Grankina, Z.A. *Zh. Neorg. Khim.* 1971, *16*, 2769.
2. Khudaibergenova, N.; Salaimankulov, K. *Zh. Neorg. Khim.* 1981, *26*, 1107.
3. Wendlandt, W.W.; Sewell, R.G. *Texas J. Sci.* 1961, *13*, 231.
4. Rard, J.A.; Spedding. F.H. *J. Chem. Eng. Data* 1982, *27*, 454.
5. Brunisholz, G.; Quinche, J.P.; Kalo, A.M. *Helv, Chim. Acta* 1964, *47*, 14.

Table 1. Estimated solubilities of $Eu(NO_3)_3$ in the hexahydrate system.

T/K	soly/mol kg^{-1}	ref
273.2	3.64	5
293.2	4.17	5
298.2	4.24	evaluators (Fig. 1)
298.2	4.33	4
323.2	5.39	5

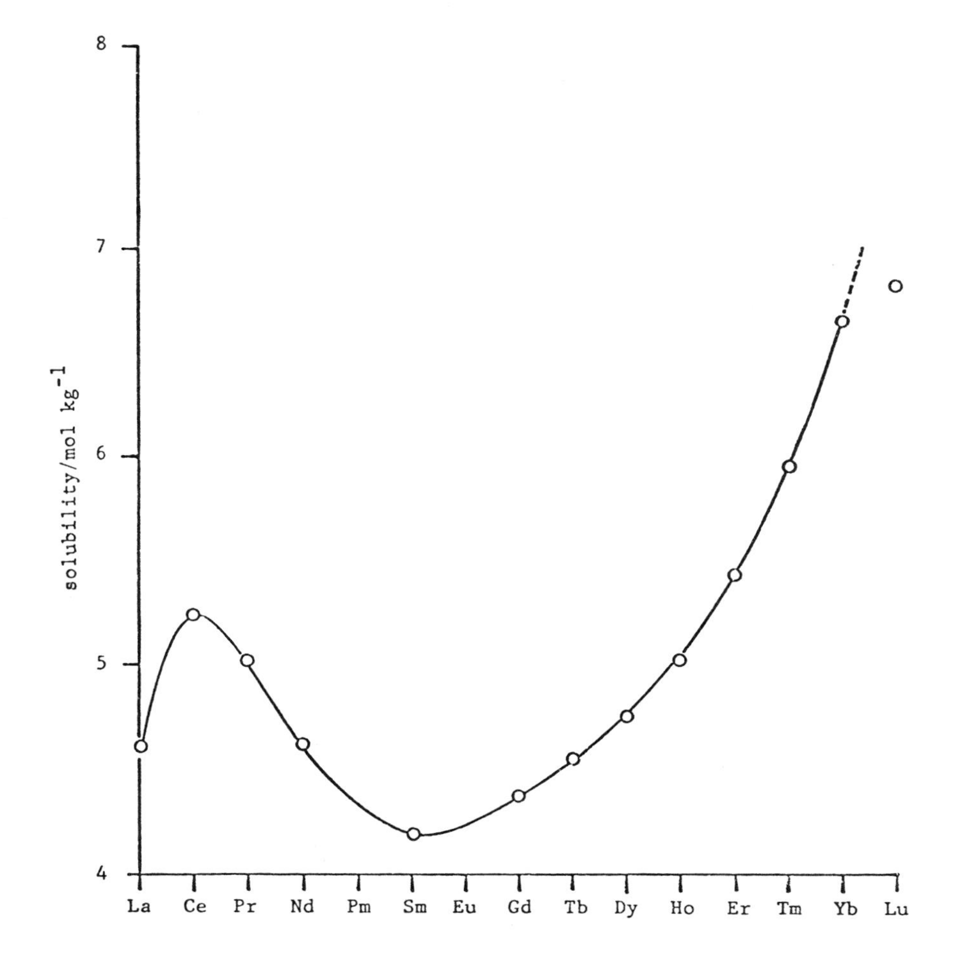

Figure 1. Recommended lanthanide nitrate solubilities at 298.2 K as a function of atomic number: solid phase is the hexahydrate except for $Lu(NO_3)_3$ where the solid phase is the pentahydrate.

COMPONENTS:

(1) Europium nitrate; $Eu(NO_3)_3$; [10138-01-9]

(2) Water ; H_2O ; [7732-18-5]

ORIGINAL MEASUREMENTS:

Mironov, K.E.; Popov, A.P.; Vorob'eva, V. Ya.; Grankina, Z.A. *Zh. Neorg. Khim.* 1971, *16,* 2769-74; *Russ. J. Inorg. Chem. Engl. Transl.* 1971, *16,* 1476-9.

VARIABLES:

Temperature

PREPARED BY:

T. Mioduski, S. Siekierski, and M. Salomon

EXPERIMENTAL VALUES:

Composition of saturated solutions as a function of temperature[a,b]

t/°C	$Eu(NO_3)_3$ mass %	mol kg^{-1}	solid phase	t/°C	$Eu(NO_3)_3$ mass %	mol kg^{-1}	solid phase
0.0	0.0	———	A	30.5	60.6	4.55	C
- 1.2	10.3	0.34	"	37.0	60.8	4.59	"
- 2.3	13.6	0.47	"	45.1	62.1	4.85	?[c]
- 2.8	14.6	0.51	"	38.1	62.7	4.97	?
- 9.9	30.2	1.28	"	47.1	64.8	5.45	C
-11.4	35.1	1.60	"	53.5	65.4	5.59	?
-18.1	39.7	1.95	"	56.8	66.5	5.87	?
-22.0	44.1	2.33	"	54.7	67.1	6.03	C
-25.8	47.0	2.62	"				
-28.1	48.6	2.80	"	59.0	67.5	6.15	D
				73.0	70.6	7.11	"
-31.6	50.3	2.99	B	74.0	71.4	7.39	?
-32.1	52.2	3.23	"	77.4	73.2	8.08	D
-20.5	53.5	3.40	"	87.0	75.8	9.27	?
- 1.8	54.7	3.57	"	77.5	78.6	10.87	?
15.4	57.0	3.92	"	89.2	83.7	15.19	D[d]
26.2	58.4	4.15	"				

a. Molalities calculated by the compilers.
b. Solid phases: A = ice ; B = $Eu(NO_3)_3 \cdot 6H_2O$

C = $Eu(NO_3)_3 \cdot 5H_2O$; D = $Eu(NO_3)_3 \cdot 4H_2O$

c. Metastable equilibra. Solid phases not specified.
d. Theoretical concentration at the congruent melting point is 13.88 mol kg^{-1} (compilers).

AUXILIARY INFORMATION

METHOD/APPARATUS/PROCEDURE:

The synthetic method was used supplemented with differential thermal analyses of the various hydrates. The solubilities were determined by visual recording of the temperatures of crystallization. The DTA studies show that the hexahydrate and pentahydrate are incongruently soluble. The tetrahydrate is congruently soluble and melted at 91.15°C.

COMMENTS AND/OR ADDITIONAL DATA:

The coordinates of the invariant points appear to have been determined graphically, and are:

t/°C	mass %	mol kg^{-1}
-36	52.0	3.21
33	60.1	4.46
60	76.4	9.58

SOURCE AND PURITY OF MATERIALS:

Europium nitrate was prepd by dissolving 99.5% pure oxide in warm nitric acid followed by evaporation to crystallization. 35.4 mass % Eu was found by EDTA titrn, 43.3 mass % NO_3 was found by the Kjeldahl method, and 22.1 mass % water was found by the Karl Fischer method. Theoretical for the pentahydrate are: 35.50 mass % Eu, 43.46 mass % NO_3, and 21.04 mass % H_2O (compilers).

Europium nitrate with less waters of hydration was prepared by drying at 100-120°C *in vacuo.*

ESTIMATED ERROR:

Nothing specified.

REFERENCES:

COMPONENTS:

(1) Europium nitrate; $Eu(NO_3)_3$; [10138-01-9]

(2) Nitric acid; HNO_3; [7697-37-2]

(3) Water ; H_2O ; [7732-18-5]

ORIGINAL MEASUREMENTS:

Mironov, K.E.; Popov, A.P.; Vorob'eva, V. Ya.; Grankina, Z.A. *Zh. Neorg. Khim.* 1971, *16*, 2769-74; *Russ. J. Inorg. Chem. Engl. Transl.* 1971, *16*, 1476-9.

VARIABLES:

Concentration of nitric acid at 25°C

PREPARED BY:

Mark Salomon

EXPERIMENTAL VALUES:

Composition of saturated solutions at the eutonic points[a]

$Eu(NO_3)_3$		HNO_3		
mass %	mol kg^{-1}	mass %	mol kg^{-1}	nature of the solid phase
9.2	0.83	58.0	28.06	$Eu(NO_3)_3 \cdot 6H_2O + Eu(NO_3)_3 \cdot 4H_2O$
12.2	1.15	56.5	28.65	$Eu(NO_3)_3 \cdot 5H_2O + Eu(NO_3)_3 \cdot 4H_2O$
20.9	3.24	60.0	49.85	$Eu(NO_3)_3 \cdot 4H_2O + Eu(NO_3)_3 \cdot 3H_2O$
13.6	2.65	71.2	74.34	$Eu(NO_3)_3 \cdot 3H_2O + Eu(NO_3)_3 \cdot H_2O$
4.6	1.02	82.0	97.11	$Eu(NO_3)_3 \cdot H_2O + Eu(NO_3)_3$

a. Molalities calculated by the compiler.

AUXILIARY INFORMATION

METHOD/APPARATUS/PROCEDURE:

The isothermal method was used. No additional information given. Presumably europium was determined by titration with EDTA, and nitric acid by titration with standard base.

Most of the data are given in a phase diagram. The only numerical data given are those for the five eutonic points given in the above table.

SOURCE AND PURITY OF MATERIALS:

Europium nitrate was prepd by dissolving 99.5% pure oxide in warm nitric acid followed by evaporation to crystallization. 35.4 mass % Eu was found by EDTA titrn, 43.3 mass % NO_3 was found by the Kjeldahl method, and 22.1 mass % water was found by the Karl Fischer method. Theoretical values for the pentahydrate are: 35.50 mass % Eu, 43.46 mass % NO_3, and 21.04 mass % water (compiler).

Source and purity of HNO_3 and water was not specified.

ESTIMATED ERROR:

Nothing specified.

REFERENCES:

COMPONENTS:

(1) Europium nitrate; $Eu(NO_3)_3$; [10138-01-9]

(2) Urea; CH_4N_2O; [57-13-6]

(3) Water ; H_2O ; [7732-18-5]

ORIGINAL MEASUREMENTS:

Khudaibergenova, N.; Sulaimankulov, K.

Zh. Neorg. Khim. 1981, *26*, 1107-9.

VARIABLES:

Composition at 30°C

PREPARED BY:

T. Mioduski and S. Siekierski

EXPERIMENTAL VALUES:

$Eu(NO_3)_3$		$CO(NH_2)_2$		
mass %	mol kg^{-1}	mass %	mol kg^{-1}	nature of the solid phase
0.0	———	57.56	22.583	$CO(NH_2)_2$
17.98	2.137	57.12	38.197	"
25.87	4.033	55.15	48.383	"
37.67	15.589	55.18	128.51	$Eu(NO_3)_3\cdot 6CO(NH_2)_2$
36.43	9.260	51.93	74.287	$Eu(NO_3)_3\cdot 4CO(NH_2)_2\cdot 2H_2O$
39.35	10.810	49.88	77.118	"
42.78	13.509	47.85	85.033	"
44.21	15.741	47.48	95.138	"
46.02	17.457	46.18	98.584	"
46.09	12.615	43.10	66.389	"
43.97	5.471	32.25	22.582	"
46.97	5.292	26.77	16.975	"
51.98	7.299	26.95	21.298	"
53.26	6.316	21.79	14.542	"
53.56	6.069	20.33	12.965	$Eu(NO_3)_3\cdot 4CO(NH_2)_2\cdot 2H_2O + Eu(NO_3)_2\cdot 2CO(NH_2)\cdot 2H_2O$
56.17	5.775	15.05	8.707	$Eu(NO_3)_3\cdot 2CO(NH_2)_2\cdot 2H_2O$
58.71	6.002	12.35	7.106	"
60.48	5.888	9.13	5.002	"
64.12	7.216	9.59	6.074	"
63.35	6.404	7.38	4.198	$Eu(NO_3)_3\cdot 6H_2O$
62.93	5.023	0.0	———	"

AUXILIARY INFORMATION

METHOD/APPARATUS/PROCEDURE:

Isothermal method employed. Equilibrium was reached within 7-9h. The liquid phase was separated using No. 3 Schott filter. Nitrogen was determined by the Kjeldahl method. Eu was determined by titration with Trilon B ($Na_2H_2(EDTA)$) using Xylenol orange indicator.

The double salts $Eu(NO_3)_3\cdot 6CO(NH_2)_2$ and $Eu(NO_3)_3\cdot 2CO(NH_2)_2\cdot 2H_2O$ are incongruently soluble, and $Eu(NO_3)_3\cdot 4CO(NH_2)_2\cdot 2H_2O$ is congruently soluble.

SOURCE AND PURITY OF MATERIALS:

Nothing specified.

ESTIMATED ERROR:

Nothing specified.

REFERENCES:

COMPONENTS:	EVALUATOR:
(1) Gadolinium nitrate; $Gd(NO_3)_3$; [10168-81-7] (2) Water ; H_2O ; [7732-18-5]	S. Siekierski, T. Mioduski Institute for Nuclear Research Warsaw, Poland and M. Salomon U.S. Army ET & DL Ft. Monmouth, NJ January 1983

CRITICAL EVALUATION:

THE BINARY SYSTEM

INTRODUCTION

Solubility data for the binary $Gd(NO_3)_3$-H_2O system have been reported in 14 publications (1-14). Only the hexahydrate (1-7) and pentahydrate (8-13) have been identified as solid phases over the temperature range of 273-323 K. In aqueous nitric acid solutions (7,8), the tetrahydrate, monohydrate and anhydrous salt are the solid phases depending upon the nitric acid concentration.

$Gd(NO_3)_3 \cdot 6H_2O$	[19598-90-4]	$Gd(NO_3)_3 \cdot 4H_2O$	[13773-30-3]
$Gd(NO_3)_3 \cdot 5H_2O$	[52788-53-1]	$Gd(NO_3)_3 \cdot H_2O$	[81201-40-3]

The hexahydrate was produced by crystallization from water (1-6), and the pentahydrate was similarly produced followed by drying over $CaCl_2$ (8-13, 15). Analysis of the waters of hydration in (2-6) showed the initial salt to be the hexahydrate within ± 0.016 water molecules. Analysis in (8-13) by Serebrennikov and Aleksennko's method (16) resulted in 21.59 mass % water which corresponds to the pentahydrate (theoretical water content is 20.79 mass % (evaluators)).

EVALUATION PROCEDURE

The data in the compilations were examined and either rejected immediately because of large obvious errors, or were analyzed by a weighted least squares method. It should be noted that only experimental solubility data were used in the least squares analyses: i.e. smoothed or extrapolated data were not used. The data were fitted to a general solubility equation based on the treatments in (17, 18) and in the INTRODUCTION to this volume:

$$Y = \ell n(m/m_o) - nM_2(m - m_o) = a + b/(T/K) + c\ \ell n(T/K) \qquad [1]$$

In eq. [1], m is the molality at temperature T, m_o is an arbitrarily selected reference molality (usually the 298.2 K value), n is the hydration number of the solid, M_2 is the molar mass of the solvent, and a, b, c are constants from which enthalpies and heat capacities of solution, ΔH_{sln} and ΔC_p, can be estimated (see INTRODUCTION). In fitting the solubility data to eq. [1], weights of 0, 1, 2 were assigned to each published value depending upon the precision of the experimental values. In this procedure, if the residual error between the observed and calculated molalities, Δm, was larger than twice the standard error of estimate for m, σ_m, the data point was either rejected or its weight factor decreased. The fitting of the data was repeated in this manner until all Δm values were equal to or less than $\pm 2\sigma_m$.

Since most authors did not report experimental errors (except in 2-6), the compilers and evaluators attempted to provide this information when possible. As discussed in previous critical evaluations, the data of highest precision are those from the isothermal studies of Spedding et al. (2-6) who reported a precision of ± 0.1% or better, and Moret (1) for which the compilers estimated a precision of around ± 0.2%. For the results reported by Afanas'ev et al. (7), the evaluators estimate an experimental precision of ± 0.5-1%. For the isothermal studies in (14, 15), experimental errors were not given, but the precision is probably around ± 0.2-0.5% (based on the number of significant figures reported by these authors). For those isothermal studies in which refractometric analyses were used (8-13), the experimental precision is estimated to be ± 1% at best as discussed in previous critical evaluations.

SOLUBILITY OF $Gd(NO_3)_3$ IN THE $Gd(NO_3)_3 \cdot 6H_2O$-H_2O SYSTEM

The solubility data for this binary system are summarized in Table 1. The results of fitting the data in Table 1 to the smoothing equation are given in Tables 2 and 3, and in Fig. 1. Table 2 gives the derived constants in eq. [1], and Table 3 gives the smoothed solubility data calculated from the smoothing equation: these results are designated as *recommended* solubility data. The hexahydrate polytherm in Fig. 1 was drawn from the smoothed solubility data, and the experimental points are included for comparison. As seen in Fig. 1, the two data points at 293.15 K and 313.15 K reported by Moret (1) do not lie on the hexahydrate polytherm, and these data points were rejected. The residual error for Moret's result of 4.005 mol kg^{-1} at 283.15 K is Δm = 0.029 mol kg^{-1} which is just

COMPONENTS:

(1) Gadolinium nitrate; $Gd(NO_3)_3$; [10168-81-7]

(2) Water ; H_2O ; [7732-18-5]

EVALUATOR: S. Siekierski, T. Mioduski
Institute for Nuclear Research
Warsaw, Poland
and
M. Salomon
U.S. Army ET & DL
Ft. Monmouth, NJ
January 1983

CRITICAL EVALUATION:

Table 1. Solubility of $Gd(NO_3)_3$ in the $Gd(NO_3)_3 \cdot 6H_2O-H_2O$ system.

T/K	soly/mol kg^{-1}	reference	initial/final weight factor
273.15	3.759	1	2/2
283.15	4.005	1	2/2
293.15	4.298	1	2/0
298.15	4.400	5	1/1
298.15	4.3766	2	2/2
298.15	4.3766	3	2/2
298.15	4.3766	4	2/2
298.15	4.3701	6	1/2
298.2	4.37	7	1/1
313.15	5.055	1	2/0
323.15	5.465	1	2/2

Table 2. Derived parameters for the smoothing equation [1].[a]

parameters	value for the hexahydrate system
a	-32.087
b	1141.2
c	4.9599
σ_a	0.002
σ_b	0.6
σ_c	0.0004
σ_Y	0.002
σ_m	0.015
ΔH_{sln}/kJ mol^{-1}	-37.8
ΔC_p/J K^{-1} mol^{-1}	165.0
congruent meltint point/K	348.5
concn at the congruent m.p./mol kg^{-1}	9.251

[a] σ_a, σ_b, σ_c are standard deviations for the constands a, b, c and σ_Y and σ_m are standard errors of estimate for the quantity Y in eq. [1] and the molality, respectively.

Table 3. Recommended and tentative solubilities for $Gd(NO_3)_3$ in the hexahydrate system.

T/K	soly/mol kg^{-1} [a]	T/K	soly/mol kg^{-1} [b]
273.15	3.76	328.15	5.78
278.15	3.87	333.15	6.16
283.15	3.98	338.15	6.64
288.15	4.10	343.15	7.31
293.15	4.23	348.15	8.72
298.15	4.38	348.50[c]	9.25
303.15	4.55		
308.15	4.74		
313.15	4.95		
318.15	5.18		
323.15	5.46		

[a] Recommended values calculated from the smoothing equation.

[b] Tentative values calculated from the smoothing equation.

[c] Melting point calculated from the smoothing equation.

COMPONENTS:	EVALUATOR:
(1) Gadolinium nitrate; $Gd(NO_3)_3$; [10168-81-7] (2) Water ; H_2O ; [7732-18-5]	S. Siekierski, T. Mioduski Institute for Nuclear Research Warsaw, Poland and M. Salomon U.S. Army ET & DL Ft. Monmouth, NJ January 1983

CRITICAL EVALUATION:

within the limit for acceptance (± $2\sigma_m$ as discussed above), and this data point was therefore not rejected although it almost lies off the smoothed line for the hexahydrate polytherm. At the 95% level of confidence and a Student's t = 2.160, the overall uncertainty in the smoothed solubilities is ± 0.008 mol kg^{-1}.

For the experimental results at 298.15 K from (2-7), the average value of the solubility is 4.378 mol kg^{-1} with a standard deviation of σ = 0.011. At the 95% level of confidence and a Student's t = 2.571, the uncertainty in the average solubility at 298.15 K is ± 0.012 mol kg^{-1}.

The value of the congruent melting point calculated from eq. [1] is 348.5 K which differs significantly from the only experimental value of 360.2 K (19). However Quill et al. (19) stated that the purity of their salt was 98% (impurities not specified), and their experimental melting point of 360.2 K is much closer to the melting point of the tetrahydrate (see below). We therefore assume that the salt used by Quill et al. was probably the tetrahydrate or a mixture of hydrates.

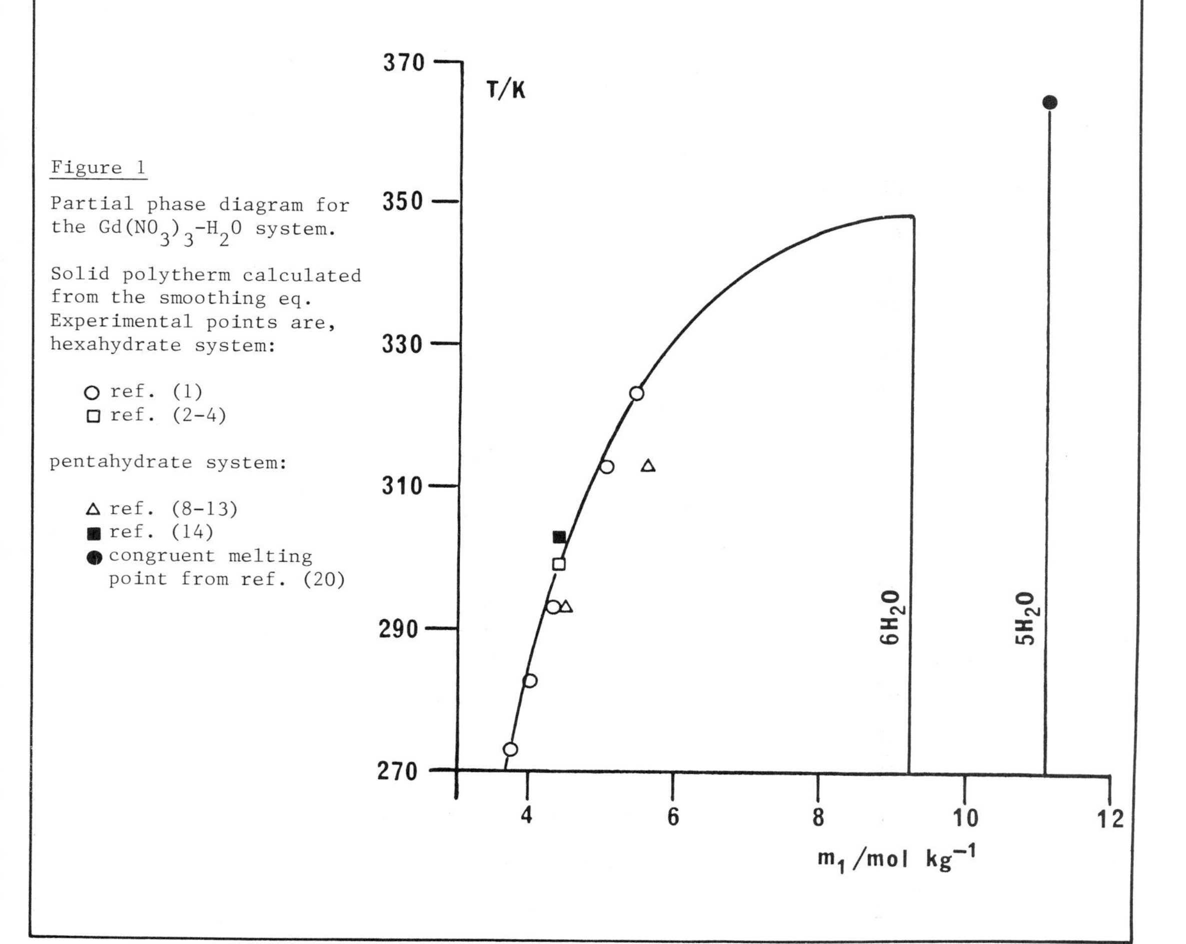

Figure 1

Partial phase diagram for the $Gd(NO_3)_3$-H_2O system.

Solid polytherm calculated from the smoothing eq. Experimental points are, hexahydrate system:

○ ref. (1)
□ ref. (2-4)

pentahydrate system:

△ ref. (8-13)
■ ref. (14)
● congruent melting point from ref. (20)

COMPONENTS:

(1) Gadolinium nitrate; $Gd(NO_3)_3$; [10168-81-7]

(2) Water ; H_2O ; [7732-18-5]

EVALUATOR:
S. Siekierski, T. Mioduski
Institute for Nuclear Research
Warsaw, Poland
and
M. Salomon
U.S. Army ET & DL
Ft. Monmouth, NJ
January 1983

CRITICAL EVALUATION:

SOLUBILITY OF $Gd(NO_3)_3$ IN THE $Gd(NO_3)_3.nH_2O-H_2O$ SYSTEM:

Zhuravlev et al. reported the same values for the solubility of gadolinium nitrate at 293.2 K and 313.2 K in the pentahydrate system in six publications (8-13). Since these values lie to the right of the hexahydrate polytherm (see Fig. 1), the pentahydrate system would have to be metastable at these temperatures. However Khudaibergenova and Sulaimankulov (14) report a solubility of 4.416 mol kg^{-1} at 303.2 K which suggests that the pentahydrate system is stable at temperatures as low as 303.2 K at least (see Fig. 1). Clearly one of these studies must be in error. We attempted to construct a polytherm for the pentahydrate system using the two data points from Zhuravlev's work and the congruent melting point of 365 K reported by Benedicks (20), but we were unable to obtain consistent results. It is possible that the pentahydrate melting point determined in 1900 by Benedicks is in error because a more recent study (21) found a congruent melting point of 363 K for the tetrahydrate. To further confuse the matter, Dvornikova et al. (22) state that the tetrahydrate melts incongruently at 364 K, but Wendlandt and Sewell (23) report a congruent melting point of 367 ± 1 K for the tetrahydrate. Because of these inconsistencies in solubility data and melting points in the lower hydrate systems, all data must be rejected for the present time.

MULTICOMPONENT SYSTEMS

TERNARY SYSTEMS WITH NITRIC ACID AT 298.2 K

Solubility data as a function of nitric acid concentration were reported by Afanas'ev et al. (7), and at several acid concentrations (32.0 and 32.8 mass % HNO_3) by Babievskaya and Perel'mam (15, 24, 25). Afanas'ev et al. found the hexahydrate, tetrahydrate, monohydrate and anhydrous salt to be the stable solid phases depending upon HNO_3 concentration: the hexahydrate is stable up to 56.4 mass % HNO_3. In (15) the authors report that the solid phase in 32.8 mass % HNO_3 is the tetrahydrate, and the solubility of $Gd(NO_3)_3$ is 26.0 mass % (1.84 molg kg^{-1}). Although not stated by the authors, this must be a metastable system since the solubility in this tetrahydrate system is greater than that in the hexahydrate system: e.g. interpolating from the data in (7) the evaluators calculate that in the hexahydrate system at a nitric acid concentration of 32.8 mass %, the solubility of $Gd(NO_3)_3$ is 23.8 mass % (1.60 mol kg^{-1}).

SYSTEMS WITH TWO SATURATING COMPONENTS

Since only one publication exists for each of these multicomponent systems, the solubility data cannot be critically evaluated, and the reader is referred directly to the compilations for further information. The papers by Perel'man and Babievskaya (24, 25) on the quaternary $Y(NO_3)_3-Gd(NO_3)_3-HNO_3-H_2O$ system were compiled and included in the $Y(NO_3)_3$ chapter.

COMPONENTS:	EVALUATOR:
(1) Gadolinium nitrate; $Gd(NO_3)_3$; [10168-81-7] (2) Water ; H_2O ; [7732-18-5]	S. Siekierski, T. Mioduski Institute for Nuclear Research Warsaw, Poland and M. Salomon U.S. Army ET & DL Ft. Monmouth, NJ January 1983

CRITICAL EVALUATION:

REFERENCES

1. Moret, R. *Thèse*. L'Université de Lausanne. 1963.
2. Spedding, F.H.; Shiers, L.E.; Rard, J.A. *J. Chem. Eng. Data* 1975, *20*, 88.
3. Rard, J.A.; Spedding, F.H. *J. Phys. Chem.* 1975, *79*, 257.
4. Spedding, F.H.; Shiers, L.E.; Brown, M.A.; Baker, J.L.; Guitierrez, L.; McDowell, L.S.; Habenschuss, A. *J. Phys. Chem.* 1975, *79*, 1087.
5. Spedding, F.H.; Derer, J.L.; Mohs, M.A.; Rard, J.A. *J. Chem. Eng. Data* 1976, *21*, 474.
6. Rard, J.A.; Shiers, L.E.; Heiser, D.J.; Spedding, F.H. *J. Chem. Eng. Data* 1977, *22*, 337.
7. Afanas'ev, Yu.A.; Azhipa, L.T.; Sal'nik, L.V. *Zh. Neorg. Khim.* 1982, *27*, 769.
8. Starikova, L.I.; Zhuravlev, E.F. *Zh. Neorg. Khim.* 1975, *20*, 2576.
9. Zhuravlev, E.F.; Starikova, L.I.; Katamonov, V.L. *Zh. Neorg. Khim.* 1975, *20*, 1113.
10. Starikova, L.I.; Zhuravlev, E.F. *Zh. Neorg. Khim.* 1975, *20*, 2294.
11. Starikova, L.I.; Zhuravlev, E.F. *Zh. Neorg. Khim.* 1975, *20*, 1676.
12. Zhuravlev, E.F.; Starikova, L.I. *Zh. Neorg. Khim.* 1975, *20*, 1406.
13. Starikova, L.I.; Zhuravlev, E.F. *Zh. Neorg. Khim.* 1980, *25*, 1723.
14. Khudaibergenova, N.; Sulaimankulov, K. *Zh. Neorg. Khim.* 1980, *25*, 2254.
15. Babievskaya, I.Z.; Perel'man, F.M. *Zh. Neorg. Khim.* 1965, *10*, 681.
16. Serebrennikov, V.V.; Alekseenko, L.A. *Kurs Khimii Redkozemel'nykh Elementov*. Tomsk. 1963, p. 352.
17. Williamson, A.T. *Trans. Faraday Soc.* 1944, *40*, 421.
18. Counioux, J.-J.; Tenu, R. *J. Chim. Phys.* 1981, *78*, 816 and 823.
19. Quill, L.L.; Robey, R.F.; Seifter, S. *Ind. Eng. Chem. Anal. Ed.* 1937, *9*, 389.
20. Benedicks, C. *Z. Anorg. Chem.* 1900, *22*, 393.
21. Babievskaya, I.Z.; Perel'man, F.M. *Russ. J. Inorg. Chem. Engl. Transl.* 1966, *11*, 971.
22. Dvornikova, L.M.; Sevost'yanov, V.P.; Ambrozhii, M.N. *Russ. J. Inorg. Chem. Engl. Transl.* 1969, *14*, 1223.
23. Wendlandt, W.W.; Sewell, R.G. *Texas J. Sci.* 1961, *13*, 231.
24. Perel'man, F.M.; Babievskaya, I.Z. *Zh. Neorg. Khim.* 1964, *9*, 1986.
25. Perel'man, F.M. *Rev. Chim. Miner.* 1970, *7*, 635.

COMPONENTS:

(1) Gadolinium nitrate; $Gd(NO_3)_3$; [10168-81-7]

(2) Water ; H_2O ; [7732-18-5]

ORIGINAL MEASUREMENTS:

Moret, R. *Thèse*. L'Université de Lausanne. 1963.

VARIABLES:

Temperature: range 0°C to 50°C

PREPARED BY:

T. Mioduski and Siekierski

EXPERIMENTAL VALUES:

Solubility of $Gd(NO_3)_3$[a]

t/°C	mass %	mol kg^{-1}	nature of the solid phase
0	56.34[b]	3.759	$Gd(NO_3)_3 \cdot 6H_2O$
10	57.89	4.005	"
20	59.60[c]	4.298	"
40	63.44	5.055	"
50	65.23[d]	5.465	"

a. Molaltites calculated by the compilers.

b. This soly reported in (1) as 1475 moles H_2O per 100 moles of $Gd(NO_3)_3$.

c. This soly reported in (1) as 1290 moles H_2O per 100 moles of $Gd(NO_3)_3$.

d. This soly reported in (1) as 1015 moles H_2O per 100 moles of $Gd(NO_3)_3$.

AUXILIARY INFORMATION

METHOD/APPARATUS/PROCEDURE:

The isothermal method was used. Gadolinium was determined by complexometric titration with the disodium salt of ethylenediamine-tetraactic acid using Xylenol orange indicator and urotropine buffer. Water was determined by difference.

SOURCE AND PURITY OF MATERIALS:

$Gd(NO_3)_3 \cdot 6H_2O$ was prepared from Gd_2O_3 of purity better than 99.7 % (obtained by the ion chromatography method). No additional details available.

ESTIMATED ERROR:

Soly: precision about ± 0.2% (compilers).

Temp: precision probably ± 0.01 K or better (compilers).

REFERENCES:

1. Brunisholz, G.; Quinche, J.P.; Kalo, A.M. *Helv. Chim. Acta* 1964, *47*, 14.

COMPONENTS:

(1) Gadolinium nitrate; $Gd(NO_3)_3$; [10168-81-7]

(2) Water ; H_2O ; [7732-18-5]

ORIGINAL MEASUREMENTS:

1. Spedding, F.H.; Shiers, L.E.; Rard, J.A. *J. Chem. Eng. Data* 1975, *20*, 88-93.
2. Rard, J.A.; Spedding, F.H. *J. Phys. Chem.* 1975, *79*, 257-62.
3. Spedding. F.H.; Shiers, L.E.; Brown, M.A.; Baker, J.L.; Guitierrez, L.; McDowell, L.S.; Habenschuss, A. *J. Phys. Chem.* 1975, *79*, 1087-96.
4. Spedding, F.H.; Derer, J.L.; Mohs, M.A.; Rard, J.A. *J. Chem. Eng. Data* 1976, *21*. 474-88.

VARIABLES:

One temperature: 25.00°C

PREPARED BY:

T. Mioduski, S. Siekierski, and M. Salomon.

EXPERIMENTAL VALUES:

The solubility of $Gd(NO_3)_3$ in water at 25.00°C has been reported by Spedding and co-workers in four publications. Source paper [4] reports the solubility to be 4.400 mol kg^{-1}, but the preferred value is given in source papers [1] -[3] as 4.3766 mol kg^{-1}.

COMMENTS AND/OR ADDITIONAL DATA:

Source paper [1] reports the relative viscosity, η_R, of a saturated solution to be 17.850. Taking the viscosity of water at 25°C to equal 0.008903 poise, the viscosity of a saturated $Gd(NO_3)_3$ solution at 25°C is 0.15892 poise (compilers calculation).

Supplementary data available in the microfilm edition to *J. Phys. Chem.* 1975, *79* and in source paper [3] enabled the compilers to provide the following additional data.

The density of the saturated solutions was calculated by the compilers from the smoothing equation, and at 25°C the value is 1.87056 kg m^{-3}. Using this density, the solubility in volume units is (based on the preferred value of 4.3766 mol kg^{-1})

$$c_{satd} = 3.2716 \text{ mol dm}^{-3}$$

Source paper [2] reports the electrolytic conductivity of the saturated solution to be (corrected for the electrolytic conductivity of the solvent) κ = 0.029735 S cm^{-1}.

The molar conductivity of the saturated solution is calculated from 1000 $\kappa/3c_{satd}$ and is

$$\Lambda\ (\tfrac{1}{3}\ Gd(NO_3)_3) = 3.030 \text{ S cm}^2 \text{ mol}^{-1}$$

AUXILIARY INFORMATION

METHOD/APPARATUS/PROCEDURE:

Isothermal method used. Solutions were prepared as described in (1) and (2). The concentration of the saturated solution was determined by both EDTA (1) and sulfate (2) methods which is said to be reliable to 0.1 % or better. In the sulfate analysis, the salt was first decomposed with HCl followed by evaporation to dryness before sulfuric acid additions were made. This eliminated the possibility of nitrate ion coprecipitation.

SOURCE AND PURITY OF MATERIALS:

$Gd(NO_3)_3 6H_2O$ was prepd by addn of HNO_3 to the oxide. The oxide was purified by an ion exchange method, and the upper limit for the impurities Ca, Fe, Si and adjacent rare earths was given as 0.15%.

In source paper [3] the salt was analysed for water of hydration and found to be within ± 0.016 water molecules of the hexahydrate.

Water was distilled form an alkaline permanganate solution.

ESTIMATED ERROR:

Soly: duplicate analyses agreed to at least ± 0.1 %.

Temp: Not specified, but probably accurate to at least ± 0.01 K as in (3) (compilers).

REFERENCES:

1. Spedding, F.G.; Cullen, P.F.; Habenschuss, A. *J. Phys. Chem.* 1974, *78*, 1106.
2. Spedding. F.H.; Pikal, M.J.; Ayers, B.O. *J. Phys. Chem.* 1966, *70*, 2440.
3. Spedding, F.H.; et. al. *J. Chem. Eng. Data.* 1975, *20*, 72.

COMPONENTS:	ORIGINAL MEASUREMENTS:
(1) Gadolinium nitrate; $Gd(NO_3)_3$; [10168-81-7] (2) Water ; H_2O ; [7732-18-5]	Rard, J.A.; Shiers, L.E.; Heiser, D.J.; Spedding, F.H. *J. Chem. Eng. Data* 1977, *22*, 337-47.
VARIABLES:	PREPARED BY:
One temperature: 25.00°C	T. Mioduski and S. Siekierski

EXPERIMENTAL VALUES:

The solubility of $Gd(NO_3)_3$ in water at 25°C was reported to be 4.3701 mol kg^{-1}.

AUXILIARY INFORMATION

METHOD/APPARATUS/PROCEDURE:

The isothermal-isopiestic method was used. The isopiestic equilibration consisted of allowing less than satd gadolinium nitrate soln to reach thermodynamic equil through the vapor phase with a reference soln. (KCl, $CaCl_2$). The soly was thus detd without sepn of the soln and solid phase. The solns were adjusted to their equivalence pH values to ensure a ratio of three nitrates to each Gd ion. Duplicate samples of the nitrate and reference soln were used and equil was approached from higher and lower concns (about 4 days). The satd soln was analysed by EDTA and gravimetric sulfate methods. The nitrate samples were evaporated with HCl before conversion to the sulfates to destroy the nitrate ions and thereby avoid their copptn. The solid phase is $Gd(NO_3)_3 \cdot 6H_2O$. The major objective of this paper was to determine the osmotic coefficients of the lanthanide nitrates.

SOURCE AND PURITY OF MATERIALS:

1. The nitrate was obtained from the oxide (purified by ion-exchange methods at the Ames Laboratory) and reagent grade HNO_3. The purity of the oxide was greater than 99.85 mass % with Ca, Fe, Si and adjacent Ln being the only significant inpurities.
2. Conductivity water distilled from alkaline $KMnO_4$ was used.

ESTIMATED ERROR: Soly: The average equil isopiestic molalities are known to at least ± 0.1%, and they differ from direct analyses values by 0.07 - 0.17%.
Temp: accuracy ± 0.01 K.

REFERENCES:

COMPONENTS:

(1) Gadolinium nitrate; $Gd(NO_3)_3$; [10168-81-7]

(2) Nitric acid; HNO_3; [7697-37-2]

(3) Water ; H_2O ; [7732-18-5]

ORIGINAL MEASUREMENTS:

Afanas'ev, Yu.A.; Azhipa, L.T.; Sal'nik, L.V.

Zh. Neorg. Khim. 1982, *27*, 769-73; *Russ. J. Inorg. Chem. Engl. Transl.* 1982. *27*,

VARIABLES:

HNO_3 concentration at 25°C

PREPARED BY:

T. Mioduski and S. Siekierski

EXPERIMENTAL VALUES:

Solubility of $Gd(NO_3)_3$ in HNO_3 solutions at 25°C[a]

$Gd(NO_3)_3$ mass %	$Gd(NO_3)_3$ mol kg^{-1}	HNO_3 mass %	HNO_3 mol kg^{-1}	nature of the solid phase
60.0	4.37	0	——	$Gd(NO_3)_3 \cdot 6H_2O$
56.0	3.93	2.5	0.96	"
50.0	3.33	6.2	2.25	"
48.1	3.33	9.8	3.69	"
39.5	2.65	17.1	6.25	"
33.8	2.26	22.6	8.23	"
25.2	1.68	31.2	11.36	"
22.3	1.54	35.5	13.35	"
14.6	1.05	44.7	17.43	"
11.3	0.88	51.3	21.77	"
7.7	0.62	56.4	24.93	$Gd(NO_3)_3 \cdot 6H_2O + Gd(NO_3)_3 \cdot 4H_2O$
6.2	0.53	59.8	27.91	$Gd(NO_3)_3 \cdot 4H_2O$
5.9	0.62	66.2	37.66	"
10.5	1.76	72.1	65.76	"
11.7	2.17	72.6	73.89	"
9.8	1.85	74.8	77.08	$Gd(NO_3)_3 \cdot 4H_2O + Gd(NO_3)_3 \cdot H_2O$
11.2	3.63	79.8	140.7	$Gd(NO_3)_3 \cdot H_2O$
8.5	3.81	85.0	207.5	$Gd(NO_3)_3 \cdot H_2O + Gd(NO_3)_3$
7.5	4.97	88.1	317.8	$Gd(NO_3)_3$
4.8	3.33	91.0	343.8	"
3.9	3.16	92.5	407.8	"
1.8	2.62	96.2	763.3	"

a. Molalities calculated by M. Salomon.

AUXILIARY INFORMATION

METHOD/APPARATUS/PROCEDURE:

The isothermal method was used. The composition of the solutions was changed by addition of 100 % HNO_3 to a saturated solution or by addition of the salt to the acid solution. Equilibrium was reached within 3-4 hours. The gadolinium content in the saturated solutions and solid phases was determined by complexometric titration using Xylenol Orange indicator. The HNO_3 content was determined by titration with NaOH using methyl red indicator. The compositions of the solid phases were determined by the Schreinemakers' method. The hydrated solid phases were separated and their infrared spectra recorded. Details are given in the source paper.

SOURCE AND PURITY OF MATERIALS:

C.p. grade gadolinium nitrate was used.

Nitric acid (source and purity not specified) was concentrated by the method recommended in the well-known Brauer's Handbook (the Russian edition was cited by the authors).

ESTIMATED ERROR:

Soly: nothing specified

Temp: precision within ± 0.1 K.

REFERENCES:

COMPONENTS:

(1) Gadolinium nitrate; $Gd(NO_3)_3$; [10168-81-7]

(2) Ammonium nitrate; NH_4NO_3; [6484-52-2]

(3) Nitric acid; HNO_3; [7697-37-2]

(4) Water ; H_2O ; [7732-18-5]

ORIGINAL MEASUREMENTS:

Babievskaya, I.Z.; Perel'man, F.M.

Zh. Neorg. Khim. 1965, *10*, 681-3; *Russ. J. Inorg. Chem. Engl. Transl.* 1965, *10*, 366-7.

VARIABLES:

Composition at 25°C

PREPARED BY:

T. Mioduski and S. Siekierski

EXPERIMENTAL VALUES:

Composition of saturated solutions at 25°C[a]

$Gd(NO_3)_3$		NH_4NO_3		HNO_3			
mass %	mol kg^{-1}	mass %	mol kg^{-1}	mass %	mol kg^{-1}	specific gravity	solid phase[c]
26.0	1.84	——	——	32.8	12.63	1.536	A
23.7	2.31	13.1	5.47	33.3	17.67	1.584	A
25.0	2.95	17.1	8.65	33.2	21.33	1.632	A
23.9	3.32	18.6	11.07	36.5	27.58	——	A
26.0	3.54	19.9	11.62	32.7	24.25	1.633	A
31.8	7.02	23.4	22.15	31.6	37.99	——	A
32.0	6.56	22.0	19.36	31.8	35.54	1.783	A+B
39.5	13.54	21.3	31.31	30.7	57.32	——	A+B
34.7	12.03	26.5	39.41	30.4	57.43	1.878	B
31.2	29.32	34.1	137.4	31.6	161.8	——	B
28.2	22.20	34.7	117.2	33.4	143.3	1.792	B
23.8	19.81	39.8	142.1	32.9	149.2	1.721	B
19.5[b]		49.7		32.4		——	B+C
15.5	11.88	50.5	166.0	30.2	126.1	1.665	C
15.0	21.85	51.3	320.5	31.7	251.5	——	C
14.2	8.44	49.4	126.0	31.5	102.0	1.623	C
9.4	2.68	47.4	58.06	33.0	51.34	1.555	C
5.1	1.02	47.0	40.22	33.3	36.20	1.480	C
——	——	48.3	33.34	33.6	29.46	1.432	C

a. Molalities calculated by M. Salomon

b. Total mass % exceeds 100%.

c. A = $Gd(NO_3)_3 \cdot 4H_2O$ B = $Gd(NO_3)_3 \cdot 2NH_4NO_3$ C = NH_4NO_3

AUXILIARY INFORMATION

METHOD/APPARATUS/PROCEDURE:

The isothermal method was used. Solutions were equilibrated for 2-3 days. Gadolinium was determined gravimetrically by precipitation as the oxalate and ignition to the oxide, and total nitrogen was determined by the Kjeldahl method. The HNO_3 content was determined by acidimetric titration (authors state that a aqueous solution of $Gd(NO_3)_3$ was neutral). Presumbly water was determined by difference.

Method of determination of the compositions of the solid phases not specified. $Gd(NO_3)_3 \cdot 4H_2O$ was isolated and studied crystal-optically. Authors report the following refractive indices for the tetrahydrate:

$n_g = 1.552$; $n_p = 1.447$

SOURCE AND PURITY OF MATERIALS:

$Gd(NO_3)_3 \cdot 5H_2O$ prepared by dissolving Gd_2O_3 in nitric acid followed by crystallization as described in (1). Analysis by the oxalate method gave 41.9 mass % Gd_2O_3 (theoretical for the pentahydrate is 41.8 mass %, authors).

No other information given.

ESTIMATED ERROR:

Soly: nothing specified.

Temp: precision ± 0.2 K.

REFERENCES:

1. Perel'man, F.M.; Babievskaya, I.Z. *Zh. Neorg. Khim.* 1962. *7*, 1479.

COMPONENTS:

(1) Gadolinium nitrate; $Gd(NO_3)_3$; [10168-81-7]

(2) 2-Aminoethanol nitrate (ethanolamine nitrate); $C_2H_8N_2O_4$; [20748-72-5]

(3) Water ; H_2O ; [7732-18-5]

ORIGINAL MEASUREMENTS:

Starikova, L.I.; Zhuravlev, E.F. *Zh. Neorg. Khim.* 1975, *20*, 2576-7; *Russ. J. Inorg. Chem. Engl. Transl.* 1975, *20*, 1428-9.

VARIABLES:

Composition at 20°C and 40°C

PREPARED BY:

T. Mioduski and S. Siekierski

EXPERIMENTAL VALUES:

20°C Isotherm[a]

$Gd(NO_3)_3$ mass %	$Gd(NO_3)_3$ mol kg^{-1}	$HOC_2H_4NH_2 \cdot HNO_3$ mass %	$HOC_2H_4NH_2 \cdot HNO_3$ mol kg^{-1}	solid phase[b]
0.0	———	86.0	49.50	A
2.5	0.52	83.5	48.06	A
6.0	1.29	80.5	48.05	A
11.0	2.47	76.0	47.11	A
15.5	3.76	72.5	48.68	A
32.0	10.97	59.5	56.41	A
52.5	10.92	33.5	19.28	B
52.7	6.88	25.0	9.03	B
54.0	5.21	15.8	4.22	B
56.0	4.41	7.0	1.52	B
60.6	4.48	0.0	———	B

40°C Isotherm[a]

$Gd(NO_3)_3$ mass %	$Gd(NO_3)_3$ mol kg^{-1}	$HOC_2H_4NH_2 \cdot HNO_3$ mass %	$HOC_2H_4NH_2 \cdot HNO_3$ mol kg^{-1}	solid phase[b]
0.0	———	95.5	171.0	A
1.5	1.09	94.5	190.4	A
2.5	1.62	93.0	166.5	A
4.0	2.33	91.0	146.7	A
6.0	4.37	90.0	181.3	A
12.0	11.65	85.0	228.3	A
56.0	10.88	29.0	15.58	B
56.5	7.48	21.5	7.88	B
58.0	5.83	13.0	3.61	B
62.0	5.56	5.5	1.36	B
65.8	5.60	0.0	———	B

a. Molalities calculated by M. Salomon.
b. Solid phases: A - $HOC_2H_4NH_2 \cdot HNO_3$; B = $Gd(NO_3)_3 \cdot 5H_2O$

AUXILIARY INFORMATION

METHOD/APPARATUS/PROCEDURE:

The solubility was studied by the method of isothermal sections (1) by measuring the refractive indices of saturated solutions along directed sections of the phase diagram. Equilibrium was checked by repeated measurements of the refractive index as a function of time. The results were used to graph the relation between the refractive indices and the composition of the components for each of the sections studied. The graphs were used to find the inflection or break points corresponding to the composition of the saturated solutions.

COMMENTS AND/OR ADDITIONAL DATA:

Between the regions of 59.5 - 33.5 mass % amine nitrate at 20°C and 85.0 - 29.0 mass % at 40°C, the mixtures of $Gd(NO_3)_3 \cdot 5H_2O$ with the ethanolamine nitrate form homogeneous liquid solutions. Due to the lack of anhydrous $Gd(NO_3)_3$, it was not possible to follow the subsequent course of the solubility lines to determine if specific interactions (i.e. compound formation) occurs.

SOURCE AND PURITY OF MATERIALS:

No information given.

ESTIMATED ERROR:

Soly: precision ± 1% at best (compilers).

Temp: precision probably ± 0.2 K (compilers).

REFERENCES:

1. Zhuravlev, E.F.; Sheveleva, A.D. *Zh. Neorg. Khim.* 1960, *5*, 2630.

COMPONENTS:

(1) Gadolinium nitrate; $Gd(NO_3)_3$; [10168-81-7]

(2) 2,2'-Iminodiethanol nitrate (diethanolamine nitrate); $C_4H_{12}N_2O_5$; [57432-67-4]

(3) Water ; H_2O ; [7732-18-5]

ORIGINAL MEASUREMENTS:

Starikova, L.I.; Zhuralev, E.F. *Zh. Neorg. Khim.* 1975, *20*, 2576-7; *Russ. J. Inorg. Chem. Engl. Transl.* 1975, *20*, 1428-9.

VARIABLES:

Composition at 20°C and 40°C

PREPARED BY:

T. Mioduski and S. Siekierski

EXPERIMENTAL VALUES:

20°C Isotherm[a]					40°C Isotherm[a]			
$Gd(NO_3)_3$		$(HOC_2H_4)_2NH \cdot HNO_3$			$Gd(NO_3)_3$		$(HOC_2H_4)_2NH \cdot HNO_3$	
mass %	mol kg^{-1}	mass %	mol kg^{-1}	solid phase[b]	mass %	mol kg^{-1}	mass %	mol kg^{-1}
0.0	——	78.5	21.71	A	0.0	——	88.5	45.77
3.5	0.49	75.5	21.38	A	1.8	0.45	86.5	43.97
8.5	1.24	71.5	21.26	A	4.5	1.14	84.0	43.44
14.5	2.28	67.0	21.54	A	9.0	2.28	79.5	41.11
20.0	3.53	63.5	22.89	A	13.5	3.58	75.5	40.82
34.0	11.01	57.0	37.66	A	28.5	11.07	64.0	50.75
51.0	11.01	35.5	15.64	B	56.2	11.06	29.0	11.65
53.3	6.99	24.5	6.56	B	58.8	7.72	19.0	5.09
56.0	5.63	15.0	3.08	B	61.5	6.64	11.5	2.53
58.0	4.76	6.5	1.09	B	64.0	6.11	5.5	1.07
60.6	4.48	0.0	——	B	65.8	5.60	0.0	——

a. Molalities caluclated by M. Salomon.
b. Solid phases: A = $(HOC_2H_4)_2NH \cdot HNO_3$; B = $Gd(NO_3)_3 \cdot 5H_2O$

AUXILIARY INFORMATION

METHOD/APPARATUS/PROCEDURE:

The solubility was studied by the method of isothermal sections (1) by measuring the refractive indices of saturated solutions along directed sections of the phase diagram. Equilibrium was checked by repeated measurements of the refractive index as a function of time. The results were used to graph the relation between the refractive indices and the composition of the components for each of the sections studied. The graphs were used to find the inflection or break points corresponding to the composition of the saturated solutions.

COMMENTS AND/OR ADDITIONAL DATA:

Between the regions of 57.0 - 35.5 mass % amine nitrate at 20°C and 64.0 - 29.0 mass % at 40°C, the mixtures of $Gd(NO_3)_3 \cdot 5H_2O$ with the ethanolamine nitrate form homogeneous liquid solutions. Due to the lack of anhydrous $Gd(NO_3)_3$, it was not possible to follow the subsequent course of the solubility lines to determine if specific interactions (i.e. compound formation) occurs.

SOURCE AND PURITY OF MATERIALS:

No information given.

ESTIMATED ERROR:

Soly: precision ± 1% at best (compilers).

Temp: precision probably ± 0.2 K (compilers).

REFERENCES:

1. Zhuravlev, E.F.; Sheveleva, A.D. *Zh. Neorg. Khim.* 1960, *5*, 2630.

COMPONENTS:

(1) Gadolinium nitrate; $Gd(NO_3)_3$; [10168-81-7]

(2) 2,2',2"-Nitrilotriethanol nitrate (triethanolamine nitrate); $C_6H_{16}N_2O_6$; [27096-29-3]

(3) Water ; H_2O ; [7732-18-5]

ORIGINAL MEASUREMENTS:

Starikova, L.I.; Zhuravlev, E.F. *Zh. Neorg. Khim.* 1975, *20*, 2576-7; *Russ. J. Inorg. Chem. Engl. Transl.* 1975, *20*, 1428-9.

VARIABLES:

Composition at 20°C and 40°C

PREPARED BY:

T. Mioduski and S. Siekierski

EXPERIMENTAL VALUES:

20°C Isotherm[a]					40°C Isotherm[a]			
$Gd(NO_3)_3$		$(HOC_2H_4)_3N{\cdot}HNO_3$			$Gd(NO_3)_3$		$(HOC_2H_4)_3N{\cdot}HNO_3$	
mass %	mol kg^{-1}	mass %	mol kg^{-1}	solid phase[b]	mass %	mol kg^{-1}	mass %	mol kg^{-1}
0.0	——	82.0	21.47	A	0.0	——	88.5	36.27
7.5	1.25	75.0	20.20	A	4.5	1.14	84.0	34.42
13.5	2.31	69.5	19.27	A	8.0	2.22	81.5	36.58
19.0	3.69	66.0	20.73	A	12.5	3.64	77.5	36.52
34.0	11.01	57.0	29.85	A	27.5	10.68	65.0	40.84
49.0	10.98	38.0	13.77	B	51.5	10.35	34.0	11.05
51.5	6.82	26.5	5.68	B	56.0	7.59	22.5	4.93
53.3	6.04	21.0	3.85	B	58.0	6.76	17.0	3.20
56.0	5.18	12.5	1.87	B	60.8	5.96	9.5	1.51
58.0	4.69	6.0	0.79	B	63.5	5.78	4.5	0.66
60.6	4.48	0.0	——	B	65.8	5.60	0.0	——

a. Molalities calculated by M. Salomon.
b. Solid phases: A = $(HOC_2H_4)_3N{\cdot}HNO_3$; B = $Gd(NO_3)_3{\cdot}5H_2O$

AUXILIARY INFORMATION

METHOD/APPARATUS/PROCEDURE:

The solubility was studied by the method of isothermal sections (1) by measuring the refractive indices of saturated solutions along directed sections of the phase diagram. Equilibrium was checked by repeated measurements of the refractive index as a function of time. The results were used to graph the relation between the refractive indices and the composition of the components for each of the sections studied. The graphs were used to find the inflection or break points corresponding to the composition of the saturated solutions.

COMMENTS AND/OR ADDITIONAL DATA:

Between the regions of 57.0 - 38.0 mass % amine nitrate at 20°C and 65.0 - 34.0 mass % at 40°C, the mixtures of $Gd(NO_3)_3{\cdot}5H_2O$ with the ethanolamine nitrate form homogeneous liquid solutions. Due to the lack of anhydrous $Gd(NO_3)_3$, it was not possible to follow the subsequent course of the solubility lines to determine if specific interactions (i.e. compound formation) occurs.

SOURCE AND PURITY OF MATERIALS:

No information given.

ESTIMATED ERROR:

Soly: precision ± 1% at best (compilers).

Temp: precision probably ± 0.2 K (compilers).

REFERENCES:

1. Zhuravlev, E.F.; Sheveleva, A.D. *Zh. Neorg. Khim.* 1960, *5*, 2630.

COMPONENTS:

(1) Gadolinium nitrate; $Gd(NO_3)_3$; [10168-81-7]

(2) Ethylenediamine dinitrate; $C_2H_{10}N_4O_6$; [20829-66-7]

(3) Water ; H_2O ; [7732-18-5]

ORIGINAL MEASUREMENTS:

Zhuravlev, E.F.; Starikova, L.I.; Katamonov, V.L. *Zh. Neorg. Khim.* 1975, *20*, 1113-6; *Russ. J. Inorg. Chem. Engl. Transl.* 1975, *20*, 626-8.

VARIABLES:

Composition at 20°C and 40°C

PREPARED BY:

T. Mioduski, S. Siekierski, and M. Salomon

EXPERIMENTAL VALUES:

20°C Isotherm[a]					40°C Isotherm[a]			
$Gd(NO_3)_3$		$H_2NC_2H_4NH_2\cdot 2HNO_3$			$Gd(NO_3)_3$		$H_2NC_2H_4NH_2\cdot 2HNO_3$	
mass %	mol kg^{-1}	mass %	mol kg^{-1}	solid phase[b]	mass %	mol kg^{-1}	mass %	mol kg^{-1}
0.0	——	43.5	4.14	A	0.0	——	60.0	8.06
9.5	0.52	37.0	3.72	A	7.2	0.51	52.0	6.85
21.5	1.25	28.5	3.06	A	17.0	1.27	44.0	6.06
36.0	2.36	19.5	2.35	A	29.3	2.42	35.5	5.42
46.7	3.55	15.0	2.10	A	38.5	3.56	30.0	5.12
51.2	4.29	14.0	2.16	A	43.0	4.32	28.0	5.19
				A	53.7	6.71	23.0	5.30
54.0	4.77	13.0	2.12	A+B	55.0	6.97	22.0	5.14
56.5	4.77	9.0	1.40	B	58.8	6.42	14.5	2.92
58.5	4.58	4.3	0.62	B	62.5	5.87	6.5	1.13
60.6	4.48	0.0	——	B	64.5	5.78	3.0	0.50
				B	65.8	5.60	0.0	——

a. Molalities calculated by the compilers.
b. Solid phases: A = $(CH_2)_2(NH_2)_2\cdot 2HNO_3$; B = $Gd(NO_3)_3\cdot 5H_2O$

AUXILIARY INFORMATION

METHOD/APPARATUS/PROCEDURE:

The solubility was studied by the method of isothermal sections (1) by measuring the refractive indices of saturated solutions along directed sections of the phase diagram. Equilibrium was checked by repeated measurements of the refractive index as a function of time. The results were used to graph the relation between the refractive indices and the composition of the components for each of the sections studied. The graphs were used to find the invlection or break points corresponding to the composition of the saturated solutions.

SOURCE AND PURITY OF MATERIALS:

C.p. grade gadolinium nitrate was twice recrystallized. Analysis for water of crystallization gave 21.59 mass % water which is close to the theoretical for the pentahydrate (theor: 20.79 mass %, compilers).

The amine nitrate was prepared by neutralizing "pure" grade amine with c.p. grade HNO_3 and evaporating the neutralized solution to crystallization. The salt was recrystallized and dried to constant weight in a desiccator over anhydrous $CaCl_2$. The salt was analysed. Doubly distilled water was used.

ESTIMATED ERROR:

Soly: precision ± 1% at best (compilers).

Temp: precision probably ± 0.2 K (compilers).

REFERENCES:

1. Zhuravlev, E.F.; Sheveleva, A.D. *Zh. Neorg. Khim.* 1960, *5*, 2630.

COMPONENTS:

(1) Gadolinium nitrate; $Gd(NO_3)_3$; [10168-81-7]

(2) Diethylenetriamine trinitrate; $C_4H_{16}N_6O_9$; [6143-55-1]

(3) Water ; H_2O ; [7732-18-5]

ORIGINAL MEASUREMENTS:

Starikova, L.I.; Zhuravlev, E.F. *Zh. Neorg. Khim.* 1975, *20*, 2294-6; *Russ. J. Inorg. Chem. Engl. Transl.* 1975, *20*, 1274-5.

VARIABLES:

Composition at 20°C and 40°C

PREPARED BY:

T. Mioduski and S. Siekierski

EXPERIMENTAL VALUES:

20°C Isotherm[a]					40°C Isotherm[a]			
$Gd(NO_3)_3$		$C_4H_{13}N_3\cdot 3HNO_3$		solid	$Gd(NO_3)_3$		$C_4H_{13}N_3\cdot 3HNO_3$	
mass %	mol kg^{-1}	mass %	mol kg^{-1}	phase[b]	mass %	mol kg^{-1}	mass %	mol kg^{-1}
0.0	——	60.5	5.24	A	0.0	——	72.5	9.02
7.0	0.49	51.0	4.16	A	5.0	0.50	66.0	7.79
18.0	1.22	39.0	3.10	A	13.0	1.22	56.0	6.18
32.0	2.39	29.0	2.54	A	25.0	2.39	44.5	4.99
42.0	3.55	23.5	2.33	A	33.3	3.57	39.5	4.97
51.0	5.31	21.0	2.57	A+B	46.5	6.95	34.0	5.97
52.0	5.17	18.7	2.18	B	47.5	6.92	32.5	5.56
55.5	4.97	12.0	1.26	B	50.0	6.47	27.5	4.18
58.0	4.65	5.7	0.54	B	57.0	6.04	15.5	1.93
60.6	4.48	0.0	——	B	60.2	5.89	10.0	1.15
				B	63.5	5.69	4.0	0.42
				B	65.8	5.60	0.0	——

a. Molalities calculated by M. Salomon.
b. Solid phases: A = $H_2NCH_2CH_2NHCH_2CH_2NH_2\cdot 3HNO_3$; B = $Gd(NO_3)_3\cdot 5H_2O$

AUXILIARY INFORMATION

METHOD/APPARATUS/PROCEDURE:

The solubility was studied by the method of isothermal sections (1) by measuring the refractive indices of saturated solutions along directed sections of the phase diagram. Equilibrium was checked by repeated measurements of the refractive index as a function of time. The results were used to graph the relation between the refractive indices and the composition of the components for each of the sections studied. The graphs were used to find the inflection or break points corresponding to the composition of the saturated solutions.

SOURCE AND PURITY OF MATERIALS:

Nothing specified for $Gd(NO_3)_3\cdot 5H_2O$.

The amine nitrate was prepared by neutralization of "pure" grade diethylenetriamine with nitric acid. The neutralized solution was evaporated to crystallization. The solid was redrystallized, dried, and analysed to check the composition of the salt. The results corresponded to the anhydrous salt.

ESTIMATED ERROR:

Soly: precision ± 1% at best (compilers).

Temp: precision probably ± 0.2 K (compilers).

REFERENCES:

1. Zhuravlev, ELF.; Sheveleva, A.D. *Zh. Neorg. Khim.* 1960, *5*, 2630.

COMPONENTS:

(1) Gadolinium nitrate; $Gd(NO_3)_3$; [10168-81-7]

(2) 1,6-Hexanediamine dinitrate (hexamethylenediamine dinitrate); $C_6H_{18}N_4O_6$; [6143-53-9]

(3) Water ; H_2O ; [7732-18-5]

ORIGINAL MEASUREMENTS:

Zhuravlev, E.F.; Starikova, L.I.; Katamanov. V.L. *Zh. Neorg. Khim.* 1975, *20*, 113-6; *Russ. J. Inorg. Chem. Engl. Transl.* 1975, *20*, 626-8.

VARIABLES:

Composition at 20°C and 40°C

PREPARED BY:

T. Mioduski, S. Siekierski, and M. Salomon

EXPERIMENTAL VALUES:

20°C Isotherm[a]					40°C Isotherm[a]			
$Gd(NO_3)_3$		$(CH_2)_6(NH_2)_2\cdot 2HNO_3$			$Gd(NO_3)_3$		$(CH_2)_6(NH_2)_2\cdot 2HNO_3$	
mass %	mol kg^{-1}	mass %	mol kg^{-1}	solid phase[b]	mass %	mol kg^{-1}	mass %	mol kg^{-1}
0.0	——	79.0	15.53	A	0.0	——	85.5	24.34
3.8	0.52	75.0	14.60	A	2.5	0.49	82.5	22.71
9.2	1.23	69.0	13.07	A	6.5	1.26	78.5	21.60
17.0	2.36	62.0	12.19	A	12.3	2.36	72.5	19.69
38.0	11.07	52.0	21.47	A	29.6	10.92	62.5	32.66
47.3	10.85	40.0	13.00	B	51.0	11.43	36.0	11.43
49.8	6.54	28.0	5.21	B	53.5	7.18	24.8	4.72
51.5	5.52	21.3	3.23	B	55.8	6.25	18.2	2.89
54.1	4.79	13.0	1.63	B	59.5	5.78	10.5	1.44
57.4	4.57	6.0	0.68	B	62.5	5.60	5.0	0.64
60.6	4.48	0.0	——	B	65.8	5.60	0.0	——

a. Molalities calculated by the compilers.
b. solid phases: A = $(CH_2)_6(NH_2)_2\cdot 2HNO_3$; B = $Gd(NO_3)_3\cdot 5H_2O$

AUXILIARY INFORMATION

METHOD/APPARATUS/PROCEDURE:

The solubility was studied by the method of isothermal sections (1) by measuring the refractive indices of saturated solutions along directed sections of the phase diagram. Equilibrium was checked by repeated measurements of the refractive indix as a function of time. The results were used to graph the relation between the refractive indices and the composition of the components for each of the sections studied. The graphs were used to find the inflection or break points corresponding to the composition of the saturated solutions.

SOURCE AND PURITY OF MATERIALS:

C.p. grade gadolinium nitrate was twice recrystallized. Analysis for water of crystallization gave 21.59 mass % water which is close to the theoretical for the pentahydrate (theor: 20.79 mass %, compilers).

The amine nitrate was prepared by neutralizing "pure" grade amine with c.p. grade HNO_3 and evaporating the neutralized solution to crystallization. The salt was recrystallized and dried to constant weight in a desiccator over anhydrous $CaCl_2$. The salt was analysed. Doubly distilled water was used.

ESTIMATED ERROR:

Soly: precision ± 1% at best (compilers).

Temp: precision probably ± 0.2 K (compilers).

REFERENCES:

1. Zhuravlev, E.F.; Sheveleva, A.D. *Zh. Neorg. Khim.* 1960, *5*, 2630.

COMPONENTS:

(1) Gadolinium nitrate; $Gd(NO_3)_3$; [10168-81-7]

(2) Pyridine nitrate; $C_5H_6N_2O_3$ [543-53-3]

(3) Water ; H_2O ; [7732-18-5]

ORIGINAL MEASUREMENTS:

Starikova, L.I.; Zhuravlev, E.F. *Zh. Neorg. Khim.* 1975, *20*, 1676-8; *Russ. J. Inorg. Chem. Engl. Transl.* 1975, *20*, 939-41.

VARIABLES:

Composition at 20°C and 40°C

PREPARED BY:

T. Mioduski and S. Siekierski

EXPERIMENTAL VALUES:

20°C Isotherm[a]

$Gd(NO_3)_3$ mass %	$Gd(NO_3)_3$ mol kg^{-1}	$C_5H_5N{\cdot}HNO_3$ mass %	$C_5H_5N{\cdot}HNO_3$ mol kg^{-1}	solid phase[b]
0.0	———	72.0	18.09	A
4.5	0.49	69.0	18.32	A
10.2	1.20	65.0	18.44	A
17.0	2.36	62.0	20.77	A
24.7	4.16	58.0	23.59	A
35.7	11.18	55.0	41.61	A
51.8	11.43	35.0	18.66	B
53.6	5.87	19.8	5.24	B
55.5	5.05	12.5	2.75	B
58.0	4.63	5.5	1.06	B
60.6	4.48	0	———	B

40°C Isotherm[a]

$Gd(NO_3)_3$ mass %	$Gd(NO_3)_3$ mol kg^{-1}	$C_5H_5N{\cdot}HNO_3$ mass %	$C_5H_5N{\cdot}HNO_3$ mol kg^{-1}
0.0	———	80.5	29.05
3.0	0.46	78.0	28.89
7.1	1.16	75.0	29.48
12.2	2.32	72.5	33.34
18.7	4.10	68.0	35.98
29.0	10.56	63.0	55.41
61.0	11.11	23.0	10.12
61.2	7.19	14.0	3.97
62.0	6.23	9.0	2.18
63.5	5.69	4.0	0.87
65.8	5.60	0.0	———

a. Molalities calculated by M. Salomon.
b. Solid phases: A = $C_5H_5N{\cdot}HNO_3$; B = $Gd(NO_3)_3{\cdot}5H_2O$

AUXILIARY INFORMATION

METHOD/APPARATUS/PROCEDURE:

The solubility was studied by the method of isothermal sections (1) by measuring the refractive indices of saturated solutions along directed sections of the phase diagram. Equilibrium was checked by repeated measurements of the refractive index as a function of time. The results were used to graph the relation between the refractive indices and the composition of the components for each of the sections studied. The graphs were used to find the inflection or break points corresponding to the composition of the saturated solutions.

SOURCE AND PURITY OF MATERIALS:

C.p. grade $Gd(NO_3)_3{\cdot}5H_2O$ was recrystallized. Pyridine nitrate, $C_5H_5N{\cdot}HNO_3$, was prepared by neutralization of pyridine with nitric acid and evaporation to crystallization. The solid was recrystallized and dried in a desiccator to constant mass.

The compositions of the compounds were checked analytically.

ESTIMATED ERROR:

Soly: precision ± 1% at best (compilers).

Temp: precision probably ± 0.2 K (compilers).

REFERENCES:

1. Zhuravlev, E.F.; Sheveleva, A.D. *Zh. Neorg. Khim.* 1960, *5*, 2630.

COMPONENTS:

(1) Gadolinium nitrate; $Gd(NO_3)_3$; [10168-81-7]

(2) Quinoline nitrate; $C_9H_8N_2O_3$; [21640-15-3]

(3) Water ; H_2O ; [7732-18-5]

ORIGINAL MEASUREMENTS:

Zhuravlev, E.F.; Starikova, L.I. *Zh. Neorg. Khim.* 1975, *20*, 1406-9; *Russ. J. Inorg. Chem. Engl. Transl.* 1975, *20*, 790-2.

VARIABLES:

Composition at 20°C and 40°C

PREPARED BY:

T. Mioduski and S. Siekierski

EXPERIMENTAL VALUES:

20°C Isotherm[a]					40°C Isotherm[a]			
$Gd(NO_3)_3$		$C_9H_7N{\cdot}HNO_3$			$Gd(NO_3)_3$		$C_9H_7N{\cdot}HNO_3$	
mass %	mol kg^{-1}	mass %	mol kg^{-1}	solid phase[b]	mass %	mol kg^{-1}	mass %	mol kg^{-1}
0.0	——	70.0	12.14	A	0.0	——	78.0	18.45
5.0	0.51	66.5	12.14	A	4.0	0.54	74.5	18.03
11.5	1.29	62.5	12.51	A	8.5	1.24	71.5	18.60
18.5	2.34	58.5	13.24	A	14.5	2.41	68.0	20.22
24.0	3.50	56.0	14.57	A	19.0	3.69	66.0	22.90
27.5	4.45	54.5	15.76	A+B	25.0	6.62	64.0	30.28
33.0	4.37	45.0	10.64	B	29.0	6.03	57.0	21.19
38.5	4.40	36.0	7.35	B	35.5	5.75	46.5	13.44
42.5	4.50	30.0	5.68	B	41.0	5.43	37.0	8.75
55.0	4.86	12.0	1.89	B	50.0	5.60	24.0	4.80
				B	61.0	5.92	9.0	1.56
58.0	5.04	8.5	1.32	B+C	63.0	6.12	7.0	1.21
59.0	4.71	4.5	0.64	C	64.0	5.83	4.0	0.65
60.0	4.66	2.5	0.35	C	65.0	5.74	2.0	0.32
60.6	4.48	0.0	——	C	65.8	5.60	0.0	——

a. Molalities calculated by M. Salomon.
b. Solid phases: A = $C_9H_7N{\cdot}HNO_3$; B = $Gd(NO_3)_3{\cdot}2C_9H_7N{\cdot}HNO_3{\cdot}2H_2O$
C = $Gd(NO_3)_3{\cdot}5H_2O$

AUXILIARY INFORMATION

METHOD/APPARATUS/PROCEDURE:

The solubility was studied by the method of isothermal sections (1) by measuring the refractive indices of saturated solutions along directed sections of the phase diagram. Equilibrium was checked by repeated measurements of the refractive index as a function of time. The results were used to graph the relation between the refractive indices and the composition of the components for each of the sections studied. The graphs were used to find the inflection or break points corresponding to the composition of the saturated solutions.

The hydrated double salt is congruently soluble over the experimental temp range. It was isolated and analysed for Gd and NO_3, and the results differ from theoretical by 0.4 - 0.5%. The compound melts at 78-79°C, has a density of 1.66 g cm^{-3} at 20°C (measured pycnometrically in benzene), and has a solubility of 79.5 mass % at 20°C and 83.3 mass % at 40°C (5.08 mol kg^{-1} at 20°C and 6.53 mol kg^{-1} at 40°C, compilers). X-ray diffraction studies are discussed in the source publication.

SOURCE AND PURITY OF MATERIALS:

Gadolinium nitrate was twice recrystallized. Analysis for water gave 21.59 mass % which corresponds to the pentahydrate.

Quinoline nitrate, $C_9H_7N{\cdot}HNO_3$, was prepared by neutralization of quinoline with c.p. grade nitric acid followed by evaporation to crystallization. The solid was recrystallized and dried to constant mass in a desiccator over anhydrous $CaCl_2$. The composition of the solid was checked analytically.

Double distilled water was used.

ESTIMATED ERROR:

Soly: precision ± 1% at best (compilers).

Temp: precision probably ± 0.2 K (compilers).

REFERENCES:

1. Zhuravlev, E.F.; Sheveleva, A.D. *Zh. Neorg. Khim.* 1960, *5*, 2630.

COMPONENTS:

(1) Gadolinium nitrate; $Gd(NO_3)_3$; [10168-81-7]

(2) 8-Methylquinoline nitrate; $C_{10}H_{10}N_2O_3$; [60491-92-1]

(3) Water ; H_2O ; [7732-]8-5]

ORIGINAL MEASUREMENTS:

Starikova, L.I.; Zhuravlev, E.F. *Zh. Neorg. Khim.* 1980, *25*, 1723-5; *Russ. J. Inorg. Chem. Engl. Transl.* 1980, *25*, 959-60.

VARIABLES:

Composition at 20°C and 40°C

PREPARED BY:

T. Mioduski and S. Siekierski

EXPERIMENTAL VALUES:

20°C Isotherm[a]					40°C Isotherm[a]			
$Gd(NO_3)_3$		$8\text{-}CH_3C_9H_6N{\cdot}HNO_3$			$Gd(NO_3)_3$		$8\text{-}CH_3C_9H_6N{\cdot}HNO_3$	
mass %	mol kg^{-1}	mass %	mol kg^{-1}	solid phase[b]	mass %	mol kg^{-1}	mass %	mol kg^{-1}
0.0	——	49.5	4.75	A	0.0	——	65.5	9.21
8.5	0.50	41.5	4.03	A	6.0	0.51	59.5	8.36
19.0	1.24	36.5	3.98	A	13.5	1.23	54.5	8.26
30.0	2.33	32.5	4.20	A	23.0	2.44	49.5	8.73
				A	28.5	3.53	48.0	9.91
36.0	3.23	31.5	4.70	A+B	30.5	3.95	47.0	10.13
36.5	3.22	30.5	4.48	B	34.0	3.81	40.0	7.46
40.0	3.19	23.5	3.12	B	38.5	3.74	31.5	5.09
45.0	3.20	14.0	1.66	B	42.0	3.71	25.0	3.67
48.5	3.18	7.0	0.76	B	48.0	3.78	15.0	1.97
54.0	3.65	3.0	0.34	B	51.0	3.67	8.5	1.02
				B	53.0	3.81	6.5	0.78
				B	58.0	4.28	2.5	0.31
59.0	4.30	1.0	0.12	B+C	63.5	5.29	1.5	0.21
60.6	4.48	0.0	——	C	65.8	5.60	0.0	——

a. Molalities calculated by M. Salomon.

b. Solid phases: A = $8\text{-}CH_3C_9H_6N{\cdot}HNO_3$; B = $Gd(NO_3)_3{\cdot}2[8\text{-}CH_3C_9H_6N{\cdot}HNO_3]$

C = $Gd(NO_3)_3{\cdot}5H_2O$

AUXILIARY INFORMATION

METHOD/APPARATUS/PROCEDURE:

The solubility was studied by the method of isothermal sections (1) by measuring the refractive indices of saturated solutions along directed sections of the phase diagram. Equilibrium was checked by repeated measurements of the refractive index as a function of time. The results were used to graph the relation between the refractive indices and the composition of the components for each of the sections studied. The graphs were used to find the invlection or break points corresponding to the composition of the saturated solutions.

COMMENTS AND/OR ADDITIONAL DATA:

The double salt is incongruently soluble at 20°C, and congruently soluble at 40°C. This salt was separated from solution at 40°C and analysed for Gd and NO_3. Its solubility at 40°C is 74.5 mass % (3.87 mol kg^{-1}, compilers). Thermal studies shows that the double salt fuses at 185°C accompanied by violent decomposition.

SOURCE AND PURITY OF MATERIALS:

C.p. grade $Gd(NO_3)_3{\cdot}5H_2O$ was recrystallized prior to use.

The amine nitrate was prepared as described previously (2).

ESTIMATED ERROR:

Soly: precision ± 1% at best (compilers).

Temp: precision probably ± 0.2 K (compilers).

REFERENCES:

1. Zhuravlev, E.F.; Sheveleva, A.D. *Zh. Neorg. Khim.* 1960, *5*, 2630.

2. Zhuravlev, E.F.; Starikova, L.M. *Zh. Neorg. Khim.* 1975, *20*, 1406.

COMPONENTS:

(1) Gadolinium nitrate; $Gd(NO_3)_3$; [10168-81-7]

(2) 8-Hydroxyquinoline nitrate (8-quinolinol nitrate); $C_9H_8N_2O_4$; [60491-93-2]

(3) Water ; H_2O ; [7732-18-5]

ORIGINAL MEASUREMENTS:

Starikova, L.I.; Zhuravlev, E.F. *Zh. Neorg. Khim.* 1980, *25*, 1723-5; *Russ. J. Inorg. Chem. Engl. Transl.* 1980, *25*, 959-60.

VARIABLES:

Composition at 20°C and 40°C

PREPARED BY:

T. Mioduski and S. Siekierski

EXPERIMENTAL VALUES:

20°C Isotherm[a]					40°C Isotherm[a]			
$Gd(NO_3)_3$		$8\text{-}HOC_9H_6N{\cdot}HNO_3$			$Gd(NO_3)_3$		$8\text{-}HOC_9H_6N{\cdot}HNO_3$	
mass %	mol kg^{-1}	mass %	mol kg^{-1}	solid phase[b]	mass %	mol kg^{-1}	mass %	mol kg^{-1}
0.0	——	23.0	1.43	A	0.0	——	47.5	4.35
13.5	0.49	7.0	0.42	A	11.0	0.53	29.0	2.32
29.0	1.24	3.0	0.21	A	27.5	1.24	8.0	0.60
44.5	2.38	1.0	0.09	A	44.0	2.40	2.5	0.22
				A+B	63.0	5.17	1.5	0.20
60.6	4.48	0.0	——	B	63.5	5.21	1.0	0.14
				B	65.8	5.60	0.0	——

a. Molalities calculated by M. Salomon.
b. Solid phases: A = $8\text{-}HOC_9H_6N{\cdot}HNO_3$; B = $Gd(NO_3)_3{\cdot}5H_2O$

AUXILIARY INFORMATION

METHOD/APPARATUS/PROCEDURE:

The solubility was studied by the method of isothermal sections (1) by measuring the refractive indices of saturated solutions along directed sections of the phase diagram. Equilibrium was checked by repeated measurements of the refractive index as a function of time. The results were used to graph the relation between the refractive indices and the composition of the components for each of the sections studied. The graphs were used to find the inflection or break points corresponding to the composition of the saturated solutions.

SOURCE AND PURITY OF MATERIALS:

C.p. grade $Gd(NO_3)_3{\cdot}5H_2O$ was recrystallized prior to use.

The amine nitrate was prepared as described previously (2).

ESTIMATED ERROR:

Soly: precision ± 1% at best (compilers).

Temp: precision probably ± 0.2 K (compilers).

REFERENCES:

1. Zhuravlev, E.F.; Sheveleva, A.D. *Zh. Neorg. Khim.* 1960, *5*, 2630.

2. Zhuravlev, E.F.; Starikova, L.M. *Zh. Neorg. Khim.* 1975, *20*, 1406.

COMPONENTS:

(1) Gadolinium nitrate; $Gd(NO_3)_3$; [10168-81-7]

(2) Piperidine nitrate; $C_5H_{12}N_2O_3$; [6091-45-8]

(3) Water ; H_2O ; [7732-18-5]

ORIGINAL MEASUREMENTS:

Starikova, L.I.; Zhuravlev, E.F. *Zh. Neorg. Khim.* 1975, *20*, 1676-8; *Russ. J. Inorg. Chem. Engl. Transl.* 1975, *20*, 939-41.

VARIABLES:

Composition at 20°C and 40°C

PREPARED BY:

T. Mioduski and S. Siekierski

EXPERIMENTAL VALUES:

20°C Isotherm[a]					40°C Isotherm[a]			
$Gd(NO_3)_3$		$C_5H_{10}NH\cdot HNO_3$		solid	$Gd(NO_3)_3$		$C_5H_{10}NH\cdot HNO_3$	
mass %	mol kg^{-1}	mass %	mol kg^{-1}	phase[b]	mass %	mol kg^{-1}	mass %	mol kg^{-1}
0	———	88.0	49.50	A	0.0	———	90.0	60.74
2.3	0.60	86.5	52.13	A	2.0	0.61	88.5	62.88
4.8	1.25	84.0	50.62	A	4.5	1.38	86.0	61.10
8.5	2.36	81.0	52.07	A	7.5	2.43	83.5	62.62
11.3	3.58	79.5	58.32	A	10.0	3.64	82.0	69.18
16.3	6.60	76.5	71.71	A+B	17.3	10.72	78.0	112.0
17.0	6.43	75.3	66.00	B	21.0	9.87	72.8	79.25
21.3	6.08	68.5	45.33	B	25.0	9.10	67.0	56.53
27.5	6.41	60.0	32.40	B	30.0	9.71	61.0	45.75
32.0	6.91	54.5	27.25	B	33.5	10.27	57.0	40.50
35.5	7.39	50.5	24.35	B	36.0	10.18	53.7	35.19
42.5	8.54	43.0	20.02	B	41.3	12.40	49.0	34.09
48.0	9.99	38.0	18.32	B+C	52.0	20.75	40.7	37.63
51.3	6.73	26.5	8.06	C	53.5	11.13	32.5	15.67
54.0	5.16	15.5	3.43	C	57.0	7.55	21.0	6.44
57.5	4.69	6.8	1.29	C	60.3	6.34	12.0	2.92
60.6	4.48	0.0	———	C	63.0	5.74	5.0	1.05
				C	65.8	5.60	0.0	———

a. Molalities calculated by M. Salomon.

b. Solid phases: A = $C_5H_{10}NH\cdot HNO_3$; B = $Gd(NO_3)_3\cdot 4C_5H_{10}NH\cdot HNO_3$

C = $Gd(NO_3)_3\cdot 5H_2O$

AUXILIARY INFORMATION

METHOD/APPARATUS/PROCEDURE:

The solubility was studied by the method of isothermal sections (1) by measuring the refractive indices of saturated solutions along directed sections of the phase diagram. Equilibrium was checked by repeated measurements of the refractive index as a function of time. The results were used to graph the relation between the refractive indices and the composition of the components for each of the sections studied. The graphs were used to find the inflection or break points corresponding to the composition of the saturated solutions.

The composition of the double salt was checked by chemical analysis: Gd was detd by the oxalate method, and nitrogen detd by pptn with nitron. The melting point of the double salt is 102°C, and its density detd pycnometrically in benezene is 1.64 g cm^{-3}. The solubility of the double salt was given as 86.5 mass % at 20°C and 90.0 mass % at 40°C (6.85 mol kg^{-1} at 20°C and 9.62 mol kg^{-1} at 40°C, compilers).

SOURCE AND PURITY OF MATERIALS:

C.p. grade $Gd(NO_3)_3\cdot 5H_2O$ was recrystallized.

Piperidine nitrate, $C_5H_{10}NH\cdot HNO_3$, was prepared by neutralization of piperidine with nitric acid and evaporation to crystallization. The salt was recrystallized and dried in a desiccator to constant mass.

The composition of the compounds were checked analytically.

ESTIMATED ERROR:

Soly: precision ± 1% at best (compilers).

Temp: precision probably ± 0.2 K (compilers).

REFERENCES:

1. Zhuravlev, E.F.; Sheveleva, A.D. *Zh. Neorg. Khim.* 1960, *5*, 2630.

COMPONENTS:

(1) Gadolinium nitrate ; $Gd(NO_3)_3$; [10168-81-7]

(2) Piperazine dinitrate; $C_4H_{12}N_4O_6$; [10308-78-8]

(3) Water ; H_2O ; [7732-18-5]

ORIGINAL MEASUREMENTS:

Zhuravlev, E.F.; Starikova, L.I. *Zh. Neorg. Khim.* 1975, *20*, 1406-9; *Russ. J. Inorg. Chem. Engl. Transl.* 1975, *20*, 790-2.

VARIABLES:

Composition at 20°C and 40°C

PREPARED BY:

T. Mioduski and S. Siekierski

EXPERIMENTAL VALUES:

20°C Isotherm[a]

$Gd(NO_3)_3$ mass %	$Gd(NO_3)_3$ mol kg^{-1}	$(CH_2CH_2)_2(NH)_2\cdot 2HNO_3$ mass %	$(CH_2CH_2)_2(NH)_2\cdot 2HNO_3$ mol kg^{-1}
0	———	24.0	1.49
12.5	0.52	17.0	1.14
26.5	1.27	12.5	0.97
41.0	2.39	9.0	0.85
51.0	3.62	8.0	0.92
56.5	4.48	6.8	0.87
57.2	4.47	5.5	0.69
59.5	4.56	2.5	0.31
60.6	4.48	0.0	———

40°C Isotherm[a]

solid phase[b]	$Gd(NO_3)_3$ mass %	$Gd(NO_3)_3$ mol kg^{-1}	$(CH_2CH_2)_2(NH)_2\cdot 2HNO_3$ mass %	$(CH_2CH_2)_2(NH)_2\cdot 2HNO_3$ mol kg^{-1}
A	0.0	———	37.0	2.77
A	10.5	0.51	29.5	2.32
A	22.5	1.24	24.5	2.18
A	37.0	2.42	18.5	1.96
A	46.2	3.61	16.5	2.08
A+B	57.5	5.68	13.0	2.08
B	60.5	5.65	8.3	1.25
B	63.5	5.69	4.0	0.58
B	65.0	5.74	2.0	0.29
B	65.8	5.60	0.0	———

a. Molalities calculated by M. Salomon.

b. Solid phases: A = $HNO_3\cdot NH_2(CH_2CH_2)_2NH_2\cdot HNO_3$; B = $Gd(NO_3)_3\cdot 5H_2O$

AUXILIARY INFORMATION

METHOD/APPARATUS/PROCEDURE:

The solubility was studied by the method of isothermal sections (1) by measuring the refractive indices of saturated solutions along directed sections of the phase diagram. Equilibrium was checked by repeated measurements of the refractive index as a function of time. The results were used to graph the relation between the refractive indices and the composition of the components for each of the sections studied. The graphs were used to find the inflection or break points corresponding to the composition of the saturated solutions.

SOURCE AND PURITY OF MATERIALS:

Gadolinium nitrate was twice recrystallized. Analysis for water gave 20.59 mass % which corresponds to the pentahydrate.

Piperazine dinitrate, $C_4H_8(NH_2)\cdot 2HNO_3$, was prepared by neutralization of piperazine with c.p. grade HNO_3 followed by evaporation to crystallization. The solid was recrystallized and dried in a desiccator over anhyrous $CaCl_2$ to constant mass. The composition of the solid was checked analytically.

Doubly distilled water was used.

ESTIMATED ERROR:

Soly: precision ± 1% at best (compilers).

Temp: precision probably ± 0.2 K (compilers).

REFERENCES:

1. Zhuravlev, E.F.; Sheveleva, A.D. *Zh. Neorg. Khim.* 1960, *5*, 2630.

COMPONENTS:

(1) Gadolinium nitrate; $Gd(NO_3)_3$; [10168-81-7]

(2) Urea; CH_4N_2O; [57-13-6]

(3) Water; H_2O; [7732-18-5]

ORIGINAL MEASUREMENTS:

Khudaibergenova, N.; Sulaimankulov, K.

Zh. Neorg. Khim. 1980, *25*, 2254-56; *Russ. J. Inorg. Chem. Engl. Transl.* 1980, *25*, 1249-50.

VARIABLES:

Composition at 30°C

PREPARED BY:

T. Mioduski and S. Siekierski

EXPERIMENTAL VALUES: Composition of saturated solutions at 30°C[a]

$Gd(NO_3)_3$		$CO(NH_2)_2$		
mass %	mol kg^{-1}	mass %	mol kg^{-1}	nature of the solid phase
0	——	57.5	22.528	$CO(NH_2)_2$
23.6	3.257	55.29	43.612	"
38.93	15.708	53.85	124.19	"
40.13	20.332	54.12	156.72	"
40.17	18.256	53.42	138.77	$CO(NH_2)_2 + Gd(NO_3)_3 \cdot 4CO(NH_2)_2$
40.26	12.721	50.52	91.248	$Gd(NO_3)_3 \cdot 4CO(NH_2)_2$
48.48	8.135	34.16	32.765	"
50.33	8.584	32.59	31.772	"
51.81	7.420	27.85	22.799	"
53.49	7.484	25.69	20.546	"
59.58	8.353	19.64	15.738	"
63.51	10.429	18.75	17.599	"
60.6	6.939	13.96	9.137	$Gd(NO_3)_3 \cdot 2CO(NH_2)_2$
61.21	7.001	13.32	8.708	"
61.65	6.682	11.47	7.105	"
63.20	5.808	5.10	2.679	"
60.89	5.081	4.20	2.003	$Gd(NO_3)_3 \cdot 5H_2O$
58.94	4.716	4.65	2.127	"
60.25	4.416	0	——	

a. Molalities calculated by M. Salomon.

AUXILIARY INFORMATION

METHOD/APPARATUS/PROCEDURE:

The isothermal method was used. Equilibrium was stated to be reached within 7 hours. The nitrogen of the urea was determined by the Kjeldahl method. Gadolinium was determined by titration with Trilon (disodium salt of ethylenediamine tetraacetic acid) using Xylenol orange indicator. The liquid phase was filtered off with a No. 3 Schott filter.

COMMENTS AND/OR ADDITIONAL DATA:

$Gd(NO_3)_3 \cdot 4CO(NH_2)_2$ is congruently soluble. It was isolated and analysed: $Gd(NO_3)_3$ content was 58.25 mass %, and urea = 41.75 mass %.

$Gd(NO_3)_3 \cdot 2CO(NH_2)_2$ is incongruently soluble.

SOURCE AND PURITY OF MATERIALS:

Nothing specified except that for the crystallization branch of the initial salt, the solid phase contained 79.22 mass % $Gd(NO_3)_3$. This corresponds to the pentahydrate, $Gd(NO_3)_3 \cdot 5H_2O$, but details on methods of analysis not specified.

ESTIMATED ERROR:

Nothing specified.

REFERENCES:

COMPONENTS:

(1) Gadolinium nitrate; $Gd(NO_3)_3$; [10168-81-7]

(2) Urea nitrate; $CH_5N_3O_4$; [124-47-0]

(3) Water ; H_2O ; [7732-18-5]

ORIGINAL MEASUREMENTS:

Starikova, L.I.; Zhuravlev, E.F. *Zh. Neorg. Khim.* 1975, *20*, 2294-6; *Russ. J. Inorg. Chem. Engl. Transl.* 1975, *20*, 1274-5.

VARIABLES:

Composition at 20°C and 40°C

PREPARED BY:

T. Mioduski and S. Siekierski

EXPERIMENTAL VALUES:

20°C Isotherm[a]					40°C Isotherm			
$Gd(NO_3)_3$		$CO(NH_2)_2 \cdot HNO_3$			$Gd(NO_3)_3$		$CO(NH_2)_2 \cdot HNO_3$	
mass %	mol kg^{-1}	mass %	mol kg^{-1}	solid phase[b]	mass %	mol kg^{-1}	mass %	mol kg^{-1}
0.0	——	16.5	1.61	A	0.0	——	24.5	2.64
13.5	0.52	11.5	1.25	A	12.0	0.50	18.0	2.09
27.5	1.21	6.5	0.80	A	26.5	1.25	11.5	1.51
43.5	2.39	3.5	0.54	A	41.5	2.32	6.5	1.02
54.5	3.57	1.0	0.18	A	52.5	3.48	3.5	0.65
60.2	4.49	0.7	0.15	A+B	65.0	5.65	1.5	0.36
60.6	4.48	0.0	——	B	65.5	5.70	1.0	0.24
				B	65.8	5.60	0.0	——

a. Molalities calculated by M. Salomon.
b. Solid phases: A = $CO(NH_2)_2 \cdot HNO_3$; B = $Gd(NO_3)_3 \cdot 5H_2O$

AUXILIARY INFORMATION

METHOD/APPARATUS/PROCEDURE:

The solubility was studied by the method of isothermal sections (1) by measuring the refractive indices of saturated solutions along directed sections of the phase diagram. Equilibrium was checked by repeated measurements of the refractive index as a function of time. The results were used to graph the relation between the refractive indices and the composition of the components for each of the sections studied. The graphs were used to find the inflection or break points corresponding to the composition of the saturated solutions.

SOURCE AND PURITY OF MATERIALS:

Nothing specified for $Gd(NO_3)_3 \cdot 5H_2O$.

The amine nitrate was prepared by neutralization of "pure" grade urea with nitric acid. The neutralized solution was evaporated to crystallization. The solid was recrystallized, dried and analysed to check the composition of the salt. The results corresponded to the anhydrous salt.

ESTIMATED ERROR:

Soly: precision ± 1% at best (compilers).

Temp: precision probably ± 0.2 K (compilers).

REFERENCES:

1. Zhuravlev, E.F.; Sheveleva, A.D. *Zh. Neorg. Khim.* 1960, *5*, 2630.

COMPONENTS:

(1) Magnesium gadolinium nitrate; $3Mg(NO_3)_2 \cdot 2Gd(NO_3)_3$; [84682-75-7]

(2) Nitric acid; HNO_3; [7697-37-2]

(3) Water ; H_2O ; [7732-18-5]

ORIGINAL MEASUREMENTS:

Jantsch, G. *Z. Anorg. Chem.* 1912, *76*, 303-23.

VARIABLES:

One temperature: 16°C

PREPARED BY:

Mark Salomon

EXPERIMENTAL VALUES:

Soly of the double salt in HNO_3 sln of density d_4^{16} = 1.325 g cm^{-3}

aliquot volume	Gd_2O_3	soly $3Mg(NO_3)_2 \cdot 2Gd(NO_3)_3$	
cm^3	g	mol dm^{-3}	
1.4638	0.1187		
1.4638	0.1200	0.2252[a]	0.2249[b]

a. Average value calculated by the author.

b. Average value calculated by the compiler using 1977 IUPAC recommended atomic masses.

ADDITIONAL DATA:

The melting point of the tetracosahydrate is 77.5°C, and the density at 0°C is 2.163 g cm^{-3}.

AUXILIARY INFORMATION

METHOD/APPARATUS/PROCEDURE:

Isothermal method used. The soly was studied in HNO_3 sln of density 1.325 g cm^{-3} at 16°C because the author did not have sufficient quantity of the rare earth to study the soly of the salt in pure water. Pulverized salt and HNO_3 sln were placed in glass-stoppered tubes and thermostated at 16°C for 24 h with periodic shaking. The solutions was then allowed to settle for 2 h, and a pipet maintained at 16°C was used to withdraw aliquots for analysis. Two analyses were performed.

Solutions were analysed by adding 2-3 g NH_4Cl and 10% NH_3 sln followed by boiling to ppt the hydroxide. The ppt was filtered, dissolved in HNO_3, reprecipitated as the hydroxide, and ignited to the oxide. Mg in the filtrate was "determined by the usual method" (no details were given).

An attempt to determine the waters of hydration by dehydration was not successful because the temperature required (120°C or higher) resulted in decomposition of the salt with the formation of basic salts. Presumably the waters of hydration were found by difference.

SOURCE AND PURITY OF MATERIALS:

"Pure" gadolinium oxide was dissolved in dil HNO_3 and $Mg(NO_3)_2$ added to give a mole ratio of Gd/Mg = 2/3. The sln was evapd and a small crystal of $Bi_2Mg_3(NO_3)_{12}$ added, and the mixt cooled to ppt the tetracosahydrate. The double nitrate was recrystd before use.

ESTIMATED ERROR:

Soly: reproducibility about ± 1-5% (compiler).

Temp: nothing specified.

REFERENCES:

COMPONENTS:

(1) Gadolinium cobalt nitrate; $2Gd(NO_3)_3 \cdot 3Co(NO_3)_2$; [84682-77-9]

(2) Nitric acid; HNO_3; [7697-37-2]

(3) Water ; H_2O ; [7732-18-5]

ORIGINAL MEASUREMENTS:

Jantsch, G. *Z. Anorg. Chem.* 1912, *76*, 303-23.

VARIABLES:

One temperature: 16°C

PREPARED BY:

Mark Salomon

EXPERIMENTAL VALUES:

Soly of the double salt in HNO_3 sln of density $d_4^{16} = 1.325$ g cm^{-3}.

aliquot volume	Gd_2O_3	Soly $2Gd(NO_3)_3 \cdot 3Co(NO_3)_2$
cm^3	g	mol dm^{-3}
1.4638	0.1435	
1.4638	0.1438	0.2706[a] , 0.2704[b]

a. Average value calculated by the author.

b. Average value calculated by the compiler using 1977 IUPAC recommended atomic masses.

ADDITIONAL DATA:

The melting point of the tetracosahydrate is 63.2°C, and the density at 0°C is 2.315 g cm^{-3}.

AUXILIARY INFORMATION

METHOD/APPARATUS/PROCEDURE:

Isothermal method used. The soly was studied in HNO_3 sln of density 1.325 g cm^{-3} at 16°C because the author did not have sufficient quantity of the rare earth to study the soly of the salt in pure water. Pulverized salt and HNO_3 sln were placed in glass-stoppered tubes and thermostated at 16°C for 24 h with periodic shaking. The solution was then allowed to settle for 2 h, and a pipet maintained at 16°C was used to withdraw aliquots for analysis. Two analyses were performed.

Solutions were analysed by adding 2-3 g NH_4Cl and 10% NH_3 sln followed by boiling to ppt the hydroxide. The ppt was filtered, dissolved in HNO_3, reprecipitated as the hydroxide, and ignited to the oxide. Co in the filtrate was "determined by the usual method" (no details were given).

An attempt to determine the waters of hydration by dehydration was not successful because the temperature required (120°C or higher) resulted in decomposition of the salt with the formation of basic salts. Presumably the waters of hydration were found by difference.

SOURCE AND PURITY OF MATERIALS:

"Pure" gadolinium oxide was dissolved in dil HNO_3 and $Co(NO_3)_2$ added to give a mole ratio of Gd/Co = 2/3. The sln was evapd and a small crystal of $Bi_2Mg_3(NO_3)_{12}$ added, and the mixt cooled to ppt the tetracosahydrate. The double nitrate was recrystd before use. The double salt was analysed gravimetrically for Gd_2O_3 and metallic Co. 0.5100 g samples of the tetracosahydrate yielded 0.1105 g oxide (21.67 mass %), and 0.0548 g Co (10.75 mass %). Theor values are 21.74 mass % Gd_2O_3 and 10.60 mass % Co (compiler).

ESTIMATED ERROR:

Soly: reproducibility about ± 1-5% (compiler).

Temp: nothing specified.

COMPONENTS:

(1) Gadolinium nickel nitrate; $2Gd(NO_3)_3 \cdot 3Ni(NO_3)_2$; [84682-79-1]

(2) Nitric acid; HNO_3; [7697-37-2]

(3) Water ; H_2O ; [7732-18-5]

ORIGINAL MEASUREMENTS:

Jantsch, C. *Z. Anorg. Chem.* 1912, *76*, 303-23.

VARIABLES:

One temperature: 16°C

PREPARED BY:

Mark Salomon

EXPERIMENTAL VALUES:

Soly of the double salt in HNO_3 sln of density d_4^{16} = 1.325 6 cm^{-3}.

aliquot volume	Gd_2O_3	Soly $2Gd(NO_3)_3 \cdot 3Ni(NO_3)_2$	
cm^3	g	$mol\ dm^{-3}$	
1.4638	0.1274		
1.4638	0.1280	0.2405[a]	0.2407[b]

a. Average value calculated by the author.

b. Average value calculated by the compiler using 1977 IUPAC recommended atomic masses.

ADDITIONAL DATA:

The melting point of the tetracosahydrate is 72.5°C, and the density at 0°C is 2.356 g cm^{-3}.

AUXILIARY INFORMATION

METHOD/APPARATUS/PROCEDURE:

Isothermal method used. The soly was studied in HNO_3 sln of density 1.325 g cm^{-3} at 16°C because the author did not have sufficient quantity of the rare earth to study the soly of the salt in pure water. Pulverized salt and HNO_3 sln were placed in glass-stoppered tubes and thermostated at 16°C for 24 h with periodic shaking. The solution was then allowed to settle for 2 h, and a pipet maintained at 16°C was used to withdraw aliquots for analysis. Two analyses were performed.

Solutions were analysed by adding 2-3 g NH_4Cl and 10% NH_3 sln followed by boiling to ppt the hydroxide. The ppt was filtered, dissolved in HNO_3, reprecipitated as the hydroxide, and ignited to the oxide. Ni in the filtrate was "determined by the usual method" (no details were given).

An attempt to determine the waters of hydration by dehydration was not successful because the temperature required (120°C or higher) resulted in decomposition of the salt with the formation of basic salts. Presumably the waters of hydration were found by difference.

SOURCE AND PURITY OF MATERIALS:

"Pure" gadolinium oxide was dissolved in dil HNO_3 and $Ni(NO_3)_2$ added to give a mole ratio of Gd/Ni = 2/3. The sln was evapd and a small crystal of $Bi_2Mg_3(NO_3)_{12}$, added, and the mixt cooled to ppt the tetracosahydrate. The double nitrate was recrystd before use.

ESTIMATED ERROR:

Soly: reproducibility about ± 1-5% (compiler).

Temp: nothing specified.

REFERENCES:

COMPONENTS:

(1) Gadolinium zinc nitrate; $2Gd(NO_3)_3 \cdot 3Zn(NO_3)_2$; [84682-81-5]

(2) Nitric acid; HNO_3; [7697-37-2]

(3) Water ; H_2O ; [7732-18-5]

ORIGINAL MEASUREMENTS:

Jantsch, G. *Z. Anorg. Chem.* 1912, *76*, 303-23.

VARIABLES:

One temperature: 16°C

PREPARED BY:

Mark Salomon

EXPERIMENTAL VALUES:

Soly of the double salt in HNO_3 sln of density d_4^{16} = 1.325 g cm^{-3}.

aliquot volume	Gd_2O_3	soly $2Gd(NO_3)_3 \cdot 3Zn(NO_3)_2$ [a]
cm^3	g	mol dm^{-3}
1.4638	0.1484	
1.4638	0.1490	0.2801

a. Average value calculated by the author.

ADDITIONAL DATA:

The melting point of the tetracosahydrate is 56.5°C, and the density at 0°C is 2.351 g cm^{-3}.

AUXILIARY INFORMATION

METHOD/APPARATUS/PROCEDURE:

Isothermal method used. The soly was studied in HNO_3 sln of density 1.325 g cm^{-3} at 16°C because the author did not have sufficient quantity of the rare earth to study the soly of the salt in pure water. Pulverized salt and HNO_3 sln were placed in glass-stoppered tubes and thermostated at 16°C for 24 h with periodic shaking. The solution was then allowed to settle for 2 h, and a pipet maintained at 16°C was used to withdraw aliquots for analysis. Two analyses were performed.

Solutions were analysed by adding 2-3 g NH_4Cl and 10% NH_3 sln followed by boiling to ppt the hydroxide. The ppt was filtered, dissolved in HNO_3, reprecipitated as the hydroxide, and ignited to the oxide. Zn in the filtrate was "determined by the usual method" (no details were given).

An attempt to determine the waters of hydration by dehydration was not successful because the temperature required (120°C or higher) resulted in decomposition of the salt with the formation of basic salts. Presumably the waters of hydration were found by difference.

SOURCE AND PURITY OF MATERIALS:

"Pure" gadolinium oxide was dissolved in dil HNO_3 and $Zn(NO_3)_2$ added to give a mole ratio of Gd/Zn = 2/3. The sln was evapd and a small crystal of $Bi_2Mg_3(NO_3)_{12}$ added, and the mixt cooled to ppt the tetracosahydrate. The double nitrate was recrystd before use. The double salt was analysed gravimetrically for Gd_2O_3 and NO. A 0.4520 g sample of the tetracosahydrate yielded 0.0972 g Gd_2O_3 (21.50 mass %): Theor value is 21.49 mass % (compiler). Additional analysis gave 21.77 mass % NO: Theor value is 21.34 mass % (compiler).

ESTIMATED ERROR:

Soly: reproducibility about ± 1-5% (compiler).

Temp: nothing specified.

COMPONENTS:	EVALUATOR:
(1) Terbium nitrate; $Tb(NO_3)_3$; [10043-27-3] (2) Water ; H_2O ; [7732-18-5]	S. Siekierski, T. Mioduski Institute for Nuclear Research Warsaw, Poland and M. Salomon U.S. Army ET & DL Ft. Monmouth, NJ January 1983

CRITICAL EVALUATION:

THE BINARY SYSTEM

INTRODUCTION

Data for the solubility in the binary terbium nitrate-water system have been reported for 298.15 K in six publications (1-6). According to Afanas'ev et al. (6), a handbook (7) states that at 293.2 K, the solubility of terbium nitrate in water is 60.6 mass % (4.46 mol kg^{-1}, evaluators): however the evaluators were unable to determine the original source of this solubility value.

In the binary system, the stable solid phase at 298.15 K is $Tb(NO_3)_3.6H_2O$, [13451-19-9]. (1-6), and presumably this is also the solid phase at 293.2 K. According to Mellor (8), Urbain and co-workers reported that the solid hexahydrate melts congruently at 362.5 K, but Ivanov-Emin et al. (9) reported a congruent melting point of 343 K for the tetrahydrate, [37131-76-3]. At least one of these melting points must be in error.

RECOMMENDED AND TENTATIVE SOLUBILITIES

For 298.15 K, the average value of the solubility in the binary system based on the results reported in (1,2,4,5) is 4.538 mol kg^{-1}. The standard deviation for this average value is σ = 0.004, and at the 95% level of confidence and a Student's t = 3.182, the uncertainty in this average value is ± 0.006 mol kg^{-1}. This solubility of 4.538 ± 0.006 mol kg^{-1} is designated as a *recommended* value. Both results from (3) and (6) are rejected because their residual errors far exceed the standard deviation of the mean.

The solubility result of 4.46 mol kg^{-1} at 293.2 K is cautiously designated as a *tentative* value. It should be noted that there is no information available on how this solubility was determined: i.e. experimental details, purity of materials, and solid phase analysis information are not available.

REFERENCES

1. Spedding, F.H.; Shiers, L.E.; Rard, J.A. *J. Chem. Eng. Data* 1975, *20*, 88.
2. Rard, J.A.; Spedding, F.H. *J. Phys. Chem.* 1975, *79*, 257.
3. Spedding, F.H.; Derer, J.L.; Mohs, M.A.; Rard, J.A. *J. Chem. Eng. Data* 1976, *21*, 474.
4. Spedding, F.H.; Shiers, L.E.; Brown. M.A.; Baker, J.L.; Guitierrez, L.; McDowell, L.S.; Habenschuss, A. *J. Phys. Chem.* 1975, *79*, 1078.
5. Rard, J.A.; Shiers, L.E.; Heiser, D.J.; Spedding, F.G. *J. Chem. Eng. Data* 1977, *22*, 337.
6. Afanas'ev, Yu.A.; Azhipa, L.T.; Shakanova, N.A. *Zh. Neorg. Khim.* 1977, *22*, 3331.
7. Kirgintsev, A.N.; Trushnikova, L.N.; Lavrent'eva, V.G. *Rastvorimost' Neorganicheskikh Veshchestv v Vode. Spravochnik. (Solubility of Inorganic Substances in Water: A Handbook).* Izd. Khimiya. Leningrad. 1972. p. 165.
8. Mellor, J.W. *A Comprehensive Treatise on Inorganic and Theoretical Chemistry.* Longmans, Green & Co. London. 1940. Vol. V.
9. Ivanov-Emin, B.N.; Odinets, Z.K.; Khaime Del'Pino; Zaitsev, B.E.; Ezhov, A.I. *Zh. Neorg. Khim.* 1975, *20*, 2952.

COMPONENTS:

(1) Terbium nitrate; $Tb(NO_3)_3$; [10043-27-3]

(2) Water ; H_2O ; [7732-18-5]

ORIGINAL MEASUREMENTS:

1. Spedding, F.H.; Shiers, L.E.; Rard, J.A. *J. Chem. Eng. Data* 1975, *20*, 88-93.
2. Rard, J.A.; Spedding, F.H. *J. Phys. Chem.* 1975, *79*, 257-62.
3. Spedding, F.H.; Derer, J.L.; Mohs, M.A.; Rard, J.A. *J. Chem. Eng. Data* 1976, *21*, 474-88.

VARIABLES:

One temperature: 25.00°C

PREPARED BY:

T. Mioduski, S. Siekierski, and M. Salomon

EXPERIMENTAL VALUES:

The solubility of $Tb(NO_3)_3$ in water at 25.00°C has been reported by Spedding and co-workers in three publications. Source paper [3] reports the solubility to be 4.738 mol kg^{-1}, but the preferred value is given in source papers [1] and [2] as 4.5395 mol kg^{-1}.

COMMENTS AND/OR ADDITIONAL DATA:

Source paper [1] reports the relative viscosity, η_R, of a saturated solution to be 20.152. Taking the viscosity of water at 25°C to equal 0.008903 poise, the viscosity of a saturated $Tb(NO_3)_3$ solution at 25°C is 0.17941 poise (compilers calculation).

Supplementary data available in the microfilm edition to *J. Phys. Chem.* 1975, *79* enabled the compilers to provide the following additional data.

The density of the saturated solutions was calculated by the compilers from the smoothing equation, and at 25°C the value is 1.89977 kg m^{-3}. Using this density, the solubility in volume units is (based on the preferred value of 4.5395 mol kg^{-1})

$$c_{satd} = 3.4030 \text{ mol dm}^{-3}$$

Source paper [2] reports the electrolytic conductivity of the saturated solution to be (corrected for the electrolytic conductivity of the solvent) κ = 0.027639 S cm^{-1}.

The molar conductivity of the saturated solution is calculated from $1000\kappa/3c_{satd}$ and is

$$\Lambda(\tfrac{1}{3}\, Tb(NO_3)_3) = 2.707 \text{ S cm}^2 \text{ mol}^{-1} \text{ (compilers' calculation)}$$

It should be noted that in the supplementary data in the microfilm edition of source paper [2], this latter quantity is given as 2.741 S cm^2 mol^{-1}.

AUXILIARY INFORMATION

METHOD/APPARATUS/PROCEDURE:

Isothermal method used. Solutions were prepared as described in (1) and (2). The concentration of the saturated solution was determined by both EDTA (1) and sulfate (2) methods which is said to be reliable to 0.1% or better. In the sulfate analysis, the salt was first decomposed with HCl followed by evaporation to dryness before sulfuric acid additions were made. This eliminated the possibility of nitrate ion coprecipitation.

SOURCE AND PURITY OF MATERIALS:

$Tb(NO_3)_3 \cdot 6H_2O$ was prepd by addn of HNO_3 to the oxide. The oxide was purified by an ion exchange method, and the upper limit for the impurities Ca, Fe, Si and adjacent rare earths was given as 0.15%.

In source paper [3] the salt was analysed for water of hydration and found to be within ± 0.016 water molecules of the hexahydrate.

Water was distilled from an alkaline permanganate solution.

ESTIMATED ERROR:

Soly: duplicate analyses agreed to at least ± 0.1%.

Temp: Not specified, but probably accurate to at least ± 0.01 K as in (3) (compilers).

REFERENCES:

1. Spedding, F.G.; Cullen. P.F.; Habenschuss, A. *J. Phys. Chem.* 1974, *78*, 1106.
2. Spedding, F.H.; Pikal, M.J.; Ayers, B.O. *J. Phys. Chem.* 1966, *70*, 2440.
3. Spedding. F.H.; et. al. *J. Chem. Eng. Data* 1975, *20*, 72.

COMPONENTS:

(1) Terbium nitrate; $Tb(NO_3)_3$; [10043-27-3]

(2) Water ; H_2O ; [7732-18-5]

ORIGINAL MEASUREMENTS:

[1] Spedding, F.H.; Shiers, L.E.; Brown, M.A.; Baker, J.L.; Guitierrez, L.; McDowell, L.S.; Habenschuss, A. *J. Phys. Chem.* 1975, *79*, 1078-96.

[2] Rard, J.A.; Shiers, L.E.; Heiser, D.J.; Spedding, F.H. *J. Chem. Eng. Data* 1977, *22*, 337-37.

VARIABLES:

One temperature: 25.00°C

PREPARED BY:

T. Mioduski, S. Siekierski, and M. Salomon

EXPERIMENTAL VALUES:

Source paper [1] reports the solubility of $Tb(NO_3)_3$ as 4.5395 mol kg^{-1}.

Source paper [2] reports the solubility of $Tb(NO_3)_3$ as 4.5320 mol kg^{-1}.

The solid phase in both studies is the hexahydrate, $Tb(NO_3)_3 \cdot 6H_2O$

AUXILIARY INFORMATION

METHOD/APPARATUS/PROCEDURE:

[1] Standard isothermal method used.
[2] Isothermal isopiestic method used in which equilibration carried out by allowing a less than satd $Tb(NO_3)_3$ sln to reach thermodynamic equilibrium through the vapor phase with a reference sln (KCl, $CaCl_2$). The soly was thus detd without sepn of the sln and solid phases. The solutions were adjusted to their equivalence pH values to insure a ratio of three nitrates to each Tb. Duplicate samples of the nitrate and reference slns were used and equil was approached from higher and lower concns (about 4d).

In both [1] and [2] the satd slns were analysed by EDTA titrn and gravimetric sulfate analysis. The methods are stated to be accurate to 0.1% or better. In the sulfate analysis, the salt was first decomposed with HCl followed by evaporation to dryness before sulfuric acid additions were made. This eliminated the possibility of nitrate ion coprecipitation.

SOURCE AND PURITY OF MATERIALS:

1. The nitrate was obtained from the oxide (purified by ion-exchange methods at the Ames Laboratory) and reagent grade HNO_3. The purity of the oxide was greater than 99.85 mass % with Ca, Fe, Si and adjacent lanthanides being the only significant impurities.

2. Conductivity water distd. from an alkaline $KMnO_4$ solution was used.

ESTIMATED ERROR:

Soly: duplicate analyses agreed to at least ± 0.1%.

Temp: precision ± 0.01 K.

REFERENCES:

COMPONENTS:

(1) Terbium nitrate ; $Tb(NO_3)_3$; [10043-27-3]

(2) Nitric acid; HNO_3; [7697-37-2]

(3) Water ; H_2O ; [7732-18-5]

ORIGINAL MEASUREMENTS:

Afanas'ev, Yu. A.; Azhipa, L.T.; Shakhanova, N.A. *Zh. Neorg. Khim.* 1977, *22*, 3331-4; *Russ. J. Inorg. Chem. Engl. Transl.* 1977, *22*, 1818-20.

VARIABLES:

Nitric acid concentration at 25°C

PREPARED BY:

T. Mioduski and S. Siekierski

EXPERIMENTAL VALUES:

Solubility of terbium nitrate in nitric acid solutions at 25°C[a]

$Tb(NO_3)_3$		HNO_3		
mass %	mol kg^{-1}	mass %	mol kg^{-1}	nature of the solid phase
57.90	3.987	0	———	$Tb(NO_3)_3 \cdot 6H_2O$
51.05	3.477	6.39	2.383	"
46.98	3.261	11.25	4.274	"
42.70	2.969	15.60	5.937	"
37.59	2.638	21.10	8.106	"
34.70	2.415	23.65	9.011	"
30.47	2.082	27.11	10.142	"
25.04	1.746	33.39	12.747	"
23.01	1.618	35.75	13.757	"
21.55	1.516	37.23	14.334	"
20.83	1.531	39.74	15.995	$Tb(NO_3)_3 \cdot 6H_2O + Tb(NO_3)_3 \cdot 4H_2O$
17.39	1.252	42.33	16.677	$Tb(NO_3)_3 \cdot 4H_2O$
16.31	1.270	46.46	19.804	"
16.63	1.331	47.16	20.669	$Tb(NO_3)_3 \cdot 4H_2O + Tb(NO_3)_3 \cdot 2H_2O$
13.83	0.962	44.49	21.412	$Tb(NO_3)_3 \cdot 2H_2O$
9.16	0.712	53.53	22.769	"
7.77	0.610	55.33	23.796	"
7.81	0.643	56.97	25.670	"
7.88	0.693	59.14	28.458	"
5.46	0.515	63.79	32.921	"

continued.........

AUXILIARY INFORMATION

METHOD/APPARATUS/PROCEDURE:

The isothermal method was used. The compositions of the saturated solutions were varied by addition of 100% nitric adic to the saturated solutions or by adding the salt to the acid solution. Equilibrium was reached in 3-4 hours.

Terbium in the liquid and solid phases was determined by complexometric titration with Xylenol Orange indicator, and the nitric acid concentrations were determined by titration with sodium hydroxide using Methyl Red indicator. The composition of the solid phases was determined by Schreinemakers' method of residues.

SOURCE AND PURITY OF MATERIALS:

C.p. grade terbium nitrate was used.

Nitric acid was concentrated to 100% by the Brauer method.

ESTIMATED ERROR:

Soly: precision about ± 0.3% (compilers).

Temp: precision ± 0.1 K.

REFERENCES:

COMPONENTS:	ORIGINAL MEASUREMENTS:
(1) Terbium nitrate; $Tb(NO_3)_3$; [10043-27-3]	Afanas'ev, Yu. A.; Azhipa, L.T.; Shakhanova, N.A. *Zh. Neorg. Khim.* 1977, *22*, 3331-4;
(2) Nitric acid; HNO_3; [7697-37-2]	*Russ. J. Inorg. Chem. Engl. Transl.* 1977, *22*, 1818-20.
(3) Water ; H_2O ; [7732-18-5]	

EXPERIMENTAL VALUES: continued ....

Solubility of terbium nitrate in nitric acid solutions at 25°C[a]

$Tb(NO_3)_3$ mass %	$Tb(NO_3)_3$ mol kg^{-1}	HNO_3 mass %	HNO_3 mol kg^{-1}	nature of the solid phase
6.24	0.624	64.75	35.421	$Tb(NO_3)_3\cdot 2H_2O + Tb(NO_3)_3\cdot H_2O$
5.41	0.598	68.37	41.381	$Tb(NO_3)_3\cdot H_2O$
5.25	0.621	70.25	45.504	"
5.69	0.712	71.14	48.726	"
6.38	0.977	74.69	62.616	"
8.04	1.542	76.84	80.650	"
9.51	2.468	79.32	112.69	$Tb(NO_3)_3\cdot H_2O + Tb(NO_3)_3\cdot nHNO_3$
8.00	2.203	81.47	122.78	$Tb(NO_3)_3\cdot nHNO_3$
6.17	2.145	85.49	162.67	"
3.69	1.917	90.73	258.04	"
2.87	2.667	94.01	478.18	"

a. Molalities calculated by M. Salomon

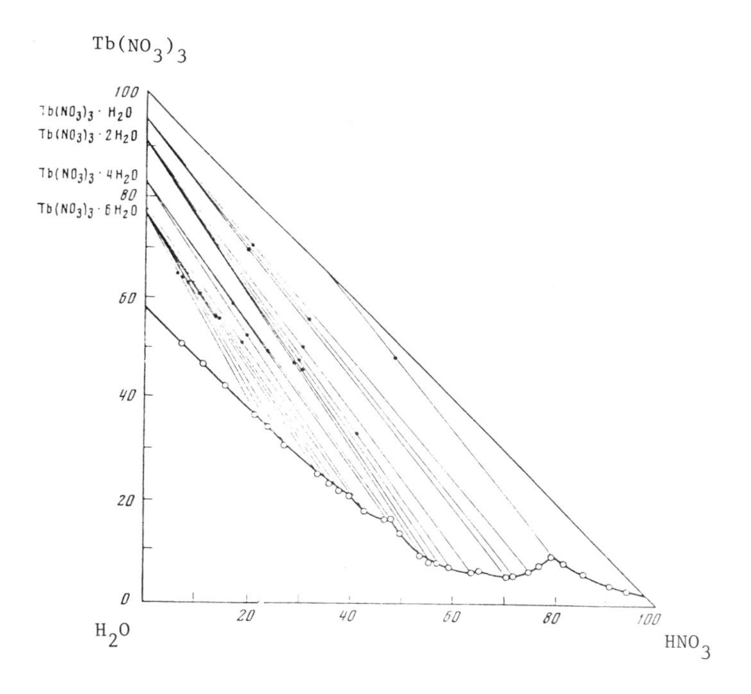

Solubility diagram of the $Tb(NO_3)_3$-HNO_3-H_2O system at 298.2 K. All compositions given in mass % units.

COMPONENTS:	EVALUATOR:
(1) Dysprosium nitrate; $Dy(NO_3)_3$; [10143-38-1] (2) Water ; H_2O ; [7732-18-5]	S. Siekierski, T. Mioduski Institute for Nuclear Research Warsaw, Poland and M. Salomon U.S. Army ET & DL Ft. Monmouth, NJ November 1982

CRITICAL EVALUATION:

THE BINARY SYSTEM

INTRODUCTION

The solubility of dysprosium nitrate in water has been reported in eight publications (1-8). Moret (1) studied the solubility as a function of temperature from 273.15 K to 323.15 K, Zhuravlev et al. (6-8) reported results for 293.2 K and 313.2 K, and Spedding et al. (2-4) and Afanas'ev et al. (5) reported results for 298.15 K. All authors specify the solid phase in equilibrium with the saturated solutions to be $Dy(NO_3)_3.6H_2O$, [35725-30-5]. For the binary system, there are no solubility data available for lower hydrate systems. However in ternary systems containing nitric acid (5) and urea (9), lower hydrates exist.

The congruent melting points of the 1:6, 1:5, and 1:4 hydrates have been reported in (10-12). According to Ivanov-Emin et al. (12), the hexahydrate melts at 358.2 K, but below the evaluators calculate a melting point of 338.9 K for this hydrate. Urbain and Jantsch (10) report a congruent melting point of 361.8 K for the pentahydrate, but Dvornikova et al. (13) state that this hydrate melts incongruently at 363 K. Wendlandt and Sewell (11) report a congruent melting point of 363 ± 1 K for the tetrahydrate.

EVALUATION PROCEDURE

The solubility data in the binary hexahydrate system given in the compilations were fitted by least squares to the general solubility equation based on the treatments in (14, 15) and in the INTRODUCTION to this volume:

$$Y = \ln(m/m_o) - nM_2(m - m_o) = a + b/(T/K) + c\ \ln(T/K) \qquad [1]$$

All terms in eq. [1] have been previously defined. The solubility data were all assigned equal weights of unity, and data were rejected for which the residual error, Δm, was greater than twice the standard error of estimate, i.e. greater than $\pm 2\sigma_m$. In previous critical evaluations we have assigned weights based upon experimental precision: e.g. the precision for the data reported by Spedding et al., by Moret and by Brunisholz et al. is around ± 0.1 to 0.2% compared to the experimental precision of around ± 1% for those results based on the method of isothermal sections. Equal weights were however assigned to all data (except those rejected values) because they appear to be of equal accuracy.

RECOMMENDED SOLUBILITIES IN THE HEXAHYDRATE SYSTEM

In fitting the solubility data to eq. [1], two values were rejected. The 298.15 K value of 4.539 mol kg^{-1} (4) and the 298.2 K value of 4.873 mol kg^{-1} (5) were rejected because their residual errors ($m_{cald} - m_{obsd}$) far exceeded $\pm 2\sigma_m$ (see Table 1). The residual errors for Moret's results at 293.15 K and 313.15 K are larger than most other Δm values (+0.05 and +0.08 mol kg^{-1}, respectively) which suggests greater inaccuracy then usually found for Moret's results. The accuracy in the results from (6-8) appear to be equal or slighty better than these two values from Moret's work.

The results of fitting the data from (1-3, 6-8) to the smoothing equation are given in Table 1. The smoothed solubilities calculated from eq. [1] are designated as *recommended*, and Table 2 gives the smoothed values at 5 K intervals up to the congruent melting point. At the 95% level of confidence and a Student's t = 2.447, the combined calculated and experimental precisions lead to an overall uncertainty of ± 0.04 mol kg^{-1} in the recommended solubilities.

The experimental melting point of 358.2 K for the hexahydrate is so divergent from the value of 338.9 K calculated from the solubility data that we believe the former to be in error due to the probable presence of lower hydrates. The melting points for the penta- and tetrahydrates are in the range of 361 K to 363 K as discussed above.

MULTICOMPONENT SYSTEMS

Since only one publication exists for the solubility of $Dy(NO_3)_3$ in aqueous solutions of nitric acid and other saturating components, these data cannot be critically evaluated. Some phase diagrams are given below, and were selected for reproduction here based mainly on clarity. The compositions of all components in the phase diagrams below are given in mass % units.

COMPONENTS:

(1) Dysprosium nitrate; $Dy(NO_3)_3$; [10143-38-1]

(2) Water ; H_2O ; [7732-18-5]

EVALUATOR:
S. Siekierski, T. Mioduski
Institute for Nuclear Research
Warsaw, Poland
and
M. Salomon
U.S. Army ET & DL
Ft. Monmouth, NJ
November 1982

CRITICAL EVALUATION:

Table 1. Derived parameters for the smoothing equation[a]

parameter	value for the hexahydrate system
a	-38.728
b	1443
c	5.9480
σ_a	0.004
σ_b	1.3
σ_c	0.0007
σ_Y	0.004
σ_m	0.05
ΔH_{sln}/kJ mol^{-1}	-47.8
ΔC_p/J K^{-1} mol^{-1}	197.8
congruent melting point/K	338.9
concn at the congr m.p./mol kg^{-1}	9.251

[a] The quantities σ_a, σ_b, σ_c are standard deviations, and σ_Y and σ_m are standard errors of estimate for Y in eq.[1], and the molality.

Table 2. Recommended solubilities calculated from the smoothing equation[a]

T/K	soly/mol kg^{-1}	T/K	soly/mol kg^{-1}
273.15	4.093	313.15	5.430
278.15	4.190	318.15	5.737
283.15	4.303	323.15	6.108
288.15	4.433	328.15	6.577
293.15	4.581	333.15	7.226
298.15	4.751	338.15	8.490
303.15	4.946	338.87[b]	9.251

[a] Although 4 significant figures are given for the recommended solubilities, it should be noted that the overall uncertainty in these values is ± 0.04 mol kg^{-1}.

[b] Congruent melting point.

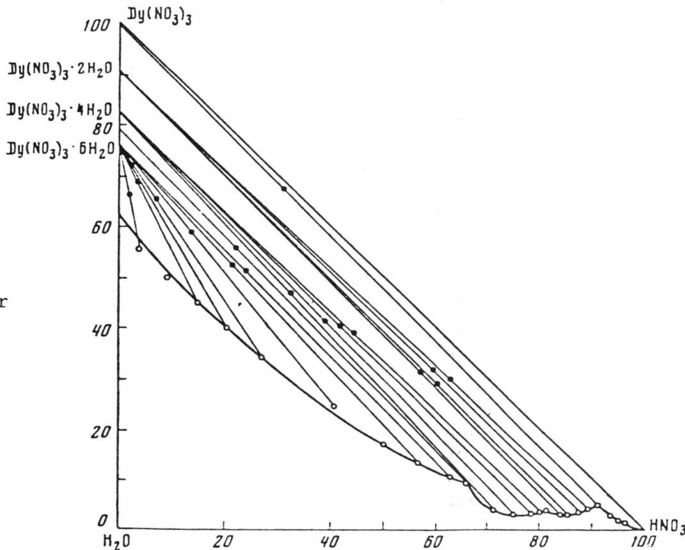

Figure 1. Solubility diagram for the $Dy(NO_3)_3$-HNO_3-H_2O system at 298.2 K: ref. (5).

COMPONENTS:

(1) Dysprosium nitrate; $Dy(NO_3)_3$; [10143-38-1]

(2) Water ; H_2O; [7732-18-5]

EVALUATOR:
S. Siekierski, T. Mioduski
Institute for Nuclear Research
Warsaw, Poland
and
M. Salomon
U.S. Army ET & DL
Ft. Monmouth, NJ
November 1982

CRITICAL EVALUATION:

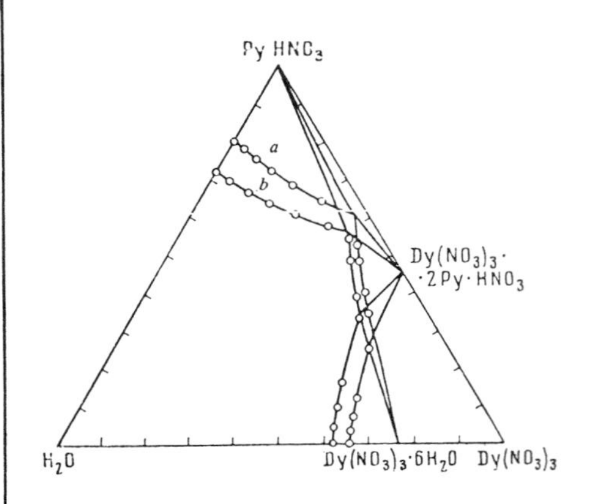

Fig. 2. Solubility isotherms in the $Dy(NO_3)_3$-pyridine nitrate-water system at 293.2 K (a), and 313.2 K (b): ref (7).

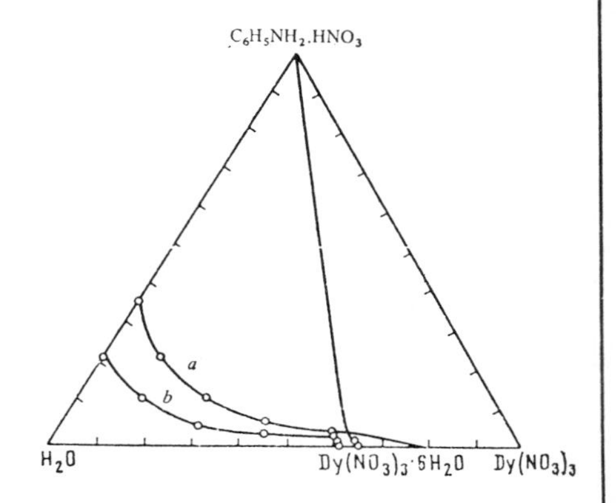

Fig. 3. Solubility isotherms in the $Dy(NO_3)_3$-aniline nitrate-water system at 293.2 K (a), and 313.2 K (b): ref. (7).

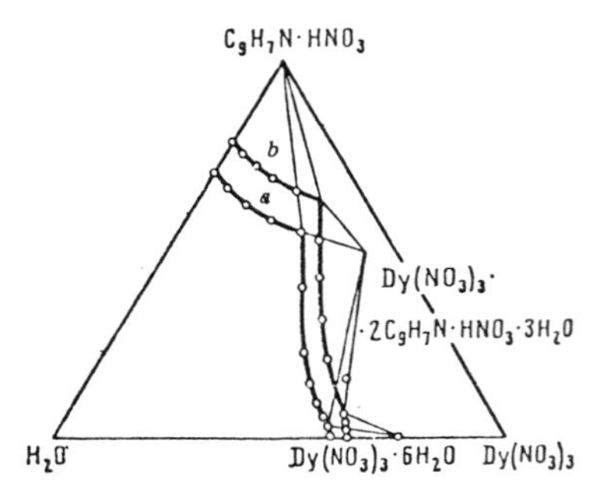

Fig. 4. Solubility isotherms in the $Dy(NO_3)_3$-quinoline nitrate-water system at 293.2 K (a), and 313.2 (b): ref. (8).

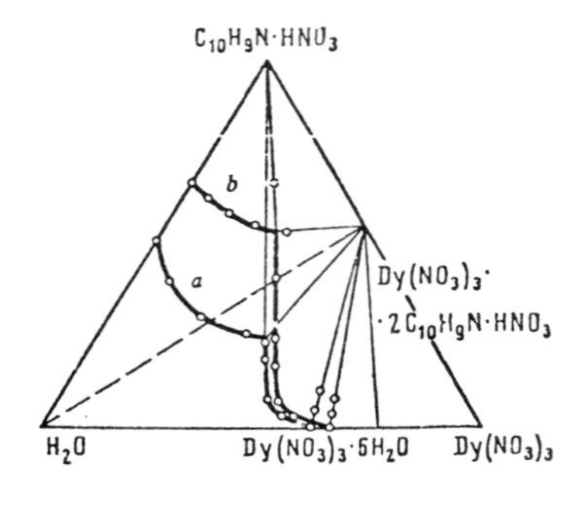

Fig. 5. Solubility isotherm in the $Dy(NO_3)_3$-8-methylquinoline nitrate-water system at 293.2 K (a), and 313.2 K (b): ref. (8).

Fig. 6. Solubility isotherms in the $Dy(NO_3)_3$-8-hydroxyquinoline nitrate-water system at 293.2 K (a), and 313.2 (b): ref. (8).

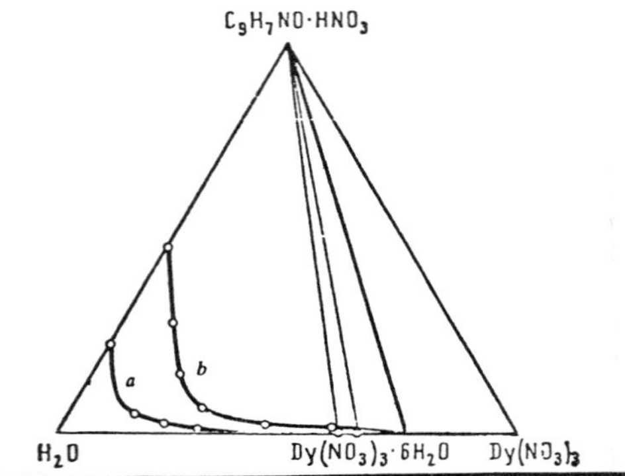

COMPONENTS:

(1) Dysprosium nitrate; $Dy(NO_3)_3$; [10143-38-1]

(2) Water ; H_2O ; [7732-18-5]

EVALUATOR:

S. Siekierski, T. Mioduski
Institute for Nuclear Research
Warsaw, Poland
and
M. Salomon
U.S. Army ET & DL
Ft. Monmouth, NJ
November 1982

CRITICAL EVALUATION:

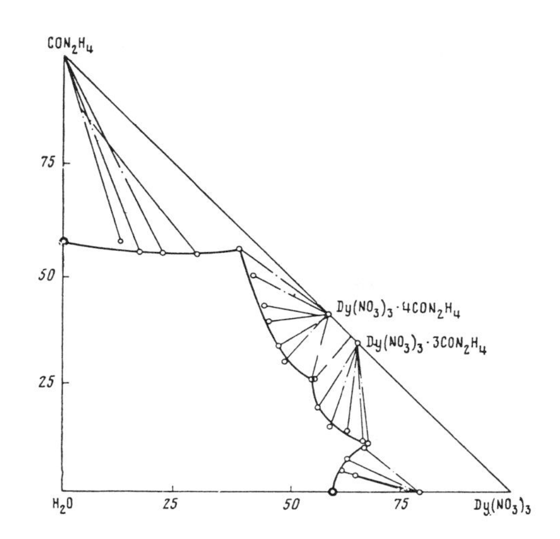

Fig. 7. Solubility isotherm for the $Dy(NO_3)_3$-urea-water system at 303.2 K: ref. (9).

REFERENCES

1. Moret, R. *Thèse*. l'Université de Lausanne. 1963.
2. Spedding, F.H.; Shiers, L.E.; Rard, J.A. *J. Chem. Eng. Data* 1975, *20*, 88.
3. Rard, J.A.; Spedding, F.H. *J. Phys. Chem.* 1975, *79*, 257.
4. Spedding. F.H.; Derer, J.L. Mohs, M.A.; Rard, J.A. *J. Chem. Eng. Data* 1976, *21*, 474.
5. Afanas'ev, Yu.A.; Azhipa, L.T.; Tarasova, N.V. *Zh. Neorg. Khim.* 1979, *24*, 550.
6. Zhuravlev, E.F.; Starikova, L.I.; Katamonov, V.L. *Zh. Neorg. Khim.* 1975, *20*, 1113.
7. Katamanov, V.L.; Zhuravlev, E.F. *Zh. Neorg. Khim.* 1976, *21*, 805.
8. Katamonov, V.L.; Zhuravlev, E.F. *Zh. Neorg. Khim.* 1976, *21*, 1610.
9. Khudaibergenova, N.; Sulaimankulov, K.; Abykeev, K. *Zh. Neorg. Khim.* 1980, *25*, 2251.
10. Urbain, G.; Jantsch, G. *Compt. Rend.* 1906, *142*, 785.
11. Wendlandt, W.W.; Sewell, R.G. *Texas J. Sci.* 1961, *13*, 231.
12. Ivanov-Emin, B.N.; Odinets, Z.K.; Khaime Del'Pino ; Zaitsev, B.E.; Ezhov, A.I. *Zh. Neorg. Khim.* 1975, *20*, 2952.
13. Dvornikova, L.M.; Sevost'yanov, V.P.; Ambrozhii, M.N. *Zh. Neorg. Khim.* 1969, *14*, 2325.
14. Williamson, A.T. *Trans. Faraday Soc.* 1944, *40*, 421.
15. Counioux, J.-J.; Tenu, R. *J. Chim. Phys.* 1981, *78*, 816 and 823.

COMPONENTS:

(1) Dysprosium nitrate; $Dy(NO_3)_3$; [10143-38-1]

(2) Water ; H_2O ; [7732-18-5]

ORIGINAL MEASUREMENTS:

Moret, R. *Thèse*. l'Université de Lausanne. 1963.

VARIABLES:

Temperature: range 0 to 50°C

PREPARED BY:

T. Mioduski and S. Siekierski

EXPERIMENTAL VALUES:

t/°C	$Dy(NO_3)_3$ mass %	mol kg^{-1} [a]	solid phase
0	58.80[b]	4.095	$Dy(NO_3)_3 \cdot 6H_2O$
10	60.03	4.309	"
20	61.73[c]	4.628	"
40	65.76	5.510	"
50	67.88[d]	6.064	"

a. Calculated by the compilers.

b. 1355 moles of H_2O per 100 moles of $Dy(NO_3)_3$ (1).

c. 1199 moles of H_2O per 100 moles of $Dy(NO_3)_3$ (1).

d. 915.3 moles of H_2O per 100 moles of $Dy(NO_3)_3$ (1).

AUXILIARY INFORMATION

METHOD/APPARATUS/PROCEDURE:

Isothermal method employed (1). Dy was determined by complexometric titration using Xylenol Orange indicator in presence of a small amount of urotropine buffer. Water was determined by difference.

COMMENTS AND/OR ADDITIONAL DATA:

Data for 0°C, 20°C, and 50°C were converted to units of moles water per 100 moles of salt and reported in reference (1).

SOURCE AND PURITY OF MATERIALS:

$Dy(NO_3)_3.6H_2O$ was prepared from Dy_2O_3 of purity higher than 99.7% (obtained by the ion exchange chromatography). No additional details available.

ESTIMATED ERROR:

Soly: precision about ± 0.2 % (compilers).

Temp: precision at least ± 0.05K (compilers).

REFERENCES:

1. Brunisholz, G.; Quinche, J.P.; Kalo, A.M. *Helv. Chim. Acta* 1964, *47*, 14.

COMPONENTS:

(1) Dysprosium nitrate; $Dy(NO_3)_3$; [10143-38-1]

(2) Water ; H_2O ; [7732-18-5]

ORIGINAL MEASUREMENTS:

1. Spedding, F.H.; Shiers, L.E.; Rard, J.A. *J. Chem. Eng. Data* 1975, *20*, 88-93.
2. Rard, J.A.; Spedding, F.H. *J. Phys. Chem.* 1975, *79*, 257-62.
3. Spedding, F.H.; Derer, J.L.; Mohs, M.A.; Rard, J.A. *J. Chem. Eng. Data* 1976, *21*, 474-88.

VARIABLES:

One temperature: 25.00°C

PREPARED BY:

T. Mioduski, S. Siekierski, and M. Salomon

EXPERIMENTAL VALUES:

The solubility of $Dy(NO_3)_3$ in water at 25.00°C has been reported by Spedding and co-workers in three publications. Source paper [3] reports the solubility to be 4.539 mol kg^{-1}, but the preferred value is given in source papers [1, 2] as 4.539 mol kg^{-1}.

COMMENTS AND/OR ADDITIONAL DATA:

Source paper [1] reports the relative viscosity, η_R, of a saturated solution to be 24.008. Taking the viscosity of water at 25°C to equal 0.008903 poise, the viscosity of a saturated $Dy(NO_3)_3$ solution at 25°C is 0.2137 poise (compilers calculation).

Supplementary data available in the microfilm edition to *J. Phys. Chem.* 1975, *79* enabled the compilers to provide the following additional data.

The density of the saturated solution was calculated by the compilers from the smoothing equation, and at 25°C the value is 1.94030 kg m^{-3}. Using this density, the solubility in volume units is (based on the preferred value of 4.7382 mol kg^{-1}).

$$c_{satd} = 3.4675 \text{ mol dm}^{-3}$$

Source paper [2] reports the electrolytic conductivity of the saturated solution to be (corrected for the electrolytic conductivity of the solvent) κ = 0.024354 S cm^{-1}.

The molar conductivity of the saturated solution is calculated from $1000\kappa/3c_{satd}$ and is

$$\Lambda(\tfrac{1}{3}Dy(NO_3)_3) = 2.341 \text{ S cm}^2 \text{ mol}^{-1}$$

AUXILIARY INFORMATION

METHOD/APPARATUS/PROCEDURE:

Isothermal method used. Solutions were prepared as described in (1) and (2). The concentration of the saturated solution was determined by both EDTA (1) and sulfate (2) methods which is said to be reliable to 0.1% or better. In the sulfate analysis, the salt was first decomposed with HCl followed by evaporation to dryness before sulfuric acid additions were made. This eliminated the possibility of nitrate ion coprecipitation.

SOURCE AND PURITY OF MATERIALS:

$Dy(NO_3)_3 \cdot 6H_2O$ prepd by addn of HNO_3 to the oxide. The oxide was purified by an ion exchange method, and the upper limit for the impurities Ca, Fe, Si and adjacent rare earths was given as 0.15%.

In source paper [3] the salt was analysed for water of dydration and found to be within ± 0.016 water molecules of the hexahydrate.

Water was distilled from an alkaline permanganate solution.

ESTIMATED ERROR:

Soly: duplicate analyses agreed to at least ± 0.1%.

Temp: Not specified, but probably accurate to at least ± 0.01 K as in (3) (compilers).

REFERENCES:

1. Spedding, F.G.; Cullen, P.F.; Habenschuss, A. *J. Phys. Chem.* 1974, *78*, 1106.
2. Spedding, F.H.; Pikal, M.J.; Ayers, B.O. *J. Phys. Chem.* 1966, *70*, 2440.
3. Spedding, F.H.; et. al. *J. Chem. Eng. Data* 1975, *20*, 72.

COMPONENTS:	ORIGINAL MEASUREMENTS:
(1) Dysprosium nitrate; $Dy(NO_3)_3$; [10143-38-1] (2) Nitric acid; HNO_3; [7697-37-2] (3) Water ; H_2O ; (7732-18-5]	Afanas'ev, Yu. A.: Azhipa, L.T.: Tarasova, N.V. *Zh. Neorg. Khim.* 1979, *24*, 550-2; *Russ. J. Inorg. Chem. Engl. Transl.* 1979, *24*, 307-8.
VARIABLES:	PREPARED BY:
Composition at 25°C	T. Mioduski and S. Siekierski

EXPERIMENTAL VALUES:

$Dy(NO_3)_3$[a] mass %	$Dy(NO_3)_3$[a] mol kg^{-1}	HNO_3[a] mass %	HNO_3[a] mol kg^{-1}	nature of the solid phase
62.94	4.873	0	———	$Dy(NO_3)_3\cdot 6H_2O$
54.5	3.797	4.31	1.661	"
49.16	3.383	9.14	3.478	"
44.83	3.198	14.95	5.899	"
40.45	2.944	20.13	8.104	"
35.10	2.661	27.05	11.342	"
29.73	2.392	34.61	15.402	"
25.46	2.147	40.51	18.892	"
17.27	1.523	50.2	24.490	"
13.25	1.270	56.81	30.112	"
9.95	1.063	63.20	37.355	"
8.81	1.002	65.97	41.512	$Dy(NO_3)_3\cdot 6H_2O + Dy(NO_3)_3\cdot 4H_2O$
3.74	0.427	71.12	44.895	$Dy(NO_3)_3\cdot 4H_2O$
3.12	0.409	75.00	54.398	"
3.15	0.493	78.52	67.981	"
4.87	0.919	79.92	83.387	"
3.98	0.786	81.5	89.076	$Dy(NO_3)_3\cdot 4H_2O + Dy(NO_3)_3\cdot 2H_2O$
2.71	0.583	83.95	99.870	$Dy(NO_3)_3\cdot 2H_2O$
2.35	0.561	85.64	113.16	"
3.60	1.165	87.53	156.60	"
4.51	1.967	88.91	214.43	"
5.00	3.560	90.97	358.23	$Dy(NO_3)_3\cdot 2H_2O + Dy(NO_3)_3$
2.65	1.980	93.51	386.45	$Dy(NO_3)_3$
1.35	1.055	94.98	410.71	"
1.20	1.252	96.05	554.29	"

a. Molalities calculated by M. Salomon

AUXILIARY INFORMATION

METHOD/APPARATUS/PROCEDURE:	SOURCE AND PURITY OF MATERIALS:
Isothermal method used. Equilibrium was established in 3-4 hours. The compositions of the saturated solutions were changed by addition of 100% nitric acid to the saturated solutions or by addition of the salt to the acid. Nitric acid was determined by titration with NaOH using methyl red indicator, and dysprosium was determined by complexometric titration using Xylenol Orange indicator. The composition of the solid phases was determined by chemical analysis and by Schreinemakers' method of residues. Data on the infrared spectra of the hydrates are discussed in the source paper.	C.p. grade $Dy(NO_3)_3\cdot 6H_2O$ was used. Nitric acid was concentrated to 100% by the Brauer method. ESTIMATED ERROR: Soly: precision ± 0.3 % (compilers). Temp: precision ± 0.1 K. REFERENCES:

COMPONENTS:

(1) Dysprosium nitrate; $Dy(NO_3)_3$; [10143-38-1]

(2) Ethylenediamine dinitrate; $C_2H_{10}N_4O_6$; [20829-66-7]

(3) Water ; H_2O ; [7732-18-5]

ORIGINAL MEASUREMENTS:

Zhuravlev, E.F.; Starikova, L.I.; Katamanov, V.L. *Zh. Neorg. Khim.* 1975, *20*, 1113-6; *Russ. J. Inorg. Chem. Engl. Transl.* 1975, *20*, 626-8.

VARIABLES:

Composition at 20°C and 40°C

PREPARED BY:

T. Mioduski and S. Siekierski

EXPERIMENTAL VALUES:

20°C Isotherm[a]					40°C Isotherm[a]			
$Dy(NO_3)_3$		$(CH_2)_2(NH_2)_2\cdot 2HNO_3$		solid	$Dy(NO_3)_3$		$(CH_2)_2(NH_2)_2\cdot 2HNO_3$	
mass %	mol kg^{-1}	mass %	mol kg^{-1}	phase[b]	mass %	mol kg^{-1}	mass %	mol kg^{-1}
0.0	———	43.5	4.137	A	0.0	———	60.0	8.059
10.0	0.512	34.0	3.262	"	7.5	0.531	52.0	6.898
22.7	1.222	24.0	2.419	"	17.5	1.255	42.5	5.709
38.1	2.382	16.0	1.873	"	30.5	2.365	32.5	4.719
52.4	4.284	12.5	1.913	"	45.0	4.304	25.0	4.477
56.2	5.152	12.5	2.146	A+B	56.5	7.540	22.0	5.498
58.4	4.843	7.0	1.087	B	60.7	6.154	11.0	2.088
60.0	4.653	3.0	0.436	"	63.2	5.703	5.0	0.845
61.2	4.526	0.0	———	"	64.2	5.532	2.5	0.403
				"	65.4	5.424	0.0	———

a. Molalities calculated by M. Salomon.
b. Solid phases A = $(CH_2)_2(NH_2)_2\cdot 2H_2O$; B = $Dy(NO_3)_3\cdot 6H_2O$

AUXILIARY INFORMATION

METHOD/APPARATUS/PROCEDURE:

Solubilities were studied by the method of isothermal sections (1) by measuring the refractive indices of saturated solutions along directed sections of the phase diagram. Equilibrium was checked by repeated measurements of the refractive index as a function of time. The results were used to graph the relation between the refractive indices and the composition of the components for each of the sections studied. The graphs were used to find the inflection or break points corresponding to the composition of the saturated solutions.

SOURCE AND PURITY OF MATERIALS:

C.p. grade dysprosium nitrate was twice recrystallized. Analysis for water of crystallization gave 24.02 mass % water which is close to the theoretical value for the hexahydrate (theor is 23.67 mass %, compilers). The amine nitrate was prepared by neutralizing "pure" grade amine with c.p. grade HNO_3 and evaporating the neutralized sln to crystallization. The salt was recrystallized and dried to const weight in a desiccator over anhydr $CaCl_2$. The salt was analysed. Doubly distilled water was used.

ESTIMATED ERROR:

Soly: precision ± 1% at best (compilers).

Temp: precision probably ± 0.2 K (compilers).

REFERENCES:

1. Zhuravlev, E.F.; Sheveleva, A.D. *Zh. Neorg. Khim.* 1960, *5*, 2630.

COMPONENTS:

(1) Dysprosium nitrate; $Dy(NO_3)_3$; [10143-38-1]

(2) 1,6-Hexanediamine dinitrate; (hexamethylenediamine dinitrate); $C_6H_{18}N_4O_6$; [6143-53-9]

(3) Water ; H_2O ; [7732-18-5]

ORIGINAL MEASUREMENTS:

Zhuravlev, E.F.; Starikova, L.I.; Katamanov, V.L. *Zh. Neorg. Khim.* 1975, *20*, 1113-6; *Russ. J. Inorg. Chem. Engl. Transl.* 1975, *20*, 626-8.

VARIABLES:

Composition at 20° and 40°C

PREPARED BY:

T. Mioduski and S. Siekierski

EXPERIMENTAL VALUES:

20°C Isotherm[a]

$Dy(NO_3)_3$ mass %	$Dy(NO_3)_3$ mol kg^{-1}	$(CH_2)_6(NH_2)_2\cdot 2HNO_3$ mass %	$(CH_2)_6(NH_2)_2\cdot 2HNO_3$ mol kg^{-1}	solid phase[b]
0.0	———	79.0	15.530	A
5.5	0.751	73.5	14.449	"
14.0	1.913	65.0	12.778	"
27.5	4.384	54.5	12.499	"
39.5	9.445	48.5	16.685	"
49.3	9.010	35.0	9.203	B
53.0	5.964	21.5	3.481	"
57.0	5.032	10.5	1.334	"
59.0	4.638	4.5	0.509	"
61.2	4.526	0.0	———	"

40°C Isotherm[a]

$Dy(NO_3)_3$ mass %	$Dy(NO_3)_3$ mol kg^{-1}	$(CH_2)_6(NH_2)_2\cdot 2HNO_3$ mass %	$(CH_2)_6(NH_2)_2\cdot 2HNO_3$ mol kg^{-1}
0.0	———	85.5	24.342
4.0	0.765	81.0	22.293
11.0	2.036	73.5	19.576
22.5	4.304	62.5	17.201
34.5	9.428	55.0	21.624
52.5	9.415	31.5	8.127
57.0	6.542	18.0	2.972
61.2	5.701	8.0	1.072
63.5	5.521	3.5	0.438
65.4	4.424	0.0	———

a. Molalities caluclated by M. Salomon.
b. Solid phases: A = $(CH_2)_6(NH_2)_2\cdot 2HNO_3$; B = $Dy(NO_3)_3\cdot 6H_2O$

AUXILIARY INFORMATION

METHOD/APPARATUS/PROCEDURE:

Solubilities were studied by the method of isothermal sections (1) by measuring the refractive indices of saturated solutions along directed sections of the phase diagram. Equilibrium was checked by repeated measurements of the refractive index as a function of time. The results were used to graph the relation between the refractive indices and the composition of the components for each of the sections studied. The graphs were used to find the inflection or break points corresponding to the composition of the saturated solutions.

SOURCE AND PURITY OF MATERIALS:

C.p. grade dysprosium nitrate was twice recrystallized. Analysis for water of crystallization gave 24.02 mass % water which is close to the theoretical value for the hexahydrate (theor is 23.67 mass %, compilers).
The amine nitrate was prepared by neutralizing "pure" grade amine with c.p. grade HNO_3 and evaporating the neutralized sln to crystallization. The salt was recrystallized and dried to const weight in a desiccator over anhydr $CaCl_2$.
Doubly distilled water was used.

ESTIMATED ERROR:

Soly: precision ± 1% at best (compilers).

Temp: precision probably ± 0.02 K (compilers).

REFERENCES:

1. Zhuravlev, E.F.; Sheveleva, A.D. *Zh. Neorg. Khim.* 1960, *5*, 2630.

COMPONENTS:

(1) Dysprosium nitrate; $Dy(NO_3)_3$; [10143-38-1]

(2) Pyridine nitrate; $C_5H_6N_2O_3$; [543-53-3]

(3) Water ; H_2O ; [7732-18-5]

ORIGINAL MEASUREMENTS:

Katamanov, V.L.; Zhuravlev, E.F. *Zh. Neorg. Khim.* 1976, *21*, 805-8; *Russ. J. Inorg. Chem. Engl. Transl.* 1976, *21*, 436-8.

VARIABLES:

Composition at 20°C and 40°C

PREPARED BY:

T. Mioduski and S. Siekierski

EXPERIMENTAL VALUES:

20°C Isotherm[a]					40°C Isotherm[a]			
$Dy(NO_3)_3$		$C_5H_5N{\cdot}HNO_3$			$Dy(NO_3)_3$		$C_5H_5N{\cdot}HNO_3$	
mass %	mol kg^{-1}	mass %	mol kg^{-1}	solid phase[b]	mass %	mol kg^{-1}	mass %	mol kg^{-1}
0.0	———	72.0	18.09	A	0.0	———	80.5	29.05
4.8	0.54	69.6	19.13	"	3.4	0.54	78.5	30.52
10.3	1.30	67.0	20.77	"	7.5	1.30	76.0	32.41
16.3	2.37	64.0	22.86	"	12.5	2.41	72.6	34.29
23.5	4.29	60.8	27.25	"	18.7	4.43	69.2	40.24
32.5	9.61	57.8	41.93	"	27.2	9.40	64.5	54.68
36.6	15.22	56.5	57.62	A+B	35.7	31.04	61.0	130.07
38.2	14.05	54.0	48.71	B	40.5	17.34	52.8	55.45
41.7	11.85	48.2	33.58	"	43.8	14.61	47.6	38.95
47.5	10.10	39.0	20.33	"	48.7	12.15	39.8	24.35
				"	52.4	10.66	33.5	16.72
50.0	9.44	34.8	16.11	B+C	56.4	10.65	28.4	13.15
50.6	8.96	33.2	14.42	C	57.5	9.02	24.2	9.31
55.5	5.63	16.2	4.03	"	61.2	6.50	11.8	3.08
57.5	5.05	9.8	2.11	"	62.8	5.97	7.0	1.63
59.5	4.74	4.5	0.88	"	64.2	5.65	3.2	0.69
61.2	4.53	0.0	———	"	65.4	5.42	0.0	———

a. Molalities calculated by

b. Solid phases: A = $C_5H_5N{\cdot}HNO_3$; B = $Dy(NO_3)_3{\cdot}2[C_5H_5N{\cdot}HNO_3]$

C = $Dy(NO_3)_3{\cdot}6H_2O$

AUXILIARY INFORMATION

METHOD/APPARATUS/PROCEDURE:

Solubilities were studied by the method of isothermal sections (1) by measuring the refractive indices of saturated solutions along directed sections of the phase diagram. Equilibrium was checked by repeated measurements of the refractive index as a function of time. The results were used to graph the relation between the refractive indices and the composition of the components for each of the sections studied. The graphs were used to find the inflection or break points corresponding to the composition of the saturated solutions.

The double nitrate was recryst and analysed for Dy and NO_3; its composition differed by 0.2 - 0.5 % from theor. The soly of this double salt is 87.0 mass % at 20°C and 88.5 mass % at 20°C (10.58 mol kg^{-1} and 12.16 mol kg^{-1}, respectively, compilers). Its density as measured in benzene is 2.23 kg m^{-3}. Thermal analyses data are presented in the source paper.

SOURCE AND PURITY OF MATERIALS:

Dysprosium nitrate hexahydrate was prepd by recrystallizing c.p. grade $Dy(NO_3)_3{\cdot}5H_2O$. Analysis of Dy confirmed the hexahydrate composition.

The amine nitrate was prepared by neutralization of pyridine with nitric acid. The salt was dried to constant mass over anhydrous $CaCl_2$, and its composition checked by analysis for NO_3 by precipitation with nitron.

ESTIMATED ERROR:

Soly: precision ± 1% at best (compilers).

Temp: precision probably ± 0.2 K (compilers).

REFERENCES:

1. Zhuravlev, E.F.; Sheveleva, A.D. *Zh. Neorg. Khim.* 1960, *5*, 2630.

COMPONENTS:

(1) Dysprosium nitrate; $Dy(NO_3)_3$; [10143-38-1]

(2) Aniline nitrate; $C_6H_8N_2O_3$; [542-15-4]

(3) Water ; H_2O ; [7732-18-5]

ORIGINAL MEASUREMENTS:

Katamanov, V.L.; Zhuravlev, E.F. *Zh. Neorg. Khim.* 1976, *21*, 805-8; *Russ. J. Inorg. Chem. Engl. Transl.* 1976, *21*, 436-8.

VARIABLES:

Composition at 20°C and 40°C

PREPARED BY:

T. Mioduski and S. Siekierski

EXPERIMENTAL VALUES:

20°C Isotherm[a]					40°C Isotherm[a]			
$Dy(NO_3)_3$		$C_6H_5NH_2\cdot HNO_3$			$Dy(NO_3)_3$		$C_6H_5NH_2\cdot HNO_3$	
mass %	mol kg^{-1}	mass %	mol kg^{-1}	solid phase[b]	mass %	mol kg^{-1}	mass %	mol kg^{-1}
0.0	———	21.5	1.75	A	0.0	———	36.0	3.60
13.4	0.51	10.5	0.88	"	11.7	0.50	21.5	2.06
28.7	1.23	4.5	0.43	"	26.6	1.23	11.5	1.19
44.0	2.36	2.5	0.30	"	42.5	2.35	5.5	0.68
59.0	4.34	2.0	0.33	"	58.0	4.30	3.3	0.55
59.2	4.38	2.0	0.33	A+B	61.3	4.93	3.0	0.54
60.0	4.44	1.2	0.20	B	62.5	5.08	2.2	0.40
61.2	4.53	0.0	———	"	63.6	5.16	1.0	0.18
				"	65.4	5.42	0.0	———

a. Molalities calculated by M. Salomon.
b. Solid phases: A = $C_6H_5NH_2\cdot HNO_3$; B = $Dy(NO_3)_3\cdot 6H_2O$

AUXILIARY INFORMATION

METHOD/APPARATUS/PROCEDURE:

Solubilities were studied by the method of isothermal sections (1) by measuring the refractive indices of saturated solutions along directed sections of the phase diagram. Equilibrium was checked by repeated measurements of the refractive index as a function of time. The results were used to graph the relation between the refractive indices and the composition of the components for each of the sections studied. The graphs were used to find the inflection or break points corresponding to the composition of the saturated solutions.

SOURCE AND PURITY OF MATERIALS:

Dysprosium nitrate hexahydrate was prepd by recrystallizing c.p. grade $Dy(NO_3)_3\cdot 5H_2O$. Analysis of Dy confirmed the hexahydrate composition.

The amine nitrate was prepared by neutralization of aniline with nitric acid. The salt was dried to constant mass over anhydrous $CaCl_2$, and its composition checked by analysis for NO_3 by precipitation with nitron.

ESTIMATED ERROR:

Soly: precision ± 1% at best (compilers).

Temp: precision probably ± 0.2 K (compilers).

REFERENCES:

1. Zhuravlev, E.F.; Sheveleva, A.D. *Zh. Neorg. Khim.* 1960, *5*, 2630.

COMPONENTS:

(1) Dysprosium nitrate; $Dy(NO_3)_3$; [10143-38-1]

(2) Quinoline nitrate; $C_9H_8N_2O_3$; [21640-15-3]

(3) Water ; H_2O ; [7732-18-5]

ORIGINAL MEASUREMENTS:

Katamanov, V.L.; Zhuravlev, E.F. *Zh. Neorg. Khim.* 1976, *21*, 1610-3; *Russ J. Inorg. Chem. Engl. Transl.* 1976, *21*, 879-81.

VARIABLES:

Composition at 20°C and 40°C

PREPARED BY:

T. Mioduski and S. Siekierski

EXPERIMENTAL VALUES:

20°C Isotherm[a]					40°C Isotherm[a]			
$Dy(NO_3)_3$		$C_9H_7N{\cdot}HNO_3$			$Dy(NO_3)_3$		$C_9H_7N{\cdot}HNO_3$	
mass %	mol kg^{-1}	mass %	mol kg^{-1}	solid phase[b]	mass %	mol kg^{-1}	mass %	mol kg^{-1}
0.0	———	70.0	12.14	A	0.0	———	78.0	18.45
5.0	0.48	65.0	11.27	"	3.7	0.50	75.0	18.32
11.8	1.20	60.0	11.07	"	8.8	1.25	71.0	18.29
19.5	2.38	57.0	12.62	"	14.0	2.26	68.2	19.94
27.5	4.15	53.5	14.65	"	20.5	4.11	65.2	23.73
28.6	4.51	53.2	15.21	A+B	27.0	7.91	63.2	33.56
29.0	4.50	52.5	14.77	B	33.0	6.11	51.5	17.29
35.5	4.03	39.2	8.06	"	38.0	5.45	42.0	10.93
42.0	3.95	27.5	4.69	"	44.0	4.89	30.2	6.09
44.6	3.83	22.0	3.43	"	50.7	4.96	20.0	3.55
49.8	3.95	14.0	2.01	"	58.2	5.17	9.5	1.53
54.0	4.17	8.8	1.23	"				
57.0	4.28	4.8	0.65	"				
59.0	4.51	3.5	0.49	B+C	60.5	5.31	6.8	1.08
59.5	4.49	2.5	0.34	C	61.0	5.30	6.0	0.95
61.2	4.53	0.0	———	"	62.8	5.35	3.5	0.54
				"	64.0	5.10	0.0	———

a. Molalities calculated by M. Salomon.

b. Solid phases: A = $C_9H_7N{\cdot}HNO_3$; B = $Dy(NO_3)_3{\cdot}2[C_9H_7N{\cdot}HNO_3]{\cdot}3H_2O$

C = $Dy(NO_3)_3{\cdot}6H_2O$

AUXILIARY INFORMATION

METHOD/APPARATUS/PROCEDURE:

Solubilities were studied by the method of isothermal sections (1) by measuring the refractive indices of saturated solutions along directed sections of the phase diagram. Equilibrium was checked by repeated measurements of the refractive index as a function of time. The results were used to graph the relation between the refractive indices and the composition of the components for each of the sections studied. The graphs were used to find the inflection or break points corresponding to the composition of the saturated solutions.

The composition of the double salt was found graphically and by chemical analysis of the isolated salt. Analyses for Dy and NO_3 differed by 0.4-0.5 % from theoretical.

The authors report data for the thermal analysis of the double nitrate.

SOURCE AND PURITY OF MATERIALS:

C.p. grade dysprosium nitrate was twice recrystallized. Analysis for water of crystallization gave 24.05 mass % which is close to the theoretical value for the hexahydrate (theor is 23.67 mass %, compilers).

The amine nitrate was prepd by neutralization of c.p. grade quinoline with c.p. grade nitric acid. The composition of the salt was checked by analysis for NO_3.

Double distilled water was used.

ESTIMATED ERROR:

Soly: precision ± 1% at best (compilers).

Temp: precision probably ± 0.2 K (compilers).

REFERENCES:

1. Zhuravlev, E.F.; Sheveleva, A.D. *Zh. Neorg. Khim.* 1960, *5*, 2630.

COMPONENTS:

(1) Dysprosium nitrate; $Dy(NO_3)_3$; [10143-38-1]

(2) 8-Methylquinoline nitrate; $C_{10}H_{10}N_2O_3$; [60491-92-1]

(3) Water ; H_2O ; [7732-18-5]

ORIGINAL MEASUREMENTS:

Katamanov. V.L.; Zhuravlev, E.F. *Zh. Neorg. Khim.* 1976, *21*, 1610-3; *Russ. J. Inorg. Chem. Engl. Transl.* 1976, *21*, 879-81.

VARIABLES:

Composition at 20°C and 40°C

PREPARED BY:

T. Mioduski and S. Siekierski

EXPERIMENTAL VALUES:

20°C Isotherm[a]					40°C Isotherm[a]			
$Dy(NO_3)_3$		$8\text{-}CH_3C_9H_6N\cdot HNO_3$			$Dy(NO_3)_3$		$8\text{-}CH_3C_9H_6N\cdot HNO_3$	
mass %	mol kg^{-1}	mass %	mol kg^{-1}	solid phase[b]	mass %	mol kg^{-1}	mass %	mol kg^{-1}
0.0	———	49.8	4.81	A	0.0	———	65.5	9.21
9.5	0.51	37.5	3.43	"	6.0	0.53	61.5	9.18
21.3	1.23	29.0	2.83	"	13.0	1.26	57.5	9.45
33.6	2.40	26.2	3.16	"	20.7	2.42	54.8	10.85
37.8	3.00	26.0	3.48	A+B	26.5	3.75	53.2	12.71
39.0	2.98	23.5	3.04	B	33.2	3.55	40.0	7.24
41.6	2.99	18.5	2.25	"	36.0	3.47	34.2	5.57
47.5	3.05	7.8	0.85	"	37.0	3.48	32.5	5.17
53.2	3.50	3.2	0.36	"	41.0	3.44	24.8	3.52
60.6	4.53	1.0	0.13	"	44.8	3.37	17.0	2.16
				"	50.0	3.35	7.2	0.82
				"	55.2	3.79	3.0	0.35
				"	60.4	4.51	1.2	0.15
60.8	4.52	0.6	0.08	B+C	65.0	5.39	0.4	0.06
61.0	4.52	0.3	0.04	C	65.2	5.41	0.2	0.03
61.2	4.53	0.0	———	"	65.4	5.42	0.0	———

a. Molalities calculated by M. Salomon.

b. Solid phases: A = $8\text{-}CH_3C_9H_6N\cdot HNO_3$; B = $Dy(NO_3)_3\cdot 2[CH_3C_9H_6N\cdot HNO_3]$

C = $Dy(NO_3)_3\cdot 6H_2O$

AUXILIARY INFORMATION

METHOD/APPARATUS/PROCEDURE:

Solubilities were studied by the method of isothermal sections (1) by measuring the refractive indices of saturated solutions along directed sections of the phase diagram. Equilibrium was checked by repeated measurements of the refractive index as a function of time. The results were used to graph the relation between the refractive indices and the composition of the components for each of the sections studied. The graphs were used to find the inflection or break points corresponding to the composition of the saturated solutions.

The composition of the double salt was found graphically and by chemical analysis of the isolated salt. Analyses for Dy and NO_3 differed by 0.4-0.5 % from theoretical.

The authors report data for the thermal analysis of the double nitrate.

SOURCE AND PURITY OF MATERIALS:

C.p. grade dysprosium nitrate was twice recrystallized. Analysis for water of crystallization gave 24.05 mass % which is close to the theoretical value for the hexahydrate (theor is 23.67 mass %, compilers).
The amine nitrate was prepd by neutralization of c.p. grade 8-methylquinoline with c.p. grade nitric acid. The composition of the salt was checked by analysis for NO_3.

Doubly distilled water was used.

ESTIMATED ERROR:

Soly: precision ± 1% at best (compilers).

Temp: precision probably ± 0.2 K (compilers).

REFERENCES:

1. Zhuravlev, E.F.; Sheveleva, A.D. *Zh. Neorg. Khim.* 1960, *5*, 2630.

COMPONENTS:	ORIGINAL MEASUREMENTS:
(1) Dysprosium nitrate; $Dy(NO_3)_3$; [10143-38-1] (2) 8-Hydroxyquinoline nitrate (8-quinolinol nitrate); $C_9H_8N_2O_4$; [60491-93-2] (3) Water ; H_2O ; [7732-18-5]	Katamanov, V.L.; Zhuravlev, E.F. *Zh. Neorg. Khim.* 1976, *21*, 1610-3; *Russ. J. Inorg. Chem. Engl. Transl.* 1976, *21*, 879-81.
VARIABLES:	PREPARED BY:
Composition at 20°C and 40°C	T. Mioduski and S. Siekierski

EXPERIMENTAL VALUES:

composition of saturated solutions at 20°C and 40°C[a]

	$Dy(NO_3)_3$		$8\text{-}HOC_9H_6N\cdot HNO_3$		
t/°C	mass %	mol kg^{-1}	mass %	mol kg^{-1}	nature of the solid phase
20	0.0	———	23.0	1.43	$8\text{-}HOC_9H_6N\cdot HNO_3$
	14.0	0.50	5.0	0.30	"
	22.0	0.83	2.0	0.13	"
	29.7	1.23	1.0	0.07	"
	45.0	2.36	0.2	0.02	"
	61.2	4.53	0.0	———	$Dy(NO_3)_3\cdot 6H_2O$
40	0.0	———	47.5	4.35	$8\text{-}HOC_9H_6N\cdot HNO_3$
	10.8	0.50	27.5	2.14	"
	19.2	0.84	15.0	1.10	"
	28.0	1.23	6.5	0.48	"
	43.5	2.31	2.5	0.22	"
	59.0	4.25	1.2	0.14	"
	64.0	5.25	1.0	0.14	$8\text{-}HOC_9H_6N\cdot HNO_3 + Dy(NO_3)_3\cdot 6H_2O$
	64.6	5.31	0.5	0.07	$Dy(NO_3)_3\cdot 6H_2O$
	65.4	5.42	0.0	———	"

a. Molalities calculated by M. Salomon.

AUXILIARY INFORMATION

METHOD/APPARATUS/PROCEDURE:

Solubilities were studied by the method of isothermal sections (1) by measuring the refractive indices of saturated solutions along directed sections of the phase diagram. Equilibrium was checked by repeated measurements of the refractive index as a function of time. The results were used to graph the relation between the refractive indices and the composition of the components for each of the sections studied. The graphs were used to find the inflection or break points corresponding to the composition of the saturated solutions.

The composition of the double salt was found graphically and by chemical analysis of the isolated salt. Analyses for Dy and NO_3 differed by 0.4-0.5 % from theoretical.

Above 45 mass % $Dy(NO_3)_3$, the solubility of 8-hydroxyquinoline nitrate is too small to be detd by the analytical method used by the authors, and it was therefore not possible to reliably determine the composition at the eutonic point at 20°C.

SOURCE AND PURITY OF MATERIALS:

C.p. grade dysprosium nitrate was twice re-crystallized. Analysis for water of crystallization gave 24.05 mass % which is close to the theoretical value for the hexahydrate (theor is 23.67 mass %, compilers).
The amine nitrate was prepd by neutralization of c.p. grade 8-quinolinol with c.p. grade nitric acid. The composition of the salt was checked by analysis for NO_3.

Doubly distilled water was used.

ESTIMATED ERROR:

Soly: precision ± 1% at best (compilers).

Temp: precision probably ± 0.2 K (compilers).

REFERENCES:

1. Zhuravlev, E.F.; Sheveleva, A.D. *Zh. Neorg. Khim.* 1960, *5*, 2630.

COMPONENTS:

(1) Dysprosium nitrate; $Dy(NO_3)_3$; [10143-38-1]

(2) Urea; CH_4N_2O; [57-13-6]

(3) Water ; H_2O ; [7732-18-5]

ORIGINAL MEASUREMENTS:

Khudaibergenova, N.; Sulaimankulov, K.; Abykeev, K. *Zh. Neorg. Khim.* 1980, *25*, 2251-3; *Russ. J. Inorg. Chem. Engl. Transl.* 1980, *25*, 1247-9.

VARIABLES:

Composition at 30°C

PREPARED BY:

T. Mioduski and S. Siekierski

EXPERIMENTAL VALUES: composition of saturated solutions at 30°C[a]

$Dy(NO_3)_3$		$CO(NH_2)_2$		
mass %	mol kg^{-1}	mass %	mol kg^{-1}	nature of the solid phase
———	———	57.50	22.528	$CO(NH_2)_2$
12.96	1.267	57.69	32.729	"
17.02	1.747	55.03	38.784	"
22.04	2.796	55.34	40.737	"
29.88	5.478	54.47	57.955	"
39.59	26.541	56.13	218.37	$Dy(NO_3)_3 \cdot 4CO(NH_2)_2$
42.61	16.130	49.81	109.42	"
44.93	10.422	42.70	57.478	"
46.14	9.049	39.23	44.650	"
48.23	7.600	33.56	30.687	"
49.39	6.803	29.78	23.806	"
55.54	8.522	25.76	22.938	"
56.00	8.912	25.97	23.984	$Dy(NO_3)_3 \cdot 3CO(NH_2)_2$
56.87	6.862	19.35	13.549	"
59.53	6.639	14.74	9.539	"
60.03	6.674	14.16	9.135	"
66.92	8.952	11.63	9.028	"
68.25	9.361	10.83	8.620	"
67.33	8.541	10.05	7.398	$Dy(NO_3)_3 \cdot 5H_2O$
63.27	6.219	7.54	4.301	"
65.45	6.058	3.55	1.907	"
62.22	5.425	4.87	2.464	"

a. Molalities calculated by M. Salomon.

AUXILIARY INFORMATION

METHOD/APPARATUS/PROCEDURE:

Isothermal method employed. Equilibrium is stated to be reached within 7 h. Nitrogen or urea was determined by the Kjeldahl method. Dysprosium was determined by titration with Trilon (Na_2H_2(EDTA)) using Xylenol Orange indicator. Liquid phase was filtered off through a Schott No. 3 filter.

The 1:4 double salt is congruently soluble, and the 1:3 double salt is incongruently soluble.

SOURCE AND PURITY OF MATERIALS:

Nothing specified.

ESTIMATED ERROR:

Nothing specified.

REFERENCES:

COMPONENTS:	EVALUATOR:
(1) Holmium nitrate; $Ho(NO_3)_3$; [10168-82-8] (2) Water ; H_2O ; [7732-18-5]	S. Siekierski, T. Mioduski Institute for Nuclear Research Warsaw, Poland and M. Salomon U.S. Army ET & DL Ft. Monmouth, NJ November 1982

CRITICAL EVALUATION:

INTRODUCTION

The solubility of holmium nitrate in water has been reported in four publications for 298.15 K (1-4), and in one publication for 303.2 K (5). The hexahydrate, $Ho(NO_3)_3.6H_2O$ [35725-31-6], was reported to be the stable solid phase in (1-4), and the pentahydrate, $Ho(NO_3)_3.5H_2O$ [14483-18-2], was reported in (5) to be the stable solid phase at 303.2 K. There are no solubility data reported for lower hydrates in pure water, but Wendlandt and Sewell (6) report that the congruent melting point of the tetrahydrate, $Ho(NO_3)_3.4H_2O$ [37131-78-5], is 357 ± 1 K.

EVALUATION

The solubility result of 3.41 mol kg^{-1} at 298.2 K (4) is rejected because it is much too low based on comparisons with the results in (1-3), and from the fact that it is incompatible with the trend of increasing solubilities with atomic number for lanthanides heavier than gadolinium (see the figures of solubility vs atomic number in the PREFACE and in the critical evaluation for the europium nitrate-water system). For the hexahydrate system, the average value from (1-3) is designated as the *recommended* solubility at 298.15 K: this recommended value including the total uncertainty is 5.02 ± 0.01 mol kg^{-1}. The total uncertainty was calculated from the experimental precision of ± 0.1% and the calculated precision in the average value (95% level of confidence and a Student's t = 4.303).

The singular value for the solubility in the pentahydrate system reported in (5) is probably of low accuracy. The result, 4.131 mol kg^{-1} at 303.2 K, is considerably lower than the recommended value at 298.15 K for the hexahydrate system. If the result from (5) were accurate to within experimental precision, then we would have to conclude that the pentahydrate solid phase is considerably more stable than the hexahydrate solid phase at temperatures well below 303.2 K. Based on the known phase relations for other binary lanthanide systems, it does not appear reasonable to assume that the pentahydrate is the stable solid phase at and below 303.2 K, and we therefore *reject* the solubility result reported in (5).

REFERENCES

1. Spedding, F.H.; Shiers, L.E.; Rard, J.A. *J. Chem. Eng. Data* 1975, *20*, 88.
2. Rard, J.A.; Spedding, F.H. *J. Phys. Chem.* 1975, *79*, 257.
3. Spedding, F.H.; Derer, J.L.; Mohs, M.A.; Rard, J.A. *J. Chem. Eng. Data* 1976, *21*, 474.
4. Afanas'ev, Yu.A.; Azhipa, L.T. *Zh. Neorg. Khim.* 1976, *21*, 2284.
5. Khudaibergenova, N.; Sulaimankulov, K.; Abykeev, K. *Zh. Neorg. Khim.* 1980, *25*, 2251.
6. Wendlandt, W.W.; Sewell, R.G. *Texas J. Sci.* 1961, *13*, 231.

COMPONENTS:

(1) Holmium nitrate; $Ho(NO_3)_3$; [10168-82-8]

(2) Water ; H_2O ; [7732-18-5]

ORIGINAL MEASUREMENTS:

1. Spedding, F.H.; Shiers, L.E.; Rard, J.A. *J. Chem. Eng. Data* 1975, *20*, 88-93.
2. Rard, J.A.; Spedding. F.H. *J. Phys. Chem.* 1975, *79*, 257-62.
3. Spedding. F.H.; Derer, J.L. ; Mohs, M.A.; Rard, J.A. *J. Chem. Eng. Data* 1976, *21*, 474-88.

VARIABLES:

One temperature: 25.00°C

PREPARED BY:

T. Mioduski, S. Siekierski, and M. Salomon

EXPERIMENTAL VALUES:

The solubility of $Ho(NO_3)_3$ in water at 25.00°C has been reported by Spedding and co-workers in three publications. Source paper [3] reports the solubility to be 5.027 mol kg^{-1}, but the preferred values are given in source papers [1] and [2] as 5.0184 mol kg^{-1} and 5.0183 mol kg^{-1}, respectively.

COMMENTS AND/OR ADDITIONAL DATA:

Source paper [1] reports the relative viscosity, η_R, of a saturated solution to be 29.773. Taking the viscosity of water at 25°C to equal 0.008903 poise, the viscosity of a saturated $Ho(NO_3)_3$ solution at 25°C is 0.26507 poise (compilers calculation).

Supplementary data available in the microfilm edition to *J. Phys. Chem.* 1975, *79*, enabled the compilers to provide the following additional data.

The density of the saturated solution was calculated by the compilers from the smoothing equation, and at 25°C the value is 1.98393 kg m^{-3}. Using this density, the solubility in volume units is (based on the preferred value of 5.0183 mol kg^{-1})

$$c_{satd} = 3.6057 \text{ mol dm}^{-3}$$

Source paper [2] reports the electrolytic conductivity of the saturated solution to be (corrected for the electrolytic conductivity of the solvent $\kappa = 0.020598$ S cm^{-1}.

The molar conductivity of the saturated solution is calculated from 1000 /3c_{satd} and is

$$\Lambda(\tfrac{1}{3}\ Ho(NO_3)_3) = 1.904 \text{ S cm}^2 \text{ mol}^{-1}$$

AUXILIARY INFORMATION

METHOD/APPARATUS/PROCEDURE:

Isothermal method used. Solutions were prepared as described in (1) and (2). The concentration of the saturated solution was determined by both EDTA (1) and sulfate (2) methods which is said to be reliable to 0.1% or better. In the sulfate analysis, the salt was first decomposed with HCl followed by evaporation to dryness before sulfuric acid additions were made. This eliminated the possibility of nitrate ion coprecipitation.

SOURCE AND PURITY OF MATERIALS:

$Ho(NO_3)_3 \cdot 6H_2O$ was prepd by addn of HNO_3 to the oxide. The oxide was purified by an ion exchange method, and the upper limit for the impurities Ca, Fe, Si and adjacent rare earths was given as 0.15%.

In source paper [3] the salt was analysed for water of hydration and found to be within ± 0.016 water molecules of the hexahydrate.

Water was distilled from an alkaline permanganate solution.

ESTIMATED ERROR:

Soly: duplicate analyses agreed to at least ± 0.1%.

Temp: not specified, but probably accurate to at least ± 0.01 K as in (3)(compilers).

REFERENCES:

1. Spedding, F.G.; Cullen, P.F.; Habenschuss, A. *J. Phys. Chem.* 1974, *78*, 1106.
2. Spedding, F.H.; Pikal, M.J.; Ayers, B.O. *J. Phys. Chem.* 1966, *70*, 2440.
3. Spedding, F.H.; et. al. *J. Chem. Eng. Data* 1975, *20*, 72.

COMPONENTS:

(1) Holmium nitrate; $Ho(NO_3)_3$; [10168-82-8]

(2) Nitric acid; HNO_3; [7697-37-2]

(3) Water ; H_2O ; [7732-18-5]

ORIGINAL MEASUREMENTS:

Afanas'ev, Yu.A.; Azhipa, L.T. *Zh. Neorg. Khim.* 1976, *21*, 2284-7; *Russ. J. Inorg. Chem. Engl. Transl.* 1976, *21*, 1257-60.

VARIABLES:

Nitric acid concentraction at 25°C

PREPARED BY:

T. Mioduski, S. Siekierski

EXPERIMENTAL VALUES:

Solubility of holmium nitrate in nitric acid solutions at 25°C[a]

$Ho(NO_3)_3$ mass %	$Ho(NO_3)_3$ mol kg^{-1}	HNO_3 mass %	HNO_3 mol kg^{-1}	nature of the solid phase
54.4	3.41	——	——	$Ho(NO_3)_3 \cdot 6H_2O$
49.5	3.13	5.4	1.93	"
46.5	3.07	10.3	3.85	"
45.2	3.02	12.2	4.62	"
40.2	2.81	19.1	7.67	"
38.8	2.75	21.0	8.42	"
35.3	2.65	26.8	11.40	"
32.4	2.76	34.1	16.41	$Ho(NO_3)_3 \cdot 6H_2O + Ho(NO_3)_3 \cdot 4H_2O$
28.2	2.30	36.9	17.05	$Ho(NO_3)_3 \cdot 4H_2O$
23.2	1.89	41.9	19.36	"
15.3	1.23	49.2	22.35	"
11.2	0.98	56.2	27.80	"
9.4	0.91	61.3	33.74	"
6.9	0.69	64.5	36.37	"
6.6	0.71	66.9	40.71	"
8.1	1.08	70.6	53.46	$Ho(NO_3)_3 \cdot 4H_2O + Ho(NO_3)_3 \cdot H_2O$
8.3	1.23	72.5	60.90	$Ho(NO_3)_3 \cdot H_2O$
7.9	1.36	75.6	73.89	$Ho(NO_3)_3 \cdot H_2O + Ho(NO_3)_3 \cdot nH_2O$

continued..........

AUXILIARY INFORMATION

METHOD/APPARATUS/PROCEDURE:

Isothermal method. Compositions of the solutions were varied by adding 100% nitric acid to a saturated solution, or by adding the salt to the acid. Equilibrium was reached in 3-4 h. Ho in both the liquid and solid phases was determined complexometrically. Nitric acid was detd by titrn with NaOH using Methyl Red indicator. Above 70 mass % nitric acid, the composition of the solid phase was determined by Schreinemakers' method of residues.

SOURCE AND PURITY OF MATERIALS:

"Pure" grade holmium nitrate was used. C.p. grade HNO_3 was concentrated by the Brauer method.

ESTIMATED ERROR:

Soly: precision about ± 0.5% (compiler).

Temp: precision ± 1 K.

REFERENCES:

COMPONENTS:	EVALUATOR:
(1) Holmium nitrate; $HO(NO_3)_3$; [10168-82-8]	Afanas'ev, Yu.A.; Azhipa, L.T. *Zh. Neorg. Khim.* 1976, *21*, 2284-7; *Russ. J. Inorg. Chem. Engl. Transl.* 1976, *21*, 1257-60.
(2) Nitric acid; HNO_3; [7697-37-2]	
(3) Water ; H_2O ; [7732-18-5]	

CRITICAL EVALUATION: continued ....

$Ho(NO_3)_3$		HNO_3		
mass %	mol kg^{-1}	mass %	mol kg^{-1}	nature of the solid phase
6.9	1.24	77.3	78.90	$Ho(NO_3)_3 \cdot nH_2O$
6.0	1.18	79.5	88.42	"
4.5	0.90	81.2	91.58	"
2.2	0.51	85.6	113.2	"
1.1	0.35	90.0	163.1	"
0.2	0.14	95.7	376.4	"

[a] Molalities calculated by M. Salomon.

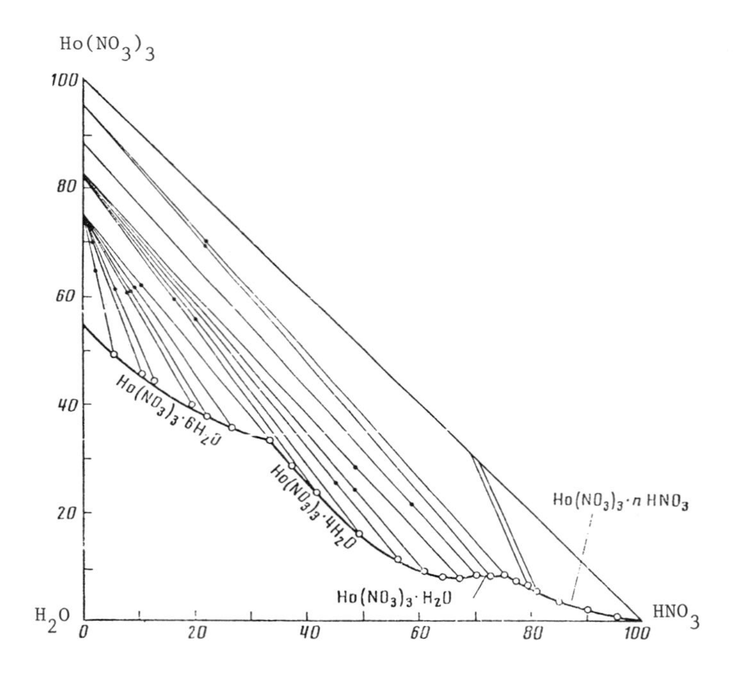

298.2 K isotherm for the $Ho(NO_3)_3$ - HNO_3 - H_2O system

Compositions given in mass % units.

COMPONENTS:

(1) Holmium nitrate; $Ho(NO_3)_3$; [10168-82-8]

(2) Urea, CH_4N_2O; [57-13-6]

(3) Water ; H_2O ; [7732-18-5]

ORIGINAL MEASUREMENTS:

Khudaibergenova, N.; Sulaimankulov, K.; Abykeev, K.

Zh. Neorg. Khim. 1980, *25*, 2251-3; *Russ. J. Inorg. Chem. Engl. Transl.* 1980, *25*, 1247-9.

VARIABLES:

Composition at 30°C

PREPARED BY:

T. Mioduski and S. Siekierski

EXPERIMENTAL VALUES:

$Ho(NO_3)_3$ [a]		$CO(NH_2)_2$ [a]		
mass %	mol kg^{-1}	mass %	mol kg^{-1}	nature of the solid phase
0.0	———	57.50	22.528	$CO(NH_2)_2$
5.31	0.405	57.37	25.597	"
24.82	3.774	56.44	50.149	"
37.53	13.435	54.51	114.03	"
40.03	15.477	52.60	118.84	"
40.52	14.414	51.47	107.00	"
41.17	17.303	52.05	127.83	$CO(NH_2)_2$ + $Ho(NO_3)_3\cdot 4CO(NH_2)_2$
41.35	12.807	49.45	89.500	$Ho(NO_3)_3\cdot 4CO(NH_2)_2$
42.03	10.636	46.71	69.074	"
44.10	8.229	40.63	44.305	"
49.07	6.577	29.67	23.238	"
53.00	7.890	27.86	24.237	$Ho(NO_3)_3\cdot 3CO(NH_2)_2$
55.42	9.289	27.58	27.014	"
55.37	7.244	22.85	17.469	"
57.35	6.098	15.85	9.848	"
58.83	6.220	14.22	8.786	"
61.27	6.725	12.77	8.191	"
62.16	6.828	11.90	7.639	"
65.14	7.039	8.49	5.361	$Ho(NO_3)_3\cdot 3CO(NH_2)_2$ + $Ho(NO_3)_3\cdot 5H_2O$
65.41	7.506	9.76	6.545	$Ho(NO_3)_3\cdot 5H_2O$
59.21	4.363	2.12	0.913	"
59.18	4.131	0.0	———	"

a. Molalities calculated by M. Salomon

AUXILIARY INFORMATION

METHOD/APPARATUS/PROCEDURE:

Isothermal method employed. Equilibrium is stated to be reached within 7 h. Nitrogen of urea was determined by the Kjeldahl method. Ho was determined by titration with Trilon (disodium salt of ethylenediamine tetraacetic acid) using Xylenol Orange indicator. The liquid phase was filtered off with a Schott No. 3 filter.

COMMENTS AND/OR ADDITIONAL DATA:

$Ho(NO_3)_3\cdot 3CO(NH_2)_2$ is incongruently soluble, and analysis yielded 66.09 mass % $Ho(NO_3)_3$ and 33.91 mass % urea. $Ho(NO_3)_3\cdot 4CO(NH_2)_2$ is congruently soluble, and analysis yielded 59.32 mass % $Ho(NO_3)_3$ and 40.68 mass % urea. Analysis of the pentahydrate solid phase yielded 79.50 mass % of the salt.

SOURCE AND PURITY OF MATERIALS:

Nothing specified.

ESTIMATED ERROR:

Soly: Nothing specified.

Temp: Precision probably ± 0.1-0.2 K (compilers).

REFERENCES:

COMPONENTS:

(1) Erbium nitrate; $Er(NO_3)_3$; [10168-80-6]

(2) Water ; H_2O ; [7732-18-5]

EVALUATOR:

S. Siekierski, T. Mioduski
Institute for Nuclear Research
Warsaw, Poland
and
M. Salomon
U.S. Army ET & DL
Ft. Monmouth, NJ
January 1983

CRITICAL EVALUATION:

INTRODUCTION

The solubility of erbium nitrate in pure water has been reported in ten publications (1-10). In the binary system, both the hexahydrate and pentahydrate have been identified as solid phases. Solid phases involving the tetrahydrate, dihydrate and anhydrous salt have been found only in solutions of high nitric acid content (> 61 mass % HNO_3, ref (5)).

$Er(NO_3)_3 \cdot 6H_2O$	[13476-05-6]	$Er(NO_3)_3 \cdot 4H_2O$	[37131-79-6]
$Er(NO_3)_3 \cdot 5H_2O$	[10031-51-3]	$Er(NO_3)_3 \cdot 2H_2O$	[71973-92-7]

Moret (1) studied the solubilities from 273.15 K to 323.15 K and found the solid phase to be the pentahydrate. Spedding et al. (2-4) determined the solubility of the hexahydrate at 298.15 K. In (4) the hydrated crystals grown from saturated solutions at 298.15 K were dried, analysed by EDTA titration, and the waters of hydration found to be within ± 0.016 water molecules of the hexahydrate. Zhuravlev et al. (6,7) reported solubilities in the pentahydrate system from 293.2 K to 323.2 K, but do not specify how the solid phases were determined. Presumably all the solid phases in (6,7) were determined by direct chemical analyses (the oxalate method), and possibly by the method of residues. Afanas'ev et al. (5) reported solubilities water and aqueous nitric acid solutions, and based on analyses by titration with Trilon and standard base reported the solid phase to be the hexahydrate at 298.2 K. Sulaimankulov et al. (8-10) reported solubilities at 303.2 K and used the method of residues to determine the compositions of the solid phases: the hexahydrate was found to be the stable solid phase at this temperature.

EVALUATION PROCEDURE

Solubilities reported as a function of temperature were fitted to the general solubility equation

$$Y = \ell n(m/m_o) - nM_2(m - m_o) = a + b/(T/K) + c\ \ell n(T/K) \qquad [1]$$

All terms in eq. [1] have been defined in previous critical evaluations. As discussed below, considerable differences are found for the solubilities reported in (1-10), and this is probably due to a combination of imprecise experimental results (5, 8-10), and uncertainties in the nature of the solid phases. A similar problem was encountered in determining the stable solid phases in the $Y(NO_3)_3$-H_2O system. From enthalpic and entropic considerations (see the PREFACE), and from ionic radius considerations (11), we have concluded that the solubility properties of $Y(NO_3)_3$ and $Er(NO_3)_3$ are almost identical. At 298.2 K the solubilities of yttrium and erbium nitrates in the hexahydrate system fall on the smoothed curve of solubility vs atomic number (see figures in the PREFACE and in the $Eu(NO_3)_3$-H_2O critical evaluation). At 323.2 K, the solubility of $Y(NO_3)_3$ in the metastable hexahydrate system is 8.481 mol kg^{-1}, but all solubilities for $Er(NO_3)_3$ in (1,6) at 323.2 K fall below 7 mol kg^{-1} thus confirming that these results correspond to solubilities in the stable pentahydrate system. We conclude that over the temperature range of 273.15 K to 323.15 K, both the hexahydrate and pentahydrate are the solid phases, that their polytherms lie very close to each other, and that the hexahydrate → pentahydrate transition temperature is very close to 298 K.

SOLUBILITY IN THE HEXAHYDRATE SYSTEM

The solubilities reported in (5, 8-10) are much lower than any of the values reported in (1-4, 6,7) at corresponding temperatures, and these data are rejected. The value of 4.325 mol kg^{-1} at 298.2 K reported by Afanas'ev et al. (5) is so low that it must be the result of failure to reach equilibrium since their experimental analyses were based on the well established method of titration with Trilon (Na_2EDTA). In fact most of the results in binary systems reported by Afanas'ev and co-workers are very low (see other critical evaluations), and we also attribute this to failure to reach equilibrium. This of course raises questions on the accuracy of the results in ternary systems reported by Afanas'ev and co-workers.

The only other data for the solubility of $Er(NO_3)_3$ in the hexahydrate system are those from Spedding's laboratory (2-4) at 298.15 K. The average solubility at 298.15 K from (2-4) is 5.442 mol kg^{-1} with an overall uncertainty of ± 0.030 mol kg^{-1} at the 95% level of confidence.

COMPONENTS:	EVALUATOR:
(1) Erbium nitrate; $Er(NO_3)_3$; [10168-80-6] (2) Water ; H_2O ; [7732-18-5]	S. Siekierski, T. Mioduski Institute for Nuclear Research Warsaw, Poland and M. Salomon U.S. Army ET & DL Ft. Monmouth, NJ January 1983

CRITICAL EVALUATION:

SOLUBILITIES IN THE PENTAHYDRATE SYSTEM

The most precise data are those of Moret, and the results of fitting these data to eq. [1] are given in Tables 1 and 2. We attempted to fit the results of Zhuravlev et al. (6,7) to eq. [1], but were unable to obtain meaningful results. The values of the constants a, b, c are trivial, and the predicted congruent melting point for the

Table 1. Derived parameters for the smoothing equation.

parameter	value in the pentahydrate system
a	-23.165
b	716.5
c	3.6473
σ_a	0.002
σ_b	0.5
σ_c	0.0003
σ_Y	0.002
σ_m	0.019
$\Delta H_{sln}/kJ\ mol^{-1}$	-23.7
$\Delta C_p/J\ K^{-1}\ mol^{-1}$	121.3
congruent melting point/K	344.6
concn at the congr m.p./mol kg^{-1}	11.102

Table 2. Tentative solubility data.

	hexahydrate system[a]	pentahydrate system[b]
T/K	m_1/mol kg^{-1}	m_1/mol kg^{-1}
273.15		4.537
278.15		4.687
283.15		4.850
288.15		5.028
293.15		5.221
298.15	5.442	5.433
303.15		5.666
308.15		5.925
313.15		6.213
318.15		6.540
323.15		6.916
328.15		7.360
333.15		7.906
338.15		8.635
343.15		9.889

[a] Average value from experimental results from (2-4).

[b] Smoothed results calculated from eq. [1] based on data from (1).

pentahydrate is 381 K which is much higher than the experimental value of 357 ± 1 K for the <u>tetrahydrate</u> reported by Wendlandt and Sewell (12). The value of the congruent melting point of the solid pentahydrate calculated from eq. [1] using Moret's solubility results is 344.6 K (see Table 1) which is, as it should be, below the melting point of the tetrahydrate. Moret's results therefore appear to be accurate as well as precise.

COMPONENTS:	EVALUATOR:
(1) Erbium nitrate; $Er(NO_3)_3$; [10168-80-6] (2) Water ; H_2O ; [7732-18-5]	S. Siekierski, T. Mioduski Institute for Nuclear Research Warsaw, Poland and M. Salomon U.S. Army ET & DL Ft. Monmouth, NJ Janaury 1983

CRITICAL EVALUATION:

The smoothed solubility value of 5.433 mol kg^{-1} at 298.15 K calculated from Moret's data is very close to the average value of 5.442 mol kg^{-1} for the solubility at 298.15 K in the hexahydrate system reported by Spedding et al. These results suggest that the hexahydrate → pentahydrate transition temperature is around 298 K.

The solubilities from (6,7) of 5.23 mol kg^{-1} at 293.2 K and 5.40 mol kg^{-1} at 298.2 K lie very close to the smoothed pentahydrate polytherm based on Moret's data, and thus appear to be accurate with the estimated (by the compilers) experimental precision of ± 1%. However above 298.2 K the solubilities from (6,7) are much lower than any of the data from (1-4), and we conclude that these results from (6,7) are in error probably due to the failure to reach equilibrium. They do not appear to be due to incorrect identification of the solid phase because the pentahydrate is the stable solid phase at 323.2 K, and solubilities in the metastable hexahydrate system would have to be much higher: e.g. at 323.2 K the solubility in the pentahydrate system is 6.916 mol kg^{-1}, and based on the similarities between $Y(NO_3)_3$ and $Er(NO_3)_3$, we would expect that the solubility of $Er(NO_3)_3$ at 323.2 K in the hexahydrate system would be about 8.5 mol kg^{-1}.

TENTATIVE SOLUBILITY VALUES

For the solubility of $Er(NO_3)_3$ in the hexahydrate system, we designate the average value from (2-4) as the *tentative* solubility at 298.15 K. All other data are *rejected*.

For solubilities in the pentahydrate system, the smoothed solubilities based on Moret's results are designated as *tentative* solubilities, and all other results are *rejected*.

All tentative solubilities are given in Table 2.

REFERENCES

1. Moret, R. *Thèse*. l'Universté de Lausanne. 1963.
2. Spedding, F.H.; Shiers, L.E.; Rard, J.A. *J. Chem. Eng. Data* 1975, *20*, 88.
3. Rard, J.A.; Spedding, F.H. *J. Phys. Chem.* 1975, *79*, 257.
4. Spedding, F.H.; Derer, J.L.; Mohs, M.A.; Rard, J.A. *J. Chem. Eng. Data* 1976, *21*, 474.
5. Afanas'ev, Yu.A.; Azhipa, L.T.; Shakhanova, N.A. *Zh. Neorg. Khim.* 1977, *22*, 3331.
6. Starikova, L.I.; Zhuravlev, E.F.; Khalfina, L.R. *Zh. Neorg. Khim.* 1979, *24*, 2803.
7. Starikova, L.I.; Zhuravlev, E.F. *Zh. Neorg. Khim.* 1980, *25*, 2007.
8. Aitimbetov, K.; Sulaimankulov, K.S.; Batyuk, A.G.; Ismailov, M. *Zh. Neorg. Khim.* 1975, *20*, 2510.
9. Isakova, S.; Sulaimankulov, K.; Aitimbetov, K.; Kozhanova, T. *Zh. Neorg. Khim.* 1980, *25*, 2271.
10. Aitimbetov, K.; Sulaimankulov, K.; Isakova, S.; Ismailov, M. *Izv. Akad. Nauk Kirg. SSSR* 1975, 46.
11. Rard, J.A.; Spedding, F.H. *J. Chem. Eng. Data* 1982, *27*, 454 (and references cited in this paper).
12. Wendlandt, W.W; Sewell, R.G. *Texas J. Sci.* 1961, *13*, 231.

COMPONENTS:

(1) Erbium nitrate; $Er(NO_3)_3$; [10168-80-6]

(2) Water ; H_2O ; [7732-18-5]

ORIGINAL MEASUREMENTS:

Moret, R. *Thèse*. l'Université de Lausanne. 1963.

VARIABLES:

Temperature: range 0 to 50°C

PREPARED BY:

T. Mioduski and S. Siekierski

EXPERIMENTAL VALUES:

t/°C	$Er(NO_3)_3$ mass %	$Er(NO_3)_3$ mol kg^{-1} [a]	solid phase
0	61.59	4.539	$Er(NO_3)_3 \cdot 5H_2O$
10	63.10	4.841	"
20	64.91	5.236	"
40	68.64	6.196	"
50	70.99	6.927	"

a. Calculated by the compilers.

AUXILIARY INFORMATION

METHOD/APPARATUS/PROCEDURE:

Isothermal method employed (1). Er was determined by complexometric titration using Xylenol Orange indicator in the presence of a small amount of urotropine buffer. Water was determined by difference.

SOURCE AND PURITY OF MATERIALS:

$Er(NO_3)_3 \cdot 5H_2O$ was prepared from Er_2O_3 of purity better than 99.7% (obtained by the ion exchange chromatography). No additional details available.

ESTIMATED ERROR:

Soly: precision about ± 0.2 % (compilers).

Temp: precision at least ± 0.05 K (compilers).

REFERENCES:

1. Brunisholz, G.; Quinche, J.P.; Kalo, A.M. *Helv. Chim. Acta* 1964, *47*, 14.

COMPONENTS:

(1) Erbium nitrate; $Er(NO_3)_3$; [10168-80-6]

(2) Water ; H_2O ; [7732-18-5]

ORIGINAL MEASUREMENTS:

1. Spedding, F.H.; Shiers, L.E.; Rard, J.A. *J. Chem. Eng. Data* 1975, *20*, 88-93.
2. Rard, J.A.; Spedding, F.H. *J. Phys. Chem.* 1975, *79*, 257-62.
3. Spedding, F.H.; Derer, J.L.; Mohs, M.A.; Rard, J.A. *J. Chem. Eng. Data* 1976, *21*, 474-88.

VARIABLES:

One temperature: 25.00°C

PREPARED BY:

T. Mioduski, S. Siekierski and M. Salomon

EXPERIMENTAL VALUES:

The solubility of $Er(NO_3)_3$ in water at 25.00°C has been reported by Spedding and co-workers in three publications. Source paper [3] reports the solubility to be 5.456 mol kg^{-1}, but the preferred value is given in source papers [1] and [2] as 5.4348 mol kg^{-1}.

COMMENTS AND/OR ADDITIONAL DATA:

Source paper [1] reports the relative viscosity, η_R, of a saturated solution to be 41.240. Taking the viscosity of water at 25°C to equal 0.008903 poise, the viscosity of a saturated $Er(NO_3)_3$ solution at 25°C is 0.3672 poise (compilers calculation).

Supplementary data available in the microfilm edition to *J. Phys. Chem.* 1975, *79* enabled the compilers to provide the following additional data.

The density of the saturated solutions was calculated by the compilers from the smoothing equation, and at 25°C the value is 2.04068 kg m^{-3}. Using this density, the solubility in volume units is (based on the preferred value of 5.4348 mol kg^{-1})

$$c_{satd} = 3.7982 \text{ mol dm}^{-3}$$

Source paper [2] reports the electrolytic conductivity of the saturated solution to be (corrected for the electrolytic conductivity of the solvent) κ = 0.015798 S cm^{-1}.

The molar conductivity of the saturated solution is calculated from $1000\kappa/3c_{satd}$ and is

$$\Lambda(\tfrac{1}{3}\, Er(NO_3)_3) = 1.386 \text{ S cm}^2 \text{ mol}^{-1}$$

AUXILIARY INFORMATION

METHOD/APPARATUS/PROCEDURE:

Isothermal method used. Solutions were prepared as described in (1) and (2). The concentration of the saturated solution was determined by both EDTA (1) and sulfate (2) methods which is said to be reliable to 0.1% or better. In the sulfate analysis, the salt was first decomposed with HCl followed by evaporation to dryness before sulfuric acid additions were made. This eliminated the possibility of nitrate ion coprecipitation.

SOURCE AND PURITY OF MATERIALS:

$Er(NO_3)_3 \cdot 6H_2O$ was prepd by addn of HNO_3 to the oxide. The oxide was purified by an ion exchange method, and the upper limit for the impurities Ca, Fe, Si and adjacent rare earths was given as 0.15%.

In source paper [3] the salt was analysed for water of hydration and found to be within ± 0.016 water molecules of the hexahydrate.

Water was distilled from an alkaline permanganate solution.

ESTIMATED ERROR:

Soly: duplicate analyses agreed to at least ± 0.1%.

Temp: not specified, but probably accurate to at least ± 0.01 K as in (3)(compilers).

REFERENCES:

1. Spedding, F.H.; Cullen, P.F.; Habenschuss, A. *J. Phys. Chem.* 1974, *78*, 1106.
2. Spedding, F.H.; Pikal, M.J.; Ayers, B.O. *J. Phys. Chem.* 1966, *70*, 2440.
3. Spedding, F.H.; et. al. *J. Chem. Eng. Data* 1975, *20*, 72.

COMPONENTS:

(1) Erbium nitrate; $Er(NO_3)_3$; [10168-80-6]

(2) Nitric acid; HNO_3; [7697-37-2]

(3) Water ; H_2O ; [7732-18-5]

ORIGINAL MEASUREMENTS:

Afanas'ev, Yu.A.; Azhipa, L.T.; Shakhanova, N.A. *Zh. Neorg. Khim.* 1977, *22*, 3331-4; *Russ. J. Inorg. Chem. Engl. Transl.* 1977, *22*, 1818-20.

VARIABLES:

Concentration of nitric acid at 25°C

PREPARED BY:

T. Mioduski and S. Siekierski

EXPERIMENTAL VALUES:

Solubility of erbium nitrate in nitric acid solutions at 25°C[a]

$Er(NO_3)_3$ mass %	$Er(NO_3)_3$ mol kg^{-1}	HNO_3 mass %	HNO_3 mol kg^{-1}	nature of the solid phase
60.44	4.325	0.0	———	$Er(NO_3)_3{\cdot}6H_2O$
55.48	4.087	6.09	2.515	"
52.26	4.126	11.89	5.263	"
45.26	3.568	18.83	8.322	"
42.46	3.354	21.71	9.616	"
38.48	2.972	24.87	10.769	"
33.74	2.576	29.19	12.496	"
27.51	2.166	36.53	16.121	"
21.86	1.847	44.63	21.136	"
16.81	1.537	52.24	26.786	"
14.91	1.457	56.13	30.759	"
12.96	1.420	61.21	37.607	$Er(NO_3)_3{\cdot}6H_2O + Er(NO_3)_3{\cdot}4H_2O$
10.88	1.220	63.87	40.143	$Er(NO_3)_3{\cdot}4H_2O$
8.66	1.023	67.37	44.604	"
6.99	0.856	69.89	47.973	"
6.18	0.810	72.23	53.093	"
5.23	0.750	75.04	60.358	"

continued ....

AUXILIARY INFORMATION

METHOD/APPARATUS/PROCEDURE:

The isothermal method was used. Equilibrium was reached in 3-4 hours. The compositions of the solutions were varied by addition of the salt to the aqueous nitric acid solutions or by addition of 100% nitric acid to the saturated solutions. Both the liquid and solid phases were analysed. Erbium was determined complexometrically by titration with Trilon using Xylenol Orange indicator, and nitric acid was determined by titration with sodium hydroxide using Methyl Red indicator.

Infrared spectra for the various hydrates are discussed in the source publication.

SOURCE AND PURITY OF MATERIALS:

C.p. grade erbium nitrate was used (presumably the hexahydrate, and used as received, compilers).

Nitric acid was concentrated to 100% by the Brauer method.

No other information given.

ESTIMATED ERROR:

Soly: precision about ± 0.3% (compilers).

Temp: precision ± 0.1 K.

REFERENCES:

COMPONENTS:	ORIGINAL MEASUREMENTS:
(1) Erbium nitrate; $Er(NO_3)_3$; [10168-80-6] (2) Nitric acid; HNO_3; [7697-37-2] (3) Water ; H_2O ; [7732-18-5]	Afanas'ev, Yu.A.; Azhipa, L.T.; Shakanova, N.A. *Zh. Neorg. Khim.* 1977, *22*, 3331-4; *Russ. J. Inorg. Chem. Engl. Transl.* 1977, *22*, 1818-20.

EXPERIMENTAL VALUES: continued ....

Solubility of erbium nitrate in nitric acid solutions at 25°C

$Er(NO_3)_3$		HNO_3		
mass %	mol kg^{-1}	mass %	mol kg^{-1}	nature of the solid phase
5.47	0.934	77.95	74.611	$Er(NO_3)_3 \cdot 4H_2O + Er(NO_3)_3 \cdot 2H_2O$
3.48	0.677	81.97	89.405	$Er(NO_3)_3 \cdot 2H_2O$
3.10	0.760	85.35	117.27	"
3.20	0.889	86.61	134.89	"
4.43	1.891	88.94	212.89	"
5.59	3.996	90.45	362.48	$Er(NO_3)_3 \cdot 2H_2O + Er(NO_3)_3$
4.20	3.112	91.98	382.12	$Er(NO_3)_3$
3.10	2.596	93.52	439.10	"
2.38	2.968	95.35	666.60	"

a. Molalities calculated by M. Salomon.

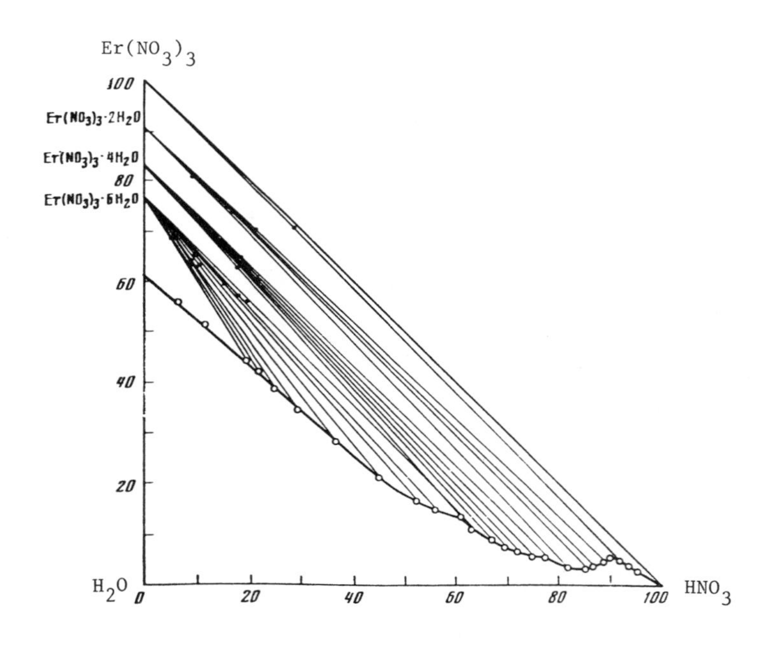

Solubility diagram of the $Er(NO_3)_3$-HNO_3-H_2O system at 298.2 K. All compositions given in mass % units.

COMPONENTS:

(1) Erbium nitrate; $Er(NO_3)_3$; [10168-80-6]

(2) Pyridine nitrate; $C_5H_6N_2O_3$; [543-53-3]

(3) Water ; H_2O ; [7732-18-5]

ORIGINAL MEASUREMENTS:

Starikova, L.I.; Zhuravlev, E.F.; Khalfina, L.R. *Zh. Neorg. Khim.* 1979, *24*, 2803-5; *Russ. J. Inorg. Chem. Engl. Transl.* 1979, *24*, 1559-60.

VARIABLES:

Composition at 25°C and 50°C

PREPARED BY:

T. Mioduski and S. Siekierski

EXPERIMENTAL VALUES:

25°C Isotherm[a]					50°C Isotherm[a]			
$Er(NO_3)_3$		$C_5H_5N{\cdot}HNO_3$			$Er(NO_3)_3$		$C_5H_5N{\cdot}HNO_3$	
mass %	mol kg^{-1}	mass %	mol kg^{-1}	solid phase[b]	mass %	mol kg^{-1}	mass %	mol kg^{-1}
0.0	———	74.5	20.56	A	0.0	———	87.0	47.09
10.0	1.26	67.5	21.11	"	5.0	1.18	83.0	48.67
17.0	2.47	63.5	22.91	"	9.5	2.56	80.0	53.61
23.5	4.16	60.5	26.61	"	14.0	4.17	76.5	56.66
34.0	12.03	58.0	51.02	"	24.0	12.35	70.5	90.20
37.0	20.95	58.0	81.62	A+B	32.0	7.88	56.5	34.57
42.5	13.37	48.5	37.92	B	44.0	15.57	48.0	42.22
48.5	10.56	38.5	20.84	"	49.5	13.34	40.0	26.81
53.0	10.00	32.0	15.01	"	53.0	12.50	35.0	20.52
54.0	10.19	31.0	14.54	B+C	58.5	13.25	29.0	16.32
59.0	6.55	15.5	4.28	C	60.0	11.32	25.0	11.73
61.5	6.00	9.5	2.31	"	64.0	7.55	12.0	3.52
63.5	5.53	4.0	0.87	"	66.0	6.92	7.0	1.82
65.5	5.37	0.0	———	"	68.0	6.87	4.0	1.01
				"	69.0	6.30	0.0	———

a. Molalities calculated by M. Salomon.
b. Solid phases: A = $C_5H_5N{\cdot}HNO_3$; B = $Er(NO_3)_3{\cdot}2[C_5H_5N{\cdot}HNO_3]$

C = $Er(NO_3)_3{\cdot}5H_2O$

AUXILIARY INFORMATION

METHOD/APPARATUS/PROCEDURE:

The method of isothermal sections was used with refractometric analyses (1). Heterogeneous and homogeneous mixtures of known composition were equilibrated until their refractive indices remained constant. The composition of the saturated solutions and the corresponding solid phases were found as inflection or "break" points on a plot of composition against refractive index.

The double salt is congruently soluble and was isolated. Analysis for Er by the oxalate method and NO_3 by pptn with nitron confirmed the composition of the double salt. Thermal and X-ray studies on the double salt are reported in the source paper.

SOURCE AND PURITY OF MATERIALS:

Commercial c.p. grade $Er(NO_3)_3{\cdot}5H_2O$ was used.

Preparation of the amine nitrate was described previously (2): i.e. neutralization of pyridine by nitric acid followed by crystallization and drying in a desiccator over anhydrous $CaCl_2$.

ESTIMATED ERROR:

Soly: precision at best is ± 1% (compilers).

Temp: precision probably ± 0.1-0.2 K (compilers).

REFERENCES:

1. Zhuravlev, E.F.; Sheveleva, A.D. *Zh. Neorg. Khim.* 1960, *5*, 2630.

2. Zhuravlev, E.F.; Starikova, L.I. *Zh. Neorg. Khim.* 1975, *20*, 1406.

COMPONENTS:

(1) Erbium nitrate; $Er(NO_3)_3$; [10168-80-6]

(2) Quinoline nitrate; $C_9H_8N_2O_3$; [21640-15-3]

(3) Water ; H_2O ; [7732-18-5]

ORIGINAL MEASUREMENTS:

Starikova, L.I.; Zhuravlev, E.F.; Khalfina, L.R. *Zh. Neorg. Khim.* 1979, *24*, 2803-5; *Russ. J. Inorg. Chem. Engl. Transl.* 1979, *24*, 1559-60.

VARIABLES:

Composition at 25°C and 50°C

PREPARED BY:

T. Mioduski and S. Siekierski

EXPERIMENTAL VALUES:

25°C Isotherm[a]					50°C Isotherm[a]			
$Er(NO_3)_3$		$C_9H_7N{\cdot}HNO_3$			$Er(NO_3)_3$		$C_9H_7N{\cdot}HNO_3$	
mass %	mol kg^{-1}	mass %	mol kg^{-1}	solid phase[b]	mass %	mol kg^{-1}	mass %	mol kg^{-1}
0.0	———	72.0	13.38	A	0.0	———	82.0	23.71
11.0	1.22	63.5	12.96	"	7.0	1.17	76.0	23.26
18.0	2.22	59.0	13.35	"	12.5	2.28	72.0	24.17
25.0	3.63	55.5	14.81	"	17.5	3.67	69.0	26.60
30.5	5.57	54.0	18.13	A+B	27.5	9.16	64.0	39.18
39.0	5.66	41.5	11.07	B	31.0	8.36	58.5	28.99
39.5	3.24	26.0	3.92	"	41.0	7.74	44.0	15.26
58.5	6.13	14.5	2.79	"	53.5	8.19	28.0	7.88
60.5	6.23	12.0	2.27	B+C	63.0	8.70	16.5	4.19
62.0	5.85	8.0	1.39	C	63.5	8.56	15.5	3.84
64.0	5.49	3.0	0.47	"	65.0	7.51	10.5	2.23
65.6	5.40	0.0	———	"	66.5	6.85	6.0	1.14
				"	68.0	6.42	2.0	0.35
				"	69.5	6.45	0.0	———

a. Molalities calculated by M. Salomon.

b. Solid phases: A = $C_9H_7N{\cdot}HNO_3$; B = $Er(NO_3)_3{\cdot}2[C_9H_7N{\cdot}HNO_3]{\cdot}4H_2O$

C = $Er(NO_3)_3{\cdot}5H_2O$

AUXILIARY INFORMATION

METHOD/APPARATUS/PROCEDURE:

The method of isothermal sections was used with refractometric analyses (1). Heterogeneous and homogeneous mixtures of known composition were equilibrated until their refractive indices remained constant. The composition of the saturated solutions and the corresponding solid phases were found as inflection or "break" points on a plot of composition against refractive index.

The double salt is congruently soluble and was isolated. Analysis for Er by the oxlate method and NO_3 by pptn with nitron confirmed the composition of the double salt. Thermal and X-ray studies on the double salt are reported in the source paper.

SOURCE AND PURITY OF MATERIALS:

Commercial c.p. grade $Er(NO_3)_3{\cdot}5H_2O$ was used.

Preparation of the amine nitrate was described previously (2): i.e. neutralization of quinoline by nitric acid followed by crystallization and drying in a desiccator over anhydrous $CaCl_2$.

ESTIMATED ERROR:

Soly: precision at best is ± 1 % (compilers).

Temp: precision probably ± 0.1-0.2 K (compilers).

REFERENCES:

1. Zhuravlev, E.F.; Sheveleva, A.D. *Zh. Neorg. Khim.* 1960, *5*, 2630.

2. Zhuravlev, E.F.; Starikova, L.I. *Zh. Neorg. Khim.* 1975, *20*, 1406.

COMPONENTS:

(1) Erbium nitrate; $Er(NO_3)_3$; [10168-80-6]

(2) 6-Methylquinoline nitrate; $C_{10}H_{10}N_2O_3$;

(3) Water ; H_2O ; [7732-18-5]

ORIGINAL MEASUREMENTS:

Starikova, L.I.; Zhuravlev, E.F.

Zh. Neorg. Khim. 1980, *25*, 2007-9; *Russ. J. Inorg. Chem. Engl. Transl.* 1980, *25*, 1114-5.

VARIABLES:

Composition at 20°C and 40°C

PREPARED BY:

T. Mioduski and S. Siekierski

EXPERIMENTAL VALUES:

20°C Isotherm[a]					40°C Isotherm[a]			
$Er(NO_3)_3$		$6\text{-}CH_3C_9H_6N\cdot HNO_3$		solid	$Er(NO_3)_3$		$6\text{-}CH_3C_9H_6N\cdot HNO_3$	
mass %	mol kg^{-1}	mass %	mol kg^{-1}	phase[b]	mass %	mol kg^{-1}	mass %	mol kg^{-1}
0.0	———	73.0	13.11	A	0.0	———	80.0	19.40
9.5	1.20	68.0	14.66	"	7.0	1.17	76.0	21.68
10.5	1.37	67.8	15.15	A+B	11.5	2.25	74.0	24.75
19.5	1.55	45.0	6.15	B	16.5	2.40	64.0	15.92
30.0	1.91	25.5	2.78	"	22.0	2.31	51.0	9.16
41.0	2.37	10.0	0.99	"	29.0	2.49	38.0	5.58
48.0	2.86	4.5	0.46	"	43.5	3.24	18.5	2.36
59.5	4.32	1.5	0.19	"	55.0	4.21	8.0	1.05
64.5	5.29	1.0	0.14	B+C	66.7	6.38	3.7	0.61
64.9	5.23	0.0	———	C	67.0	6.43	3.5	0.58
				"	67.3	6.15	1.7	0.27
				"	67.7	5.93	0.0	———

a. Molalities calculated by M. Salomon.

b. Solid phases: A = $6\text{-}CH_3C_9H_6N\cdot HNO_3$; B = $Er(NO_3)_3\cdot 4[CH_3C_9H_6N\cdot HNO_3]$

C = $Er(NO_3)_3\cdot 5H_2O$

AUXILIARY INFORMATION

METHOD/APPARATUS/PROCEDURE:

The method of isothermal sections was used with refractometric analyses (1). Heterogeneous and homogeneous mixtures of known composition were equilibrated until their refractive indices remained constant. The composition of the saturated solutions and the corresponding solid phases were found as inflection or "break" points on a plot of composition against refractive index.

The composition of the double salt was detd graphically and was confirmed by chem anal for the Er content (the oxalate method) and for the NO_3 content (pptn with nitron). The double salt is congruenty soluble, and its solubility in water is 63.0 mass % at 20°C, and 72.5 mass % at 40°C (1.45 mol kg^{-1} and 2.24 mol kg^{-1}, respectively: compilers).

Thermal studies are presented in the source paper.

SOURCE AND PURITY OF MATERIALS:

1. C.p. grade $Er(NO_3)_3\cdot 5H_2O$ recrystd. prior to use.

2. 6-Methylquinoline nitrate was obtained as previously described (2).

ESTIMATED ERROR:

Nothing specified.

REFERENCES:

1. Zhuravlev, E.F.; Sheveleva, A.D. *Zh. Neorg. Khim.* 1960, *5*, 2630.
2. Zhuravlev, E.F.; Sheveleva, A.D. *Zh. Neorg. Khim.* 1975, *20*, 1406.

COMPONENTS:

(1) Erbium nitrate; $Er(NO_3)_3$; [10168-80-6]

(2) Urea; CH_4N_2O; [57-13-6]

(3) Water ; H_2O ; [7732-18-5]

ORIGINAL MEASUREMENTS:

Aitimbetov, K.; Sulaimankulov, K.S.; Batyuk, A.G.; Ismailov, M. *Zh. Neorg. Khim.* 1975, *20*, 2510-3; *Russ. J. Inorg. Chem. Engl. Transl.* 1975, *20*, 1391-2.

VARIABLES:

Composition at 30°C

PREPARED BY:

T. Mioduski and S. Siekierski

EXPERIMENTAL VALUES:

Composition of saturated solutions at 30°C[a]

$Er(NO_3)_3$		$CO(NH_2)_2$		
mass %	mol kg^{-1}	mass %	mol kg^{-1}	nature of the solid phase
———	———	57.56	22.583	$CO(NH_2)_2$
3.33	0.232	56.08	23.006	"
7.83	0.597	55.06	24.705	"
11.15	0.906	54.00	25.801	"
15.86	1.455	53.29	28.763	"
19.80	1.993	52.08	30.839	"
25.26	3.014	51.02	35.815	"
35.14	6.413	49.35	52.981	"
37.54	7.623	48.52	57.957	"
39.80	9.349	48.15	66.536	$CO(NH_2)_2 + Er(NO_3)_3 \cdot 4CO(NH_2)_2$
41.50	9.765	46.47	64.321	$Er(NO_3)_3 \cdot 4CO(NH_2)_2$
41.05	8.931	45.94	58.797	"
41.72	7.076	41.59	41.493	"
44.34	6.069	34.98	28.165	"
46.52	5.475	29.43	20.376	"
48.74	5.588	26.57	17.919	"
51.20	6.340	25.94	18.895	$CO(NH_2)_2 + Er(NO_3)_3 \cdot 3CO(NH_2)_2 \cdot 2H_2O$

continued ....

AUXILIARY INFORMATION

METHOD/APPARATUS/PROCEDURE:

The isothermal method was used. Equilibrium was reached after 7-8 hours. The liquid phase was separated from the solid phase using a Schott No. 3 filter. Erbium was determined by titration with Trilon. Nitrogen of urea was determined by the Kjeldahl method. The composition of the solid phases was determined by Schreinemakers' method of residues.

SOURCE AND PURITY OF MATERIALS:

Nothing specified.

ESTIMATED ERROR:

Nothing specified.

REFERENCES:

COMPONENTS:	ORIGINAL MEASUREMENTS:
(1) Erbium nitrate; $Er(NO_3)_3$; [10168-80-6] (2) Urea; CH_4N_2O; [57-13-6] (3) Water ; H_2O ; [7732-18-5]	Aitimbetov, K.; Sulaimankulov, K.S.; Batyuk, A.G.; Ismailov, M. *Zh. Neorg. Khim.* 1975, *20*, 2510-3; *Russ. J. Inorg. Chem. Engl. Transl.* 1975, *20*, 1391-2.

EXPERIMENTAL VALUES: continued ....

Composition of saturated solutions at 30°C[a]

$Er(NO_3)_3$		$CO(NH_2)_2$		
mass %	mol kg^{-1}	mass %	mol kg^{-1}	nature of the solid phase
51.96	6.136	24.07	16.721	$Er(NO_3)_3 \cdot 3CO(NH_2)_2 \cdot 2H_2O$
54.00	6.011	20.57	13.469	"
55.51	5.947	18.07	11.389	"
59.96	6.395	13.50	8.470	"
63.43	7.100	11.28	7.427	"
65.46	7.350	9.33	6.162	"
64.54	6.702	8.20	5.009	$Er(NO_3)_3 \cdot 6H_2O$
65.84	6.341	4.77	2.702	"
65.00	5.257	——	——	"

a. Molalities calculated by M. Salomon.

COMPONENTS:

(1) Erbium nitrate; $Er(NO_3)_3$; [10168-80-6]

(2) Acetylurea; $C_3H_6N_2O_2$; [591-07-1]

(3) Water ; H_2O ; [7732-18-5]

ORIGINAL MEASUREMENTS:

Isakova, S.; Sulaimankulov, K.; Aitimbetov, K.; Kozhanova, T. *Zh. Neorg. Khim.* 1980, *25*, 2271-3; *Russ. J. Inorg. Chem. Engl. Transl.* 1980, *25*, 1257-8.

VARIABLES:

Composition at 30°C

PREPARED BY:

T. Mioduski and S. Siekierski

EXPERIMENTAL VALUES:

Composition of saturated solutions at 30°C[a]

$Er(NO_3)_3$ mass %	$Er(NO_3)_3$ mol kg^{-1}	$CH_3CONHCONH_2$ mass %	$CH_3CONHCONH_2$ mol kg^{-1}	nature of the solid phase
0	———	4.01	0.409	$CH_3CONHCONH_2$
0.02	0.001	4.28	0.438	"
16.83	0.600	3.80	0.469	"
23.58	0.924	4.17	0.565	"
30.84	1.337	3.85	0.577	"
39.18	1.963	4.33	0.751	"
49.53	3.010	3.89	0.818	"
54.10	3.636	3.78	0.879	"
61.12	4.974	4.10	1.155	$CH_3CONHCONH_2 + Er(NO_3)_3 \cdot 6H_2O$
62.17	5.216	4.09	1.187	"
61.18	4.955	3.87	1.085	$Er(NO_3)_3 \cdot 6H_2O$
62.64	5.042	2.19	0.610	"
65.18	5.299	0	———	"

a. Molalities calculated by M. Salomon

AUXILIARY INFORMATION

METHOD/APPARATUS/PROCEDURE:

Isothermal method employed. Equilibrium is stated to be reached within 7-8 h. Saturated solutions and solid phases were analysed for the Er content by titrn with Trilon (Na_2H_2(EDTA)) using Xylenol Orange indicator, and for nitrogen of acetylurea by the Kjeldahl method.

SOURCE AND PURITY OF MATERIALS:

Nothing specified

ESTIMATED ERROR:

Nothing specified.

REFERENCES:

COMPONENTS:

(1) Erbium nitrate; $Er(NO_3)_3$; [10168-80-6]

(2) Acetamide; C_2H_5NO; [60-35-5]

(3) Water ; H_2O ; [7732-18-5]

ORIGINAL MEASUREMENTS:

Aitimbetov, K.; Sulaimankulov, K.; Isakova, S.; Ismailov, M. *Izv. Akad. Nauk Kirg. SSR* 1975, 46-8.

VARIABLES:

Composition at 30°C

PREPARED BY:

T. Mioduski and S. Siekierski

EXPERIMENTAL VALUES:

Composition of saturated solutions at 30°C[a]

$Er(NO_3)_3$		CH_3CONH_2		nature of the solid phase
mass %	mol kg^{-1}	mass %	mol kg^{-1}	
———	———	78.08	60.304	CH_3CONH_2
4.08	0.554	75.06	60.917	"
13.57	2.631	71.83	83.291	"
20.72	3.523	62.63	63.682	"
36.48	9.816	53.00	85.291	"
42.55	14.529	49.16	100.39	"
46.21	21.063	47.58	129.71	"
50.62	30.358	44.66	160.19	$CH_3CONH_2 + Er(NO_3)_3 \cdot 3CH_3CONH_2$
51.42	40.886	45.02	214.09	$Er(NO_3)_3 \cdot 3CH_3CONH_2$
51.27	14.931	39.01	67.945	"
51.13	11.729	36.53	50.116	"
53.89	7.416	25.54	21.020	"
59.02	7.040	17.25	12.307	"
62.21	7.590	14.59	10.647	"
62.77	6.357	9.28	5.621	$Er(NO_3)_3 \cdot 3CH_3CONH_2 + Er(NO_3)_3 \cdot 6H_2O$
63.49	6.117	7.13	4.108	$Er(NO_3)_3 \cdot 6H_2O$
63.23	5.280	2.73	1.358	"
65.18[b]	5.299	———	———	"

a. Molalities calculated by M. Salomon.

b. In the text, the authors report the soly in water at 30°C as 65.08 mass % (5.275 mol kg^{-1}).

AUXILIARY INFORMATION

METHOD/APPARATUS/PROCEDURE:

Isothermal method employed. Equilibrium is stated to be reached in 8-9 h. Er was determined by titration with Trilon (Na_2H_2(EDTA)) and nitrogen of acetamide by the Kjeldahl method.

SOURCE AND PURITY OF MATERIALS:

Nothing specified.

ESTIMATED ERROR:

Nothing specified.

REFERENCES:

COMPONENTS:

(1) Erbium nitrate; $Er(NO_3)_3$; [10168-80-6]
(2) Diethyl ether; $C_4H_{10}O$; [60-29-7]

ORIGINAL MEASUREMENTS:

Wells, R.C. *J. Wash. Acad. Sci.* 1930, *20*, 146-8.

VARIABLES:

Room temperature (about 20°C)

PREPARED BY:

T. Mioduski, S. Siekierski, M. Salomon

EXPERIMENTAL VALUES:

Experiment 1. This experiment involves the hydrated nitrate as the initial solid, and which the compilers assume to be the hexahydrate.

Authors report the solubility as 0.162 g Er_2O_3 in 10 ml ether.

This is equivalent to a $Er(NO_3)_3$ soly of 0.0847 mol dm^{-3} (compilers).

Experiment 2. This experiment involves erbium nitrate dehydrated as described in the METHOD/APPARUTUS/PROCEDURE box below.

Authors report the solubility as 0.0190 g Er_2O_3 in 10 ml ether.

This is equivalent to a $Er(NO_3)_3$ soly of 0.0993 mol dm^{-3} (compilers).

AUXILIARY INFORMATION

METHOD/APPARATUS/PROCEDURE:

The isothermal method was used. The soly of erbium nitrate was determined in two experiments in which the nature of the initial solid phase differs.

Experiment 1. A few grams of erbium nitrate (presumably the hexahydrate, compilers) was added to about 20 ml of ether in small stoppered flasks. The flasks were periodically agitated and permitted to stand at about 20°C overnight. A 10 ml sample was removed, filtered, the solvent evaporated and the salt ignited to the oxide and weighed.

Experiment 2. The remaining salt in the flask was freed from ether, dissolved in water and a few drops of HNO_3 added. The solution was evaporated to dryness and heated to 150°C. The solubility in ether was then determined again with this "dehydrated" salt.

SOURCE AND PURITY OF MATERIALS:

Nothing specified.

ESTIMATED ERROR:

Soly: precision probably around ± 10% (compilers).
Temp: precision probably ± 4 K (compilers).

REFERENCES:

COMPONENTS:	EVALUATOR:
(1) Thulium nitrate; $Tm(NO_3)_3$; [14985-19-4] (2) Water ; H_2O ; [7732-18-5]	S. Siekierski, T. Mioduski Institute for Nuclear Research Warsaw, Poland and M. Salomon U.S. Army ET & DL Ft. Monmouth, NJ January 1983

CRITICAL EVALUATION:

INTRODUCTION AND EVALUATION PROCEDURE

The solubility of thulium nitrate in water has been reported in four publications at 298.15 K (1-4), and in one publication at 303.2 K (5). The solid phases identified in the binary system are the hexa- and pentahydrates, and in nitric acid solutions (4), the dihydrate and anhydrous salt are stable solid phases depending upon nitric acid concentration.

$Tm(NO_3)_3 \cdot 6H_2O$	[35725-33-8]	$Tm(NO_3)_3 \cdot 4H_2O$	[37131-80-9]
$Tm(NO_3)_3 \cdot 5H_2O$	[36548-87-5]	$Tm(NO_3)_3 \cdot 2H_2O$	[81201-55-0]

For solubilities in the hexahydrate system, the data reported by Spedding et al (1-3) are considered to be the most accurate since previous results from Spedding's laboratory have proved accurate and precise. The results of Afanas'ev et al. for Dy, Ho, Er, Tm, and Yb nitrates in the hexahydrate system at 298.2 K are all very low and highly improbable. In the $Er(NO_3)_3$-H_2O critical evaluation we concluded that the results of Afanas'ev et al. are in error, probably due to the failure to reach equilibrium.

For the solubilities in the metastable pentahydrate system at 298.15 K, Afanas'ev's result (4) of 5.40 mol/kg is lower than that for the stable hexahydrate system reported in (1-3), and again we reject Afanas'ev's result.

We note that there is considerable difficulty in identification of the solid phases. Afanas'ev et al. (4) report that while the pentahydrate is metastable at 298.2 K, it does not easily convert to the stable hexahydrate (see the compilation for this paper). Ivanov-Emin et al. (6) report a congruent melting point of 335 K for the tetrahydrate which may actually correspond to the hexahydrate since Wendlandt and Sewell (7) report that the tetrahydrate melts congruently at 363 ± 1 K.

TENTATIVE SOLUBILITIES

For solubilities in the hexahydrate system, we consider the results from (1-3) to be the most accurate as stated above. The difference between the highest and lowest value reported by Spedding et al. is significantly greater than the differences reported by these authors for other lanthanide nitrates. Since it is not very probable that this relatively large difference is the result of an analytical error typical just for the determination of thulium, the observed scatter of data may be due to difficulties in establishing true equilibrium, perhaps due to the possibility that the hexahydrate → pentahydrate transition temperature is close to 298.2 K. If so, then the lower results from (1-3) should be preferred, and thus the *tentative* solubility of $Tm(NO_3)_3$ at 298.15 K in the hexahydrate system is 5.95 mol kg^{-1} with an overall uncertainty of about -0.02 to + 0.05 mol kg^{-1}.

For the solubility of $Tm(NO_3)_3$ at 303.2 K in the pentahydrate system, the value of 6.692 mol kg^{-1} reported in (5) cannot be evaluated at this time. We cautiously designate this result as a *tentative* value, and note that this value suggests that the pentahydrate system is still metastable at 303.2 K.

REFERENCES

1. Rard, J.A.; Shiers, L.E.; Heiser, D.J.; Spedding, F.H. *J. Chem. Eng. Data* 1977, *22*, 337.
2. Rard, J.A.; Spedding, F.H. *J. Phys. Chem.* 1975, *79*, 257.
3. Spedding, F.H.; Derer, J.L.; Mohs, M.A.; Rard, J.A. *J. Chem. Eng. Data* 1976, *21*, 474.
4. Afanas'ev, Yu.A.; Azhipa, L.T.; Linnik, N.V. *Zh. Neorg. Khim.* 1976, *21*, 1661.
5. Khudaibergenova, N.; Sulaimankulov, K. *Zh. Neorg. Khim.* 1981, *26*, 1156.
6. Ivanov-Emin, B.N.; Odinets, Z.K.; Del'Pino, Kh.; Zaitsev, B.E. *Zh. Neorg. Khim.* 1976, *21*, 873.
7. Wendlandt, W.W.; Sewell, R.G. *Texas J. Sci.* 1961, *13*, 231.

COMPONENTS:

(1) Thulium nitrate; $Tm(NO_3)_3$; [14985-19-4]

(2) Water ; H_2O ; [7732-18-5]

ORIGINAL MEASUREMENTS:

1. Rard, J.A.; Shiers, L.E.; Heiser, D.J.; Spedding, F.H. *J. Chem. Eng. Data* 1977 *22*, 337-47.
2. Rard, J.A.; Spedding. F.H. *J. Phys. Chem.* 1975, *79*, 257-62.
3. Spedding, F.H.; Derer, J.L.; Mohs, M.A.; Rard, J.A. *J. Chem. Eng. Data* 1976, *21*, 474-88.

VARIABLES:

One temperature: 25.00°C

PREPARED BY:

T. Mioduski, S. Siekierski and M. Salomon

EXPERIMENTAL VALUES:

Source paper [1] reports the solubility of $Tm(NO_3)_3$ as 5.9526 mol kg^{-1}

Source paper [2] reports the solubility of $Tm(NO_3)_3$ as 5.9483 mol kg^{-1}

Source paper [3] reports the solubility of $Tm(NO_3)_3$ as 6.028 mol kg^{-1}

AUXILIARY INFORMATION

METHOD/APPARATUS/PROCEDURE:

[2] and [3] Standard isothermal method used. Solutions prepared as in (1) and (2).
[1] Isothermal isopiestic method used in which equilibration carried out by allowing a less than satd $Tm(NO_3)_3$ sln to reach thermodynamic equilibrium through the vapor phase with a reference sln (KCl, $CaCl_2$). The soly was thus detd without sepn of the sln and solid phases. The solutions were adjusted to their equivalence pH values to insure a ratio of three nitrates to each Tm. Duplicate samples of the nitrate and reference slns were used and equil was approached from higher and lower concns (about 4 days).

In [1], [2] and [3] the satd slns were analysed by EDTA titrn and gravimetric sulfate analysis. The methods are stated to be accurate to 0.1% or better. In the sulfate analysis, the salt was first decomposed with HCl followed by evaporation to dryness before sulfuric acid additions were made. This eliminated the possibility of nitrate ion coprecipitation.

SOURCE AND PURITY OF MATERIALS:

$Tm(NO_3)_3 \cdot 6H_2O$ was prepd by addn of HNO_3 to the oxide. The oxide was purified by an ion exchange method, and the upper limit for the impurities Ca, Fe, Si and adjacent rare earths was given as 0.15%.

In source paper [3] the salt was analysed for water of hydration and found to be within ± 0.016 water molecules of the hexahydrate.

Water was distilled from an alkaline permanganate solution.

ESTIMATED ERROR:

Soly: duplicate analyses agreed to at least ± 0.1%.

Temp: not specified, but probably accurate to at least ± 0.01 K as in (3)(compilers).

REFERENCES:

1. Spedding, F.G.; Cullen, P.F.; Habenschuss, A. *J. Phys. Chem.* 1974, *78*, 1106.
2. Spedding, F.H.; Pikal, M.J.; Ayers, B.O. *J. Phys. Chem.* 1966, *70*, 2440.
3. Spedding, F.H.; et. al *J. Chem. Eng. Data* 1975, *20*, 72.

COMPONENTS:

(1) Thulium nitrate; $Tm(NO_3)_3$; [14985-19-4]

(2) Nitric acid; HNO_3; [7697-37-2]

(3) Water ; H_2O ; [7732-18-5]

ORIGINAL MEASUREMENTS:

Afanas'ev, Yu. A.; Azhipa, L.T.; Linnik, N.V. *Zh. Neorg. Khim.* 1976, *21*, 1661-3; *Russ. J. Inorg. Chem. Engl. Transl.* 1976, *21*, 909-10.

VARIABLES:

Concentration of HNO_3 at 25°C

PREPARED BY:

T. Mioduski and S. Siekierski

EXPERIMENTAL VALUES:

Solubility of thulium nitrate in nitric acid solutions at 25°C[a]

$Tm(NO_3)_3$ mass %	$Tm(NO_3)_3$ mol kg^{-1}	HNO_3 mass %	HNO_3 mol kg^{-1}	nature of the solid phase
65.7	5.40	0	———	$Tm(NO_3)_3\cdot5H_2O$
56.1	4.52	8.9	4.04	"
47.6	3.68	16.0	6.98	"
42.6	3.69	24.9	12.16	"
35.0	2.99	32.0	15.39	"
34.6	3.19	34.8	18.05	$Tm(NO_3)_3\cdot5H_2O + Tm(NO_3)_3\cdot4H_2O$
51.2	2.96	0	———	$Tm(NO_3)_3\cdot6H_2O$
44.5	2.61	7.5	2.48	"
38.1	2.46	18.2	6.61	"
35.9	2.52	24.0	9.50	"
31.0	2.40	32.6	14.21	"
29.0	2.29	35.4	15.78	"
29.2	2.44	37.1	17.47	"
28.7	2.55	39.6	19.82	$Tm(NO_3)_3\cdot6H_2O + Tm(NO_3)_3\cdot4H_2O$

continued..........

AUXILIARY INFORMATION

METHOD/APPARATUS/PROCEDURE:

The isothermal method was used. Compositions of the solutions were varied by adding 100% nitric acid to the saturated solutions, or by adding the salt to the acid solutions. Equilibrium was reached in 3-4 hours. The concentration of Tm in the liquid and the residue was detd complexometrically, and the concentration of HNO_3 was detd by titrn with NaOH using Methyl Red indicator. The composition of the solid phase was detd by Schreinemakers' method of residues.

The pentahydrate is metastable, and attempts to crystallize the hexahydrate by seeding an equilibrated solution failed. The hexahydrate could be produced by heating a solution in equilibrium with the pentahydrate followed by slow cooling.

SOURCE AND PURITY OF MATERIALS:

C.p. grade thulium nitrate was used.

C.p. grade nitric acid was concentrated to 100% by the Brauer method.

No other information given.

ESTIMATED ERROR:

Soly: precision probably about ± 0.5%, but accuracy is very poor (see critical evaluation).

Temp: precision ± 0.1 K.

REFERENCES:

COMPONENTS:	ORIGINAL MEASUREMENTS:
(1) Thulium nitrate; $Tm(NO_3)_3$; [14985-19-4]	Afanas'ev, Yu.A.; Azhipa, L.T.; Linnik, N.V.
(2) Nitric acid; HNO_3; [7697-37-2]	*Zh. Neorg. Khim.* <u>1976</u>, *21*, 1661-3; *Russ. J. Inorg. Chem. Engl. Transl.* <u>1976</u>, *21*, 909-10.
(3) Water ; H_2O ; [7732-18-5]	

EXPERIMENTAL VALUE: continued ....

Solubility of thulium nitrate in nitric acid solutions at 25°C[a]

$Tm(NO_3)_3$		HNO_3		
mass %	mol kg^{-1}	mass %	mol kg^{-1}	nature of the solid phase
23.4	2.07	44.7	22.24	$Tm(NO_3)_3\cdot4H_2O$
22.5	2.06	46.7	24.06	"
21.4	1.96	47.9	24.76	"
21.9	2.21	50.2	28.55	$Tm(NO_3)_3\cdot4H_2O + Tm(NO_3)_3\cdot2H_2O$
14.0	1.31	55.9	29.47	$Tm(NO_3)_3\cdot2H_2O$
10.8	1.04	59.9	32.44	"
7.6	0.75	63.8	35.40	"
7.4	0.86	68.4	44.86	$Tm(NO_3)_3\cdot2H_2O + Tm(NO_3)_3$
4.0	0.48	72.6	49.24	$Tm(NO_3)_3$
2.3	0.36	79.8	70.75	$Tm(NO_3)_3\cdot nNO_2$
1.4	0.28	84.7	96.70	"
0.3	0.12	92.5	203.9	"

a. Molalities calculated by M. Salomon.

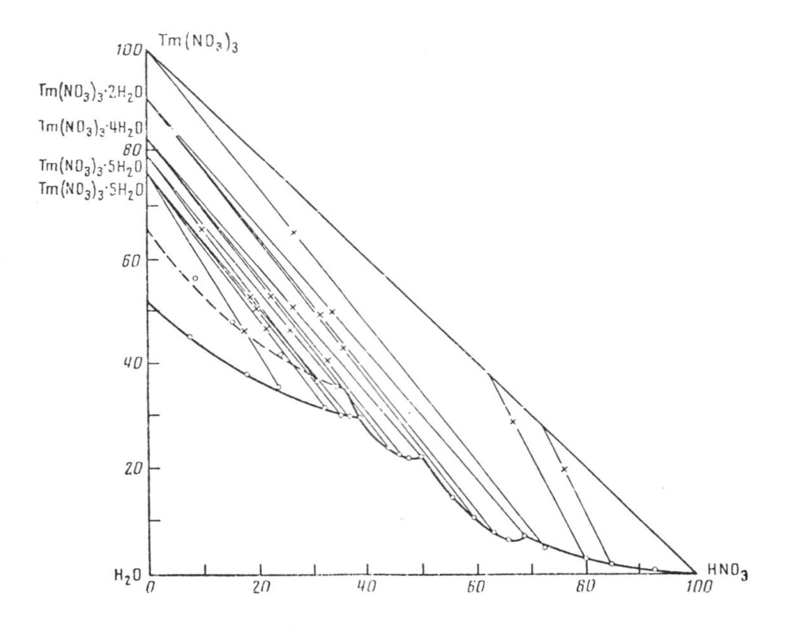

Solubility diagram for the $Tm(NO_3)_3 - HNO_3 - H_2O$ system at 298.2 K. All compositions given in mass % units.

COMPONENTS:	ORIGINAL MEASUREMENTS:
(1) Thulium nitrate; $Tm(NO_3)_3$; [14985-19-4] (2) Urea; CH_4N_2O; [57-13-6] (3) Water ; H_2O ; [7732-18-5]	Khudaibergenova, N.; Sulaimankulov, K. *Zh. Neorg. Khim.* 1981, *26*, 1156-9; *Russ. J. Inorg. Chem. Engl. Transl.* 1981, *26*, 627-8.

VARIABLES:	PREPARED BY:
Composition at 30°C	M. Salomon

EXPERIMENTAL VALUES:

Composition of saturated solutions[a]

$Tm(NO_3)_3$ mass %	$Tm(NO_3)_3$ mol kg^{-1}	$CO(NH_2)_2$ mass %	$CO(NH_2)_2$ mol kg^{-1}	nature of the solid phase
——	——	57.56	22.583	$CO(NH_2)_2$
18.36	1.932	54.86	34.111	"
25.81	3.718	54.63	46.506	"
25.76	3.329	52.44	40.054	"
35.72	8.168	51.96	70.227	"
37.72	10.327	51.99	84.130	"
39.35	13.980	52.72	110.70	$CO(NH_2)_2$ + $Tm(NO_3)_3\cdot4CO(NH_2)_2\cdot2H_2O$
37.63	7.556	48.34	57.371	$Tm(NO_3)_3\cdot4CO(NH_2)_2\cdot2H_2O$
38.47	6.799	45.59	47.624	"
38.92	5.417	40.84	33.599	"
43.54	5.674	34.84	26.833	"
52.83	6.128	22.88	15.685	"
56.29	5.792	16.33	9.931	$Tm(NO_3)_3\cdot3CO(NH_2)_2$
61.20	6.492	12.24	7.674	"
67.20	8.407	10.28	7.601	"
70.11	9.044	8.05	6.137	$Tm(NO_3)_3\cdot3CO(NH_2)_2$ + $Tm(NO_3)_3\cdot5H_2O$
69.12	7.172	3.73	2.288	$Tm(NO_3)_3\cdot5H_2O$
70.37	6.691	——	——	"

a. Molalities calculated by the compiler.

AUXILIARY INFORMATION

METHOD/APPARATUS/PROCEDURE:

Isothermal method used. Equilibrium was reached after 7-9 h. After the liquid and solid phases had been separated, their nitrogen content was determined by the Kjeldahl method, and thulium was determined as described previously (1).

COMMENTS AND/OR ADDITIONAL DATA:

The complex $Tm(NO_3)_3\cdot3CO(NH_2)_2$ is incongruently soluble, and $Tm(NO_3)_3\cdot4CO(NH_2)_2\cdot2H_2O$ is congruently soluble. The compositions of these salts were confirmed by chemical analyses.

SOURCE AND PURITY OF MATERIALS:

"Chemically pure" grade area and crystalline yttrium nitrate hydrate were used. No other information given.

ESTIMATED ERROR:

Soly: accuracy for the $Tm(NO_3)_3$-H_2O binary system very poor (see critical evaluation.

Temp: nothing specified.

REFERENCES:

1. Khudaibergenova, N.; Sulaimankulov, K. *Zh. Neorg. Khim.* 1979, *24*, 2005.

COMPONENTS:	EVALUATOR:
(1) Ytterbium nitrate; $Yb(NO_3)_3$; [13768-67-7] (2) Water ; H_2O ; [7732-18-5]	S. Siekierski, T. Mioduski Institute for Nuclear Research Warsaw, Poland and M. Salomon U.S. Army ET & DL Ft. Monmouth, NJ January 1983

CRITICAL EVALUATION:

INTRODUCTION AND EVALUATION PROCEDURE

Solubility data in the binary $Yb(NO_3)_3$-H_2O system have been reported in six publications (1-6). The solid phases reported in this binary system are the hexahydrate (2-5), the pentahydrate (1), and the tetrahydrate (6).

$Yb(NO_3)_3 \cdot 6H_2O$	[13839-85-5]	$Yb(NO_3)_3 \cdot 4H_2O$	[10035-00-4]
$Yb(NO_3)_3 \cdot 5H_2O$	[35725-34-9]		

Moret (1) studied the solubilities from 273.15 K to 313.15 K in the pentahydrate system, Afanas'ev and Azhipa (5) reported the solubility at 298.2 K in the hexahydrate system, and Khudaibergenova and Sulaimankulov (6) reported the solubility at 303.2 K in the tetrahydrate system. Spedding and co-workers (2-4) reported the solubility at 298.15 K, but did not identify the nature of the solid phase.

Marsh (7) states that the stable solid phase at "room temperature" is the pentahydrate, and that when the pentahydrate is heated to 321 K, there is a transition point at which water is liberated and a new hydrate is formed which does not dissolve in the water until heated to near 373 K. Based on Moret's solubility results for the pentahydrate system, below we calculate a congruent melting point of 325.8 K for the pentahydrate which is close to Marsh's experimental value of 321 K (Marsh also refers to this as an "indefinite" melting point). Wendlandt and Sewell (8) state that the unspecified hydrate $Yb(NO_3)_3 \cdot nH_2O$ does not fuse, but Ivanov-Emin et al.(9) report a congruent melting point of 333 K for the tetrahydrate. The only direct analyses of the solid phase for the hexahydrate system is that reported in the rejected paper (5). Although Spedding et al. (2-4) did not specify the nature of the stable solid phase, the implication is that it is the hexahydrate. This is based on the fact that the solubility of $Yb(NO_3)_3$ at 298.15 K reported in (2-4) falls on the smoothed curve of solubility vs atomic number for the hexahydrate system (see the figures in the PREFACE and in the $Eu(NO_3)_3$-H_2O critical evaluation). Moreover, at 298.2 K the stable solid phase in the saturated $Tm(NO_3)_3$-H_2O system is the hexahydrate, and in the saturated $Lu(NO_3)_3$-H_2O system the stable solid phase is the pentahydrate. It thus appears that the transition from stable hexahydrate to stable pentahydrate occurs at atomic number 70 (i.e. at Yb in the figure of solubility vs atomic number), and that this transition temperature must be very close to 298.2 K. We therefore conclude that at 298.15 K the stable solid phase is either the hexahydrate, or the eutonic hexahydrate-pentahydrate mixture. Considering the results from (1,7) we also conclude that the pentahydrate does not readily convert to the more stable hexahydrate at temperatures at least down to 273 K. This type of behavior has been observed for other lanthanide systems such as $Nd(NO_3)_3$ and $Sm(NO_3)_3$ where the solubilities in the metastable pentahydrate systems were readily measured at temperatures as low as 240 K.

For solubilities in the pentahydrate system, Moret's results (1) were fitted to the general solubility equation

$$Y = \ln(m/m_o) - nM_2(m - m_o) = a + b/(T/K) = c \ln(T/K) \quad [1]$$

where all terms in eq. [1] have been previously defined. At 298.15 K the smoothed value for the solubility in the pentahydrate system is 6.623 mol kg^{-1} which is slightly lower than the value of 6.650 mol kg^{-1} reported by Spedding et al. (2-4). Since we estimate the total uncertainty in Spedding's results to be ± 0.007 mol kg^{-1} or better, and the uncertainty in the smoothed value based on Moret's results as ± 0.07 mol kg^{-1}, it is difficult to attribute much significance to the difference between these two results: i.e. we would have to conclude that both solubility values are essentially identical and either correspond to the same solid phase (the pentahydrate), or to the eutonic mixture. If, on the other hand, we assume that the solid phase in Spedding's studies is the hexahydrate, then we would have to conclude that the difference between Spedding's result and the smoothed result at 298.15 K is significant. Since we concluded above that the hexahydrate is the stable solid phase at 298.15 K, then we also conclude that the difference between the two results is indeed significant. This is the basis upon which we selected the tentative solubilities (see below).

COMPONENTS:	EVALUATOR:
(1) Ytterbium nitrate; $Yb(NO_3)_3$; [13768-67-7] (2) Water ; H_2O ; [7732-18-5]	S. Siekierski, T. Mioduski Institute for Nuclear Research Warsaw, Poland and M. Salomon U.S. Army ET & DL Ft. Monmouth, NJ January 1983

CRITICAL EVALUATION:

Afanas'ev and Azhipa (5) reported a solubility of 4.16 mol kg^{-1} at 298.2 K in the hexahydrate system which is so low that it can immediately be rejected. At 303.2 K, Khudaibergenova and Sulaimankulov (6) reported the solubility to be 5.825 mol kg^{-1}, and that the solid phase (by direct chemical analysis) is the tetrahydrate. Since the tetrahydrate is probably highly metastable at this temperature, this low value for the solubility probably represents failure to attain equilibrium, and this result is also rejected.

TENTATIVE SOLUBILITIES

For the solubility of $Yb(NO_3)_3$ in the hexahydrate system at 298.15 K, the average value of 6.650 mol kg^{-1} from (2-4) is designated at a *tentative* value. It is not designated as a recommended value because of the lack of experimental data confirming the composition of the solid phase. Based on the experimental precision, we estimate an overall uncertainty of at least ± 0.007 mol kg^{-1} for this tentative solubility.

For the solubility of $Yb(NO_3)_3$ in the pentahydrate system as a function of temperature, smoothed *tentative* solubility data were obtained by fitting Moret's results (1) to the smoothing equation [1]. The derived parameters for the fitting of these data to the smoothing equation are given in Table 1. AT the 95% level of confidence and a Student's t = 4.303, the total uncertainty in the smoothed solubility values is ± 0.07 mol kg^{-1}.

All tentative solubility data are given in Table 2.

REFERENCES

1. Moret, R. *Thèse*. l'Université de Lausanne. 1963.
2. Spedding, F.H.; Shiers, L.E.; Rard, J.A. *J. Chem. Eng. Data* 1975, *20*, 88.
3. Rard, J.A.; Spedding, F.H. *J. Phys. Chem.* 1975, *79*, 257.
4. Spedding, F.H.; Derer, J.L.; Mohs, M.A.; Rard, J.A. *J. Chem. Eng. Data* 1976, *21*, 474.
5. Afanas'ev, Yu.A.; Azhipa, L.T. *Zh. Neorg. Khim.* 1976, *21*, 2284.
6. Khudaibergenova, N.; Sulaimankulof, K. *Zh. Neorg. Khim.* 1980, *25*, 2254.
7. Marsh, J.K. *J. Chem. Soc.* 1941, 561.
8. Wendlandt, W.W.; Sewell, R.G. *Texas J. Sci.* 1961, *13*, 231.
9. Ivanov-Emin, B.N.; Odinets, Z.K.; Del'Pino, Kh.; Zaitsev, B.E. *Zh. Neorg. Khim.* 1976, *21*, 837.

COMPONENTS:

(1) Ytterbium nitrate; $Yb(NO_3)_3$; [13768-67-7]

(2) Water ; H_2O ; [7732-18-5]

EVALUATOR:
S. Siekierski, T. Mioduski
Institute for Nuclear Research
Warsaw, Poland
and
M. Salomon
U.S. Army ET & DL
Ft. Monmouth, NJ
January 1983

CRITICAL EVALUATION:

Table 1. Derived parameters for the smoothing equation.[a]

parameter	value in the pentahydrate system
a	-17.917
b	492.4
c	2.8580
σ_a	0.003
σ_b	0.8
σ_c	0.0005
σ_Y	0.003
σ_m	0.037
ΔH_{sln}/mol kg^{-1}	-16.3
ΔC_p/J K^{-1} mol^{-1}	95.0
congruent melting point/K	325.8
concn at the congr m.p./mol kg^{-1}	11.102

[a] σ_a, σ_b, σ_c are standard deviation for the derived parameters a, b, c, and σ_Y and σ_m are the standard errors of estimate for Y in eq. [1] and the molality, respectively.

Table 2. Tentative solubilities.

	hexahydrate system[a]	pentahydrate system[b]
T/K	m_1/mol kg^{-1}	m_1/mol kg^{-1}
273.15		5.346
278.15		5.554
283.15		5.782
288.15		6.032
293.15		6.311
298.15	6.650	6.623
303.15		6.979
308.15		7.394
313.15		7.897
318.15		8.547
323.15		9.550

[a] Average value based on results from (2-4).

[b] Smoothed values based on results from (1).

COMPONENTS:

(1) Ytterbium nitrate; $Yb(NO_3)_3$; [13768-67-7]

(2) Water ; H_2O ; [7732-18-5]

ORIGINAL MEASUREMENTS:

Moret, R. *Thèse*. l'Université de Lausanne. 1963.

VARIABLES:

Temperature: range 0°C to 40°C

PREPARED BY:

T. Mioduski and S. Siekierski

EXPERIMENTAL VALUES:

Solubility[a]

t/°C	mass %	moles of water per 100 moles salt	mol kg^{-1}	solid phase
0	65.76	1035	5.349	$Yb(NO_3)_3 \cdot 5H_2O$
10	67.51		5.787	"
15	68.28		5.995	"
20	69.50	875	6.346	"
40	73.91	703	7.890	"

a. Molalities calculated by compilers from mass % values.

AUXILIARY INFORMATION

METHOD/APPARATUS/PROCEDURE:

The isothermal method was used as described in (1). Yb was determined by complexometric titration using Xylenol Orange indicator in the presence of a small amount of urotropine buffer. Water was determined by difference.

SOURCE AND PURITY OF MATERIALS:

Ytterbium nitrate was prepared from Yb_2O_3 of purity better than 99.7% (obtained by the ion exchange chromatographic method).

ESTIMATED ERROR:

Soly: precision about ± 0.1 % (compilers).

Temp: precision at least ± 0.05 K (compilers).

REFERENCES:

1. Brunisholz, G.; Quinche, J.P.; Kalo, A.M. *Helv. Chim. Acta* 1964, *47*, 14.

COMPONENTS:

(1) Ytterbium nitrate; $Yb(NO_3)_3$; [13768-67-7]

(2) Water ; H_2O ; [7732-18-5]

ORIGINAL MEASUREMENTS:

1. Spedding, F.H.; Shiers, L.E.; Rard, J.A. *J. Chem. Eng. Data* 1975, *20*, 88-93.
2. Rard, J.A.; Spedding, F.H. *J. Phys. Chem.* 1975, *79*, 257-62.
3. Spedding, F.H.; Derer, J.L.; Mohs, M.A.; Rard, J.A. *J. Chem. Eng. Data* 1976, *21*, 474-88.

VARIABLES:

One temperature: 25.00°C

PREPARED BY:

T. Mioduski, S. Siekierski, and M. Salomon

EXPERIMENTAL VALUES:

The solubility of $Yb(NO_3)_3$ in water at 25.00°C has been reported by Spedding and co-workers in three publications. Source paper [3] reports the solubility to be 6.650 mol kg^{-1}, and in source papers [1] and [2] as 6.6500 mol kg^{-1}.

COMMENTS AND/OR ADDITIONAL DATA:

Source paper [1] reports the relative viscosity, η_R, of a saturated solution to be 124.32. Taking the viscosity of water at 25°C to equal 0.008903 poise, the viscosity of a saturated $Yb(NO_3)_3$ solution at 25°C is 1.1068 poise (compilers calculation).

Supplementary data available in the microfilm edition to *J. Phys. Chem.* 1975, *79* enabled the compilers to provide the following additional data.

The density of the saturated solution was calculated by the compilers from the smoothing equation, and at 25°C the value is 2.19776 kg m^{-3}. Using this density, the solubility in volume units is

$$c_{satd} = 4.3141 \text{ mol dm}^{-3}$$

Source paper [2] reports the electrolytic conductivity of the saturated solution to be (corrected for the electrolytic conductivity of the solvent) κ = 0.006548 S cm^{-1}.

The molar conductivity of the saturated solution is calculated from $1000\kappa/3c_{satd}$ and is

$$\Lambda(\tfrac{1}{3}\ Yb(NO_3)_3) = 0.506 \text{ S cm}^2 \text{ mol}^{-1}$$

AUXILIARY INFORMATION

METHOD/APPARATUS/PROCEDURE:

Isothermal method used. Solutions were prepared as described in (1) and (2). The concentration of the saturated solution was determined by both EDTA (1) and sulfate (2) methods which is said to be reliable to 0.1% or better. In the sulfate analysis, the salt was first decomposed with HCl followed by evaporation to dryness before sulfuric acid additions were made. This eliminated the possibility of nitrate ion coprecipitation.

The composition of the solid phase was not specified in the source papers, but we assume it to be the hexahydrate (see the critical evaluation).

SOURCE AND PURITY OF MATERIALS:

$Yb(NO_3)_3 \cdot nH_2O$ was prepd by addn of HNO_3 to the oxide. The oxide was purified by an ion exchange method, and the upper limit for the impurities Ca, Fe, Si and adjacent rare earths was given as 0.15%.

Water was distilled from an alkaline permanganate soltuion.

ESTIMATED ERROR:

Soly: duplicate analyses agreed to at least ± 0.1%.

Temp: Not specified, but probably accurate to at least ± 0.01 K as in (3) (compilers).

REFERENCES:

1. Spedding, F.G.; Cullen, P.F.; Habenschuss, A. *J. Phys. Chem.* 1974, *78*, 1106.
2. Spedding, F.H.; Pikal, M.J.; Ayers, B.O. *J. Phys. Chem.* 1966, *70*, 2440.
3. Spedding, F.H.; et. al. *J. Chem. Eng. Data* 1975, *20*, 72.

COMPONENTS:

(1) Ytterbium nitrate; $Yb(NO_3)_3$; [13768-67-7]

(2) Nitric acid; HNO_3; [7697-37-2]

(3) Water ; H_2O ; [7732-18-5]

ORIGINAL MEASUREMENTS:

Afanas'ev, Yu.A.; Azhipa, L.T. *Zh. Neorg. Khim.* 1976, *21*, 2284-7; *Russ. J. Inorg. Chem. Eng. Transl.* 1976, *21*, 1257-60.

VARIABLES:

Concentration of HNO_3 at 25°C

PREPARED BY:

T. Mioduski, S. Siekierski and M. Salomon

EXPERIMENTAL VALUES:

$Yb(NO_3)_3$ [a]		HNO_3 [a]		
mass %	mol kg^{-1}	mass %	mol kg^{-1}	nature of the solid phase
59.9	4.16	0	———	$Yb(NO_3)_3 \cdot 6H_2O$
56.2	3.90	3.7	1.46	"
54.0	3.80	6.4	2.56	"
51.9	3.72	9.2	3.75	"
46.8	3.41	15.0	6.23	"
42.5	3.14	19.8	8.33	"
39.2	2.90	23.2	9.79	"
36.2	2.87	28.7	12.98	"
30.5	2.46	35.0	16.10	"
24.7	2.18	43.7	21.95	"
16.8	1.50	52.1	26.59	"
12.7	1.26	59.3	33.61	$Yb(NO_3)_3 \cdot 6H_2O + Yb(NO_3)_3 \cdot 4H_2O$
10.3	1.02	61.6	34.79	$Yb(NO_3)_3 \cdot 4H_2O$
8.0	0.82	64.7	37.61	"
7.5	0.85	67.8	43.56	"
5.5	0.70	72.7	52.92	"
5.8	0.82	74.5	60.02	$Yb(NO_3)_3 \cdot 4H_2O + Yb(NO_3)_3$
6.0	0.92	75.8	66.09	$Yb(NO_3)_3$
6.0	1.06	78.2	78.55	$Yb(NO_3)_3 \cdot nHNO_3$
4.0	0.80	82.0	92.95	"
1.4	0.33	86.8	116.7	"
1.2	0.48	91.9	211.4	"

a. Molalities calculated by M. Salomon.

AUXILIARY INFORMATION

METHOD/APPARATUS/PROCEDURE:

The isothermal method was used. Compositions of the saturated solutions were varied by adding 100% nitric acid to the saturated solutions, or by adding the salt to acid solutions. The concentration of Yb in the liquid and the residue was detd complexometrically, and the concentration of HNO_3 was detd by titrn with NaOH using Methyl Red indicator. The composition of the solid phase was detd by Schreinemakers' method of residues.

One data point was omitted from the above table: $Yb(NO_3)_3$ = 97.1 mass % and HNO_3 = 0.3 mass %. It is highly probable that this is a typographical error, and that the mass % values should be reversed.

Data on the infrared spectra of the hexahydrate and tetrahydrate are presented in the source publication.

SOURCE AND PURITY OF MATERIALS:

C.p. grade ytterbium nitrate was used.

C.p. grade nitric acid was concentrated to 100% by the Brauer method.

No other information given.

ESTIMATED ERROR:

Soly: precision probably about ± 0.5%, but accuracy is very poor (see critical evaluation).

Temp: precision ± 0.1 K.

REFERENCES:

COMPONENTS:

(1) Ytterbium nitrate; $Yb(NO_3)_3$; [13768-67-7]

(2) Urea; CH_4N_2O; [57-13-6]

(3) Water ; H_2O ; [7732-18-5]

ORIGINAL MEASUREMENTS:

Khudaibergenova, N.; Sulaimankulov, K.

Zh. Neorg. Khim. 1980, *25*, 2254-56; *Russ. J. Inorg. Chem. Engl. Transl.* 1980, *25*, 1249-50.

VARIABLES:

Composition at 30°C

PREPARED BY:

T. Mioduski and S. Siekierski

EXPERIMENTAL VALUES:

Composition of saturated solutions at 30°C[a]

$Yb(NO_3)_3$ mass %	$Yb(NO_3)_3$ mol kg^{-1}	$CO(NH_2)_2$ mass %	$CO(NH_2)_2$ mol kg^{-1}	nature of the solid phase
——	——	57.50	22.528	$CO(NH_2)_2$
12.50	1.112	56.20	29.898	"
15.50	1.438	54.49	30.234	"
24.23	3.015	53.39	39.723	"
28.47	4.398	53.50	49.409	"
33.05	6.813	53.44	65.865	"
35.87	9.597	53.72	85.927	"
36.03	8.232	51.78	70.730	$Yb(NO_3)_3 \cdot 4CO(NH_2)_2$
38.09	7.497	47.76	56.206	"
37.57	6.610	46.60	49.017	"
38.56	5.649	42.43	37.165	"
45.19	4.545	27.12	16.308	"
45.72	4.267	24.44	13.638	"
51.62	4.964	19.42	11.166	"
63.41	7.609	13.38	9.599	"
66.88	7.873	9.46	6.658	$Yb(NO_3)_3 \cdot CO(NH_2)_2$
70.07	8.455	6.85	4.942	"
71.72	8.107	3.64	2.460	"
74.53	8.998	2.40	1.732	$Yb(NO_3)_3 \cdot CO(NH_2)_2$ + $Yb(NO_3)_3 \cdot 4H_2O$
67.65	5.824	——	——	$Yb(NO_3)_3 \cdot 4H_2O$

a. Molalities calculated by M. Salomon.

AUXILIARY INFORMATION

METHOD/APPARATUS/PROCEDURE:

The isothermal method was used. Equilibrium was stated to be reached within 7 hours. The nitrogem of the urea was determined by the Kjeldahl method. Ytterbium was determined by titration with Trilon (disodium salt of ethylene diamine tetraacetic acid) using Xylenol Orange indicator. The liquid phase was filtered off with a No. 3 Schott filter.

COMMENTS AND/OR ADDITIONAL DATA:

$Yb(NO_3)_3 \cdot 4CO(NH_2)_2$ is congruently soluble. It was isolated and analysed: $Yb(NO_3)_3$ content was 59.99 mass %, and urea = 30.01 mass %.
$Yb(NO_3)_3 \cdot CO(NH_2)_2$ is incongruently soluble.

Analysis of the solid phase for the binary system resulted in 83.30 mass % $Yb(NO_3)_3$. This corresponds to the tetrahydrate (theor for the tetrahydrate is 83.30 mass % (compilers).

SOURCE AND PURITY OF MATERIALS:

Nothing specified.

ESTIMATED ERROR:

Nothing specified.

REFERENCES:

COMPONENTS:	EVALUATOR:
(1) Lutetium nitrate; $Lu(NO_3)_3$; [10099-67-9] (2) Water ; H_2O ; [7732-18-5]	S. Siekierski, T. Mioduski Institute for Nuclear Research Warsaw, Poland and M. Salomon U.S. Army ET & DL Ft. Monmouth, NJ January 1983

CRITICAL EVALUATION:

The solubility of lutetium nitrate in pure water at 298.15 K and 303.2 K has been reported in five publications (1-5). The solid phases reported are

$Lu(NO_3)_3 \cdot 6H_2O$ [36549-50-5] $Lu(NO_3)_3 \cdot 4H_2O$ [17836-45-2]

$Lu(NO_3)_3 \cdot 5H_2O$ [34767-08-3]

Marsh (6) reported that the pentahydrate is the stable solid phase at "room temperature," and that it is easily dehydrated. Marsh also reported the existence of $Lu(NO_3)_3 \cdot 4.5H_2O$. Molodkin et al. (7) reported that the trihydrate can be produced by desiccation over concentrated sulfuric acid.

Afanas'ev et al. (4) analysed the solid phase in saturated solutions at 298.2 K by complexometric titration, and reported it to be the hexahydrate. However their solubility value is so low (4.12 mol kg^{-1}) that we must reject the results in (4). Spedding et al. (3) analysed the solid phase in equilibrium with saturated solutions at 298.15 K by titration with EDTA, and the hydrate was found to be within ± 0.016 water molecules of the pentahydrate. We therefore conclude that the stable solid phase at 298.14 K is the pentahydrate. This conclusion is also supported by the fact that there is a sudden break for $Lu(NO_3)_3$ to a lower solubility in the plot of solubility vs atomic number at 298.15 K as seen in the figure in the $Eu(NO_3)_3$-H_2O critical evaluation.

Wendlandt and Sewell (8) reported that an unspecified hydrate, $Lu(NO_3)_3 \cdot nH_2O$, failed to undergo fusion, but Molodkin et al. (7) reported that the trihydrate melts congruently at 328 K.

TENTATIVE SOLUBILITIES

At 298.15 K, the *tentative* solubility of $Lu(NO_3)_3$ in the pentahydrate system is obtained from (1-2) and is 6.822 ± 0.007 mol kg^{-1}. The uncertainty in this tentative value is based upon the experimental precision reported in (1-2). For the metastable hexahydrate system, the solubility of $Lu(NO_3)_3$ is *estimated* as 7.60 ± 0.10 mol kg^{-1} (extrapolated value from the figure of solubility vs atomic number).

At 303.2 K the solubility of $Lu(NO_3)_3$ in the tetrahydrate system is 7.47 mol kg^{-1} (5) which is lower than the estimated value of 7.60 mol kg^{-1} for the metastable hexahydrate system at 298.2 K. We therefore designate this experimental solubility in the tetrahydrate system as *doubtful*.

REFERENCES

1. Spedding, F.H.; Shiers, L.E.; Rard, J.A. *J. Chem. Eng. Data* 1975, *20*, 88.
2. Rard, J.A.; Spedding, F.H. *J. Phys. Chem.* 1975, *79*, 257.
3. Spedding, F.H.; Derer, J.L.; Mohs, M.A.; Rard, J.A. *J. Chem. Eng. Data* 1976, *21*, 474.
4. Afanas'ev, Yu.A.; Azhipa, L.T.; Sal'nik, L.V. *Zh. Neorg. Khim.* 1982, *27*, 769.
5. Khudaibergenova, N.; Sulaimankulov, K.S. *Zh. Neorg. Khim.* 1979, *24*, 2005.
6. Marsh, J.K. *J. Chem. Soc.* 1941, 561.
7. Molodkin, A.K.; Odinets, Z.K.; Chuvelev, A.V. *Zh. Neorg. Khim.* 1977, *22*, 1520.
8. Wendlandt, W.W.; Sewell, R.G. *Texas J. Sci.* 1961, *13*, 231.

COMPONENTS:

(1) Lutetium nitrate; $Lu(NO_3)_3$; [10099-67-9]

(2) Water ; H_2O ; [7732-18-5]

ORIGINAL MEASUREMENTS:

1. Spedding, F.H.; Shiers, L.E.; Rard, J.A. *J. Chem. Eng. Data* 1975, *20*, 88-93.
2. Rard, J.A.; Spedding. F.H. *J. Phys. Chem.* 1975, *79*, 257-62.
3. Spedding, F.H.; Derer, J.L.; Mohs, M.A.; Rard, J.A. *J. Chem. Eng. Data* 1976, *21*, 474-88.

VARIABLES:

One temperature: 25.00°C

PREPARED BY:

T. Mioduski, S. Siekierski, and M. Salomon

EXPERIMENTAL VALUES:

The solubility of $Lu(NO_3)_3$ in water at 25.00°C has been reported by Spedding and co-workers in three publications. Source paper [3] reports the solubility to be 6.792 mol kg^{-1}, but the preferred value is given in source paper [1] and [2] as 6.8219 mol kg^{-1}.

COMMENTS AND/OR ADDITIONAL DATA:

Source paper [1] reports the relative viscosity, η_R, of a saturated solution to be 149.92. Taking the viscosity of water at 25°C to equal 0.008903 poise, the viscosity of a saturated $Lu(NO_3)_3$ solution is 25°C is 1.3347 poise (compilers calculation).

Supplementary data available in the microfilm edition to *J. Phys. Chem.* 1975, *79* enabled the compilers to provide the following additional data.

The density of the saturated solutions was calculated by the compilers from the smoothing equation, and at 25°C the value is 2.22323 kg m^{-3}. Using this density, the solubility in volume units is (based on the preferred value of 6.8219 mol kg^{-1})

$$c_{satd} = 4.3801 \text{ mol dm}^{-3}$$

Source paper [2] reports the electrolytic conductivity of the saturated solution to be (corrected for the electrolytic conductivity of the solvent) $\kappa = 0.005660$ S cm^{-1}.

The molar conductivity of the saturated solution is calculated from $1000\kappa/3c_{satd}$ and is

$$\Lambda(\tfrac{1}{3}\, Lu(NO_3)_3) = 0.431 \text{ S cm}^2 \text{ mol}^{-1}$$

AUXILIARY INFORMATION

METHOD/APPARATUS/PROCEDURE:

Isothermal method used. Solutions were prepared as described in (1) and (2). The concentration of the saturated solution was determined by both EDTA (1) and sulfate (2) methods which is said to be reliable to 0.1% or better. In the sulfate analysis, the salt was first decomposed with HCl followed by evaporation to dryness before sulfuric acid additions were made. This eliminated the possibility of nitrate ion coprecipitation.

SOURCE AND PURITY OF MATERIALS:

$Lu(NO_3)_3 \cdot 5H_2O$ was prepd by addn of HNO_3 to the oxide. The oxide was purified by an ion exchange method, and the upper limit for the impurities Ca, Fe, Si and adjacent rare earths was given as 0.15%.

In source paper [3] the salt was analysed for water of hydration and found to be within ± 0.016 water molecules of the pentahydrate.

Water was distilled from an alkaline permanganate solution.

ESTIMATED ERROR:

Soly: duplicate analyses agreed to at least ± 0.1%.

Temp: Not specified, but probably accurate to at least ± 0.01 K as in (3) (compilers).

REFERENCES:

1. Spedding, F.G.; Cullen, P.F.; Habenschuss, A. *J. Phys. Chem.* 1974, *78*, 1106.
2. Spedding, F.H. Pikal, M.J.; Ayers, B.O. *J. Phys. Chem.* 1966, *70*, 2440.
3. Spedding, F.H.; et. al. *J. Chem. Eng. Data* 1975, *20*, 72.

COMPONENTS:

(1) Lutetium nitrate; $Lu(NO_3)_3$; [10099-67-9]

(2) Nitric acid; HNO_3; [7697-37-2]

(3) Water ; H_2O ; [7732-18-5]

ORIGINAL MEASUREMENTS:

Afanas'ev, Yu. A.; Azhipa, L.T.; Sal'nik, L.V. *Zh. Neorg. Khim.* 1982, *27*, 769-73; *Russ. J. Inorg. Chem. Engl. Transl.* 1982, *27*, 431-4.

VARIABLES:

HNO_3 concentration at 25°C

PREPARED BY:

T. Mioduski and S. Siekierski

EXPERIMENTAL VALUES:

Solubility of $Lu(NO_3)_3$ in HNO_3 solutions at 25°C

$Lu(NO_3)_3$ mass %	$Lu(NO_3)_3$ mol kg^{-1}	HNO_3 mass %	HNO_3 mol kg^{-1}	solid phase[b]
59.8	4.12	0	———	A
51.6	3.55	8.1	3.19	"
49.3	3.40	10.5	4.15	"
46.8	3.19	12.5	4.87	"
42.1	2.85	17.0	6.60	"
36.4	2.41	21.7	8.22	"
32.5	2.12	25.0	9.34	"
29.0	1.88	28.3	10.52	"
28.4	1.87	29.5	11.12	"
27.7	1.90	31.9	12.53	"
27.5	1.98	34.0	14.01	A+B
22.0	1.54	38.5	15.47	B
19.2	1.38	42.2	17.35	"
19.1	1.43	43.8	18.74	"
19.0	1.46	45.0	18.94	"
18.9	1.68	50.0	25.51	B+C
13.7	1.18	54.1	26.66	C
11.8	1.06	57.5	29.72	"
10.0	0.96	61.0	33.38	"
9.4	0.96	63.4	36.99	"
6.9	0.79	68.9	45.18	"
7.4	1.02	72.5	57.24	C+D
6.1	0.85	74.0	59.01	D
4.2	0.68	78.8	73.56	"
3.7	0.74	82.5	94.87	"
3.8	0.98	85.5	126.8	"
3.2	0.84	86.2	129.1	D+E
2.3	0.62	87.4	134.7	E
1.5	0.40	88.1	134.4	"
0.7	0.19	89.0	137.1	"

a. Molalities calculated by M. Salomon.

b. Solid phases: A = $Lu(NO_3)_3 \cdot 6H_2O$; B = $Lu(NO_3)_3 \cdot 4H_2O$; C = $Lu(NO_3)_3 \cdot 2H_2O$; D = $Lu(NO_3)_3 \cdot H_2O$; E = $Lu(NO_3)_3$

AUXILIARY INFORMATION

METHOD/APPARATUS/PROCEDURE:

The isothermal method was used. The composition of the solutions was changed by addition of 100% HNO_3 to a saturated solution or by addition of the salt to the acid solution. Equilibrium was reached within 3-4 hours. The lutetium content in the saturated solutions and solid phases was determined by complexometric titration using Xylenol Orange indicator. The HNO_3 content was determined by titration with NaOH with Methyl Orange indicator. The compositions of the solid phases were determined by the Schreinemakers' method. The hydrated solid phases were separated and their infrared spectra recorded. Details are given in the source paper.

SOURCE AND PURITY OF MATERIALS:

C.p. grade lutetium nitrate was used.

Nitric acid (source and purity not specified) was concentrated by method recommended in the well-known Brauer's Handbook (the Russian edition was cited by the authors).

ESTIMATED ERROR:

Soly: nothing specified.

Temp: precision within ± 0.1 K.

REFERENCES:

COMPONENTS:

(1) Lutetium nitrate; $Lu(NO_3)_3$; [10099-67-9]

(2) Urea; CH_4N_2O; [57-13-6]

(3) Water ; H_2O ; [7732-18-5]

ORIGINAL MEASUREMENTS:

Khudaibergenova, N.; Sulaimankulov, K.S.
Zh. Neorg, Khim. 1979, *24*, 2005-8;
Russ. J. Inorg. Chem. Engl. Transl. 1979, *24*, 1112-4.

VARIABLES:

Composition at 30°C

PREPARED BY:

T. Mioduski and S. Siekierski

EXPERIMENTAL VALUES:

$Lu(NO_3)_3$ [a]		$CO(NH_2)_2$ [a]		
mass %	mol kg^{-1}	mass %	mol kg^{-1}	nature of the solid phase
0	———	55.62	20.868	$CO(NH_2)_2$
10.73	0.820	53.03	24.366	"
15.94	1.419	52.95	28.341	"
19.76	2.009	52.99	32.380	"
27.64	4.045	53.43	46.998	"
28.92	4.330	52.58	47.325	"
35.58	11.596	55.92	109.55	$Lu(NO_3)_3 \cdot 4CO(NH_2)_2$
35.67	9.032	53.39	81.262	"
35.31	5.372	46.48	42.501	"
36.06	4.351	40.98	29.720	"
37.80	4.322	37.97	26.093	"
41.93	6.198	39.33	34.946	"
38.38	3.268	29.09	14.890	"
43.44	3.762	24.57	12.789	"
46.58	3.882	20.18	10.109	"
50.15	5.209	23.18	14.472	"
52.80	4.797	16.71	9.126	"
53.36	4.500	13.79	6.990	"
54.61	4.747	13.52	7.064	"
58.73	5.560	12.01	6.835	"
64.86	7.382	10.80	7.388	"
69.29	8.685	8.61	6.487	"
71.15	8.710	6.22	4.577	$Lu(NO_3)_3 \cdot 4CO(NH_2)_2 + Lu(NO_3)_3 \cdot 4H_2O$
71.55	7.872	3.27	2.162	$Lu(NO_3)_3 \cdot 4H_2O$
72.95	7.471	0	———	"

a. Molalities calculated by M. Salomon.

AUXILIARY INFORMATION

METHOD/APPARATUS/PROCEDURE:

The isothermal method was used. Equil was attained within 7-8 h. To separate the phases, a Schott filter No. 3 was employed. Nitrogen was detd by the Kjeldahl method, and Lu detd by titrn with Trilon in the presence of Xylenol Orange indicator. The compilers assume that compn of the solid phases was found by the Schreinemakers' residue method. The double salt is congruently soluble. It was isolated and analysed: $Lu(NO_3)_3$ = 60 mass %, and $CO(NH_2)_2$ = 40 mass %.

SOURCE AND PURITY OF MATERIALS:

Nothing specified.

ESTIMATED ERROR:

Nothing specified.

REFERENCES:

SYSTEM INDEX

Underlined page numbers refer to evaluation text and those not underlined to compiled tables. All compounds are listed as in Chemical Abstracts. Systems are listed with solvent last, for example, cerium nitrate + water system occurs as Nitric acid, cerium salt + water but not as water + nitric acid, cerium salt. Ternary mixtures involving a salt and water as components are listed as salt (aqueous) + other component and these occur after salt + other components. For example, Nitric acid, cerium salt + ethane, 1,1'-oxybis- comes before Nitric acid, cerium salt + water which comes before Nitric acid, cerium salt (aqueous) + Nitric acid, barium salt.

S

T

U

REGISTRY NUMBER INDEX

Underlined numbers refer to evaluations. Other numbers refer to compiled tables.

AUTHOR INDEX

Underlined numbers refer to evaluations. Other numbers refer to compiled tables.